Fundamental Identities

$$\sin \theta \csc \theta = 1 \qquad \sin^2 \theta + \cos^2 \theta = 1$$
$$\cos \theta \sec \theta = 1 \qquad 1 + \tan^2 \theta = \sec^2 \theta$$
$$\tan \theta \cot \theta = 1 \qquad 1 + \cot^2 \theta = \csc^2 \theta$$
$$\tan \theta = \frac{\sin \theta}{\cos \theta} \qquad \cot \theta = \frac{\cos \theta}{\sin \theta}$$

Cofunction Identities

$$\cos \left(\frac{\pi}{2} - \theta \right) = \sin \theta \qquad \sin \left(\frac{\pi}{2} - \theta \right) = \cos \theta$$

$$\cot \left(\frac{\pi}{2} - \theta \right) = \tan \theta \qquad \tan \left(\frac{\pi}{2} - \theta \right) = \cot \theta$$

$$\csc \left(\frac{\pi}{2} - \theta \right) = \sec \theta \qquad \sec \left(\frac{\pi}{2} - \theta \right) = \csc \theta$$

Negative-Angle Identities

$$\sin (-\theta) = -\sin \theta \qquad \csc (-\theta) = -\csc \theta$$
$$\cos (-\theta) = \cos \theta \qquad \sec (-\theta) = \sec \theta$$
$$\tan (-\theta) = -\tan \theta \qquad \cot (-\theta) = -\cot \theta$$

Double-Angle Identities

$$\sin 2\theta = 2 \sin \theta \cos \theta$$
$$\cos 2\theta = \cos^2 \theta - \sin^2 \theta$$
$$= 2 \cos^2 \theta - 1$$
$$= 1 - 2 \sin^2 \theta$$
$$\tan 2\theta = \frac{2 \tan \theta}{1 - \tan^2 \theta}$$

Half-Angle Identities

$$\sin \frac{\theta}{2} = \pm \sqrt{\frac{1 - \cos \theta}{2}} \qquad \tan \frac{\theta}{2} = \pm \sqrt{\frac{1 - \cos \theta}{1 + \cos \theta}}$$

$$\cos \frac{\theta}{2} = \pm \sqrt{\frac{1 + \cos \theta}{2}} \qquad = \frac{\sin \theta}{1 + \cos \theta}$$

$$= \frac{1 - \cos \theta}{\sin \theta}$$

Binomial Coefficients

$$\binom{n}{r} = \frac{n!}{(n - r)! \, r!}$$

Binomial Theorem

$$(a + b)^n = a^n + \binom{n}{1} a^{n-1} b + \binom{n}{2} a^{n-2} b^2 + \cdots + \binom{n}{r} a^{n-r} b^r + \cdots + \binom{n}{n-2} a^2 b^{n-2} + \binom{n}{n-1} ab^{n-1} + b^n$$

Reduction Identity

$$a \sin \theta + b \cos \theta = \sqrt{a^2 + b^2} \sin (\theta + \alpha),$$

where

$$\cos \alpha = \frac{a}{\sqrt{a^2 + b^2}} \qquad \text{and} \qquad \sin \alpha = \frac{b}{\sqrt{a^2 + b^2}}$$

Identities for Sum and Difference of Two Angles

$$\cos (\alpha - \beta) = \cos \alpha \cos \beta + \sin \alpha \sin \beta$$
$$\cos (\alpha + \beta) = \cos \alpha \cos \beta - \sin \alpha \sin \beta$$
$$\sin (\alpha + \beta) = \sin \alpha \cos \beta + \cos \alpha \sin \beta$$
$$\sin (\alpha - \beta) = \sin \alpha \cos \beta - \cos \alpha \sin \beta$$
$$\tan (\alpha + \beta) = \frac{\tan \alpha + \tan \beta}{1 - \tan \alpha \tan \beta}$$
$$\tan (\alpha - \beta) = \frac{\tan \alpha - \tan \beta}{1 + \tan \alpha \tan \beta}$$

Product Identities

$$2 \sin \alpha \cos \beta = \sin (\alpha + \beta) + \sin (\alpha - \beta)$$
$$2 \cos \alpha \sin \beta = \sin (\alpha + \beta) - \sin (\alpha - \beta)$$
$$2 \cos \alpha \cos \beta = \cos (\alpha + \beta) + \cos (\alpha - \beta)$$
$$2 \sin \alpha \sin \beta = \cos (\alpha - \beta) - \cos (\alpha + \beta)$$

Sum Identities

$$\sin x + \sin y = 2 \sin \left(\frac{x + y}{2} \right) \cos \left(\frac{x - y}{2} \right)$$

$$\sin x - \sin y = 2 \cos \left(\frac{x + y}{2} \right) \sin \left(\frac{x - y}{2} \right)$$

$$\cos x + \cos y = 2 \cos \left(\frac{x + y}{2} \right) \cos \left(\frac{x - y}{2} \right)$$

$$\cos x - \cos y = -2 \sin \left(\frac{x + y}{2} \right) \sin \left(\frac{x - y}{2} \right)$$

Vectors

Vector addition:

$$(u_1 \mathbf{i} + u_2 \mathbf{j}) + (v_1 \mathbf{i} + v_2 \mathbf{j}) = (u_1 + v_1)\mathbf{i} + (u_2 + v_2)\mathbf{j}$$

Scalar multiplication: $a(v_1 \mathbf{i} + v_2 \mathbf{j}) = (av_1)\mathbf{i} + (av_2)\mathbf{j}$

Magnitude: $|v_1 \mathbf{i} + v_2 \mathbf{j}| = \sqrt{v_1^2 + v_2^2}$

Precalculus
FUNCTIONS AND GRAPHS

Precalculus
FUNCTIONS AND GRAPHS

Linda Gilbert
Jimmie Gilbert

University of South Carolina at Spartanburg

ADDISON-WESLEY PUBLISHING COMPANY

Reading, Massachusetts ■ Menlo Park, California ■ New York
Don Mills, Ontario ■ Wokingham, England ■ Amsterdam ■ Bonn
Sydney ■ Singapore ■ Tokyo ■ Madrid ■ San Juan

Executive Editor: *David F. Pallai*
Production Supervisor: *Karen Garrison*
Text Designer: *Piñeiro Design Associates*
Copyeditor: *Jacqueline Dormitzer*
Manufacturing Supervisor: *Roy Logan*
Cover Designer: *Marshall Henrichs*

Library of Congress Cataloging-in-Publication Data

Gilbert, Jimmie, 1934–
    Precalculus : functions and graphs / by Jimmie Gilbert and Linda Gilbert.
        p.   cm.
    Includes index.
    ISBN 0-201-51020-0
    1. Functions.   2. Algebra—Graphic methods.   I. Gilbert, Linda.
    II. Title.
    QA331.3.G55   1990
    515—dc20   89-31751
                                                                                    CIP

ABCDEFGHIJ-MU-9987654321089

To our children:
    Donna, Lisa, Martin, Dan, Beckie and Matt;
and our grandchildren:
    James, Beth, Jill, Mary, and Adrienne.

Preface

Purpose

This book is written as a text for today's college precalculus course. It fills the needs of students majoring in mathematics, engineering, science, business, or other technical areas. The knowledge and maturity that an average student has after two courses in algebra is adequate for the study of this text.

Content Features

The following features reflect the fact that preparation for the calculus is emphasized throughout the text.

- Section 1.7, **"Algebraic Skills for the Calculus,"** is devoted exclusively to skills that are necessary for success in the calculus. (See pages 45–52.)
- Informal but mathematically correct material on **limits** is included in the treatment of asymptotes (pages 114–115 and 117–120) and also in the discussion of infinite geometric series in Section 9.3.

- All of Section 4.2 is devoted to the **natural exponential function.**
- The entire Chapter 10 is devoted to **analytic geometry** that is repeated in a briefer form in most calculus courses. It includes a section on the **three-dimensional** coordinate system.
- Seven **Drawing Tips** are included in Section 10.8 as an aid to sketching planes and spheres in a three-dimensional coordinate system.

Pedagogical Features

The following features are intended to make the study of this book a successful and rewarding experience for the student:

- **critical thinking problems** at the end of each chapter with solutions in the answer section;
- opening paragraphs for each chapter that convey the significance and applications of the material in that chapter;
- a format that highlights key equations, rules, and procedures for problem solving;
- an abundance and variety of applications in the examples as well as the exercises;
- an early introduction of the trigonometric functions of an angle in standard position, with trigonometric functions of real numbers in the next section;
- a large number of examples and exercises, with the exercises odd-even paired and arranged in order of increasing difficulty;
- a sampling of problems that require the use of a calculator, labeled to make them conspicuous;
- a chapter review at the end of each chapter that includes a list of **Key Words and Phrases** and a **Summary of Important Concepts and Formulas,** along with a generous supply of review problems for that chapter.

The exercises requiring trigonometric function values or logarithms may be worked either with a calculator or with tables. Explanations are included in the body of the text for the calculator approach, and the use of tables is explained in the appendix. Tables are provided for common logarithms and trigonometric functions in degrees and minutes, in decimal degrees, and in radians or real numbers. The choice of computational methods is left to the instructor.

Some students need to learn the procedure of linear interpolation for use with tables unrelated to logarithms or trigonometry. For this reason, interpolation is included with table evaluation of logarithms in the appendix.

Each critical thinking problem is followed by a worked-out solution that contains one or more errors. The challenge for the student is to think critically and find the errors. The errors are marked and corrected in the answer section of the book. It is hoped that these critical thinking problems will help the student recognize and avoid the common pitfalls of each type of problem.

Supplements

Instructor's Solutions Manual
Complete worked-out solutions for every exercise in the text.

Student's Solutions Manual
Worked-out solutions to odd problems.

Printed Test Bank
Includes multiple choice, open-ended essay problems, and True/False questions.

Answer Book
Answers to odd and even questions.

Computerized Testing—Test Edit
Contains over 1000 problems easily accessed by computer in either multiple-choice or open-ended format. Features include editing of problems, option to leave space for working problems on the text, scrambling problem and answer orders, and printed answer keys (for IBM-PC®).

Transparency Masters
One hundred twenty printed transparency masters covering key definitions, theorems, and graphs from the text.

Precalculus Acetate Package
Twenty acetates to be used with overhead projectors. Includes key figures to illustrate precalculus concepts, four-color transparencies with overlays. Available to adoptions of 500 copies or more.

Master Grapher/3D Grapher
A powerful, interactive graphing utility for functions, polar equations, parametric equations, conic equations, and functions of two variables. Available for IBM-PC®, Apple®, and Macintosh®.

x   PREFACE

Acknowledgments

We wish to acknowledge with thanks the helpful suggestions made in reviews by the following persons:

Professor Dennis Bertholf, *Oklahoma State University*

Professor Steven Blasberg, *West Valley College*

Professor Gary Carlson, *Utah Valley Community College*

Professor Christopher Ennis, *Carleton College*

Dr. Maher Shawer, *Indiana University of Pennsylvania*

Professor Marcia Siderow, *California State University, Northridge*

Professor Lynn Tooley, *Bellevue Community College*

We also express our gratitude to David Pallai, Executive Editor at Addison-Wesley, for the inspiration and guidance that he gave us in writing this book. It is a much better book because of his influence.

Linda Gilbert

Jimmie Gilbert

Contents

1 Fundamentals

1.1 The Real Numbers   2
1.2 Polynomials and Rational Expressions   9
1.3 Radicals and Rational Exponents   20
1.4 Complex Numbers   27
1.5 Solutions of Equations with Real Coefficients   32
1.6 Nonlinear Inequalities in One Real Variable   40
1.7 Algebraic Skills for the Calculus   45
Chapter Review   52

2 Functions and Graphs

2.1 Functions and Relations   60
2.2 The Algebra of Functions   72
2.3 Linear Functions   76
2.4 Quadratic Functions   86
2.5 Inverse Relations and Functions   92
2.6 Symmetry and Translations   98

2.7 Graphs of Polynomial Functions   107
2.8 Graphs of Rational Functions   112
Chapter Review   121

3 ▌Theory of Polynomials

3.1 The Factor and Remainder Theorems   126
3.2 The Fundamental Theorem of Algebra and Descartes' Rule of Signs   132
3.3 Rational Zeros   139
3.4 Approximation of Real Zeros   147
Chapter Review   150

4 ▌Exponential and Logarithmic Functions

4.1 Exponential Functions   156
4.2 The Natural Exponential Function   161
4.3 Logarithmic Functions   165
4.4 Logarithmic Equations, Exponential Equations, and Natural Logarithms   171
Chapter Review   178

5 ▌The Trigonometric Functions

5.1 Angles, Triangles, and Arc Length   182
5.2 Trigonometric Functions of Angles   192
5.3 Trigonometric Functions of Real Numbers   202
5.4 Some Fundamental Properties   207
5.5 Values of Trigonometric Functions   211
5.6 Right Triangles   216
5.7 Sinusoidal Graphs   226
5.8 Other Trigonometric Graphs   239
Chapter Review   250

6 Analytic Trigonometry

6.1 Verifying Identities 258
6.2 Cosine of the Sum or Difference of Two Angles 266
6.3 Identities for the Sine and Tangent 273
6.4 Double-Angle and Half-Angle Identities 278
6.5 Product and Sum Identities 285
6.6 Trigonometric Equations 288
6.7 Inverse Trigonometric Functions 295
Chapter Review 303

7 Additional Topics in Trigonometry

7.1 The Law of Sines 310
7.2 The Law of Cosines 318
7.3 The Area of a Triangle 325
7.4 Vectors 330
7.5 Trigonometric Form of Complex Numbers 341
7.6 Powers and Roots of Complex Numbers 347
Chapter Review 351

8 Systems of Equations and Inequalities

8.1 Systems of Equations 358
8.2 Matrix Algebra 366
8.3 Solution of Linear Systems by Matrix Methods 376
8.4 Inverses of Matrices 385
8.5 Partial Fractions 390
8.6 The Definition of a Determinant 396
8.7 Evaluation of Determinants 402
8.8 Cramer's Rule 407
8.9 Systems of Inequalities 412
8.10 Linear Programming 417
Chapter Review 423

9 Further Topics in Algebra

9.1 Sequences and Series   430
9.2 Arithmetic Sequences   434
9.3 Geometric Sequences   440
9.4 Mathematical Induction   449
9.5 The Binomial Theorem   455
Chapter Review   461

10 Analytic Geometry

10.1 The Parabola   466
10.2 The Ellipse   472
10.3 The Hyperbola   479
10.4 Rotation of Axes   486
10.5 Polar Coordinates   493
10.6 Polar Equations of Conics   500
10.7 Parametric Equations   506
10.8 Three-Dimensional Rectangular Coordinates   513
Chapter Review   522

Appendix

Table Evaluation of Logarithms   A-2
Computations with Logarithms   A-7
Table Evaluation of Trigonometric Functions   A-11

Tables

Trigonometric Functions, Degrees and Minutes or Radians   A-17
Trigonometric Functions, Decimal Degrees   A-23
Common Logarithms   A-35
**Answers to Odd-Numbered Exercises, Review Problems,
and Critical Thinking Problems   A-37**
Index   I-1

Precalculus
FUNCTIONS AND GRAPHS

Chapter 1

Fundamentals

This chapter reviews the most important ideas from high school algebra and concludes with a section on algebraic skills for the calculus. The examples and exercises in that section are taken directly from applications of the formulas of the calculus, and the skills emphasized there are precisely those needed in the calculus.

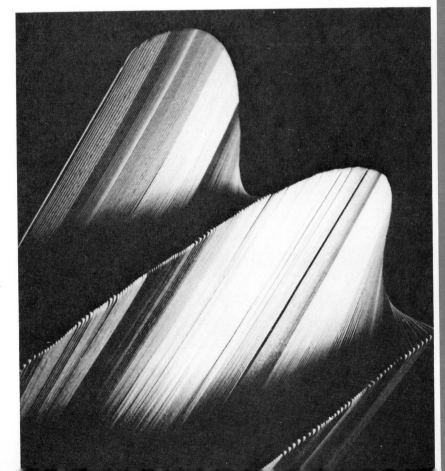

Much of our work in algebra involves sets of real numbers, and it is frequently convenient to use some of the terminology and notation for sets.

The word **set** is used to refer to a collection of objects where it is possible to determine whether or not a certain object is in the set. The individual objects in the set are called the **members** of the set, or the **elements** of the set.

A standard way to indicate a set, even when all the elements cannot be listed, is to use braces. Thus the notation

$$E = \{2, 4, 6, 8, \ldots\}$$

indicates that E is the set of all positive even integers. Another standard way to describe a set is to use braces and indicate the property that qualifies objects for membership in the set. With this notation, the same set E would be indicated as

$$E = \{x : x \text{ is a positive even integer}\}.$$

This notation is called **set-builder notation,** and is read as "E is the set of all x such that x is a positive even integer." The colon is taken as shorthand for "such that."

Other shorthand symbols are also used with sets. For example, we write $x \in S$ as shorthand for the phrase "x is an element of S," or "x is in S." If x is not an element of S, we write $x \notin S$. If every element of the set A is an element of the set B, then A is called a **subset** of B, and we write $A \subseteq B$. We write $A \not\subseteq B$ to indicate that A is not a subset of B. The notation $B \supseteq A$ is used interchangeably with $A \subseteq B$. We write $A = B$ to mean that A and B are composed of exactly the same elements.

Two operations on sets are frequently used in mathematics. One of these operations is that of forming the union of two sets. If A and B are sets, the **union** of A and B is the set $A \cup B$ given by

$$A \cup B = \{x : x \in A \quad \text{or} \quad x \in B\}.$$

That is, $A \cup B$ consists of all those elements x that are either an element of A, or an element of B, or an element of both A and B.

The other operation frequently used is that of forming the intersection of sets. For sets A and B, the **intersection** of A and B is the set $A \cap B$ given by

$$A \cap B = \{x : x \in A \quad \text{and} \quad x \in B\}.$$

Thus $A \cap B$ consists of those elements that are in both A and B.

The empty set is the set that has no members. The **empty set** is denoted by $\varnothing$ or $\{\,\}$, and is regarded as a subset of every set.

$(-\infty, 0) \cup (0, \infty)$

$x > 1$

$(1, +\infty)$

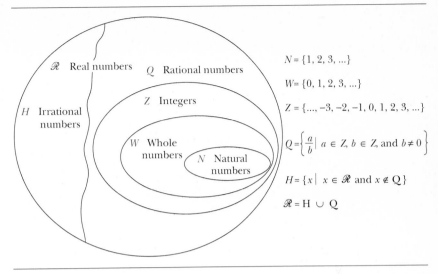

Real numbers $\mathcal{R}$ Rational numbers Q

$\mathcal{R}$ Real numbers Q Rational numbers

Z Integers

H Irrational numbers

W Whole numbers N Natural numbers

$N = \{1, 2, 3, ...\}$

$W = \{0, 1, 2, 3, ...\}$

$Z = \{..., -3, -2, -1, 0, 1, 2, 3, ...\}$

$Q = \left\{ \dfrac{a}{b} \,\middle|\, a \in Z,\, b \in Z, \text{ and } b \neq 0 \right\}$

$H = \{ x \mid x \in \mathcal{R} \text{ and } x \notin Q \}$

$\mathcal{R} = H \cup Q$

FIGURE 1.1

Throughout this book the set of real numbers is denoted by $\mathcal{R}$. Several subsets of $\mathcal{R}$ have special names. Those that we are concerned with now and their relationships to each other are shown in Figure 1.1.

One of the most useful aids in working with real numbers is the association between real numbers and points on a straight line. We start with a horizontal line that extends indefinitely in each direction, as indicated by the arrowheads in Figure 1.2. A point is chosen, labeled 0, and referred to as the **origin.** A unit of measure is chosen, and points successively one unit apart are located on the line in both directions from the origin. It is conventional to label the points so located to the right of 0 in succession with the positive integers 1, 2, 3, . . . , and to label those to the left of 0 in succession with the negative integers -1, -2, -3, ... Since any rational number can be represented as the quotient of two integers, rational numbers are then located on the line by using appropriate portions of the chosen unit of measure. For example, $5/2 = 2.5$ is located two and one-half units to the right of 0, $-7/4 = -1.75$ is located one and three-fourths units to the left of 0, and so on. An irrational number is characterized by the fact that its decimal representation is nonterminating and nonrepeating. Thus irrational numbers can be located to any desired degree of accuracy by

FIGURE 1.2

using their decimal approximations. In Figure 1.2, $\sqrt{2}$ is located by using $\sqrt{2} \approx 1.41$, and $-\pi$ is located by using $-\pi \approx -3.14$. (The symbol $\approx$ means "approximately equals.")

The association that has thus been established between real numbers and the points on the line is a one-to-one correspondence: Each real number corresponds to exactly one point on the line, and each point on the line corresponds to one real number. After that correspondence between real numbers and points on the line has been made, the line is commonly referred to as a *number line.*

Ordering between real numbers is indicated by using the inequality symbols $>$ and $<$, read "greater than" and "less than," respectively.

On the number line, the geometric representation of $a > b$ is that a *lies to the right of b.* In particular, $a > 0$ means that a lies to the right of 0, or that *a is positive.*

On the number line, $a < b$ means that a *lies to the left of b.* In particular, $a < 0$ means that a lies to the left of 0, or that *a is negative.* Whether we write $a < b$ or $b > a$ is often a matter of emphasis or of personal preference.

The symbols $<$ and $>$ are often used in combination with equality to form compound statements. For example, $x \leq a$ means that x is less than or equal to a, and $x \geq a$ indicates that x is greater than or equal to a.

Statements using the symbols $<$, $>$, $\leq$, and $\geq$ are called **inequalities.** At times it is convenient to use inequalities that involve two of these symbols. Such inequalities are called **compound inequalities.** As an example, suppose we wish to indicate that x lies between 2 and 4 on the number line. This can be indicated by $2 < x < 4$. The compound inequality $x > 3$ or $x \leq -1$ indicates that x lies either to the right of 3, or at -1, or to the left of -1.

The use of parentheses and brackets leads to a very compact and useful notation for the following types of sets involving inequalities:

The letters a and b represent real numbers, with $a < b$.

$(a, b) = \{x : a < x < b\}$ $(a, \infty) = \{x : x > a\}$

$[a, b] = \{x : a \leq x \leq b\}$ $[a, \infty) = \{x : x \geq a\}$

$(a, b] = \{x : a < x \leq b\}$ $(-\infty, a] = \{x : x \leq a\}$

$[a, b) = \{x : a \leq x < b\}$ $(-\infty, a) = \{x : x < a\}$

The symbols ∞ and $-\infty$ are read "infinity" and "negative infinity," respectively. They are purely notational and do not represent real numbers. A set of any one of these types is called an **interval,** and the compact notation is referred to as the **interval notation.** The notation $(-\infty, \infty)$ is sometimes used to denote the set $\mathcal{R}$ of all real numbers.

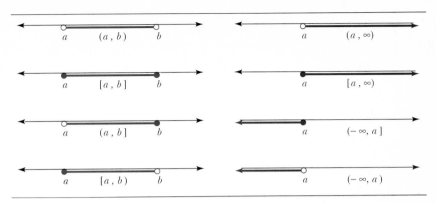

FIGURE 1.3

The **graph** of an interval consists of all points on the number line corresponding to numbers in the interval. The points associated with a and b are called the **endpoints** of the interval. A solid dot at the endpoint indicates that it is included in the graph, whereas an open dot indicates that the endpoint is not included. Figure 1.3 illustrates the graphs of each type of interval.

In describing distances between points on the number line, the concept of absolute value is essential.

Definition 1.1

The **absolute value** of a real number a is denoted by $|a|$. It is defined by

$$|a| = \begin{cases} a & \text{if } a \geq 0, \\ -a & \text{if } a < 0. \end{cases}$$

Note that $|a|$ is never negative. If $a \neq 0$, then $|a|$ is a positive number. It is also worth noting that in the case where $|a| = -a$, the value $-a$ signifies a positive number. That is, when a is a negative number, then $-a$ is a positive number. For example, if $a = -3$, then $-a = -(-3) = 3$, and $|a| = 3$.

Example 1 Express the following without absolute value symbols.

a) $|2x - 4|$, if $x > 2$ b) $|x - 1|$, if $x < 1$

Solution

a) If $x > 2$, then $x - 2 > 0$ and

$$2x - 4 = 2(x - 2) > 2(0) = 0.$$

Thus, since $2x - 4 > 0$, we have

$$|2x - 4| = 2x - 4.$$

b)   If $x < 1$, then $x - 1 < 0$ and

$$|x - 1| = -(x - 1) = 1 - x. \qquad \square$$

The next three examples illustrate the solution of certain types of equations that involve absolute values.

Example 2   Solve $|x - 5| = 4$.

Solution
In order for $|x - 5| = 4$, then

$$x - 5 = \pm 4$$

$$x = 5 \pm 4,$$

$$x = \begin{cases} 5 + 4 = 9, \\ 5 - 4 = 1. \end{cases}$$

Thus the solutions are $x = 9$ and $x = 1$. $\square$

Example 3   Solve $|3x - 7| = -4$.

Solution
Now $|3x - 7|$ is either positive or zero and cannot be negative. Hence the equation has no solution. In other words, the solution set is $\varnothing$. $\square$

Example 4   Solve $|w - 2| = |3w + 1|$.

Solution
The solutions to the given equation occur when the expressions within the absolute value symbols are equal or when they are opposites of each other.

$$w - 2 = 3w + 1 \qquad \text{or} \qquad w - 2 = -(3w + 1)$$
$$-2w = 3 \qquad\qquad\qquad\qquad 4w = 1$$

$$w = -\frac{3}{2} \qquad\qquad\qquad\qquad w = \frac{1}{4}$$

Thus the solutions to $|w - 2| = |3w + 1|$ are $w = -\frac{3}{2}$ and $w = \frac{1}{4}$. $\square$

The geometric interpretation of $|a|$ on the number line is that $|a|$ denotes the distance between 0 and a. When we think of absolute value in terms of distance on the number line, the following theorem seems fairly evident. A proof using the definition of absolute value involves consideration of several cases and is omitted in this book.

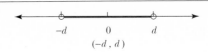

FIGURE 1.4 Absolute value inequality: $|x| < d$

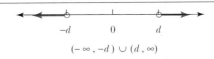

FIGURE 1.5 Absolute value inequality: $|x| > d$

Theorem 1.2

Let x and d be real numbers with $d > 0$. Then

a) $-|x| \le x \le |x|$;

b) $|x| < d$ if and only if $-d < x < d$;

c) $|x| > d$ if and only if either $x > d$ or $x < -d$.

   The graphs of parts (b) and (c) of Theorem 1.2 are given in Figures 1.4 and 1.5.

   Theorem 1.2 proves useful in solving absolute value inequalities. This is demonstrated in the following examples.

Example 5   Solve $|3x - 1| < 4$.

Solution
We must solve the equivalent inequality

$$-4 < 3x - 1 < 4.$$

Adding 1 to all members, we have

$$-3 < 3x < 5.$$

Dividing by 3 then gives

$$-1 < x < \frac{5}{3}.$$

The solution set is $(-1, \frac{5}{3})$, as shown in Figure 1.6.   ❏

FIGURE 1.6

Example 6   Solve $|9 - 2x| > 5$.

Solution
The inequality $|9 - 2x| > 5$ is equivalent to

$$9 - 2x > 5 \quad \text{or} \quad 9 - 2x < -5.$$

Solving these inequalities separately, we get

$$-2x > -4 \qquad -2x < -14$$
$$x < 2, \qquad\quad x > 7.$$

The solution set for the compound inequality is the union of these two solution sets:

$$\{x : x < 2 \quad \text{or} \quad x > 7\} = (-\infty, 2) \cup (7, \infty).$$

The solution set is graphed in Figure 1.7.   ❏

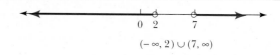

$$(-\infty, 2) \cup (7, \infty)$$

FIGURE 1.7

There are special cases with inequalities involving absolute value. An illustration is given in the next example.

Example 7   Solve $|x - 5| < -1$.

Solution
Since the absolute value of a real number is either zero or a positive number, $|x - 5| < -1$ is impossible for any x, and the solution set is $\varnothing$.   ❏

Exercises 1.1

In Exercises 1–16, use interval notation to write the solution set of the given inequality, and draw the graph of the solution set.

1. $-2 > x \geq -7$
2. $-1 \leq x \leq 3$
3. $-4 < x \leq 0$
4. $-2 < x < 2$
5. $3 \geq x > 1$
6. $2 \geq x \geq 0$
7. $x > 1$ or $x < -1$
8. $x \leq 1$ or $x > 3$
9. $x > -1$ or $x < 1$
10. $x > 2$ or $x \leq 0$
11. $-2 < x \leq 0$ or $x > 1$
12. $3 > x \geq 1$ or $x < 0$
13. $2 \geq x > -1$ or $x < -3$
14. $3 > x > 1$ or $x > 5$
15. $-1 \leq x \leq 2$ or $x < -4$
16. $1 \leq x \leq 3$ or $-3 < x < -1$

Express each of the following without using absolute value symbols. (See Example 1.)

17. $|y - 4|$, if $y > 4$ **18.** $|x - 5|$, if $x < 5$

19. $|y - 4|$, if $y < 4$ **20.** $|x - 5|$, if $x > 5$

21. $|2x - 10|$, if $x < 5$ **22.** $|3x - 12|$, if $x < 4$

23. $|a - b|$, if $a < b$ **24.** $|2a - b|$, if $a > b/2$

25. $|2a - b|$, if $a < b/2$ **26.** $|a - 2b|$, if $a < 2b$

27. $\left|\dfrac{x}{5}\right|$, if $x < 0$ **28.** $\dfrac{|x|}{5}$, if $x > 0$

29. $|-a^2|$, where $a^2 = a \cdot a$

30. $|a^2|$, where $a^2 = a \cdot a$

31. $|-5 - 2a|$, if $a < -3$

32. $|3a - 14|$, if $a < 4$

33. $|3a - 14|$, if $a > 6$ **34.** $|15 - 6a|$, if $a > 3$

Solve the following equations. (See Examples 2–4.)

35. $|x - 3| = 6$ **36.** $|x - 2| = 4$

37. $|2r + 7| = 11$ **38.** $|3t + 4| = 13$

39. $|2z - 7| = 0$ **40.** $|4y - 9| = 0$

41. $|4w + 13| = -3$ **42.** $|3x + 5| = -2$

43. $|3y + 8| = |2 - y|$ **44.** $|8q + 5| = |q + 4|$

45. $\dfrac{|2x - 5|}{|x + 2|} = 1$ **46.** $\dfrac{|4p + 3|}{|p - 5|} = 1$

47. $\dfrac{|9 - 4t|}{|t + 2|} = 0$ **48.** $\dfrac{|3 - 2t|}{|t + 9|} = 0$

49. $|x| + 3x - 9 = 0$ **50.** $|x + 5| = 3x - 2$

51. $|x - 1| = 5 + 3x$ **52.** $|x - 6| = 2x - 3$

53. $|x - 3| = 4 - 3x$ **54.** $|2x - 5| = x - 7$

Solve the following inequalities and graph the solution set on a number line. (See Examples 5–7.)

55. $|x| < 2$ **56.** $|x| < 4$

57. $|x| > 3$ **58.** $|x| > 1$

59. $|x| \le 4$ **60.** $|x| \le 1$

61. $|x - 2| < 1$ **62.** $|x - 1| < 3$

63. $|x - 4| \le 6$ **64.** $|x - 5| \le 2$

65. $|x + 1| < 2$ **66.** $|x + 3| < 1$

67. $|x - 3| \ge 4$ **68.** $|x - 1| \ge 5$

69. $|x + 6| \ge 2$ **70.** $|x + 3| \ge 1$

71. $|2x - 5| < 3$ **72.** $|10x - 3| < 12$

73. $|7 - x| \le 2$ **74.** $|2 - x| \le 5$

75. $|8x + 5| < 25$ **76.** $|6x + 4| < 18$

77. $|2x - 4| \ge 3$ **78.** $|x + 4| \ge 4$

79. $|3 - 4x| > 2$ **80.** $|1 - 3x| > 2$

81. $|5x + 3| > 7$ **82.** $|3x + 2| > 7$

83. $|4 - 5x| < 0$ **84.** $|7x - 2| < -3$

85. $|3x - 6| \le 0$ **86.** $|2x + 7| \le 0$

87. $|1 - x| > 0$ **88.** $|3x + 1| > 0$

89. $|2x + 5| > -5$ **90.** $|4x + 7| \ge -2$

1.2 Polynomials and Rational Expressions

We begin this section with a brief review of the properties of integral exponents. From earlier courses recall examples such as

$$x^2 = x \cdot x,$$

$$3^4 = 3 \cdot 3 \cdot 3 \cdot 3,$$

$$2^0 = 1, \quad \text{and}$$

$$4^{-1} = \frac{1}{4}.$$

These examples illustrate the general definition of integral exponents.

> ### Definition 1.3
>
> If a is a nonzero real number and n is a positive integer, then
>
> $$a^n = \underbrace{a \cdot a \cdot a \cdot \ldots \cdot a}_{n \text{ factors of } a}, \qquad a^0 = 1, \quad \text{and} \quad a^{-n} = \frac{1}{a^n}.$$
>
> An expression of the form a^n is called an **exponential,** a is called the **base,** and n is called the **exponent,** or **power.**

The special case for $a = 0$ and n any integer can be summarized by

$$0^n = \begin{cases} 0, \text{ for } n > 0, \\ \text{undefined, for } n \le 0. \end{cases}$$

The following laws govern some of the operations involving exponentials:

> ### Laws of Exponents
>
> Let a and b be real numbers. The following properties hold for all integers m and n for which the quantities involved are defined.
>
> **1.** $a^m \cdot a^n = a^{m+n}$ Product Rule
>
> **2.** $(a^m)^n = a^{mn}$ Power Rule
>
> **3.** $\dfrac{a^m}{a^n} = a^{m-n}$, for $a \ne 0$ Quotient Rule
>
> **4.** $(ab)^n = a^n b^n$
>
> **5.** $\left(\dfrac{a}{b}\right)^n = \dfrac{a^n}{b^n}$, for $b \ne 0$

To avoid cumbersome instructions, we adhere to the following convention.

To simplify an exponential expression is to rewrite it so that in each term, each variable occurs only once, and all negative and zero exponents have been eliminated.

     Example 1   Use the Laws of Exponents to simplify each of the following:

a) $[(x^{-1}y^2)^3]^{-2}$ b) $\left(\dfrac{(-r^2 t^{-3})^{-1} s^3}{2r^2 s^{-2}}\right)^{-2}$

Solution

a) $[(x^{-1}y^2)^3]^{-2} = (x^{-3}y^6)^{-2} = x^6 y^{-12} = \dfrac{x^6}{y^{12}}$

b) $\left(\dfrac{(-r^2 t^{-3})^{-1} s^3}{2r^2 s^{-2}}\right)^{-2} = \left(\dfrac{-r^{-2} t^3 s^3}{2r^2 s^{-2}}\right)^{-2}$

$\qquad\qquad = \dfrac{(-1)^{-2} r^4 t^{-6} s^{-6}}{2^{-2} r^{-4} s^4}$

$\qquad\qquad = \dfrac{4r^8}{s^{10} t^6}$ ❏

An **algebraic expression** is an expression involving numbers and variables that represents a result obtained by performing the operations of addition, subtraction, multiplication, division, or extraction of roots. In this section we assume that all variables represent real numbers.

A **polynomial** is an algebraic expression in which variables have only nonnegative integral exponents and no variable is in a denominator.

If only one variable is involved in a polynomial, we can give a more precise description.

Definition 1.4

A **polynomial in the variable** x is an algebraic expression of the form

$$a_n x^n + a_{n-1} x^{n-1} + \cdots + a_1 x + a_0,$$

where n is a nonnegative integer, and the coefficients $a_0, a_1, a_2, \ldots,$ a_n are constants. If $n = 0$, the polynomial has only one term, a_0.

One distinguishing feature of a polynomial is the number of terms that it contains. A polynomial that contains only one term is called a **monomial.** One with two terms is called a **binomial,** and one with three terms is a **trinomial.**

Another important feature of a polynomial is its *degree*. The *degree of a nonzero term* in a polynomial is the sum of the exponents on the variables that are involved in that term. The **degree of a nonzero polynomial** is the largest degree that occurs among the terms of the polynomial. The coefficient in the term of highest degree is called the **leading coefficient,** and a **monic polynomial** is one in which the leading coefficient is 1.

Nonzero constants are polynomials of degree 0, and the real number 0 is the only polynomial that has no degree defined.

A polynomial in x is in a **simplest form** if all like terms have been combined and written either in decreasing powers of x or in increasing powers of x.

Example 2   Perform the indicated operations and write the results in simplest form.

a) $(4x^3 - 8x + 7) + (5x^3 + 3x - 6)$
b) $(6x^6 + 9x^2 - 5x) - (11x^2 - 3x - 8)$
c) $(2x^2 - 5x - 3)(4x^2 + 3x)$

Solution

a) $(4x^3 - 8x + 7) + (5x^3 + 3x - 6) = (4 + 5)x^3$
$$+ (-8 + 3)x + (7 - 6)$$
$$= 9x^3 - 5x + 1$$

b) $(6x^6 + 9x^2 - 5x) - (11x^2 - 3x - 8) = 6x^6 + (9 - 11)x^2$
$$+ [-5 - (-3)]x - (-8)$$
$$= 6x^6 - 2x^2 - 2x + 8$$

c) The distributive property can be used to systematically write out all the products involved.

$(2x^2 - 5x - 3)(4x^2 + 3x) = (2x^2 - 5x - 3)(4x^2) + (2x^2 - 5x - 3)(3x)$
$$= 8x^4 - 20x^3 - 12x^2 + 6x^3 - 15x^2 - 9x$$
$$= 8x^4 - 14x^3 - 27x^2 - 9x \qquad \square$$

Certain types of products are important enough to deserve special attention. These are stated in the following list of *special product rules*. Each of them can be verified by direct multiplication.

Special Product Rules

Let x and y represent arbitrary real numbers. Then the following special product rules are valid.

Difference of Two Squares: $(x - y)(x + y) = x^2 - y^2$

Square of a Binomial: $\begin{cases} (x + y)^2 = x^2 + 2xy + y^2 \\ (x - y)^2 = x^2 - 2xy + y^2 \end{cases}$

Cube of a Binomial: $\begin{cases} (x + y)^3 = x^3 + 3x^2y + 3xy^2 + y^3 \\ (x - y)^3 = x^3 - 3x^2y + 3xy^2 - y^3 \end{cases}$

Sum of Two Cubes: $(x + y)(x^2 - xy + y^2) = x^3 + y^3$

Difference of Two Cubes: $(x - y)(x^2 + xy + y^2) = x^3 - y^3$

Example 3   One of the special product rules is illustrated in each of the following products.

a) $(3x - y^2)(3x + y^2) = (3x)^2 - (y^2)^2$ Difference of
                                                                              Two Squares

$$= 9x^2 - y^4$$

b) $(5x - 2y^3)^2 = (5x)^2 - 2(5x)(2y^3) + (2y^3)^2$ Square of a Binomial

$$= 25x^2 - 20xy^3 + 4y^6$$

c) $(p + 2q)^3 = p^3 + 3p^2(2q) + 3p(2q)^2 + (2q)^3$ Cube of a Binomial

$$= p^3 + 6p^2q + 12pq^2 + 8q^3$$

d) $(u - 2v)(u^2 + 2uv + 4v^2) = u^3 - (2v)^3$ Difference of
                                                                                      Two Cubes

$$= u^3 - 8v^3$$ ❏

In a product of polynomials, the polynomials being multiplied are called the *factors* of the product. To factor a given polynomial is to write the polynomial as a product of two or more factors. In this sense, factoring is the reverse process of multiplication.

In this section we consider only polynomials that have integers as coefficients, and we require that all factors be polynomials whose coefficients are integers. This requirement does not allow factorizations such as

$$x^2 - 1 = (2x - 2)(\tfrac{1}{2}x + \tfrac{1}{2})   \text{or}   x - y = (\sqrt{x} - \sqrt{y})(\sqrt{x} + \sqrt{y}).$$

A polynomial with integral coefficients is **completely factored** when it is written as a product of polynomials with integral coefficients, and none of the nonconstant polynomial factors can be written as a product of two polynomials.

Since factoring is the reverse process of multiplication, the special product rules can be regarded as patterns for factoring. With this in mind, we list the following formulas.

Formulas for Factoring

Common Factor:	$ax + ay = a(x + y)$
Difference of Two Squares:	$x^2 - y^2 = (x + y)(x - y)$
Square of a Binomial:	$\begin{cases} x^2 + 2xy + y^2 = (x + y)^2 \\ x^2 - 2xy + y^2 = (x - y)^2 \end{cases}$
Sum of Two Cubes:	$x^3 + y^3 = (x + y)(x^2 - xy + y^2)$
Difference of Two Cubes:	$x^3 - y^3 = (x - y)(x^2 + xy + y^2)$
Quadratic Trinomial:	$x^2 + (a + b)x + ab = (x + a)(x + b)$

The use of these formulas is illustrated in the following example.

Example 4

a) $4x^3 - 100xy^2 = 4x(x^2 - 25y^2)$ Common Factor

$= 4x(x + 5y)(x - 5y)$ Difference of
Two Squares

b) $4x^2 + 12xy + 9y^2 = (2x)^2 + 2(2x)(3y) + (3y)^2$

$= (2x + 3y)^2$ Square of a
Binomial

c) $x^3 + 8 = x^3 + 2^3$

$= (x + 2)(x^2 - 2x + 4)$ Sum of Two Cubes

d) $2a^4b - 2ab^4 = 2ab(a^3 - b^3)$ Common Factor

$= 2ab(a - b)(a^2 + ab + b^2)$ Difference of
Two Cubes

e) $x^2 + x - 2 = (x - 1)(x + 2)$ Quadratic
Trinomial   ❏

In some instances a factorization of the type we are considering may not be possible. The polynomial $x^2 - 2$, for example, cannot be factored as a product of polynomials with integers as coefficients (except trivial factorizations, with one factor 1 or -1). The polynomial $x^2 + y^2$ is another example. A polynomial with integral coefficients that cannot be factored as a product of polynomials, both different from 1 and -1, and both with integral coefficients, is called a **prime polynomial,** or an **irreducible polynomial.**[1]

If a polynomial has more than three terms, it is sometimes necessary to group two or more terms together and factor out a common factor in each group. This method is called **factoring by grouping.**

Example 5   Factor $4hx - 4bh - 5cx + 5bc$.

Solution
The first two terms contain the common factor $4h$, and the last two terms contain the common factor $-5c$. Thus we can write

$$4hx - 4bh = 4h(x - b)$$

$$-5cx + 5bc = -5c(x - b),$$

and we note that $(x - b)$ is common to each group. Hence the polynomial

---

1. In Section 1.7 and Chapter 3 other factorizations of polynomials are considered.

can be factored by grouping as follows:

$$4hx - 4bh - 5cx + 5bc = 4h(x - b) - 5c(x - b)$$

$$= (x - b)(4h - 5c). \qquad \square$$

A **rational expression** is a quotient of polynomials. One of the basic formulas needed in working with rational expressions states that

$$\frac{ad}{bd} = \frac{a}{b}, \quad \text{if } b \neq 0 \text{ and } d \neq 0.$$

This property is sometimes called the **Fundamental Principle of Fractions.** It states that any nonzero factor that occurs in both numerator and denominator may be divided out, or removed. Reading from right to left, it also states that numerator and denominator may be multiplied by the same nonzero quantity.

A rational expression is in **lowest terms** if the only common factors in the numerator and denominator are 1 and -1. The Fundamental Principle of Fractions justifies the procedure used to reduce a rational expression to lowest terms.

Example 6   Reduce $\dfrac{x^2 + 4xy + 3y^2}{x^2 + 2xy - 3y^2}$ to lowest terms.

Solution

$$\frac{x^2 + 4xy + 3y^2}{x^2 + 2xy - 3y^2} = \frac{(x + y)(x + 3y)}{(x - y)(x + 3y)} = \frac{x + y}{x - y} \qquad \square$$

When the reduction in Example 6 is made, all values of x and y for which $x + 3y = 0$ or $x - y = 0$ must be excluded. That is, the equality

$$\frac{x^2 + 4xy + 3y^2}{x^2 + 2xy - 3y^2} = \frac{x + y}{x - y}$$

holds, subject to the conditions that $x \neq -3y$ and $x \neq y$.

To avoid continually having to note restrictions in our work, *in the remainder of this chapter we assume that all necessary restrictions on the variables are made to avoid zero denominators.*

Multiplication and division of rational expressions follow the rules

$$\frac{a}{b} \cdot \frac{c}{d} = \frac{ac}{bd} \quad \text{and} \quad \frac{\dfrac{a}{b}}{\dfrac{c}{d}} = \frac{a}{b} \div \frac{c}{d} = \frac{a}{b} \cdot \frac{d}{c} = \frac{ad}{bc}.$$

When using these rules, we factor all numerators and denominators completely and reduce all results to lowest terms.

The following examples demonstrate the procedures.

Example 7

$$\frac{2x^2 + 5x - 3}{4x^2 - 4x + 1} \cdot \frac{x^2 - 4}{x^2 + x - 6} = \frac{(2x - 1)(x + 3)(x - 2)(x + 2)}{(2x - 1)^2(x + 3)(x - 2)}$$

$$= \frac{x + 2}{2x - 1}$$

Example 8
$$\frac{9y^2 - 4}{y^2 - 9} \div \frac{3y^2 + 2y}{y - 3} = \frac{9y^2 - 4}{y^2 - 9} \cdot \frac{y - 3}{3y^2 + 2y}$$

$$= \frac{(3y - 2)(3y + 2)(y - 3)}{(y + 3)(y - 3)(y)(3y + 2)}$$

$$= \frac{3y - 2}{y(y + 3)}$$

Example 9
$$\frac{\dfrac{4x^2 + 4x + 1}{9x^2 + 3x - 2}}{\dfrac{4x^2 - 1}{6x^2 + x - 2}} = \frac{4x^2 + 4x + 1}{9x^2 + 3x - 2} \div \frac{4x^2 - 1}{6x^2 + x - 2}$$

$$= \frac{4x^2 + 4x + 1}{9x^2 + 3x - 2} \cdot \frac{6x^2 + x - 2}{4x^2 - 1}$$

$$= \frac{(2x + 1)^2(3x + 2)(2x - 1)}{(3x - 1)(3x + 2)(2x - 1)(2x + 1)}$$

$$= \frac{2x + 1}{3x - 1}$$

The quotient of rational expressions in Example 9 is a simple instance of a *complex fraction*. A **complex fraction** is a fraction that has fractions in its numerator or its denominator.

For addition and subtraction of rational expressions that have the same denominator, we use the rules

$$\frac{a}{c} + \frac{b}{c} = \frac{a + b}{c} \quad \text{and} \quad \frac{a}{c} - \frac{b}{c} = \frac{a - b}{c}.$$

When the fractions involved have different denominators, the Fundamental Principle of Fractions is used to convert the fractions to fractions with the same denominator (that is, a *common denominator*). The best choice of common denominator is the *least common denominator*.

> **Procedure for Finding the Least Common Denominator**
>
> 1. Factor each denominator completely (into prime factors).
> 2. Form the product of all the different prime factors, with each factor raised to the highest power with which it appears in any of the given denominators.

This procedure is illustrated in the next example.

Example 10

$$\frac{2}{y-3} - \frac{12}{y^2-9} = \frac{2(y+3)}{(y-3)(y+3)} - \frac{12}{(y-3)(y+3)}$$

$$= \frac{2y+6-12}{(y-3)(y+3)}$$

$$= \frac{2y-6}{(y-3)(y+3)}$$

$$= \frac{2\overset{1}{\cancel{(y-3)}}}{\underset{1}{\cancel{(y-3)}}(y+3)}$$

$$= \frac{2}{y+3} \qquad \qquad \Box$$

The least common denominator can also be used to simplify complex fractions.

Example 11   Express

$$\frac{x^{-2} + y^{-1}}{(xy)^{-2}}$$

as a rational expression in lowest terms.

Solution
When we eliminate the negative exponents, the resulting fraction is a complex fraction:

$$\frac{x^{-2} + y^{-1}}{(xy)^{-2}} = \frac{\dfrac{1}{x^2} + \dfrac{1}{y}}{\dfrac{1}{(xy)^2}}.$$

We could rewrite this complex fraction as a quotient of two fractions and then invert the divisor and multiply. Or, we can simplify the complex fraction by multiplying its numerator and denominator by the least common denominator of all the fractions internal to the complex fraction:

$$\frac{x^{-2} + y^{-1}}{(xy)^{-2}} = \frac{\left(\dfrac{1}{x^2} + \dfrac{1}{y}\right)x^2 y^2}{\left(\dfrac{1}{(xy)^2}\right)x^2 y^2}$$

$$= \frac{y^2 + x^2 y}{1}$$

$$= y(y + x^2). \qquad \square$$

Exercises 1.2

Simplify by using the Laws of Exponents. (See Example 1.)

1. $(4pq^2)^3$

2. $(-2r^2 s^3)^4$

3. $\left(\dfrac{2x}{7}\right)^2$

4. $\left(\dfrac{2r}{3s}\right)^3$

5. $2y^{-3}$

6. $3z^{-2}$

7. $\dfrac{(6xy^4)^0}{2x^2 y^{-3}}$

8. $\dfrac{(2y^2 z)^3}{(16y^3)^0}$

9. $(4xy^2)^{-3}(x^{-2}y^3)^2$

10. $(3a^{-2}b^3)^{-1}(a^3 b)^{-2}$

11. $\dfrac{3p^{-3}q^{-1}}{8^{-1}pq^3}$

12. $\dfrac{5m^{-2}n^3}{4^{-2}m^4 n^{-2}}$

13. $\left(\dfrac{(-x^2)^3 z}{(2x^4 y)^2}\right)^3$

14. $\left(\dfrac{(a^2 b)^3}{(-ab^3)^2}\right)^2$

15. $\left(\dfrac{(ab)^{-1}c^2}{(ac^{-2})^{-1}b^2}\right)^{-2}$

16. $\left(\dfrac{-2x^{-3}z}{(xy)^{-2}z^{-4}}\right)^{-3}$

In Exercises 17–20, m and n denote positive integers. Simplify by using the Laws of Exponents.

17. $\left(\dfrac{3x^m}{y^n}\right)^2$

18. $\left(\dfrac{a^m b^n}{c^2}\right)^2$

19. $\dfrac{x^{2n-2}}{x^{n-1}}$

20. $\dfrac{x^{n+4}}{x^{2n-1}} \cdot \dfrac{x^n}{x^2}$

Scientific and technical work often involves very large numbers or very small numbers. In these cases it is help- ful to write the numbers in scientific notation. In **scientific notation** a positive number N is written in the form

$$N = a \times 10^n,$$

where $1 \leq a < 10$ and n is an integer.

Write the following numbers in scientific notation.

21. 1020

22. 40,300

23. 41,610,000

24. 940,200,000

25. 0.0035

26. 0.00073

Write the following numbers in decimal notation.

27. 8.22×10^3

28. 1.06×10^4

29. 6.6×10^{-1}

30. 7.9×10^{-2}

31. 2.87×10^{-5}

32. 5.1×10^{-4}

Perform the indicated operations and write the results in simplest form. (See Examples 2 and 3.)

33. $(8r^4 - 3r^5 + 4) - (r^2 + 2r^4 - r^5)$

34. $(7p^5 - 4p^3 + p) - (5p^4 - 2p^5 + 4p)$

35. $(-3 + 8z + 4z^3) - (-5z^2 + 3z^3 - 12)$

36. $(2y^4 - 5y^2 - 7y^3 + 8) - (1 + 7y - 4y^2 - 6y^4)$

37. $(5q + 3)(8q - 1)$

38. $(11r + 4)(r - 3)$

39. $(5a - b)(2a + 3b)$

40. $(8m + 3n)(2m - 5n)$

41. $(x - 4)(2x^3 + 3x^2 - 1)$

42. $(x^2 - 3)(x^3 - x + 5)$

43. $(2x^2 + 3x - 1)(x^2 - 2x + 4)$

44. $(9x^2 - 2x - 5)(3x^2 - x - 2)$

45. $(2r + s)(2r - s)$ **46.** $(4p + 3q)(4p - 3q)$

47. $(a + 3bc^2)^2$ **48.** $(m^2 - 2np)^2$

49. $(2x^3 - y)(2x^3 + y)$ **50.** $(p - 3q^2)(p + 3q^2)$

51. $(a - 2b)(a^2 + 2ab + 4b^2)$

52. $(2x + 1)(4x^2 - 2x + 1)$

53. $(x - 2y)^3$ **54.** $(2m + n)^3$

Factor each expression completely. (See Examples 4 and 5.)

55. $ax^2 - 9ay^4$ **56.** $x^4 - 81y^4$

57. $(4x - 3y)^2 - 25$ **58.** $(2x + y)^2 - 9z^2$

59. $4a^4 + 12a^2b^2 + 9b^4$ **60.** $4x^4 + 28x^2y^2 + 49y^4$

61. $3x^2 + 7ax - 6a^2$ **62.** $45a^2 - 8ay - 4y^2$

63. $8x^3 + 27$ **64.** $64x^3 + 27y^3$

65. $ax^2 + bx^2 + ad^2 + bd^2$

66. $2ax^2 - bx^2 + 6a - 3b$

67. $25x^2 + 10x + 1 - y^2$

68. $16x^2 + 8x + 1 - 4y^2$

Perform the indicated operations and reduce the results to lowest terms. (See Examples 6–10.)

69. $\dfrac{z^2 - 5z - 14}{z^2 - 4} \cdot \dfrac{z^2 - 4z + 4}{6 - 3z}$

70. $\dfrac{3v^2 - 4v - 4}{3v^2 + 4v - 4} \cdot \dfrac{3v^2 - 11v + 6}{3v^2 - 7v - 6}$

71. $\dfrac{w + 2}{w^2 + 7w + 12} \div \dfrac{w^2 + 4w + 4}{w^2 + 5w + 6}$

72. $\dfrac{t^2 + 5t - 14}{t^2 + 10t + 21} \div \dfrac{t^2 + 5t + 6}{t^2 + 7t + 12}$

73. $\dfrac{\dfrac{3w^2 + w - 2}{4w^2 - w - 5}}{\dfrac{3w^2 - 11w + 6}{4w^2 - 7w - 15}}$ **74.** $\dfrac{\dfrac{2p^2 - 7p + 5}{3p^2 + p - 4}}{\dfrac{2p^2 - p - 10}{3p^2 + 2p - 8}}$

75. $\dfrac{9x^2 - 4y^2}{2x + 2y} \cdot \dfrac{x^3 + y^3}{3x + 2y}$

76. $\dfrac{16 - 9z^2}{z^3 + 8} \cdot \dfrac{z^2 - 2z + 4}{8 + 10z + 3z^2}$

77. $\dfrac{\dfrac{25 - 4x^2}{x^3 - 1}}{\dfrac{15 - x - 2x^2}{x^3 + x^2 + x}}$ **78.** $\dfrac{\dfrac{a^3 - b^3}{3a + 4b}}{\dfrac{3a - 3b}{9a^2 - 16b^2}}$

79. $\dfrac{w^3 - 4w^2 + 9w - 36}{w^4 - 81} \cdot \dfrac{2aw^2 - aw - 15a}{w^3 - 64}$

80. $\dfrac{2x + y}{3abx + 3by} \cdot \dfrac{a^3x^3 + y^3}{ax + 3ay} \cdot \dfrac{x^2 + 6xy + 9y^2}{4x^2 - y^2}$

81. $\dfrac{7}{3y^2 + 11y + 6} - \dfrac{11}{6y^2 + 19y + 10}$

82. $\dfrac{t - 12}{2t^2 - 13t + 15} - \dfrac{t - 15}{2t^2 - 15t + 18}$

83. $1 + \dfrac{u}{u + 1} - \dfrac{2u^2}{u^2 - 1}$

84. $t - \dfrac{4}{2 - t} - \dfrac{t^3}{t^2 - 4}$

85. $(25c^2 - 16d^2) \div \left(1 + \dfrac{5c}{4d}\right)$

86. $(16a^2 - 9b^2) \div \left(1 - \dfrac{3b}{4a}\right)$

87. $\left(\dfrac{1}{y + 4} + \dfrac{2}{y - 4}\right)\left(\dfrac{y - 4}{y + 12}\right)$

88. $\left(\dfrac{2}{x - 3} - \dfrac{1}{x + 3}\right)\left(\dfrac{x + 3}{x + 9}\right)$

89. $\left(x - 3 - \dfrac{4}{x}\right) \div \left(1 - \dfrac{2}{x} - \dfrac{8}{x^2}\right)$

90. $\left(\dfrac{2}{p} + \dfrac{9}{p^2} + \dfrac{9}{p^3}\right) \div \left(2 - \dfrac{1}{p} - \dfrac{6}{p^2}\right)$

91. $\dfrac{2x + 2y}{x^2 + 2xy - 3y^2} - \dfrac{x - 2y}{x^2 + xy - 6y^2}$

92. $\dfrac{2x + y}{4x^2 - 4xy - 3y^2} + \dfrac{x - 2y}{2x^2 - 5xy + 3y^2}$

In Exercises 93–100, express the given fraction as a rational expression in lowest terms. (See Examples 9 and 11.)

93. $\dfrac{\dfrac{1}{a} - \dfrac{1}{b}}{\dfrac{1}{a} + \dfrac{1}{b}}$ **94.** $\dfrac{\dfrac{1}{a} + \dfrac{1}{b}}{\dfrac{1}{a^2} - \dfrac{1}{b^2}}$

95. $\dfrac{\dfrac{1}{x-1} + \dfrac{1}{x+1}}{\dfrac{1}{x-1} - \dfrac{1}{x+1}}$

96. $\dfrac{\dfrac{1}{x} - \dfrac{1}{x-2}}{\dfrac{1}{x} + \dfrac{1}{x-2}}$

99. $\dfrac{x^{-1} - y^{-1}}{x^{-2} - y^{-2}}$

100. $\dfrac{9x - x^{-1}}{3 + x^{-1}}$

97. $\dfrac{x+2 - \dfrac{6}{x+3}}{x+4 - \dfrac{2}{x+3}}$

98. $\dfrac{x-1 + \dfrac{1}{x-3}}{x+2 + \dfrac{4}{x-3}}$

1.3 | Radicals and Rational Exponents

If a is a *positive* real number and b is a number such that $b^2 = a$, then b is called a *square root* of a. Since $(3)^2 = 9$ and $(-3)^2 = 9$, both 3 and -3 are square roots of 9. The positive square root, 3, is called the *principal square root* of 9, and the symbol $\sqrt{9}$ is used to denote it. We write

$$\sqrt{9} = 3 \quad \text{and} \quad -\sqrt{9} = -3.$$

A *negative* real number a does not have a square root in the set of real numbers, because the square of a real number is never negative. If $a = 0$, its only square root is 0. For any *nonnegative* real number a, the **principal square root** of a is defined to be the *nonnegative* number $\sqrt{a}$ such that

$$(\sqrt{a})^2 = a.$$

The symbol $\sqrt{}$ is called the **radical sign.**

Other roots of numbers are defined in a similar manner. A **cube root** of a is a number b such that $b^3 = a$, and a **fourth root** of a is a number b such that $b^4 = a$. In general, if n is a positive integer, then b is an ***n*th root** of a if $b^n = a$. An nth root is sometimes referred to as a root of **order n.**

With cube roots, the situation is simpler than it is with square roots. Every real number a has exactly one real cube root. This cube root of a is denoted by $\sqrt[3]{a}$ and is called the **principal cube root** of a. For example,

$$\sqrt[3]{64} = 4 \quad \text{since} \quad (4)^3 = 64;$$
$$\sqrt[3]{-27} = -3 \quad \text{since} \quad (-3)^3 = -27.$$

In fact, for any odd n, every real number has exactly one real nth root called the **principal nth root.**

The situation for the 4th roots, 6th roots, and other roots of even order is much the same as for square roots. For n even, every positive number has two real nth roots, with the positive root designated as the principal nth root. But a real nth root (n even) of a negative number does not exist in the real numbers.

This discussion leads us to the following definition.

Definition 1.5

Let n be a positive integer. The **principal nth root** of a real number a is the real number $\sqrt[n]{a}$ defined by the following statements.

1. If n is even and $a \geq 0$, then $\sqrt[n]{a}$ is the *nonnegative number* such that
$$(\sqrt[n]{a})^n = a.$$

2. If n is odd, then $\sqrt[n]{a}$ is the real number such that
$$(\sqrt[n]{a})^n = a.$$

The symbol $\sqrt[n]{a}$ is called a **radical,** a is called the **radicand,** and n is called the **index,** or **order,** of the radical.

Note that $\sqrt[n]{a}$ is left undefined for the present when n is even and a is negative. This point comes up again in the next section, in the study of complex numbers.

Since a^2 is always nonnegative, then $\sqrt{a^2}$ is always defined and designates the nonnegative number that yields a^2 when it is squared. Since a may be negative, $\sqrt{a^2}$ is not always the same as a. For example, if $a = -1$, then
$$\sqrt{a^2} = \sqrt{(-1)^2} = \sqrt{1} = 1 \neq a,$$

whereas

$$\sqrt{a^2} = \sqrt{(-1)^2} = 1 = |-1| = |a|,$$

This property of radicals can be extended for any even positive integer n and any real a: $\sqrt[n]{a^n} = |a|$. Whenever $a \geq 0$, this reduces to $\sqrt[n]{a^n} = a$.

Example 1   Simplify each radical.

a) $\sqrt[5]{7^5}$ b) $\sqrt[4]{x^4}$ c) $\sqrt[3]{(-y)^3}$ d) $\sqrt{x^8}$

Solution

a) $\sqrt[5]{7^5} = 7$

b) $\sqrt[4]{x^4} = |x|$

c) $\sqrt[3]{(-y)^3} = -y$

d) $\sqrt{x^8} = \sqrt{(x^4)^2} = x^4$, since $x^4 \geq 0$ for any x.   ❑

The following list contains properties of radicals that are useful in changing the form of expressions involving radicals.

Basic Properties of Radicals

Let n and m be positive integers, and let a and b be real numbers. Whenever each indicated root is defined, then:

1. $(\sqrt[n]{a})^n = a$;

2. $\sqrt[n]{a^n} = \begin{cases} a & \text{if } n \text{ is odd} \\ |a| & \text{if } n \text{ is even} \end{cases}$ (if $a \geq 0$, $\sqrt[n]{a^n} = a$);

3. $\sqrt[n]{ab} = \sqrt[n]{a}\ \sqrt[n]{b}$;

4. $\sqrt[n]{\dfrac{a}{b}} = \dfrac{\sqrt[n]{a}}{\sqrt[n]{b}}$, $b \neq 0$;

5. $\sqrt[m]{\sqrt[n]{a}} = \sqrt[mn]{a}$.

Example 2   Find the value of each given expression or simplify it by using the Basic Properties of Radicals. Variables represent nonzero real numbers.

a) $(\sqrt[4]{5})^4$ b) $\sqrt[3]{2x}\ \sqrt[3]{4x^2}$ c) $\sqrt[5]{\dfrac{-32a^5}{b^{10}}}$ d) $\sqrt[4]{16a^4}$ e) $\sqrt[3]{\sqrt{7}}$

Solution

a) $(\sqrt[4]{5})^4 = 5$

b) $\sqrt[3]{2x}\ \sqrt[3]{4x^2} = \sqrt[3]{8x^3} = \sqrt[3]{2^3}\ \sqrt[3]{x^3} = 2x$

c) $\sqrt[5]{\dfrac{-32a^5}{b^{10}}} = \dfrac{\sqrt[5]{-32a^5}}{\sqrt[5]{b^{10}}} = \dfrac{\sqrt[5]{(-2)^5}\ \sqrt[5]{a^5}}{\sqrt[5]{(b^2)^5}} = \dfrac{-2a}{b^2}$

d) $\sqrt[4]{16a^4} = \sqrt[4]{(2)^4}\ \sqrt[4]{a^4} = 2|a|$

e) $\sqrt[3]{\sqrt{7}} = \sqrt[6]{7}$ since $3 \cdot 2 = 6$ ❑

In the remainder of this section, *unless otherwise indicated, we make the assumption that all variables represent positive numbers*. This ensures that all indicated roots are defined and that property 2 occurs only in the form $\sqrt[n]{a^n} = a$. Also, we use the phrase "simplify the radical" to mean "write the expression in simplest radical form." An expression involving radicals is in simplest radical form if the following conditions are satisfied.

Simplest Radical Form

---

1. The expression contains no zero or negative exponents.
2. The radicand contains no factor to a power greater than or equal to the index of the radical.
3. The denominator contains no radicals, and the radicand contains no fractions.
4. The index is as small as possible.

The next example illustrates a process called **rationalizing the denominator.** This process is used to remove radicals from the denominator of an expression. This is accomplished by multiplying the numerator and denominator of a fraction by a factor chosen to force factors of the form $\sqrt[n]{a^n}$ into the denominator.

Example 3   Write each of the following in simplest radical form.

a) $\sqrt{\dfrac{x}{y}}$ b) $\dfrac{1}{\sqrt[3]{x^2 y}}$ c) $\dfrac{1}{1 + \sqrt{x}}$

Solution

a) $\sqrt{\dfrac{x}{y}} = \sqrt{\dfrac{x \cdot y}{y \cdot y}} = \sqrt{\dfrac{xy}{y^2}} = \dfrac{\sqrt{xy}}{\sqrt{y^2}} = \dfrac{\sqrt{xy}}{y}$

b) $\dfrac{1}{\sqrt[3]{x^2 y}} = \dfrac{1 \cdot \sqrt[3]{xy^2}}{\sqrt[3]{x^2 y} \cdot \sqrt[3]{xy^2}} = \dfrac{\sqrt[3]{xy^2}}{\sqrt[3]{x^3 y^3}} = \dfrac{\sqrt[3]{xy^2}}{xy}$

c) The radical $\sqrt{x}$ in the denominator $1 + \sqrt{x}$ will be eliminated by multiplying numerator and denominator by $1 - \sqrt{x}$ since

$$(1 + \sqrt{x})(1 - \sqrt{x}) = 1 - (\sqrt{x})^2 = 1 - x.$$

Thus we have

$$\frac{1}{1 + \sqrt{x}} = \frac{1 \cdot (1 - \sqrt{x})}{(1 + \sqrt{x})(1 - \sqrt{x})} = \frac{1 - \sqrt{x}}{1 - x}.$$ ❏

The expression $1 - \sqrt{x}$ is called the *conjugate* of $1 + \sqrt{x}$. In general, the two binomials $a + b$ and $a - b$ are **conjugates** of each other. In Section 1.6 we shall use conjugates to rationalize numerators of rational expressions.

The concept of simplest radical form is not as important today as it was before the widespread use of calculators. For example, the form $1/\sqrt{x}$ is easier to use than $\sqrt{x}/x$ in computational work with a calculator.

Sums involving radicals can be combined by using the distributive property if the radical is a common factor of each term in the sum.

Example 4   Combine $\sqrt{98x^3y} - \sqrt{50xy^5}$ by using the distributive property.

Solution

We must first simplify each radical, and then use the distributive property if there is a factor common to both terms:

$$\sqrt{98x^3y} - \sqrt{50xy^5} = \sqrt{49x^2 \cdot 2xy} - \sqrt{25y^4 \cdot 2xy}$$
$$= 7x\sqrt{2xy} - 5y^2\sqrt{2xy}$$
$$= (7x - 5y^2)\sqrt{2xy}. \qquad \Box$$

In Section 1.2 we defined a^n for any integer n. In this section we consider rational exponents, and in Chapter 4 we complete the story by studying a^n for any real number n.

At all stages of this development, the Laws of Exponents hold. We first consider expressions of the form $a^{1/n}$, for n a positive integer. The Power Rule leads us to write

$$(a^{1/n})^n = a^{n/n} = a^1,$$

and $a^{1/n}$ is an nth root of a. Hence we make the following definition, which ties together the notions of rational exponents and radicals.

Definition 1.6

If a is a real number and n is a positive integer such that $\sqrt[n]{a}$ exists, then

$$a^{1/n} = \sqrt[n]{a}.$$

This definition allows us to evaluate $16^{1/2}$ by using radicals:

$$16^{1/2} = \sqrt{16} = 4.$$

We next consider exponents of the form m/n. Again using the Power Rule, we write $a^{m/n}$ in two ways:

$$a^{m/n} = \begin{cases} (a^m)^{1/n} = \sqrt[n]{a^m} \\ (a^{1/n})^m = (\sqrt[n]{a})^m, \end{cases}$$

whenever the indicated roots are defined. Note that $a^{m/n}$ is undefined when n is even and $a < 0$, or when $m \le 0$ and $a = 0$. The form $a^{m/n}$ is called the **exponential form** of either of the **radical forms** $\sqrt[n]{a^m}$ or $(\sqrt[n]{a})^m$.

The radical forms can be used for evaluation purposes, although the exponential form lends itself more appropriately to the use of the calculator. For example, we write

$$32^{-3/5} = \frac{1}{32^{3/5}} = \frac{1}{(\sqrt[5]{32})^3} = \frac{1}{2^3} = \frac{1}{8}.$$

Alternatively we "punch" the calculator and use the $\boxed{y^x}$ key to find $32^{-3/5} = 0.125$.

Quite often, the exponential form is handier to use than the equivalent radical form. We illustrate this type of conversion in the next example.

Example 5

a) $\sqrt[3]{a^2} = a^{2/3}$

b) $\sqrt[4]{y^3} = y^{3/4}$

c) $\sqrt{x^2 + y^2} = (x^2 + y^2)^{1/2}$

We caution the reader against a common mistake by noting

$$(x^2 + y^2)^{1/2} \neq (x^2)^{1/2} + (y^2)^{1/2} = x + y. \qquad \square$$

It can be shown that the Laws of Exponents hold for rational exponents as long as the radicals involved are defined. Although we do not prove them here, we will illustrate their use in the following example.

Example 6   Use the Laws of Exponents to simplify.

a) $\dfrac{x^{1/2}x^{3/4}}{x^{1/3}}$ b) $(a^{1/4}b^{1/5})^{20}$ c) $(r^{-1/3}s^{2/3})^{-3/2}$

Solution

a) $\dfrac{x^{1/2}x^{3/4}}{x^{1/3}} = x^{1/2 + 3/4 - 1/3} = x^{(6 + 9 - 4)/12} = x^{11/12}$

b) $(a^{1/4}b^{1/5})^{20} = (a^{1/4})^{20}(b^{1/5})^{20} = a^{20/4}b^{20/5} = a^5 b^4$

c) $(r^{-1/3}s^{2/3})^{-3/2} = (r^{-1/3})^{-3/2}(s^{2/3})^{-3/2} = r^{1/2}s^{-1} = \dfrac{r^{1/2}}{s} \qquad \square$

Rational exponents can be used to simplify radicals—in particular, to reduce the index of a radical. Recall that one of the conditions necessary for a radical to be in simplest form is that the index be as small as possible.

Example 7   Rewrite the radical

$$\sqrt[10]{x^5 y^{15} z^5}$$

as a radical with index 2 and simplify.

Solution

$$\sqrt[10]{x^5 y^{15} z^5} = (x^5 y^{15} z^5)^{1/10} = x^{5/10} y^{15/10} z^{5/10}$$

$$= x^{1/2} y^{3/2} z^{1/2} = (xy^3 z)^{1/2} = \sqrt{xy^3 z}$$

$$= y\sqrt{xyz} \qquad \square$$

Exercises 1.3

(Throughout this exercise set, unless otherwise stated, *all variables represent positive real numbers.*)

Simplify by using the Basic Properties of Radicals. (See Examples 1 and 2.)

1. $\sqrt{16x^2 y^4}$ **2.** $\sqrt{z^8 y^{10}}$

3. $\sqrt[3]{\sqrt[4]{y}}$ **4.** $\sqrt[3]{\sqrt[3]{z}}$

5. $\sqrt{xy}\,\sqrt{xy^3}$ **6.** $\sqrt[3]{ab^4}\,\sqrt[3]{a^2 b^2}$

7. $\sqrt[3]{2c^4 d^5}\,\sqrt[3]{4c^{-1}d}$ **8.** $\sqrt{r^{-3}s^9}\,\sqrt{r^5 s^7}$

9. $\sqrt{\dfrac{x^8 y^{10}}{x^2 y^4}}$ **10.** $\sqrt{\dfrac{98a^9 b^7}{2ab}}$

11. $\sqrt[3]{\dfrac{-p^{10}q^2 r}{p^4 q^{-1} r^{-2}}}$ **12.** $\sqrt[3]{\dfrac{-56a^{17}b^{13}}{7a^{23}b^{16}}}$

In Exercises 13–20, assume that variables represent arbitrary real numbers. Use absolute values as needed to simplify the radicals. (See Examples 1 and 2.)

13. $\sqrt{25a^2}$ **14.** $\sqrt[4]{x^4 y^4}$

15. $-\sqrt[4]{(a-b)^4}$ **16.** $-\sqrt{(cb)^2}$

17. $\sqrt{x^2 - 2x + 1}$ **18.** $\sqrt{a^4 + 4a^2 + 4}$

19. $\sqrt{a^{10}b^8}$ **20.** $\sqrt[4]{a^{12}b^8}$

Write in simplest radical form. All variables represent positive real numbers. (See Example 3.)

21. $\sqrt{250x^2 y^5}$ **22.** $\sqrt[3]{250x^2 y^5}$

23. $\sqrt[3]{(r-s)^5}$ **24.** $\sqrt{(x+y)^3}$

25. $\sqrt[4]{2a^{-8}}$ **26.** $\sqrt[3]{4a^{-6}}$

27. $\sqrt{a^9 b^{-2} c^0}$ **28.** $\sqrt{a^7 b^0 c^{-4}}$

29. $\sqrt{\dfrac{48(x+y)^2}{(x+y)^{-3}}}$ **30.** $\sqrt{\dfrac{50(a+b)^{-4}}{(a+b)^2}}$

31. $\sqrt{\dfrac{x^2}{2}}$ **32.** $\sqrt[3]{\dfrac{x^3}{9}}$

33. $\sqrt{\dfrac{3a}{2b^3}}$ **34.** $\sqrt{\dfrac{x^5}{z^5}}$

35. $\sqrt[3]{\dfrac{1}{5a^5}}$ **36.** $\sqrt[3]{\dfrac{xy}{2z^2}}$

37. $\sqrt[5]{\dfrac{3a^2 b^6}{c^3 d^5}}$ **38.** $\sqrt[5]{\dfrac{ax^6}{27y^7}}$

39. $\dfrac{1}{\sqrt{x} - \sqrt{y}}$ **40.** $\dfrac{\sqrt{x}}{\sqrt{x} + 2}$

41. $\dfrac{1}{\sqrt{2} - \sqrt{3}}$ **42.** $\dfrac{3}{\sqrt{2} + \sqrt{7}}$

43. $\sqrt{\dfrac{x}{2} - \dfrac{2}{x}}$ **44.** $\sqrt{\dfrac{3}{x} - \dfrac{x}{2}}$

45. $\dfrac{1}{1 + \sqrt{1 + x}}$ **46.** $\dfrac{1}{1 - \sqrt{1 - x}}$

Use the distributive property to perform the indicated operations. Write the results in simplest radical form. (See Examples 3 and 4.)

47. $(2 - \sqrt{3})(1 + 2\sqrt{3})$ **48.** $(\sqrt{7} - 2)(\sqrt{7} + 3)$

49. $(\sqrt{2} - \sqrt{3})(2\sqrt{2} + \sqrt{3})$

50. $(\sqrt{5} + \sqrt{2})(\sqrt{5} + 3\sqrt{2})$

51. $(\sqrt{x} - 1)^2$ **52.** $(\sqrt{x} + 2)(\sqrt{x} - 1)$

53. $(\sqrt{x} - \sqrt{y})(\sqrt{x} + 2\sqrt{y})$

54. $(\sqrt{x} - \sqrt{y})^2$

55. $2\sqrt{75} - 3\sqrt{48} + 6\sqrt{12}$

56. $4\sqrt{98} + 3\sqrt{72} - 10\sqrt{8}$

57. $\sqrt{8ab^2} - \sqrt{32ab^2} - b\sqrt{50a}$

58. $2\sqrt{48a^3} - 4\sqrt{108a}$

59. $3\sqrt[3]{81a^4 b^5} + 2\sqrt[3]{375ab^8}$

60. $6\sqrt[3]{-16y^8} - 2\sqrt[3]{-54x^6 y^5}$

Use the Laws of Exponents to simplify. (See Example 6.)

61. $x^{2/3}x^{7/3}$

62. $r^{1/5}r^{4/5}$

63. $(x^{1/3}y^{5/6})^{12}$

64. $(a^{4/9}b^{2/3})^{18}$

65. $(a^{-1/2}b^{3/2})^{-4/3}$

66. $(t^{20/9}r^{-10/3})^{-9/10}$

67. $[(x^2y^{-2/3})^{-3}]^0$

68. $[2c^9d^{5/4})^0]^{-3}$

69. $[(4x^{1/3})^{-2}y]^3$

70. $[4x^2(2y^{-1/4})^2]^{-1/2}$

71. $x^{1/3}(x^{2/3} - x^{-1/3})$

72. $y^{2/5}(y^{3/5} + y^{-2/5})$

73. $\dfrac{(16x^{4/3})(4x^{1/3})}{32x^{3/2}}$

74. $\dfrac{(2x^{3/2})(9x^{3/2})}{36x^{5/2}}$

75. $\dfrac{(9x)^{1/2}(27y)^{2/3}}{(8xy)^{1/3}}$

76. $\dfrac{(16x^3)^{1/4}(-27y^2)^{2/3}}{(81xy^3)^{1/4}}$

77. $\left(\dfrac{x^{3/2}y^{-5/4}}{x^{-5/2}y^{3/4}}\right)^{1/2}$

78. $\left(\dfrac{a^{-2/3}b^{-4/5}}{a^{-5/3}b^{-7/5}}\right)^{1/3}$

79. $\dfrac{(x^{3/4}y^{3/2}z)^{1/3}}{(x^0y^{-1/2}z^{1/2})^{-1/2}}$

80. $\dfrac{(r^{1/5}s^{2/5}t^{-3/5})^{-1}}{(rs^{-2})^{1/5}(r^3s)^{-2/5}}$

Rewrite in simplest radical form with the index as small as possible. (See Example 7.)

81. $\sqrt[6]{x^4}$

82. $\sqrt[8]{a^6}$

83. $\sqrt[6]{25x^4y^8}$

84. $\sqrt[6]{27x^9y^3}$

85. $\sqrt{\sqrt[3]{x^2}}$

86. $\sqrt{\sqrt[4]{x^6}}$

87. $\sqrt{x\sqrt[3]{x}}$

88. $\sqrt[4]{y\sqrt{y^3}}$

89. $\sqrt{x}\,\sqrt[3]{x^2}$

90. $\sqrt[4]{y}\,\sqrt[5]{y}$

91. $\dfrac{\sqrt[3]{a^3b^5}}{\sqrt{a^2b^3}}$

92. $\dfrac{\sqrt[6]{a^5b^8}}{\sqrt[3]{ab^4}}$

Express the results of the following in scientific notation rounded to four digits.

93. $(13.91)^{1.8}$

94. $(127.1)^{0.9}$

95. $(1.421)^{-0.15}$

96. $(0.1815)^{-2.8}$

97. $(2724)^{0.35}(4.150)^{-3.2}$

98. $(5.820)^{2.2}(81.13)^{-1.5}$

99. $\dfrac{(0.01723)^{-1.4}}{(99.21)^{0.051}}$

100. $\dfrac{(0.4189)^{-0.78}}{(81.92)^{-1.2}}$

1.4 Complex Numbers

In Definition 1.5, $\sqrt[n]{a}$ is not defined when a is negative and n is even. This is a consequence of a fundamental deficiency in the set of real numbers: *A negative real number does not have a root of even order in the set of real numbers*, since the square of a real number is never negative.

This situation is very unsatisfactory to a mathematician. This inadequacy of the real numbers is the primary reason for the construction of the system of complex numbers, which begins with the introduction of a number i such that $i^2 = -1$. The formal definition is as follows.

Definition 1.7

The number i is, by definition, a number such that

$$i^2 = -1.$$

That is, $i = \sqrt{-1}$. A **complex number** is a number of the form

$$a + bi,$$

where a and b are real numbers. The real number a is called the **real part** of the complex number, and the real number b is called the **imaginary part** of the complex number. The set of all complex numbers is denoted by $\mathscr{C}$.

Example 1   Examples of complex numbers are

$$3 + 7i, \quad \pi + 4i, \quad 1 + (-2)i, \quad -5 + (-\tfrac{3}{7})i, \quad 0 + 3i, \quad 4 + 0i. \quad \square$$

If b is negative, as it is in $1 + (-2)i$, we drop the $+$ sign and simply write $1 - 2i$. In this set $\mathscr{C}$, we identify each real number a as being the same as $a + 0i$. For example, we write 4 for $4 + 0i$. Numbers of the form $0 + bi$ are called **imaginary numbers** and are written as bi instead of $0 + bi$. The list of complex numbers in Example 1 can be written as

$$3 + 7i, \quad \pi + 4i, \quad 1 - 2i, \quad -5 - \tfrac{3}{7}i, \quad 3i, \quad 4.$$

Two complex numbers are said to be **equal** if they have equal real parts and equal imaginary parts. That is, for two complex numbers $a + bi$ and $c + di$,

$$a + bi = c + di \quad \text{if and only if} \quad a = c \text{ and } b = d.$$

The operations of addition, subtraction, and multiplication are performed on complex numbers in the same way that they are performed on polynomials, except that results of multiplication are simplified by using $i^2 = -1$. As a consequence of performing the operations in this way, the same basic properties of these operations that hold for real numbers also hold for complex numbers.

Any sum, difference, or product of complex numbers can be expressed in the form $a + bi$, with a and b real numbers. This form is called the **standard form** for complex numbers. These ideas are illustrated in the next example.

Example 2

a) $(2 - 3i) + (7 + i) = (2 + 7) + (-3 + 1)i$

$$= 9 - 2i$$

b) $(9 - 5i) - (6 + 4i) = (9 - 6) + (-5 - 4)i$

$$= 3 - 9i$$

c) $(3 + 7i)(2 - 5i) = 3(2 - 5i) + 7i(2 - 5i)$

$$= (3)(2) + (3)(-5i) + (7i)(2) + (7i)(-5i)$$

$$= 6 - 15i + 14i - 35i^2$$

$$= 6 - i - 35(-1)$$

$$= 41 - i$$

d) $(5 + 3i)(5 - 3i) = 25 - 15i + 15i - 9i^2$

$$= 25 - 9(-1)$$

$$= 34 \quad\quad\quad\quad \square$$

In part (d) of Example 2, the product had the form $(a + bi)(a - bi)$ and turned out to be a positive real number. The following multiplication shows this was no accident.

$$(a + bi)(a - bi) = a^2 - abi + abi - b^2i^2$$
$$= a^2 - b^2(-1)$$
$$= a^2 + b^2$$

If $a + bi \neq 0$, then $a^2 + b^2$ is a positive real number. In connection with this result, we make the following definition.

Definition 1.8

For any complex number $z = a + bi$, the **conjugate** of z is the complex number

$$\bar{z} = a - bi.$$

Example 3   Some examples of conjugates follow.

a) If $z = 3 + 4i$, then $\bar{z} = 3 - 4i$. That is,

$$\overline{3 + 4i} = 3 - 4i.$$

b) The conjugate of $2 - 3i$ is $2 + 3i$. That is,

$$\overline{2 - 3i} = 2 + 3i.$$

c) $\bar{4} = 4$ ❑

The next example illustrates how a quotient of two complex numbers can be found by multiplying the numerator and denominator by the conjugate of the denominator.

Example 4   Perform the following divisions and express each result in standard form.

a) $\dfrac{6 - 2i}{3 + i}$

b) $\dfrac{1}{2 - 3i}$

Solution

a) $\dfrac{6 - 2i}{3 + i} = \dfrac{(6 - 2i)(3 - i)}{(3 + i)(3 - i)}$ b) $\dfrac{1}{2 - 3i} = \dfrac{(1)(2 + 3i)}{(2 - 3i)(2 + 3i)}$

$\quad\quad\quad = \dfrac{16 - 12i}{9 + 1}$ $\quad\quad\quad = \dfrac{2 + 3i}{4 + 9}$

$\quad\quad\quad = \dfrac{(2)(8 - 6i)}{(2)(5)}$ $\quad\quad\quad = \dfrac{2}{13} + \dfrac{3}{13}i$

$\quad\quad\quad = \dfrac{8 - 6i}{5}$

$\quad\quad\quad = \dfrac{8}{5} - \dfrac{6}{5}i$ $\quad\quad\quad\quad\quad\quad\quad\quad\quad\quad\quad\square$

According to Definition 1.7, -1 has a square root in the set of complex numbers since

$$i^2 = -1.$$

We find that -1 also has $-i$ as another square root since

$$(-i)^2 = (-i)(-i) = +i^2 = -1.$$

Thus i and $-i$ are both square roots of -1. Similarly, $2i$ and $-2i$ are square roots of -4. Every negative real number is of the form $-a$, where a is a positive real number, and $-a$ has two square roots, $i\sqrt{a}$ and $-i\sqrt{a}$. The square root of $-a$ that has the positive imaginary part is designated as the **principal square root** and is denoted by $\sqrt{-a}$. Thus

$$\sqrt{-1} = i, \quad \sqrt{-4} = 2i, \quad \sqrt{-25} = 5i, \quad \sqrt{-7} = i\sqrt{7},$$

and so on. We usually write, and accept as standard form, $i\sqrt{7}$ instead of $\sqrt{7}i$. This is done to avoid confusion between the two different numbers $\sqrt{7}i$ and $\sqrt{7i}$.

The property $\sqrt{a}\,\sqrt{b} = \sqrt{ab}$, where $a > 0$ and $b > 0$, *does not carry over to principal square roots of negative numbers:*

$$\sqrt{-4}\,\sqrt{-9} = (2i)(3i) = 6i^2 = -6,$$

and

$$\sqrt{(-4)(-9)} = \sqrt{36} = 6 \neq \sqrt{-4}\,\sqrt{-9}.$$

Computations involving square roots of negative numbers are best handled by writing all numbers involved in standard form $a + bi$ and then performing the computations. This procedure is followed in Example 5.

Example 5

a) $-\sqrt{-36} = -6i$

b) $\sqrt{-36} \sqrt{-4} = (6i)(2i) = 12i^2 = -12$

c) $\dfrac{\sqrt{-36}}{\sqrt{-9}} = \dfrac{6i}{3i} = 2$ ❑

Using the fact that $i^2 = -1$, any power of the complex number i can be reduced to one of the values 1, -1, i, or $-i$. The technique is demonstrated in Example 6.

Example 6 Write i^{47} in standard form.

Solution

$$i^{47} = i^{46} \cdot i$$
$$= (i^2)^{23} \cdot i$$
$$= (-1)^{23} \cdot i$$
$$= -i$$ ❑

Exercises 1.4

Evaluate each expression. (See Example 5.)

1. $\sqrt{-16}$

2. $\sqrt{-25}$

3. $-\sqrt{-49}$

4. $-\sqrt{-64}$

5. $\sqrt{-25} \sqrt{-9}$

6. $\sqrt{-16} \sqrt{-36}$

7. $\dfrac{\sqrt{-45}}{\sqrt{-5}}$

8. $\dfrac{\sqrt{-75}}{\sqrt{-3}}$

Perform the indicated operations and write the results in standard form. (See Examples 2 and 4.)

9. $(3 + 2i) + (6 - 3i)$

10. $(7 - i) - (3 - 6i)$

11. $(65 - 3i) - (50 - 70i)$

12. $(64 + 32i) + (-59 - 75i)$

13. $(2 - 5i) - (6 - 3i)$

14. $(6 - 19i) + (32 - 7i)$

15. $(2 - 3i) - (7 - 6i)$

16. $(4 - 7i) + (16 - 5i)$

17. $(3 - 7i)(2 - i)$

18. $(-8 - 2i)(5 - 3i)$

19. $(2 + 3i)(3 - i)$

20. $(11 - 5i)(2 - 3i)$

21. $(6 + 4i)(-7 + 3i)$

22. $(2 - 3i)(2 + 3i)$

23. $(6 - 3i)^2$

24. $(3 - 2i)^2$

25. $i(12 - 4i)(3 + i)$

26. $i(2 - 5i)(3 - 7i)$

27. $-i(-2 + 3i)(4 - i)$

28. $-2i(1 + 5i)(2 - 5i)$

29. $(6 - 7i) \div 3$

30. $(3 - 9i) \div 9$

31. $\dfrac{2}{i}$

32. $\dfrac{5}{i}$

33. $\dfrac{4}{3i}$

34. $\dfrac{7}{2i}$

35. $\dfrac{6 - i}{-3i}$

36. $\dfrac{4 + i}{-2i}$

37. $\dfrac{1}{3 + 4i}$

38. $\dfrac{1}{4 - 5i}$

39. $\dfrac{1}{5 - 12i}$

40. $\dfrac{1}{15 + 8i}$

41. $6 \div (3 - i)$

42. $4 \div (2 - i)$

43. $\dfrac{5 - 3i}{6 + i}$

44. $\dfrac{9 - i}{3 + 2i}$

45. $\dfrac{4 - 7i}{1 - 2i}$ **46.** $\dfrac{7 - 5i}{3 - 7i}$

47. $(4 - 3i) \div (3 + 4i)$ **48.** $(5 - 10i) \div (2 + i)$

49. $(-1 + 4i) \div (1 - 3i)$

50. $(-2 - 3i) \div (-3 + i)$

Write each power of i in standard form. (See Example 6.)

51. i^{72} **52.** i^{56} **53.** i^{29} **54.** i^{13}

55. i^{91} **56.** i^{39} **57.** $\dfrac{1}{i^{15}}$ **58.** $\dfrac{1}{i^{95}}$

59. i^{-62} **60.** i^{-78} **61.** i^{915} **62.** i^{231}

63. For any two complex numbers $z_1 = a_1 + b_1 i$ and $z_2 = a_2 + b_2 i$, in standard form, prove that

$$\overline{z_1 + z_2} = \bar{z}_1 + \bar{z}_2.$$

64. With the same notation as in Problem 63, prove that $\overline{z_1 \cdot z_2} = \bar{z}_1 \cdot \bar{z}_2$.

1.5 Solutions of Equations with Real Coefficients

In this section we demonstrate some techniques for solving certain types of nonlinear equations with coefficients that are real numbers. Prior experience in solving linear and quadratic equations is assumed here.

One of the simplest nonlinear equations is of the form

$$x^2 = a,$$

where a is a real number. The two solutions are the numbers whose square is a.

We write $x = \pm\sqrt{a}$ to indicate the solutions of $x^2 = a$, and we say that these solutions are obtained by "taking the square root of both sides." If $a < 0$, the values of $\pm\sqrt{a}$ are not real numbers. For example, $\pm\sqrt{-16} = \pm 4i$.

Example 1   Solve $(2x + 3)^2 = 4$.

Solution
We use the method of taking the square root of both sides.

$2x + 3 = \pm 2$ Equating square roots

$2x = -3 \pm 2$ Subtracting 3 from both sides

$x = -\dfrac{3}{2} \pm 1$ Dividing both sides by 2

$x = \begin{cases} -\dfrac{3}{2} + 1 = -\dfrac{1}{2} & \text{Using the } + \text{ sign} \\[2mm] -\dfrac{3}{2} - 1 = -\dfrac{5}{2} & \text{Using the } - \text{ sign} \end{cases}$

The solutions are $x = -\frac{1}{2}$ and $x = -\frac{5}{2}$.   ❑

Another method used in solving nonlinear equations is that of factoring. The method of factoring relies on this fact: *The product of two factors is zero if and only if at least one of the factors is zero.* That is,

$$xy = 0 \quad \text{if and only if} \quad \text{either } x = 0 \text{ or } y = 0.$$

Example 2   Solve the cubic equation $x^3 + 3x^2 - x - 3 = 0$.

Solution
Factoring by grouping leads to a product of three linear factors equal to zero.

$$x^3 + 3x^2 - x - 3 = 0$$
$$x^2(x + 3) - 1(x + 3) = 0$$
$$(x^2 - 1)(x + 3) = 0$$
$$(x - 1)(x + 1)(x + 3) = 0$$

This product can be zero only if at least one of the linear factors is zero. Thus

$$x - 1 = 0 \qquad \text{or} \qquad x + 1 = 0 \qquad \text{or} \qquad x + 3 = 0$$
$$x = 1, \qquad\qquad\qquad x = -1, \qquad\qquad\qquad x = -3,$$

and the solutions are 1, -1, and -3. ❏

If the quadratic expression in the equation

$$ax^2 + bx + c = 0,$$

where $a \neq 0$, cannot be factored easily, then the Quadratic Formula can be used.

The Quadratic Formula

The solutions to the quadratic equation $ax^2 + bx + c = 0$, $a \neq 0$, are given by

$$x = \frac{-b \pm \sqrt{b^2 - 4ac}}{2a}.$$

The expression $b^2 - 4ac$, which occurs under the radical in the quadratic formula, is called the **discriminant**. The value of $b^2 - 4ac$ will be either positive, negative, or zero. We can classify the roots of any quadratic equation by simply evaluating the discriminant as follows.

1. If $b^2 - 4ac = 0$, then the two roots will be equal real numbers.
2. If $b^2 - 4ac > 0$, then the two roots will be distinct (unequal) real numbers.
3. If $b^2 - 4ac < 0$, then the two roots will be distinct complex numbers that are conjugates of each other.

Example 3   Use the discriminant to determine the type of roots of the following equations:

a) $2x^2 - x - 15 = 0$ b) $4x^2 - 20x + 25 = 0$
c) $x^2 + 4x + 13 = 0$

Solution

a)   Since
$$b^2 - 4ac = (-1)^2 - 4(2)(-15) = 121 > 0,$$
the equation has two distinct real roots.

b)   Since
$$b^2 - 4ac = (-20)^2 - 4(4)(25) = 0,$$
the two roots are equal real numbers.

c)   The discriminant
$$b^2 - 4ac = 4^2 - 4(1)(13) = -36 < 0,$$
and the two roots are complex numbers that are conjugates of each other.   ❑

A combination of the method of factoring and the Quadratic Formula is sometimes useful, as illustrated in the next example.

Example 4   Solve $x^3 - 8 = 0$.

Solution
The key tools for solving this equation are the factorization of the difference of two cubes and the quadratic formula.
$$x^3 - 8 = 0$$
$$(x - 2)(x^2 + 2x + 4) = 0$$
$$x - 2 = 0 \quad \text{or} \quad x^2 + 2x + 4 = 0$$
$$x = 2 \quad \text{or} \qquad x = \frac{-2 \pm \sqrt{2^2 - 4(1)(4)}}{2(1)}$$
$$= \frac{-2 \pm \sqrt{-12}}{2}$$

$$= \frac{-2 \pm 2i\sqrt{3}}{2}$$

$$= -1 \pm i\sqrt{3}$$

The solution set is $\{2, -1 \pm i\sqrt{3}\}$. ❏

Sometimes a simple substitution of variables will transform an equation into one that is quadratic.

Example 5 Solve $4z^4 = 13z^2 - 9$.

Solution
This equation is quadratic in the variable z^2. Hence we make the substitution $u = z^2$. Since $u^2 = z^4$, we have a quadratic equation in u.

$$4u^2 = 13u - 9$$

$$4u^2 - 13u + 9 = 0$$

$$(4u - 9)(u - 1) = 0$$

$$4u - 9 = 0 \quad \text{or} \quad u - 1 = 0$$

$$u = \frac{9}{4} \quad \text{or} \quad u = 1$$

To solve for z, we make the substitution $u = z^2$ again.

$$z^2 = \frac{9}{4} \quad \text{or} \quad z^2 = 1$$

$$z = \pm\frac{3}{2} \quad \text{or} \quad z = \pm 1$$

Hence the solution set is $\{\pm\frac{3}{2}, \pm 1\}$. ❏

We next consider equations in one real variable that contain a radical in one or both members. The standard method for solving an equation of this type is to raise both sides of the equation to a power that will eliminate one or more of the radicals, simplify the resulting equation, and solve for the variable by repeating the process, if necessary.

In some instances this procedure leads to solutions that do not satisfy the original equation. Such solutions are called **extraneous solutions,** or **extraneous roots.** To determine whether or not a solution is extraneous, we must check the solution in the original equation (or any equivalent equation, before both sides are raised to a power). Consider, for example, the equation $x = -3$. Squaring both sides yields $x^2 = 9$, whose solutions are $x = 3$ and $x = -3$. But $x = 3$ does not satisfy the original equation. That is, $x = 3$ is an extraneous root.

Example 6   Solve $\sqrt{8x + 1} - 4 = 1 - 2x$.

Solution
Squaring both sides will not eliminate the radical. To eliminate the radical, we must first isolate it on one side of the equation.

$$\sqrt{8x + 1} = 5 - 2x \qquad \text{Adding 4 to both sides}$$

$$8x + 1 = 25 - 20x + 4x^2 \qquad \text{Squaring both sides}$$

$$0 = 24 - 28x + 4x^2 \qquad \text{Rewriting to force 0 on one side}$$

$$0 = 6 - 7x + x^2 \qquad \text{Dividing both sides by 4}$$

$$0 = (1 - x)(6 - x) \qquad \text{Factoring}$$

The solutions here are $x = 1$ and $x = 6$. Since one or both solutions might be extraneous, they need to be checked in the original equation.

Check: $x = 1$
Since

$$\sqrt{8(1) + 1} - 4 = 1 - 2(1),$$

the value $x = 1$ is a solution to the original equation.

Check: $x = 6$
Since

$$\sqrt{8(6) + 1} - 4 \neq 1 - 2(6),$$

then $x = 6$ is an extraneous solution, and the solution set to the original equation is {1}. ❏

In some equations involving radicals, it is necessary to apply the procedure of raising both sides of the equation to a power more than once in order to eliminate all the radicals. In such cases it is important that the checking procedure be performed in the original equation (or any equivalent equation, before both sides are raised to an even power).

We next consider an equation involving fractional exponents. Because of the relationship between rational exponents and radicals, the solution techniques are the same.

Example 7   Solve $(3x + 1)^{1/2} = (x + 4)^{1/2} + 1$.

Solution
Although squaring both sides will not eliminate both fractional exponents, one of them will be eliminated.

$$3x + 1 = x + 4 + 2(x + 4)^{1/2} + 1$$

$$2x - 4 = 2(x + 4)^{1/2} \qquad \text{Isolating } 2(x + 4)^{1/2}$$

$$x - 2 = (x + 4)^{1/2} \qquad \text{Dividing by 2}$$

$$x^2 - 4x + 4 = x + 4 \qquad \text{Squaring both sides}$$

$$x^2 - 5x = 0 \qquad \text{Simplifying}$$

$$x(x - 5) = 0 \qquad \text{Factoring}$$

The solutions to the quadratic equation are $x = 0$ and $x = 5$. These must be checked in the original equation.

Check: $x = 0$
Since

$$[3(0) + 1]^{1/2} \neq (0 + 4)^{1/2} + 1,$$

then $x = 0$ is an extraneous solution.

Check: $x = 5$
Since

$$[3(5) + 1]^{1/2} = (5 + 4)^{1/2} + 1,$$

then 5 is a solution, and {5} is the solution set for the original equation. ❏

As a final example, we solve an applied problem involving uniform motion. First we recall the following principle: If an object moves at a constant rate of speed r for a time t, then it will have moved a distance d that is equal to the product of the rate and the time:

$$d = rt.$$

In using this formula, the units in which the quantities are expressed must be in agreement. For example, if d is measured in miles and t in hours, then r must be in miles per hour.

Example 8 Matt rode his bike to the store 3 miles from home. If he traveled 3 miles per hour faster going to the store than he did returning and his total travel time was 27 minutes, how fast did he go each way?

Solution
Let r represent Matt's speed returning from the store. Then $r + 3$ is his speed going to the store. His travel time going is $3/(r + 3)$, and his travel time returning is $3/r$. Since the total travel time is 27 minutes

(9/20 hour), we must solve the following equation.

$$\frac{3}{r + 3} + \frac{3}{r} = \frac{9}{20}$$

$$\frac{1}{r + 3} + \frac{1}{r} = \frac{3}{20}$$

$$20r + 20(r + 3) = 3r(r + 3)$$

$$20r + 20r + 60 = 3r^2 + 9r$$

$$0 = 3r^2 - 31r - 60$$

$$0 = (3r + 5)(r - 12)$$

$$3r + 5 = 0 \qquad \text{or} \qquad r - 12 = 0$$

$$r = -5/3 \qquad \text{or} \qquad r = 12$$

The value $r = -5/3$ does not make sense as a solution. Thus Matt's rate returning was 12 miles per hour, and his rate going was 15 miles per hour. ❑

Exercises 1.5

Solve by factoring.

1. $4x^2 - 25 = 0$ **2.** $9x^2 - 16 = 0$

3. $r^2 + 3r = 0$ **4.** $u^2 - 5u = 0$

5. $y^2 + 3y - 10 = 0$ **6.** $x^2 + 2x - 15 = 0$

7. $3t^2 - 10t + 3 = 0$ **8.** $2y^2 - 5y + 2 = 0$

9. $49x^3 + 7x^2 = 2x$ **10.** $2x^3 + 5x^2 = 3x$

11. $2x^3 - 5x^2 - 2x + 5 = 0$

12. $3x^3 + x^2 - 6x - 2 = 0$

Solve by using the Quadratic Formula.

13. $x^2 + 3x - 28 = 0$ **14.** $x^2 - 16x + 64 = 0$

15. $4x^2 - 8x + 3 = 0$ **16.** $6x^2 - 11x - 35 = 0$

17. $x^2 + x = 0$ **18.** $4x^2 = -7x$

19. $-x^2 - 4x + 12 = 0$ **20.** $-x^2 - 2x + 1 = 0$

21. $16x^2 - 25 = 0$ **22.** $2x^2 = 1$

23. $x^2 + 5x + 5 = 0$ **24.** $3x^2 + 8x + 3 = 0$

25. $x^2 - 2x + 4 = 0$ **26.** $x^2 + 3x + 9 = 0$

▦ Use the Quadratic Formula to approximate the solutions of the following equations, rounded to two decimal places.

27. $0.26x^2 + 1.02x + 0.82 = 0$

28. $91.31x^2 + 21.81x - 11.24 = 0$

29. $1.01x^2 + 3.72x + 1.82 = 0$

30. $6.51x^2 - 9.15x + 2.11 = 0$

Solve by any method.

31. $7 - 15x + 2x^2 = 0$ **32.** $3z^2 = 11z + 4$

33. $x^2 + 2x - 1 = 0$ **34.** $y^2 + 5y + 5 = 0$

35. $-27y^2 + 3y + 2 = 0$ **36.** $15x^2 - x - 28 = 0$

37. $2z^2 + 2z - 5 = 0$ **38.** $2y^2 = 6y + 1$

39. $4x^2 - 8x + 5 = 0$ **40.** $4t^2 - 16t + 17 = 0$

Use the discriminant to determine the type of roots of each of the following equations. Do not evaluate the roots. (See Example 3.)

41. $x^2 - 4 = 0$ **42.** $4x^2 + 7x - 15 = 0$

43. $2x^2 - 5x + 7 = 0$ **44.** $-x^2 - x - 2 = 0$

45. $49x^2 + 14x + 1 = 0$ **46.** $-x^2 + 2x - 1 = 0$

▦ **47.** $5.09x^2 + 9.21x + 1.82 = 0$

48. $61.28x^2 + 45.19x + 21.54 = 0$

49. $27.81x^2 - 15.25x + 11.84 = 0$

50. $8.880x^2 - 27.21x + 15.62 = 0$

Determine the solution set for each of the following. (See Examples 4 and 5.)

51. $2x - 1 = \dfrac{-5(3x + 2)}{2x + 1}$

52. $2(x + 1) = 13 - \dfrac{4}{x - 1}$

53. $x^3 - 27 = 0$ **54.** $x^3 + 8 = 0$

55. $8x^4 - 27x = 0$ **56.** $64x^4 + 27x = 0$

57. $27z^6 + 215z^3 - 8 = 0$

58. $z^6 + 16z^3 + 64 = 0$

Find the solution set of the given equation in the real numbers. (See Examples 6 and 7.)

59. $2y^{2/3} + y^{1/3} - 1 = 0$

60. $y^{2/3} - y^{1/3} - 12 = 0$

61. $\sqrt{x + 10} = x - 2$ **62.** $\sqrt{4x + 1} = x - 5$

63. $\sqrt{10 + x} - x = 10$ **64.** $\sqrt{x - 1} + 2 = 2x$

65. $(9 - x)^{1/2} + (x + 8)^{1/2} = 3$

66. $(2x + 3)^{1/2} - (2x)^{1/2} = 1$

67. $\sqrt[3]{3x - 1} + 1 = x$

68. $\sqrt{5 + \sqrt{x + 1}} + 1 = \sqrt{x + 1}$

69. For two consecutive integers whose product is 462.

70. The sum of the squares of four consecutive integers is 174. Find the integers.

71. The length of a rectangular field is twice its width, and the area is 1800 square meters. Find the dimensions of the field.

72. Suppose the perimeter of a rectangular field is 480 meters, and the area is 10,800 square meters. Determine the dimensions of the field.

73. A positive number minus its square is -182. Determine the number.

74. If an object is thrown vertically upward with an initial velocity of v_0 feet per second, then the distance s in feet that the object will be above the earth at time t, in seconds, is given by the equation $s = v_0 t - 16t^2$.
   a) Find t if $v_0 = 128$ feet per second and $s = 192$ feet.
   b) Find t if $v_0 = 128$ feet per second and $s = 0$ feet.

75. A farmer is plowing a rectangular field that is 100 meters wide and 200 meters long, as sketched in the figure. How wide a strip must she plow around the field so that 52% of the field is plowed?

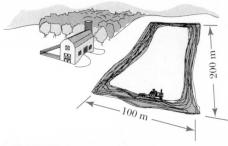

EXERCISE 75

76. An open box is to be made from a rectangular piece of tin that is 12 centimeters wide and 14 centimeters long by cutting equal squares from the four corners and turning up the sides, as shown in the accompanying figure. How large a square must be cut from each corner if the area of the base is to be 60 square centimeters?

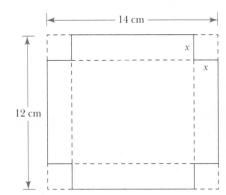

EXERCISE 76

77. A car traveled 660 miles at a uniform rate. If the rate had been 5 miles per hour more, the trip would have taken 1 hour less. Find the rate of the car.

78. Adrienne drove her boat 12 miles upstream, as shown below, and returned making the round-trip in 54 minutes. If the rate of the current was 3 miles

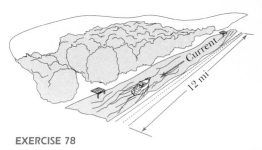

EXERCISE 78

per hour, what was the speed of her boat in still water?

79. A plane traveled 800 miles at a uniform rate of speed. If the rate had been 40 miles per hour more, the trip would have taken 1 hour less. Find the rate of the plane.

80. Beth flew her airplane against the wind a one-way distance of 540 miles to Anaheim. If the speed of her plane in still air was 150 miles per hour and her round-trip travel time was 7.5 hours, what was the speed of the wind? Assume the wind was blowing uniformly with the same speed and direction during the entire flight.

1.6 Nonlinear Inequalities in One Real Variable

In this section our work is restricted to the set of real numbers. A **quadratic inequality** in the real variable x is an inequality of the form

$$ax^2 + bx + c > 0,$$

with $a \neq 0$, or a similar statement with one of the other inequality symbols: $\geq$, $<$, or $\leq$. The **solution set** of the quadratic inequality is the set of all real numbers x for which the statement of the inequality is true.

One way to determine the solution set for a given inequality is first to write the quadratic expression as a product of linear factors and then to construct a diagram showing the signs of each factor. Such a diagram is called a **sign graph.** We illustrate this technique in the first example.

Example 1   Solve the inequality $x^2 - 2x - 3 \leq 0$.

Solution
Factoring the quadratic expression, we have

$$(x - 3)(x + 1) \leq 0.$$

The solution set for the inequality consists of all real numbers x for which the product $(x - 3)(x + 1)$ is zero or negative. The product is zero at $x = 3$ and at $x = -1$. The sign of the product depends on the sign of each of the factors. Considering each factor individually, we know that

$$x - 3 > 0 \quad \text{when} \quad x > 3,$$
$$x - 3 = 0 \quad \text{when} \quad x = 3,$$
$$x - 3 < 0 \quad \text{when} \quad x < 3;$$

and

$$x + 1 > 0 \quad \text{when} \quad x > -1,$$
$$x + 1 = 0 \quad \text{when} \quad x = -1,$$
$$x + 1 < 0 \quad \text{when} \quad x < -1.$$

Sign graph for $x^2 - 2x - 3 \le 0$

Factors $\begin{cases} x - 3. \\ x + 1 \end{cases}$

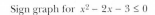

$$\begin{array}{cccccccccccc}
x - 3. & - & - & - & - & - & - & - & 0 & + & + \\
x + 1 & - & - & - & - & 0 & + & + & + & + & + & + \\
\text{Product} & + & + & + & + & 0 & - & - & - & 0 & + & + \\
\end{array}$$

$-1 \le x \le 3$

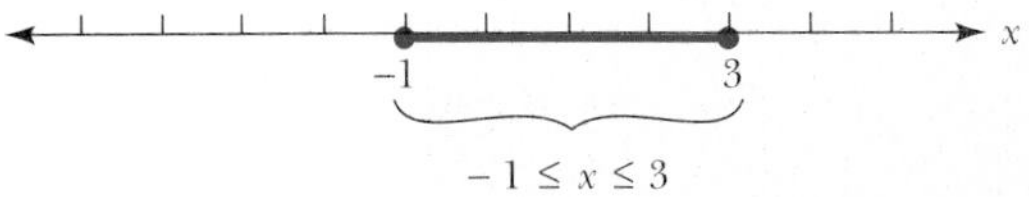

FIGURE 1.8

We can summarize this information on a sign graph, which is constructed by first locating $x = 3$ and $x = -1$ on the number line in Figure 1.8. To indicate the sign of each factor, we place $+$ signs in the interval where each factor is positive, $-$ signs where each factor is negative, and 0 where each factor assumes the value zero. Finally, we consider the product of the two factors. Since quantities with like signs multiply to yield a positive number and quantities with opposite signs multiply to yield a negative number, we can determine the sign of the product. Thus the solution set for the inequality

$$x^2 - 2x - 3 \le 0$$

is the set of real numbers in the interval where the product is negative or zero. From the sign graph we see that the solution set is

$$\{x : -1 \le x \le 3\} = [-1, 3]. \qquad \square$$

The sign graph method extends easily to products that contain more than two linear factors and also to quotients made up of linear factors.

Example 2 Solve the inequality $\dfrac{2}{w - 1} > \dfrac{1}{w + 2}$.

Solution

To use the sign graph method, we must compare a product or quotient to *zero*. Thus we first obtain an inequality with one member zero by subtracting $1/(w + 2)$ from both sides:

$$\frac{2}{w - 1} - \frac{1}{w + 2} > 0 \quad \text{or} \quad \frac{w + 5}{(w - 1)(w + 2)} > 0.$$

The sign of the quotient depends on the sign of each linear factor used in forming the quotient. The signs of the linear factors are displayed on

Sign graph for $\dfrac{w+5}{(w-1)(w+2)} > 0$

Factors $\begin{cases} w+5 \\ w-1 \\ w+2 \end{cases}$

$w+5$	$-$	$-$	$-$	$-$	0	$+$	$+$	$+$	$+$	$+$	$+$	$+$	$+$	$+$	$+$
$w-1$	$-$	$-$	$-$	$-$	$-$	$-$	$-$	$-$	$-$	$-$	0	$+$	$+$	$+$	$+$
$w+2$	$-$	$-$	$-$	$-$	$-$	$-$	$-$	0	$+$	$+$	$+$	$+$	$+$	$+$	$+$
Quotient	$-$	$-$	$-$	$-$	0	$+$	$+$	U	$-$	$-$	U	$+$	$+$	$+$	$+$

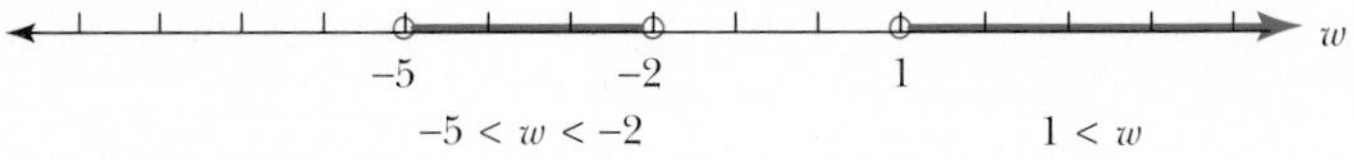

$$-5 < w < -2 \qquad\qquad\qquad 1 < w$$

with points at -5, -2, 1 on the w-axis.

FIGURE 1.9

the first three lines of the sign graph in Figure 1.9. The last line indicates the sign of the quotient in each interval, and U indicates that the quotient is undefined at that point. Therefore the solution set for the inequality

$$\frac{2}{w-1} > \frac{1}{w+2} \text{ is given by } (-5, -2) \cup (1, \infty).$$ ❏

At this point it is easy to see that the solution of an inequality in x amounts to determining the values of x that make an expression positive, negative, or zero, depending on which symbol of inequality is involved. These values of x can be determined without drawing a sign graph as we did in the last two examples.

We must first determine the linear factors that are involved in the expression and then find the points where the linear factors are zero. These points separate the real numbers into intervals over which the expression has the same sign (positive or negative). This is so because the expression cannot change sign unless one of its factors changes sign. We can determine the sign of the expression in each interval by testing a value of x from that region. Systematically testing all the intervals leads to the solution set of the given inequality. This method is called the **algebraic method.**

Example 3   Use the algebraic method to solve the inequality

$$\frac{x+2}{x-1} \ge 2.$$

Solution
We begin by transforming the given inequality into one in which one member is zero.

$$\frac{x + 2}{x - 1} - 2 \geq 0 \qquad \text{Subtracting 2}$$

$$\frac{x + 2 - 2(x - 1)}{x - 1} \geq 0 \qquad \text{Adding fractions}$$

$$\frac{-x + 4}{x - 1} \geq 0 \qquad \text{Simplifying}$$

We now reason that the value of the quotient

$$\frac{-x + 4}{x - 1}$$

does not change sign without taking on the value of zero or becoming undefined because of a zero denominator. The quotient is zero at $x = 4$ and it is undefined at $x = 1$. These values of x separate the real numbers into the intervals

$$(-\infty, 1), \quad (1, 4), \quad \text{and} \quad (4, \infty).$$

Selecting a test point in each interval, we obtain the results recorded in Figure 1.10.

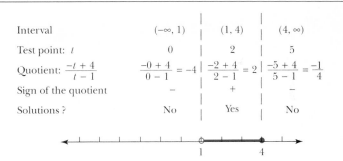

Interval	$(-\infty, 1)$	$(1, 4)$	$(4, \infty)$
Test point: t	0	2	5
Quotient: $\dfrac{-t + 4}{t - 1}$	$\dfrac{-0 + 4}{0 - 1} = -4$	$\dfrac{-2 + 4}{2 - 1} = 2$	$\dfrac{-5 + 4}{5 - 1} = \dfrac{-1}{4}$
Sign of the quotient	$-$	$+$	$-$
Solutions ?	No	Yes	No

FIGURE 1.10

Keeping in mind that the quotient $(-x + 4)/(x - 1)$ is zero at $x = 4$ and undefined at $x = 1$, we find that the solution set is

$$\{x : 1 < x \leq 4\} = (1, 4],$$

as graphed in Figure 1.10. ❏

It is important to notice in Example 3 that *we did not multiply both sides by $x - 1$*. This is not a valid step, because $x - 1$ may be positive or negative. It is possible to consider the two cases (1) when $x - 1 > 0$ and

(2) when $x - 1 < 0$, and solve the inequality by considering these cases separately. However, this method is somewhat tedious and inefficient, and we do not go into it here.

Example 4   In the World Series playoffs, Slugger Sam hits a foul ball. The ball leaves his bat at a height of 3 feet and is forced vertically upward with initial velocity of 80 feet per second. At t seconds after leaving the bat, the height h of the ball is given in feet by

$$h = 3 + 80t - 16t^2.$$

At what times will the ball be above the lights if the lights are 67 feet high?

Solution
The ball will be above the lights when the height h is greater than 67. Thus we need to solve the quadratic inequality

$$3 + 80t - 16t^2 > 67, \qquad t \geq 0.$$

We rewrite this inequality so that one side is 0 and factor the other side.

$$-64 + 80t - 16t^2 > 0 \qquad \text{Subtracting 67}$$

$$4 - 5t + t^2 < 0 \qquad \text{Dividing by } -16 \text{ and reversing the inequality}$$

$$(4 - t)(1 - t) < 0 \qquad \text{Factoring}$$

The chart in Figure 1.11 indicates that the ball will be above the lights between the first and fourth seconds after leaving the bat. Notice that only the nonnegative portion of the number line is used in Figure 1.11 because of the restriction that $t \geq 0$.   ❑

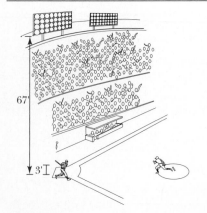

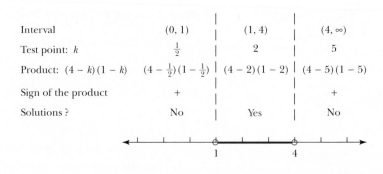

Interval	$(0, 1)$	$(1, 4)$	$(4, \infty)$
Test point: k	$\frac{1}{2}$	2	5
Product: $(4 - k)(1 - k)$	$(4 - \frac{1}{2})(1 - \frac{1}{2})$	$(4 - 2)(1 - 2)$	$(4 - 5)(1 - 5)$
Sign of the product	$+$		$+$
Solutions ?	No	Yes	No

FIGURE 1.11

Exercises 1.6

Solve the following quadratic inequalities.

1. $x^2 - x - 12 > 0$
2. $x^2 + 7x + 10 > 0$
3. $x^2 - 2x - 3 \leq 0$
4. $x^2 - 8x + 12 \leq 0$
5. $-x^2 + 4x - 3 \leq 0$
6. $9 - 8x - x^2 \leq 0$
7. $-2x^2 + 6 < 0$
8. $x^2 - 4 < 0$
9. $x^2 - 9x > 0$
10. $6x + x^2 > 0$
11. $8x - 2x^2 \geq 0$
12. $4x - x^2 \geq 0$
13. $-2x^2 + 4 < 0$
14. $-3x^2 + 9 < 0$
15. $2x - x^2 \leq 0$
16. $6x - 2x^2 \leq 0$
17. $x^2 + 6x + 16 < 8$
18. $x^2 + 4x + 6 \geq 3$
19. $z^2 + 2z \leq 15$
20. $w^2 + 2w > 99$
21. $3w^2 + 13w - 10 > 0$
22. $4s^2 + 3s - 1 \geq 0$
23. $-15x^2 + 28x - 12 > 0$
24. $-2z^2 + 11z - 15 > 0$
25. $3z^2 > 4z$
26. $6w^2 < w$
27. $r - 6r^2 > -35$
28. $4x^2 > 2 - 7x$
29. $21 - 4x^2 > -5x$
30. $2x^2 < 12 - 5x$
31. $-4s^2 - 4s + 1 \geq 0$
32. $w^2 + 8w + 13 \geq 0$

Solve the following inequalities. (See Examples 2 and 3.)

33. $(x - 1)(x - 2)(x - 3) > 0$
34. $(x - 1)(x - 2)(x - 3)(x - 4) \leq 0$
35. $\dfrac{x + 1}{2x - 1} > 3$
36. $\dfrac{2x - 1}{x} < 0$

37. $\dfrac{3x - 1}{x + 2} < 1$
38. $\dfrac{3w + 2}{w} \leq 2$
39. $\dfrac{w + 1}{w + 3} \geq 2$
40. $\dfrac{1}{w - 1} > \dfrac{1}{3}$
41. $\dfrac{7 - z}{(z - 2)(z - 3)} < 0$
42. $\dfrac{x - 5}{(3x - 1)(2x - 3)} < 0$
43. $\dfrac{2}{w + 2} \geq \dfrac{1}{2w + 1}$
44. $\dfrac{1}{r - 3} \leq \dfrac{3}{r + 1}$
45. $\dfrac{5}{2w + 3} \geq \dfrac{-5}{w}$
46. $\dfrac{3}{x + 1} < \dfrac{2}{x + 2}$

For the following exercises see Example 4.

47. The length of a rectangular field is twice its width. For what dimensions is the area at least 1800 square feet?

48. Suppose the perimeter of a rectangular field is 480 meters. For what dimensions is the area at least 10,800 square meters?

49. Suppose an object is thrown vertically upward with initial velocity 128 feet per second. If s is the height above ground in feet and t is the time elapsed in seconds, then $s = 128t - 16t^2$. At what times will the height of the object be 192 feet or more?

50. A farmer is plowing a rectangular field that is 100 meters wide and 200 meters long. How wide a strip must he plow around the field so that at least 52% of the field is plowed?

1.7 Algebraic Skills for the Calculus

Up to this point this chapter has been a review of algebra. Now we elaborate on some algebra-based techniques that are predominant in the calculus.

Rewriting absolute value expressions

In Section 1.1 we studied the concept of absolute value. In the calculus, however, it is often advantageous to work with expressions that are free of absolute value symbols. For example, we can rewrite $|x + 2|$ without the absolute value symbol by considering the following two cases.

Case 1: If $x + 2 \geq 0$, then $|x + 2| = x + 2$.

Case 2: If $x + 2 < 0$, then $|x + 2| = -(x + 2) = -x - 2$.

We can write this more compactly as

$$|x + 2| = \begin{cases} x + 2 & \text{if } x \geq -2, \\ -x - 2 & \text{if } x < -2. \end{cases}$$

Example 1 Rewrite $\dfrac{|x|}{x}$ without the absolute value symbol.

Solution
We first note that this expression is not defined whenever $x = 0$.

If $x > 0$, then $|x| = x$ and $\dfrac{|x|}{x} = \dfrac{x}{x} = 1$.

If $x < 0$, then $|x| = -x$ and $\dfrac{|x|}{x} = \dfrac{-x}{x} = -1$.

More compactly, we write

$$\frac{|x|}{x} = \begin{cases} 1 & \text{if } x > 0, \\ -1 & \text{if } x < 0, \\ \text{undefined} & \text{if } x = 0. \end{cases} \qquad \square$$

Using special factorizations

In Section 1.2 we considered only factorizations in which the coefficients were integers. In this light, $x^2 - 2$ is considered to be a prime polynomial. If that restriction is lifted, we can consider $x^2 - 2$ to be the difference of two squares and write

$$x^2 - 2 = x^2 - (\sqrt{2})^2 = (x - \sqrt{2})(x + \sqrt{2}).$$

Similarly, $x - a$ is considered to be a prime polynomial. However, it is sometimes desirable in a calculus problem to change the form of $x - a$ by considering it to be the difference of two squares or even possibly the difference of two cubes. In that sense, then, $x - a$ can indeed be factored.

Difference of two squares: $x - a = (\sqrt{x})^2 - (\sqrt{a})^2$, $x \geq 0$ and $a \geq 0$

$$= (\sqrt{x} - \sqrt{a})(\sqrt{x} + \sqrt{a})$$

Difference of two cubes: $x - a = (\sqrt[3]{x})^3 - (\sqrt[3]{a})^3$

$$= (\sqrt[3]{x} - \sqrt[3]{a})(\sqrt[3]{x^2} + \sqrt[3]{xa} + \sqrt[3]{a^2})$$

A factorization using negative exponents can be helpful in rewriting a quantity as a rational expression. This approach may seem to be somewhat messy, but at times it is very effective.

Example 2   Rewrite $2(x - 1)^{-3} - 3(2x + 5)(x - 1)^{-2}$ as a rational expression.

Solution
In the expression

$$2(x - 1)^{-3} - 3(2x + 5)(x - 1)^{-2},$$

$x - 1$ is a common factor. The least power of $x - 1$ is -3, and we can factor out $(x - 1)^{-3}$.

$$
\begin{aligned}
2(x - 1)^{-3} - 3(2x + 5)(x - 1)^{-2} &= (x - 1)^{-3}[2 - 3(2x + 5)(x - 1)^1] \\
&= (x - 1)^{-3}[2 - 3(2x^2 + 3x - 5)] \\
&= (x - 1)^{-3}(2 - 6x^2 - 9x + 15) \\
&= \frac{17 - 9x - 6x^2}{(x - 1)^3} \qquad \square
\end{aligned}
$$

With a little practice, this factorization technique is more efficient than the use of a least common denominator.

Separating fractions

In Section 1.2 we also saw that fractions with the same denominator can be combined. For example,

$$\frac{4x}{x^2 + 1} + \frac{3}{x^2 + 1} = \frac{4x + 3}{x^2 + 1}.$$

We can reverse the process and separate a fraction into a sum or difference of two or more fractions with the same denominator.

Example 3   Separate $\dfrac{8x - 3}{x^2 + x + 1}$ into two fractions.

Solution

$$\frac{8x - 3}{x^2 + x + 1} = \frac{8x}{x^2 + x + 1} - \frac{3}{x^2 + x + 1} \qquad \square$$

Notice that the solution to Example 3 is not unique, since another possible separation is

$$\frac{8x - 3}{x^2 + x + 1} = \frac{4(2x + 1)}{x^2 + x + 1} - \frac{7}{x^2 + x + 1}.$$

A more elaborate procedure for separating fractions into sums or differences of simpler fractions is presented in Section 8.5.

Rationalizing the numerator

In Section 1.3 rationalizing the denominator was used to remove radicals from the denominator of a rational expression. Historically, before the calculator became commonplace, rationalizing the denominator was used mainly for computational reasons. For example, without the advantage of a calculator

$$\frac{\sqrt{2}}{2} \approx \frac{1.414}{2}$$

is easier to compute than is

$$\frac{1}{\sqrt{2}} \approx \frac{1}{1.414}.$$

However, some situations in the calculus demand that a rational expression be free of radicals in the numerator. We illustrate some methods of rationalizing the numerator in the next examples.

Example 4　Rationalize the numerator of $\dfrac{\sqrt{x} + 1}{x - 1}$.

Solution
We multiply numerator and denominator of the rational expression by the conjugate of the *numerator*.

$$\frac{\sqrt{x} + 1}{x - 1} \cdot \frac{\sqrt{x} - 1}{\sqrt{x} - 1} = \frac{x - 1}{(x - 1)(\sqrt{x} - 1)} = \frac{1}{\sqrt{x} - 1}. \qquad \square$$

Alternatively, in Example 4 we can use factorization to write

$$\frac{\sqrt{x} + 1}{x - 1} = \frac{\sqrt{x} + 1}{(\sqrt{x} - 1)(\sqrt{x} + 1)} = \frac{1}{\sqrt{x} - 1}.$$

Example 5　Rationalize the numerator of

$$\frac{2(x - 1)^{1/2} - x(x - 1)^{-1/2}}{x - 1}.$$

Solution

Although we could multiply numerator and denominator by the conjugate of the numerator, a quicker way of clearing the numerator of fractional exponents is to multiply by $(x - 1)^{1/2}$.

$$\frac{2(x - 1)^{1/2} - x(x - 1)^{-1/2}}{x - 1} = \frac{[2(x - 1)^{1/2} - x(x - 1)^{-1/2}]}{x - 1} \cdot \frac{(x - 1)^{1/2}}{(x - 1)^{1/2}}$$

$$= \frac{2(x - 1)^{1} - x(x - 1)^{0}}{(x - 1)^{3/2}}$$

$$= \frac{2x - 2 - x}{(x - 1)^{3/2}}$$

$$= \frac{x - 2}{(x - 1)^{3/2}}. \qquad \square$$

Making *u*-substitutions

In Section 1.4 we solved certain types of equations by using a substitution. We next incorporate substitutions with the method of completing the square.

Some formulas in the calculus contain expressions of the form

$$u^2 + a^2.$$

In order to apply these formulas, the method of **completing the square** is used to rewrite a quadratic trinomial $ax^2 + bx + c$ $(a \neq 0)$ in the form of the sum or difference of two squares. Consider, for example,

$$x^2 + 4x + 13.$$

We separate the constant, 13, away from the variable terms, $x^2 + 4x$:

$$x^2 + 4x + (\ ) + 13 - (\ ).$$

The idea is to fill in the parentheses so that the first three terms become a perfect square trinomial, $x^2 + 2kx + k^2$. In other words, we wish to add (and at the same time subtract) the constant k^2 so that the first three terms will factor as $(x + k)^2$. In our example, $2k = 4$ and $k = \frac{1}{2}(4) = 2$. Thus $k^2 = 2^2 = 4$. Adding (and subtracting) 4 yields

$$x^2 + 4x + 13 = x^2 + 4x + (4) + 13 - (4)$$

$$= (x + 2)^2 + 9$$

$$= (x + 2)^2 + 3^2.$$

Finally, if we let $u = x + 2$ and $a = 3$, we have the desired form:

$$(x + 2)^2 + 3^2 = u^2 + a^2.$$

The next example illustrates the situation where the leading coefficient is different from 1.

Example 6   Rewrite the following fraction so that it contains an expression of the form $c(u^2 + a^2)$.

$$\frac{5}{2x^2 + 12x + 3}$$

Solution

$$\frac{5}{2x^2 + 12x + 3} = \frac{5}{2[x^2 + 6x + (\ ) + \frac{3}{2} - (\ )]} \qquad \begin{array}{l}\text{Factoring 2 out of}\\\text{the denominator}\end{array}$$

$$= \frac{5}{2[x^2 + 6x + (9) + \frac{3}{2} - (9)]} \qquad \begin{array}{l}\text{Adding and subtracting}\\ [\frac{1}{2}(6)]^2 = (3)^2 = 9\end{array}$$

$$= \frac{5}{2[(x + 3)^2 - \frac{15}{2}]} \qquad \text{Rewriting}$$

$$= \frac{5}{2(u^2 - a^2)} \qquad \begin{array}{l}\text{Letting } u = x + 3 \text{ and}\\ a = \sqrt{\frac{15}{2}}\end{array} \qquad \square$$

Solving equations that involve negative exponents

Some of the formulas in differential calculus may give a result that involves negative exponents in an algebraic expression. It then becomes important to find when the expression is zero and also to find when the expression is undefined. An illustration is given in Example 7.

Example 7   Find the values of x for which the following expression is (a) zero or (b) undefined:

$$\frac{x(x + 1)^{-2/3}}{3} + (x + 1)^{1/3}.$$

Solution
The negative exponent can be eliminated by the same technique used in Example 5. In this case we multiply by $(x + 1)^{2/3}/(x + 1)^{2/3}$.

$$\frac{(x + 1)^{2/3}}{(x + 1)^{2/3}} \cdot \left[\frac{x(x + 1)^{-2/3}}{3} + (x + 1)^{1/3}\right] = \frac{x(x + 1)^0}{3(x + 1)^{2/3}} + \frac{x + 1}{(x + 1)^{2/3}}$$

$$= \frac{x(1) + 3(x + 1)}{3(x + 1)^{2/3}}$$

$$= \frac{4x + 3}{3(x + 1)^{2/3}}$$

a) From the last fraction, it is clear that the expression is zero when $4x + 3 = 0$; that is, when $x = -3/4$.

b) The last fraction also shows that the expression is undefined when $3(x + 1)^{2/3} = 0$; that is, when $x = -1$.   ❑

This entire chapter has been devoted to reviewing and extending some of the knowledge and skills that are necessary for the successful study of the calculus. Although much of the chapter is essentially review material, its importance cannot be overemphasized.

Exercises 1.7

Rewrite each expression without absolute value symbols. (See Example 1.)

1. $|x| - x$ **2.** $|x| + x$ **3.** $\dfrac{x}{|x|}$ **4.** $x|x|$

5. $\dfrac{|2x|}{x}$ **6.** $\dfrac{2x - |x|}{x}$ **7.** $\dfrac{|x - 2|}{2 - x}$ **8.** $\dfrac{|3 - x|}{x - 3}$

9. $\dfrac{|4 - x|}{4 - x}$ **10.** $\dfrac{|1 + x|}{1 + x}$

11. $\dfrac{|1 + x| - |1|}{x}$ **12.** $\dfrac{|2 + x| - |2|}{x}$

A restriction on the distance, $|x - a|$, between x and some constant a determines a corresponding restriction on any multiple of that distance. Fill in the blank in each of the following statements so as to make a true statement.[2]

13. If $|x - 2| < \delta$ then $|9x - 18| <$ _____ .

14. If $|x - 1| < \delta$ then $|4x - 4| <$ _____ .

15. If $|x + 3| < \delta$ then $|2(x - 4) + 14| <$ _____ .

16. If $|x + 5| < \delta$ then $|3(x - 1) + 18| <$ _____ .

17. If $|x - 1| < \delta$ then $|4 - 4x| <$ _____ .

18. If $|x - 2| < \delta$ then $|6 - 3x| <$ _____ .

19. If $|x - 1| < \delta$ then $|2 + 7x - 9| <$ _____ .

20. If $|x + 2| < \delta$ then $\left|\frac{4}{9}(x + 5) - \frac{12}{9}\right| <$ _____ .

Factor as the difference of two squares. Assume variables represent positive numbers.

21. $x^2 - 3$ **22.** $x^2 - 5$

23. $x - 2$ **24.** $x - h$

Factor as the sum or difference of two cubes.

25. $x^3 - 2$ **26.** $x^3 + 4$

27. $x + 1$ **28.** $x - 8$

29. $x + 2$ **30.** $x - 3$

31. $x - h$ **32.** $x + h$

Use factoring to rewrite each of the following as a quotient that is free of negative exponents. (See Example 2.)

33. $(2x - 1)^{-2} - 4(x + 1)(2x - 1)^{-3}$

34. $4x(x^2 + 2)(x - 1)^{-2} - 2(x^2 + 2)^2(x - 1)^{-3}$

35. $-2(x^2 + x)^{-3}(2x + 1)(x + 1)^{-3} - 3(x + 1)^{-4}(x^2 + x)^{-2}$

36. $-5(x^2 - 2x)^{-6}(2x - 2)(x - 1)^{-3} - 3(x - 1)^{-4}(x^2 - 2x)^{-5}$

37. $3x^2(2x - 1)^{1/2} + x^3(2x - 1)^{-1/2}$

38. $(x^2 + 1)^{-1/2} - x(x + 2)(x^2 + 1)^{-3/2}$

39. $6x(x + 2)^{-1/2} - \frac{3}{2}x^2(x + 2)^{-3/2}$

40. $\frac{1}{2}(x + 5)^{-1/2}(3x - 2)^{-1/2} - \frac{3}{2}(3x - 2)^{-3/2}(x + 5)^{1/2}$

Separate each of the following into two or more fractions. (See Example 3.)

41. $\dfrac{4x - 1}{x^{1/2}}$ **42.** $\dfrac{x + 7}{x^{3/2}}$

43. $\dfrac{4x - 3}{x^2 + x + 1}$ **44.** $\dfrac{2x + 7}{x^2 + 3x + 5}$

45. $\dfrac{x - 3}{x^2 + 1}$ **46.** $\dfrac{9x - 5}{x^2 + 4}$

47. $\dfrac{x^2 + x + 2}{x^3 + 3x - 1}$ **48.** $\dfrac{x^5 - 4x + 9}{x^4 + x^2 + 1}$

2. The symbol δ is the Greek letter *delta*.

Rationalize the numerator and simplify the result. (See Examples 4 and 5.)

49. $\dfrac{\sqrt{x}}{x}$ **50.** $\dfrac{\sqrt[3]{x}}{x}$

51. $\dfrac{\sqrt{x}+2}{x-4}$ **52.** $\dfrac{\sqrt{x}+5}{x-25}$

53. $\dfrac{\sqrt{x}-\sqrt{a}}{x-a}$ **54.** $\dfrac{\sqrt{x}-a}{x-a^2}$

55. $\dfrac{\sqrt{x+h}-\sqrt{x}}{h}$ **56.** $\dfrac{\sqrt{x-1+h}-\sqrt{x-1}}{h}$

57. $\dfrac{(x-2)^{1/2}(2x)-\frac{1}{2}x^2(x-2)^{-1/2}}{x-2}$

58. $\dfrac{5(x^2+1)^{1/2}-5x^2(x^2+1)^{-1/2}}{x^2+1}$

59. $\dfrac{\frac{1}{2}(x+3)^{1/2}(x-2)^{-1/2}-\frac{1}{2}(x-2)^{1/2}(x+3)^{-1/2}}{x+3}$

60. $\dfrac{(x+2)^{1/2}(\frac{3}{2})x^{1/2}-x^{3/2}(\frac{1}{2})(x+2)^{-1/2}}{x+2}$

61. $\dfrac{2x^{5/4}-\frac{1}{4}x^{-3/4}(x^2+1)}{x^{1/2}}$

62. $\dfrac{2(x-1)^{1/4}-\frac{1}{2}x(x-1)^{-3/4}}{(x-1)^{1/2}}$

Rewrite each of the following so that it contains an expression of the form $c(u^2+a^2)$. (See Example 6.)

63. $x^2+6x+10$ **64.** $x^2-8x+25$

65. $3x^2+12x-15$ **66.** $5x^2-10x+25$

67. $2x^2-x-4$ **68.** $3x^2-4x+9$

69. $\dfrac{1}{x^2-4x+7}$ **70.** $\dfrac{1}{x^2+10x+10}$

71. $\dfrac{1}{2x^2+2x+1}$ **72.** $\dfrac{1}{2x^2+4x+10}$

73. $\dfrac{3}{2x^2+4x+5}$ **74.** $\dfrac{7}{2x^2+10x+3}$

Find the values of x for which the given expression is (a) zero or (b) undefined. (See Example 7.)

75. $(x^2-2x)(x+4)^{-2/3}$

76. $(x^2+4x)(3x-6)^{-1/3}$

77. $(x^2-9)(x+2)^{-2}$ **78.** $(x^2-4)(x+6)^{-3}$

79. $4x(4x^2+9)^{-1/2}+(4x^2+9)^{1/2}$

80. $(x^2+9x)(x^2+4)^{-1/2}+(x^2+4)^{1/2}$

81. $x^{1/3}+\dfrac{x^{-2/3}(x-4)}{3}$ **82.** $x^{5/3}+\dfrac{x^{-1/3}(x^2-4)}{3}$

83. $\dfrac{3x(x-1)^{-2/5}}{5}+(x-1)^{3/5}$

84. $\dfrac{x^2(x-9)^{-2/3}}{3}+2x(x-9)^{1/3}$

85. $(5-x^2)^{1/3}-\dfrac{2x^2(5-x^2)^{-2/3}}{3}$

86. $x(2x-5)^{1/3}+\dfrac{x^2(2x-5)^{-2/3}}{3}$

CHAPTER REVIEW

Key Words and Phrases

Subset	Binomial	Simplest radical form
Union	Trinomial	Rationalizing the denominator
Intersection	Degree of a nonzero polynomial	Conjugate of a binomial
Empty set	Leading coefficient	Exponential form
Infinity	Monic polynomial	Radical form
Negative infinity	Prime polynomial	Complex number in standard
Interval	Fundamental Principle of	form
Interval notation	Fractions	Conjugate of a complex number
Absolute value	Rational expression in lowest	Quadratic Formula
Exponential	terms	Discriminant
Algebraic expression	Complex fraction	Extraneous solutions
Polynomial	Least common denominator	Rationalizing the numerator
Monomial	Principal nth root	Completing the square

Summary of Important Concepts and Formulas

Interval Notation ($a < b$)

$(a, b) = \{x : a < x < b\}$ $(a, \infty) = \{x : x > a\}$

$[a, b] = \{x : a \leq x \leq b\}$ $[a, \infty) = \{x : x \geq a\}$

$(a, b] = \{x : a < x \leq b\}$ $(-\infty, a] = \{x : x \leq a\}$

$[a, b) = \{x : a \leq x < b\}$ $(-\infty, a) = \{x : x < a\}$

$(-\infty, \infty) = \mathcal{R}$

Absolute Value

$$|a| = \begin{cases} a & \text{if } a \geq 0 \\ -a & \text{if } a < 0 \end{cases}$$

Absolute Value Inequalities

$|x| < d$ if and only if $-d < x < d$

$|x| > d$ if and only if   either $x > d$ or $x < -d$

Laws of Exponents

$a^m \cdot a^n = a^{m+n}$ Product Rule

$(a^m)^n = a^{mn}$ Power Rule

$\dfrac{a^m}{a^n} = a^{m-n}$ Quotient Rule

$(ab)^n = a^n b^n$

$\left(\dfrac{a}{b}\right)^n = \dfrac{a^n}{b^n}$

Fundamental Principle of Fractions

$$\frac{ad}{bd} = \frac{a}{b}, \quad b \neq 0, \quad d \neq 0$$

Basic Properties of Radicals

$(\sqrt[n]{a})^n = a$

$\sqrt[n]{a^n} = \begin{cases} a & \text{if } n \text{ is odd} \\ |a| & \text{if } n \text{ is even} \end{cases}$ (if $a \geq 0$, $\sqrt[n]{a^n} = a$)

$\sqrt[n]{ab} = \sqrt[n]{a}\,\sqrt[n]{b}$

$\sqrt[n]{\dfrac{a}{b}} = \dfrac{\sqrt[n]{a}}{\sqrt[n]{b}}, \quad b \neq 0$

$\sqrt[m]{\sqrt[n]{a}} = \sqrt[mn]{a}$

Formulas for Factoring and Special Product Rules

Common factor: $ax + ay = a(x + y)$

Difference of two squares: $x^2 - y^2 = (x + y)(x - y)$

Square of a binomial: $x^2 \pm 2xy + y^2 = (x \pm y)^2$

Cube of a binomial: $\begin{cases} x^3 + 3x^2y + 3xy^2 + y^3 = (x + y)^3 \\ x^3 - 3x^2y + 3xy^2 - y^3 = (x - y)^3 \end{cases}$

Sum of two cubes: $x^3 + y^3 = (x + y)(x^2 - xy + y^2)$

Difference of two cubes: $x^3 - y^3 = (x - y)(x^2 + xy + y^2)$

Quadratic trinomial: $x^2 + (a + b)x + ab = (x + a)(x + b)$

Arithmetic of Rational Expressions

$$\frac{a}{b} + \frac{c}{b} = \frac{a + c}{b} \qquad \frac{a}{b} \cdot \frac{c}{d} = \frac{ac}{bd} \qquad \frac{a}{b} \div \frac{c}{d} = \frac{a}{b} \cdot \frac{d}{c} = \frac{ad}{bc}$$

Rational Exponents

$$a^{m/n} = \begin{cases} \sqrt[n]{a^m} \\ (\sqrt[n]{a})^m \end{cases}$$

Equality of Complex Numbers

$a + bi = c + di$ if and only if $a = c$ and $b = d$

Solving Quadratic Equations by Taking Square Roots

$x^2 = a$ if and only if $x = \pm\sqrt{a}$

Solving Equations by Factoring

$xy = 0$ if and only if $x = 0$ or $y = 0$

Quadratic Formula

$ax^2 + bx + c = 0$ ($a \neq 0$)   if and only if

$$x = \frac{-b \pm \sqrt{b^2 - 4ac}}{2a}$$

Review Problems for Chapter 1

Use interval notation to write the solution set of the given inequality, and draw the graph of the solution set.

1. $-3 \le x < 2$ **2.** $-1 < x \le 4$

3. $x < -2$ or $0 < x \le 3$

4. $x \le 1$ or $3 \le x < 5$

Express each of the following without using absolute value symbols.

5. $|3x - 6|$, if $x \le 2$ **6.** $|2x - 8|$, if $x \le 4$

7. $|5x - 15|$, if $x > 3$ **8.** $|2x - 6|$, if $x > 3$

Solve the following equations.

9. $|3x + 7| = |x - 5|$ **10.** $|2x + 3| = |x - 7|$

11. $|2x + 3| = x + 11$ **12.** $|3x + 7| = x + 15$

Solve the following inequalities and graph the solution set on a number line.

13. $|x - 1| < 2$ **14.** $|x - 2| < 1$

15. $|x + 2| \ge 3$ **16.** $|x + 1| \ge 4$

17. $|2x - 5| > 3$ **18.** $|3x - 2| > 7$

19. $|4x + 3| \le -1$ **20.** $|6x + 1| \le -3$

Simplify by using the Laws of Exponents.

21. $[(uv^{-1})^3]^{-2}$ **22.** $[(p^{-1}q^{-1})^{-2}]^{-1}$

23. $\left(\dfrac{x^2}{y^{-1}z^3}\right)^{-3}$ **24.** $\left(\dfrac{yz^{-2}}{2x^{-1}w^2}\right)^{-2}$

Perform the indicated operations and write the results in simplest form.

25. $(1 + x)[1 + x(1 - x^2)]$

26. $1 + x[1 + x(1 - x^2)]$

27. $(y^2 - 2)^2$ **28.** $(x^2 + 3)^2$

29. $(p - q)(p^2 + pq + q^2)$ **30.** $(y + 3)(y^2 - 3y + 9)$

Factor each expression completely.

31. $-x^6 - x^3$ **32.** $-64y - 8yx^3$

33. $9x^2 - y^2 + 2yz - z^2$ **34.** $16a^2 - 9b^2 + 6b - 1$

Perform the indicated operations and reduce the results to lowest terms.

35. $\dfrac{x^3 - 3x^2 + 4x - 12}{x^4 - 16} \cdot \dfrac{3x^2y - xy - 10y}{x^3 - 27}$

36. $\dfrac{3x^2 + 7x - 6}{x^3 + 3x^2 + 9x} \cdot \dfrac{6x^3 + 54x}{2x^4 - 162} \div \dfrac{27x^2 - 12}{x^3 - 27}$

37. $\dfrac{1}{2z^2 - 7z + 3} + \dfrac{1}{2z^2 + 3z - 2}$

38. $\dfrac{3b + 8}{3b^2 + b - 2} - \dfrac{3b - 10}{3b^2 - 8b + 4}$

Express the given fraction as a rational expression in lowest terms.

39. $\dfrac{(ab)^{-2}}{a^{-2} - b^{-2}}$ **40.** $\dfrac{x^{-1} + y^{-1}}{x^{-1} - y^{-1}}$

Write in simplest radical form. All variables represent positive real numbers.

41. $\sqrt[4]{\dfrac{a^5b^2}{ab^{10}}}$ **42.** $\sqrt[4]{\dfrac{243x^{14}y^2}{3x^2y^{10}}}$

43. $\dfrac{\sqrt{75x^5y^9}}{\sqrt{3xy}}$ **44.** $\dfrac{\sqrt[3]{250a^8b^5}}{\sqrt[3]{2a^{-1}b^2}}$

45. $\sqrt{a^6b^9}\,\sqrt{a^3b^{11}}$ **46.** $\sqrt[3]{r^5z^2}\,\sqrt[3]{r^4z^2}$

47. $\sqrt[4]{\dfrac{p^7}{q^5}}$ **48.** $\sqrt[4]{\dfrac{2x^5}{25y}}$

49. $\dfrac{3\sqrt{x}}{2\sqrt{x} - 1}$ **50.** $\dfrac{\sqrt{a}}{2\sqrt{a} + 3}$

Use the distributive property to perform the indicated operations. Write the results in simplest radical form.

51. $3x\sqrt[3]{40xy} - 20\sqrt[3]{135xy^4}$

52. $\sqrt[5]{64rs^8} + \sqrt[5]{-2r^{11}s^3}$

Use the Laws of Exponents to simplify.

53. $[(a^{2/3}b^{-1/4})^{-2}]^{-6}$ **54.** $[(x^{1/3}y^{-2/3})^{-1/2}]^6$

55. $(a^{1/2} - b^{1/2})(a^{1/2} + b^{1/2})$

56. $(a^{-3/2} + b^{-3/2})(a^{-3/2} - b^{-3/2})$

57. $\dfrac{ab^0c^{2/5}}{a^{3/4}b^{-1/3}c^{1/5}}$ **58.** $\dfrac{x^{2/3}y^{7/4}z}{x^0y^{3/4}z^{1/2}}$

Rewrite in simplest radical form with the index as small as possible.

59. $\sqrt[3]{r^2s}\,\sqrt{r^5s}$ **60.** $\sqrt[5]{x^2y}\,\sqrt[3]{xy^4}$

Express each of the following in terms of i.

61. $\sqrt{-9}$

62. $\sqrt{-25}$

63. $-\sqrt{-8}$

64. $-\sqrt{-32}$

Perform the indicated operations and write each result in standard form.

65. $(1 + i) - (6 + 5i)$

66. $(11 + 2i) - (-8 + 7i)$

67. $(-3 - i)(2 - 3i)$

68. $(1 - i)(-2 - 7i)$

69. $(1 + 2i)^2$

70. $(3 - 4i)^2$

71. $(\sqrt{2} + i\sqrt{2})^4$

72. $(\sqrt{2} - i\sqrt{2})^4$

73. $\dfrac{2 - 3i}{-1 + 4i}$

74. $\dfrac{3 - i}{-2 + 5i}$

75. $\dfrac{1 + i}{-2i}$

76. $\dfrac{-4 + 5i}{-3i}$

77. i^{11}

78. i^{25}

79. $\dfrac{1}{i^7}$

80. $\dfrac{1}{i^{10}}$

Solve by any method.

81. $x^2 + 6x + 9 = 0$

82. $16z^2 - 40z + 25 = 0$

83. $4x^2 = -7x - 2$

84. $2x^2 + 7x = 4$

85. $x^2 + 3x + 1 = 0$

86. $-5x^2 + 2x + 1 = 0$

87. $4t^2 - 5t - 6 = 0$

88. $6w^2 - w - 12 = 0$

89. $3x^2 + 5x - 1 = 0$

90. $2x^2 + 3x - 1 = 0$

91. $x^4 + x^3 - 27x - 27 = 0$

92. $x^5 - x^3 + x^2 - 1 = 0$

Find the solution set of each equation in the real numbers.

93. $x = 8\sqrt{x} - 15$

94. $(x - 3)^{1/2} - 5(x - 3)^{1/4} + 6 = 0$

95. $\sqrt{x + 8} - 1 = 2x$

96. $4 + \sqrt{x + 2} = x$

97. $\sqrt{2x - 3} - \sqrt{x + 2} = 1$

98. $\sqrt{x - 2} = 3 + \sqrt{x + 1}$

99. $\sqrt[3]{x^2 + 2x} = -1$

100. $\sqrt{2x + 2} + \sqrt{2x + 6} = 4$

In Problems 101–104, solve the given inequality.

101. $x^2 + 3x + 2 > 6$

102. $w^2 + 6w + 8 < 0$

103. $\dfrac{1}{x + 2} \geq \dfrac{1}{2}$

104. $\dfrac{3x - 1}{x + 1} > 2$

Rewrite each expression without absolute value symbols.

105. $\dfrac{|2x| - 2x}{x}$

106. $\dfrac{2x + |x|}{x}$

107. $\dfrac{|x - 2| - |2|}{x}$

108. $\dfrac{|x - 3| - |3|}{x}$

Factor as the difference of two squares. Assume variables represent positive numbers.

109. $x - 4$

110. $x - 9$

111. $x^2 - a$

112. $x^2 - h$

Use factoring to rewrite each of the following as a quotient that is free of negative exponents.

113. $(3x + 2)^{-2} - 6(x + 1)(3x + 2)^{-3}$

114. $(2x + 1)^{-3} - 6(x - 2)(2x + 1)^{-4}$

115. $3(x^2 + 1)^2(6x - 1)^{-1/2} + 4x(x^2 + 1)(6x - 1)^{1/2}$

116. $3(x^2 - x)^2(2x - 1)(8x + 3)^{-1/2} - 4(x^2 - x)^3(8x + 3)^{-3/2}$

Rationalize the numerator and simplify the result.

117. $\dfrac{\sqrt{3x + 3h} - \sqrt{3x}}{h}$

118. $\dfrac{\sqrt{2x + 2h + 1} - \sqrt{2x + 1}}{h}$

119. $\dfrac{x^2(2x + 1)^{-1/2} + 2x(2x + 1)^{1/2}}{(2x + 1)^{3/2}}$

120. $\dfrac{2x^2(4x - 3)^{-1/2} + 2x(4x - 3)^{1/2}}{(4x - 3)^{3/2}}$

Rewrite each of the following so that it contains an expression of the form $c(u^2 \pm a^2)$.

121. $2x^2 + 5x + 1$

122. $4x^2 - 10x + 3$

123. $\dfrac{5}{2x^2 - 2x + 1}$

124. $\dfrac{3}{2x^2 + 4x - 1}$

Find the values of x for which the given expression is (a) zero or (b) undefined.

125. $2x^2(6x + 1)^{-2/3} + 2x(6x + 1)^{1/3}$

126. $x^2(3x - 2)^{-2/3} + 2x(3x - 2)^{1/3}$

127. $2x^2(3x + 4)^{-1/3} + 2x(3x + 4)^{2/3}$

128. $6x^2(9x - 1)^{-1/3} + 2x(9x - 1)^{2/3}$

▲ ▲ CRITICAL THINKING: FIND THE ERRORS

Each of the following nonsolutions has at least one error. Can you find them?

Problem 1 Draw the graph of $|x| \leq 2$.

Nonsolution The graph is shown in Figure 1.12.

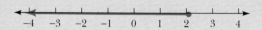

FIGURE 1.12

Problem 2 Solve $|4x - 1| = -3$.

Nonsolution

$$4x - 1 = \pm 3$$
$$4x = 1 \pm 3$$
$$x = \frac{1 \pm 3}{4}$$
$$x = \begin{cases} \dfrac{1 + 3}{4} = 1 \\ \dfrac{1 - 3}{4} = -\dfrac{1}{2} \end{cases}$$

Problem 3 Simplify $(x^{-2}y)^{-3}$.

Nonsolution

$$(x^{-2}y)^{-3} = \frac{1}{(x^2y)^3} = \frac{1}{x^5y^3}$$

Problem 4 Factor by grouping: $r^2 - s^2 - r - s$.

Nonsolution

$$\begin{aligned} r^2 - s^2 - r - s &= (r^2 - s^2) - (r - s) \\ &= (r - s)(r + s) - (r - s) \\ &= (r - s)(r + s - 1) \end{aligned}$$

Problem 5 Perform the subtraction and write the result in lowest terms.

$$\frac{p - 3}{p + 1} - \frac{2p - 1}{p + 2}$$

Nonsolution

$$\frac{p-3}{p+1} - \frac{2p-1}{p+2} = \frac{p-3-2p+1}{(p+1)(p+2)}$$

$$= \frac{-p-2}{(p+1)(p+2)}$$

$$= \frac{-(p+2)}{(p+1)(p+2)}$$

$$= \frac{-1}{p+1}$$

Problem 6 Evaluate $(-27)^{2/3}$.

Nonsolution $\quad (-27)^{2/3} = [(-27)^{1/3}]^2 = (-9)^2 = 81$

Problem 7 If possible, combine $2\sqrt{3} - 4\sqrt{2}$ by using the distributive property.

Nonsolution $\quad 2\sqrt{3} - 4\sqrt{2} = (2-4)(\sqrt{3} - \sqrt{2})$

$$= (-2)(\sqrt{1})$$

$$= -2$$

Problem 8 Change $\sqrt[3]{8x^2y^3}$ to simplest radical form.

Nonsolution $\quad \sqrt[3]{8x^2y^3} = \sqrt{2 \cdot 4 \cdot x^2 \cdot y \cdot y^2} = 2xy\sqrt{2y}$

Problem 9 Evaluate $\sqrt{-25}\ \sqrt{-4}$.

Nonsolution $\quad \sqrt{-25}\ \sqrt{-4} = \sqrt{100} = 10$

Problem 10 Write $\dfrac{1}{2+i}$ in standard form.

Nonsolution $\quad \dfrac{1}{2+i} = \dfrac{1}{2+i} \cdot \dfrac{2-i}{2-i} = \dfrac{2-i}{4-1} = \dfrac{2}{3} - \dfrac{1}{3}i$

Problem 11 Solve by factoring: $6x^2 - x = 1$.

Nonsolution $\quad x(6x-1) = 1$

$$x = 1 \quad \text{or} \quad 6x - 1 = 1$$

$$6x = 2$$

$$x = \frac{1}{3}$$

The solution set is $\{1, 1/3\}$.

Solve the equation: $2 - \sqrt{1 - x} = 2x$.

Nonsolution

$$4 + 1 - x = 4x^2$$

$$0 = 4x^2 + x - 5$$

$$0 = (4x + 5)(x - 1)$$

$$4x + 5 = 0 \qquad \text{or} \qquad x - 1 = 0$$

$$x = -\frac{5}{4} \qquad\qquad\qquad x = 1$$

Since $x = -5/4$ is an extraneous solution, the solution set is $\{1\}$.

Solve the following inequality: $x^2 + 6x + 8 < 3$.

Nonsolution

$$(x + 4)(x + 2) < 3$$

$$x + 4 < 3 \qquad \text{or} \qquad x + 2 < 3$$

$$x < -1 \qquad \text{or} \qquad x < 1$$

The solution set is $\{x : x < -1\} \cup \{x : x < 1\} = (-\infty, 1)$.

Factor $x - 3$ as the difference of two squares.

Nonsolution

$$x - 3 = (\sqrt{x} - 3)(\sqrt{x} + 3)$$

Rewrite $2x^2 + 6x - 7$ in the form $c(u^2 + a^2)$.

Nonsolution

$$2x^2 + 6x - 7 = 2x^2 + 6x + (\ ) - 7 - (\ )$$

$$= 2x^2 + 6x + 9 - 7 - 9$$

$$= (2x + 3)^2 - 16$$

$$= 2\left(x + \frac{3}{2}\right)^2 - 4^2$$

$$= 2(u^2 - a^2), \text{ where } u = x + \frac{3}{2} \text{ and } a = 4$$

Chapter 2

Functions and Graphs

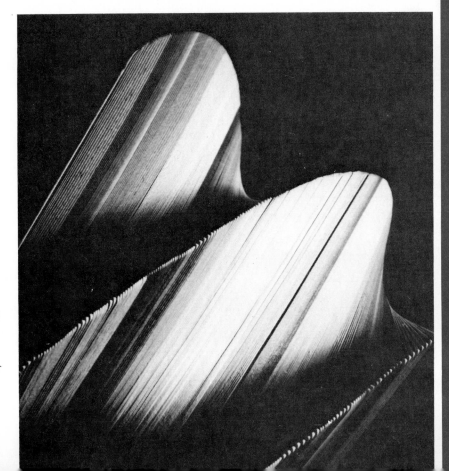

The examples and exercises of this chapter illustrate the use of functions in manufacturing design, dosages of medicine, business profits, wildlife management, and other areas. Many of these applications involve the formulation of functions that are analyzed in calculus courses. Skill with this type of formulation and a knowledge of the material on graphing in this chapter are both essential for successful study of the calculus.

2.1 Functions and Relations

The concept of a function is one of the most important concepts in college-level mathematics. Strangely enough, two quite different formulations of the function concept are commonly used. Some mathematicians prefer one formulation because it is natural and intuitive, while others prefer the second formulation because it is more precise and rigorous. In both, functions are denoted by letters such as f, g, h, and r. (Throughout this chapter we restrict our attention to real variables.)

Historically, a **function** f was first defined as a **correspondence** of a certain type between the elements of two sets. In this correspondence a rule of association exists between the elements of a first set D and those of a second set. The association must be such that for each element x in D there is one and only one associated element y in the second set. This discussion is summarized in the following definition.

Definition 2.1

Let D be a nonempty set of real numbers. A **function** f with **domain** D is a correspondence such that for each element x in D there is one and only one associated element y in a second set of real numbers.

To indicate that y is associated with x by the function f, we write $y = f(x)$ and say that y is the **value of f** at x. The notation $f(x)$ *does not*

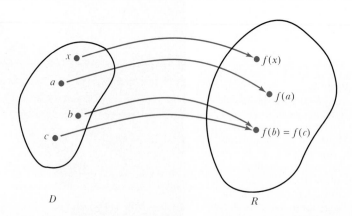

FIGURE 2.1

indicate multiplication. It is read "f of x" or "f at x." The set R of all values of $f(x)$ is called the **range** of f.

With Definition 2.1, it is natural to think of starting with a value of x in D and using the correspondence to obtain a value $y = f(x)$ in R. In this way, the value y is assigned to x. Thinking like this, we call x the **independent variable** and y the **dependent variable,** since the value obtained for y depends on the value selected for x. The diagram in Figure 2.1 illustrates this intuitive definition of a function as a correspondence between the elements of two sets.

In Figure 2.1, it is indicated that $f(b) = f(c)$, with $b \neq c$ in D. This may happen with a function, as illustrated in Example 1. However, it is important to keep in mind that there is a *unique value* $f(x)$ for each x in D.

Example 1 Consider the function f with domain $D = \{-2, 1, 2\}$ and $f(x) = x^2$. The values of f are given by

$$f(-2) = (-2)^2 = 4, \qquad f(1) = 1^2 = 1, \qquad f(2) = 2^2 = 4.$$

We note that $f(-2) = f(2)$ with $-2 \neq 2$, as indicated in Figure 2.2. ❑

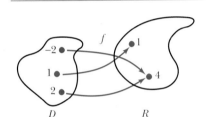

FIGURE 2.2

$D = x$

The second formulation of the definition of a function has become widely used in the last half of the twentieth century. This formulation involves the idea of ordered pairs of real numbers.

By an **ordered pair of real numbers,** we mean a pairing (a, b) of real numbers a and b, where a distinction is made between the pair (a, b) and the pair (b, a), if a and b are not equal. That is, a first position and a second position are designated in the pair of numbers. It is traditional to use parentheses to indicate ordered pairs, where the number on the left is called the **first entry,** or **first component,** and the number on the right is called the **second entry,** or **second component.** Two ordered pairs (a, b) and (c, d) are **equal** if and only if $a = c$ and $b = d$.

Definition 2.2

A **relation** is a nonempty set of ordered pairs (x, y) of real numbers x and y. The **domain** D of a relation is the set of all first-entry elements, or the set of all x-values, that occur in the relation. The **range** of a relation is the set of all y-values that occur in the relation. A **function** is a relation in which no two distinct ordered pairs have the same x-value.

Example 2 The set

$$g = \{(5, -2), (5, 2), (-5, -2), (-5, 2)\}$$

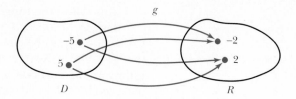

FIGURE 2.3

is a relation with domain $D = \{-5, 5\}$ and range $R = \{-2, 2\}$. We can describe g by the rule:

$$(x, y) \in g \quad \text{if and only if} \quad |x| = y^2 + 1 \quad \text{and} \quad x \in \{-5, 5\}.$$

The relation g can be visualized as in Figure 2.3. ❏

    In our work here, we have little use for relations with domains that have only two or three elements. Our relations are usually determined by a rule relating the entries x and y of the ordered pairs (x, y), and we make the following convention regarding these rules.

> Unless otherwise specified, the domain of a relation described by a certain rule is the set of all real numbers x that yield a real number y in the rule for the relation.

This convention is illustrated in the next example.

Example 3   In each of the following relations, list three sample ordered pairs in the relation and state the domain D and range R of the relation.

a) $h = \{(x, y) : y = x^2\}$ b) $r = \{(x, y) : y = \sqrt{x}\}$

Solution

a)   Three ordered pairs in the relation h are $(1, 1)$, $(2, 4)$, and $(-7, 49)$. The domain of h is the set of all real numbers, since any real number x has a square that is a real number. However, the range consists only of nonnegative real numbers since the rule gives $y = x^2$ and the square of a real number is never negative That is, h has domain[1]

$$D = \mathcal{R}$$

---

1. Recall that $\mathcal{R}$ denotes the set of all real numbers.

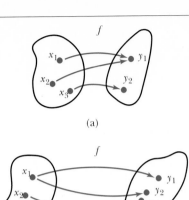

(a)

(b)

FIGURE 2.4

and range

$$R = \{y : y \geq 0\} = [0, \infty).$$

b) Three ordered pairs in r are $(1, 1)$, $(2, \sqrt{2})$, and $(4, 2)$. In this case, the only real numbers that are in the domain are nonnegative real numbers, for they are the only ones that have a square root in the real numbers. The range consists of all nonnegative real numbers, since $\sqrt{x}$ denotes the nonnegative square root of x. That is, r has domain

$$D = [0, \infty)$$

and range

$$R = [0, \infty).$$ ❏

The diagrams in Figure 2.4 illustrate the difference between a function and a relation that is not a function. In part (a) we see that for each x-value there is a unique y-value. Hence the relation f is a function. But the relation f in part (b) is not a function since the x-value, x_1, is associated with two different y-values, y_1 and y_2 ($y_1 \neq y_2$).

Example 4   Which of the relations g, h, and r in the preceding examples are functions?

Solution

a) The relation g in Example 2 contains both of the pairs $(5, -2)$ and $(5, 2)$. This means there are two y-values, -2 and 2, for $x = 5$, so g is not a function.

b) The relation h in Example 3 has the rule $y = x^2$, which clearly gives exactly one y-value for each x. Thus h is a function.

c) The relation r in Example 3 is also a function since $y = \sqrt{x}$ gives a unique value when it is defined.   ❏

In Example 3 we could simply write

$$h(x) = x^2$$

and

$$r(x) = \sqrt{x}$$

to describe the functions h and r. This kind of notation is emphasized in the next two examples.

Example 5   For $F(x) = x^2 - 1$ and $G(x) = \sqrt{x - 1}$, find the values of $F(2)$, $F(-3)$, $G(1)$, and $G(14)$.

Solution
We have

$$F(2) = (2)^2 - 1 = 3;$$

$$F(-3) = (-3)^2 - 1 = 8;$$

$$G(1) = \sqrt{1 - 1} = \sqrt{0} = 0;$$

$$G(14) = \sqrt{14 - 1} = \sqrt{13}.$$

Using the ordered pair notation, we could give the same information in this form: $(2, 3) \in F$, $(-3, 8) \in F$, $(1, 0) \in G$, $(14, \sqrt{13}) \in G$.   ❑

Example 6   Suppose $f(x) = 2x^2 - 3$. Evaluate

a) $f(-x)$,   b) $f(x + h)$.

Solution

a) $f(-x) = 2(-x)^2 - 3 = 2x^2 - 3 = f(x)$

b) $f(x + h) = 2(x + h)^2 - 3 = 2(x^2 + 2xh + h^2) - 3$

$$= 2x^2 + 4xh + 2h^2 - 3$$ ❑

The usual rectangular coordinate system in a plane is often very helpful in working with relations and functions. It can be used to provide a "picture" or graph of a relation. With this purpose in mind, we briefly review the construction of a **rectangular coordinate system**[2] in a plane.

To construct a rectangular coordinate system in a plane, we begin with a horizontal line, called the **x-axis,** and a vertical line, called the **y-axis,** which intersect at a point O, called the **origin.** (See Figure 2.5.)

On each of these lines, we set up a one-to-one correspondence between the points on the line and the real numbers, as was described in Section 1.1. The correspondence on the x-axis is set up so that 0 is at the origin, with the positive direction to the right, indicated by an arrowhead, and the negative direction to the left. Similarly, the correspondence on the y-axis is set up so that 0 is at the origin, with the

---

2. The rectangular coordinate system is also frequently referred to as the **Cartesian coordinate system.** This is in honor of René Descartes (1596–1650), the French mathematician who is credited with inventing the system.

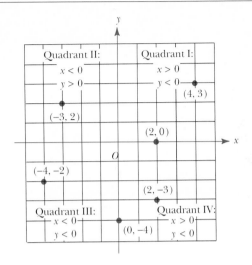

FIGURE 2.5

positive direction upward, indicated by an arrowhead, and the negative direction downward.

We can now set up a one-to-one correspondence between points in the plane and ordered pairs (x, y) of real numbers. To describe this correspondence, let P be a point in the plane. We let x denote the directed horizontal distance from the y-axis to the point P, so that x is positive if P is to the right of the y-axis, x is negative if P is to the left of the y-axis, and x is zero if P is on the y-axis. Similarly, we let y denote the directed vertical distance from the x-axis to the point P, where y is positive in the upward direction, negative in the downward direction, and zero on the x-axis. The ordered pair (x, y) is then assigned to the point P. The first entry, x, is called the **abscissa**, or **x-coordinate,** of P, and the second entry, y, is called the **ordinate**, or **y-coordinate,** of P. The ordered pair (x, y) is referred to as the **coordinates** of P.

Conversely, each ordered pair (x, y) determines a unique point P that has the pair (x, y) as coordinates. The point P is located by simply using x as the directed horizontal distance from the y-axis, and y as the directed vertical distance from the x-axis. Several points are located by their coordinates in Figure 2.5.

The axes in a Cartesian (or rectangular) coordinate system separate the plane into four regions that are called **quadrants** and numbered I, II, III, and IV, as shown in Figure 2.5. The signs of the coordinates in the ordered pair (x, y) are determined by the quadrant in which the point with coordinates (x, y) lies, as indicated in Figure 2.5. The points on the coordinate axes are not in any quadrant.

As mentioned before, a rectangular coordinate system can be used to obtain a picture of a relation. This picture is formed by sketching the graph of the relation.

Definition 2.3

The **graph** of a relation is the set of all points whose coordinates (x, y) are members of the relation.

The sketch of the graph is made by plotting several points on the graph of the relation and then drawing a curve through these points. This is illustrated in the following examples.

Example 7   Sketch the graph of the function $f(x) = x^2$.

Solution

We first assign several sample values to x and compute the corresponding values of $f(x)$. The resulting ordered pairs are recorded in the table of Figure 2.6. We then locate the points corresponding to these coordinates, and sketch the graph as well as possible from these points. The graph provides a picture of the behavior of the function, and it clearly shows that the domain D of f is the set of all real numbers, and the range R of f is the set of all nonnegative real numbers.

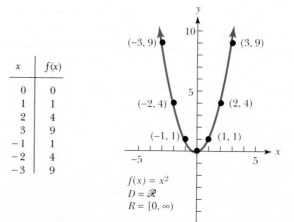

x	$f(x)$
0	0
1	1
2	4
3	9
-1	1
-2	4
-3	9

$f(x) = x^2$
$D = \mathcal{R}$
$R = [0, \infty)$

FIGURE 2.6

Example 8   Sketch the graph of the relation $g = \{(x, y) : x + 1 = |y|\}$.

Solution

Following the same procedure as in Example 7, we obtain the table given in Figure 2.7. It soon becomes evident that g is not a function, since two values of y are sometimes obtained from a given value of x. Also, we discover that if x is assigned a value less than -1, then there is no y that satisfies the equation, since an absolute value cannot be negative. When the points corresponding to the coordinates in the table are plotted, it appears that they lie along two half-lines that have a common endpoint at $(-1, 0)$, and the graph is drawn accordingly. It is clear that g has domain $D = [-1, \infty)$ and range $R = \mathcal{R}$.

x	y
0	1
0	-1
1	2
1	-2
2	3
2	-3
-1	0
-2	Undefined

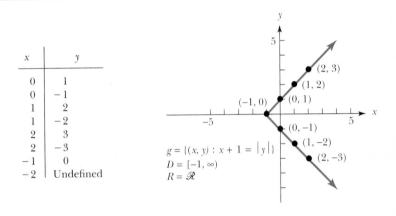

$g = \{(x, y) : x + 1 = |y|\}$
$D = [-1, \infty)$
$R = \mathcal{R}$

FIGURE 2.7

The graph of f in Figure 2.6 makes it easy to see that f is a function because there is exactly one y-value for each x-value. Similarly, the graph of g in Figure 2.7 shows that there are two y-values for any $x > -1$, so that g is not a function. A graphical description of this situation is that some vertical lines intersect the graph of g at more than one point, and that no vertical line crosses the graph of f at more than one point. This illustrates the **Vertical Line Test** for functions, which is stated below.

Vertical Line Test

If any vertical line intersects the graph of a relation at two or more points, the relation is *not* a function. On the other hand, if no vertical line intersects the graph in more than one point, the relation is a function.

In most cases, the rule defining a relation is simply an equation in x and y which determines the ordered pairs (x, y) that belong to the relation. In such a case it is common to refer to the graph of the relation as the **graph of the equation.** As examples, the graph of the equation $y = x^2$ is given in Figure 2.6, and the graph of the equation $x + 1 = |y|$ is given in Figure 2.7.

For any real number x, the notation $[x]$ denotes the *greatest integer less than or equal to x.* Some values of $[x]$ are given by

$$[2] = 2, \quad [4.12] = 4, \quad [\sqrt{10}] = 3, \quad [-5.9] = -6, \quad [-\sqrt{10}] = -4.$$

A more precise definition of $[x]$ is provided by the following statement.

If n is an integer such that $n \leq x < n + 1$, then $[x] = n$.

The **greatest integer function** is the function defined by $f(x) = [x]$.

Example 9 Sketch the graph of the greatest integer function
$$f(x) = [x].$$

Solution
To sketch the graph, we assign some values to n in the statement above that defined $[x]$. For $n = 1$, the statement says that if $1 \leq x < 2$, then $[x] = 1$. This and some similar statements are tabulated in Figure 2.8. From the table, we see that the graph consists of horizontal line segments over intervals of length 1, as shown in Figure 2.8.

x	y
$0 \leq x < 1$	0
$1 \leq x < 2$	1
$2 \leq x < 3$	2
$-1 \leq x < 0$	-1

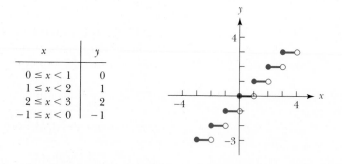

FIGURE 2.8

Because of the resemblance of its graph to stair steps, the greatest integer function is called a **step function.** The next example illustrates another type of step function.

Example 10   Suppose the dosage of a certain pediatric drug depends on the weight of the child in the following manner: 30 milligrams of the drug plus an additional 10 milligrams for each 5 pounds or portion of 5 pounds of body weight above 20 pounds. Sketch a graph of the dosage function for children weighing from 10 through 40 pounds.

Solution
Let x represent the body weight of a child, and $D(x)$ the corresponding dosage. The domain of the dosage function is the interval $[10, 40]$. For a child weighing from 10 through 20 pounds, the dosage is 30 milligrams. For children weighing more than 20 pounds through 25 pounds, an additional 10 milligrams of drug should be administered. Following this type of reasoning, we complete the chart and sketch the graph in Figure 2.9.

Weight x	Dosage $D(x)$
$10 \leq x \leq 20$	30
$20 < x \leq 25$	$30 + 1(10) = 40$
$25 < x \leq 30$	$30 + 2(10) = 50$
$30 < x \leq 35$	$30 + 3(10) = 60$
$35 < x \leq 40$	$30 + 4(10) = 70$

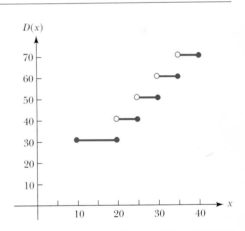

FIGURE 2.9

Exercises 2.1

Give the domain D and range R of each function. (See Example 3.)

1. $f(x) = 3x$

2. $g(x) = 4x - 3$

3. $h(x) = -\sqrt{x - 4}$

4. $F(x) = \sqrt{x + 2}$

5. $G(x) = x^2 - 3$

6. $p(x) = 2(x - 1)^2 + 1$

7. $y = \dfrac{1}{x - 1}$

8. $y = \dfrac{2}{x - 3}$

9. $y = |4 - x|$

10. $y = |5x - 7|$

11. $y = |x - 1| + 3$

12. $y = |x + 2| - 1$

Which of the following relations are functions? (See Example 4.)

13. $g = \{(1, 2), (2, 1), (3, 2)\}$

14. $g = \{(-1, 0), (0, -1), (1, 0)\}$

15. $h = \{(2, 1), (1, 2), (2, 3)\}$

16. $h = \{(0, -1), (-1, 0), (0, 1)\}$

17. $p = \{(x, y) : x = 3\}$ **18.** $p = \{(x, y) : x = -2\}$

19. $h = \{(x, y) : x = y + 1\}$

20. $h = \{(x, y) : x = y - 3\}$

21. $r = \{(x, y) : y = |x|\}$

22. $r = \{(x, y) : y = |x + 1|\}$

If $f(x) = 4x - 2$ and $g(x) = x^2$, find each of the following. (See Examples 5 and 6.)

23. $f(3)$ **24.** $f(-5)$

25. $g(4)$ **26.** $g(-7)$

27. $f(b)$ **28.** $g(a)$

29. $f(-x)$ **30.** $g(-x)$

31. $g(x + 1)$ **32.** $f(x + 1)$

If $f(x) = 2x$ and $g(x) = x^2 - x$, find each of the following.

33. $f(3)$ **34.** $f(-2)$

35. $g(-3)$ **36.** $g(2)$

37. $f(0)$ **38.** $g(0)$

39. $f(-x)$ **40.** $g(-x)$

41. $f(x + h)$ **42.** $g(x + h)$

43. $g(x + h) - g(x)$ **44.** $f(x + h) - f(x)$

Find $\dfrac{f(x + h) - f(x)}{h}$ in lowest terms for each of the following.

45. $f(x) = 3x - 4$ **46.** $f(x) = 4x + 5$

47. $f(x) = 2x^2 + 1$ **48.** $f(x) = 3x^2 - 1$

49. $f(x) = 3x^2 - 2x + 4$ **50.** $f(x) = 2x^2 - 3x + 5$

51. $f(x) = 2x^3 + 1$ **52.** $f(x) = x^3 + 2x$

Sketch the graph of the following equations. (See Examples 7–9.)

53. $y = |x|$ **54.** $y = -|x|$

55. $y = |x| + 1$ **56.** $y = |x| - 1$

57. $y = |x + 1|$ **58.** $y = |x - 1|$

59. $y = x^2 - 4$ **60.** $y = x^2 + 1$

61. $y = [x + 1]$ **62.** $y = [x - 1]$

63. $y = [2x]$ **64.** $y = [3x]$

State the domain D and range R of the relation whose graph is given in Exercises 65–72. In each case determine if the relation is a function.

65.
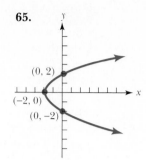

66.
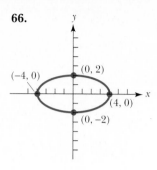

67.
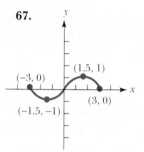

68.
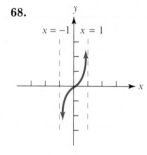

69.
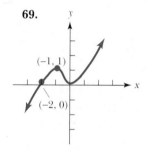

70.
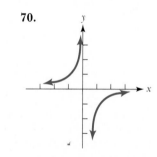

71.
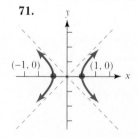

72.
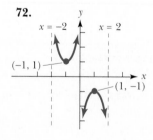

73. Suppose a delivery service charges for delivery of items based on mileage in the following way: $10 plus an additional $0.50 per mile or fraction of a mile. Sketch the graph of the delivery charges for mileage from 0 through 10 miles.

74. Postage rates for a first-class letter are determined as follows: 25¢ for the first ounce plus an additional 20¢ for each ounce or fraction of an ounce above 1 ounce and less than 12 ounces. Graph the function representing first-class postage rates for letters between 0 and 12 ounces.

75. The members of the Big Woods Hunting Club have 120 feet of fencing to build a dog pen. They plan to build a rectangular pen with one cross fence as shown in the figure below.
   a) Using the fact that there is a total of 120 feet of fencing, express the length l in terms of the width w.
   b) Use the results of part (a) to express the area A of the pen as a function of the width w.

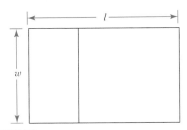

EXERCISE 75

76. Suppose a political campaign poster is to contain 600 square inches of printing with 6-inch margins at the top and bottom and 4-inch margins along the sides, as shown in the figure.
   a) Using the fact that the printed area totals 600 square inches, express the width w of the printed area in terms of its length l.
   b) Using the results of part (a), express the total area A of the poster as a function of l.

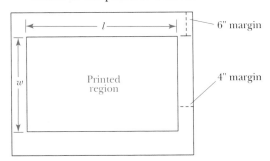

EXERCISE 76

77. An open box (without a top) is to be made from a piece of cardboard 18 inches square by cutting equal square pieces from each corner and folding up the sides as shown. Express the volume V of the box as a function of x.

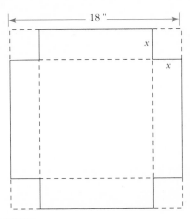

EXERCISE 77

78. The closed box with a square bottom and top in the following figure is to be made by using 400 square inches of plastic.
   a) Express the height h of the box in terms of the width w of the bottom.
   b) Write the volume V of the box as a function of w.

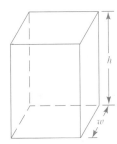

EXERCISE 78

79. A rectangular sheet of tin is rolled into a cylinder as shown in the figure.
   a) If the perimeter of the sheet is 32 inches, express the width x of the sheet in terms of the length y.
   b) Express the area A of the circular end of the

Circle: $\begin{cases} \text{Circumference} = 2\pi r \\ \text{Area} = \pi r^2 \end{cases}$

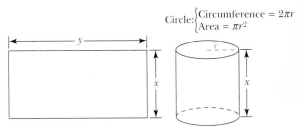

Cylinder:
Volume = (area of the circular end) × (height of the cylinder)

EXERCISE 79

cylinder in terms of y, the length of the sheet.

c) Express the volume V of the cylinder as a function of y.

80. The piece of wire in the following figure is 20 inches long and is to be cut into two pieces. The piece of length x is bent to form a square and the other to form an equilateral triangle. Find a function f that represents the sum of the areas of the square and the triangle in terms of the length x. [*Hint:* The area A of an equilateral triangle with

side a is given by $A = \sqrt{3}\, a^2/4$.]

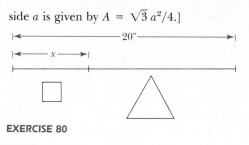

EXERCISE 80

2.2   **The Algebra of Functions**

Given two functions f and g, it is natural to consider creating new functions by using the operations of addition, subtraction, multiplication, and division with f and g. For example, if $f(x) = x^2$ and $g(x) = x - 2$, it is natural to define the sum $f + g$ by

$$(f + g)(x) = x^2 + x - 2 = f(x) + g(x);$$

the difference $f - g$ by

$$(f - g)(x) = x^2 - (x - 2) = f(x) - g(x);$$

the product $f \cdot g$ by

$$(f \cdot g)(x) = x^2(x - 2) = f(x) \cdot g(x);$$

and the quotient f/g by

$$\left(\frac{f}{g}\right)(x) = \frac{x^2}{x - 2} = \frac{f(x)}{g(x)}, \quad \text{if } g(x) \neq 0.$$

Care must be taken when forming these new functions. Consider the next example.

Example 1   Suppose $f(x) = -x$ and $g(x) = \sqrt{x}$. Describe $f + g$ and evaluate $f + g$ at $x = -4$.

Solution
The sum $f + g$ is

$$(f + g)(x) = -x + \sqrt{x}.$$

Now suppose $x = -4$. Then

$$(f + g)(-4) = -(-4) + \sqrt{-4} = 4 + \sqrt{-4},$$

which is not defined in the real numbers. This is because -4 is not in the domain of g. Any element in the domain of $f + g$ has to be an element of both the domain of f and the domain of g. Thus a complete description of $f + g$ requires that the domain be fully described. Hence the sum $f + g$ is

$$(f + g)(x) = -x + \sqrt{x}, \quad x \geq 0.$$ ❏

Example 2 Let $f(x) = \sqrt{x - 4}$ and $g(x) = \sqrt{5 - x}$. Describe completely $f + g$, $f - g$, $f \cdot g$, and f/g.

Solution
The domain of f is $D_f = [4, \infty)$, and the domain of g is $D_g = (-\infty, 5]$. Since $D_f \cap D_g = [4, 5]$, we have

$$(f + g)(x) = \sqrt{x - 4} + \sqrt{5 - x}, \quad x \in [4, 5];$$

$$(f - g)(x) = \sqrt{x - 4} - \sqrt{5 - x}, \quad x \in [4, 5];$$

$$(f \cdot g)(x) = \sqrt{x - 4} \, \sqrt{5 - x} = \sqrt{(x - 4)(5 - x)}, \quad x \in [4, 5];$$

$$\left(\frac{f}{g}\right)(x) = \frac{\sqrt{x - 4}}{\sqrt{5 - x}} = \sqrt{\frac{x - 4}{5 - x}}, \quad x \in [4, 5).$$

Note that 5 had to be excluded from the domain of f/g since f/g is not defined whenever $g(x) = 0$. ❏

We summarize the results of the preceding discussion in the following definition.

Definition 2.4

Suppose f and g are real-valued functions whose domains are D_f and D_g, respectively. The functions $f + g$, $f - g$, $f \cdot g$, and f/g are defined by the following equations.

$$(f + g)(x) = f(x) + g(x)$$

$$(f - g)(x) = f(x) - g(x)$$

$$(f \cdot g)(x) = f(x) \cdot g(x)$$

$$\left(\frac{f}{g}\right)(x) = f(x)/g(x), \quad g(x) \neq 0$$

The domain of each of these functions is $D_f \cap D_g$, with the additional condition that f/g is not defined when $g(x) = 0$.

Whether a student graduates is a function of passing each required course. Whether he passes each course is a function of what grade he earns on his examinations. What grade he earns on his examinations is a function of how much he studies. How much he studies is certainly a

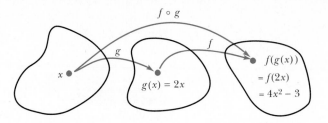

FIGURE 2.10

function of some other variable, and on, and on, and on. This sort of chain of events has a mathematical counterpart called the composition of functions.

Consider the diagram in Figure 2.10 of the two functions $f(x) = x^2 - 3$ and $g(x) = 2x$. For any x in the domain of g, we can write

$$f(g(x)) = f(2x) = (2x)^2 - 3 = 4x^2 - 3.$$

definition of g definition of f

Thus the functions g and f have been used to obtain yet another function. We call this new function the composition (or composite) function and denote it by $f \circ g$. For this example, we have

$$(f \circ g)(x) = f(g(x)) = 4x^2 - 3$$

for any real x.

Definition 2.5

If f and g are functions, the **composition** (or **composite**) **function** $f \circ g$ is defined by

$$(f \circ g)(x) = f(g(x)).$$

The domain of $f \circ g$ is the set of all x in the domain of g such that f is defined at $g(x)$.

Example 3   Let $f(x) = 2x - 1$ and $g(x) = x^2 + 1$. Determine $(f \circ g)(2)$ and $(g \circ f)(2)$.

Solution
Since $g(2) = 5$ and $f(2) = 3$, we have

$$(f \circ g)(2) = f(g(2)) = f(5) = 2 \cdot 5 - 1 = 9$$

and

$$(g \circ f)(2) = g(f(2)) = g(3) = 3^2 + 1 = 10. \qquad \Box$$

Note in Example 3 that $(f \circ g)(x) \neq (g \circ f)(x)$, which is usually the case. In Section 2.5 we shall examine the situation when $f \circ g$ and $g \circ f$ are equal, and in particular when $(f \circ g)(x) = (g \circ f)(x) = x$.

Example 4   Let $f(x) = 1/x$ and $g(x) = x - 3$. Determine:

a) $(f \circ g)(x)$ and the domain of $f \circ g$,

b) $(g \circ f)(x)$ and the domain of $g \circ f$.

Solution

a) $(f \circ g)(x) = f(g(x)) = f(x - 3) = \dfrac{1}{x - 3}$

The domain of $f \circ g$ is $\{x : x \neq 3\} = (-\infty, 3) \cup (3, \infty)$.

b) $(g \circ f)(x) = g(f(x)) = g\left(\dfrac{1}{x}\right) = \dfrac{1}{x} - 3$

The domain of $g \circ f$ is $\{x : x \neq 0\} = (-\infty, 0) \cup (0, \infty)$.   ❑

Exercises 2.2

Suppose $f(x) = x + 2$, $g(x) = 2x - 3$, and $h(x) = x^2$. Evaluate each of the following. (See Examples 1 and 3.)

1. $(f + g)(3)$
2. $(f + g)(2)$
3. $(h - g)(-1)$
4. $(g - h)(0)$
5. $(f \cdot h)(-3)$
6. $(h \cdot g)(-2)$
7. $(f \cdot f)(0)$
8. $(g \cdot g)(1)$
9. $(f/g)(1)$
10. $(g/h)(1)$
11. $(h/g)(0)$
12. $(f/h)(-2)$
13. $(f - g + h)(1)$
14. $(f + g - h)(1)$
15. $((f \cdot g) - h)(2)$
16. $(f - (g \cdot h))(-1)$
17. $(f \circ g)(3)$
18. $(g \circ f)(3)$
19. $(h \circ g)(0)$
20. $(h \circ f)(-3)$
21. $(f \circ f)(-1)$
22. $(g \circ g)(0)$
23. $((f \circ g) \circ h)(-2)$
24. $((f \circ g) \circ h)(1)$

For the given functions, determine $f + g$, $f - g$, $f \cdot g$, and f/g, and state their domains. (See Example 2.)

25. $f(x) = x - 2$, $g(x) = 2x + 1$
26. $f(x) = 3x$, $g(x) = x - 2$
27. $f(x) = 4x - 1$, $g(x) = x(x - 1)$

28. $f(x) = x - 3$, $g(x) = (x - 1)(x - 2)$
29. $f(x) = \sqrt{x}$, $g(x) = 2x$
30. $f(x) = \sqrt{x - 1}$, $g(x) = 4x$
31. $f(x) = 2x + 1$, $g(x) = \sqrt{x + 2}$
32. $f(x) = 3x - 4$, $g(x) = \sqrt{x - 2}$
33. $f(x) = \sqrt{x - 2}$, $g(x) = \sqrt{3 - x}$
34. $f(x) = \sqrt{x + 1}$, $g(x) = \sqrt{4 - x}$
35. $f(x) = \sqrt{x + 1}$, $g(x) = \sqrt{x + 6}$
36. $f(x) = \sqrt{x - 2}$, $g(x) = \sqrt{x - 4}$
37. $f(x) = x - 4$, $g(x) = x^2 - 5x + 6$
38. $f(x) = 2x + 7$, $g(x) = x^2 - 4x - 8$
39. $f(x) = 6x^2 + 2x - 5$, $g(x) = x^2 + 3x - 4$
40. $f(x) = 5x^2 - 7$, $g(x) = x^2 - 4$

For the given functions, determine $f \circ g$ and $g \circ f$, and state their domains. (See Example 4.)

41. $f(x) = x + 1$, $g(x) = x^2$
42. $f(x) = 2x - 1$, $g(x) = x^3$
43. $f(x) = x^{10}$, $g(x) = x - 1$
44. $f(x) = x^{40}$, $g(x) = 2x + 5$

45. $f(x) = 2x^2 + x, \quad g(x) = x + 3$
46. $f(x) = x - 1, \quad g(x) = 3x^2 + 4x - 1$
47. $f(x) = \sqrt{x}, \quad g(x) = x - 3$
48. $f(x) = \sqrt{x - 1}, \quad g(x) = 2x$
49. $f(x) = 1/x, \quad g(x) = 1 - x$
50. $f(x) = 1/(x + 2), \quad g(x) = x - 1$
51. $f(x) = \sqrt{x + 2}, \quad g(x) = x - 3$
52. $f(x) = \sqrt{x + 4}, \quad g(x) = x - 5$
53. $f(x) = 3x - 5, \quad g(x) = (x + 5)/3$
54. $f(x) = 7x - 1, \quad g(x) = (x + 1)/7$
55. $f(x) = 1/(x + 1), \quad g(x) = (1 - x)/x$
56. $f(x) = 2/(x - 1), \quad g(x) = (x + 2)/x$

57. $f(x) = \sqrt{x - 1}, \quad g(x) = x^2 + 1$
58. $f(x) = \sqrt[3]{x + 2}, \quad g(x) = x^3 - 2$

Determine two functions f and g such that $(f \circ g)(x)$ is the given function.

59. $(f \circ g)(x) = x^3 - 1$　**60.** $(f \circ g)(x) = (x - 1)^3$
61. $(f \circ g)(x) = \sqrt{x + 3}$　**62.** $(f \circ g)(x) = \sqrt{x} + 3$
63. $(f \circ g)(x) = (2x - 9)^{50}$　**64.** $(f \circ g)(x) = (4x + 7)^{23}$
65. $(f \circ g)(x) = \left(\dfrac{1}{x + 3}\right)^3$
66. $(f \circ g)(x) = \left(\dfrac{1}{1 - x}\right)^4$

2.3 | Linear Functions

One of the simplest and most useful types of functions is the linear function. A linear function is a special case of a linear relation, as stated in the following definition.

Definition 2.6

If A, B, and C are real numbers, with not both A and B zero, the set

$$\{(x, y) : Ax + By = C\}$$

is called a **linear relation.** A **linear function** is a linear relation that is a function.

Linear relations get their name from the fact that their graphs are always straight lines. That is, *the graph of any equation of the form*

$$Ax + By = C,$$

with not both A and B zero, is a **straight line.** For this reason, an equation of this type is called a **linear equation.** If there is a point where the graph crosses the y-axis, the y-coordinate of that point is called a **y-intercept** of the graph. Similarly, the x-coordinate of a point where the graph crosses the x-axis is called an **x-intercept.**

Example 1　Find the x- and y-intercepts, if they exist, and sketch the graph of the equation

$$2x + 3y = 6.$$

Solution

To find the y-intercept, we let $x = 0$ and obtain $3y = 6$. Thus 2 is the y-intercept. Similarly, $y = 0$ gives $2x = 6$, and 3 is the x-intercept. Plotting the points $(0, 2)$ and $(3, 0)$, we draw the graph as in Figure 2.11. The point $(2, \frac{2}{3})$ is included as a check on the line drawn through the other two points.

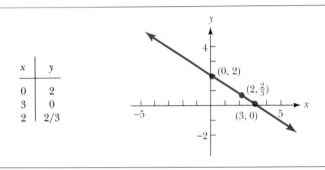

FIGURE 2.11   ❏

Two special cases of the equation $Ax + By = C$ are worth noting. One of these is the equation

$$y = c.$$

The graph of $y = c$ consists of all points with coordinates of the form (x, c), where x may be any real number. Thus the graph of $y = c$, where c is a constant, is a **horizontal straight line.** Similarly, the graph of the equation

$$x = c,$$

where c is a constant, is a **vertical straight line.** As examples, the graphs of $y = 4$ and $x = -2$ are drawn in Figure 2.12.

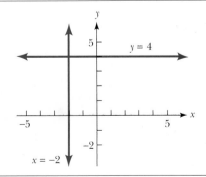

FIGURE 2.12

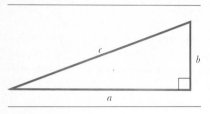

FIGURE 2.13

The *Distance Formula* and the concept of the *slope* of a line are extremely useful in the study of analytic geometry and the differential calculus, and in many other areas as well. We turn our attention first to the Distance Formula.

Let a, b, and c represent the lengths of the sides of a right triangle, with c the length of the hypotenuse. (See Figure 2.13.) The **Pythagorean Theorem** states the following relationship between the lengths of the sides.

$$c^2 = a^2 + b^2 \quad \text{or} \quad c = \sqrt{a^2 + b^2}$$

Suppose now that we have a right triangle drawn as in Figure 2.14. The coordinates of the vertices of the right triangle are (x_1, y_1), (x_2, y_1), and (x_2, y_2), as indicated in the figure. For the figure as drawn, the length of the side parallel to the x-axis is $x_2 - x_1$, and the length of the side parallel to the y-axis is $y_2 - y_1$. The Pythagorean Theorem can be used to determine the length d of the hypotenuse:

$$d = \sqrt{(x_2 - x_1)^2 + (y_2 - y_1)^2}.$$

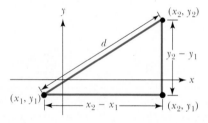

FIGURE 2.14

The length d of the hypotenuse represents the distance between the points with coordinates (x_1, y_1) and (x_2, y_2). This equation holds independently of the quadrants in which the points lie, and independently of the points' orientation to each other. This result is known as the *Distance Formula*.

Theorem 2.7	The Distance Formula

The distance d between the points with coordinates (x_1, y_1) and (x_2, y_2) is given by

$$d = \sqrt{(x_2 - x_1)^2 + (y_2 - y_1)^2}.$$

Example 2 The distance between the points $(-1, 3)$ and $(7, -3)$ is given by

$$d = \sqrt{[7 - (-1)]^2 + (-3 - 3)^2}$$
$$= \sqrt{64 + 36}$$
$$= 10.$$

In using the distance formula, we chose $(x_1, y_1) = (-1, 3)$, and $(x_2, y_2) = (7, -3)$. The same result would have been obtained, however, if we had chosen $(x_1, y_1) = (7, -3)$ and $(x_2, y_2) = (-1, 3)$. ❏

We begin our study of the slope of a line with the following definition.

Definition 2.8

The **slope** m of the line through two distinct points (x_1, y_1) and (x_2, y_2) is given by

$$m = \frac{y_2 - y_1}{x_2 - x_1}.$$

Two special cases should be noted. If the line is horizontal, then $y_1 = y_2$ and $x_1 \neq x_2$ in the definition, and $m = 0$. That is, **a horizontal line has slope 0.** If the line is vertical, then $x_1 = x_2$ and $y_1 \neq y_2$ in the definition, and $m = (y_2 - y_1)/0$ is undefined, since division by zero is impossible. This means that **the slope of a vertical line is undefined.**

The slope of a line that is not horizontal or vertical is a real number $m \neq 0$, and m is *independent of the choice of the points* (x_1, y_1) and (x_2, y_2). In Figure 2.15 another pair of points (x_1', y_1') and (x_2', y_2') on the same

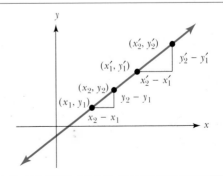

FIGURE 2.15

line is indicated, and we see that

$$\frac{y_2' - y_1'}{x_2' - x_1'} = \frac{y_2 - y_1}{x_2 - x_1}.$$

This follows from the fact that the ratios of corresponding sides of similar triangles are always equal.

Example 3   Find the slope of the line $2x - y = 4$.

Solution
To find the slope, we locate two points on the line and use the formula in Definition 2.8. If we let $x = 1$ and solve for y in the equation of the line, we get $y = -2$. That is, $(1, -2)$ is on the graph. For $x = 3$, we get $y = 2$, and $(3, 2)$ is on the graph. Thus the slope of the line is

$$m = \frac{2 - (-2)}{3 - 1} = \frac{4}{2} = 2.$$

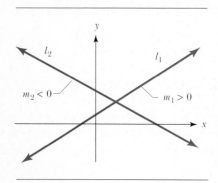

FIGURE 2.16

If the graph of a line slants upward to the right, it indicates that y increases as x increases along the line, and the slope of the line is **positive.** However, if the graph slants downward to the right, then y decreases as x increases along the line, and the slope of the line is **negative.** These situations are pictured in Figure 2.16, where line l_1 has slope $m_1 > 0$ and line l_2 has slope $m_2 < 0$.

A given straight line would be the graph of many different equations, because the linear equation $Ax + By = C$ can be written in many equivalent ways. The form $Ax + By = C$ is called a **standard form** of the equation of a straight line, but other forms are sometimes more useful. Two of these other forms are presented in Theorems 2.9 and 2.10.

Suppose that a line has slope m and that b is the y-intercept. Then $(0, b)$ is on the line, and any other point (x, y) is on the line if and only if

$$\frac{y - b}{x - 0} = m.$$

Solving for y, we have

$$y = mx + b,$$

and this is an equation for the line. This result is stated formally in the following theorem.

> ### Theorem 2.9   Slope-Intercept Form
>
> An equation of the straight line that has slope equal to m and y-intercept equal to b is
>
> $$y = mx + b.$$
>
> This form of the equation of the line is called the **slope-intercept form.**

Another special form for the equation of a straight line is the point-slope form. This is described in the next theorem.

> ### Theorem 2.10   Point-Slope Form
>
> An equation of the straight line that has slope m and passes through the point (x_1, y_1) is
>
> $$y - y_1 = m(x - x_1). \tag{1}$$
>
> This form of the equation of the line is called the **point-slope form.**

To see why Theorem 2.10 is true, suppose that m is the slope of the line and (x_1, y_1) is a point on the line. If (x, y) is any other point on the line, then

$$\frac{y - y_1}{x - x_1} = m$$

by the definition of the slope. When both sides of this equation are multiplied by $x - x_1$, we have

$$y - y_1 = m(x - x_1).$$

This is the equation stated in the theorem. We note that this last equation is satisfied when $x = x_1$ and $y = y_1$, so every point on the line, including (x_1, y_1), satisfies Eq. (1).

Example 4   Find an equation of the straight line that has slope -2 and passes through the point $(-1, 4)$, and draw the graph.

Solution
Using $m = -2$ and $(x_1, y_1) = (-1, 4)$ in the point-slope form of the equation, we have

$$y - 4 = -2(x + 1)$$

as an equation of the line. The graph is drawn in Figure 2.17.   ❑

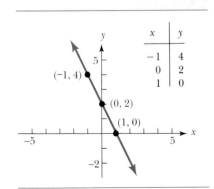

x	y
-1	4
0	2
1	0

FIGURE 2.17

Example 5   Find an equation of the straight line passing through the points $(-1, 3)$ and $(2, -4)$.

Solution

Here we are given two distinct points, and any two distinct points determine a unique line. We can find the slope of the line by using these two points in the formula from Definition 2.8. We get

$$m = \frac{-4 - 3}{2 - (-1)} = -\frac{7}{3}.$$

Now we can use the point-slope form with this slope and either of the points $(-1, 3)$ or $(2, -4)$ to write an equation of the line. Using $(-1, 3)$, we have

$$y - 3 = -\frac{7}{3}(x + 1),$$

and using $(2, -4)$, we have

$$y + 4 = -\frac{7}{3}(x - 2).$$

Both of these equations reduce to $7x + 3y = 2$.   ❏

It is easy to see that **two nonvertical lines are parallel if and only if they have the same slope.** Slopes can also be used to determine whether or not two nonvertical lines are perpendicular to each other. If line l_1 has slope m_1 and line l_2 has slope m_2, then **l_1 and l_2 are perpendicular to each other if and only if $m_1 m_2 = -1$.** In other words, l_1 and l_2 *intersect at right angles if and only if the slope of one line is the negative reciprocal of the slope of the other line.*

Example 6   Find an equation of the line that is perpendicular to the line $-3x + y = 7$ and has -5 as the y-intercept.

Solution

In order for a line to be perpendicular to $-3x + y = 7$, it must have slope equal to $-\frac{1}{3}$. Using $m = -\frac{1}{3}$ and $b = -5$ in the slope-intercept form, we obtain

$$y = -\frac{1}{3}x - 5$$

as an equation of the line.   ❏

Some functions have graphs that are composed of parts of lines instead of one complete line. Such graphs are sometimes called **broken-line graphs.** A function with a graph of this type is given in the next example.

Example 7   Draw the graph of the function f defined by

$$f(x) = \begin{cases} x + 1 & \text{if } x \le 2, \\ 2 - \dfrac{x}{2} & \text{if } x > 2. \end{cases}$$

Solution

For $x \le 2$, the graph of f coincides with the line $y = x + 1$. For $x > 2$, it coincides with the line $y = 2 - x/2$. A table of values and the graph are shown in Figure 2.18. Note that the point $(2, 1)$ is plotted as an open circle to show that it is not on the graph of f.

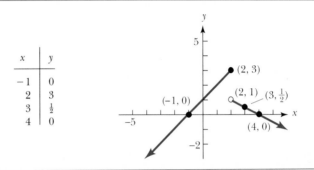

x	y
-1	0
2	3
3	$\frac{1}{2}$
4	0

FIGURE 2.18

Most businesses use equipment that loses its value over a period of time. Such a loss in value is called **depreciation.** We define the net cost N to be the difference between the original cost C and the scrap value S:

$$N = C - S.$$

Suppose that an item has a lifespan of n years. One of the simplest methods for computing depreciation is based on the assumption that an item depreciates $1/n$ of its net cost each year, so that after n years it is totally depreciated. This method is called **linear,** or **straight-line, depreciation.**

Example 8   Suppose a bagging machine cost $42,000, has a lifespan of 5 years, and can be scrapped for $12,000. What is the amount of the annual straight-line depreciation? Determine the linear function f that gives the undepreciated value of the machine during its lifespan and sketch its graph.

Solution

The net cost N is the difference between the original cost and the scrap value. Thus

$$N = \$42,000 - \$12,000 = \$30,000.$$

Over 5 years the bagging machine will depreciate

$$\frac{\$30,000}{5} = \$6000 \text{ annually.}$$

To determine the linear function that gives the straight-line depreciation, we first let x represent the number of years after the purchase of the bagging machine and then let $f(x)$ be the corresponding undepreciated value of the machine. Hence x is restricted to the interval $[0, 5]$. Setting $y = f(x)$, we note that when $x = 0$, $y = \$42,000$, and when $x = 5$, $y = \$12,000$. The straight line in Figure 2.19 drawn through these two points has equation

$$y = -6000x + 42,000, \qquad x \in [0, 5].$$

Thus the linear function f that gives the undepreciated value for year x is

$$f(x) = -6000x + 42,000 \quad \text{for} \quad x \in [0, 5]. \qquad \square$$

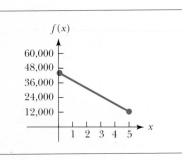

FIGURE 2.19

Exercises 2.3

Graph each linear relation and find the x-intercepts and y-intercepts, if they exist. (See Example 1.)

1. $y = 2x - 3$
2. $2x - 3y = 6$
3. $4x = 5$
4. $5x + 2y = 10$
5. $2x + 5y = 7$
6. $y - x = 5$
7. $x + y = 3$
8. $y = -3$

Compute the distance between each pair of points and find the slope of the line through the points, if it exists. (See Examples 2 and 3.)

9. $(2, -1)$ and $(-1, 3)$
10. $(0, 4)$ and $(3, 7)$
11. $(3, -1)$ and $(3, 3)$
12. $(-2, 3)$ and $(-2, 5)$
13. $(-4, 5)$ and $(2, 5)$
14. $(2, 7)$ and $(-3, 7)$

If it exists, find the slope of the line that has the given equation.

15. $3x - 4y = 6$
16. $4x - 2y = 20$
17. $x + 4 = 0$
18. $y = -2$
19. $2y = 7x - 5$
20. $3y = 5 - 2x$

Suppose the two given points lie on a straight line. Find the missing coordinate if the line has the given slope m.

21. $(1, y)$, $(2, 3)$, $m = 5$
22. $(-3, y)$, $(1, -5)$, $m = -3/5$

Suppose the three given points lie on a straight line. Find the missing coordinate.

23. $(0, 0)$, $(1, 2)$, $(3, y)$
24. $(1, -1)$, $(-2, 8)$, $(x, -7)$

Find an equation of the straight line that satisfies the given conditions and draw the graph. (See Examples 4–6.)

25. Through $(2, 2)$ and $(-1, 3)$
26. Through $(3, 4)$ and $(4, 3)$
27. x-intercept 7, slope 4
28. x-intercept 2, y-intercept -1
29. y-intercept 3, through $(-1, -1)$
30. y-intercept 3, no x-intercept
31. y-intercept -7, slope 0
32. Through $(3, 4)$, slope does not exist
33. Through $(-5, -3)$, perpendicular to $5x = 6y - 1$
34. Through $(-7, 2)$, perpendicular to $-2x = -3y + 7$
35. y-intercept -1, parallel to $5x - 7y = 5$

36. x-intercept 5, parallel to $9x - 7 = 3y$

Find an equation, in standard form, of the straight line whose graph is given.

37.

38.

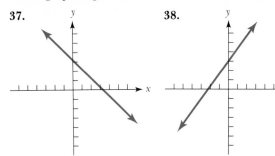

39.

40.

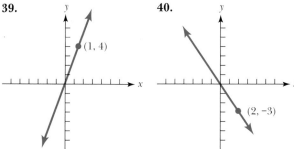

(1, 4)

(2, −3)

41.

42.

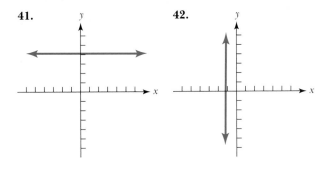

Graph each of the following functions. (See Example 7.)

43. $f(x) = \begin{cases} 1 - 2x & \text{if } x < 0 \\ 1 + 2x & \text{if } x \geq 0 \end{cases}$

44. $f(x) = \begin{cases} 2 - 3x & \text{if } x < 1 \\ -1 & \text{if } x \geq 1 \end{cases}$

45. $f(x) = \begin{cases} 2 + x & \text{if } x \leq 2 \\ 2 - x & \text{if } x > 2 \end{cases}$

46. $f(x) = \begin{cases} -5 - 3x & \text{if } x \leq -2 \\ 1 - x & \text{if } x > -2 \end{cases}$

Determine if the four given points are the vertices of a parallelogram.

47. $(-1, 3), (-2, 0), (2, 0), (3, 3)$

48. $(-1, 1), (-2, -1), (1, -4), (0, -6)$

49. $(-2, 8), (-6, 4), (-4, 1), (2, 2)$

50. $(2, 4), (3, 8), (1, -8), (4, 4)$

Determine if the three given points are the vertices of a right triangle.

51. $(-1, 5), (1, 2), (-2, 0)$

52. $(-3, 3), (1, 4), (-1, -5)$

53. $(0, 6), (-3, 0), (3, 0)$

54. $(-2, 3), (-7, 0), (3, 1)$

55. Show that the point

$$\left(\frac{x_1 + x_2}{2}, \frac{y_1 + y_2}{2} \right)$$

is the **midpoint** of the line segment joining the points (x_1, y_1) and (x_2, y_2).

56. Use the midpoint formula in Exercise 55 to find the midpoint of the line segment joining each of the following pairs of points.
   a. $(2, 7)$ and $(-3, 1)$
   b. $(-1, 1)$ and $(10, -4)$
   c. $(-\pi, 0)$ and $(3\pi, 0)$
   d. $(\pi/4, 0)$ and $(9\pi/4, 0)$

57. Suppose a plumber charges $20 for a house call plus $18 per hour while he is there. Write out the linear function f that gives the plumber's charges for one house call in terms of the number of hours worked.

58. A florist purchases a delivery van for $20,000. Suppose it costs an additional $1.50 per mile to operate and maintain the van. Write out the linear function whose value is the total expense of buying and operating the delivery van in terms of mileage.

59. A new car is purchased for $18,000. After 6 years of use its trade-in value is $3,000. Write out the linear function f for the undepreciated value of the car in terms of the number of years since its purchase. Assume the car depreciates linearly and note any restrictions on the independent variable.

60. A company purchases a copy machine for $2500 and anticipates using it for 4 years. If the trade-in value of the machine is $1100 and the company plans to depreciate it linearly, write out the function f that gives the undepreciated value of the machine.

61. A typical slope for a household stairway is 7/9. Referring to the figure below, determine the height of the risers if the treads are 12 inches wide.

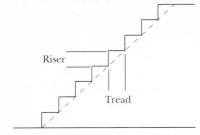

EXERCISE 61

62. If the upper floor is 9′4″ above the lower floor in the figure above, how much horizontal distance must be allowed for the stairway?

63. The slope of a railroad or a highway is called its **grade** and is usually written as a percent. For example, a 5% grade means that the slope is 5/100. Suppose the grade of a straight roadway is 4%. How far has a truck traveled along the road if its

Courtesy of the Massachusetts Department of Public Works
Photo by Susan Van Etten

altitude has changed from 3402 feet above sea level to 3349 feet above sea level?

64. Suppose the steepest incline on a roller coaster has a 30% grade (see Exercise 63). On this incline the passengers in the front seat are 9 feet above those in the rear seat. How long is the roller coaster train?

2.4   Quadratic Functions

Although they are not as simple as the linear functions, the quadratic functions are fully as useful and important. It has often been found that a quadratic function can be used to satisfactorily represent a quantity in real life.

> **Definition 2.11**
>
> A function f is a **quadratic function** if there are real numbers a, b, and c, with $a \neq 0$, such that the function value for f is given by
> $$f(x) = ax^2 + bx + c.$$

That is, a quadratic function is a function that has a defining equation of the form

$$y = ax^2 + bx + c,$$

with $a \neq 0$.

Example 1   Which of the following equations define a quadratic function?

a) $y = 2x^2$ b) $2x + x^2 = y$ c) $y^2 = 4x$
d) $x^3 + 4x = y$ e) $y = 2x + 7$

Solution
The equations in parts (a) and (b) can be written in the form $y = ax^2 + bx + c$, with $a \neq 0$, so each of them defines a quadratic function. The relation defined by $y^2 = 4x$ in part (c) is not a function, and so is not a quadratic function. The equations in parts (d) and (e) cannot be written in the required form, so they do not define quadratic functions. ❏

The graph of a quadratic function is called a **parabola.** We have seen the graph of the quadratic function $y = x^2$ in Figure 2.6 (Example 7 of Section 2.1). Another example is provided below.

Example 2   Sketch the graph of $y = 2x^2 - 12x + 17$.

Solution
One technique that can be used to great advantage in graphing a quadratic function is to complete the square on the x-terms. We first factor the coefficient of x^2 away from the x-terms:

$$y = 2x^2 - 12x + 17$$
$$= 2(x^2 - 6x) + 17.$$

To complete the square on $x^2 - 6x$, we add $[\frac{1}{2}(-6)]^2 = 9$ inside the parentheses. In doing this, we are actually adding $2(9) = 18$, so we must subtract 18 from 17 to have an equivalent equation:

$$y = 2(x^2 - 6x + 9) + 17 - 18$$
$$= 2(x - 3)^2 - 1.$$

This form of the equation gives some important information about the graph. Since $2(x - 3)^2 \geq 0$ for all x, the smallest possible value for y is -1, and this occurs when $2(x - 3)^2 = 0$, that is, when $x = 3$. This means that *the lowest point on the graph is at* $(3, -1)$.

We observe now that $(x - 3)^2 = |x - 3|^2$ is the square of the distance from x to 3; this means we get the same value for y when we move a certain distance to the left from $x = 3$ as we get when we move the same distance to the right from $x = 3$. In other words, *the graph is symmetric, or balanced, with respect to the line* $x = 3$.

We now assign some values to x and compute the corresponding y-values. These are recorded in the table in Figure 2.20, and a sketch of the graph is shown there. The extreme point on the graph is called the *vertex.* ❏

x	y
1	7
2	1
3	-1
4	1
5	7

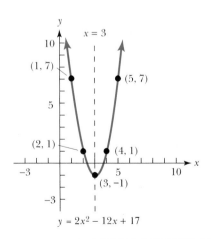

$y = 2x^2 - 12x + 17$

FIGURE 2.20

As mentioned just before Example 2, the graph of a quadratic function is called a *parabola*. The parabolas in Figures 2.6 and 2.20 are typical as to shape. Parabolas always have a bullet shape, with an extreme point (the **vertex**) and a **line of symmetry** through the vertex. However, the vertex may be the highest point on the graph instead of the lowest point. The graph opens downward and has a highest point when $a < 0$ in $y = ax^2 + bx + c$. A parabola of this type is shown in Example 3.

The discussion given in Example 2 can be extended to the general case.

If the equation $y = ax^2 + bx + c$ is rewritten as

$$y = a(x - h)^2 + k, \qquad a \neq 0,$$

then the parabola has the following properties:

1. The vertex is at (h, k);
2. The line $x = h$ is the axis of symmetry;
3. The graph opens upward if $a > 0$, downward if $a < 0$.

Example 3   Sketch the graph of $y = -2x^2 + 8x - 7$.

Solution

We first rewrite the equation, completing the square on the *x*-terms to obtain the form $y = a(x - h)^2 + k$.

$$y = -2(x^2 - 4x) - 7 = -2(x^2 - 4x + 4) - 7 + 8$$
$$= -2(x - 2)^2 + 1$$

We observe that $a = -2$, $h = 2$, and $k = 1$. The vertex is at $(2, 1)$, the line $x = 2$ is the axis of symmetry, and the parabola opens downward since $a < 0$. With this much knowledge of the graph, it is sufficient to plot five points and then sketch the graph. (See Figure 2.21.) In selecting these points, it is efficient to make use of symmetry. ❑

It is possible to obtain formulas that give the vertex and the axis of symmetry for

$$y = ax^2 + bx + c, \qquad a \neq 0$$

in terms of a, b, and c. All we need do is complete the square in the general case. We have

$$y = a\left(x^2 + \frac{b}{a}x\right) + c = a\left(x^2 + \frac{b}{a}x + \frac{b^2}{4a^2}\right) + c - \frac{b^2}{4a}$$

$$= a\left(x + \frac{b}{2a}\right)^2 + \frac{4ac - b^2}{4a}.$$

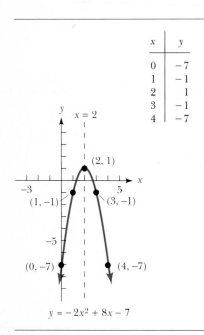

x	y
0	−7
1	−1
2	1
3	−1
4	−7

$y = -2x^2 + 8x - 7$

FIGURE 2.21

Careful examination of this equation leads to the statements made in Theorem 2.12.

Theorem 2.12

The graph of $y = ax^2 + bx + c$, $a \neq 0$, is a parabola that

1. Has vertex at $\left(-\dfrac{b}{2a}, \dfrac{4ac - b^2}{4a} \right)$;

2. Has the line $x = -\dfrac{b}{2a}$ as the axis of symmetry;

3. Opens upward if $a > 0$, downward if $a < 0$.

It is actually not necessary to memorize the y-coordinate of the vertex, because it can be readily obtained by substituting the value $x = -b/(2a)$ into the equation $y = ax^2 + bx + c$.

In many cases it is useful to find the intercepts for a parabola $y = ax^2 + bx + c$. The y-intercept is easy: When $x = 0$, $y = c$. In looking for the x-intercepts, we are at once confronted with the problem of solving

$$ax^2 + bx + c = 0.$$

Recall from Section 1.5 that this equation may have two equal real roots, two unequal real roots, or no real roots, depending on the discriminant $b^2 - 4ac$. The graphical interpretation of this situation is that the parabola may be tangent to the x-axis at the vertex, it may cross the x-axis in two places, or it may not cross the x-axis at all. All three cases are illustrated in Figure 2.22.

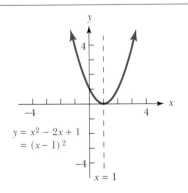

$b^2 - 4ac = 0$
One x-intercept

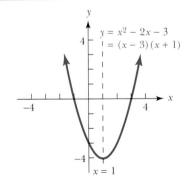

$b^2 - 4ac = 16 > 0$
Two x-intercepts

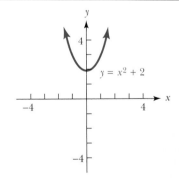

$b^2 - 4ac = -8 < 0$
No x-intercepts

FIGURE 2.22

Since the y-coordinate of the vertex of a parabola represents the maximum (or minimum) function value of $y = f(x) = ax^2 + bx + c$, the parabola is often useful in solving applied problems in which it is necessary to determine the maximum (or minimum) value of some quadratic function. We illustrate the technique in the next example.

Example 4   Farmer Fred wishes to fence off part of his pasture in a rectangle for his prize-winning bull. One side is bordered by a straight river and he needs no fence there. Using 120 yards of fencing, what should the dimensions of the rectangle be if Fred wants to maximize the grazing area for his bull?

Solution

Suppose we let the dimensions of the rectangle in Figure 2.23(a) be length $= l$ and width $= w$. Since there is a total of 120 yards of fencing, we can express l in terms of w as $l = 120 - 2w$. The area A of the rectangle is

$$A = lw$$
$$= (120 - 2w)w$$
$$= -2w^2 + 120w.$$

But this is the equation of the parabola in Figure 2.23(b). Since the parabola opens downward, the second coordinate of the vertex represents the maximum area. Hence we see that the area is a maximum of 1800 square yards when the dimensions of the rectangle are

$$w = 30 \text{ yards} \quad \text{and} \quad l = 120 - 2(30) = 60 \text{ yards.} \qquad \square$$

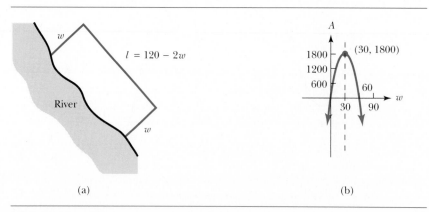

(a)   (b)

FIGURE 2.23

Exercises 2.4

Graph each of the following parabolas. Locate the vertex and the axis of symmetry, and plot at least two other points on each graph. (See Examples 2 and 3.)

1. $y = x^2 - 4$

2. $y = -2x^2 + 6$

3. $y = \frac{2}{3}(x - 2)^2 - 1$

4. $y = -\frac{4}{5}(x + 1)^2 + 2$

5. $y = 4x - x^2$

6. $y = x^2 - 9x$

7. $y = x^2 + 2x + 4$

8. $y = x^2 - 2x + 3$

9. $y = x^2 + 8x + 13$

10. $y = 5 + 6x + x^2$

11. $y = -x^2 - 2x - 2$

12. $y = -x^2 + 4x - 3$

13. $y = -x^2 - 4x - 1$

14. $y = -x^2 + 6x - 6$

15. $y = 3x^2 + 12x + 17$

16. $y = 2x^2 + 8x + 9$

17. $y = -2x^2 + 4x - 5$

18. $y = -4x^2 - 4x + 1$

Find the maximum or minimum value of the given function and state whether that value represents a maximum or minimum. (See Example 4.)

19. $f(x) = x^2 - 8x + 5$

20. $f(x) = x^2 + 6x + 10$

21. $f(x) = x^2 + 10x - 20$

22. $f(x) = x^2 + 4x - 9$

23. $f(x) = -x^2 + 6x + 3$

24. $f(x) = -x^2 + 8x - 6$

25. $f(x) = -2x^2 + 4x - 3$

26. $f(x) = -2x^2 + 8x - 11$

27. $f(x) = -9 - 6x - x^2$

28. $f(x) = -16 - 8x - x^2$

29. $f(x) = 3x^2 + 9x + 14$

30. $f(x) = 2x^2 - 5x + 7$

31. Sketch the graph of each of the following parabolas on the same coordinate system. Describe the effect of k on the graph of $y = x^2 + k$.
a) $y = x^2$
b) $y = x^2 + 2$
c) $y = x^2 - 2$
d) $y = x^2 + 4$

32. Sketch the graph of each of the following parabolas on the same coordinate system. Describe the effect of a on the graph of $y = ax^2$ when $a > 0$.
a) $y = x^2$
b) $y = 2x^2$
c) $y = 3x^2$
d) $y = \frac{1}{2}x^2$

33. Sketch the graph of each of the following parabolas on the same coordinate system. Describe the effect of a on the graph of $y = ax^2$ when $a < 0$.
a) $y = x^2$
b) $y = -x^2$
c) $y = 2x^2$
d) $y = -2x^2$

34. Sketch the graph of each of the following parabolas on the same coordinate system. Describe the effect of h on the graph of $y = (x - h)^2$.
a) $y = x^2$
b) $y = (x - 1)^2$
c) $y = (x + 1)^2$
d) $y = (x - 2)^2$

Sketch the graph of each of the following functions.

35. $f(x) = \begin{cases} x^2 & \text{if } x \geq 0 \\ 1 - x & \text{if } x < 0 \end{cases}$

36. $f(x) = \begin{cases} 4 - x^2 & \text{if } x \geq 0 \\ x + 2 & \text{if } x < 0 \end{cases}$

37. $f(x) = \begin{cases} 2 - x & \text{if } x > 1 \\ x^2 - 1 & \text{if } x \leq 1 \end{cases}$

38. $f(x) = \begin{cases} 1 - x^2 & \text{if } x > -2 \\ 2x + 4 & \text{if } x \leq -2 \end{cases}$

39. $f(x) = \begin{cases} (x - 1)^2 + 1 & \text{if } x > 1 \\ 1 & \text{if } -1 \leq x \leq 1 \\ (x + 1)^2 + 1 & \text{if } x < -1 \end{cases}$

40. $f(x) = \begin{cases} x - 1 & \text{if } x \geq 2 \\ 1 & \text{if } -1 < x < 2 \\ 1 - x^2 & \text{if } x \leq -1 \end{cases}$

41. Find two positive numbers whose sum is 54 and whose product is a maximum.

42. Find two positive numbers whose sum is 92 and whose product is a maximum.

43. Find the dimensions of a rectangle with maximum area whose perimeter is 48 feet.

44. Find the dimensions of a rectangle with maximum area whose perimeter is 88 feet.

45. The cost C in dollars of producing x units of a certain item is given by $C = 0.001x^2 - 0.5x + 800$. Find the production level x for which the cost will be a minimum. What is the minimum cost?

46. The cost C in dollars of producing x units of a certain item is given by $C = 0.002x^2 - 3x + 9000$. Find the production level x that yields a minimum cost. What is the minimum cost?

47. If a rock is thrown upward from the ground with an initial velocity of 64 feet per second, the distance s in feet of the rock from the ground after t seconds is $s = -16t^2 + 64t$. When does the rock reach its maximum height, and what is that maximum height?

48. Do Exercise 47 if the initial velocity is 128 feet per second and s is given by $s = -16t^2 + 128t$.

49. A rectangular dog pen is fenced off along the wall of a house, with no fence needed along the wall, as indicated below. If 40 feet of fence is available, find the dimensions of the dog pen with maximum area.

EXERCISE 49

50. A rectangular dog pen is to be made in the corner of a barn. There is 16 feet of fencing material available for the remaining two sides of the pen. Determine the dimensions of the pen with maximum area.

51. Suppose the window in the figure is made by placing a semicircle of radius r atop a rectangle with height h. Determine the dimensions r and h that

yield the maximum area if the outside perimeter of the window is 20 feet.

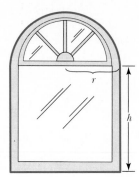

EXERCISE 51

52. The mirror shown below is constructed by joining two semicircular pieces of radius r to a rectangular piece of width w. Determine the value of r that yields the maximum area of the rectangular piece if the outside perimeter of the mirror is 40 feet.

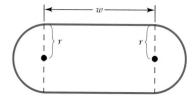

EXERCISE 52

2.5   Inverse Relations and Functions

As stated in Definition 2.2, a *relation* is a nonempty set of ordered pairs (x, y) of real numbers x and y. Every relation has an *inverse relation*, as described in the following definition.

> **Definition 2.13**
>
> For any relation g, the **inverse** of g is the relation g^{-1} defined by
>
> $$g^{-1} = \{(b, a) : (a, b) \text{ is in } g\}.$$

The relation g^{-1} is obtained from g by interchanging the components a and b in all ordered pairs (a, b) in g. It follows from this that the domain of g^{-1} is the range of g, and the range of g^{-1} is the domain of g.

Example 1   If
$$g = \{(1, 0), (2, 3), (-1, 4)\},$$
then
$$g^{-1} = \{(0, 1), (3, 2), (4, -1)\}. \qquad ❏$$

The notation g^{-1} is read as "the inverse of g" or as "g-inverse." This should not be confused with the reciprocal of a function[3] g, which is defined by $[g(x)]^{-1} = 1/g(x)$. That is,

$$g^{-1}(x) \neq \frac{1}{g(x)}$$

even if g is a function.

In most cases, a relation g is described by a defining equation in x and y. In such cases the inverse g^{-1} may be obtained by interchanging x and y in the defining equation for g. It is customary to solve for y in the equation that results from the interchange, if this can be done with a reasonable amount of work. An illustration is given in Example 2.

Example 2   Find a defining equation for the inverse of the relation
$$g = \{(x, y) : y = 3x + 6\}$$
and solve the equation for y. Draw the graphs of g and g^{-1} on the same coordinate system.

Solution
The defining equation for g is $y = 3x + 6$. Interchanging x and y in this equation, we obtain
$$x = 3y + 6$$
as a defining equation for g^{-1}. Solving for y in the last equation, we get
$$y = \frac{1}{3}x - 2.$$

The graphs of g and g^{-1} are straight lines, as shown in Figure 2.24. The line $y = x$ is drawn in the figure as a dashed line, and the graphs of g and g^{-1} are symmetric about the line $y = x$. That is, the graph of g^{-1} is the reflection (mirror image) of the graph of g through the line $y = x$. ❏

---

3. As noted in Definition 2.2, a *function* is a special type of relation in which no two different y-values are paired with the same x-value.

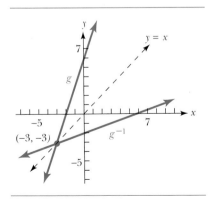

FIGURE 2.24

If a and b are any real numbers, the points (a, b) and (b, a) are located symmetrically with respect to the line $y = x$. For this reason, the graph of the inverse of a relation is always the reflection of the graph of the original relation through the line $y = x$.

In Examples 1 and 2, it happens that both g and g^{-1} are functions. Our next example shows that either g or g^{-1} may be a function while the other is not.

Example 3   For each of the following relations, find a defining equation for the inverse relation, draw the graphs of the relation and its inverse, and decide if the relation or its inverse is a function.

a) $g = \{(x, y) : y = x^2 + 1\}$ b) $h = \{(x, y) : |y| = x - 2\}$

Solution

a)   A defining equation for g^{-1} is

$$x = y^2 + 1.$$

The graph of g is a parabola, as shown in Figure 2.25, and the graph of g^{-1} is obtained by reflecting the parabola through the line $y = x$. Since no vertical line intersects the graph of g at more than one point, g is a function. However, any vertical line to the right of $x = 1$ intersects the graph of g^{-1} at more than one point, so g^{-1} is not a function.

b)   A defining equation for h^{-1} is

$$|x| = y - 2.$$

The graphs of h and h^{-1} are shown in Figure 2.26. The Vertical Line Test shows that h is not a function and that h^{-1} is a function. ❏

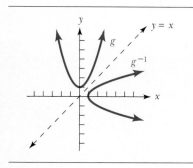

FIGURE 2.25

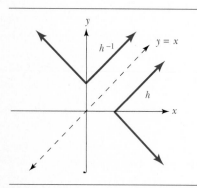

FIGURE 2.26

The situation where both a relation f and its inverse f^{-1} are functions is of special interest. In this case, a pairing takes place between the elements in the domain D and the range R of f, with the following properties.

1. **For each $a \in D$, there is exactly one $b = f(a)$ in R that is paired with a by f.**
2. **For each $b \in R$, there is exactly one $a = f^{-1}(b)$ in D that has b paired with it by f.**

This pairing is diagrammed in Figure 2.27.

Stating these properties another way, we can say that (1) different y-values in the equation $y = f(x)$ require different x-values, and (2) different x-values in $y = f(x)$ require different y-values. Because of this one-to-one pairing when both f and f^{-1} are functions, f is called a **one-**

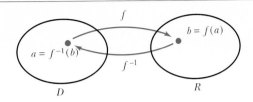

FIGURE 2.27

to-one function, or a **one-to-one correspondence,** from D to R. Corresponding to the Vertical Line Test for functions, we can state the following Horizontal Line Test for one-to-one functions.

Horizontal Line Test

If any horizontal line intersects the graph of the function f at more than one point, then f is *not* a one-to-one function. If no horizontal line intersects the graph of the function f at more than one point, then f is a one-to-one function.

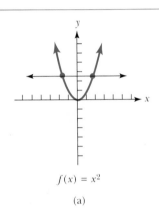

$f(x) = x^2$

(a)

Example 4   The function with defining equation

$$f(x) = x^2$$

is *not* a one-to-one function, since any horizontal line above the x-axis cuts the graph in more than one place. This is shown in Figure 2.28(a). The graph of the function defined by $f(x) = x^3$ is shown in Figure 2.28(b). The Horizontal Line Test shows that $f(x) = x^3$ defines a one-to-one function. ❑

When f is a one-to-one function, we refer to f^{-1} as the **inverse function** of f. It follows from properties (1) and (2) (stated just after Example 3) that *for a one-to-one function f,*

$$f^{-1}(f(x)) = x \text{ for all } x \text{ in the domain of } f,$$

and

$$f(f^{-1}(x)) = x \text{ for all } x \text{ in the domain of } f^{-1}.$$

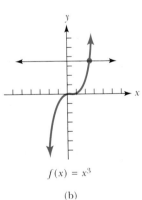

$f(x) = x^3$

(b)

FIGURE 2.28

Example 5   Each of the following equations defines a function f. Find an expression for $f(x)$ and also for $f^{-1}(x)$ if f is one-to-one. If f is one-to-one, verify each of the equations $f^{-1}(f(x)) = x$ and $f(f^{-1}(x)) = x$.

a) $x^2 - y + 3 = 0$ b) $3x - 2y = 6$

Solution

a) Solving for y in the defining equation, we get $y = x^2 + 3$, so
$$f(x) = x^2 + 3.$$

Since $f(1) = 4$ and $f(-1) = 4$, f is not one-to-one. (The graph of $y = x^2 + 3$ would also show that f is not one-to-one.)

b) Solving for y, we obtain $y = (3/2)x - 3$, so
$$f(x) = \frac{3}{2}x - 3.$$

The graph of f is a nonhorizontal straight line, so f is one-to-one. A defining equation for f^{-1} is
$$3y - 2x = 6.$$

Solving for y in this equation, we get $y = (2/3)x + 2$, so
$$f^{-1}(x) = \frac{2}{3}x + 2.$$

Verifying the equation $f^{-1}(f(x)) = x$, we have
$$f^{-1}(f(x)) = \frac{2}{3}f(x) + 2$$
$$= \frac{2}{3}\left(\frac{3}{2}x - 3\right) + 2$$
$$= x - 2 + 2 = x.$$

Also,
$$f(f^{-1}(x)) = \frac{3}{2}f^{-1}(x) - 3$$
$$= \frac{3}{2}\left(\frac{3}{2}x + 2\right) - 3$$
$$= x + 3 - 3 = x,$$

and the equation $f(f^{-1}(x)) = x$ is verified.   ❑

The equations $f^{-1}(f(x)) = x$ and $f(f^{-1}(x)) = x$ are sometimes used to define the inverse of a function in the following way. Let f be a function with domain D and range R. If g is a function with domain R and range D such that
$$f(g(x)) = x \text{ for all } x \in R \qquad \text{and} \qquad g(f(x)) = x \text{ for all } x \in D,$$
then $g = f^{-1}$ and $f = g^{-1}$.

Example 6   Determine if the following functions f and g are inverse functions of each other.

$$f(x) = \frac{x}{x-1}, \qquad g(x) = \frac{x-1}{x}$$

Solution

Both the equations $f(g(x)) = x$ and $g(f(x)) = x$ must hold in order for f and g to be inverse functions of each other. However,

$$f(g(x)) = \frac{g(x)}{g(x)-1} = \frac{\dfrac{x-1}{x}}{\dfrac{x-1}{x}-1} \cdot \frac{x}{x} = \frac{x-1}{x-1-x}$$

$$= -x + 1$$

$$\neq x.$$

Since $f(g(x)) \neq x$, f and g are not inverse functions of each other.   ❏

Exercises 2.5

Find the inverse of each of the following relations. Also determine if each relation and/or its inverse is a function. (See Examples 1–3.)

1. $g = \{(0, 3), (1, 2), (-1, 2)\}$ $(3, 0)$

2. $g = \{(1, 5), (0, 2), (0, -2)\}$

3. $f = \{(x, y) : y = 3x - 2\}$

4. $f = \{(x, y) : x - y = 5\}$ **5.** $p = \{(x, y) : y = |x|\}$

6. $p = \{(x, y) : y = |x + 1|\}$

7. $g = \{(x, y) : x = 2y^2\}$ **8.** $g = \{(x, y) : y = 3x^2\}$

9. $f = \{(x, y) : y = x^2 + 2x\}$

10. $f = \{(x, y) : y = 2x^2 - 8x\}$

11. $g = \{(x, y) : x^2 + y^2 = 4\}$

12. $g = \{(x, y) : 4x^2 + 9y^2 = 36\}$

13. $h = \{(x, y) : y = -\sqrt{4 - x^2}\}$

14. $h = \{(x, y) : y = 2\sqrt{9 - x^2}/3\}$

Each of the following equations defines a relation g. In each problem, (a) find a defining equation for the inverse g^{-1}, (b) draw the graphs of g and g^{-1}, and (c) decide if the relation or its inverse is a function. (See Examples 2 and 3.)

15. $y = x + 3$ **16.** $y = 4 - x$

17. $4x - 3y = 12$ **18.** $2x - 3y = 12$

19. $y = -1 - x^2$ **20.** $y = -3 - x^2$

In Exercises 21–24, use the Horizontal Line Test to decide whether or not the given graph is the graph of a one-to-one function. (See Example 4.)

21.
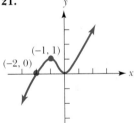

22.
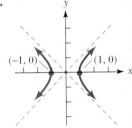

23.
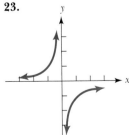

24.
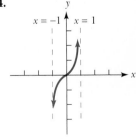

Each of the following equations defines a function f. Find an expression for $f(x)$ and also for $f^{-1}(x)$ if f is one-to-one. If f is one-to-one, verify each of the equations $f^{-1}(f(x)) = x$ and $f(f^{-1}(x)) = x$. (See Example 5.)

25. $2x - y = 4$ **26.** $3x - y = 9$

27. $y = x^2 + 2$ **28.** $y = x^2 - 1$

29. $y = (x - 1)^2$ **30.** $y = (x + 2)^2$

31. $y = x^3 - 1$ **32.** $y = x^3 + 1$

Which of the following pairs of functions f and g are inverse functions of each other? (See Example 6.)

33. $f(x) = -\frac{3}{4}x + 3$, $g(x) = -\frac{4}{3}x + 4$

34. $f(x) = -\frac{3}{5}x + 3$, $g(x) = -\frac{5}{3}x + 5$

35. $f(x) = 3x - 9$, $g(x) = \dfrac{x + 9}{3}$

36. $f(x) = \dfrac{4x - 10}{5}$, $g(x) = \dfrac{5x + 10}{4}$

37. $f(x) = \dfrac{2x + 1}{x}$, $g(x) = \dfrac{1}{x - 2}$

38. $f(x) = \dfrac{x - 1}{2x}$, $g(x) = \dfrac{1}{1 - 2x}$

39. $f(x) = \dfrac{2x}{x + 1}$, $g(x) = \dfrac{x + 1}{2x}$

40. $f(x) = \dfrac{x}{2x - 1}$, $g(x) = \dfrac{2x - 1}{x}$

41. $f(x) = \sqrt{x - 1}$, $g(x) = x^2 + 1$ where $x \geq 0$

42. $f(x) = \sqrt{x^2 - 4}$ where $x \geq 2$,
      $g(x) = \sqrt{x^2 + 4}$ where $x \geq 0$

43. $f(x) = \sqrt{x + 2}$, $g(x) = x^2 - 2$

44. $f(x) = \sqrt{2x - 6}$, $g(x) = \frac{1}{2}x^2 + 3$

45. Prove that the points (a, b) and (b, a) are symmetric to each other with respect to the line $y = x$. That is, prove that the line $y = x$ is the perpendicular bisector of the line segment joining (a, b) and (b, a).

46. Prove that $(f^{-1})^{-1} = f$, for any relation f.

2.6   Symmetry and Translations

In Section 2.4, we saw that a parabola is symmetric with respect to its axis of symmetry. The parabola $y = x^2$ is sketched in Figure 2.29, and we see that its axis of symmetry is the line with equation $x = 0$, the y-axis. Therefore we say that this parabola is symmetric with respect to the y-axis. Note that whenever the point (x, y) is on the graph, so is $(-x, y)$.

Other types of symmetry can be considered. For example, a graph is said to be symmetric with respect to the x-axis if the graph below the x-axis is a mirror image of the graph above the x-axis. That is, whenever (x, y) lies on the graph of the relation, so does $(x, -y)$. This means that whenever (x, y) satisfies the equation of the relation, so does $(x, -y)$. The graph in Figure 2.30 is symmetric with respect to the x-axis.

A graph is said to be symmetric with respect to the origin if, whenever (x, y) lies on the graph of the relation, $(-x, -y)$ does also. Thus the ordered pair $(-x, -y)$ satisfies the equation of the relation whenever (x, y) does. The curve in Figure 2.31 is an example of a graph that is symmetric with respect to the origin.

Knowledge about symmetry is a very useful tool in graphing an equation, as illustrated in the following example.

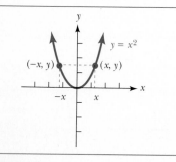

FIGURE 2.29

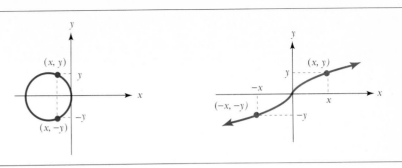

FIGURE 2.30   **FIGURE 2.31**

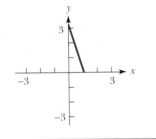

FIGURE 2.32

Example 1   The graph in Figure 2.32 is a portion of the graph of a relation. Complete the graph so that it is symmetric with respect to the

a) *y*-axis,   b) *x*-axis,   c) origin.

Solution
The graphs are completed in the corresponding parts of Figure 2.33.

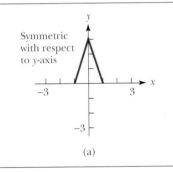

Symmetric
with respect
to *y*-axis

(a)

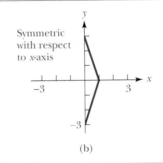

Symmetric
with respect
to *x*-axis

(b)

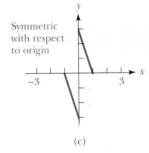

Symmetric
with respect
to origin

(c)

FIGURE 2.33   ❏

When the defining equation is given for a graph, the following tests for symmetry can be utilized.

Tests for Symmetry	
The graph of an equation is symmetric with respect to the	**if the following substitution yields an equivalent equation**
y-axis	$-x$ for x
x-axis	$-y$ for y
origin	$-x$ for x and $-y$ for y

Example 2   Test each of the following equations for symmetry and sketch its graph.

a) $y = x^4$ b) $y^2 = x$ c) $y = x^3$

Solution

a)   Replacing $-x$ for x in $y = x^4$ yields

$$y = (-x)^4 = x^4,$$

which is an equivalent equation. Thus the graph of $y = x^4$ is symmetric with respect to the y-axis. The graph is not symmetric with respect to the x-axis or to the origin, since replacing $-y$ for y yields $-y = x^4$, and replacing both $-x$ for x and $-y$ for y yields $-y = x^4$. To sketch the graph in Figure 2.34(a), we plot a few points for positive values of x and use symmetry to obtain the portion of the graph in the second quadrant.

b)   Replacing $-y$ for y in $y^2 = x$ yields the equivalent equation

$$(-y)^2 = x$$

or

$$y^2 = x.$$

Thus the graph is symmetric with respect to the x-axis. The other tests for symmetry fail. The graph in Figure 2.34(b) is sketched by plotting a few points in the first quadrant and using symmetry to complete the graph.

c)   The graph of $y = x^3$ is symmetric with respect to the origin since replacing $-x$ for x and $-y$ for y yields

$$-y = (-x)^3 = -x^3$$

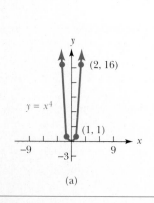
(a)

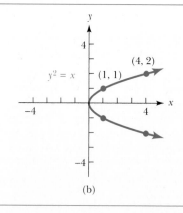
(b)

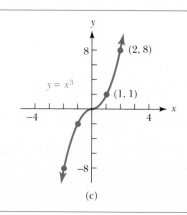
(c)

FIGURE 2.34

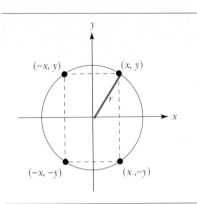

FIGURE 2.35

or

$$y = x^3.$$

The graph in Figure 2.34(c) is sketched by plotting points in the first quadrant and using symmetry to obtain the portion in the third quadrant. ❑

There are some graphs that are symmetric with respect to the *x*-axis, the *y*-axis, and the origin. One such example is a circle of radius $r > 0$, if it is positioned in a coordinate system with its center at the origin, as shown in Figure 2.35. Hence the equation for this circle must satisfy all three tests for symmetry. To derive its equation, we first formally define the circle in terms of the notion of distance and then use the Distance Formula.

Definition 2.14

A **circle** is the set of all points equally distant from a given point, called the **center** of the circle. The distance from the center of the circle to any point on the circle is called the **radius** of the circle.

The distance between the point (x, y) and the origin $(0, 0)$ is given by

$$d = \sqrt{(x - 0)^2 + (y - 0)^2} = \sqrt{x^2 + y^2}.$$

If we square both sides of $d = \sqrt{x^2 + y^2}$, we see that the set of all points (x, y) whose distance from $(0, 0)$ is r is given by

$$\{(x, y) : x^2 + y^2 = r^2\}.$$

Hence an equation for a circle with center at $(0, 0)$ and radius r is

$$x^2 + y^2 = r^2.$$

It is easy to see that this equation satisfies all three tests for symmetry.
Other aids for graphing are described in the next examples.

Example 3   Sketch the graphs of $y = x^2$, $y = 2x^2$, and $y = \frac{1}{2}x^2$ on the same coordinate system and compare their shapes.

Solution
Each of the graphs is sketched in Figure 2.36. Points on the graph of $y = 2x^2$ can be obtained by multiplying each *y*-coordinate on the graph of $y = x^2$ by 2. Similarly, if each *y*-coordinate on the graph of $y = x^2$ is multiplied by $\frac{1}{2}$, the resulting point is on the graph of $y = \frac{1}{2}x^2$. A change in graphs, such as we see here, is called a *stretching* of the function $y = x^2$. ❑

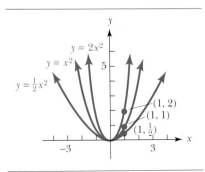

FIGURE 2.36

2   FUNCTIONS AND GRAPHS

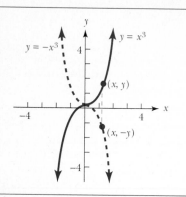

FIGURE 2.37

Stretching of $y = f(x)$

To graph $y = cf(x)$, $c > 0$, multiply each y-coordinate on the graph of $y = f(x)$ by c.

Example 4   Sketch the graphs of $y = x^3$ and $y = -x^3$ on the same coordinate system.

Solution

The multiplier -1 in $y = -x^3$ multiplies the function values of $y = x^3$ by -1. Thus for every point (x, y) on the graph of $y = x^3$, there is a corresponding point $(x, -y)$ on the graph of $y = -x^3$. (See Figure 2.37.) We call the graph of $y = -x^3$ a *reflection* through the x-axis of the graph of $y = x^3$.   ❑

Reflection of $y = f(x)$

To graph $y = -f(x)$, reflect the graph of $y = f(x)$ through the x-axis.

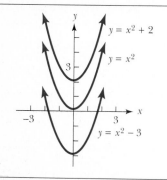

FIGURE 2.38

Example 5   Sketch the graphs of the three parabolas $y = x^2$, $y = x^2 + 2$, and $y = x^2 - 3$ on the same coordinate system.

Solution

The graph of the parabola $y = x^2 + 2$ can be obtained by shifting the graph of $y = x^2$ upward two units. That is, if 2 is added to the y-coordinate of each point on the parabola $y = x^2$, the result is the y-coordinate of a point on $y = x^2 + 2$. Similarly, if the graph of $y = x^2$ is shifted downward three units, the result is the graph of $y = x^2 - 3$. These parabolas are sketched in Figure 2.38.   ❑

Vertical Translation of $y = f(x)$

To graph $y = f(x) + c$:

1. If $c > 0$, shift the graph of $y = f(x)$ upward c units;
2. If $c < 0$, shift the graph of $y = f(x)$ downward $|c|$ units.

Another type of translation is illustrated in the next example.

Example 6   Sketch the parabolas $y = x^2$ and $y = (x - 2)^2$ on the same coordinate system.

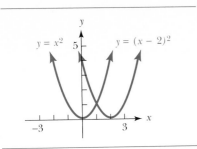

FIGURE 2.39

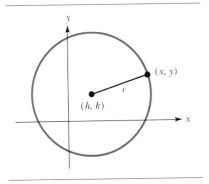

FIGURE 2.40

Solution

In Figure 2.39, we see that the graph of $y = (x - 2)^2$ can be obtained by shifting (or translating horizontally) the graph of $y = x^2$ two units to the right. ❏

Horizontal Translation of $y = f(x)$

To graph $y = f(x - c)$:

1. If $c > 0$, shift the graph of $y = f(x)$ to the right c units;
2. If $c < 0$, shift the graph of $y = f(x)$ to the left $|c|$ units.

The concept of translation can be extended to the graphs of equations in general. For example, if a circle of radius $r > 0$ centered at the origin is translated h units horizontally and k units vertically, the center of the translated circle will be located at (h, k). Any point (x, y) on the circle is r units away from the center. The Distance Formula gives

$$r = \sqrt{(x - h)^2 + (y - k)^2},$$

or

$$r^2 = (x - h)^2 + (y - k)^2. \tag{2}$$

That is, a point (x, y) is on the circle with center (h, k) and radius r if and only if its coordinates (x, y) satisfy Equation (2). This important result is stated in the following theorem and is represented graphically in Figure 2.40.

Theorem 2.15

The **standard equation** for a circle with center (h, k) and radius r is

$$(x - h)^2 + (y - k)^2 = r^2.$$

Example 7 Determine whether or not the graph of each equation below is a circle. If it is, find the center and radius, and draw the graph.

a) $x^2 + 2x + y^2 - 6y = 0$ b) $x^2 + y^2 = 10x - 8y - 41$
c) $2x^2 + 2y^2 - 4x + 10 = 0$

Solution

a) We must complete the squares on the variables x and y in order to change the equation into the standard form. The given equation

$$x^2 + 2x + y^2 - 6y = 0$$

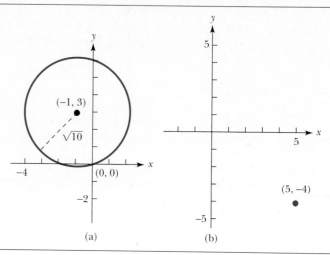

FIGURE 2.41

is equivalent to

$$x^2 + 2x + 1 + y^2 - 6y + 9 = 1 + 9.$$

Expressing in standard form, we have

$$[x - (-1)]^2 + (y - 3)^2 = (\sqrt{10})^2,$$

which is the equation of the circle shown in Figure 2.41(a) with center $(-1, 3)$ and radius $\sqrt{10}$.

b) Completing the squares on the variables x and y, and rewriting in standard form, we have

$$x^2 - 10x + 25 + y^2 + 8y + 16 = -41 + 25 + 16$$

and

$$(x - 5)^2 + [y - (-4)]^2 = 0. \qquad (3)$$

Since the square of a real number cannot be negative, this equation requires that both $x - 5 = 0$ and $y + 4 = 0$. Thus the graph of Eq. (3) consists of the single point $(5, -4)$ in Figure 2.41(b). For this reason, the graph of Eq. (3) is referred to as a **point circle,** or a **degenerate circle.**

c) Dividing both sides of the given equation by 2 and rearranging yields

$$x^2 - 2x + y^2 = -5.$$

Completing the square on x gives

$$x^2 - 2x + 1 + y^2 = -5 + 1$$

or

$$(x - 1)^2 + (y - 0)^2 = -4.$$

Since the sum of the squares of two real numbers can never be negative, there are no values of x and y that satisfy this equation. Hence the equation

$$2x^2 + 2y^2 - 4x + 10 = 0$$

does not represent a circle. The set of solutions to this equation is the empty set $\varnothing$.　　　　❏

Exercises 2.6

In Exercises 1–8, a portion of a graph is given. Use it to construct three additional graphs: (a) one that is symmetric with respect to the y-axis, (b) one that is symmetric with respect to the x-axis, and (c) one that is symmetric with respect to the origin. (See Example 1.)

1.
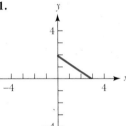

2.
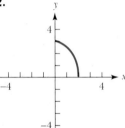

3.
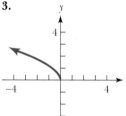

4.
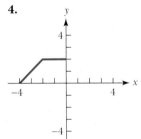

5.
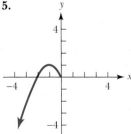

6.
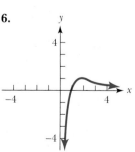

7.
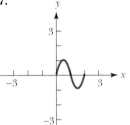

8.
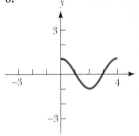

Test each of the following equations for symmetry. (See Example 2.)

9. $y = |x|$　　　　　　　　**10.** $|y| = x$

11. $|y| = |x|$　　　　　　　**12.** $y = 2x^4$

13. $y^2 = x + 1$　　　　　　**14.** $y^4 = x$

15. $y = x$　　　　　　　　　**16.** $y = x^5$

17. $x^2 + y^2 = 25$　　　　　**18.** $4x^2 + y^2 = 4$

19. $xy = 9$　　　　　　　　　**20.** $4 = xy$

21. $y = x^4 - 4x^2$　　　　　**22.** $y = x^5 - 4x^3$

(See Examples 3–6.)

23. Suppose the graph in Exercise 1 is the graph of $y = f(x)$. Sketch the graph of (a) $y = 2f(x)$ and (b) $y = f(x) - 2$.

24. Suppose the graph in Exercise 2 is the graph of $y = f(x)$. Sketch the graph of (a) $y = -f(x)$ and (b) $y = f(x) + 1$.

25. Suppose the graph in Exercise 3 is the graph of $y = f(x)$. Sketch the graph of (a) $y = f(x - 1)$ and (b) $y = -f(x)$.

26. Suppose the graph in Exercise 4 is the graph of $y = f(x)$. Sketch the graph of (a) $y = -2f(x)$ and (b) $y = f(x - 2)$.

27. Suppose the graph in Exercise 5 is the graph of $y = f(x)$. Sketch the graph of (a) $y = 3f(x)$ and (b) $y = f(x + 1)$.

28. Suppose the graph in Exercise 6 is the graph of $y = f(x)$. Sketch the graph of (a) $y = f(x + 3)$ and (b) $y = 3f(x)$.

29. Suppose the graph in Exercise 7 is the graph of $y = f(x)$. Sketch the graph of (a) $y = f(x + 2)$ and (b) $y = 3f(x + 2)$.

30. Suppose the graph in Exercise 8 is the graph of $y = f(x)$. Sketch the graph of (a) $y = -2f(x)$ and (b) $y = -2f(x) + 1$.

Sketch the graphs of each of the following by using either one of—or a combination of—a stretching, a reflection, or a translation of a familiar function. (See Examples 3–6.)

31. $y = \frac{1}{2}x^3$

32. $y = \frac{1}{4}x^4$

33. $y = -2|x|$

34. $y = -x^2 + 1$

35. $y = |x + 1|$

36. $y = |x - 2|$

37. $y = |x| + 1$

38. $y = |x| - 2$

39. $y = (x - 1)^3$

40. $y = (x + 2)^3$

41. $y = x^3 - 1$

42. $y = x^3 + 2$

43. $y = (x + 1)^3 - 2$

44. $y = (x - 1)^3 + 2$

Determine whether or not the graph of each equation is a circle. If it is, find the center and radius. (See Example 7.)

45. $x^2 + y^2 = 16$

46. $(x + 1)^2 + y^2 = 1$

47. $(x - 3)^2 + (y + 5)^2 = 25$

48. $(x - 4)^2 + (y + 1)^2 = 16$

49. $x^2 + 2x + y^2 + 6y = 0$

50. $x^2 - 2x + y^2 + 4y = 4$

51. $x^2 + y^2 = 3x$

52. $x^2 + y^2 = 8y$

53. $x^2 + y^2 = 10x - 6y - 36$

54. $x^2 + y^2 - 3x + 3y + 12 = 0$

55. $x^2 + y^2 - 4x + 16y + 68 = 0$

56. $x^2 + y^2 = 12x + 2y - 37$

57. $2x^2 + 2y^2 + 4x + 8y = 8$

58. $7x^2 + 7y^2 = x$

Write an equation of the circle with the given properties.

59. Center $(0, 1)$, radius 2

60. Center $(-1, 3)$, radius 1

61. Center $(2, -2)$, through $(5, 1)$

62. Center $(-3, 0)$, through $(3, -8)$

63. Center $(-3, -4)$, through $(-7, -1)$

64. Center $(10, -3)$, through $(6, -6)$

Write an equation of the given circle with endpoints of a diameter as shown.

65.

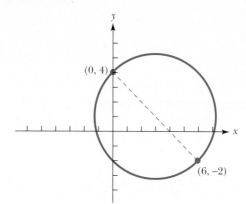

66.

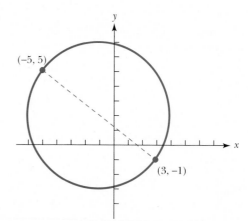

67.

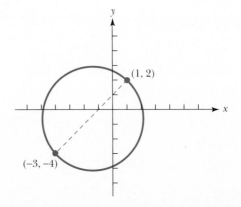

68.

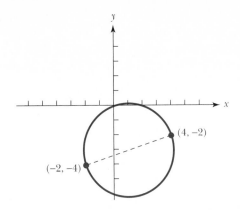

$(4, -2)$

$(-2, -4)$

Graph the following relations. (See Example 7.)

69. $x^2 + y^2 = 4$

70. $(x - 2)^2 + (y - 1)^2 = 9$

71. $(x + 1)^2 + (y + 2)^2 = 16$

72. $x^2 + (y - 4)^2 = 1$

73. $x^2 + y^2 - 4x = 21$

74. $x^2 + y^2 + 6x - 12y + 9 = 0$

75. $x^2 + y^2 = 10x$

76. $x^2 + y^2 = -4y$

2.7 | Graphs of Polynomial Functions

We restrict our attention in this section to polynomial functions defined by $y = P(x)$, where x is a real variable and $P(x)$ is a polynomial with real coefficients.

A **zero** of the polynomial $P(x)$ is a real number r such that $P(r) = 0$. If r is a zero of $P(x)$, then the point $(r, 0)$ lies on the graph of the equation $y = P(x)$. If the polynomial $P(x)$ can be written in factored form

$$P(x) = (x - r_1)^{k_1}(x - r_2)^{k_2} \cdots (x - r_m)^{k_m},$$

where the numbers $r_1, r_2, \ldots, r_m$ are distinct, then each of $r_1, r_2, \ldots, r_m$ is a zero of the polynomial $P(x)$. The exponent k_i counts the number of times $(x - r_i)$ occurs as a factor of $P(x)$. We say that $x - r_i$ is a factor of **multiplicity k_i** and that r_i is a zero of $P(x)$ of **multiplicity k_i**.

Example 1 State the zeros (and their multiplicities) of the following polynomials.

a) $P(x) = (x - 1)^2(x + 2)$ b) $P(x) = 2x^3 + 12x^2 + 18x$

Solution

a) The zeros of $P(x)$ are 1 with multiplicity 2 and -2 with multiplicity 1.

b) The factored form of $P(x) = 2x^3 + 12x^2 + 18x$ is

$$P(x) = 2x(x + 3)^2.$$

Thus 0 is a zero of $P(x)$ with multiplicity 1, and -3 is a zero of $P(x)$ with multiplicity 2. ❏

In Section 2.3 we saw that the graph of a first-degree polynomial function is always a straight line. Then in Section 2.4 we saw that the graph of a second-degree polynomial function is always a parabola. In this section we study the graphs of polynomial functions that have degree greater than 2.

The number of possible shapes of the graph increases as the degree of the polynomial function increases. Even for third-degree polynomials there are several possibilities. Some (not all) of these are indicated in Figure 2.42, where attention is called to the number of real zeros of the polynomial.

The graphs shown in Figure 2.42 represent third-degree polynomials

$$P(x) = a_3x^3 + a_2x^2 + a_1x + a_0$$

with a leading coefficient a_3 which is positive. Similar graphs could be drawn for the case where $a_3 < 0$. These graphs would simply correspond to reflections through the x-axis, since that is the effect of multiplying a function by -1.

The point here is that there is no single characteristic shape, such as a parabola, for the graphs of higher-degree polynomials. As the degree increases, the number of possible shapes increases. However, some general remarks can be made. First, the graph of a polynomial function is a continuous, smooth curve (no breaks, gaps, or sharp corners). It usually happens that the number of turning points on the graph is one less than the degree (note the exceptions in Figure 2.42).

For maximum efficiency in graphing a polynomial function, it is necessary to use calculus. For certain simpler functions, however, a rea-

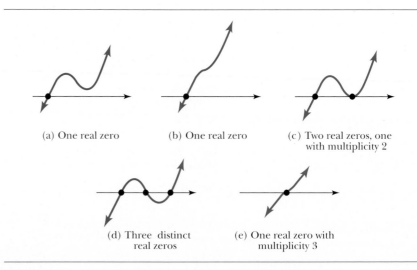

(a) One real zero

(b) One real zero

(c) Two real zeros, one with multiplicity 2

(d) Three distinct real zeros

(e) One real zero with multiplicity 3

FIGURE 2.42

sonably good graph can be drawn by plotting suitably selected points on the curve that have integral x-coordinates and connecting these points with a smooth curve. Knowledge of the zeros of the polynomial, of symmetry, and even of the algebraic sign of the polynomial between the zeros should be utilized whenever possible. A complete analysis of the sign of the polynomial between its zeros can be carried out by using the techniques from Section 1.6 for solving nonlinear inequalities.

Example 2 Sketch the graph of P if $P(x) = (x + 1)(x - 2)(x + 3)$.

Solution

The zeros of the polynomial are $x = -1$, $x = 2$, and $x = -3$, and they separate the coordinate plane into four vertical regions. We can determine the sign of $P(x)$ in each region by choosing test points. If $P(x)$ is positive for one value of x in a given interval, then it is positive for all x in that interval. Similarly, if $P(x)$ is negative for one value of x in a given interval, then it is negative for all x in that interval. The test points, along with the zeros, are used to sketch the graph of $y = P(x)$ in Figure 2.43. Additional points can be plotted, as desired. Determining those points where the curve turns requires the use of calculus.

	x	y
Test point	-4	-18
Zero	-3	0
Test point	-2	4
Zero	-1	0
Test point	0	-6
Additional point	1	-8
Zero	2	0
Test point	3	24

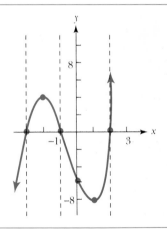

FIGURE 2.43

It is easy to see that if $x - r$ is a factor of $P(x)$ with *even* multiplicity, then $P(x)$ does not change sign as x increases through r, and the graph of $y = P(x)$ is tangent to the x-axis at $x = r$. However, if $x - r$ is a factor of $P(x)$ with *odd* multiplicity, then $P(x)$ does change sign as x increases through r, and the graph of $y = P(x)$ crosses the x-axis at $x = r$.

x	y
0	-4
1/2	$-9/8$
1	0
3/2	1/8
2	0
3	2

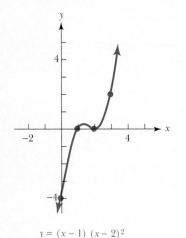

$y = (x - 1)(x - 2)^2$

FIGURE 2.44

These remarks are illustrated in the following examples.

Example 3   Sketch the graph of P if $P(x) = (x - 1)(x - 2)^2$.

Solution
Since $x - 2$ is a factor with *even* multiplicity, the graph is tangent to the x-axis at $x = 2$. Also, since $x - 1$ is a factor with *odd* multiplicity, the graph crosses the x-axis at $x = 1$. A table of values and the graph are shown in Figure 2.44. ❏

Symmetry and translation play key roles in graphing the polynomial in the next example.

Example 4   Sketch the graph of $y = (x - 2)^2(x + 2)^2 + 1$.

Solution
Let $P(x) = (x - 2)^2(x + 2)^2$. Then y can be written as

$$y = P(x) + 1.$$

We can graph $y = P(x) + 1$ by translating the graph of $y = P(x)$ upward one unit. We note that the graph of $y = P(x)$ is symmetric with respect to the y-axis and is tangent to the x-axis at ± 2. Plotting a few points for nonnegative values of x and using symmetry, we graph $y = P(x)$ as the dashed curve in Figure 2.45. The graph of $y = (x - 2)^2(x + 2)^2 + 1$ is shown as the solid curve.

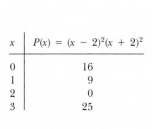

x	$P(x) = (x - 2)^2(x + 2)^2$
0	16
1	9
2	0
3	25

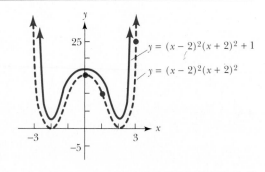

$y = (x - 2)^2(x + 2)^2 + 1$

$y = (x - 2)^2(x + 2)^2$

FIGURE 2.45   ❏

Example 5   Acme Anchor Company can sell up to 8000 anchors per day, but it can produce at most 4000 anchors per day with the single machine that it owns. It is considering the purchase of a second machine that could double production. An expert predicts that with two machines in operation, the company's profit P, in thousands of dollars per day, will be given by

$$P(x) = x(x - 4)(x - 6),$$

where x is the daily production in thousands of anchors.

a) Find the nonnegative values of x for which $P(x) > 0$.
b) Sketch the graph of $P(x)$ for $0 \leq x \leq 8$.

Solution

a) We see by inspection that the zeros of P are 0, 4, and 6. Testing a point in each interval, we find that $P(2) = 16$, $P(5) = -5$, and $P(7) = 21$. Thus $P(x) > 0$ for $x \in (0, 4) \cup (6, \infty)$.

b) We use our test points and the additional function values in the table of Figure 2.46 to draw the graph shown in Figure 2.46.

x	$P(x)$
0	0
1	15
2	16
3	9
4	0
5	-5
6	0
7	21
8	64

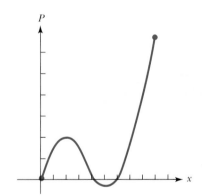

FIGURE 2.46

Exercises 2.7

Sketch the graph of the polynomial function defined by each equation.

1. $P(x) = (x - 1)(x + 2)(x - 3)$
2. $P(x) = (x + 1)(x - 2)(x - 4)$
3. $P(x) = (x - 1)^2(x + 4)^2$
4. $P(x) = (x + 3)^2(x - 2)^2$
5. $P(x) = (-3x + 2)(2x - 1)(x + 3)$

6. $P(x) = (x + 2)(-2x + 3)(2x + 1)$
7. $P(x) = -2(x + 1)^2(1 - x)$
8. $P(x) = (x - 2)^2(x + 3)$
9. $P(x) = x^2(x + 2)(x - 2)$
10. $P(x) = x^2(x + 1)(x - 1)$
11. $P(x) = (x - 1)^2(x + 1)^2$
12. $P(x) = (x - 3)^2(x + 3)^2$

13. $P(x) = (2 - x)(x^2 - 3x + 2)$

14. $P(x) = (1 - x)(x^2 - 3x + 2)$

15. $P(x) = (2x - 3)^2(3x - 4)(2x + 1)$

16. $P(x) = (2x + 1)^2(4x - 3)(2x - 5)$

17. $P(x) = x^4 - 8x^2$

18. $P(x) = x^4 - 27x^2$

19. $P(x) = x^3 - 2x^2 - 15x$

20. $P(x) = x^3 + 4x^2 + 4x$

21. $P(x) = x(x - 1)^2 - 3$

22. $P(x) = (x - 1)(x + 2)^2 - 2$

23. $P(x) = (x - 2)^2(x + 3) + 4$

24. $P(x) = (x + 1)^2(x - 2) + 2$

25. $P(x) = -2(x^2 + 1)(x - 1)(x + 2)$

26. $P(x) = -2(x^2 + 2)(x^2 - 1)$

27. $P(x) = (x^2 + x + 1)(x - 2)$

28. $P(x) = (x^2 + x + 2)(x + 1)$

29. $P(x) = (x - 4)(x^3 + 8)$

30. $P(x) = (x + 3)(x^3 - 8)$

31. Acme Boat Company has found that the revenue from producing x canoes is given in dollars by $R(x) = 8x^2 - 0.02x^3$.
    a) Find the nonnegative values of x for which $R(x) \geq 0$.
    b) Sketch the graph of $R(x)$ for $x \geq 0$.

32. The demand $D(x)$ for Pedalex cars, in thousands of cars, is given by $D(x) = x(2 + x)(8 - x)$, where x is the price in thousands of dollars.
    a) Find the nonnegative values of x for which $D(x) \geq 0$.
    b) Sketch the graph of D for $x \geq 0$.

33. An open box is to be made from a rectangular piece of tin by cutting equal squares from each corner and bending up the sides as shown below. If the piece of tin measures 8 inches by 12 inches, the volume of the resulting box is given by $V(x) = x(8 - 2x)(12 - 2x)$. Find the nonnegative values of x for which $V(x) \geq 0$.

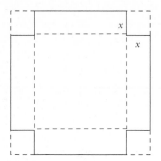

EXERCISE 33

34. If a right circular cylinder is inscribed in a sphere as shown in the accompanying figure, the volume of the cylinder is $V(h) = \pi h(256 - h^2)/4$, where h is the altitude of the cylinder. Find the nonnegative values of h for which $V(h) \geq 0$.

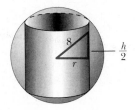

EXERCISE 34

2.8 | Graphs of Rational Functions

In Section 1.2 a rational expression was defined as a quotient of polynomials. Similarly, a **rational function** is a function defined by an equation of the form

$$f(x) = \frac{P(x)}{Q(x)},$$

where $P(x)$ and $Q(x)$ are polynomials in the variable x, and $Q(x)$ is not the zero polynomial. In this section we consider only those rational functions that are quotients of polynomials with real coefficients, and our goal is to learn to sketch the graph of such a function with some degree of efficiency. If such a rational function f is defined by

$$y = f(x) = \frac{P(x)}{Q(x)},$$

it is understood that the domain of f is the set of all real numbers x such that $Q(x) \neq 0$.

In very simple cases the graph can be sketched satisfactorily by plotting several points. In some cases, symmetry and translation can be helpful. Consider the following example.

Example 1   Sketch the graph of the function f defined by

$$f(x) = \frac{2}{x - 1}.$$

Solution

The domain of the function is the set of all real numbers x such that $x \neq 1$. This means that there is no point on the graph where $x = 1$, and the graph is separated into two parts: those points (x, y) with $x > 1$, and those with $x < 1$. We concentrate first on the points with $x > 1$. We note that the corresponding y-coordinates are positive. Using the values $x = 2, 5, 9$ in succession, we obtain the first three sets of coordinates in the following table. Then using $x = \frac{3}{2}, 1.2, 1.01$, we obtain the next three sets.

x	2	5	9	$\frac{3}{2}$	1.2	1.01
f(x)	2	$\frac{1}{2}$	$\frac{1}{4}$	4	10	200

After locating the points

$$A(2, 2), \quad B\left(5, \frac{1}{2}\right), \quad C\left(9, \frac{1}{4}\right),$$

we find that, as x continues to increase, the corresponding y-values are positive but steadily decreasing toward 0. We see that there is no x-intercept since $2/(x - 1) = 0$ has no solution. After locating the points

$$D\left(\frac{3}{2}, 4\right), \quad E(1.2, 10), \quad F(1.01, 200),$$

we find that, as x moves closer to 1 from the right, the corresponding y-values increase steadily without bound. Thus the points in the table lead us to draw the right-hand part of the curve, as shown in Figure 2.47.

In similar fashion, we obtain the following table for some values of $x < 1$. We note that whenever $x < 1$, we have $y < 0$. In making this

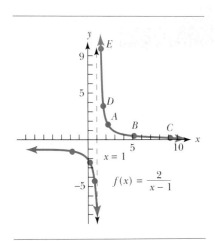

$$f(x) = \frac{2}{x - 1}$$

FIGURE 2.47

table, we use $x = 0$ to find that the y-intercept is -2, and we observe that no symmetry is detected by the tests in Section 2.6.

x	0	-2	-10	$\frac{1}{2}$	0.9	0.99
$f(x)$	-2	$-\frac{2}{3}$	$-\frac{2}{11}$	-4	-20	-200

After plotting these points, we draw the left-hand part of the curve, as shown in Figure 2.47. The dashed line $x = 1$ is used as a guide in drawing the curve and is *not* a part of the graph of the function.   ❑

The line $x = 1$ is useful in drawing the graph in Example 1 because the points on the graph are closer to the line when the chosen values of x are closer to 1. A line such as this is called a **vertical asymptote** of the graph. From Figure 2.47, it is clear that the graph has the same sort of relationship to the x-axis. As $|x|$ increases without bound, the points on the graph come closer and closer to the line $y = 0$. A line such as this is called a **horizontal asymptote** of the curve.

In describing the graph of a function near an asymptote, some of the language and notation from the calculus is very helpful. To indicate that x increases without bound, we write $x \rightarrow \infty$ and read this as "x approaches infinity." Similarly, we write

$$\frac{2}{x-1} \rightarrow 0 \quad \text{as} \quad x \rightarrow \infty$$

and read this as

"$\dfrac{2}{x-1}$ approaches 0 as x approaches ∞."

This is also written in a more formal way as

$$\lim_{x \to \infty} \frac{2}{x-1} = 0,$$

and this symbolism is read as

"the limit of $\dfrac{2}{x-1}$, as x approaches ∞, is 0."

We write $x \rightarrow -\infty$ to indicate that x decreases without bound. With this type of notation, there are other limit statements appropriate to describe the behavior of f in Figure 2.47:

$$\frac{2}{x-1} \rightarrow 0 \quad \text{as} \quad x \rightarrow -\infty; \text{ that is,} \quad \lim_{x \to -\infty} \frac{2}{x-1} = 0;$$

$$\frac{2}{x - 1} \to \infty \quad \text{as} \quad x \to 1^+; \text{ that is,} \quad \lim_{x \to 1^+} \frac{2}{x - 1} = \infty;$$

$$\frac{2}{x - 1} \to -\infty \quad \text{as} \quad x \to 1^-; \text{ that is,} \quad \lim_{x \to 1^-} \frac{2}{x - 1} = -\infty.$$

The notation $x \to 1^+$ indicates that x approaches 1 from the right, and $x \to 1^-$ indicates that x approaches 1 from the left. The superscripts $^+$ and $^-$ are not exponents in these notations, but they indicate "right side" and "left side," respectively. The superscript $^+$ indicates that x is greater than 1, and the superscript $^-$ indicates that x is less than 1.

It is no accident that the vertical asymptote in Example 1 occurred at a zero of the denominator of the function. In the general case, suppose that

$$y = \frac{P(x)}{Q(x)}$$

and that $x = a$ is a value of x for which $Q(a) = 0$ and $P(a) \neq 0$. As x takes on values very close to a, $Q(x)$ is very close to 0, and $P(x)$ is very close to $P(a)$. This means that the quotient $P(x)/Q(x)$ grows larger numerically (that is, in absolute value) as x gets closer to a, and

$$|y| \to \infty \quad \text{as} \quad x \to a^+ \quad \text{or} \quad x \to a^-.$$

A detailed discussion of this situation calls for a precise formulation of the concept of a limit. The concept of a limit is the fundamental concept of the calculus, and limits are treated there in a very rigorous fashion. However, our intuitive discussion should make the following theorem seem plausible.

Theorem 2.16

Let f be a rational function defined by

$$f(x) = \frac{P(x)}{Q(x)} = \frac{a_n x^n + a_{n-1} x^{n-1} + \cdots + a_1 x + a_0}{b_m x^m + b_{m-1} x^{m-1} + \cdots + b_1 x + b_0},$$

where $P(x)$ and $Q(x)$ are polynomials with real coefficients, and $Q(x)$ is not the zero polynomial. The horizontal and vertical asymptotes of the graph of f may be found by the following rules.

Vertical Asymptotes

If a is a real number such that $Q(a) = 0$ and $P(a) \neq 0$, then the line $x = a$ is a vertical asymptote of the graph of f.

Horizontal Asymptotes

a) If $n < m$, then $y = 0$ is a horizontal asymptote.
b) If $n = m$, then $y = a_n/b_m$ is a horizontal asymptote.
c) If $n > m$, there are no horizontal asymptotes.

A systematic routine for using Theorem 2.16 and other information is given in the following steps.

Procedure for Graphing a Rational Function $f(x) = \dfrac{P(x)}{Q(x)}$

---

1. Locate all asymptotes of the graph.
2. Locate the x-intercepts and y-intercepts, if there are any.
3. Note the sign of $f(x)$ in the intervals determined by the zeros of $P(x)$ and $Q(x)$.
4. Note any symmetry detected by the tests in Section 2.6.
5. Plot a few points on either side of each vertical asymptote and check to see if the graph crosses a horizontal asymptote.
6. Sketch the graph, using the points plotted and the asymptotes as guides. The graph will be a smooth curve, except for breaks at the asymptotes.

Example 2　　Sketch the graph of

$$f(x) = \frac{2x^2 - 1}{x^2 - 3x},$$

locating all vertical or horizontal asymptotes.

Solution
To find vertical asymptotes, we set the denominator equal to 0:

$$x^2 - 3x = 0,$$

or

$$x(x - 3) = 0.$$

Since $P(x) = 2x^2 - 1$ is not zero at $x = 0$ or $x = 3$, the lines $x = 0$ and $x = 3$ are vertical asymptotes. The numerator and denominator have the same degree, so $y = 2$ is a horizontal asymptote. Setting $y = 2$ in

$$y = \frac{2x^2 - 1}{x^2 - 3x},$$

we find that $x = 1/6 \approx 0.2$, and the graph crosses the horizontal asymptote there. Since 0 is not in the domain, there is no y-intercept. The x-intercepts are given by $x = \pm 1/\sqrt{2}$.

We rewrite the rational function in factored form

$$f(x) = \frac{(\sqrt{2}x - 1)(\sqrt{2}x + 1)}{x(x - 3)}$$

$\sqrt{2}x - 1$	$-$		$-$	$-$	$-$	$-$	0	$+$		$+$	$+$
$\sqrt{2}x + 1$	$-$		0	$+$	$+$	$+$	$+$	$+$		$+$	$+$
x	$-$		$-$	$-$	0	$+$	$+$	$+$		$+$	$+$
$x - 3$	$-$		$-$	$-$	$-$	$-$	$-$	$-$		0	$+$
$f(x) = \dfrac{(\sqrt{2}x - 1)(\sqrt{2}x + 1)}{x(x - 3)}$	$+$		0	$-$	U	$+$	0	$-$		U	$+$

```
        -2   -1/√2  0  1/√2      2       3       4
             x-intercept  x-intercept

             Vertical asymptote     Vertical asymptote
```

FIGURE 2.48

and use the sign graph in Figure 2.48 to analyze the sign of $y = f(x)$ between the x-intercepts and the vertical asymptotes.

No symmetry is detected by the tests in Section 2.6. The following table of values is for the graph in Figure 2.49, with coordinates rounded to the nearest tenth. We use these values in sketching the graph, and we also use the facts that

$$f(x) \to 2^- \quad \text{as} \quad x \to -\infty \quad \text{and} \quad f(x) \to 2^+ \quad \text{as} \quad x \to \infty;$$

$$f(x) \to -\infty \quad \text{as} \quad x \to 0^- \quad \text{and} \quad f(x) \to \infty \quad \text{as} \quad x \to 0^+;$$

$$f(x) \to -\infty \quad \text{as} \quad x \to 3^- \quad \text{and} \quad f(x) \to \infty \quad \text{as} \quad x \to 3^+.$$

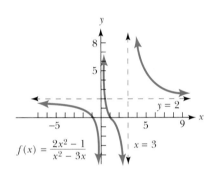

$$f(x) = \frac{2x^2 - 1}{x^2 - 3x}$$

FIGURE 2.49

x	-3	-2	-1	-0.7	-0.5	0	0.2	0.7	1	2	3	4	5	8
$f(x)$	1	0.7	0.2	0	-0.3	U	1.6	0	-0.5	-3.5	U	7.8	4.9	3.2

❑

Under certain conditions, the graph of a rational function may have an asymptote that is of a type different from those described up to this point. If the degree of $A(x)$ is greater than or equal to the degree of $B(x)$, the equation

$$f(x) = \frac{A(x)}{B(x)} = Q(x) + \frac{R(x)}{B(x)},$$

where degree $R(x) <$ degree $B(x)$, means that $y = Q(x)$ is an asymptote since $R(x)/B(x) \to 0$ as $|x| \to \infty$. In particular, if the degree of $A(x)$ is exactly 1 more than the degree of $B(x)$, and

$$f(x) = \frac{A(x)}{B(x)} = ax + b + \frac{R(x)}{B(x)},$$

then the graph of f has the line $y = ax + b$ as an **oblique asymptote** when $x \to \infty$ and when $x \to -\infty$.

Example 3   Sketch the graph of

$$f(x) = \frac{x^2}{x - 1}$$

and locate all asymptotes of the graph.

Solution

According to Theorem 2.16, the graph has a vertical asymptote at $x = 1$, and it has no horizontal asymptote. The only point where the graph crosses a coordinate axis is at the origin $(0, 0)$. To find the oblique asymptote, we divide x^2 by $x - 1$ and obtain

$$f(x) = x + 1 + \frac{1}{x - 1}.$$

From this equation, it can be seen that for $|x|$ very large, the value of $1/(x - 1)$ is near zero, and points (x, y) on the graph of f are close to the line

$$y = x + 1.$$

That is, $y = x + 1$ is an oblique asymptote.

We analyze the sign of $y = f(x)$ in the following manner. Since x^2 is always nonnegative, then the sign of y depends only on the sign of $x - 1$. Thus $y > 0$ when $x > 1$, and $y \leq 0$ when $x < 1$.

The tests in Section 2.6 do not detect any symmetry. A table of values, rounded to the nearest tenth, is given below.

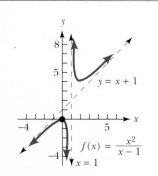

$y = x + 1$

$f(x) = \dfrac{x^2}{x - 1}$

$x = 1$

FIGURE 2.50

	-1	-2	-3	-4	0	0.5	2	2.5	3	4
$f(x)$	-0.5	-1.3	-2.2	-3.2	0	-0.5	4	4.2	4.5	5.3

The graph is shown in Figure 2.50.   ❏

In Theorem 2.16 and all the examples to this point, we have considered only rational functions $f(x) = P(x)/Q(x)$, where $P(a)$ was *not* zero if $Q(a) = 0$. As our last example in this section, we consider a case where both $P(a) = 0$ and $Q(a) = 0$ for some value of a. In any case such as this, there must be a break in the graph of f at $x = a$, because

$$f(a) = \frac{P(a)}{Q(a)} = \frac{0}{0} \text{ is undefined.}$$

Our example shows that this situation is not as difficult to handle as it might seem.

Example 4   Sketch the graph of

$$f(x) = \frac{2x - 10}{x^2 - 6x + 5}$$

and locate all asymptotes of the graph.

Solution
When $P(x) = 2x - 10$ and $Q(x) = x^2 - 6x + 5$ are written in factored form, we have

$$f(x) = \frac{2(x - 5)}{(x - 1)(x - 5)}.$$

In this form we see that

$$f(5) = \frac{0}{0} \text{ is undefined,}$$

and that

$$f(x) = \frac{2}{x - 1} \quad \text{for all} \quad x \neq 5.$$

Thus the graph of f is *almost* the same as the graph of $y = 2/(x - 1)$ in Example 1 of this section. The only difference is that the point $(5, \frac{1}{2})$ is on the graph of $y = 2/(x - 1)$, but $(5, \frac{1}{2})$ is *not* on the graph of

$$f(x) = \frac{2(x - 5)}{(x - 1)(x - 5)}.$$

Thus the graph of f has a "hole" in it, and this is indicated in Figure 2.51 where the graph of f is drawn with an open dot at $(5, \frac{1}{2})$.   ❑

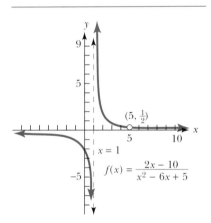

FIGURE 2.51

Exercises 2.8

In Exercises 1–16, find the value of the given limit as a real number, as ∞, or as $-\infty$.

1. $\lim\limits_{x \to 3^+} \dfrac{2}{x - 3}$

2. $\lim\limits_{x \to 2^+} \dfrac{5}{x - 2}$

3. $\lim\limits_{x \to 3^-} \dfrac{2}{x - 3}$

4. $\lim\limits_{x \to 2^-} \dfrac{5}{x - 2}$

5. $\lim\limits_{x \to \infty} \dfrac{2}{x - 3}$

6. $\lim\limits_{x \to \infty} \dfrac{5}{x - 2}$

7. $\lim\limits_{x \to -\infty} \dfrac{2}{x - 3}$

8. $\lim\limits_{x \to -\infty} \dfrac{5}{x - 2}$

9. $\lim\limits_{x \to -2^+} \dfrac{3}{x^2 - 4}$

10. $\lim\limits_{x \to -3^+} \dfrac{2}{x^2 - 9}$

11. $\lim\limits_{x \to -2^-} \dfrac{3}{x^2 - 4}$

12. $\lim\limits_{x \to -3^-} \dfrac{2}{x^2 - 9}$

13. $\lim\limits_{x \to \infty} \dfrac{3}{x^2 - 4}$

14. $\lim\limits_{x \to \infty} \dfrac{2}{x^2 - 9}$

15. $\lim\limits_{x \to -\infty} \dfrac{3}{x^2 - 4}$ **16.** $\lim\limits_{x \to -\infty} \dfrac{2}{x^2 - 9}$

Sketch the graphs of the rational functions defined by the following equations. Locate all vertical, horizontal, and oblique asymptotes.

17. $f(x) = \dfrac{2}{x + 4}$ **18.** $f(x) = -\dfrac{3}{2 - x}$

19. $f(x) = \dfrac{2x + 1}{x - 3}$ **20.** $f(x) = \dfrac{3x - 4}{2x + 1}$

21. $f(x) = \dfrac{5 - 4x}{3x + 5}$ **22.** $f(x) = \dfrac{3x + 5}{5 - 4x}$

23. $f(x) = \dfrac{1}{x^2 - 3x}$ **24.** $f(x) = \dfrac{1}{x^2 - x}$

25. $f(x) = \dfrac{1}{(x + 1)^2}$ **26.** $f(x) = \dfrac{1}{x^2}$

27. $f(x) = \dfrac{1}{x^2 + 4}$ **28.** $f(x) = \dfrac{1}{x^2 + 1}$

29. $f(x) = \dfrac{2x}{(x + 2)^2}$

30. $f(x) = \dfrac{2x}{(x - 1)(x + 2)}$

31. $f(x) = \dfrac{1}{x^2(x + 1)}$ **32.** $f(x) = \dfrac{1}{x^2(x - 2)}$

33. $f(x) = \dfrac{4}{(x + 2)(x - 3)}$

34. $f(x) = \dfrac{2}{(x - 1)(x + 3)}$

35. $f(x) = \dfrac{x^2 - x - 2}{x^2 - x}$ **36.** $f(x) = \dfrac{2x^2 + 5x - 3}{x^2 + 2x}$

37. $f(x) = \dfrac{x^2 - 3}{2x^2 - x - 3}$

38. $f(x) = \dfrac{x^2 - 9}{2x^2 - 5x - 3}$

39. $f(x) = \dfrac{2x^2}{x + 1}$ **40.** $f(x) = \dfrac{3x^2 - 1}{x}$

41. $f(x) = \dfrac{3x - 4}{x^2 - 3x + 2}$ **42.** $f(x) = \dfrac{2x + 1}{x^2 - 2x - 3}$

43. $f(x) = \dfrac{x^2 - 1}{x^2(x + 2)}$ **44.** $f(x) = \dfrac{x^2 - 4}{x^2(x + 3)}$

45. $f(x) = \dfrac{x^2 + 4}{x - 2}$ **46.** $f(x) = \dfrac{x^2 + 9}{x - 3}$

47. $f(x) = \dfrac{x^2 - 9}{x^3}$ **48.** $f(x) = \dfrac{x^2 - 4}{x^3}$

49. $f(x) = \dfrac{x^2 - 4}{x^2 - 1}$ **50.** $f(x) = \dfrac{x^2 - 4}{x^2 - 9}$

51. $f(x) = \dfrac{x^2 - 4}{x + 2}$ **52.** $f(x) = \dfrac{x^2 - 9}{x - 3}$

53. $f(x) = \dfrac{6x^2 + x - 12}{3x - 4}$

54. $f(x) = \dfrac{4x^2 - 7x - 15}{x - 3}$

55. $f(x) = \dfrac{x^3 - 27}{x - 3}$ **56.** $f(x) = \dfrac{x^3 + 8}{x + 2}$

57. If A denotes the adult dosage of a medication in milligrams, a formula that is commonly used to find the dosage $D(x)$ for a child x years old is

$$D(x) = \dfrac{Ax}{x + 12}.$$

For the medication phenylpropanolamine hydrochloride, $A = 25$. Sketch the graph of D for $A = 25$ and $x > 0$.

58. In a restocking program, 60 deer are released in a wildlife management area. The expected population $P(x)$ of deer after x years is

$$P(x) = \dfrac{60 + 15x}{1 + 0.05x}.$$

Sketch the graph of P for $x \geq 0$ and estimate the limiting value of the population as time increases without bound.

59. The cost $C(x)$, in thousands of dollars, of removing x percent of a certain oil spill off the Texas coast is approximated by

$$C(x) = \dfrac{15x}{100 - x}.$$

Sketch the graph of C for $0 \leq x < 100$.

60. Biologists estimate the number S of fish spawning this season and use this to predict the number of breeding fish, or "recruits," that will return next year. The number R of recruits is estimated from S by a formula of the form

$$R(S) = \dfrac{S}{aS + b},$$

where a and b are positive constants. Find the limiting value of R for extremely large values of S.

CHAPTER REVIEW

Key Words and Phrases

Function
Domain
Range
Independent variable
Dependent variable
Relation
Graph
Vertical Line Test
Composite function
Linear function
Intercepts
Pythagorean Theorem
Distance Formula

Slope of a line
Standard form of the equation of
  a straight line
Slope-intercept form
Point-slope form
Quadratic function
Parabola
Vertex of a parabola
Axis of symmetry of a parabola
Inverse relation
One-to-one function
Horizontal Line Test

Inverse function
Symmetry
Circle
Center
Radius
Stretching
Reflection
Translation
Multiplicity of a factor
Vertical asymptote
Horizontal asymptote
Oblique asymptote

Summary of Important Concepts and Formulas

Greatest Integer Function

If $n \leq x < n + 1$, then $[x] = n$.

Composite Function

$$(f \circ g)(x) = f(g(x))$$

Vertical Line Test

If some vertical line intersects the graph of a relation at two or more points, the relation is *not* a function. If no vertical line intersects the graph in more than one point, the relation is a function.

Distance Formula

$$d = \sqrt{(x_2 - x_1)^2 + (y_2 - y_1)^2}$$

Midpoint Formula

$$\left(\frac{x_1 + x_2}{2}, \frac{y_1 + y_2}{2}\right)$$

Slope of a Line

$$m = \frac{y_2 - y_1}{x_2 - x_1}$$

Slope-Intercept Form

$$y = mx + b$$

Point-Slope Form

$$y - y_1 = m(x - x_1)$$

Parallel Lines

$$m_1 = m_2$$

Perpendicular Lines

$$m_1 m_2 = -1$$

A parabola with equation $y = a(x - h)^2 + k$ has vertex (h, k), has $x = h$ as the axis of symmetry, and opens upward if $a > 0$, downward if $a < 0$.

If a parabola has equation $y = ax^2 + bx + c$, the x-coordinate of the vertex is $-b/(2a)$, the line $x = -b/(2a)$ is the axis of symmetry, and it opens upward if $a > 0$, downward if $a < 0$.

Horizontal Line Test

If any horizontal line intersects the graph of the function f at more than one point, then f is *not* a one-to-one function. If no horizontal line intersects the graph of the function f at more than one point, then f is a one-to-one function.

Inverse Function Equations

$$f^{-1}(f(x)) = x \text{ for all } x \text{ in the domain of } f$$

$$f(f^{-1}(x)) = x \text{ for all } x \text{ in the domain of } f^{-1}$$

Tests for Symmetry

The graph of an equation is symmetric with respect to the	if the following substitution yields an equivalent equation
y-axis	$-x$ for x
x-axis	$-y$ for y
origin	$-x$ for x and $-y$ for y

Standard Equation for a Circle

$$(x - h)^2 + (y - k)^2 = r^2$$

Stretching of $y = f(x)$
To graph $y = cf(x)$, $c > 0$, multiply each y-coordinate on the graph of $y = f(x)$ by c.

Reflection of $y = f(x)$
To graph $y = -f(x)$, reflect the graph of $y = f(x)$ through the x-axis.

Vertical Translation of $y = f(x)$
To graph $y = f(x) + c$, shift the graph of $y = f(x)$ upward c units if $c > 0$, or downward $|c|$ units if $c < 0$.

Horizontal Translation of $y = f(x)$
To graph $y = f(x - c)$, shift the graph of $y = f(x)$ to the right c units if $c > 0$, or to the left $|c|$ units if $c < 0$.

Procedure for Graphing a Rational Function
$$f(x) = \frac{P(x)}{Q(x)}$$
a. Locate all asymptotes of the graph.
b. Locate the x-intercepts and y-intercepts, if there are any.
c. Note the sign of $f(x)$ in the intervals determined by the zeros of $P(x)$ and $Q(x)$.
d. Note any symmetry detected by the tests in Section 2.6.
e. Plot a few points on either side of each vertical asymptote and check to see if the graph crosses a horizontal asymptote.
f. Sketch the graph, using the points plotted and the asymptotes as guides. The graph will be a smooth curve, except for breaks at the asymptotes.

Review Problems for Chapter 2

1. Find the domain D and the range R of the function defined by $y = \sqrt{x - 9}$.

If $f(x) = x^2 - 2x$ and $g(x) = \sqrt{x} + 3$, find each of the following.

2. a) $f(4)$ b) $f(g(9))$ c) $(g \circ f)(4)$

3. a) $(f \circ g)(a)$ b) $f(x + h) - f(x)$

Sketch the graph of the given function.

4. $f(x) = |x + 2|$ **5.** $f(x) = [x/2]$

Let $f(x) = \sqrt{x - 3}$ and $g(x) = x - 4$. Determine the following functions and state their domains.

6. $f + g$ **7.** $f - g$

8. $f \cdot g$ **9.** f/g

10. $f \circ g$

11. Graph each of the following linear relations, and find the x-intercepts and y-intercepts, if they exist.
a) $3x - 2y = 12$ b) $y = 4$

12. Compute the slope of the line joining the points $(-2, 4)$ and $(-7, -8)$. Also compute the distance between the points.

Find an equation, in standard form, of the straight line that satisfies the given conditions.

13. Through $(3, -1)$ and $(2, 7)$

14. Through $(-2, 3)$, parallel to $3x - 4y = 10$

15. Through $(-1, 4)$, perpendicular to $3x = 6y + 1$

16. Find the value of y if the line through $(-1, 1)$ and $(3, y)$ has slope 2.

17. Graph the following function.
$$f(x) = \begin{cases} 1 - x & \text{if } x < 0 \\ 3x + 1 & \text{if } x \geq 0 \end{cases}$$

In Problems 18 and 19, graph the given parabolas. Locate the vertex and axis of symmetry, and plot at least two other points on each graph.

18. $y = 3(x - 1)^2 + 2$ **19.** $y = -x^2 + 4x - 3$

Find the maximum or minimum value of the given function and state whether that value represents a maximum or a minimum.

20. $f(x) = 2x^2 - 8x + 13$ **21.** $f(x) = 1 - 6x - x^2$

22. The profit P, in dollars, from producing x units of a certain item is given by $P = 0.8x - 0.001x^2$. Find the production level x for which the profit is a maximum, and find the maximum profit.

23. Sketch the graph of the following function.
$$f(x) = \begin{cases} 4 - x^2 & \text{if } x < 2 \\ 4 - x & \text{if } x \geq 2 \end{cases}$$

Each of the following equations defines a relation g. In each problem, (a) find a defining equation for the inverse g^{-1}, (b) draw the graphs of g and g^{-1}, and (c) decide if g or g^{-1} is a function.

24. $3x + 2y = 12$ **25.** $x = y^2 + 1$

Decide if the following pairs of functions f and g are inverse functions of each other.

26. $f(x) = 2x - 1$, $g(x) = \dfrac{1}{2x - 1}$

27. $f(x) = \frac{2}{3}x + 6$, $g(x) = \frac{3}{2}x - 9$

Each of the following equations defines a function f. Find an expression for $f(x)$ and also for $f^{-1}(x)$ if f is one-to-one. If f is one-to-one, verify each of the equations $f^{-1}(f(x)) = x$ and $f(f^{-1}(x)) = x$.

28. $2x - 3y = 6$ **29.** $y = x^2 - 4$

The graph of $y = f(x)$ is given in the accompanying figure. Sketch the graphs of each of the following functions.

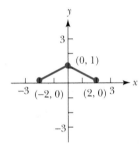

PROBLEMS 30–33

30. $y = 2f(x)$ **31.** $y = f(x) - 2$

32. $y = -f(x)$ **33.** $y = f(x - 2)$

Test the given equation for symmetry with respect to the x-axis, the y-axis, and the origin, and sketch its graph.

34. $y = |x| - 1$ **35.** $|y| = x^2$

36. Determine whether or not the graph of the following equation is a circle. If it is, find the center and the radius.

$$x^2 - 2x + y^2 = 12y - 38$$

37. Write an equation of the circle with center $(2, -3)$ and radius 4.

38. Write an equation of the circle that has a diameter with $(-2, 3)$ and $(6, -3)$ as endpoints.

In Problems 39 and 40, sketch the graph of the polynomial function defined by the given equation.

39. $P(x) = x(x + 2)(x - 3)$

40. $P(x) = (x - 1)^2(x + 2)$

In Problems 41–43, find the value of the given limit as a real number, as ∞, or as $-\infty$.

41. $\displaystyle\lim_{x \to -4^+} \frac{2}{x + 4}$ **42.** $\displaystyle\lim_{x \to -4^-} \frac{2}{x + 4}$

43. $\displaystyle\lim_{x \to -\infty} \frac{2}{x + 4}$

Sketch the graph of the rational functions defined by the following equations. Locate all vertical, horizontal, and oblique asymptotes.

44. $f(x) = \dfrac{2x - 4}{x + 1}$ **45.** $f(x) = \dfrac{2x}{(x - 2)^2}$

Each of the following nonsolutions has at least one error. Can you find them?

Problem 1 If $f(x) = x^2 - 1$ and $g(x) = \sqrt{x - 3}$, find the value of $f(g(7))$.

Nonsolution
$$f(g(7)) = (7^2 - 1)(\sqrt{7 - 3})$$
$$= (48)(2) = 96$$

Problem 2 Write the equation, in standard form, of the line with x-intercept 3 and y-intercept -2.

Nonsolution
$$m = \frac{-2}{3} = -\frac{2}{3}$$

$$y - (-2) = -\frac{2}{3}(x - 3)$$

$$3(y + 2) = -2(x - 3)$$

$$3y + 6 = -2x + 6$$

$$2x + 3y = 0$$

Problem 3 Find the distance between the points $(4, 6)$ and $(1, 2)$.

Nonsolution
$$d = \sqrt{(4 - 1)^2 + (6 - 2)^2}$$
$$= \sqrt{3^2 + 4^2} = 3 + 4 = 7$$

Problem 4 Find the inverse f^{-1} of the function defined by $y = \sqrt{x + 1}$.

Nonsolution
$$f = \{(x, y) : y = \sqrt{x + 1}\}$$

so

$$f^{-1} = \left\{(x, y) : y = \frac{1}{\sqrt{x + 1}}\right\}$$

Problem 5 If the graph of the following equation is a circle, find the center and the radius.

$$x^2 + 2x + y^2 = 11 - 4y$$

Nonsolution
$$x^2 + 2x + y^2 + 4y = 11$$
$$x^2 + 2x + 1 + y^2 + 4y + 4 = 11 + 1 + 4$$
$$(x + 1)^2 + (y + 2)^2 = 16$$

The center is at $(1, 2)$, and the radius is 16.

Chapter 3

Theory of Polynomials

The polynomial functions studied in this chapter are used extensively in mathematical models of production costs, consumer demands, biological processes, and many other scientific studies. The chapter concludes with a section that deals with the approximation of real zeros. Approximation techniques steadily gain in importance with the growing use of computers and calculators.

3.1 The Factor and Remainder Theorems

In this chapter the domain of a variable is the set $\mathscr{C}$ of all complex numbers, unless otherwise stated. For instance, we consider quadratic equations that involve complex numbers. When the coefficients a, b, and c in

$$ax^2 + bx + c = 0, \qquad a \neq 0,$$

are allowed to be complex numbers, the Quadratic Formula still yields the solutions to the equation.

Example 1   Solve the quadratic equation $2ix^2 - 3x + 3i = 0$.

Solution
We use $a = 2i$, $b = -3$, and $c = 3i$ in the Quadratic Formula to obtain

$$x = \frac{3 \pm \sqrt{9 - 4(2i)(3i)}}{4i}$$

$$= \frac{3 \pm \sqrt{9 - 24i^2}}{4i}$$

$$= \frac{3 \pm \sqrt{33}}{4i}$$

$$= \frac{(3 \pm \sqrt{33})(-i)}{(4i)(-i)}$$

$$= \frac{-3 \pm \sqrt{33}}{4} i.$$

The solutions are $x = (-3 + \sqrt{33})i/4$ and $x = (-3 - \sqrt{33})i/4$. ❏

We recall that when we divide a polynomial $P(x)$ by a nonzero polynomial $D(x)$, there is a quotient $Q(x)$ and remainder $R(x)$ such that

$$\frac{P(x)}{D(x)} = Q(x) + \frac{R(x)}{D(x)},$$

or

$$P(x) = D(x)Q(x) + R(x),$$

where the remainder $R(x)$ is either 0 or has degree less than the degree of $D(x)$. This is valid whether the coefficients of $P(x)$ and $Q(x)$ are real or complex numbers. In particular, when the divisor is a linear polynomial of the form $x - c$, the remainder is a constant:

$$P(x) = (x - c) \cdot Q(x) + r,$$

where r is a constant. This equality of polynomials holds for all values of x and, in particular, for $x = c$. If we assign the value c to x, we obtain

$$P(c) = (c - c) \cdot Q(c) + r$$

$$= 0 \cdot Q(c) + r$$

$$= 0 + r$$

$$= r.$$

This means that the remainder r is the same as the value of $P(x)$ when $x = c$. This important result is known as the *Remainder Theorem*.

Theorem 3.1   **Remainder Theorem**

If a real or complex polynomial $P(x)$ is divided by $x - c$, for c real or complex, the remainder is $P(c)$.

Example 2   Use the Remainder Theorem to find the remainder in each of the following divisions.

a) $(x^2 + 3x + 2) \div (x - 2)$

b) $\dfrac{x^{40} - 2x^{10} + 1}{x + 1}$

c) $(x^3 - x^2 - x + 2) \div (x + i)$

Solution

a) By the Remainder Theorem, when $P(x) = x^2 + 3x + 2$ is divided by $x - 2$, the remainder is

$$r = P(2) = (2)^2 + 3(2) + 2 = 12.$$

b) With $P(x) = x^{40} - 2x^{10} + 1$ and $x - c = x + 1 = x - (-1)$, the remainder is

$$r = P(-1) = (-1)^{40} - 2(-1)^{10} + 1 = 1 - 2 + 1 = 0.$$

c) With $P(x) = x^3 - x^2 - x + 2$ and $x + i$ as divisor, the remainder is

$$r = P(-i)$$

$$= (-i)^3 - (-i)^2 - (-i) + 2$$

$$= i + 1 + i + 2$$

$$= 3 + 2i.$$

The Remainder Theorem can also be used in the other direction: The value of $P(c)$ can be found by obtaining the remainder when $P(x)$ is divided by $x - c$.

Example 3　Use the Remainder Theorem to find the value of $P(2)$ if

$$P(x) = x^3 - 3x^2 - x + 4.$$

Solution
We divide $P(x)$ by $x - 2$ as follows:

$$
\begin{array}{r}
x^2 - x \quad - 3 \quad \text{Quotient} \\
x - 2 \overline{)\ x^3 - 3x^2 - \ \ x + 4} \quad \text{Dividend} \\
\underline{x^3 - 2x^2} \\
-\ x^2 - \ \ x + 4 \\
\underline{-\ x^2 + 2x} \\
-\ 3x + 4 \\
\underline{-\ 3x + 6} \\
-\ 2 \quad \text{Remainder}
\end{array}
$$

Divisor appears before $x - 2$.

Thus $P(2) = -2$, by the Remainder Theorem. ❑

Many people prefer a shortened form that allows more rapid division of a polynomial by a binomial of the form $x - c$. This procedure is called **synthetic division,** or **detached coefficients.** To understand why the synthetic division procedure works, we will rework the last example with the goal of streamlining the procedure in mind.

$$
\begin{array}{r}
x^2 - x - 3 \\
x - 2 \overline{)\ x^3 - 3x^2 - \ \ x + 4} \\
\underline{x^3 - 2x^2} \\
-\ x^2 - \ \ x + 4 \\
\underline{-\ x^2 + 2x} \\
-\ 3x + 4 \\
\underline{-\ 3x + 6} \\
-\ 2
\end{array}
$$

We eliminate the terms in color and write this more compact form.

$$
\begin{array}{r}
x^2 - x - 3 \\
x - 2 \overline{)\ x^3 - 3x^2 - \ \ x + 4} \\
-\ 2x^2 + 2x + 6 \\
\hline
x^3 - \ \ x^2 - 3x - 2
\end{array}
$$

If we go a step further and eliminate the variables and omit the quotient, we have the following:

$$1 - 2 \overline{)1 - 3 - 1 \quad 4}$$
$$\overline{1 - 1 - 3 - 2}$$

As final refinements, we drop the coefficient of x in the divisor and change the sign of the constant in the divisor $x - 2$ to 2, so that we can **add** at each stage rather than subtract. This yields the routine known as synthetic division.

Divisor $x - 2 \rightarrow 2 \overline{)1 - 3 \quad -1 \quad 4} \leftarrow$ Dividend $x^3 - 3x^2 - x + 4$

Quotient $= x^2 - x - 3$ Remainder $= -2$

The mechanics are indicated by the arrows shown in the diagram constructed by using the coefficients of the dividend $P(x)$ and the constant c in the divisor $x - c$. The first step is to bring down the leading coefficient of $P(x)$. Then repeat the following two steps until all the coefficients of $P(x)$ are exhausted.

1. Multiply each number in the bottom row by c to obtain the next number in the second row.

2. Add the numbers in the first and second rows to obtain the next number in the bottom row.

We demonstrate the synthetic division procedure with one more example. This example illustrates the fact that if any coefficients in the dividend are zero, these zeros *must* be recorded in the synthetic division procedure.

Example 4   Divide $2x^4 - 12x^2 - 5$ by $x + 3$.

Solution

$$-3 \overline{)2 \quad\quad 0 \quad -12 \quad\quad 0 \quad -5}$$
$$\quad\quad\quad -6 \quad\quad 18 \quad -18 \quad\quad 54$$
$$\overline{2 \quad -6 \quad\quad 6 \quad -18 \quad\quad 49}$$

Hence

$$\frac{2x^4 - 12x^2 - 5}{x + 3} = 2x^3 - 6x^2 + 6x - 18 + \frac{49}{x + 3}.$$

Note that $x + 3 = x - (-3)$, and -3 was recorded in the divisor position. Also, 0 was recorded as the coefficient of x^3 and of x in the dividend.   ❏

For an arbitrary complex number c, the Remainder Theorem states that

$$P(x) = (x - c)Q(x) + P(c).$$

Now $x - c$ is a factor of $P(x)$ if and only if the remainder is 0 when $P(x)$ is divided by $x - c$, and therefore the equation above shows that $x - c$ is a factor of $P(x)$ if and only if $P(c) = 0$. This establishes the following Factor Theorem.

Theorem 3.2   **Factor Theorem**

A polynomial $P(x)$ has a factor $x - c$ if and only if $P(c) = 0$.

A **zero** of the polynomial $P(x)$ is a complex number c such that $P(c) = 0$. The Factor Theorem states that $x - c$ is a factor of $P(x)$ if and only if c is a zero of $P(x)$.

Example 5   Use the Factor Theorem to decide whether or not -2 is a zero of $P(x) = x^4 - x^3 - 5x^2 + 3x + 2$.

Solution

By the Factor Theorem, -2 is a zero of $P(x) = x^4 - x^3 - 5x^2 + 3x + 2$ if and only if $x + 2$ is a factor of $P(x)$. Synthetic division can be used to check whether or not the remainder is 0 when $P(x)$ is divided by $x + 2 = x - (-2)$.

$$
\begin{array}{r|rrrrr}
-2 & 1 & -1 & -5 & 3 & 2 \\
   &   & -2 & 6 & -2 & -2 \\
\hline
   & 1 & -3 & 1 & 1 & 0
\end{array}
$$

Thus $x + 2$ is a factor of $P(x)$, and -2 is a zero of $P(x)$.   ❑

Sometimes it is easier to compute $P(c)$ than to divide $P(x)$ by $x - c$. In these cases the Factor Theorem is useful in deciding whether or not $x - c$ is a factor of $P(x)$.

Example 6   Use the Factor Theorem to decide whether or not $x + 2$ is a factor of $P(x) = x^4 - 3x + 2$.

Solution
By direct computation,

$$P(-2) = (-2)^4 - 3(-2) + 2$$
$$= 16 + 6 + 2 = 24.$$

Thus $P(-2) \neq 0$, and $x + 2$ is not a factor of $x^4 - 3x + 2$.   ❑

In certain instances the Factor Theorem can be useful in solving equations. This is illustrated in the next example.

Example 7 Given that 1 is a zero of $P(x) = x^3 - 3x^2 + x + 1$, solve the equation

$$x^3 - 3x^2 + x + 1 = 0.$$

Solution

Since it is given that 1 is a zero, we know that $x - 1$ is a factor of $P(x)$. To find the quotient, we divide by $x - 1$.

$$
\begin{array}{r|rrrr}
1 & 1 & -3 & 1 & 1 \\
  &   & 1 & -2 & -1 \\
\hline
  & 1 & -2 & -1 & 0
\end{array}
$$

Thus the quotient is $x^2 - 2x - 1$, and the given equation is equivalent to

$$(x - 1)(x^2 - 2x - 1) = 0.$$

Since a product is zero if and only if one of the factors is zero, the other solutions of the given equation are the same as the solutions to

$$x^2 - 2x - 1 = 0.$$

These solutions are easily found to be

$$x = \frac{2 \pm \sqrt{4 + 4}}{2}$$

$$= \frac{2 \pm 2\sqrt{2}}{2}$$

$$= 1 \pm \sqrt{2}.$$

Thus the solution set for the equation $x^3 - 3x^2 + x + 1 = 0$ is

$$\{1, 1 + \sqrt{2}, 1 - \sqrt{2}\}.$$

Exercises 3.1

Solve each quadratic equation and write the solutions in standard form $a + bi$. (See Example 1.)

1. $x^2 - 3ix + 4 = 0$
2. $x^2 + ix + 2 = 0$
3. $2ix^2 + 5x - 2i = 0$
4. $2ix^2 - 3x + 3i = 0$
5. $2ix^2 + 3x + 2i = 0$
6. $3ix^2 - x + 4i = 0$
7. $(1 + i)x^2 + ix + 1 - i = 0$
8. $(1 - i)x^2 - ix + 1 + i = 0$

In Exercises 9–16, use synthetic division and leave the results in the form

$$\frac{\text{dividend}}{\text{divisor}} = \text{quotient} + \frac{\text{remainder}}{\text{divisor}}.$$

(See Example 4.)

9. $(2x^3 - 4x^2 + 5x - 1) \div (x - 2)$
10. $(5x^3 - 10x^2 - 11x + 8) \div (x - 3)$

11. $(-3x^3 + 2x - 75) \div (x + 3)$

12. $(x^4 - 7x^2 - 6x) \div (x + 2)$

13. $(x^5 - x^4 - 7x^3 + 24) \div (x - 3)$

14. $(4x^5 + 30x^2 + 3) \div (x + 2)$

15. $(x^4 - a^4) \div (x + a)$ **16.** $(x^5 - a^5) \div (x - a)$

Use the Remainder Theorem to find the remainder when $P(x)$ is divided by $D(x)$. (See Example 2.)

17. $P(x) = x^4 - 2x^3 + 2x^2 - 1, \quad D(x) = x - 2$

18. $P(x) = 2x^4 - 3x^3 + 4x - 5, \quad D(x) = x - 1$

19. $P(x) = 2x^3 - 2x^2 + 4, \quad D(x) = x - \sqrt{2}$

20. $P(x) = x^3 + 2x^2 - 5x + 1, \quad D(x) = x + 3$

21. $P(x) = x^2 - 3x + 1, \quad D(x) = x - i$

22. $P(x) = x^3 + x^2 + x + 1, \quad D(x) = x - 2i$

23. $P(x) = x^{1023} - 3x^{15} + 7, \quad D(x) = x + 1$

24. $P(x) = x^{95} - 17x^4 + 9, \quad D(x) = x - 1$

Use the Remainder Theorem and synthetic division to find $P(c)$.

25. $P(x) = 4x^3 - 2x + 6, \quad c = \sqrt{2}$

26. $P(x) = 5x^3 + 2x^2 - \sqrt{3}, \quad c = \sqrt{3}$

27. $P(x) = 3x^4 - 2x^2 + x - 5, \quad c = 1 - i$

28. $P(x) = 2x^4 - 4x^2 + 2x - 3, \quad c = 1 + i$

29. $P(x) = x^4 + 2x^2 - 3x, \quad c = 1 + 2i$

30. $P(x) = x^4 - 3x^2 + 4x, \quad c = 2 - i$

Use the Factor Theorem to decide whether or not the given number c is a zero of $P(x)$. (See Example 5.)

31. $P(x) = x^3 - x^2 - x - 5, \quad c = 2$

32. $P(x) = x^3 - 2x^2 - 2x + 7, \quad c = 3$

33. $P(x) = x^3 - 3x^2 + 4x - 2, \quad c = 1 + i$

34. $P(x) = x^3 - 3x^2 + x - 3, \quad c = i$

Use the Factor Theorem to determine whether or not $D(x)$ is a factor of $P(x)$. (See Example 6.)

35. $P(x) = x^2 - 3x + 4, \quad D(x) = x + 1$

36. $P(x) = x^3 - 4x^2 + 9, \quad D(x) = x - 3$

37. $P(x) = x^{19} - x^{17} + x^2 - 1, \quad D(x) = x - 1$

38. $P(x) = x^{30} - x^5 - 2, \quad D(x) = x + 1$

39. $P(x) = x^4 - 3x^2 + 7x - 1, \quad D(x) = x + 2$

40. $P(x) = 7x^3 - 3x^2 - 19, \quad D(x) = x - 3$

41. $P(x) = x^3 + 5x^2 + x + 5, \quad D(x) = x - i$

42. $P(x) = x^2 + x + 4, \quad D(x) = x - (1 - 2i)$

43. $P(x) = x^2 - (3 - i)x + 8 + i, \quad D(x) = x - (1 + 2i)$

44. $P(x) = x^3 - 3ix + i + 7, \quad D(x) = x - (1 - i)$

(See Example 7.)

45. Given that 3 is a zero of $P(x) = x^3 - 3x^2 + x - 3$, solve the equation
$$x^3 - 3x^2 + x - 3 = 0.$$

46. Given that -1 is a zero of $P(x) = x^3 - x^2 + 2$, solve the equation
$$x^3 - x^2 + 2 = 0.$$

47. Determine k so that $P(x) = x^3 - kx^2 + 3x + 7k$ is divisible by $x + 2$.

48. Determine k so that $P(x) = x^4 + kx^3 - 3kx + 9$ is divisible by $x - 3$.

3.2 | **The Fundamental Theorem of Algebra and Descartes' Rule of Signs**

We recall from the last section that a **zero** of a polynomial $P(x)$ is a complex number c such that $P(c) = 0$. If c is a real number and $P(c) = 0$, then c is called a **real zero** of $P(x)$. We have seen that any second-degree polynomial $P(x) = ax^2 + bx + c$ has two zeros in the complex numbers, and the zeros are given by the quadratic formula.

In this chapter we are concerned primarily with the problem of finding the zeros of a given polynomial
$$P(x) = a_n x^n + a_{n-1} x^{n-1} + \cdots + a_1 x + a_0.$$

Carl Friedrich Gauss

This is the same problem as that of solving the polynomial equation

$$a_n x^n + a_{n-1} x^{n-1} + \cdots + a_1 x + a_0 = 0.$$

The solutions to an equation are frequently referred to as the **roots** of the equation. We are especially interested in finding the real roots of a polynomial equation that has real coefficients.

Since the problem of finding the roots of a polynomial equation of degree 2 is completely resolved by the Quadratic Formula, it might be expected that similar formulas exist for the roots of polynomial equations with higher degree. This is true, up to a point. In the period 1500–1550, the Italian mathematicians Tartaglia, Cardan, and Ferrari obtained formulas that could be used to solve the general equations of the third and fourth degrees. For over 200 years afterward, mathematicians searched for similar formulas for equations of degree greater than 4. It was not until 1824 that a Norwegian mathematician, N. H. Abel, proved that it is impossible to express the roots of a general equation of degree greater than 4 by a formula involving only the four fundamental operations and the extraction of roots.

Thus the problem we are dealing with in this chapter is far from simple. Our development begins with the following theorem, which was first proved in 1799 by the German mathematician C. F. Gauss (1777–1855).

Theorem 3.3   The Fundamental Theorem of Algebra

Let

$$P(x) = a_n x^n + a_{n-1} x^{n-1} + \cdots + a_1 x + a_0$$

denote a polynomial of degree $n \geq 1$ with coefficients that are real or complex numbers. Then $P(x)$ has a zero in the complex numbers.

In other words, the conclusion of Theorem 3.3 states that there is a complex number r_1, such that

$$P(r_1) = 0.$$

This complex number r_1 may be a real number; that is, r_1 may be a real zero of $P(x)$. In any case, $x - r_1$ is a factor of $P(x)$, by the Factor Theorem. Thus we can write

$$P(x) = (x - r_1)Q_1(x),$$

where $Q_1(x)$ has degree $n - 1$. If the quotient $Q_1(x)$ has degree ≥ 1, then by Theorem 3.3, $Q_1(x)$ has a zero r_2 and a corresponding factor $x - r_2$. That is,

$$Q_1(x) = (x - r_2)Q_2(x)$$

and

$$P(x) = (x - r_1)(x - r_2)Q_2(x).$$

If the quotient $Q_2(x)$ has degree ≥ 1, the procedure can be repeated again, obtaining a factor $x - r_3$ of $P(x)$. Each time another factor is obtained, the degree of the new quotient is one less than the degree of the previous quotient. After n applications of the Fundamental Theorem and the Factor Theorem, we arrive at the factorization

$$P(x) = (x - r_1)(x - r_2) \cdots (x - r_n)(a_n).$$

The last quotient must be a_n, since this is the coefficient of x^n in $P(x)$ in the beginning. Thus we have the following theorem.

Theorem 3.4

Let $P(x)$ be a polynomial of degree $n \geq 1$ with coefficients that are real or complex numbers. Then $P(x)$ can be factored as

$$P(x) = a_n(x - r_1)(x - r_2) \cdots (x - r_n),$$

where $r_1, r_2, \ldots, r_n$ are n complex numbers that are zeros of $P(x)$ and a_n is the leading coefficient of $P(x)$.

The zeros $r_1, r_2, \ldots, r_n$ in Theorem 3.4 are not necessarily distinct. That is, a given factor $x - r$ may be repeated in the factorization of $P(x)$. If the factor $x - r$ is of **multiplicity k**, then r is a zero of multiplicity k. Thus a polynomial of degree $n \geq 1$ has exactly n zeros in the complex numbers if a zero of multiplicity k is counted as a zero k times.

Example 1   Find a polynomial $P(x)$ of least degree that has 2, -1, and $2i$ as zeros.

Solution
From Theorem 3.4, we know that $P(x)$ must have degree 3 and must factor as

$$P(x) = a_3(x - 2)(x + 1)(x - 2i),$$

where a_3 is the leading coefficient in $P(x)$. The choice $a_3 = 1$ makes $P(x)$ a monic polynomial:

$$P(x) = (x - 2)(x + 1)(x - 2i)$$
$$= x^3 - (1 + 2i)x^2 - (2 - 2i)x + 4i. \qquad \square$$

Example 2   State the factors $x - r$ and the zeros of the polynomial

$$P(x) = 3(x - 2)^3(x + 1)^2(x - 3).$$

Solution

According to our preceding discussion, $P(x)$ has six factors of the form $x - r$ and six zeros, counting repetitions. The six factors are

$$x - 2, \quad \text{with multiplicity 3,}$$

$$x - (-1), \quad \text{with multiplicity 2, and}$$

$$x - 3.$$

The six zeros are $2, 2, 2, -1, -1$, and 3. ❏

Although Theorem 3.4 assures us that a polynomial of degree $n \geq 1$ has exactly n zeros in the complex numbers, it gives us no help at all in finding these zeros. As was indicated earlier, we are especially interested in finding the real zeros of a polynomial that has real coefficients. The next theorem is often useful for this purpose when it is used in combination with results that we shall present in the next two sections. This theorem is known as *Descartes' Rule of Signs*.

Descartes' Rule of Signs allows us to make certain predictions about the number of positive real zeros, or about the number of negative real zeros, of a polynomial that has *real coefficients*. The predictions are based on the number of "variations in sign" that occur when the terms of the polynomial are arranged in the usual order of descending powers of x:

$$P(x) = a_n x^n + a_{n-1} x^{n-1} + \cdots + a_1 x + a_0.$$

After any powers of x with zero coefficients are deleted, a *variation in sign* is said to occur when two consecutive coefficients are opposite in sign. For example,

$$P(x) = x^4 - 3x^2 + 7x + 11$$

has two variations in sign, as indicated by the arrows.

Theorem 3.5 **Descartes' Rule of Signs**

Let

$$P(x) = a_n x^n + a_{n-1} x^{n-1} + \cdots + a_1 x + a_0$$

be a polynomial with real coefficients. The number of positive real zeros of $P(x)$ is either equal to the number of variations in sign occurring in $P(x)$, or is less than this number by an even positive integer. The number of negative real zeros of $P(x)$ is either equal to the number of variations in sign occurring in $P(-x)$, or is less than this number by an even positive integer.

The proof of Descartes' Rule of Signs, like the proof of the Fundamental Theorem of Algebra, is much beyond the level of this text and so is not included. Some applications of the rule are given in Example 3, but its full usefulness is not illustrated until Section 3.3.

Example 3   Use Descartes' Rule of Signs to discuss the nature of the zeros of each polynomial. That is, describe the possibilities as to the number of positive real zeros, the number of negative real zeros, and the number of complex zeros that are not real.

a) $P(x) = x^3 - x^2 - 3$

b) $P(x) = x^4 - 3x^2 - 7x + 11$

Solution

a) Counting the number of variations of sign in

$$P(x) = x^3 - x^2 - 3,$$

we see there is only one variation, so $P(x)$ has one positive real zero. In

$$P(-x) = (-x)^3 - (-x)^2 - 3$$
$$= -x^3 - x^2 - 3,$$

there are no variations in sign, so $P(x)$ has no negative real zeros. Thus we know that $P(x)$ has one positive real zero and therefore two complex zeros that are not real.

b) There are two variations of sign in

$$P(x) = x^4 - 3x^2 - 7x + 11$$

and also two variations of sign in

$$P(-x) = x^4 - 3x^2 + 7x + 11.$$

Thus any one of the following is a possibility as to the nature of the zeros of $P(x)$:

i. Two positive zeros, two negative zeros;

ii. Two positive zeros, two nonreal complex zeros;

iii. Two negative zeros, two nonreal complex zeros;

iv. Four nonreal complex zeros.   ❏

There is one more basic fact we will need concerning the zeros of a polynomial that has real coefficients. This fact, which is stated in Theorem 3.6, involves the conjugates of the zeros of $P(x)$.

We recall from Section 1.4 that if $z = a + bi$ is a complex number in standard form, the conjugate of z is the complex number $\bar{z} = a - bi$. Exercises 63 and 64 in Exercises 1.4 can be extended to state that

$$\overline{z_1 + z_2 + \cdots + z_n} = \bar{z}_1 + \bar{z}_2 + \cdots + \bar{z}_n$$

and

$$\overline{z_1 \cdot z_2 \cdots \cdots z_n} = \bar{z}_1 \cdot \bar{z}_2 \cdots \cdots \bar{z}_n.$$

In words, the conjugate of a sum is the sum of the conjugates, and the conjugate of a product is the product of the conjugates. A special case of the last equation is

$$\overline{(z^n)} = (\bar{z})^n.$$

These facts about conjugates have an interesting implication concerning polynomials with real coefficients. Let

$$P(x) = a_n x^n + a_{n-1} x^{n-1} + \cdots + a_1 x + a_0$$

represent a polynomial that has real coefficients. For any complex number z,

$$
\begin{aligned}
\overline{P(z)} &= \overline{a_n z^n + a_{n-1} z^{n-1} + \cdots + a_1 z + a_0} \\
&= \overline{a_n z^n} + \overline{a_{n-1} z^{n-1}} + \cdots + \overline{a_1 z} + \overline{a_0} \\
&= \overline{a_n}\,\overline{(z^n)} + \overline{a_{n-1}}\,\overline{(z^{n-1})} + \cdots + \overline{a_1}\,\overline{z} + \overline{a_0} \\
&= \overline{a_n}\,(\bar{z})^n + \overline{a_{n-1}}\,(\bar{z})^{n-1} + \cdots + \overline{a_1}\,\overline{z} + \overline{a_0} \\
&= a_n (\bar{z})^n + a_{n-1} (\bar{z})^{n-1} + \cdots + a_1 \bar{z} + a_0,
\end{aligned}
$$

where the last equality follows from the fact that $\overline{a_i} = a_i$ because each a_i is real. Now the last expression above is the same as that obtained when $P(x)$ is evaluated at $\bar{z}$. That is,

$$\overline{P(z)} = P(\bar{z})$$

for any complex number z. The main use that we have for this result is when z is a zero of $P(x)$. For when $P(z) = 0$, then $\overline{P(z)} = \overline{0} = 0$, and therefore $P(\bar{z}) = 0$. That is, if z is a zero of $P(x)$, then $\bar{z}$ is also a zero. This is stated in the following theorem.

Theorem 3.6

Let

$$P(x) = a_n x^n + a_{n-1} x^{n-1} + \cdots + a_1 x + a_0$$

be a polynomial with real coefficients. If $z = a + bi$ is a zero of $P(x)$, then $\bar{z} = a - bi$ is also a zero of $P(x)$.

This means that, if $P(x)$ is a polynomial with real coefficients, the complex zeros of $P(x)$ always occur in conjugate pairs.

Example 4   Given that $1 - i$ is a zero of $P(x) = x^3 - 4x^2 + 6x - 4$, find all zeros of $P(x)$.

Solution

Since $1 - i$ is a zero of $P(x)$, $x - (1 - i)$ is a factor of $P(x)$. We use synthetic division to divide $P(x)$ by $x - (1 - i)$.

$$
\begin{array}{r|rrrr}
1 - i & 1 & -4 & 6 & -4 \\
      &   & 1 - i & -4 + 2i & 4 \\
\hline
      & 1 & -3 - i & 2 + 2i & 0
\end{array}
$$

Thus

$$P(x) = [x - (1 - i)][x^2 + (-3 - i)x + (2 + 2i)].$$

By Theorem 3.6, $\overline{1 - i} = 1 + i$ is also a zero of $P(x)$. Since $1 + i$ is not a zero of the factor $x - (1 - i)$, it must be a zero of the quotient. We have

$$
\begin{array}{r|rrr}
1 + i & 1 & -3 - i & 2 + 2i \\
      &   & 1 + i & -2 - 2i \\
\hline
      & 1 & -2 & 0
\end{array}
$$

and the new quotient is $x - 2$. This means that

$$P(x) = [x - (1 - i)][x - (1 + i)](x - 2),$$

so the zeros of $P(x)$ are $1 - i$, $1 + i$, and 2.   ❏

Example 5   Find a polynomial $Q(x)$ of least degree with *real* coefficients that has $3i$ and 4 as zeros.

Solution

Since $Q(x)$ is to have *real* coefficients and is to have $3i$ as a zero, it must also have $\overline{3i} = -3i$ as a zero. The product

$$
\begin{aligned}
Q(x) &= (x - 3i)(x + 3i)(x - 4) \\
&= (x^2 + 9)(x - 4) \\
&= x^3 - 4x^2 + 9x - 36
\end{aligned}
$$

is a monic polynomial with the required properties.   ❏

In connection with Example 5, we note that the monic polynomial of least degree that has $3i$ and 4 as zeros is

$$
\begin{aligned}
P(x) &= (x - 3i)(x - 4) \\
&= x^2 - (4 + 3i)x + 12i,
\end{aligned}
$$

but this polynomial does not have the real coefficients that were required in Example 5.

Exercises 3.2

Find (a) a polynomial $P(x)$ of least degree that has the given numbers as zeros and (b) a polynomial $Q(x)$ of least degree with real coefficients that has the given numbers as zeros. (See Examples 1 and 5.)

1. $3, -5$

2. $2, -2$

3. $2i$

4. $-3i$

5. $3, 2 - i$

6. $2, 1 + i$

7. $3, 1 - i, 3 + 2i$

8. $1 - 2i, 3i, -2$

Use Descartes' Rule of Signs to discuss the nature of the zeros of each polynomial. (See Example 3.)

9. $P(x) = 2x^4 + 3x^3 - 2x + 1$

10. $P(x) = 4x^4 - 7x^3 + 3x + 2$

11. $P(x) = 2x^4 - x^3 + 3x - 1$

12. $P(x) = 4x^4 - 2x^2 + 5x - 1$

13. $P(x) = x^6 + x^3 + 2x + 3$

14. $P(x) = 4x^4 + 2x^3 + 4x - 2$

15. $P(x) = x^4 + 3x^3 + 12x + 8$

16. $P(x) = 3x^3 + 19x^2 + 30x + 8$

17. $P(x) = 4x^4 + 1$

18. $P(x) = 2x^7 - 3$

19. $P(x) = x^6 + 4x^4 + x^2 + 5$

20. $P(x) = x^5 + 3x^3 + 7x$

In Exercises 21–36, some of the zeros of the polynomial are given. Find the other zeros. (See Example 4.)

21. $P(x) = x^2 + 9$; $-3i$ is a zero.

22. $P(x) = x^2 + 2x + 2$; $-1 + i$ is a zero.

23. $Q(x) = x^3 + x + 10$; -2 and $1 + 2i$ are zeros.

24. $Q(x) = x^3 + x^2 - x + 15$; -3 and $1 + 2i$ are zeros.

25. $P(x) = x^4 + 20x^2 + 64$; $-2i$ and $4i$ are zeros.

26. $P(x) = x^4 + 17x^2 + 16$; $-i$ and $4i$ are zeros.

27. $Q(x) = x^3 + 4x^2 + 4x + 16$; $-2i$ is a zero.

28. $Q(x) = x^3 - 5x^2 + 9x - 45$; $3i$ is a zero.

29. $P(x) = x^3 - 2x^2 - 3x + 10$; $2 + i$ is a zero.

30. $P(x) = x^3 + x^2 - 4x + 6$; $1 - i$ is a zero.

31. $Q(x) = x^4 + x^3 + 10x^2 + 9x + 9$; $-3i$ is a zero.

32. $Q(x) = x^4 + 3x^3 + 6x^2 + 12x + 8$; $2i$ is a zero.

33. $P(x) = x^4 - 2x^3 + x^2 + 2x - 2$; $1 + i$ is a zero.

34. $P(x) = x^4 - 7x^3 + 19x^2 - 23x + 10$; $2 + i$ is a zero.

35. $P(x) = x^5 - 6x^4 + 16x^3 - 24x^2 + 20x - 8$; $1 + i$ is a zero of multiplicity 2.

36. $P(x) = x^5 - 9x^4 + 34x^3 - 66x^2 + 65x - 25$; $2 - i$ is a zero of multiplicity 2.

37. Prove that if all the coefficients of $P(x)$ are positive, then $P(x)$ has no positive real zeros.

38. Prove that if $P(x)$ has no odd powers of x and all its coefficients are of the same sign, then $P(x)$ has no real zeros different from 0.

39. Prove that every polynomial of odd degree with real coefficients has at least one real zero.

3.3 | Rational Zeros

In the practical applications of algebra, it is common to encounter a problem that calls for finding the rational numbers[1] that are zeros of a polynomial $P(x)$ with integral coefficients. In such cases the following theorem is fundamental to the solution of the problem.

---

1. Recall that a rational number is a quotient p/q of integers p and q, with $q \neq 0$.

Theorem 3.7

Let

$$P(x) = a_n x^n + a_{n-1} x^{n-1} + \cdots + a_1 x + a_0$$

be a polynomial in which all coefficients are integers, and let p/q denote a rational number that has been reduced to lowest terms. If p/q is a zero of $P(x)$, then p is a factor of a_0 and q is a factor of a_n.

That is, if p/q is a zero of $P(x)$ which is written so that the greatest common divisor of p and q is 1, then p must be a factor of the constant term, and q must be a factor of the leading coefficient of $P(x)$.

To see why the theorem is true, suppose that p/q, in lowest terms, is a zero of $P(x)$. Then

$$a_n \left(\frac{p}{q}\right)^n + a_{n-1} \left(\frac{p}{q}\right)^{n-1} + \cdots + a_1 \left(\frac{p}{q}\right) + a_0 = 0.$$

Multiplying both sides by q^n, we have

$$a_n p^n + a_{n-1} p^{n-1} q + \cdots + a_1 p q^{n-1} + a_0 q^n = 0.$$

Subtracting $a_0 q^n$ from both sides yields

$$a_n p^n + a_{n-1} p^{n-1} q + \cdots + a_1 p q^{n-1} = -a_0 q^n$$

and

$$p(a_n p^{n-1} + a_{n-1} p^{n-2} q + \cdots + a_1 q^{n-1}) = -a_0 q^n.$$

This equation shows that p is a factor of $a_0 q^n$. Since p/q is in lowest terms, the greatest common divisor of p and q is 1, and this means that p is a factor of a_0. Similarly, the equation

$$a_{n-1} p^{n-1} q + \cdots + a_1 p q^{n-1} + a_0 q^n = -a_n p^n$$

can be used to show that q is a factor of a_n.

The special case where $P(x)$ is a monic polynomial is important enough to designate as a corollary to Theorem 3.7.

Corollary 3.8

Let

$$P(x) = x^n + a_{n-1} x^{n-1} + \cdots + a_1 x + a_0$$

be a monic polynomial with integral coefficients. Then any rational zero of $P(x)$ is an integral factor of a_0.

Example 1   Find all rational zeros of $P(x) = 2x^3 - x^2 - 8x - 5$.

Solution

By Theorem 3.7, any rational zero of $P(x)$ has the form p/q, where p is a factor of -5 and q is a factor of 2. This means that

$$p \in \{\pm 1, \pm 5\}$$

and

$$q \in \{\pm 1, \pm 2\}.$$

Thus any rational zero p/q of $P(x)$ is included in the list

$$\pm 1, \quad \pm 5, \quad \pm \frac{1}{2}, \quad \pm \frac{5}{2}.$$

It is good practice to list the possible zeros in order from left to right and to test them systematically. We rewrite the possibilities as

$$-5, \quad -\frac{5}{2}, \quad -1, \quad -\frac{1}{2}, \frac{1}{2}, \quad 1, \quad \frac{5}{2}, \quad 5.$$

Descartes' Rule of Signs indicates there is one positive zero of

$$P(x) = 2x^3 - x^2 - 8x - 5,$$

but this positive zero is not necessarily rational. Using synthetic division and testing the positive possibilities in order, we have the following:

$$
\frac{1}{2} \begin{array}{|rrrr} 2 & -1 & -8 & -5 \\ & 1 & 0 & -4 \\ \hline 2 & 0 & -8 & -9 \end{array}
\qquad
1 \begin{array}{|rrrr} 2 & -1 & -8 & -5 \\ & 2 & 1 & -7 \\ \hline 2 & 1 & -7 & -12 \end{array}
\qquad
\frac{5}{2} \begin{array}{|rrrr} 2 & -1 & -8 & -5 \\ & 5 & 10 & 5 \\ \hline 2 & 4 & 2 & 0 \end{array}
$$

Thus $\frac{5}{2}$ is a rational zero of $P(x)$.

Before continuing, let us observe that the three synthetic divisions above could have been tabulated as follows, where the bottom row of each synthetic division is written to the right of the possibility that is being tested.

	2	**−1**	**−8**	**−5**
$\frac{1}{2}$	2	0	−8	−9
1	2	1	−7	−12
$\frac{5}{2}$	2	4	2	0

A table such as this is efficient when several divisions are to be performed in succession. With a little practice, the necessary arithmetic can be done mentally.

It would be straightforward to continue testing the remaining possible rational zeros, which are $5, -\frac{1}{2}, -1, -\frac{5}{2}, -5$. However, we know from the last division performed that $x - \frac{5}{2}$ is a factor of $P(x)$, and

$$P(x) = \left(x - \frac{5}{2}\right)(2x^2 + 4x + 2).$$

Any remaining zeros of $P(x)$ are zeros of the quotient $2x^2 + 4x + 2 = 2(x^2 + 2x + 1)$. We need only solve

$$x^2 + 2x + 1 = 0$$

to finish the problem. We have

$$(x + 1)^2 = 0,$$

so the other zero of $P(x)$ is -1 with a multiplicity of 2. That is, the rational zeros of $P(x)$ are given by

$$\left\{\frac{5}{2}, -1, -1\right\}.$$

As Example 1 shows, the testing for rational zeros may lead to a situation where the quotient is a quadratic polynomial. In such a case, it is easier to find the zeros of the quotient than to continue checking rational zero possibilities. Sometimes this method will produce all the zeros of $P(x)$, not just the rational zeros.

Example 2   Find all zeros of $P(x) = x^3 + x^2 - x + 2$.

Solution
According to Corollary 3.8, any rational zero of $P(x)$ is a factor of 2. Thus the possible rational zeros are

$$-2, \quad -1, \quad 1, \quad 2.$$

There are two variations of sign in

$$P(x) = x^3 + x^2 - x + 2$$

and one in

$$P(-x) = -x^3 + x^2 + x + 2.$$

Thus the number of positive zeros is either 2 or 0, and there is one negative zero. Using synthetic division, we obtain the following table.

	1	1	−1	2
1	1	2	1	3
2	1	3	5	12
−1	1	0	−1	3
−2	1	−1	1	0

The last division shows that -2 is a zero and that

$$P(x) = (x + 2)(x^2 - x + 1).$$

To find the other zeros of $P(x)$, we set

$$x^2 - x + 1 = 0$$

and obtain

$$x = \frac{1 \pm i\sqrt{3}}{2}.$$

The zeros of $P(x)$ are given by $\left\{ -2, \dfrac{1 + i\sqrt{3}}{2}, \dfrac{1 - i\sqrt{3}}{2} \right\}.$ ❏

It is important to note that a polynomial may have a large set of possible rational zeros and yet in fact not have any rational zeros. For example, the polynomial

$$P(x) = 4x^4 + 7x^2 + 3$$

has, as possible rational zeros, the set

$$\left\{ \pm 1, \ \pm 3, \ \pm\frac{1}{2}, \ \pm\frac{3}{2}, \ \pm\frac{1}{4}, \ \pm\frac{3}{4} \right\}.$$

But Descartes' Rule of Signs shows that $P(x)$ has no real zeros and certainly no rational zeros.

When there is a large number of possible rational zeros, the following theorem can be extremely useful. We accept the theorem without proof.

Theorem 3.9   **The Upper and Lower Bound Theorem**

Let

$$P(x) = a_n x^n + a_{n-1}x^{n-1} + \cdots + a_1 x + a_0$$

be a polynomial with real coefficients and $a_n > 0$, and suppose that synthetic division is used to divide $P(x)$ by $x - r$, where r is real. The last row in the synthetic division can be used in the following manner.

1. If $r > 0$ and all numbers in the last row are nonnegative, then $P(x)$ has no zero greater than r.

2. If $r < 0$ and the numbers in the last row alternate in sign (with 0 written as $+0$ or -0), then $P(x)$ has no zero less than r.

In case $P(x)$ has no zero less than the number a, we say that a is a **lower bound** for the zeros. Similarly, if $P(x)$ has no zeros greater than

b, then b is called an **upper bound** for the zeros. Theorem 3.9 can frequently be used to obtain a positive upper bound and a negative lower bound for the zeros of $P(x)$.

Example 3   Find all rational zeros of

$$P(x) = 2x^4 + x^3 - 8x^2 + x - 10.$$

Solution

Any rational zero of $P(x)$ has the form p/q, where p is a factor of -10 and q is a factor of 2. That is,

$$p \in \{\pm 1, \pm 2, \pm 5, \pm 10\}$$

and

$$q \in \{\pm 1, \pm 2\}.$$

This gives the following set of possible rational zeros:

$$\left\{ \pm 1, \pm 2, \pm 5, \pm 10, \pm \frac{1}{2}, \pm \frac{5}{2} \right\}.$$

Since there are three variations of sign in

$$P(x) = 2x^4 + x^3 - 8x^2 + x - 10,$$

the number of positive zeros is either 3 or 1. We arrange the positive possibilities in order of size,

$$\frac{1}{2}, \quad 1, \quad 2, \quad \frac{5}{2}, \quad 5, \quad 10,$$

and begin a systematic check.

	2	1	-8	1	-10
$\frac{1}{2}$	2	2	-7	$-\frac{5}{2}$	$-\frac{45}{4}$
1	2	3	-5	-4	-14
2	2	5	2	5	0

The last division shows two things: 2 is a rational zero of $P(x)$, and 2 is an upper bound of the zeros of $P(x)$. There is no need to try the other positive possibilities for rational zeros. Since

$$P(x) = (x - 2)(2x^3 + 5x^2 + 2x + 5),$$

we concentrate now on the negative zeros of

$$2x^3 + 5x^2 + 2x + 5.$$

Systematically checking the possibilities

$$-\frac{1}{2}, \quad -1, \quad -2, \quad -\frac{5}{2}, \quad -5$$

(notice that -10 is no longer a possibility), we obtain the following table.

	2	5	2	5
$-\frac{1}{2}$	2	4	0	5
-1	2	3	-1	6
-2	2	1	0	5
$-\frac{5}{2}$	2	-0	2	-0

From the last row of the table, we see that $-\frac{5}{2}$ is a rational zero, and $-\frac{5}{2}$ is a lower bound of the zeros. Since $-\frac{5}{2}$ is a lower bound for the zeros, there is no need to test -5. We have

$$P(x) = (x - 2)\left(x + \frac{5}{2}\right)(2x^2 + 2)$$

$$= 2(x - 2)\left(x + \frac{5}{2}\right)(x^2 + 1).$$

The complete set of zeros of $P(x)$ is $\{2, -\frac{5}{2}, i, -i\}$, and the rational zeros of $P(x)$ are given by $\{2, -\frac{5}{2}\}$. ❏

Example 4   Kracmore Pottery Company can produce ceramic flower pots and make a profit $P(x)$, in thousands of dollars, given by $P(x) = x^3 - 2x^2 - 13x - 2$, where x is the number of pots produced in thousands. To stay in business, the company needs to turn a profit of \$8000 in March. How many pots should be produced in March to make \$8000?

Solution
To find the required production, we set $P(x) = 8$ and solve for x:

$$x^3 - 2x^2 - 13x - 2 = 8.$$

This equation is equivalent to

$$x^3 - 2x^2 - 13x - 10 = 0.$$

Since negative values of x are not meaningful here, we consider only the possible rational solutions that are positive. They are 1, 2, 5, and 10. Testing these possibilities, we find that neither 1 nor 2 is a solution, but that 5 is a solution.

$$
\begin{array}{r|rrrr}
5 & 1 & -2 & -13 & -10 \\
  &   & 5 & 15 & 10 \\
\hline
  & 1 & 3 & 2   & 0
\end{array}
$$

We have

$$x^3 - 2x^2 - 13x - 10 = (x - 5)(x^2 + 3x + 2).$$

Since

$$x^2 + 3x + 2 = (x + 1)(x + 2),$$

5 is the only positive solution to our equation. Hence the company needs to produce 5000 pots in March to make an $8000 profit.   ❑

Exercises 3.3

In Exercises 1–8, (a) find the smallest positive integer that Theorem 3.9 detects as an upper bound for the zeros of the given polynomial, and (b) find the negative integer nearest 0 that Theorem 3.9 detects as a lower bound for the zeros of the given polynomial.

1. $2x^3 + x^2 - 10x - 4$
2. $3x^3 - 2x^2 - 21x + 15$
3. $2x^3 - 3x^2 + 8x - 13$
4. $3x^3 - 7x^2 + 15x - 35$
5. $x^4 + x^3 + 2x^2 + 4x - 9$
6. $x^4 - x^3 - 4x^2 - 2x - 13$
7. $x^4 + x^3 - 5x^2 + x - 5$
8. $x^4 + x^3 - 11x^2 + x - 10$

Find all zeros of the given polynomial. (See Examples 1 and 2.)

9. $x^3 - 3x^2 + 4x - 12$ 10. $x^3 + 2x^2 + 6x + 12$
11. $2x^3 + 7x^2 + 2x - 3$ 12. $2x^3 - 3x^2 - 7x - 6$
13. $3x^3 - 5x^2 - 4$ 14. $2x^3 + x^2 - 2x - 6$

Find all solutions to the given equations.

15. $x^4 - x^3 - 2x^2 + 6x - 4 = 0$
16. $x^4 + x^3 - 2x^2 - 6x - 4 = 0$
17. $2x^4 - x^3 - 13x^2 + 5x + 15 = 0$
18. $2x^4 + x^3 - 13x^2 - 5x + 15 = 0$
19. $2x^4 + 5x^3 - 7x^2 - 10x + 6 = 0$
20. $2x^4 - x^3 - x^2 - x - 3 = 0$

Find all rational zeros of the given polynomial. (See Example 3.)

21. $x^3 + 7x - 6$ 22. $x^3 - 2x^2 + 10$
23. $x^4 + x^3 + 2x^2 + 4x - 8$
24. $x^4 + 3x^3 + 6x^2 + 12x + 8$
25. $3x^3 + 8x^2 + 3x - 2$ 26. $2x^3 + 9x^2 + 7x - 6$
27. $2x^3 + 17x^2 + 38x + 15$
28. $2x^3 - 9x^2 - 8x + 15$

29. $2x^4 - x^3 - 4x^2 - x - 6$
30. $2x^4 - x^3 - x^2 - x - 3$
31. $2x^4 + 3x^3 + 2x^2 + 11x + 12$
32. $3x^4 - 2x^3 + 3x^2 + 16x - 12$
33. $2x^5 - x^4 - 8x^3 + 3x^2 + 5x - 6$
34. $3x^5 + 4x^4 - 7x^3 - x^2 + 8x - 4$
35. $3x^4 + 5x^2 + 6$ 36. $5x^6 + 2x^2 + 4$
37. Show that $\sqrt{3}$ is irrational by applying Theorem 3.7 to $P(x) = x^2 - 3$.
38. Show that $\sqrt{2}$ is irrational by applying Theorem 3.7 to $P(x) = x^2 - 2$.
39. An open box is to be made from a square piece of tin that measures 3 meters on each side by cutting equal squares from the four corners and bending up the sides. By using calculus, it can be shown that the largest possible volume for such a box is 2 cubic meters. How large a square must be cut from each corner to have a volume of 2 cubic meters?
40. At time $t = 0$, a bactericide is introduced into a medium in which bacteria are growing. The num-

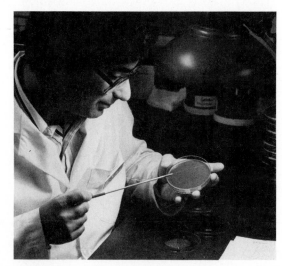

Photo by David Witbeck/The Picture Cube

ber $P(t)$ of thousands of bacteria present t hours later is approximated by $P(t) = 1000 + 10t - t^2$. How long will it be before $P(t) = 800$?

41. A rectangular box is to be made with an open top and vertical sides. The bottom is to be a square with sides of length x meters, and only 12 square meters of material is available to make the box. By using calculus, it can be shown that the largest possible volume for such a box is 4 cubic meters. Find the value of x that yields this maximum volume.

42. A package is to be mailed that has the shape of the rectangular parallelepiped with a square base

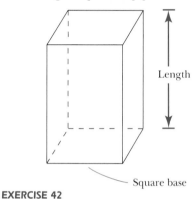

Length

Square base

EXERCISE 42

shown in the figure. It can be mailed by parcel post only if the sum of its length and girth (perimeter of the base) is no more than 9 feet. By using calculus, it can be shown that the largest possible volume for such a package is $\frac{27}{4}$ cubic feet. Find the dimensions that yield this maximum volume.

43. If a right circular cylinder is inscribed in a sphere of radius 3, the volume of the cylinder is given by $V(h) = \frac{\pi}{4}(36h - h^3)$, where h is the altitude of the cylinder. Find two values of h for which $V(h) = \frac{55\pi}{4}$.

44. The demand $D(x)$ for Pedalex cars, in thousands of cars, is given by $D(x) = x(2 + x)(8 - x)$, where x is the price in thousands of dollars. Show that there are two values of x for which the demand is 96,000 ($D = 96$).

45. Acme Boat Company has found that the cost $C(x)$ of producing x canoes is given in dollars by $C(x) = x^3 - 20x^2 + 500x$. How many canoes can be produced at a cost of $1347?

46. Ace Gravel Company can produce t tons of gravel at a cost of $C(t) = t^3 - 5t^2 + 9t$, where $C(t)$ is in hundreds of dollars. How many tons of gravel can be produced at a cost of $9000 [C(t) = 90]$?

3.4 | Approximation of Real Zeros

In the preceding section we saw how the rational zeros of a polynomial with integral coefficients can be found. The other real zeros (that is, the irrational zeros) of such a polynomial can be found to any desired degree of accuracy by using the Location Theorem, which is stated below.

Theorem 3.10   **The Location Theorem**

Let $P(x)$ be a polynomial with real coefficients. If a and b are real numbers such that $P(a)$ and $P(b)$ have opposite signs, then $P(x)$ has at least one zero between a and b.

The essence of this theorem is that the values of $P(x)$ cannot change from positive to negative, or from negative to positive, without assuming

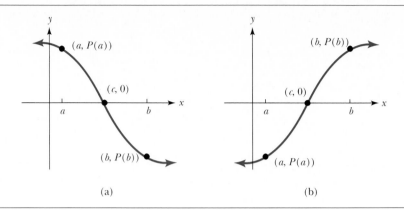

(a)　　　　　　　　　　　　　　(b)

FIGURE 3.1

the value 0 between them. The sketches in Figure 3.1 show how the graph of $y = P(x)$ might look over the interval from $x = a$ to $x = b$. A point c where $P(c) = 0$ is indicated in each case.

Theorem 3.10 is the basis of our method for approximating the real zeros of a polynomial with real coefficients to any desired degree of accuracy. The fundamental idea is to locate the zeros of $P(x)$ in intervals having length small enough to give the desired accuracy. The method is illustrated in the following example.

Example 1　Find the positive real zeros of the polynomial

$$P(x) = x^4 + 4x^2 - 6x - 5,$$

correct to the nearest tenth.

Solution
Since there is one variation in sign in $P(x)$, the polynomial has one positive real zero, by Descartes' Rule of Signs. The set of possible positive rational zeros is

$$\{1, 5\},$$

by Corollary 3.8. Using synthetic division, we construct the following table.

	1	0	4	-6	-5
1	1	1	5	-1	-6
5	1	5	29	139	690

There are no positive rational zeros of $P(x)$, but the work above shows that $P(1) = -6$ and $P(5) = 690$, so the positive real zero is between 1 and 5, by the Location Theorem. The values $P(1) = -6$ and $P(5) = 690$ indicate that the zero is probably much closer to 1 than 5, so we try 2 next in synthetic division.

$$
\begin{array}{r|rrrr}
2 & 1 & 0 & 4 & -6 & -5 \\
  &   & 2 & 4 & 16 & 20 \\
\hline
  & 1 & 2 & 8 & 10 & 15
\end{array}
$$

We have $P(1) = -6$ and $P(2) = 15$, so the zero is between 1 and 2. We next evaluate $P(1.5)$, since this will locate the zero in one half of the interval or the other.

$$
\begin{array}{r|rrrrr}
1.5 & 1 & 0 & 4 & -6 & -5 \\
    &   & 1.5 & 2.25 & 9.375 & 5.0625 \\
\hline
    & 1 & 1.5 & 6.25 & 3.375 & 0.0625
\end{array}
$$

Since $P(1.5) = 0.0625$ is positive, the zero is between 1 and 1.5, probably nearer 1.5. We next find $P(1.4)$.

$$
\begin{array}{r|rrrrr}
1.4 & 1 & 0 & 4 & -6 & -5 \\
    &   & 1.4 & 1.96 & 8.344 & 3.2816 \\
\hline
    & 1 & 1.4 & 5.96 & 2.344 & -1.7184
\end{array}
$$

Thus the zero is between 1.4 and 1.5, probably closer to 1.5. Computing $P(1.45)$, we have the following:

$$
\begin{array}{r|rrrrr}
1.45 & 1 & 0 & 4 & -6 & -5 \\
     &   & 1.45 & 2.1025 & 8.8486 & 4.1305 \\
\hline
     & 1 & 1.45 & 6.1025 & 2.8486 & -0.8695
\end{array}
$$

Since $P(1.5) = 0.0625$ and $P(1.45) = -0.8695$, the positive zero of $P(x)$ is between 1.45 and 1.5. To the nearest tenth, then, the zero is 1.5. ❏

The real (rational or irrational) zeros of a polynomial with real (not necessarily rational) coefficients can be approximated to the nearest hundredth, or to any needed degree of accuracy, by the method used in Example 1. The numerical calculations, of course, become more and more tedious as accuracy increases. However, any inexpensive calculator can relieve the burden of hand computations.

Exercises 3.4

Use the Location Theorem to approximate, to the nearest tenth, the zero of the given polynomial that is in the indicated interval.

1. $2x^3 - 11x^2 + 15x - 1$, between 2 and 3

2. $x^3 - 3x^2 + x + 1$, between 0 and -1
3. $x^3 + x^2 - 9x + 4$, between 0 and 1
4. $x^3 - 9x + 7$, between 2 and 3
5. $2x^4 - 5x^3 + 6x^2 - 22x + 7$, between 2 and 3

6. $x^4 + 2x^3 - 3x^2 + 7x - 4$,   between 0 and 1

Each of the following polynomials has exactly one positive real zero. Find the value of the positive zero, correct to the nearest tenth.

7. $-x^3 - 3x^2 - x + 1$ **8.** $-x^3 - x^2 + 3x + 2$

9. $2x^3 + 2x^2 - 3x - 2$ **10.** $2x^3 - x^2 - 4x - 1$

11. $3x^4 + 6x^3 - 2x^2 - 10x - 5$

12. $4x^4 + 12x^3 + 5x^2 - 9x - 8$

Each of the following polynomials has exactly one real zero. Find the value of the zero, correct to the nearest tenth.

13. $x^3 + x^2 + x - 1$ **14.** $x^3 + x^2 + x - 2$

15. $4x^3 + 4x^2 + 2x + 1$ **16.** $x^3 + 2x^2 + 2x + 2$

17. $x^3 + 2x^2 + x - 5$ **18.** $x^3 - 2x^2 + 2x - 3$

19. $x^3 + 3x^2 + 3x - 10$ **20.** $x^3 + x^2 - x + 1$

Find all real zeros of the given polynomial, correct to the nearest tenth.

21. $x^3 - x^2 - 3x + 1$ **22.** $x^3 - 3x + 1$

23. $2x^3 + 2x^2 - 2x - 1$ **24.** $3x^3 - 7x^2 - 6x + 8$

25. $x^3 - x^2 - 15x - 17$ **26.** $x^3 + 2x^2 - 14x - 32$

27. $2x^4 - 8x^3 + x^2 + 4$ **28.** $x^4 - 2x^3 - 2x^2 - 3$

CHAPTER REVIEW

Key Words and Phrases

Remainder Theorem	Fundamental Theorem of Algebra	Rational zeros
Synthetic division	Factor of multiplicity k	Lower bound for the zeros
Factor Theorem	Zero of multiplicity k	Upper bound for the zeros
Zero of a polynomial	Variation in sign	Location Theorem
Roots of an equation	Descartes' Rule of Signs	

Summary of Important Theorems

Theorem 3.1 Remainder Theorem
If a real or complex polynomial $P(x)$ is divided by $x - c$, for c real or complex, the remainder is $P(c)$.

Theorem 3.2 Factor Theorem
A polynomial $P(x)$ has a factor $x - c$ if and only if $P(c) = 0$.

Theorem 3.3 The Fundamental Theorem of Algebra
Let

$$P(x) = a_n x^n + a_{n-1} x^{n-1} + \cdots + a_1 x + a_0$$

denote a polynomial of degree $n \geq 1$ with coefficients that are real or complex numbers. Then $P(x)$ has a zero in the complex numbers.

Theorem 3.4
Let $P(x)$ be a polynomial of degree $n \geq 1$, with coefficients that are real or complex numbers. Then $P(x)$ can be factored as

$$P(x) = a_n(x - r_1)(x - r_2) \cdots (x - r_n),$$

where $r_1, r_2, \ldots, r_n$ are n complex numbers that are zeros of $P(x)$ and a_n is the leading coefficient of $P(x)$.

Theorem 3.5 Descartes' Rule of Signs
Let

$$P(x) = a_n x^n + a_{n-1} x^{n-1} + \cdots + a_1 x + a_0$$

be a polynomial with real coefficients. The number of positive real zeros of $P(x)$ is either equal to the number of variations in sign occurring in $P(x)$, or is less than this number by an even positive integer. The number of negative real zeros of $P(x)$ is either equal to the number of variations in sign occurring in $P(-x)$, or is less than this number by an even positive integer.

Theorem 3.6
Let

$$P(x) = a_n x^n + a_{n-1} x^{n-1} + \cdots + a_1 x + a_0$$

be a polynomial with real coefficients. If $z = a + bi$ is a zero of $P(x)$, then $\bar{z} = a - bi$ is also a zero of $P(x)$.

Theorem 3.7
Let

$$P(x) = a_n x^n + a_{n-1} x^{n-1} + \cdots + a_1 x + a_0$$

be a polynomial in which all coefficients are integers, and let p/q denote a rational number that has been reduced to lowest terms. If p/q is a zero of $P(x)$, then p is a factor of a_0 and q is a factor of a_n.

Corollary 3.8
Let

$$P(x) = x^n + a_{n-1}x^{n-1} + \cdots + a_1x + a_0$$

be a monic polynomial with integral coefficients. Then any rational zero of $P(x)$ is an integral factor of a_0.

Theorem 3.9
Let

$$P(x) = a_nx^n + a_{n-1}x^{n-1} + \cdots + a_1x + a_0$$

be a polynomial with real coefficients and $a_n > 0$, and suppose that synthetic division is used to divide $P(x)$ by $x - r$, where r is real. The last row in the synthetic division can be used in the following manner.

1. If $r > 0$ and all numbers in the last row are non-negative, then $P(x)$ has no zero greater than r.
2. If $r < 0$ and the numbers in the last row alternate in sign (with 0 written as $+0$ or -0), then $P(x)$ has no zero less than r.

Theorem 3.10 The Location Theorem
Let $P(x)$ be a polynomial with real coefficients. If a and b are real numbers such that $P(a)$ and $P(b)$ have opposite signs, then $P(x)$ has at least one zero between a and b.

Review Problems for Chapter 3

Solve the given equations.

1. $2x^2 - 3ix + 3 = 0$
2. $ix^2 - x + 2i = 0$
3. $(1 + i)x^2 + x - 1 + i = 0$

In Problems 4–6, use synthetic division and write the results in the form

$$\frac{\text{dividend}}{\text{divisor}} = \text{quotient} + \frac{\text{remainder}}{\text{divisor}}.$$

4. $(2x^3 - x^2 - 4x - 30) \div (x - 3)$
5. $(3x^3 + 5x^2 + 7) \div (x + 2)$
6. $(2x^4 + 6x^2 - 2x + 1) \div (x + 1)$

Use the Remainder Theorem and synthetic division to find $P(c)$.

7. $P(x) = 4x^3 + 5x^2 - 3x - 40, \quad c = 2$
8. $P(x) = 5x^4 - 2x^3 - 4x + 1, \quad c = -2$
9. $P(x) = 2x^4 - 3x^2 + 4x, \quad c = 1 + i$

Use the Factor Theorem and synthetic division to decide whether or not the given number c is a zero of $P(x)$.

10. $P(x) = x^3 - 2x^2 + 2x - 4, \quad c = 2$
11. $P(x) = x^3 - 7x + 6, \quad c = -3$
12. $P(x) = 3x^3 - 14x + 8, \quad c = 4$

Use the Factor Theorem to determine whether or not $D(x)$ is a factor of $P(x)$.

13. $P(x) = 2x^3 - 5x^2 + x + 10, \quad D(x) = x - 2$
14. $P(x) = x^4 + 4x^3 + x^2 + 18, \quad D(x) = x + 3$
15. $P(x) = 2x^3 - 4x^2 + 2x - 4, \quad D(x) = x + i$
16. Find a polynomial of least degree that has 4 and $2i$ as zeros.
17. Find a polynomial of least degree with real coefficients that has -1, -2, and $3i$ as zeros.

In Problems 18–21, use Descartes' Rule of Signs to discuss the nature of the zeros of the given polynomial.

18. $P(x) = 5x^3 - x^2 + 4x + 9$
19. $P(x) = 2x^4 + x^3 - 4x - 3$
20. $P(x) = 5x^4 - 2x^3 - 7x + 4$
21. $P(x) = 3x^4 - 5x^3 - 7x^2 - 4x + 6$

In Problems 22–25, some of the zeros of the polynomial are given. Find the other zeros.

22. $P(x) = x^3 - 5x^2 + 8x - 6; \quad 3$ and $1 - i$ are zeros.
23. $P(x) = x^3 - 2x^2 - 3x + 10; \quad -2$ is a zero.
24. $P(x) = x^3 + x^2 - 4x + 6; \quad 1 - i$ is a zero.
25. $P(x) = x^4 - x^3 - x^2 - x - 2; \quad -i$ is a zero.

26. a) Find a positive integer, as small as possible, that Theorem 3.9 detects as an upper bound for the zeros of the polynomial $2x^3 + 4x^2 - 3x - 6$.

b) Find a negative integer, as large as possible, that Theorem 3.9 detects as a lower bound for the zeros of the polynomial in part (a).

Find all rational zeros of the given polynomial.

27. $P(x) = 2x^3 + 4x^2 - 3x - 6$

28. $P(x) = 5x^6 + 8x^4 + 6$

29. $P(x) = 3x^3 + 2x^2 - 7x + 2$

In Problems 30–33, find all solutions of the given equation.

30. $x^3 - 6x^2 + 3x + 10 = 0$

31. $x^3 + 3x^2 - 10x - 24 = 0$

32. $x^3 - x^2 - x - 2 = 0$

33. $x^4 - x^3 - 12x^2 - 4x + 16 = 0$

In Problems 34–37, find all zeros of the given polynomial.

34. $P(x) = 3x^3 - 7x^2 + 8x - 2$

35. $P(x) = 6x^3 + 11x^2 + x - 4$

36. $P(x) = 9x^3 + 27x^2 + 8x - 20$

37. $P(x) = 3x^3 + 19x^2 + 30x + 8$

38. The following polynomial has one negative real zero. Find the value of the zero, correct to the nearest tenth.

$$P(x) = 2x^4 + x^2 - 4x - 3$$

39. The following polynomial has exactly one real zero. Find the value of the zero, correct to the nearest tenth.

$$P(x) = x^3 + x^2 - 1$$

40. Find all real zeros of the following polynomial, correct to the nearest tenth.

$$P(x) = 2x^3 + 5x + 2$$

Each of the following nonsolutions has at least one error. Can you find them?

Problem 1

Use the Factor Theorem to determine whether or not the number $c = 2$ is a zero of the polynomial $P(x) = 2x^3 - 7x^2 + 6$.

Nonsolution

$$2 \overline{)\begin{array}{rrr} 2 & -7 & 6 \\ & 4 & -6 \\ \hline 2 & -3 & 0 \end{array}}$$

Since the remainder is zero, $x - 2$ is a factor of $P(x)$ and 2 is a zero of $P(x)$.

Problem 2

Find a polynomial $P(x)$ of least degree with real coefficients that has -2 and $3i$ as zeros.

Nonsolution

$$P(x) = (x + 2)(x - 3i)$$
$$= x^2 + (2 - 3i)x - 6i$$

Problem 3

Use Descartes' Rule of Signs to discuss the nature of the zeros of the polynomial $P(x) = x^3 + x^2 - x + 15$.

Nonsolution

There are two variations in sign in

$$P(x) = x^3 + x^2 - x + 15$$

and one variation in sign in

$$P(-x) = -x^3 + x^2 + x + 15.$$

Therefore $P(x)$ has two positive zeros and one negative zero.

Chapter 4

Exponential and Logarithmic Functions

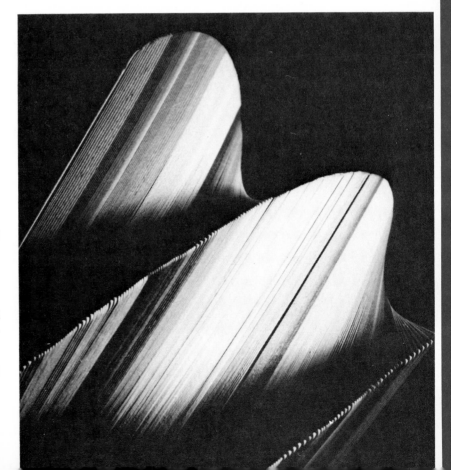

The applications of exponential and logarithmic functions that appear in this chapter illustrate the importance of these functions. An example in Section 4.2 concerns one of the most famous applications in recent years, the dating of the Shroud of Turin by carbon-14 tests in 1988. Many other examples and exercises show the usefulness of these functions in a variety of fields, including engineering and business as well as the biological sciences.

4.1   Exponential Functions

In this chapter we consider only real variables and two types of functions: exponential functions and logarithmic functions. These functions are indispensable in working with problems that involve population growth, decay of radioactive materials, and other processes that occur in nature.

In an exponential function, the function value is obtained by raising a fixed number, called the **base,** to a variable power. Recall that for $a \neq 0$, a^x was defined for integral values of x in Section 1.2. In Section 1.3 this definition was extended to include values $x = m/n$, which are rational numbers. However, if a is negative, there may be no real number $a^{m/n}$. [Such an example is provided by $(-4)^{1/2}$.] For this reason, *we restrict our attention in this chapter to the case where a is positive.* Once this restriction is made, a^x is defined for all rational values of x. It is possible to extend the definition of a^x to include irrational values of x, but a rigorous treatment of this extension requires a degree of mathematical sophistication that is beyond this text. A complete treatment of this sort of topic belongs to the area of mathematics known as *analysis.*

To develop some intuitive feeling for the situation involving an irrational exponent, we consider a specific example, say $2^{\sqrt{5}}$. The reasoning we use is this: If $\sqrt{5}$ is approximated by a rational number m/n, then $2^{\sqrt{5}}$ should be approximated by $2^{m/n}$. More specifically, if $\sqrt{5}$ is approximated by successively closer rational numbers such as

$$2.2, \quad 2.23, \quad 2.236,$$

then successively closer approximations to $2^{\sqrt{5}}$ should be provided by

$$2^{2.2}, \quad 2^{2.23}, \quad 2^{2.236}.$$

Each of these approximations is meaningful, because the exponents are rational. For example,

$$2^{2.2} = 2^{22/10} = 2^{11/5} = (\sqrt[5]{2})^{11}.$$

This intuitive procedure is justified in more advanced courses. We accept it here without justification. In the same spirit, we accept the following generalized statement of the Laws of Exponents.

Laws of Exponents

If a, b, x, and y are real numbers with a and b positive, then

a) $a^x \cdot a^y = a^{x+y}$ Product Rule

b) $(a^x)^y = a^{xy}$ Power Rule

c) $\dfrac{a^x}{a^y} = a^{x-y}$ Quotient Rule

d) $(ab)^x = a^x b^x$

e) $\left(\dfrac{a}{b}\right)^x = \dfrac{a^x}{b^x}$

Definition 4.1

If a is a positive real number and $a \neq 1$, the function f defined by

$$f(x) = a^x$$

is an **exponential function with base a**.

This definition excludes values of a that are not positive, so that a^x will be a real number for all real numbers x, and it excludes $a = 1$ because $1^x = 1$ defines a constant function.

Some graphs of exponential functions are shown in the next two examples.

Example 1   Sketch the graph of the exponential function

$$f(x) = 3^x.$$

Solution
After tabulating the function values displayed in Figure 4.1, we plot the corresponding points and draw a smooth curve, as shown in the figure. Notice that the x-axis is an asymptote for the graph: As x decreases without bound, 3^x approaches 0. Similarly, as x increases without bound, so does 3^x.   ❑

Example 2   Sketch the graph of the exponential function

$$f(x) = \left(\dfrac{1}{2}\right)^x.$$

Solution
We follow the same procedure as in Example 1 and obtain the results shown in Figure 4.2. Note that the x-axis is an asymptote in this case also: As x increases without bound, $(\tfrac{1}{2})^x$ approaches 0.   ❑

The graphs in Figures 4.1 and 4.2 illustrate the two typical graphs for exponential functions. If $a > 1$, the graph of $f(x) = a^x$ resembles the graph in Figure 4.1. If $0 < a < 1$, the graph of $f(x) = a^x$ resembles the graph in Figure 4.2. This is illustrated in Figure 4.3. Note that for any

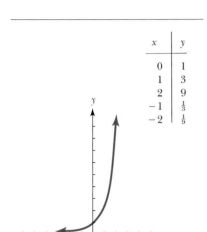

x	y
0	1
1	3
2	9
-1	$\frac{1}{3}$
-2	$\frac{1}{9}$

$f(x) = 3^x$

FIGURE 4.1

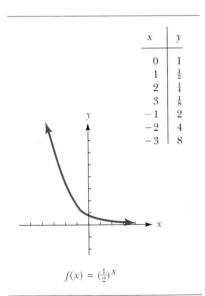

x	y
0	1
1	$\frac{1}{2}$
2	$\frac{1}{4}$
3	$\frac{1}{8}$
-1	2
-2	4
-3	8

$f(x) = (\frac{1}{2})^x$

FIGURE 4.2

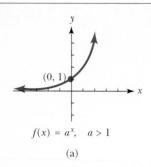

$f(x) = a^x, \quad a > 1$

(a)

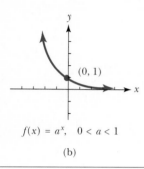

$f(x) = a^x, \quad 0 < a < 1$

(b)

FIGURE 4.3

exponential function, the domain is the set of all real numbers and the range is the set of all positive real numbers.

Figure 4.3 shows graphically how an exponential function is one-to-one, and for a positive and $a \neq 1$, $a^u = a^v$ if and only if $u = v$. If both sides of an equation can be written as a power of the same base, this one-to-one property can often be used to solve the equation.

Example 3   Solve the following equations.

a) $2^x = 32$ b) $9^x = \frac{1}{27}$

Solution

a)   Since $32 = 2^5$, we have

$$2^x = 2^5,$$

and therefore $x = 5$, by the one-to-one property.

b)   Since $9 = 3^2$ and $27 = 3^3$, we can write both sides of the equation as a power of 3.

$$(3^2)^x = \frac{1}{3^3}$$

$$3^{2x} = 3^{-3}$$

$$2x = -3$$

$$x = -\frac{3}{2}$$

The solution is $x = -\frac{3}{2}$.   ❑

An important application of exponential functions occurs in the formula for computing the value of an investment, or *original principal P*, when interest is added to the principal at the end of certain periods of time, so the accumulated interest also earns interest in the next period of time. Interest rates are usually stated at an annual rate r, and interest is converted to principal, or *compounded*, a specified number n times per year. The *rate per period* is then r/n, and the original investment P, together with accumulated interest, has the value

$$A = P\left(1 + \frac{r}{n}\right)^{tn}$$

after t years. The total amount A is called the *compound amount*.

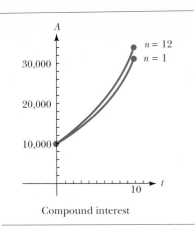

Compound interest

FIGURE 4.4

Example 4 Suppose $10,000 is invested at an annual rate of 12% for a period of 10 years. Find the compound amount if interest is compounded

a) annually,   b) daily.

Solution
From the given information, $P = \$10,000$, $r = 0.12$, and $t = 10$. In each part we use the formula

$$A = P\left(1 + \frac{r}{n}\right)^{tn}$$

and compute the exponential values with a calculator.

a) We have $n = 1$ and
$$A = 10,000(1 + 0.12)^{10} = \$31,058.48.$$

b) With interest compounded daily, we have $n = 365$ and
$$A = 10,000\left(1 + \frac{.12}{365}\right)^{3650} = \$33,194.62.$$

The amounts obtained here illustrate how the frequency of compounding can affect the value of an investment. This is shown graphically in Figure 4.4. ❑

The following example illustrates an application of exponential functions in the natural sciences. Other applications are included in the exercises for this section.

Example 5 Archaeologists are able to estimate the age of organic remains by measuring the amount of radioactive carbon ^{14}C in the organism's remains. If an organism contained 600 milligrams of ^{14}C at its death, the amount $A(t)$ in its remains t years after death is given by

$$A(t) = 600 \cdot 2^{-0.000175t}.$$

Find the amount of ^{14}C in the remains 10,000 years after death.

Solution
We have $t = 10,000$ and
$$A(10,000) = 600 \cdot 2^{-1.75}$$
$$= 600(0.2973018)$$
$$= 178.$$

Archaeologist taking measurements of a fossil jawbone. *Photo By K. Cannon-Bonventre/ Anthrophoto*

There will be approximately 178 milligrams of ^{14}C remaining from the original 600 milligrams. ❑

Exercises 4.1

Solve each of the following equations. (See Example 3.)

1. $2^x = 8$ **2.** $3^x = 81$

3. $2^x = \frac{1}{16}$ **4.** $3^x = \frac{1}{81}$

5. $4^x = 32$ **6.** $27^x = 81$

7. $27^x = \frac{1}{9}$ **8.** $16^x = \frac{1}{32}$

9. $(\frac{1}{3})^{2x+1} = 27$ **10.** $(\frac{1}{5})^{x-1} = 25$

11. $25^{x+1} = \frac{1}{125}$ **12.** $8^{x+1} = \frac{1}{4}$

13. $4^{x-2} = 8^{1+3x}$ **14.** $9^{x+2} = 27^{4-x}$

15. $(\frac{1}{5})^{2x} = 25^{x+3}$ **16.** $(625)^{-x} = 25^{2x-3}$

17. $\dfrac{1}{27^{1+x}} = 81^{3x+5}$ **18.** $16^{x-1} = \dfrac{1}{64^{4+x}}$

Sketch the graphs of the functions defined by the following equations. (See Examples 1 and 2.)

19. $f(x) = 4^x$ **20.** $f(x) = 2^x$

21. $f(x) = 10^x$ **22.** $f(x) = 5^x$

23. $f(x) = (\frac{1}{3})^x$ **24.** $f(x) = (\frac{1}{4})^x$

25. $f(x) = 2^{-x}$ **26.** $f(x) = 3^{-x}$

27. $f(x) = -3^x$ **28.** $f(x) = -(\frac{1}{2})^x$

29. $f(x) = 3^{x-1}$ **30.** $f(x) = 2^{x+1}$

31. $f(x) = 2^x - 1$ **32.** $f(x) = 3^x + 1$

33. $f(x) = 2 + 3^{-x}$ **34.** $f(x) = 3 - 2^x$

35. $f(x) = 2^{|x|}$ **36.** $f(x) = 3^{|x|}$

37. $f(x) = \dfrac{2^x + 2^{-x}}{2}$ **38.** $f(x) = \dfrac{2^x - 2^{-x}}{2}$

39. Find the compound amount if $20,000 is deposited at 12% compounded quarterly for 8 years.

40. Find the compound amount if $15,000 is deposited for 12 years in an account that pays interest at 8% compounded semiannually.

41. Find the amount of interest earned if $42,000 is deposited at 9% compounded monthly for 5 years.

42. Find the amount of interest earned if $50,000 is deposited at 10% compounded quarterly for 6 years.

43. If an item has initial cost C and depreciates at a fixed annual rate r, its value A after t years is given by $A = C(1 - r)^t$. Find the value after 4 years of a copier that cost $5000 new and depreciates at 10% annually.

44. With A and C as defined in Exercise 43, the amount of depreciation is $D = C - A$. Find the amount of depreciation after 3 years on a $10,000 computer that depreciates at 7% annually.

45. The number of cholera bacteria in a laboratory culture is doubling every 2 hours. The culture started with 2000 bacteria at time $t = 0$, and the number $N(t)$ of bacteria after t hours is given by $N(t) = 2000 \cdot 2^{t/2}$. Find the number of bacteria present after 24 hours.

46. In processing raw sugar, the sugar's molecular structure changes in a step that is called *inversion*. If the inversion process starts with 6000 pounds of sugar, the amount of raw sugar remaining after t hours is given by $S(t) = 6000 \cdot 2^{-0.0322t}$ pounds. Find the amount of sugar remaining after 8 hours.

47. A typing teacher has found that the number of words per minute that an average student can type after t months in typing class is given by $W(t) = 50 - 45(2)^{-0.361t}$. Find the number of words per minute that an average student can type after 4 months of typing class.

48. If a bowl of boiling soup (temperature 212°F) is placed in a refrigerator where the temperature is 35°F, the Fahrenheit temperature of the soup after t minutes is given by $F(t) = 35 + 177(2)^{-0.0256t}$. Find the temperature of the soup after 1 hour.

49. When it is open, the spring-loaded valve on a water tank allows $\frac{1}{2}$ of the water in the tank to flow out in 1 hour. If the tank contains 1000 gallons of water when the valve is opened, find a formula for the amount $A(t)$ of water in the tank t hours later.

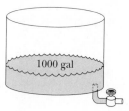

EXERCISE 49

50. The bacteria population in a laboratory culture doubles every day. If the culture starts with 10,000 bacteria, find a formula for the number of bacteria $B(x)$ at the end of x days.

4.2 The Natural Exponential Function

Two exponential functions are especially useful in applications. One of these is the function $f(x) = 10^x$. The other is

$$f(x) = e^x,$$

where e is an irrational number with value

$$e \approx 2.71828,$$

correct to five decimal places. The number e is frequently called the *natural number e*, although it is not an element of the set of natural numbers: $N = \{1, 2, 3, \ldots\}$. Many calculus formulas are simpler when the base e is used, so e is the "natural" choice in those situations.

Example 1 Sketch the graph of the exponential function f where $f(x) = e^x$.

x	y
0	1
1	2.7
2	7.4
3	20.1
-1	0.4
-2	0.1

Solution
For sketching the graph, we use a calculator to obtain the function values shown in Figure 4.5. These values have been rounded to the nearest tenth. ❏

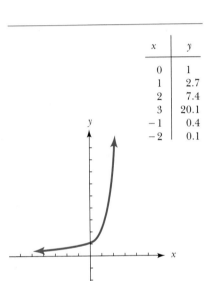

$f(x) = e^x$

FIGURE 4.5

The number e is useful in connection with many physical situations. One such situation is given in Example 2.

Example 2 The measure of the average atmospheric pressure P in inches of mercury at an altitude x miles above sea level is given approximately by

$$P(x) = 30e^{-0.21x}.$$

The swinging bridge on Grandfather Mountain in North Carolina (see photo on the next page) is at an altitude of 1 mile. Find the atmospheric pressure there.

Solution
With $x = 1$ in the given formula, we have

$$P(1) = 30e^{-0.21}.$$

Using a calculator, we compute the value of $e^{-0.21}$ as 0.8106. Therefore

$$P(1) = 30(0.8106)$$

$$= 24.32 \text{ inches of mercury.} \qquad ❏$$

Photo by Hugh Morton

The compound interest formula

$$A = P\left(1 + \frac{r}{n}\right)^{tn}$$

that we used in Section 4.1 has an interesting connection with the number e. If we take $r = 1$ and $t = 1$, the compound amount A is given by

$$A = P\left(1 + \frac{1}{n}\right)^{n}$$

(This corresponds to taking 100% as an interest rate and a time of 1 year.) As the number n of periods per year increases without bound, the factor

$$\left(1 + \frac{1}{n}\right)^{n}$$

approaches $e \approx 2.718281828$ as a limiting value. This can be observed in the values shown in Table 4.1.

n	1	10	100	10,000	1,000,000
$\left(1 + \dfrac{1}{n}\right)^{n}$	2	2.593743	2.704814	2.718146	2.718281

TABLE 4.1

The result in the preceding paragraph can be generalized: For an arbitrary interest rate r and time of t years, the factor

$$\left(1 + \frac{r}{n}\right)^{nt}$$

approaches e^{rt} as the number of periods n increases without bound. The compound amount A thus approaches Pe^{rt}. The compound amount is sometimes computed from the formula

$$A = Pe^{rt}.$$

We then say that interest is being **compounded continuously.**

Example 3   Suppose \$10,000 is invested at an annual rate of 12% for a period of 10 years with interest compounded continuously. Find the value of the investment at the end of 10 years.

Solution
The value of the investment is given by

$$A = 10{,}000e^{(0.12)(10)} = 10{,}000e^{1.2} = \$33{,}201.17.$$

It is worth noting that this value differs from the one obtained by compounding annually in part (a) of Example 4 in the last section by \$2142.69, but it differs from that obtained by compounding daily by only \$6.55.   ❑

Example 4   For centuries it was thought that the Shroud of Turin might be the burial cloth of Jesus Christ. In 1988 the archbishop of Turin announced that carbon-14 dating tests had concluded that the shroud was made between 1260 and 1390.[1] If A_0 was the amount of radioactive ^{14}C in the shroud when it was made, the amount $A(t)$ remaining t years later is given approximately by $A(t) = A_0e^{-0.000124t}$. Find the percentage of ^{14}C remaining in the shroud in 1988 if it was made in 1260.

Solution
The number of years from 1260 to 1988 is 728. With $t = 728$, we find that

$$A(728) = A_0e^{(-0.000124)(728)} = A_0e^{-0.0903} = A_0(0.914).$$

Hence approximately 91.4% of A_0 remained in the shroud in 1988.   ❑

---

1. The accuracy of this conclusion was reported to be 95%; that is, the range of years computed in this manner includes the actual date 95% of the time.

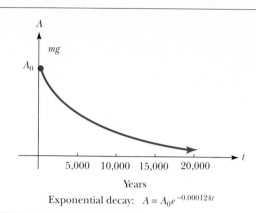

Exponential decay: $A = A_0 e^{-0.000124t}$

FIGURE 4.6

If f is a constant multiple of an exponential function of t and if $f(t)$ increases as t increases, then f is called an **exponential growth function.** When P and r are positive constants, the function $A(t) = Pe^{rt}$ used in Example 3 provides an illustration of exponential growth. On the other hand, if $f(t)$ *decreases* as t increases, then f is called an **exponential decay function.** Exponential decay functions appear in both Examples 2 and 4. Figure 4.6 shows the graph of the function $A(t) = A_0 e^{-0.000124t}$ that was used in Example 4.

For an arbitrary function f, a number c is a **zero** of f if $f(c) = 0$. The next example illustrates a technique that is sometimes used to find the zeros of a function that contains products of exponentials and polynomials. This technique is another of those algebraic skills needed in the calculus.

Example 5   Find the zeros of f where $f(x) = 2xe^{2x} + e^{2x}$.

Solution
To find the zeros of f, we must solve the equation $f(x) = 0$.

$$2xe^{2x} + e^{2x} = 0 \qquad \text{Setting } f(x) = 0$$

$$e^{2x}(2x + 1) = 0 \qquad \text{Factoring}$$

Since $e^{2x} > 0$ for all x, then it must be that

$$2x + 1 = 0$$

$$x = -\frac{1}{2}.$$

Thus $x = -\frac{1}{2}$ is the only zero of f.   ❑

Exercises 4.2

Use a calculator to obtain function values, and sketch the graphs of the following functions. Round the coordinates to the nearest tenth. (See Example 1.)

1. $f(x) = e^{0.1x}$
2. $f(x) = e^{0.2x}$
3. $f(x) = e^{-0.2x}$
4. $f(x) = e^{-0.1x}$
5. $f(x) = e^x + 1$
6. $f(x) = e^{-x} + 1$
7. $f(x) = e^{x-1}$
8. $f(x) = e^{-x+1}$
9. $f(x) = \dfrac{e^x + e^{-x}}{2}$
10. $f(x) = \dfrac{e^x - e^{-x}}{2}$

Find all the real zeros of each of the following functions. (See Example 5.)

11. $f(x) = -xe^{-x} + e^{-x}$
12. $f(x) = 3xe^{3x} + e^{3x}$
13. $f(x) = 2x^2e^{2x} + 2xe^{2x}$
14. $f(x) = -x^3e^{-x} + 3x^2e^{-x}$
15. $f(x) = (x^2 - 3)e^x + 2xe^x$
16. $f(x) = xe^{x^2+1}(2x) + e^{x^2+1}$

17. Find the compound amount if \$20,000 is deposited at 12% compounded continuously for 8 years.

18. Find the compound amount if \$15,000 is deposited for 12 years in an account that pays interest at 8% compounded continuously.

19. Find the interest earned if \$42,000 is deposited at 9% compounded continuously for 5 years.

20. Find the interest earned if \$50,000 is deposited at 10% compounded continuously for 6 years.

21. (See Example 4.) Find the percentage of ^{14}C remaining in the Shroud of Turin in 1988 if it was made in 1390.

22. The atmospheric pressure $P(x)$, in inches of mercury, at an altitude x miles above sea level is approximated by $P(x) = 30e^{-0.21x}$. Find the atmospheric pressure at the top of Mount Everest, which is 5.5 miles above sea level.

23. We say that the *half-life* of radium is 1690 years because it takes 1690 years for half of a given amount of radium to decay into another substance. Starting with 100 milligrams of radium, the number $A(x)$ of milligrams that will be left after x years is $A(x) = 100e^{-0.000411x}$. How much radium will remain after 1000 years?

24. The radioactive carbon atom ^{14}C is formed in the upper atmosphere and has a half-life of about 5600 years. If an organism contains 400 milligrams of ^{14}C at its death, the amount $A(x)$ of ^{14}C remaining x years later is $A(x) = 400e^{-0.000124x}$ milligrams. Find the amount of ^{14}C remaining in the organism after 1000 years.

25. While Route I-85 was being moved to another location in Spartanburg County, South Carolina, a fossilized bone was discovered and estimated to be 3000 years old. Using the formula $A(t) = A_0e^{-0.000124t}$ from Example 4, find the percentage of ^{14}C left in the bone when it was discovered.

26. The half-life of polonium is 140 days. Starting with A_0 milligrams of polonium, the amount $A(x)$ remaining after x days is $A(x) = A_0e^{-0.00495x}$ milligrams. Find the percentage of A_0 remaining after 100 days.

27. As part of a restocking program, 60 white-tail deer were imported from Michigan and released in the Jackson–Bienville Wildlife Management Area in Louisiana. The number $N(x)$ of these deer expected to be alive after x years is approximated by $N(x) = 60e^{-0.163x}$. Find the approximate number of deer expected to survive to the end of the second year.

28. If a bowl of soup with temperature 90°C is placed in a room at 20°C, its temperature $T(x)$ after x minutes is approximated by $T(x) = 20 + 70e^{-0.056x}$. Find its temperature after 10 minutes.

4.3 | Logarithmic Functions

We noted in Section 4.1 that any exponential function $f(x) = a^x$ is a one-to-one function with domain the set of all real numbers and range the set of all positive real numbers. It follows from the results of Section

2.5 that the exponential function has an inverse function with domain the set of all positive numbers and range the set of all real numbers.

The inverse of the exponential function $f(x) = a^x$ is called the **logarithmic function with base a**. The value of this logarithmic function at x is called the **logarithm of x to the base a**, and is abbreviated **$\log_a x$**.

In Section 2.5 we saw that a defining equation for f^{-1} may be obtained by interchanging x and y in the defining equation for f. The defining equation for the logarithmic function is $x = a^y$, and the defining statement is

$$y = \log_a x \quad \text{if and only if} \quad x = a^y.$$

In these equations we must keep in mind that x and a are positive and $a \neq 1$. Rewording the boxed statement, we can say that **the exponent y to which a must be raised in order to obtain x is $y = \log_a x$**.

The defining statement for the logarithmic function sets up an equivalence between the logarithmic statement $L = \log_a N$ and the exponential statement $N = a^L$. A statement in either of these forms can be changed to an equivalent statement in the other form.

Example 1   Change the following statements to logarithmic form.

a) $16 = 2^4$ b) $\frac{1}{9} = (27)^{-2/3}$

Solution

a) $16 = 2^4$ implies that $\log_2 16 = 4$.

b) $\frac{1}{9} = (27)^{-2/3}$ implies that $\log_{27} \frac{1}{9} = -\frac{2}{3}$.   ❑

Example 2   Change the following statements to exponential form.

a) $\log_3 81 = 4$ b) $\log_{16} \frac{1}{64} = -\frac{3}{2}$

Solution

a) $\log_3 81 = 4$ implies that $81 = 3^4$.

b) $\log_{16} \frac{1}{64} = -\frac{3}{2}$ implies that $\frac{1}{64} = (16)^{-3/2}$.   ❑

From Chapter 2 we know that $f^{-1}(f(x)) = x$ for all x in the domain of f and that $f(f^{-1}(x)) = x$ for all x in the domain of f^{-1}. For the exponential function $f(x) = a^x$, these equations read as

$$\log_a a^x = x$$

for all real numbers x, and

$$a^{\log_a x} = x$$

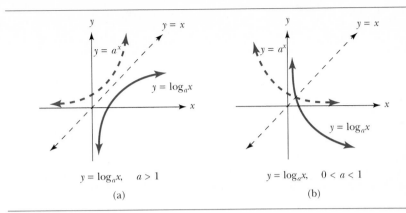

FIGURE 4.7

for all positive numbers x. In the first equation we get

$$\log_a a = 1$$

with $x = 1$, and we get

$$\log_a 1 = 0$$

with $x = 0$.

Since the logarithmic function $y = \log_a x$ is the inverse of the exponential function $y = a^x$, the graph of $y = \log_a x$ can be obtained by reflecting the graph of $y = a^x$ through the line $y = x$. This is illustrated in Figure 4.7.

To graph a logarithmic function, keep Figure 4.7 in mind and plot a few strategic points using the equation $\log_a a^x = x$.

x	y
$1 = 2^0$	0
$2 = 2^1$	1
$4 = 2^2$	2
$8 = 2^3$	3
$\frac{1}{2} = 2^{-1}$	-1
$\frac{1}{4} = 2^{-2}$	-2

Example 3   Sketch the graph of $f(x) = \log_2 x$.

Solution
We replace x with some convenient powers of 2 and use the equality

$$y = \log_2 x = \log_2 2^n = n$$

to obtain the points tabulated in Figure 4.8. After plotting these points, we join them with a smooth curve that has the typical shape shown in Figure 4.7(a). ❏

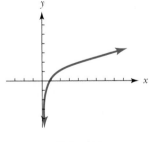

$$f(x) = \log_2 x$$

FIGURE 4.8

If only one of the variables in the equation $y = \log_a x$ is unknown, the equation can often be solved by changing it to the equivalent exponential statement. The following example illustrates the technique.

Example 4   Solve for the unknown variable.

a) $y = \log_{3/2} \frac{9}{4}$ b) $\log_4 x = -\frac{3}{2}$ c) $\log_a 64 = -3$

Solution

a) $y = \log_{3/2} \frac{9}{4}$ is equivalent to $(\frac{3}{2})^y = \frac{9}{4}$. Since $(\frac{9}{4}) = (\frac{3}{2})^2$, we have

$$\left(\frac{3}{2}\right)^y = \left(\frac{3}{2}\right)^2,$$

and therefore $y = 2$.

b) $\log_4 x = -\frac{3}{2}$ is equivalent to

$$x = (4)^{-3/2} = \frac{1}{4^{3/2}} = \frac{1}{(\sqrt{4})^3} = \frac{1}{8}.$$

c) $\log_a 64 = -3$ is equivalent to $a^{-3} = 64$. Therefore

$$\frac{1}{a^3} = 64$$

$$a^3 = \frac{1}{64}$$

$$a = \sqrt[3]{\frac{1}{64}} = \frac{1}{4}. \qquad \square$$

Since the exponential and logarithmic functions with base a are inverse functions, the Laws of Exponents have some important implications for logarithms. These are stated in the following properties of logarithms.

Properties of Logarithms

Let a, u, v, and r be real numbers with u, v, and a positive and $a \neq 1$. Then

a) $\log_a u = \log_a v$ if and only if $u = v$,
b) $\log_a (uv) = \log_a u + \log_a v$,
c) $\log_a (u/v) = \log_a u - \log_a v$,
d) $\log_a u^r = r \log_a u$.

Numerical computations can be made by using these properties with a table of logarithms, but calculators have virtually eliminated this use of logarithms.

Property (a) follows from the fact that the logarithmic function is the inverse of the exponential function. More explicitly,

$$\log_a u = \log_a v \quad \text{implies that} \quad a^{\log_a u} = a^{\log_a v},$$

and therefore

$$u = v.$$

The Laws of Exponents can be used to derive properties (b), (c), and (d). We shall prove property (b) here and leave the other proofs as exercises.

Let $x = \log_a u$ and $y = \log_a v$. Then $u = a^x$ and $v = a^y$. By the product rule in the Laws of Exponents,

$$uv = a^x a^y = a^{x+y},$$

and therefore

$$\log_a uv = x + y$$

$$= \log_a u + \log_a v.$$

Logarithmic functions are defined only for positive real numbers, and **we assume for the remainder of this section that all variables are appropriately restricted so that the functions involved are defined.**

The properties of logarithms can be used to expand a single logarithmic expression into terms that involve simpler logarithmic expressions or to change an expanded expression to a single logarithm. This is illustrated in the following example.

Example 5

a) Express $\log_a \sqrt[3]{\dfrac{x^3 y^5}{z^6}}$ in terms of the logarithms of x, y, and z.

b) Write the following expression as a single logarithm.

$$4 \log_a x + 3 \log_a 4 - 2 \log_a y - 3 \log_a z$$

Solution

a) $\log_a \sqrt[3]{\dfrac{x^3 y^5}{z^6}} = \log_a \left(\dfrac{x^3 y^5}{z^6}\right)^{1/3} = \tfrac{1}{3}\log_a \left(\dfrac{x^3 y^5}{z^6}\right)$

$$= \tfrac{1}{3}(\log_a x^3 y^5 - \log_a z^6)$$

$$= \tfrac{1}{3}(\log_a x^3 + \log_a y^5 - \log_a z^6)$$

$$= \tfrac{1}{3}(3 \log_a x + 5 \log_a y - 6 \log_a z)$$

$$= \log_a x + \tfrac{5}{3}\log_a y - 2 \log_a z$$

b) $4 = \log_a x + 3 \log_a 4 - 2 \log_a y - 3 \log_a z$

$\qquad = \log_a x^4 + \log_a 4^3 - \log_a y^2 - \log_a z^3$

$\qquad = \log_a x^4 + \log_a 64 - (\log_a y^2 + \log_a z^3)$

$\qquad = \log_a 64x^4 - \log_a y^2 z^3$

$\qquad = \log_a \dfrac{64x^4}{y^2 z^3}$ ❑

One of the most important uses of logarithms is to solve for a variable that occurs in an exponent. We illustrate this use in the next example.

Example 6   Starting with 400 milligrams of ^{14}C in a dead organism, the amount remaining after x years is $A = 400e^{-0.000124x}$ milligrams. Use logarithms to solve for x in terms of A.

Solution
Since the variable x is in the exponent, our first step is to take the logarithm of each side of the equation, using the base e.

$$\log_e A = \log_e (400e^{-0.000124x})$$

$$\log_e A = \log_e 400 + \log_e e^{-0.000124x}$$

$$\log_e A = \log_e 400 - 0.000124x$$

Solving for x in this linear equation, we get

$$x = \frac{\log_e 400 - \log_e A}{0.000124}.$$

         ❑

Exercises 4.3

Change the following statements to logarithmic form. (See Example 1.)

1. $2^3 = 8$ **2.** $7^2 = 49$

3. $3^{-4} = \frac{1}{81}$ **4.** $4^{-2} = \frac{1}{16}$

5. $5^0 = 1$ **6.** $3^0 = 1$

7. $(64)^{-2/3} = \frac{1}{16}$ **8.** $4^{-3/2} = \frac{1}{8}$

Change the following statements to exponential form. (See Example 2.)

9. $\log_5 5 = 1$ **10.** $\log_7 1 = 0$

11. $\log_3 81 = 4$ **12.** $\log_2 8 = 3$

13. $\log_2 \frac{1}{16} = -4$ **14.** $\log_3 \frac{1}{27} = -3$

15. $\log_{3/5} \frac{25}{9} = -2$ **16.** $\log_{5/2} \frac{4}{25} = -2$

Sketch the graphs of the functions defined by the following equations. In Exercises 23 and 24, round coordinates to the nearest tenth. (See Example 3.)

17. $f(x) = \log_3 x$ **18.** $f(x) = \log_4 x$

19. $f(x) = \log_{1/2} x$ **20.** $f(x) = \log_{1/3} x$

21. $f(x) = \log_2 (x + 3)$ **22.** $f(x) = \log_3 (x - 1)$

23. $f(x) = \log_e x$ **24.** $f(x) = \log_{1.6} x$

Write each of the following expressions as a single logarithm. (See part b of Example 5.)

63. $\log_a x + 3 \log_a y$ **64.** $4 \log_a x + 2 \log_a z$

Solve for the unknown variable. (See Example 4.)

25. $y = \log_2 1$ **26.** $y = \log_6 6$

65. $2 \log_a x + \log_a y + 2 \log_a z$

27. $y = \log_2 32$ **28.** $y = \log_3 27$

66. $3 \log_a x + 2 \log_a y + \log_a z$

29. $y = \log_3 \frac{1}{9}$ **30.** $y = \log_2 \frac{1}{8}$

67. $\frac{1}{2} \log_a y - \frac{2}{3} \log_a z$ **68.** $2 \log_a x - \frac{1}{2} \log_a y$

31. $\log_4 x = 3$ **32.** $\log_5 x = 2$

69. $3 \log_a x + 2 \log_a z - 3 \log_a 4 - 2 \log_a y$

33. $\log_3 x = -2$ **34.** $\log_2 x = -3$

70. $2 \log_a x + 4 \log_a 2 - \log_a y - 2 \log_a z$

35. $\log_8 x = -\frac{2}{3}$ **36.** $\log_4 x = -\frac{1}{2}$

71. $\frac{1}{2}(3 \log_a x - \log_a y + 2 \log_a z)$

37. $\log_a 64 = 3$ **38.** $\log_a 36 = 2$

72. $\frac{1}{2}(4 \log_a x - 2 \log_a y - 3 \log_a z)$

39. $\log_a 16 = -2$ **40.** $\log_a 8 = -3$

73. $\frac{3}{2} \log_a 9x^4 y^6 - 4 \log_a xz + \frac{1}{3} \log_a x^3 z^6$

41. $\log_a 3 = -\frac{1}{3}$ **42.** $\log_a 2 = -\frac{1}{3}$

74. $\frac{2}{3} \log_a 8x^3 y^2 + \frac{1}{2} \log_a x^4 z^4 - \frac{1}{3} \log_a yz^6$

43. $3^{\log_3 x} = 7$ **44.** $5^{\log_5 4x} = 12$

45. $\log_4 4^{x^2} = 1$ **46.** $\log_{10} 10^{5x+3} = 4$

75. Starting with 100 milligrams of radium, the number A of milligrams remaining after x years is given by $A = 100e^{-0.000411x}$. Solve for x in terms of A by using logarithms to the base e.

47. $\log_2 (x - 3) = 0$ **48.** $\log_e (x + 7) = 0$

49. $\log_6 (5x + 4) = 1$ **50.** $\log_5 (3x - 1) = 1$

76. Starting with 800 milligrams of radioactive polonium, the amount A remaining after x days is $A = 800e^{-0.00495x}$ milligrams. Use logarithms to the base e to solve for x in terms of A.

Express each of the following in terms of the logarithms of x, y, and z. (See part a of Example 5.)

51. $\log_a x^3 y^4$ **52.** $\log_a x^2 y^5$

77. The number N of bacteria in a given culture after t hours is given by $N = 2000 \cdot 2^{t/2}$. Solve for t in terms of N by using logarithms to the base 2.

53. $\log_a x^2 y \sqrt{z}$ **54.** $\log_a xy^3 \sqrt[3]{z}$

55. $\log_a \frac{xy}{z}$ **56.** $\log_a \frac{x}{yz}$

78. At a certain stage in refining raw sugar, the amount S of raw sugar remaining after t hours is given by $S = 6000 \cdot 2^{-0.0322t}$ pounds. Solve for t in terms of S by using logarithms to the base 2.

57. $\log_a \frac{x^2}{y \sqrt[3]{z}}$ **58.** $\log_a \frac{x^4 \sqrt{y}}{z^3}$

79. Prove property (c) of logarithms:
$$\log_a (u/v) = \log_a u - \log_a v.$$

59. $\log_a \sqrt[3]{x^4 y^3 z^6}$ **60.** $\log_a \sqrt[4]{xy^4 z^5}$

80. Prove property (d) of logarithms:
$$\log_a u^r = r \log_a u.$$

61. $\log_a \sqrt[3]{\dfrac{x^6}{y^5 z^3}}$ **62.** $\log_a \dfrac{\sqrt[3]{x^3 y^5}}{\sqrt{z^3}}$

4.4   Logarithmic Equations, Exponential Equations, and Natural Logarithms

In technical work it is sometimes necessary to solve equations that contain logarithmic functions or exponential functions. Numerical values of logarithms are required to obtain satisfactory solutions to these types of equations.

Before hand-held calculators came into common use, logarithms were frequently used in performing numerical computations. Since our number system uses the base 10, logarithms to base 10 are the most practical for computational purposes, and they are referred to as **common logarithms.** For the remainder of this book we adopt the convention that

$$\log x = \log_{10} x.$$

That is, the base is understood to be 10 unless it is specified otherwise. We also use the phrase *logarithm of x* to mean "logarithm of *x* to the base 10."

The `log` button on most scientific calculators is used for computations with common logarithms. Tables can also be used, and a table is provided at the end of the Appendix to this book. The Appendix gives instructions on using the table and also on using logarithms to perform computations.

An equation that contains logarithmic functions is called a **logarithmic equation.** Many logarithmic equations can be solved by using the properties and the definition of a logarithm. Such an equation is solved in Example 1.

Example 1   Find the solution set for the equation

$$\log (2x + 1) + \log (3x - 4) = 1.$$

Solution
Using property (b) of logarithms, the left side can be written as a single logarithm. The 1 on the right side can be replaced by log 10.

$$\log (2x + 1)(3x - 4) = \log 10$$

By the one-to-one property of a logarithmic function, this is equivalent to

$$(2x + 1)(3x - 4) = 10.$$

Therefore

$$6x^2 - 5x - 4 = 10$$
$$6x^2 - 5x - 14 = 0,$$

and

$$(x - 2)(6x + 7) = 0.$$

This leads to

$$x - 2 = 0 \quad \text{or} \quad 6x + 7 = 0,$$
$$x = 2 \quad \text{or} \quad x = -\frac{7}{6}.$$

We must check to see if the logarithms in the original equation are defined at these values. With $x = -\frac{7}{6}$, the left member is

$$\log\left(-\frac{4}{3}\right) + \log\left(-\frac{15}{2}\right).$$

This quantity is not defined because it involves logarithms of negative numbers. The value $x = -\frac{7}{6}$ must be rejected. With $x = 2$, all logarithms in the original equation are defined, and $x = 2$ is a solution. Thus the solution set is $\{2\}$.   ❏

An equation that contains an exponential function is called an **exponential equation**. In many cases an exponential equation can be solved by converting it to a logarithmic equation and solving the logarithmic equation.

Example 2   Solve the equation $2^{2x+1} = 3^{x+4}$.

Solution
Taking the logarithm of each side of the equation, we have

$$(2x + 1) \log 2 = (x + 4) \log 3,$$

or

$$2x \log 2 + \log 2 = x \log 3 + 4 \log 3.$$

Solving for x, we have

$$(2 \log 2 - \log 3)x = 4 \log 3 - \log 2$$

and

$$x = \frac{4 \log 3 - \log 2}{2 \log 2 - \log 3} = 12.87.$$ ❏

Example 3   If an original investment of \$1000 grows to a value of \$1500 in 5 years when interest is compounded quarterly at an annual rate r, find r.

Solution
Using the formula for compound amount from Example 4 of Section 4.1 with $A = \$1500$, $P = \$1000$, $n = 4$, and $t = 5$, we have

$$1500 = 1000\left(1 + \frac{r}{4}\right)^{20}.$$

This gives

$$1.5 = (1 + 0.25r)^{20}.$$

Taking the logarithm of each side, we have

$$\log 1.5 = 20 \log (1 + 0.25r)$$

and

$$\log (1 + 0.25r) = \frac{\log 1.5}{20}$$

$$= \frac{0.1761}{20}$$

$$= 0.0088.$$

Since INV **log** $0.0088 = 1.02$, we have

$$1 + 0.25r = 1.02$$

and

$$r = \frac{0.02}{0.25}$$

$$= 0.08.$$

That is, the annual interest rate is 8%.   ❏

Example 4   Logarithms to the base 10 are used in the db ("d, b") scale for measuring loudness in decibels. If the intensity of the sound in watts per square meter is I, the decibel level L is given by

$$L = 10 \log (I \times 10^{12})\text{db}.$$

For ordinary conversation, L is about 65 db, and the threshold of pain is about 120 db. Find the effect of doubling the power I on an audio amplifier.

Solution
If we replace I by $2I$ in the formula for L, we obtain

$$L = 10 \log (2I \times 10^{12})$$

$$= 10[\log 2 + \log (I \times 10^{12})]$$

$$= 10 \log 2 + 10 \log (I \times 10^{12})$$

$$= 3 + 10 \log (I \times 10^{12}).$$

Thus doubling the power increases L by only about 3 decibels.   ❏

Although logarithms to base 10 are adequate for many purposes, other logarithms are used more extensively in scientific and technical work. There are logarithms to the base e. The notation "ln x" is commonly used to denote logarithms to the base e and "ln x" is read "the natural logarithm of x." We write:

$$\ln x = \log_e x.$$

Many of the formulas used in calculus have a particularly simple form when expressed using logarithms to the base e, and many physical situations can be described with these logarithms. Most books of mathematical tables have a table of natural logarithms, and scientific calculators have a button marked **ln x** that will yield the natural logarithm of any positive number.

When exponential equations involving the base e occur, natural logarithms are useful in solving these equations. An illustration is given in the next example.

Example 5   The measure of the atmospheric pressure P in inches of mercury at an altitude x miles above sea level is given approximately by

$$P = 30e^{-0.21x}.$$

If the pressure at a mountain peak is 15 inches of mercury, find the altitude of the peak.

Solution
We put $P = 15$ and solve for x.

$$30e^{-0.21x} = 15$$

$$e^{-0.21x} = \frac{1}{2}$$

$$\ln(e^{-0.21x}) = \ln(2^{-1})$$

$$-0.21x = -\ln 2$$

$$x = \frac{\ln 2}{0.21}$$

$$= 3.30$$

The peak is approximately 3.30 miles above sea level.   ❏

It is possible to use either common logarithms or natural logarithms to calculate the value of logarithms to any desired base. To see how this is done, we consider the general problem of changing logarithms from one base to another.

To obtain an equation relating $\log_a N$ and $\log_b N$, let

$$L = \log_b N.$$

Then

$$N = b^L$$

and

$$\log_a N = \log_a b^L$$

$$= L \log_a b.$$

Solving for L, we get

$$L = \frac{\log_a N}{\log_a b}.$$

Since $L = \log_b N$ from our first equation, we have the following change-of-base formula.

Change-of-Base Formula

$$\log_b N = \frac{\log_a N}{\log_a b}$$

The special cases in which $a = 10$ or $a = e$ in the change-of-base formula give the formulas

$$\log_b N = \frac{\log N}{\log b} \quad \text{or} \quad \log_b N = \frac{\ln N}{\ln b}.$$

Either of these formulas will yield the same results. In Example 6 we use logarithms to base 10.

Example 6   Find a three-digit value for $\log_5 19$.

Solution
Using the change-of-base formula, we obtain

$$\log_5 19 = \frac{\log 19}{\log 5} = 1.83.$$

❏

Exercises 4.4

Find the solution set for the following equations. (See Example 1.)

1. $\log (2x + 1) - \log 5 = \log x$
2. $\log 2x - \log 5 = 2$
3. $\log x^2 - \log 9 = \log x$
4. $\log x^2 - \log 4 = \log x$
5. $\log_5 (4x - 1) - \log_5 3 = \log_5 (x + 2)$
6. $\log_3 (9x - 6) - \log_3 4 = \log_3 (x + 1)$
7. $\log_2 x - \log_2 (x - 1) = 3$
8. $\log_3 x - \log_3 (x - 1) = 2$
9. $\log x + \log (3x - 7) = 1$
10. $\log (3x - 16) + \log (x - 1) = 1$
11. $\log 8x + \log (x + 2) = 1$
12. $\log 5x + \log (6x + 1) = 1$
13. $\log 2x + \log (13x - 1) = 2$
14. $\log (2 - 6x) + \log (8 + x) = 2$
15. $\log_4 x + \log_4 (x + 6) = 2$
16. $\log_4 (x - 2) + \log_4 (x + 1) = 1$
17. $e^{\ln x^2} = 4$
18. $e^{\ln(6x - 5)} = 7$
19. $\ln e^{4x} = 8$
20. $\ln e^{9x + 8} = 10$

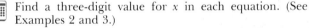

Find a three-digit value for x in each equation. (See Examples 2 and 3.)

21. $3^x = 17$
22. $2^x = 5$
23. $4^{1-x} = 19$
24. $7^{2-x} = 6$
25. $9^{2x^2-1} = 11$
26. $5^{x^2+2} = 39$
27. $3^x = 2^{4x+3}$
28. $5^x = 6^{x-1}$
29. $5^{x-3} = 9^{x+4}$
30. $7^{1-x} = 6^{3-x}$
31. $9^{x+2} = 7^{3x-1}$
32. $5^{2+3x} = 8^{4x-1}$
33. $(1 + x)^3 = 3.12$
34. $(1 + x)^5 = 1.34$
35. $(1 + 0.06)^x = 3.17$
36. $60(1 + 0.02)^{4x} = 130$

Find a three-digit value for each of the following logarithms. (See Example 6.)

37. $\log_3 7$
38. $\log_4 8$
39. $\log_6 43$
40. $\log_8 51$
41. $\log_2 28.6$
42. $\log_5 41.3$
43. $\log_7 132$
44. $\log_9 158$

Use natural logarithms to solve the following equations and round the answers to three digits. (See Example 5.)

45. $e^x = 4.2$
46. $e^x = 26$
47. $e^{-x} = 5.3$
48. $e^{-x} = 6.1$
49. $e^{x+1} = 67.1$
50. $e^{x+1} = 24.7$
51. $e^{2x+1} = 114$
52. $e^{2x+1} = 92.5$

53. Find the compound amount A of $P = \$1000$ compounded quarterly for 5 years at an annual rate of 12%.

54. If $P = \$1000$ amounts to $1250 in 2 years and interest is compounded semiannually, find the annual interest rate r.

55. How long will it take an original principal P to double if it is invested at 14% compounded semiannually?

56. How long will it take an original principal P to triple if it is invested at 12% compounded monthly?

57. If money earns 10% compounded quarterly, how much should be invested today to accumulate to $10,000 in 6 years?

58. If money earns 8% compounded monthly, how much should be invested today to accumulate to $5000 in 4 years?

59. How long will it take an original investment to double if interest is at 10% compounded continuously?

60. How long will it take an original investment to triple if interest is at 9% compounded continuously?

61. If a colony of bacteria starts with 2000 bacteria and increases by 15% of its population each day, the population P after t days is given approximately by $P = 2000e^{0.14t}$. How long will it take the population to reach 8000?

62. A colony of bacteria increases by 20% of its population each day. Starting with 3000 bacteria, the population P after x days is approximately $P = 3000e^{0.18x}$. How long will it take the population to reach 6000?

63. How long will it take the population in Exercise 62 to reach 9000?

64. The atmospheric pressure P at an altitude x miles above sea level is given approximately by $P = 30e^{-0.21x}$, where P is measured in inches of mercury. Find the altitude of a mountain peak if the pressure there is 16 inches of mercury.

65. If an organism contains 40 milligrams of the radioactive carbon atom ^{14}C at its death, the amount A of ^{14}C remaining x years later is $A = 40e^{-0.000124x}$ milligrams. How long after its death will 10 milligrams of ^{14}C remain?

66. If an organism contains 10 milligrams of the radioactive carbon atom ^{14}C at its death, the amount of ^{14}C remaining x years later is $A = 10e^{-0.000124x}$ milligrams. How long has it been since death when 4 milligrams of ^{14}C remain?

67. Starting with 100 milligrams of radium, the amount A of radium remaining after x years of radioactive decay is $A = 100e^{-0.000411x}$ milligrams. How long will it be until half the radium remains?

68. The pH of a chemical solution is given by pH $= -\log (H_3O^+)$, where (H_3O^+) is the concentration in moles per liter of the hydronium ion. Find the hydronium ion concentration (H_3O^+) if the pH is 6.8.

69. Using the decibel formula $L = 10 \log (I \times 10^{12})$ db, find the effect on L of tripling the power I on an audio amplifier.

70. Using the formula in Exercise 69, find the effect on L of increasing the power from I to $5I$.

CHAPTER REVIEW

Key Words and Phrases

Exponential function	Logarithmic function	Exponential equation
Base	Common logarithm	Natural logarithm
Natural number e	Logarithmic equation	

Summary of Important Concepts and Formulas

Laws of Exponents

$$a^x \cdot a^y = a^{x+y} \qquad \text{Product Rule}$$

$$(a^x)^y = a^{xy} \qquad \text{Power Rule}$$

$$\frac{a^x}{a^y} = a^{x-y} \qquad \text{Quotient Rule}$$

$$(ab)^x = a^x \cdot b^x$$

$$\left(\frac{a}{b}\right)^x = \frac{a^x}{b^x}$$

Change-of-Base Formula

$$\log_b N = \frac{\log_a N}{\log_a b}$$

One-to-One Property

$$a^u = a^v \quad \text{if and only if} \quad u = v.$$

Logarithmic Function

$$y = \log_a x \quad \text{if and only if} \quad x = a^y.$$

Properties of Logarithms

$$\log_a a^x = x \qquad \log_a u = \log_a v \quad \text{if and only if} \quad u = v.$$

$$a^{\log_a x} = x \qquad \log_a(uv) = \log_a u + \log_a v$$

$$\log_a a = 1 \qquad \log_a (u/v) = \log_a u - \log_a v$$

$$\log_a 1 = 0 \qquad \log_a u^r = r \log_a u$$

Review Problems for Chapter 4

1. If $f(x) = (2/3)^x$, find the following function values.
   a. $f(-2)$ b. $f(\tfrac{1}{2})$

Solve for x.

2. $(81)^x = \tfrac{1}{27}$

3. $8^x = \tfrac{1}{32}$

4. $4^{x+3} = 32^{2x-5}$

5. $10^{-2x} = \dfrac{1}{1000^{x+3}}$

Sketch the graphs of the functions defined by the following equations.

6. $f(x) = 2^{x+2}$

7. $f(x) = 3^{-x}$

8. $f(x) = e^x - 2$

9. $f(x) = \ln x + 1$

10. $f(x) = \log_3 x$

11. $f(x) = \log_2 (x - 2)$

Find the value of the unknown variable.

12. $\log_a 64 = \frac{3}{2}$

13. $y = \log_{16} \frac{1}{32}$

14. $\log_4 x = -\dfrac{5}{2}$

15. $\log_3 3^{4x} = 18$

16. $e^{\ln e^3} = y$

17. $\log 10^x = -3$

Expand each of the following into terms involving the logarithms of x, y, and z.

18. $\log_a \left(\dfrac{x^2 z}{y^3} \right)$

19. $\log_a \dfrac{\sqrt[3]{x^2 y}}{z^4}$

Write each of the following as a single logarithm.

20. $2 \log_a x^2 z + \log_a 9xy^3 - 2 \log_a 4yz^3$

21. $2 \log_a x^2 yz + 3 \log_a 2xyz^2 - \log_a 4x^3 y^2 z^4$

Find a three-digit value for x in each equation.

22. $x = \log_3 17$

23. $e^x = 4.7$

24. $5^{x+1} = 192$

25. $5^{2x-1} = 8^{x+1}$

Find the solution set for each of the following equations.

26. $\log x + \log (3x + 1) = 1$

27. $\log (6 + 3x) + \log (1 - 3x) = 1$

28. $\log_8 (x + 5) + \log_8 (3x - 1) = 2$

29. Find all the zeros of the function f where $f(x) = -2x^3 e^{-2x} + 3x^2 e^{-2x}$.

30. The atmospheric pressure P, in inches of mercury, at an altitude x miles above sea level is approximated by $P = 30e^{-0.21x}$. Find the atmospheric pressure at the top of Mount Mitchell, which is about 1.27 miles above sea level.

31. How long will it take an original principal P to increase by 50% if it is invested at 10% compounded monthly?

32. An anthill containing 130,000 ants is sprayed with an insecticide. If 23% of the ant population dies each hour, the population P after t hours is approximately $P = 130,000e^{-0.26t}$. How long will it take the population to reach 10,000 ants?

33. After t months in typing class, the number of words per minute that an average student can type is given by $W(t) = 50 - 45(2)^{-0.361t}$. Find the length of time until the average student can type 42 words per minute.

34. According to *Newton's Law of Cooling*, the temperature T of an object placed in a medium with constant temperature T_0 changes at a rate proportional to $T - T_0$. If a bowl of soup with temperature 90°C is placed in a room at 20°C, its temperature $T(x)$ after x minutes is $T(x) = 20 + 70e^{-0.056x}$ degrees Celsius. Find how long it will take the soup to cool to 37°C (body temperature).

35. The loss of value of equipment or buildings in business is called *depreciation*. If an item has initial cost C and depreciates at 8% per year, its value after t years is $A = C(1 - 0.08)^t$. Find the length of time it takes for the item to lose 65% of its original value.

36. When 60 deer are released in a wildlife management area, the number $N(x)$ of these deer expected to be alive at the end of x years is $N(x) = 60e^{-0.163x}$. Find the expected length of time it takes for 40 of the deer to die.

37. Starting with A_0 milligrams of polonium, the amount $A(x)$ remaining after x days is $A(x) = A_0 e^{-0.000495x}$ milligrams. Find the number of days until $0.37A_0$ milligrams of polonium remain.

Each of the following nonsolutions has at least one error. Can you find them?

Problem 1   Solve $\log_2 x - \log_2 (x + 1) = 3$.

Nonsolution

$$\log_2 \frac{x}{x + 1} = 3$$

$$\frac{x}{x + 1} = 8$$

$$x = 8x + 8$$

$$-7x = 8$$

$$x = -\frac{8}{7}$$

The solution set is $\{-\frac{8}{7}\}$.

Problem 2   Solve $\log (2x + 3) + \log (3x - 1) = 1$.

Nonsolution

$$\log [(2x + 3) + (3x - 1)] = 1$$

$$\log (5x + 2) = 1$$

$$5x + 2 = 10$$

$$5x = 8$$

$$x = \frac{8}{5}$$

The solution is $x = \frac{8}{5}$.

Problem 3   Solve $\log_3 x + \log_3 (2x + 3) = 2$.

Nonsolution

$$x(2x + 3) = 2$$

$$2x^2 + 3x - 2 = 0$$

$$(2x - 1)(x + 2) = 0$$

$$2x - 1 = 0 \quad \text{or} \quad x + 2 = 0$$

$$x = \frac{1}{2} \quad \text{or} \quad x = -2$$

When $x = -2$, the left member of the original equation contains logarithms of negative numbers. This value must be rejected. The only solution is $x = \frac{1}{2}$.

Chapter 5

The Trigonometric Functions

This chapter presents many interesting and practical problems that can be solved by using the trigonometric functions. One such problem that received a great deal of publicity in 1988 is related in Section 5.6. In that year a surveyor in Louisville, Kentucky, used the tangent function to prove that the Louisville Falls Fountain on the Ohio River is the world's largest floating fountain. The rich abundance of applications of trigonometry is evident throughout the chapter.

5.1   Angles, Triangles, and Arc Length

Much of trigonometry involves working with triangles. In fact, the word *trigonometry* means "triangle measurement" and comes from the Greek words *tri* ("three"), *gonia* ("angle"), and *metron* ("measure"). In order to work with triangles, some definitions and terminology concerning angles are essential.

Any point on a straight line separates the line into two parts, called **half-lines.** A half-line together with its endpoint is called a **ray.** An **angle** consists of two rays that have a common endpoint. The two rays are called the **sides** of the angle.

In trigonometry we usually think of an angle as being formed by revolving, or rotating, a ray about its endpoint. The endpoint is called the **vertex** of the angle. We label the initial side and terminal side of the angle according to the direction of rotation (see Figure 5.1). An angle is called a **positive angle** if the direction of rotation is counterclockwise, and a **negative angle** if the direction of rotation is clockwise. The direction of rotation is usually indicated by an arrow.

Angles are commonly denoted by the Greek letters α, β, γ, θ, and so on, or by capital letters A, B, C, and so on. If an angle has its vertex at point O, its initial side passing through point A, and its terminal side passing through point P, then the angle might be labeled "angle at O" or "angle AOP." The symbol $\sphericalangle$ is used to denote angles, but often this symbol is omitted if the context is clear.

The **measure** of an angle is given by stating the amount of rotation used to revolve from the initial position of the ray to the terminal position. We write $A = B$ to mean "the measure of angle A equals the measure of angle B." There is more than one unit for measuring angles,

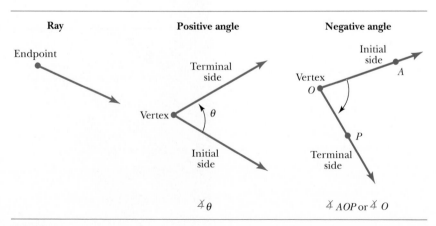

FIGURE 5.1

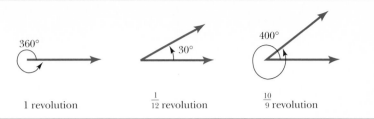

FIGURE 5.2

just as there is more than one unit of linear measure (feet, miles, meters, kilometers, and so on). The commonly used units of measure for angles are **revolution, degree,** and **radian.**

A measure of 1 *revolution* is, by definition, the amount of rotation needed for one full turn of a ray about its endpoint, so that the initial and terminal sides of the angle coincide. Historically, 1 revolution is taken to be an angle of *degree measure* 360, written 360°. Thus an angle measuring 1° is $\frac{1}{360}$ part of 1 revolution, an angle measuring 30° is $\frac{30}{360}$ or $\frac{1}{12}$ of a revolution, one measuring 400° is $\frac{400}{360}$ or $\frac{10}{9}$ of a revolution, and so on (see Figure 5.2).

A **right angle** measures 90°, and a **straight angle** measures 180°. A positive angle measuring less than 90° is called an **acute angle,** and one measuring more than 90° but less than 180° is called an **obtuse angle.** If the sum of two positive angles is 90°, the angles are called **complementary angles.** If the sum of two positive angles is 180°, the angles are called **supplementary angles.**

Just as a kilometer can be subdivided into meters and further subdivided into centimeters, or a yard can be subdivided into feet and then inches, a degree can be subdivided into units called **minutes,** denoted by ′, and **seconds,** denoted by ″. There are 60 minutes in 1 degree and 60 seconds in 1 minute.

$$1° = 60' = 3600'', \qquad 1' = 60''$$

Example 1   Find the angle A that is complementary to the angle $B = 27°20'14''$.

Solution
We know that the sum of A and B must be 90°. We rewrite 90° as 89°60′ and then as 89°59′60″ so that we can subtract B from 90°.

$$
\begin{array}{llcl}
90°: & 90° & \xrightarrow{\text{Rewriting}} & 89°59'60'' \\
B: & \underline{27°20'14''} & & \underline{27°20'14''} \\
A: & & & 62°39'46''
\end{array}
$$

Thus angles $A = 62°39'46''$ and $B = 27°20'14''$ are complementary. ❑

Many calculators use only decimals to denote fractional parts of a degree. Thus it is desirable to become acquainted with the procedure for transforming fractional parts of a degree to minutes and seconds, and vice versa.

Example 2 Transform $23.24°$ to degree–minute–second format.

Solution

$$23.24° = 23° + 0.24°$$

$$= 23° + 0.24\,(60)' \qquad \text{Transforming decimal degrees to minutes}$$

$$= 23° + 14.4'$$

$$= 23° + 14' + 0.4'$$

$$= 23° + 14' + 0.4\,(60)'' \qquad \text{Transforming decimal minutes to seconds}$$

$$= 23°14'24'' \qquad\qquad\qquad ❏$$

Example 3 Convert $143°27'15''$ to decimal–degree format.

Solution

$$143°27'15'' = 143° + 27' + 15''$$

$$= 143° + \frac{27°}{60} + \frac{15}{3600}^{°} \qquad \text{Converting minutes and seconds to degrees}$$

$$= 143° + 0.45000\ldots° + 0.00416\ldots°$$

$$= 143.4542° \qquad\qquad\qquad ❏$$

Technically the last = symbol should be $\approx$ for "approximately equals." Here and elsewhere we use = for convenience in dealing with approximations.

The sum of the three angles in any triangle is $180°$. Consequently the two acute angles in a *right triangle* are complementary angles. An **oblique triangle** is a triangle that is not a right triangle. An **isosceles triangle** is one that has two equal sides (and hence two equal angles). If all three sides are equal (or all three angles are equal), the triangle is called **equilateral.**

Two triangles are said to be **similar** if two angles of one triangle are equal to two angles of the other triangle. The sides of similar triangles satisfy the following property:

The corresponding sides of similar triangles are proportional.

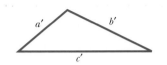

FIGURE 5.3

If the sides of similar triangles are labeled as in Figure 5.3, then

$$\frac{a}{a'} = \frac{b}{b'} = \frac{c}{c'}.$$

This property is frequently useful in practical applications. (See Exercises 69–74 at the end of this section.)

Degree measure is commonly used in fields such as surveying and navigation, but radian measure is more conventional in applications that require the use of the calculus.

In order to formulate the definition of a radian, we consider angles that have vertex at the center of a circle. Such angles are called **central angles.** Since an angle measuring $1°$ is $\frac{1}{360}$ of a revolution, a central angle of $1°$ subtends an arc with length $\frac{1}{360}$ of the circumference C of the circle. The radian unit of measure is defined in terms of the length of the intercepted arc.

Definition 5.1

A central angle has measure 1 **radian** if it intercepts an arc with length equal to the radius of the circle.

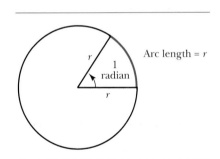

FIGURE 5.4

A central angle of measure 1 radian is drawn in Figure 5.4.

It follows from Definition 5.1 that an angle has measure 2 radians if it intercepts an arc that has length $2r$ when placed with its vertex at the center of a circle having radius r. Similarly, a central angle of 3 radians intercepts an arc of length $3r$. The number of radians in 1 revolution is the number of times the radius r can be measured off along the circumference of the circle. Since $C = 2\pi r$, we conclude that there are 2π radians in 1 revolution and that

$$2\pi \text{ radians} = 1 \text{ revolution} = 360°.$$

Dividing by 2, we obtain the **basic equation** relating radians and degrees:

$$\pi \text{ radians} = 180°.$$

Dividing both sides of the basic equation first by π and then by 180, we get

$$1 \text{ radian} = \frac{180°}{\pi} \approx 57.2958° \approx 57°17'45'',$$

$$1° = \frac{\pi}{180} \text{ radians} \approx 0.0174533 \text{ radians}.$$

These equations lead to the following conversion procedures.

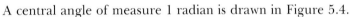

1. To change radians to degrees, multiply by $180/\pi$.
2. To change degrees to radians, multiply by $\pi/180$.

Example 4   Convert each of the following to degree measure.

a) $-\dfrac{\pi}{3}$ radians   b) $\dfrac{3\pi}{4}$ radians

Solution

In each part, we multiply by $\dfrac{180}{\pi}$.

a) $-\dfrac{\pi}{3}$ radians $= -\left(\dfrac{\pi}{3} \cdot \dfrac{180}{\pi}\right)^{\circ} = -60°$

b) $\dfrac{3\pi}{4}$ radians $= \left(\dfrac{3\pi}{4} \cdot \dfrac{180}{\pi}\right)^{\circ} = 135°$ ❏

Example 5   Convert each of the following to radian measure in terms of π.

a)   210°   b)   3°56′15″

Solution

In each part, we multiply by $\dfrac{\pi}{180}$.

a) $210° = \left(210 \cdot \dfrac{\pi}{180}\right)$ radians $= \dfrac{7\pi}{6}$ radians

b) $3°56′15″ = 3° + \dfrac{56°}{60} + \dfrac{15}{3600}^{\circ} = 3° + \dfrac{56°}{60} + \dfrac{1}{240}^{\circ} = \dfrac{945°}{240}$

     $= \left(\dfrac{945}{240} \cdot \dfrac{\pi}{180}\right)$ radians $= \dfrac{7\pi}{320}$ radians   ❏

Radian measures are frequently left in terms of π, as they were in the preceding examples. **It is customary to omit the word *radian* when radian measure is being used.**[1] Thus we would write equations such as

$$30° = \dfrac{\pi}{6}, \quad 45° = \dfrac{\pi}{4}, \quad 60° = \dfrac{\pi}{3}, \quad 90° = \dfrac{\pi}{2}.$$

It is sometimes necessary to work with radian measures that are expressed as decimal numbers instead of multiples of π. In such cases we may have to convert a decimal radian measure to degrees, or vice versa. Some calculators have a conversion key that makes it possible to convert from either type of measure to the other. If such a calculator is

---

1. The word *radian* is sometimes abbreviated "rad."

not available, then we must multiply with the appropriate multiplier: $180/\pi$ or $\pi/180$. If accuracy to the nearest second (or $0.0001°$) is desired, then we must use an approximation for π at least as accurate as 3.14159. Even with this value, a conversion to decimal degrees may be off in the fourth decimal place.

Some illustrations of conversions going both ways are given in the following example.

Example 6   In making the following conversions, use 3.14159 as an approximation for π.

a)   Convert 2.05 radians to decimal degrees, rounding to four decimal places.

b)   Convert $14°35'22''$ to decimal radians, rounding to four decimal places.

Solution

a) 2.05 radians $= 2.05\,(180°/\pi) = 117.4564°$

b)   We first change $14°35'22''$ to decimal degrees, then multiply by $\pi/180$.

$$14°35'22'' = 14° + \frac{35°}{60} + \frac{22\ ^\circ}{3600} = 14° + 0.58333° + 0.00611°$$

$$= 14.5894° = 14.5894\left(\frac{\pi}{180}\right)\text{ radians} = 0.2546\text{ radians}$$

❏

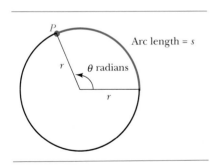

FIGURE 5.5

Consider a central angle of measure θ radians in a circle with radius r. As in Figure 5.5, let s denote the length of the arc intercepted by θ on the circle, and assume that r and s are measured in the same linear units (for instance, both in feet or both in miles).

We have seen that $s = r$ when $\theta = 1$ radian. It is not hard to see, then, that $s = 2r$ when $\theta = 2$ radians, $s = r/2$ when $\theta = \frac{1}{2}$ radian, and generally that

$$s = r\theta.$$

There are two important facts to keep in mind when using this formula.

  1.   **θ is the radian measure of the central angle.**
  2.   **s and r are measured in the same linear units.**

Example 7   A bicycle has wheels that are 28 inches in diameter. How far does the bicycle move as the wheels roll through an angle of $15°$?

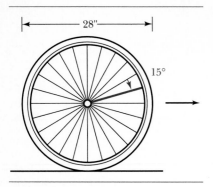

FIGURE 5.6

Solution

The distance that the bicycle moves is the same as the arc length intercepted on a wheel by a central angle of 15° (see Figure 5.6). The radius of a wheel is

$$r = \tfrac{1}{2}(28) = 14 \text{ inches,}$$

and

$$\theta = 15° = 15\left(\frac{\pi}{180}\right) = \frac{\pi}{12} \text{ radians.}$$

Thus the bicycle moves

$$s = 14\left(\frac{\pi}{12}\right) = \frac{7\pi}{6} = 3.7 \text{ inches,}$$

rounded to two digits.   ❏

If an object moves in a circular path, two speeds are involved: the rate at which distance is traveled along the circle, and the rate at which the object revolves about the center of the circle.

In Figure 5.5, suppose that the particle P has traveled at a constant rate along the arc length s intercepted by the central angle θ, and that t is the time that has elapsed during this motion. The ratio $v = s/t$ is called the **linear speed** of the particle, since it gives the rate at which distance is covered along the circular path. The ratio $\omega = \theta/t$ is called the **angular speed** of the particle, since it describes the rate at which the particle is revolving about the center of the circle.

If θ is the measure of the central angle *in radians*, then $s = r\theta$, where s and r are measured *in the same units*. When both sides of this equation are divided by t, we get

$$\frac{s}{t} = r \cdot \frac{\theta}{t},$$

or

$$v = r\omega.$$

This equation relates linear speed and angular speed. It must be kept in mind that

1. ω is angular speed in radians per unit time;
2. The time units used in v and ω must be the same;
3. The linear units used in v and r must be the same.

Example 8   A car has wheels 28 inches in diameter and is traveling at 45 miles per hour. Find the angular speed of the wheels in radians per minute, accurate to four digits.

Solution
The radius of a wheel is half the diameter.

$$r = \tfrac{1}{2}(28) = 14 \text{ inches}$$

We are given $v = 45$ miles per hour, and we need the same linear units in r and v. A good plan would be to change to feet in both cases, and also to change time units to minutes since we want ω in radians per minute.

$$r = 14 \text{ inches} = \frac{14}{12} \text{ feet} = \frac{7}{6} \text{ feet}$$

$$v = \frac{45 \text{ miles}}{1 \text{ hour}} \cdot \frac{5280 \text{ feet}}{1 \text{ mile}} \cdot \frac{1 \text{ hour}}{60 \text{ minutes}}$$

$$= \frac{45(5280) \text{ feet}}{60 \text{ minutes}}$$

$$= 3960 \text{ feet per minute}$$

Solving $v = r\omega$ for ω, we obtain $\omega = v/r$ and

$$\omega = \frac{3960}{\frac{7}{6}} = 3394 \text{ radians per minute,}$$

rounded to four digits.   ❏

Exercises 5.1

Find the angle θ such that θ and the given angle are complementary. (See Example 1.)

1. $42°$
2. $77°$
3. $14°25'$
4. $47°50'$
5. $18°42'14''$
6. $49°17'52''$
7. $18.27°$
8. $36.2582°$

Convert the given angle to degree–minute–second format. (See Example 2.)

9. $14.27°$
10. $-108.45°$
11. $-729.5055°$
12. $271.9340°$

Convert the given angle to decimal–degree format. (See Example 3.)

13. $-18°50'$
14. $-193°45'$
15. $53°14'26''$
16. $169°50'18''$

Convert each of the following radian measures to degree measure. (See Example 4.)

17. $\dfrac{\pi}{6}$
18. $\dfrac{\pi}{9}$
19. $-\dfrac{\pi}{12}$
20. $-\dfrac{\pi}{5}$
21. $-\dfrac{7\pi}{5}$
22. $-\dfrac{3\pi}{10}$
23. $\dfrac{11\pi}{4}$
24. $\dfrac{7\pi}{30}$

Convert each of the following to radian measure in terms of π. (See Example 5.)

25. $-90°$
26. $-60°$
27. $120°$
28. $135°$
29. $520°$
30. $1260°$
31. $28°7'30''$
32. $14°3'45''$

Convert the following radian measures to decimal degrees. Use 3.14159 as an approximation for π and round answers to three decimal places. (See Example 6.)

33. 3 radians   **34.** 4 radians

35. 0.513 radians   **36.** 1.67 radians

 Convert each of the following to decimal radians. Use $\pi = 3.14159$ and round answers to four decimal places. (See Example 6.)

37. 18°30′   **38.** 23°40′

39. 72°14′20″   **40.** 144°11′45″

Convert each of the following to radian measure.

41. 2 revolutions   **42.** 4 revolutions

43. $\frac{3}{2}$ revolutions   **44.** $\frac{9}{4}$ revolutions

Convert each of the following to revolutions.

45. $\frac{\pi}{36}$ radians   **46.** $\frac{\pi}{16}$ radians

47. $\frac{2\pi}{3}$ radians   **48.** $\frac{3\pi}{4}$ radians

In Exercises 49–54, s is the length of the arc intercepted by the central angle θ in a circle of radius r. Find the exact value of the missing variable. (See Example 7.)

49. $r = 213$ m, $\theta = \dfrac{5\pi}{3}$

50. $r = 246$ m, $\theta = \dfrac{5\pi}{6}$

51. $r = 1.8$ ft, $\theta = 210°$

52. $r = 6.3$ ft, $\theta = 120°$

53. $s = 9$ cm, $\theta = 2$

54. $s = 13$ cm, $\theta = 4$

For the following exercises, see Example 7.

55. A wagon has wheels that are 3.6 feet in diameter. How far does the wagon move as the wheels turn through 72°?

56. A wheel on a truck has a radius of 1.5 feet. How far will the truck move if the wheel turns through 36°?

57. How far does the tip of the 18-inch pendulum shown travel as it swings through an angle of 10°?

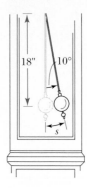

EXERCISE 57

58. The minute hand on a tower clock is 2.0 feet in length. How far does its tip travel in 5 minutes?

59. Madison, Wisconsin, is almost due north of Poplarville, Mississippi. The latitude of Madison is 43°N, and that of Poplarville is 31°N (see the accompanying figure). Use 4000 miles as the radius of the earth, and find the distance between the two cities, rounded to the nearest 10 miles.

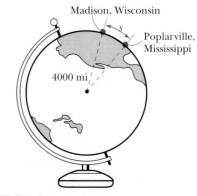

EXERCISE 59

60. (See Exercise 59.) Memphis, Tennessee, is almost due north of New Orleans, Louisiana. The latitude of Memphis is 35°N and that of New Orleans is 30°N. Find the distance between the cities, rounded to the nearest 10 miles.

For the following exercises, see Example 8.

61. A seat on the ferris wheel shown in the figure is located 20 feet from the axle. If the wheel turns at the rate of 18 degrees per second, find the linear speed of the seat in miles per hour.

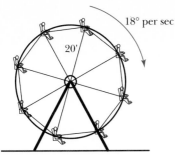

18° per sec

20'

EXERCISE 61

62. The weight in the figure below is being raised by a rope that passes over a pulley with a diameter of 10 inches. If the pulley turns through 240 degrees per second, find the rate at which the weight is rising in feet per minute.

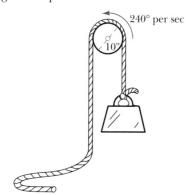

240° per sec

10"

EXERCISE 62

63. The front and rear wheels of a bicycle have diameters of 20 inches and 30 inches, respectively. If the bike is moving so that the angular speed of the rear wheel is 60 revolutions per minute, find the angular speed of the front wheel in radians per second.

64. The gear having radius 3 inches in the figure is meshed with another gear having radius 7 inches. If the larger gear is turning at 84 revolutions per minute, find the angular speed of the smaller gear in radians per second.

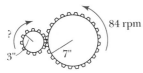

? 3"   7"   84 rpm

EXERCISE 64

65. Find the diameter d of the pulley shown below that is driven at 360 revolutions per minute by a belt moving at 40 feet per second.

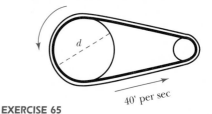

d

40' per sec

EXERCISE 65

66. The wheels on a truck turn at the rate of $30/\pi$ revolutions per second when the truck is traveling at 90 feet per second. Find the diameter of a wheel.

67. Quito, Ecuador, is located on the equator of the earth. Take 4000 miles as the radius of the earth and find the linear speed, to the nearest 10 miles per hour, of a person in Quito due to the rotation of the earth.

68. (See Exercise 67.) What must be the linear speed of a satellite if it is to stay in an orbit that keeps it 300 miles directly above Quito, Ecuador?

Use similar triangles to solve the following problems.

69. If the 4-foot fence post shown in the figure casts a 1.8-foot shadow at the same time a pine tree casts a 36-foot shadow, how tall is the pine tree?

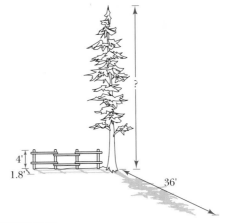

?

4'
1.8'   36'

EXERCISE 69

70. If a 4-foot boy casts a 7-foot shadow at the same time a vertical cliff casts a 91-foot shadow, how high is the cliff?

71. Sarah and Joey are positioned as indicated below, so they see each other's reflections in the mirror. How far is Joey standing from the wall?

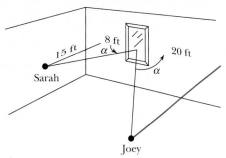

EXERCISE 71

72. Matt places a mirror on the pavement in such a position that he sees the top of a flagpole. The mirror is lying 3.1 feet from Matt's feet and 12.4 feet from the base of the flagpole. How tall is the flagpole if Matt's eye level is 5.2 feet high?

73. To find the distance across a slough, a surveyor drove stakes at points A, B, C, D, and E (see the accompanying figure) so that the line segment joining points A and E was parallel to the line segment joining points D and B. He measured and found the distance from A to C to be 42 yards, from B to D to be 15 yards, and from B to C to be 18 yards. Find the distance AE across the slough.

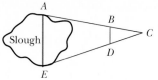

EXERCISE 73

74. To find the distance across a sinkhole, a surveyor drove stakes at points A, B, C, D, and E (see the figure) so that the line segment joining points A and E was parallel to the line segment joining points B and D. She measured and found the distance from A to C to be 215 meters, from B to D to be 60 meters, and from B to C to be 50 meters. Find the distance AE across the sinkhole.

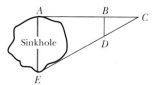

EXERCISE 74

5.2 | Trigonometric Functions of Angles

The following terminology plays an important role in much of the subsequent material.

Definition 5.2

An angle is in **standard position** if its vertex is placed at the origin of the coordinate system and its initial side lies along the positive x-axis.

An angle in standard position is called a **first, second, third,** or **fourth quadrant angle,** depending on whether its terminal side lies in the first, second, third, or fourth quadrant, respectively.

For example, $150° = 5\pi/6$ is a second quadrant angle, and $-20° = -\pi/9$ is a fourth quadrant angle.

> **Definition 5.3**
>
> Two angles are called **coterminal angles,** if, when placed in standard position, their terminal sides coincide.

For a given angle θ, in degrees, the expression

$$\theta + n \cdot 360°, \qquad n \text{ any integer,}$$

characterizes all angles that are coterminal to θ. If θ is measured in radians, the corresponding expression for coterminal angles is

$$\theta + 2\pi n, \qquad n \text{ any integer.}$$

Suppose now that θ is an angle in standard position, as shown in Figure 5.7. Let (x, y) be a point, other than the origin, on the terminal side of θ. Then the distance r from the origin to (x, y) is always positive and is given by

$$r = \sqrt{x^2 + y^2},$$

Six possible ratios can be formed by using the three values x, y, and r. They are

$$\frac{y}{r}, \quad \frac{x}{r}, \quad \frac{y}{x}, \quad \frac{x}{y}, \quad \frac{r}{x}, \quad \text{and} \quad \frac{r}{y}.$$

These six ratios are given the special names sine θ, cosine θ, tangent θ, cotangent θ, secant θ, and cosecant θ, with respective abbreviations sin θ, cos θ, tan θ, cot θ, sec θ, and csc θ. This is formalized in the following definition.

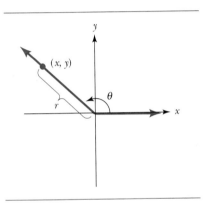

FIGURE 5.7

> **Definition 5.4**
>
> Let θ be an angle in standard position, and let (x, y) be a point, other than the origin, on the terminal side of θ. With $r = \sqrt{x^2 + y^2}$, the **trigonometric functions** of θ are defined as follows:
>
> $$\sin \theta = \frac{y}{r}, \qquad \cot \theta = \frac{x}{y},$$
>
> $$\cos \theta = \frac{x}{r}, \qquad \sec \theta = \frac{r}{x},$$
>
> $$\tan \theta = \frac{y}{x}, \qquad \csc \theta = \frac{r}{y}.$$

The values of the six trigonometric functions of an angle θ in standard position are independent of the choice of the point (x, y) on the terminal side of θ. This can be shown by using similar triangles. Hence the value of $\sin \theta = y/r$ is *unique* for a given θ, and we are justified in using the term *function*. Similar remarks can be made concerning the other trigonometric functions.

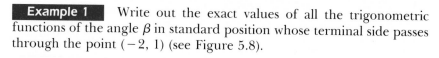

Example 1   Write out the exact values of all the trigonometric functions of the angle β in standard position whose terminal side passes through the point $(-2, 1)$ (see Figure 5.8).

Solution

With $x = -2$ and $y = 1$, we find $r = \sqrt{(-2)^2 + 1^2} = \sqrt{5}$. Then the exact values of the trigonometric functions of β are

$$\sin \beta = \frac{y}{r} = \frac{1}{\sqrt{5}} = \frac{\sqrt{5}}{5}, \qquad \cot \beta = \frac{x}{y} = \frac{-2}{1} = -2,$$

$$\cos \beta = \frac{x}{r} = \frac{-2}{\sqrt{5}} = \frac{-2\sqrt{5}}{5}, \qquad \sec \beta = \frac{r}{x} = \frac{\sqrt{5}}{-2} = -\frac{\sqrt{5}}{2},$$

$$\tan \beta = \frac{y}{x} = \frac{1}{-2} = -\frac{1}{2}, \qquad \csc \beta = \frac{r}{y} = \frac{\sqrt{5}}{1} = \sqrt{5}. \qquad \square$$

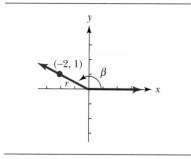

FIGURE 5.8

The signs of the trigonometric function values of θ depend on the quadrant in which θ terminates. Since r is *always* positive, the sign of $\sin \theta$ and $\csc \theta$ depends only on the sign of y. Hence $\sin \theta$ and $\csc \theta$ are both positive whenever y is positive—that is, whenever θ is in quadrants I or II. Sin θ and csc θ are both negative in quadrants III or IV.

The signs of $\cos \theta$ and $\sec \theta$ depend only on the sign of x. Since x is positive in quadrants I or IV, $\cos \theta$ and $\sec \theta$ are positive there and negative in quadrants II or III.

The remaining two functions, tangent and cotangent, depend on both x and y. We see that $\tan \theta$ and $\cot \theta$ are positive whenever x and y have the same sign and negative whenever x and y have opposite signs. Thus whenever θ is a first or third quadrant angle, $\tan \theta$ and $\cot \theta$ are positive; whenever θ is a second or fourth quadrant angle, $\tan \theta$ and $\cot \theta$ are negative. These results are summarized in Figure 5.9.

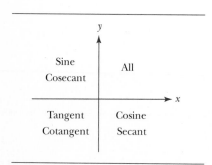

FIGURE 5.9

Example 2   If $\sin \theta < 0$ and $\cos \theta > 0$, in which quadrant does θ terminate?

Solution

Sin θ is negative in quadrants III or IV; cos θ is positive in quadrants I or IV. Thus for $\sin \theta < 0$ and $\cos \theta > 0$, θ must be a fourth quadrant angle. $\qquad \square$

The next example illustrates how the values of all the trigonometric functions of an angle θ can be determined if one function value is known, along with the quadrant in which θ terminates.

Example 3   Determine the values of all the trigonometric functions of the fourth quadrant angle α if $\cos \alpha = \frac{8}{17}$.

Solution

We must determine the values x, y, and r where (x, y) is a point on the terminal side of α and r is the distance from the origin to (x, y). Since

$$\cos \alpha = \frac{x}{r} = \frac{8}{17},$$

we choose x and r so that their ratio is 8 to 17. The value of r must be positive. Also x must be positive since α terminates in the fourth quadrant. Hence we choose $x = 8$ and $r = 17$. To determine y, we use $r^2 = x^2 + y^2$. Thus

$$289 = 64 + y^2$$

$$225 = y^2$$

$$\pm 15 = y.$$

The value of y is chosen to be negative since α terminates in quadrant IV. Now with $x = 8$, $y = -15$, and $r = 17$, we write out the values of all the other trigonometric functions.

$$\sin \alpha = -\tfrac{15}{17} \qquad \sec \alpha = \tfrac{17}{8}$$

$$\tan \alpha = -\tfrac{15}{8} \qquad \csc \alpha = -\tfrac{17}{15}$$

$$\cot \alpha = -\tfrac{8}{15}$$

Recall that the values of the trigonometric functions are independent of the choice of the point on the terminal side of the angle. Thus in Example 3 we could have used the values $x = 16$ and $r = 34$, since the ratio of x to r is

$$\frac{x}{r} = \frac{16}{34} = \frac{8}{17}.$$

Later in this chapter we shall find values of trigonometric functions by using a calculator or a table of values. However, function values for certain special angles can be found by using the definitions. These special angles are used so much in future work that it is worthwhile to learn how to find their function values without using tables or a calculator.

Among the special angles are those whose terminal sides lie along one of the coordinate axes. Angles such as these, whose terminal sides do not lie in any quadrant, are called **quadrantal angles.**

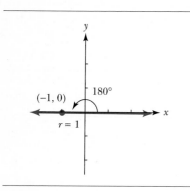

FIGURE 5.10

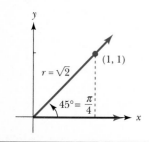

FIGURE 5.11

Example 4   Determine the values of the trigonometric functions of 180°.

Solution

Figure 5.10 shows 180° placed in standard position, with the point $(-1, 0)$ on its terminal side. With $x = -1$, $y = 0$, we find $r = \sqrt{(-1)^2 + 0^2} = 1$. Then, remembering that division by 0 is undefined, we have

$$\sin 180° = \frac{0}{1} = 0, \qquad \cot 180° = \frac{-1}{0} = \text{undefined},$$

$$\cos 180° = \frac{-1}{1} = -1, \qquad \sec 180° = \frac{1}{-1} = -1,$$

$$\tan 180° = \frac{0}{-1} \doteq 0, \qquad \csc 180° = \frac{1}{0} = \text{undefined}. \qquad \square$$

The values of the trigonometric functions of any quadrantal angle are found by using the same procedure as in Example 4.

We use some knowledge of geometry to determine the values of the trigonometric functions of the special angles 30°, 45°, and 60°. In the right triangle in Figure 5.11 with 45° acute angles and legs of length 1, the hypotenuse has length $c = \sqrt{1^2 + 1^2} = \sqrt{2}$. If we place this triangle in a coordinate system with one of the 45° angles in standard position, then we find that the point $(1, 1)$ lies on the terminal side of a 45° angle. Using $r = \sqrt{2}$, we write out the values of the trigonometric functions of $45° = \pi/4$.

$$\sin \frac{\pi}{4} = \sin 45° = \frac{1}{\sqrt{2}} = \frac{\sqrt{2}}{2} \qquad \cot \frac{\pi}{4} = \cot 45° = 1$$

$$\cos \frac{\pi}{4} = \cos 45° = \frac{1}{\sqrt{2}} = \frac{\sqrt{2}}{2} \qquad \sec \frac{\pi}{4} = \sec 45° = \sqrt{2}$$

$$\tan \frac{\pi}{4} = \tan 45° = \frac{1}{1} = 1 \qquad \csc \frac{\pi}{4} = \csc 45° = \sqrt{2}$$

To obtain the values of the trigonometric functions of 30°, we place a 30°–60° right triangle in the coordinate system as shown in Figure 5.12(a), with the 30° angle in standard position. If the length of the hypotenuse is 2, then the length of the shorter side is 1, and the length of the remaining side is $\sqrt{3}$. With this information we can write out the values of the trigonometric functions of 30°. Similarly, in Figure 5.12(b) we position the 60° angle in standard position to obtain its function values.

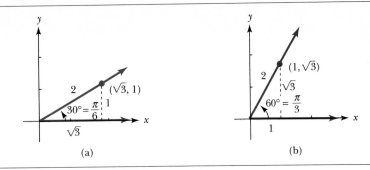

FIGURE 5.12

$$\sin \frac{\pi}{6} = \sin 30° = \frac{1}{2}$$

$$\cos \frac{\pi}{6} = \cos 30° = \frac{\sqrt{3}}{2}$$

$$\tan \frac{\pi}{6} = \tan 30° = \frac{1}{\sqrt{3}} = \frac{\sqrt{3}}{3}$$

$$\cot \frac{\pi}{6} = \cot 30° = \sqrt{3}$$

$$\sec \frac{\pi}{6} = \sec 30° = \frac{2}{\sqrt{3}} = \frac{2\sqrt{3}}{3}$$

$$\csc \frac{\pi}{6} = \csc 30° = 2$$

(a)

$$\sin \frac{\pi}{3} = \sin 60° = \frac{\sqrt{3}}{2}$$

$$\cos \frac{\pi}{3} = \cos 60° = \frac{1}{2}$$

$$\tan \frac{\pi}{3} = \tan 60° = \sqrt{3}$$

$$\cot \frac{\pi}{3} = \cot 60° = \frac{1}{\sqrt{3}} = \frac{\sqrt{3}}{3}$$

$$\sec \frac{\pi}{3} = \sec 60° = 2$$

$$\csc \frac{\pi}{3} = \csc 60° = \frac{2}{\sqrt{3}} = \frac{2\sqrt{3}}{3}$$

(b)

The next definition is invaluable in determining the values of the trigonometric functions of many angles.

Definition 5.5

Suppose θ is an angle in standard position and is not a quadrantal angle. The **related** (or **reference**) **angle** θ' for the angle θ is the positive acute angle that the terminal side of θ makes with the x-axis.

Example 5   Determine the related angle θ' for the given angle θ.

a) $\theta = 115°$ b) $\theta = \frac{4\pi}{3}$ c) $\theta = 290°$ d) $\theta = -\frac{5\pi}{3}$

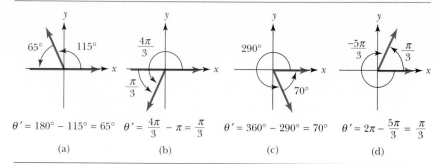

$$\theta' = 180° - 115° = 65° \quad \theta' = \frac{4\pi}{3} - \pi = \frac{\pi}{3} \qquad \theta' = 360° - 290° = 70° \qquad \theta' = 2\pi - \frac{5\pi}{3} = \frac{\pi}{3}$$

(a)　　　　　　　　(b)　　　　　　　　(c)　　　　　　　　(d)

FIGURE 5.13

Solution
First visualize the situation with a sketch such as those shown in Figure 5.13. Notice that the related angle θ' is *always* drawn between the terminal side of θ and the x-axis, *never* the y-axis. ❑

Example 6　　Write out the exact values of the trigonometric functions of $5\pi/4$.

Solution
The related angle for $5\pi/4$ is $\pi/4 = 45°$, one of the special angles. When we place a 45° right triangle as shown in Figure 5.14, we see that $(-1, -1)$ is a point on the terminal side of $5\pi/4$ with $r = \sqrt{2}$. Using the definition of the trigonometric functions, we can write out their values.

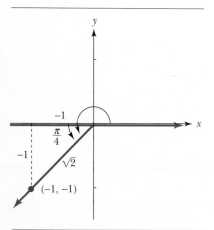

FIGURE 5.14

$$\sin\frac{5\pi}{4} = \frac{-1}{\sqrt{2}} = -\frac{\sqrt{2}}{2} \qquad \cot\frac{5\pi}{4} = \frac{-1}{-1} = 1$$

$$\cos\frac{5\pi}{4} = \frac{-1}{\sqrt{2}} = -\frac{\sqrt{2}}{2} \qquad \sec\frac{5\pi}{4} = \frac{\sqrt{2}}{-1} = -\sqrt{2}$$

$$\tan\frac{5\pi}{4} = \frac{-1}{-1} = 1 \qquad \csc\frac{5\pi}{4} = \frac{\sqrt{2}}{-1} = -\sqrt{2}$$

❑

　　When one of the special triangles is positioned with hypotenuse coincident to the terminal side of θ and one side lying along the x-axis, it is called the **related** (or **reference**) **triangle.** The related angle θ' for θ is the angle in the related triangle between the hypotenuse and the side along the x-axis.
　　The values of the trigonometric functions of an angle θ can be found by considering the values of the trigonometric functions of the related angle. Let θ_1, θ_2, θ_3, and θ_4 be angles in quadrants I, II, III, and IV,

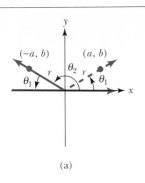

(a)

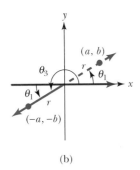

(b)

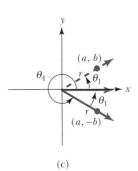

(c)

FIGURE 5.15

respectively, each having related angle the same measure as θ_1 (see Figure 5.15).

Let (a, b) be a point on the terminal side of θ_1, with $r = \sqrt{a^2 + b^2}$. Then:

1. The point $(-a, b)$ lies on the terminal side of θ_2, r units away from the origin;

2. The point $(-a, -b)$ lies on the terminal side of θ_3, r units away from the origin;

3. The point $(a, -b)$ lies on the terminal side of θ_4, r units away from the origin.

The sine of each of the angles θ_2, θ_3, and θ_4 can be expressed in terms of $\sin \theta_1$.

$$\sin \theta_1 = \frac{b}{r}$$

$$\sin \theta_2 = \frac{b}{r} = \sin \theta_1$$

$$\sin \theta_3 = \frac{-b}{r} = -\frac{b}{r} = -\sin \theta_1$$

$$\sin \theta_4 = \frac{-b}{r} = -\frac{b}{r} = -\sin \theta_1$$

The values of the sine of each angle θ_1, θ_2, θ_3, and θ_4 are the same except for sign. Each sign depends on the quadrant in which the corresponding angle terminates. As before, the sine function assumes positive values for angles in quadrants I or II and negative values in quadrants III or IV.

Similar remarks can be made for each of the other trigonometric functions, and we have the following result.

Related Angle Theorem

The value of any trigonometric function of an angle θ is equal to the value of the corresponding trigonometric function of the related angle, except possibly for the sign. The sign depends on the quadrant θ is in and can be determined by using the diagram in Figure 5.9.

Example 7   Write each of the following in terms of the same trigonometric function of a related angle.

a) $\sin 261°$ b) $\cos \dfrac{35\pi}{6}$ c) $\tan (-218°30')$

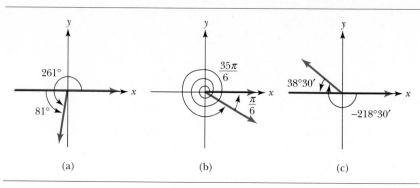

(a)   (b)   (c)

FIGURE 5.16

Solution

The related angle for each angle is shown in the various parts of Figure 5.16.

a) Since the sine function assumes negative values in quadrant III,

$$\sin 261° = -\sin 81°.$$

b) The cosine function values are positive in quadrant IV. So

$$\cos \frac{35\pi}{6} = \cos \frac{\pi}{6}.$$

c) The tangent function values are negative in quadrant II. Thus

$$\tan(-218°30') = -\tan 38°30'. \qquad \square$$

Exercises 5.2 _____

Determine the exact values of the six trigonometric functions of an angle θ in standard position if the terminal side of θ goes through the given point. (See Example 1.)

1.

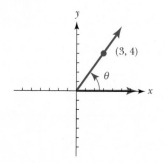

2.

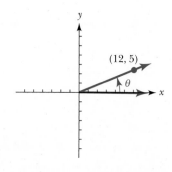

3.

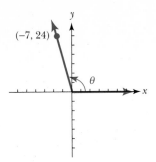

4.

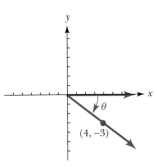

5.

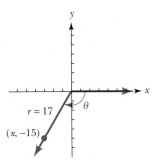

6.

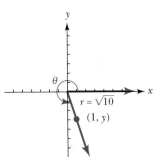

7. $(-\sqrt{2}, -7)$

8. $(\sqrt{10}, -2)$

9. $(x, 1), \quad x > 0$

10. $(1, y), \quad y > 0$

11. $(x, \sqrt{1 - x^2}), \quad x > 0$

12. $(\sqrt{1 - y^2}, y), \quad y > 0$

Use a calculator to determine the approximate values of the six trigonometric functions of an angle α in standard position if the terminal side of α passes through the given point. Round your answers to three decimal places.

13. $(-1.271, 3.122)$

14. $(-0.511, -0.698)$

15. $(10.211, -35.513)$

16. $(-44.821, 13.514)$

Determine the quadrant(s) in which the angle lies if the following conditions are satisfied. (See Example 2.)

17. $\sin \alpha > 0$ and $\sec \alpha > 0$

18. $\cos \beta < 0$ and $\tan \beta < 0$

19. $\cot \gamma > 0$ and $\sin \gamma > 0$

20. $\sin \theta < 0$ and $\tan \theta < 0$

21. $\tan \alpha < 0$ and $\csc \alpha > 0$

22. $\cot \theta > 0$ and $\sec \theta < 0$

23. $\sin \alpha < 0$ and $\csc \alpha < 0$

24. $\sec \alpha > 0$ and $\cos \alpha > 0$

25. $\tan \beta < 0$

26. $\sec \alpha < 0$

Determine the exact values of all the trigonometric functions of the angle that terminates in the given quadrant. (See Example 3.)

27. $\sin \alpha = 3/5$, Q I

28. $\tan \beta = 4/3$, Q I

29. $\cos A = -7/25$, Q II

30. $\cot B = -8/15$, Q II

31. $\tan \theta = 1/3$, Q III

32. $\cos \alpha = -2/5$, Q III

33. $\tan \alpha = -2/3$, Q IV

34. $\csc \alpha = -3$, Q IV

35. $\cos \theta = -0.4$, Q III

36. $\cot \theta = -1.3$, Q IV

37. $\sin \theta = y$, Q I

38. $\cos \theta = x$, Q I

39. $\sec \theta = 1/x$, Q II

40. $\tan \theta = y$, Q IV

Determine the values of the trigonometric functions of each of the following. (See Example 4.)

41. $360°$

42. $-90°$

43. $-\dfrac{5\pi}{2}$

44. 7π

Determine the related angle of each of the given angles. (See Example 5.)

45. $247°$

46. $223°$

47. $\dfrac{7\pi}{5}$

48. $\dfrac{7\pi}{10}$

49. $-\dfrac{8\pi}{9}$

50. $-423°$

Write each of the following in terms of the same trigonometric function of a related angle. (See Example 7.)

51. $\sin \dfrac{16\pi}{9}$ **52.** $\sin \dfrac{13\pi}{8}$

53. $\cos(-73°)$ **54.** $\cos 140°$

55. $\tan \dfrac{7\pi}{9}$ **56.** $\tan\left(-\dfrac{13\pi}{5}\right)$

57. $\csc(-20°30')$ **58.** $\cot 83°20'$

Write the exact values of the trigonometric functions of each angle. (See Example 6.)

59. $120°$ **60.** $135°$

61. $\dfrac{5\pi}{6}$ **62.** $\dfrac{7\pi}{6}$

63. $-60°$ **64.** $-135°$

65. $-\dfrac{5\pi}{4}$ **66.** $-\dfrac{4\pi}{3}$

Find the exact value of each of the following expressions. The notation $\sin^2 \theta$ means $(\sin \theta)^2$ and $\cos^2 \theta$ means $(\cos \theta)^2$.

67. $\cos^2 \dfrac{\pi}{3} - \sin^2 \dfrac{\pi}{3}$

68. $3 \tan 180° - 5 \csc 270°$

69. $2 \sec 0 - 3 \csc \dfrac{\pi}{2}$ **70.** $\sin \dfrac{3\pi}{4} + \cos \dfrac{5\pi}{4}$

71. $\sec \dfrac{7\pi}{6} + 6 \cot \dfrac{4\pi}{3}$

72. $5 \cot 150° + 6 \tan 300°$

Determine whether each of the following is true or false.

73. $\sin(30° + 60°) = \sin 30° + \sin 60°$

74. $\cos \dfrac{\pi}{3} = \cos^2 \dfrac{\pi}{6} - \sin^2 \dfrac{\pi}{6}$

75. $\cos \dfrac{2\pi}{3} = \tfrac{1}{2} \cos \dfrac{4\pi}{3}$ **76.** $\sin 150° = \tfrac{1}{2} \sin 300°$

5.3 Trigonometric Functions of Real Numbers

When the trigonometric functions were defined in Section 5.2, we emphasized that they were *functions of an angle*. There are many practical applications for the trigonometric functions defined in this way. But this situation is in strong contrast to the study of algebra, where only functions of numbers are considered.

In order that the methods of calculus and other more advanced courses may be applied to the trigonometric functions, the definitions of the trigonometric functions must be reformulated so they are *functions of real numbers* rather than functions of angles. Such a reformulation is the main goal of this section. The method that we use to achieve this goal utilizes a **unit circle,** that is, a circle that has radius 1 unit in length.

Consider a unit circle in a coordinate plane with center at the origin, as shown in each part of Figure 5.17. With each real number t, we associate a point $P(t) = (x, y)$ *on the unit circle* that is located by the following two-part rule.

1. If $t \geq 0$, $P(t)$ is at a distance t units from $(1, 0)$ along an arc of the circle in a counterclockwise direction.

2. If $t < 0$, $P(t)$ is at a distance $|t|$ units from $(1, 0)$ along an arc of the circle in a clockwise direction.

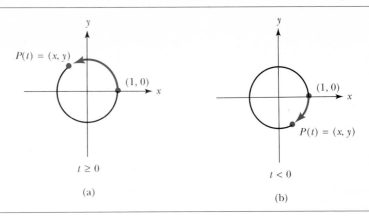

FIGURE 5.17

Figure 5.17 (a) indicates how $P(t)$ is located for $t \geq 0$, and Figure 5.17(b) indicates how $P(t)$ is located for $t < 0$. The rule may be summarized by saying that t is the directed arc length along the circle, with t positive in the counterclockwise direction and negative in the clockwise direction.

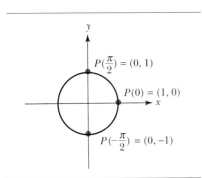

FIGURE 5.18

Example 1   Find the rectangular coordinates of each of the following points on the unit circle.

a) $P(0)$ b) $P(\pi/2)$ c) $P(-\pi/2)$

Solution
All three points are graphed in Figure 5.18.

a) With $t = 0$ in the first part of the rule, we see that $P(0) = (1, 0)$.

b) Since the unit circle has radius $r = 1$, its circumference is $C = 2\pi r = 2\pi$. The distance $t = \pi/2$ reaches one-fourth of the way around the circle since $\pi/2 = (1/4)(2\pi)$. Thus $P(\pi/2)$ is located on the y-axis at $(0, 1)$.

c) By the same reasoning as in part (b), $P(-\pi/2)$ is located one-fourth of the way around the circle in a clockwise direction, at $(0, -1)$. ❏

As pointed out in Example 1, the circumference of the unit circle is 2π. This means that the number $t + 2\pi$ corresponds to the same point on the unit circle as t. That is,

$$P(t + 2\pi) = P(t),$$

since adding 2π to the distance results in one additional full turn counterclockwise about the circle. Similarly,

$$P(t - 2\pi) = P(t).$$

These results generalize easily to any integral multiple of 2π, and

$$P(t) = P(t + 2\pi n)$$

for an arbitrary integer n (including negative n) and every real number t.

Example 2   Determine which quadrant contains the given point $P(t)$.

a) $P(13\pi/5)$ b) $P(12)$

Solution

In each part our first step is to locate the given value of t between multiples of 2π. After this is done, we can find a number between 0 and 2π that differs from t by a multiple of 2π and that locates the same point on the unit circle.

a) Since

$$\frac{13\pi}{5} = 2\pi + \frac{3}{5}\,\pi,$$

we have

$$P\left(\frac{13\pi}{5}\right) = P\left(\frac{3}{5}\,\pi + 2\pi\right) = P\left(\frac{3}{5}\,\pi\right).$$

Since $\frac{1}{2}\pi < \frac{3}{5}\pi < \pi$, $P(13\pi/5)$ lies in quadrant II.

b) Using $\pi = 3.1416$, we compute successive multiples of 2π and find that

$$2\pi = 6.2832, \qquad 4\pi = 12.5664.$$

This locates 12 between 2π and 4π. We can write 12 as

$$12 = 5.7168 + 6.2832$$
$$= 5.7168 + 2\pi,$$

so

$$P(12) = P(5.7168).$$

Therefore $P(12)$ lies in the fourth quadrant, since

$$\frac{3\pi}{2} < 5.7168 < 2\pi.$$

        ❏

The pairing that we have made between the real numbers t and the corresponding points $P(t)$ on the unit circle is frequently called the **wrap-**

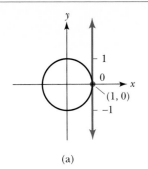

(a)

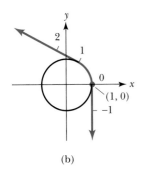

(b)

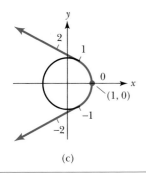

(c)

FIGURE 5.19

ping function. This name comes about because the association that we have made between numbers and points on the circle is the same as would occur if a number line were placed with its origin at (1, 0) and wrapped around the unit circle, as indicated in Figure 5.19. The positive part of the number line is wrapped in a counterclockwise direction, whereas the negative part is wrapped in a clockwise direction.

The wrapping function can be used in the following manner to define trigonometric functions of real numbers.

Definition 5.6

For any real number t, let $P(t) = (x, y)$ be the point on the unit circle that corresponds to t with the wrapping function. Using the same abbreviations as in Definition 5.4, the six trigonometric functions of t are defined as follows:

$$\sin t = y, \qquad \cot t = \frac{x}{y},$$

$$\cos t = x, \qquad \sec t = \frac{1}{x},$$

$$\tan t = \frac{y}{x}, \qquad \csc t = \frac{1}{y}.$$

Definition 5.6 reformulates the definitions of the trigonometric functions so that they appear as *functions of the real number t*, with no reference whatever to an angle.[2] This accomplishes the main goal of this section.

A natural question arises immediately, however. What, if any, relation exists between these new functions and the original trigonometric functions in Definition 5.4? This question is answered in the following two paragraphs.

Consider a point $P(t) = (x, y)$ on the unit circle that is located by the arc with directed length t, and let θ be the radian measure of the central angle in the unit circle that intercepts this arc. This is shown in Figure 5.20. Since $r = 1$ in the unit circle, it follows from the arc length formula $s = r\theta$ that

$$t = \text{arc length} = r\theta = 1 \cdot \theta = \theta.$$

That is, the measure of the central angle in radians is equal to the directed length of the intercepted arc on the unit circle.

---

2. The trigonometric functions defined in this way are sometimes called circular functions. However, the term *circular functions* has been used with different meanings in various texts.

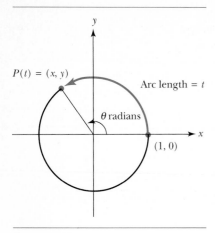

FIGURE 5.20

Referring to Figure 5.20 and using Definitions 5.4 and 5.6, we have

$$\sin \theta = \frac{y}{r} = \frac{y}{1} = \sin t,$$

$$\cos \theta = \frac{x}{r} = \frac{x}{1} = \cos t,$$

$$\tan \theta = \frac{y}{x} = \tan t.$$

In similar fashion, we find that $\csc \theta = \csc t$, $\sec \theta = \sec t$, and $\cot \theta = \cot t$. Thus we arrive at the following conclusion:

> Any trigonometric function of a real number t is equal to the same function of an angle with measure t radians.

For example, the value of $\sin t$ that Definition 5.6 gives for a real number t is the same as the value of $\sin t$ that Definition 5.4 gives for an angle of t radians.

Example 3 Find the rectangular coordinates of the point $P(-5\pi/6)$ on the unit circle.

Solution
According to Definition 5.6, the coordinates (x, y) of $P(t)$ are given by $x = \cos t$ and $y = \sin t$, for any value of t. For $P(-5\pi/6)$, we have

$$x = \cos\left(-\frac{5\pi}{6}\right) = -\frac{\sqrt{3}}{2},$$

$$y = \sin\left(-\frac{5\pi}{6}\right) = -\frac{1}{2}.$$ ❑

Exercises 5.3

Find the rectangular coordinates of each of the following points on the unit circle. (See Examples 1 and 3.)

1. $P(\pi)$ **2.** $P(3\pi/2)$ **3.** $P(-3\pi/2)$
4. $P(2\pi)$ **5.** $P(\pi/6)$ **6.** $P(-\pi/3)$
7. $P(-\pi/4)$ **8.** $P(\pi/4)$ **9.** $P(5\pi/6)$
10. $P(4\pi/3)$ **11.** $P(5\pi/3)$ **12.** $P(3\pi/4)$
13. $P(-3\pi/4)$ **14.** $P(-2\pi/3)$ **15.** $P(-7\pi/3)$
16. $P(-5\pi/3)$ **17.** $P(13\pi/4)$ **18.** $P(10\pi/3)$

19. $P(23\pi/6)$ **20.** $P(19\pi/6)$

In Exercises 21–28, determine the quadrant that contains the given point $P(t)$. (See Example 2.)

21. $P(17\pi/5)$ **22.** $P(22\pi/7)$
23. $P(-43\pi/9)$ **24.** $P(-37\pi/7)$
25. $P(10)$ **26.** $P(14)$
27. $P(-20)$ **28.** $P(-17)$

Find the exact value of the given trigonometric function.

29. $\tan 3\pi$

30. $\sec(-\pi)$

31. $\sin \dfrac{7\pi}{4}$

32. $\cos \dfrac{11\pi}{6}$

33. $\cot \dfrac{2\pi}{3}$

34. $\csc\left(-\dfrac{5\pi}{6}\right)$

35. $\cos \dfrac{7\pi}{6}$

36. $\tan \dfrac{11\pi}{6}$

Find the exact value of each of the following expressions. The notation $\sin^2 t$ means $(\sin t)^2$ and $\cos^2 t$ means $(\cos t)^2$.

37. $\sin^2 \dfrac{\pi}{2} + \cos^2 \dfrac{\pi}{2}$

38. $2 \sin \dfrac{\pi}{6} \cos \dfrac{\pi}{6}$

39. $2 \cos^2 \dfrac{\pi}{4} - 1$

40. $\cos \dfrac{11\pi}{6} - 4 \sin \dfrac{2\pi}{3}$

Determine whether each of the following is true or false.

41. $2 \sin \dfrac{\pi}{4} = \sin \dfrac{\pi}{2}$

42. $\cos \dfrac{2\pi}{3} = 2 \cos^2 \dfrac{\pi}{3} - 1$

43. $\tan \dfrac{2\pi}{3} = \dfrac{2 \tan \dfrac{\pi}{3}}{1 - \tan^2 \dfrac{\pi}{3}}$

44. $\tan \dfrac{\pi}{3} = \dfrac{1 - \cos \dfrac{2\pi}{3}}{\sin \dfrac{2\pi}{3}}$

45. Use the figure below to show that $\tan(-t) = -\tan t$ for $0 \le t < 2\pi$.

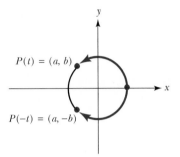

$P(t) = (a, b)$

$P(-t) = (a, -b)$

EXERCISE 45

46. a) Use the figure accompanying Exercise 45 to show that $\sin(-t) = -\sin t$ for $0 \le t < 2\pi$.
   b) Use the result in part (a) and the fact that $P(t) = P(t + 2\pi n)$ to show that $\sin(-t) = -\sin t$ for any real number t.

47. a) Use the figure given for Exercise 45 to show that $\cos(-t) = \cos t$ for $0 \le t < 2\pi$.
   b) Use the result in part (a) and the fact that $P(t) = P(t + 2\pi n)$ to show that $\cos(-t) = \cos t$ for any real number t.

5.4 | Some Fundamental Properties

In this section we obtain several useful results that are consequences of the definitions of the trigonometric functions. To derive the first of these results, we notice that the sine and cosecant functions are both defined in terms of y and r. When $y \ne 0$,

$$\sin \theta \csc \theta = \frac{y}{r} \cdot \frac{r}{y} = 1.$$

The equation $\sin \theta \csc \theta = 1$ is true for all values of θ for which $\sin \theta$ and $\csc \theta$ are defined. Equations that are always true whenever both sides are defined are called **identities.** We call the identity

$$\sin \theta \csc \theta = 1$$

a Reciprocal Identity, and we say that $\sin \theta$ and $\csc \theta$ are reciprocals.[3] Other forms of this relationship are

$$\sin \theta = \frac{1}{\csc \theta} \quad \text{and} \quad \csc \theta = \frac{1}{\sin \theta}.$$

Similarly, $\cos \theta$ and $\sec \theta$ are reciprocals, as are $\tan \theta$ and $\cot \theta$. We record these results, labeling them the Reciprocal Identities. It is understood that the equations hold only when each side is defined.

Reciprocal Identities

$$\sin \theta \, \csc \theta = 1 \qquad \cos \theta \, \sec \theta = 1 \qquad \tan \theta \, \cot \theta = 1$$

$$\sin \theta = \frac{1}{\csc \theta} \qquad \cos \theta = \frac{1}{\sec \theta} \qquad \tan \theta = \frac{1}{\cot \theta}$$

$$\csc \theta = \frac{1}{\sin \theta} \qquad \sec \theta = \frac{1}{\cos \theta} \qquad \cot \theta = \frac{1}{\tan \theta}$$

Two additional identities can be derived directly from the definitions of the trigonometric functions. Consider the quotients formed using $\sin \theta$ and $\cos \theta$.

$$\text{If } \cos \theta \neq 0, \qquad \frac{\sin \theta}{\cos \theta} = \frac{\dfrac{y}{r}}{\dfrac{x}{r}} = \frac{y}{x} = \tan \theta.$$

$$\text{If } \sin \theta \neq 0, \qquad \frac{\cos \theta}{\sin \theta} = \frac{\dfrac{x}{r}}{\dfrac{y}{r}} = \frac{x}{y} = \cot \theta.$$

These relationships are true for all θ, as long as all denominators are different from zero. We call these relationships the Quotient Identities.

Quotient Identities

$$\tan \theta = \frac{\sin \theta}{\cos \theta}, \qquad \cot \theta = \frac{\cos \theta}{\sin \theta}$$

---

3. If $a \neq 0$, the *reciprocal* of a is $1/a$.

The next example illustrates how useful the Reciprocal and Quotient Identities can be.

Example 1   Suppose $\sin \theta = -1/5$ and $\cos \theta = -2\sqrt{6}/5$. Use identities to determine the values of the remaining trigonometric functions of θ.

Solution

Using the Quotient Identities for $\tan \theta$ and $\cot \theta$, we have

$$\tan \theta = \frac{\sin \theta}{\cos \theta} = \frac{-\dfrac{1}{5}}{-\dfrac{2\sqrt{6}}{5}} = \frac{1}{2\sqrt{6}} = \frac{1\sqrt{6}}{2\sqrt{6}\,\sqrt{6}} = \frac{\sqrt{6}}{12},$$

$$\cot \theta = \frac{\cos \theta}{\sin \theta} = \frac{-\dfrac{2\sqrt{6}}{5}}{-\dfrac{1}{5}} = 2\sqrt{6}.$$

The Reciprocal Identities are useful for obtaining the values of $\sec \theta$ and $\csc \theta$.

$$\sec \theta = \frac{1}{\cos \theta} = \frac{5}{-2\sqrt{6}} = -\frac{5\sqrt{6}}{2\sqrt{6}\,\sqrt{6}} = -\frac{5\sqrt{6}}{12}$$

$$\csc \theta = \frac{1}{\sin \theta} = -5$$

Next we turn our attention to the ranges of the sine and cosine functions. Let θ be an angle in standard position, with (x, y) the point where the terminal side intersects the unit circle as shown in Figure 5.21. Then $\sin \theta = y$, and $\cos \theta = x$.

Since any point on the unit circle has x- and y-coordinates that satisfy

$$-1 \le x \le 1 \quad \text{and} \quad -1 \le y \le 1,$$

we have

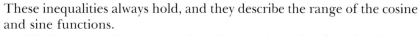

$$-1 \le \cos \theta \le 1 \quad \text{and} \quad -1 \le \sin \theta \le 1.$$

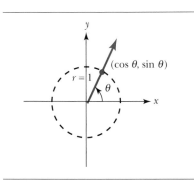

FIGURE 5.21

These inequalities always hold, and they describe the range of the cosine and sine functions.

We have seen that $\sec \theta$ and $\csc \theta$ are reciprocals of $\cos \theta$ and $\sin \theta$, respectively. It follows then that the ranges of the secant and cosecant functions can be described by the following inequalities.

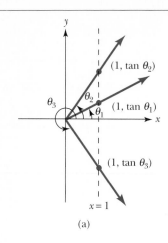

(a)

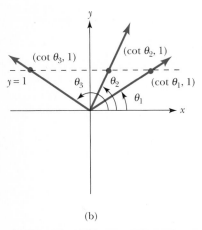

(b)

FIGURE 5.22

$$\text{sec } \theta \leq -1 \quad \text{or} \quad \text{sec } \theta \geq 1$$

$$\text{csc } \theta \leq -1 \quad \text{or} \quad \text{csc } \theta \geq 1$$

To study the range of the tangent function, we consider the angle in standard position, choosing the point (x, y) on the terminal side with $x = 1$, as shown in Figure 5.22(a). Then

$$\tan \theta = \frac{y}{x} = \frac{y}{1} = y,$$

and we see that **the tangent function ranges over all real numbers.** In Figure 5.22(b) we choose (x, y) with $y = 1$. Then

$$\cot \theta = \frac{x}{y} = \frac{x}{1} = x,$$

and we see that **the cotangent function also ranges over all real numbers.**

Example 2   Determine whether each of the following statements is possible or impossible.

a) $\sin \theta = 2$ b) $\cos \theta = 1$ and $\sec \theta = -1$

c) $\sec \theta = -3994$ d) $\csc \theta = 0$

Solution

a)   Impossible. The value of $\sin \theta$ cannot be larger than 1.

b)   Impossible. Since $\cos \theta$ and $\sec \theta$ are reciprocals, their product must be 1, not -1.

c)   Possible. The value -3994 lies in the range of the secant function.

d)   Impossible. The value 0 lies outside the range of the cosecant function.   ❑

Exercises 5.4

 Use identities and a calculator to determine the indicated function value. Round each result to four decimal places. (See Example 1.)

1. $\sin t = 0.2193$, $\csc t = ?$

2. $\sec \theta = 4.2711$, $\cos \theta = ?$

3. $\tan u = 2.1370$, $\cot u = ?$

4. $\cot \alpha = 1.5131$, $\tan \alpha = ?$

5. $\cos t = -0.9132$, $\sec t = ?$

6. $\sin \beta = -0.3196$, $\csc \beta = ?$

7. $\csc v = -1.4576$, $\sin v = ?$

8. $\tan \theta = -0.8352$, $\cot \theta = ?$

9. $\sin \beta = 0.4722$, $\cos \beta = 0.8815$, $\tan \beta = ?$

10. $\sin t = -0.5798$, $\cos t = -0.8148$, $\tan t = ?$

11. $\sin \theta = -0.7813$, $\cos \theta = 0.6242$, $\cot \theta = ?$

12. $\sin \alpha = 0.1391$, $\cos \alpha = -0.9903$, $\cot \alpha = ?$

13. $\sin A = 0.1521$, $\tan A = 0.1539$, $\cos A = ?$
14. $\cos B = -0.6731$, $\tan B = 1.0987$, $\sin B = ?$
15. $\sin D = 0.1216$, $\cot D = -8.1627$, $\cos D = ?$
16. $\cos C = -0.4561$, $\cot C = 0.5125$, $\sin C = ?$

Use identities to determine the exact values of the remaining trigonometric functions of the angle satisfying the given conditions. (See Example 1.)

17. $\sin \theta = -\frac{3}{5}$, $\cos \theta = \frac{4}{5}$
18. $\sin s = \frac{12}{13}$, $\cos s = -\frac{5}{13}$
19. $\tan \theta = \frac{12}{5}$, $\sec \theta = \frac{13}{5}$
20. $\cot t = \frac{8}{15}$, $\sin t = \frac{15}{17}$
21. $\csc \alpha = \frac{17}{8}$, $\tan \alpha = -\frac{8}{15}$
22. $\csc \theta = -\frac{5}{4}$, $\tan \theta = -\frac{4}{3}$
23. $\sin v = -\frac{7}{25}$, $\sec v = -\frac{25}{24}$
24. $\tan \beta = \frac{24}{7}$, $\sec \beta = -\frac{25}{7}$
25. $\cos t = -\frac{1}{3}$, $\tan t = -2\sqrt{2}$
26. $\sin A = \sqrt{21}/5$, $\tan A = -\sqrt{21}/2$
27. $\csc s = -\sqrt{2}$, $\tan s = -1$
28. $\cot B = -\sqrt{3}$, $\sin B = \frac{1}{2}$
29. $\sin \alpha = -\sqrt{7}/3$, $\cos \alpha = -\sqrt{2}/3$
30. $\sin u = -\sqrt{11}/4$, $\cos u = -\sqrt{5}/4$
31. $\cot \theta = -\sqrt{10}/2$, $\csc \theta = \sqrt{14}/2$
32. $\sec \theta = \frac{3}{2}$, $\sin \theta = -\sqrt{5}/3$

Determine whether each of the following statements is possible or impossible. (See Example 2.)

33. $\cos \alpha = 1.1$
34. $\sin t = 2.3$
35. $\tan s = 0$
36. $\sec \beta = 0$
37. $\csc \gamma = -1.1$
38. $\cot u = -1$
39. $\sec \theta = -0.99$
40. $\cos \alpha = -0.33333$
41. $\sin u = \frac{5}{4}$
42. $\cos \alpha = -\frac{10}{3}$
43. $\sec t = -\frac{1}{2}$
44. $\csc v = -\frac{3}{4}$
45. $\sin \alpha > 0$ and $\csc \alpha < 0$
46. $\cos s < 0$ and $\sec s > 0$
47. $\tan \theta < 0$ and $\cot \theta > 0$
48. $\csc \alpha > 0$ and $\sin \alpha < 0$
49. $\cos t = 3$ and $\sec t = \frac{1}{3}$
50. $\tan \alpha = 2$ and $\cot \alpha = -2$
51. $\sin \beta = \frac{2}{3}$ and $\csc \beta = -\frac{3}{2}$
52. $\cos v = -\frac{1}{4}$ and $\sec v = -4$

5.5   Values of Trigonometric Functions

Values of trigonometric functions for angles other than the special ones considered in Section 5.2 must be found by using a table or a calculator. A calculator is easier to use, if that option is available.

Scientific calculators have buttons labeled sin , cos , and tan that will yield values of the corresponding trigonometric functions of an angle. With many calculators the angle may be entered in either degrees or radian measure, but most calculators require that an angle be entered in decimal degrees instead of in degrees, minutes, and seconds. The following illustrations should clarify the procedures to be employed.

Example 1   Use a calculator to find the value of each of the following, and round the results to four digits.[4]

a) $\sin 39.7$ b) $\cos 14°23'$ c) $\csc(-214.4°)$

---

4. By a four-digit number, we mean that there are four digits when the number is written in scientific notation. That is, zeros used only to place the decimal are not counted.

Solution

a) To find sin 39.7, we set the calculator for radian measure, enter 39.7, and press the sin button. The value obtained is 0.9089275, which rounds to give

$$\sin 39.7 = 0.9089.$$

b) To find cos 14°23′, we first change 14°23′ to decimal degrees:

$$14°23′ = \left(14 + \frac{23}{60}\right)^\circ = 14.383333°.$$

With this angle entered and the calculator set for degrees, we press the cos button, which gives a displayed value of 0.96865546. Rounding gives

$$\cos 14°23′ = 0.9687.$$

c) To find csc (−214.4°), we use the fact that csc $\theta = 1/\sin \theta$. We set the calculator for degrees, enter −214.4°, and press the sin button to obtain

$$\sin (−214.4°) = 0.564967.$$

Pressing the 1/x button then yields

$$\csc (−214.4°) = \frac{1}{\sin (−214.4°)} = 1.770.$$

Values of trigonometric functions may also be found by using a table. Two such tables are found in the Appendix to this book. Table I gives values of all six trigonometric functions of θ at increments of 0.0029, from 0 to 1.5708. The increment 0.0029 in radian measure corresponds to 10 minutes in degree measure, and the interval from 0 to 1.5708 radians corresponds to the interval from 0° to 90°. Table II gives values of sin θ, cos θ, tan θ, and cot θ for θ in decimal degrees at increments of 0.1° from 0° to 90°. **The use of trigonometric function tables is described in Section A.3 of the Appendix.**

In using a calculator to find a value of the angle from a given function value, the procedure varies from one type of calculator to another. With any type, however, it is necessary to have a value of one of the functions sine, cosine, or tangent. If a function value different from these is given, one of the reciprocal identities can be used to obtain the value of either sine, cosine, or tangent.

We consider first the case in which a positive number is given as one of the values sin θ, cos θ, or tan θ, and we want to find a first quadrant angle θ in degrees or radians. First, the calculator is set for the desired measure and the number is entered into the calculator. Depending on the type of calculator in use, one of the following routines is then performed.

1. Press either (a) the INV button, (b) the arc button, or (c) the 2nd function button, and then one of the sin , cos , or tan buttons, whichever corresponds to the given function value.
2. Press one of the $\sin^{-1}$, $\cos^{-1}$, or $\tan^{-1}$ buttons, whichever corresponds to the given function value.

The value of the angle in the selected measure will then be displayed by the calculator.

Example 2   Use a calculator to find a value of θ in the indicated measure that satisfies the given equation. Round the value obtained to the nearest tenth of a degree or nearest thousandth of a radian.

a) $\cos \theta = 0.8951$, degrees   b) $\cot \theta = 0.6060$, degrees
c) $\csc \theta = 1.093$, radians

Solution

a)   With the calculator set for degrees, enter the number 0.8951. If part (a) in the first of the preceding routines is being used, press the INV button and then the cos button. The displayed value will be 26.478723, and the value of θ will be

$$\theta = 26.5°,$$

rounded to the nearest tenth of a degree.

b)   With the calculator set for degrees, enter the number 0.6060 and press the 1/x button. If part (b) in the first of the preceding routines is being used, next press the arc button, followed by the tan button. The result will be 58.784137, which rounds to the nearest tenth of a degree as

$$\theta = 58.8°.$$

c)   With the calculator set for radians, enter 1.093 and press the 1/x button. If the second routine described earlier is being used, next press the $\sin^{-1}$ button. The calculator then displays 1.1552925, which rounds to the nearest thousandth of a radian as

$$\theta = 1.155. \qquad ❑$$

When a negative number is given as a value of a trigonometric function of θ, we use the related angle θ' to find θ. **The negative number is not entered into a calculator.** The reason for this restriction will become clear in Chapter 6.

Example 3   If $180° \leq \theta \leq 270°$ and $\cos \theta = -0.9641$, find θ to the nearest tenth of a degree.

Solution
We first find the related angle θ' in degrees.

$$\cos \theta' = 0.9641$$

$$\theta' = 15.4°$$

Since $180° \le \theta \le 270°$, we have

$$\theta = 180° + \theta' = 195.4°.$$ ❏

Example 4   If $\tan \theta = -1.419$, $0 \le \theta < 2\pi$, and θ lies in Q II, find the value of θ in radians, rounded to four decimal places.

Solution
This time, we find the related angle θ' in radians.

$$\tan \theta' = 1.419$$

$$\theta' = 0.9569$$

Since $\pi/2 < \theta < \pi$, we have

$$\theta = \pi - \theta' = 2.1847.$$ ❏

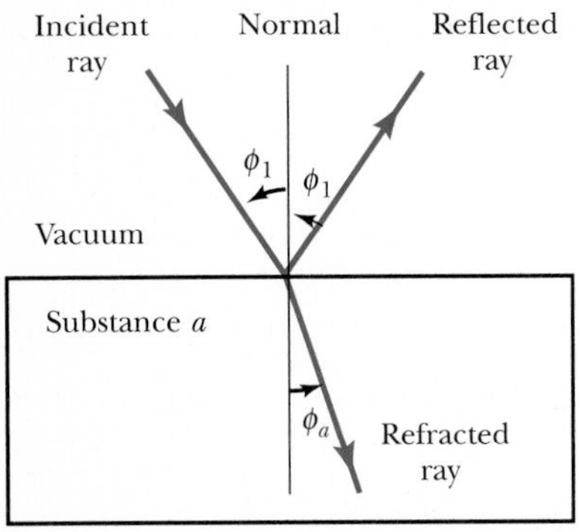

Example 5   Consider a beam of yellow sodium light that travels in a vacuum and makes an angle of incidence ϕ_1 with the normal to the surface of substance a, as shown in Figure 5.23. The beam is in part reflected and in part refracted, and the angle ϕ_a in Figure 5.23 is called the *angle of refraction* of the substance.

One form of **Snell's law** in physics states that

$$\frac{\sin \phi_1}{\sin \phi_a} = n_a,$$

where n_a is a constant, called the *index of refraction* of the substance a. Table 5.1 gives the index of refraction for several substances. We shall find the angle of refraction ϕ_a of water at 20°C when the angle of incidence ϕ_1 is 45°.

Solution
We have $\phi_1 = 45°$, and Table 5.1 gives $n_a = 1.333$ for water at 20° C. According to Snell's law,

$$\frac{\sin 45°}{\sin \phi_a} = 1.333,$$

and

$$\sin \phi_a = \frac{\sin 45°}{1.333}.$$

FIGURE 5.23

Index of Refraction for Yellow Sodium Light

Substance	Index of refraction
Ice	1.309
Water at 20°C	1.333
Ethyl alcohol	1.360
Rock salt	1.544
Diamond	2.417

TABLE 5.1

Using a calculator or Tables I and II in the Appendix, we get

$$\sin \phi_a = 0.5305$$

and

$$\phi_a = 32°.$$

Although Snell's law applies to light traveling in a vacuum, ordinary atmospheric pressure does not have enough effect to change the result that $\phi_a = 32°$ for water at 20°C when $\phi_1 = 45°$. ❑

Exercises 5.5

Find the values of the following trigonometric functions by using either a calculator or Tables I and II, as instructed by the teacher. In either case, give answers correct to four digits. (See Example 1 or Section A.3 in the Appendix.)

1. $\sin 31.3°$ 2. $\sin 42.2°$
3. $\cos 58.7°$ 4. $\cos 76.5°$
5. $\tan 63.2°$ 6. $\tan 81.8°$
7. $\cot 115.6°$ 8. $\cot 295.6°$
9. $\cos 32°20'$ 10. $\cos 57°40'$
11. $\tan 108°40'$ 12. $\tan 260°50'$
13. $\sec (-112°50')$ 14. $\sec (-278°30')$
15. $\sin (-401°30')$ 16. $\sin (-526°20')$
17. $\sin 0.3054$ 18. $\csc 0.4422$
19. $\sec 5.6258$ 20. $\cos 3.7816$
21. $\tan 2.8158$ 22. $\cot 3.2667$
23. $\cos 5.1720$ 24. $\sin 5.4745$

In Exercises 25–36, θ has the given function value and lies in the given quadrant, with $0° \le \theta < 360°$. According to the teacher's instructions, use either a calculator or Table II to find θ in decimal degrees to the nearest tenth of a degree. (See Examples 2 and 3.)

25. $\sin \theta = 0.5299$, Q I   26. $\sin \theta = 0.6587$, Q I
27. $\cot \theta = 1.6842$, Q I   28. $\cot \theta = 1.2482$, Q I
29. $\cos \theta = 0.5793$, Q I   30. $\cos \theta = 0.6756$, Q I
31. $\tan \theta = -0.6924$, Q II
32. $\tan \theta = -0.4813$, Q II
33. $\cos \theta = -0.8171$, Q III
34. $\cos \theta = -0.8616$, Q III

35. $\sin \theta = -0.4478$, Q IV
36. $\cot \theta = -8.028$, Q IV

In Exercises 37–48, θ has the given function value and lies in the given quadrant, with $0 \le \theta < 2\pi$. According to the teacher's instructions, use either a calculator or Table I to find θ in radians, rounded to four decimal places. (See Examples 2 and 4.)

37. $\tan \theta = 1.767$, Q I   38. $\tan \theta = 0.3121$, Q I
39. $\sec \theta = 1.231$, Q I   40. $\sin \theta = 0.9051$, Q I
41. $\cos \theta = -0.8732$, Q III
42. $\tan \theta = 1.756$, Q III
43. $\sin \theta = -0.7214$, Q IV
44. $\cot \theta = -0.9827$, Q IV
45. $\cos \theta = -0.5901$, Q II
46. $\cos \theta = -0.5446$, Q II
47. $\csc \theta = -1.244$, Q III
48. $\sec \theta = -1.448$, Q II

In Exercises 49–60, θ has the given function value and lies in the given quadrant, with $0° \le \theta < 360°$. According to the teacher's instructions, use either a calculator or Table I to find θ in degrees and minutes, correct to the nearest 10 minutes.

49. $\cos \theta = 0.9013$, Q I   50. $\cos \theta = 0.6450$, Q I
51. $\sin \theta = 0.9250$, Q I   52. $\sin \theta = 0.7698$, Q I
53. $\cot \theta = 1.473$, Q III   54. $\cot \theta = 2.066$, Q III
55. $\sec \theta = -1.111$, Q II
56. $\sec \theta = -1.023$, Q II
57. $\csc \theta = -1.485$, Q III

58. $\csc \theta = -1.951$, Q III

59. $\tan \theta = -1.626$, Q IV

60. $\tan \theta = -2.590$, Q IV

(See Example 5.)

61. Assume that a beam of yellow sodium light travels in a vacuum and makes an angle of incidence of 30° with the normal to a flat surface of rock salt. Use either a calculator or Tables I and II to find the angle of refraction to the nearest degree.

62. Assume that a beam of yellow sodium light travels in a vacuum and makes an angle of incidence of 60° with the normal to a flat surface of diamond. Use either a calculator or Tables I and II to find the angle of refraction to the nearest degree.

5.6	Right Triangles

Numerous practical problems that involve right triangles can be solved by using the trigonometric functions. One problem of this type received much publicity in 1988 when a surveyor in Louisville, Kentucky, used the tangent function to prove that the Louisville Falls Fountain on the Ohio River is the world's largest floating fountain. A diagram of the procedure employed is shown in Figure 5.24, and the calculation of the height of the fountain is given in Example 3 of this section.

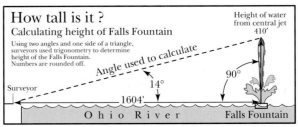

FIGURE 5.24

We adopt conventional notation by labeling the right angle in a right triangle C and the two acute angles A and B. The hypotenuse is always labeled c; side a is opposite angle A; and side b is opposite angle B. Thus side b is adjacent to angle A, and side a is adjacent to angle B. The six trigonometric functions of an acute angle in a right triangle can be restated in terms of the sides of the triangle. We first place the triangle ABC in Figure 5.25(a) in a coordinate system in Figure 5.25(b) so that angle A is in standard position.

The point with coordinates $x = b$, $y = a$ is located on the terminal side of A and at the vertex of B. The distance from the origin to this point is $r = c$. From Definition 5.4, it follows that

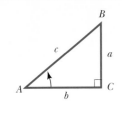

(a)

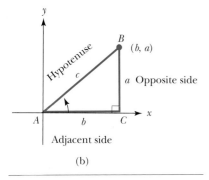

(b)

FIGURE 5.25

$$\sin A = \frac{a}{c} = \frac{\text{opposite side}}{\text{hypotenuse}}, \qquad \cot A = \frac{b}{a} = \frac{\text{adjacent side}}{\text{opposite side}},$$

$$\cos A = \frac{b}{c} = \frac{\text{adjacent side}}{\text{hypotenuse}}, \qquad \sec A = \frac{c}{b} = \frac{\text{hypotenuse}}{\text{adjacent side}},$$

$$\tan A = \frac{a}{b} = \frac{\text{opposite side}}{\text{adjacent side}}, \qquad \csc A = \frac{c}{a} = \frac{\text{hypotenuse}}{\text{opposite side}}.$$

Although we have formulated the definitions for angle A, the same relationships hold for angle B:

$$\sin B = \frac{b}{c} = \frac{\text{opp}}{\text{hyp}}, \quad \cos B = \frac{a}{c} = \frac{\text{adj}}{\text{hyp}}, \quad \text{and so on.}$$

In connection with these relationships, we note that A and B are complementary angles, and we can write

$$\sin A = \frac{a}{c} = \cos (90° - A), \qquad \cos A = \frac{b}{c} = \sin (90° - A),$$

$$\tan A = \frac{a}{b} = \cot (90° - A), \qquad \cot A = \frac{b}{a} = \tan (90° - A),$$

$$\sec A = \frac{c}{b} = \csc (90° - A), \qquad \csc A = \frac{c}{a} = \sec (90° - A).$$

Because of these relations, we say that sine and cosine are *complementary functions,* or *cofunctions.* Similarly, tangent and cotangent are cofunctions, and secant and cosecant are cofunctions. In Chapter 6 we shall see that any function of an angle equals the cofunction of the complementary angle, and that this is true for all angles (positive, negative, or zero) as long as the functions involved are defined. This is the origin of the cofunctions' names. For example, *cosine* is a shortened form of *complementary sine.*

To **solve** a triangle means to find the value of all six parts of the triangle: the three sides and the three angles. With the convention that $C = 90°$ in effect, all parts of a triangle are determined when one side and one other quantity are specified.

In a calculation involving approximate numbers, the digits known to be correct are called **significant digits.** The **number of significant digits** in a number is the total when the digits are counted from left to right, starting with the first nonzero digit and ending with the last digit in the number. The placing of the decimal in a number has nothing to do with the number of significant digits, but there may be confusion when zeros are present.

Zeros between nonzero digits are always significant, and **zeros used only to place the decimal are not significant.** Thus each of the numbers 405.8 and 0.004295 has four significant digits.

Zeros at the right end of a number are sometimes significant and sometimes not. **Zeros that are at the right end of a number and to the right of the decimal are significant.** For example, writing a measurement as 43.60 meters indicates that the measurement is accurate to the nearest hundredth of a meter, and that the true value lies between 43.595 meters and 43.605 meters.

In a measurement such as 6400 kilometers, the zeros may or may not be significant. In the absence of additional information or some prior agreement, this could represent a measurement to the nearest kilometer, to the nearest 10 kilometers, or to the nearest 100 kilometers. Such uncertainty can be avoided by writing the number in scientific notation:

6.4×10^3 kilometers indicates accuracy to nearest 100 kilometers;

6.40×10^3 kilometers indicates accuracy to nearest 10 kilometers;

6.400×10^3 kilometers indicates accuracy to nearest kilometer.

Rather than use scientific notation, we shall adopt the convention in this text that **all zeros in whole numbers are significant unless it is explicitly stated otherwise.**

In solving triangles, we need a guide as to how the accuracy in measuring sides corresponds to the accuracy in measuring angles. Such a guide is given in the following table.

Number of significant digits in side measures	Angle accuracy to the nearest
2	1°
3	10′, or 0.1°
4	1′, or 0.01°

We follow this guide in the examples as well as in the exercises throughout this chapter.

Example 1   Solve the right triangle that has $c = 33$, $B = 22°$.

Solution

It may be helpful to sketch the triangle, as shown in Figure 5.26.

As a first step, we find angle A to be 68°. To find side b, we may use either of the equations

$$\sin B = \frac{b}{c} \quad \text{or} \quad \csc B = \frac{c}{b},$$

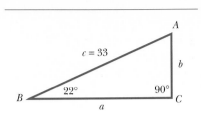

FIGURE 5.26

since both B and c are known. Solving the first equation for b, we get

$$b = c \sin B$$
$$= 33 \sin 22°$$
$$= 33(0.3746)$$
$$\mathbf{b = 12.}$$

Solving now for a, we use the equation

$$\cos B = \frac{a}{c}.$$

Multiplying by c,

$$a = c \cos B$$
$$= 33 \cos 22°$$
$$= 33(0.9272)$$
$$\mathbf{a = 31.}$$

All parts of the triangle are now known:

$$\mathbf{a = 31,} \qquad \mathbf{A = 68°}$$
$$\mathbf{b = 12,} \qquad B = 22°,$$
$$c = 33, \qquad C = 90°. \qquad \square$$

In most cases there is more than one correct procedure for solving a triangle. Occasionally two correct procedures will yield results that differ by 1 or 2 in the rightmost significant digit. This is nothing to worry about, however. It merely reflects the fact that we are working with approximate numbers.

Example 2   A guy wire on a vertical utility pole is anchored at a point on level ground 18 feet from the base of the pole. If the wire makes an angle of 63° with the ground, how long is the wire?

Solution
Let c be the length of the wire, as shown in Figure 5.27.

From the figure we see that

$$\sec A = \frac{c}{b}$$
$$c = b \sec A = 18 \sec 63° = 18(2.203) = 40.$$

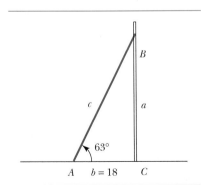

FIGURE 5.27

The guy wire is 40 feet in length. $\square$

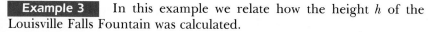

FIGURE 5.28

Whenever an observer looks at an object, an acute angle is formed in a vertical plane between the horizontal direction and the line of sight. This acute angle is called an **angle of elevation** if the line of sight is above the horizontal, and an **angle of depression** if the line of sight is below the horizontal. This is illustrated in Figure 5.28.

Example 3   In this example we relate how the height h of the Louisville Falls Fountain was calculated.

Solution

Louisville Falls Fountain
© 1988 *The Courier-Journal, reprinted with permission.*

Roger Basham, the Louisville Water Company surveyor, obtained a measurement of the distance along a horizontal line between the fountain and his instrument on River Road. This distance was approximately 1604 feet, as shown in Figure 5.29. Since the central stream of water in the fountain was vertical, it formed a 90° angle with the horizontal line.

With his instrument, Basham measured the angle of elevation of the top of the fountain stream as 14°38′ (see Figure 5.29). He then calculated the distance a in Figure 5.29 as follows:

$$\tan 14°38' = \frac{a}{1604}$$

$$a = 1604 \tan 14°38'$$

$$= 1604\,(0.2611)$$

$$= 418.8.$$

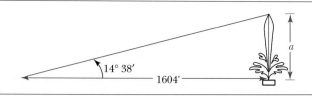

FIGURE 5.29

Since the opening of the fountain's central jet was 8.8 feet above the horizontal line, he subtracted to obtain

$$h = a - 8.8 = 410.$$

Thus the height of the fountain was found to be 410 feet. ❏

In surveying and navigation, two methods are commonly used to describe the direction, or **bearing,** of a line. The method ordinarily used in surveying is to state the acute angle that measures the variation from one of the directions north or south and to indicate whether this variation is to the east or to the west. As examples:

N 40° E indicates a bearing 40° to the east of north;

N 70° W indicates a bearing 70° to the west of north;

S 60° E indicates a bearing 60° to the east of south;

S 30° W indicates a bearing 30° to the west of south.

All these directions are pictured in Figure 5.30.

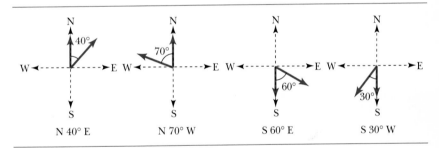

FIGURE 5.30

In all cases, either N or S comes first, and this tells the direction from which the acute angle is measured. Either E or W comes last, and this tells on which side of the north–south line the acute angle lies.

Example 4   The distance from Vicksburg to Laurel is 114 miles, on a bearing S 60.0° E. From Laurel, the distance to Meridian is 53.0 miles, on a bearing N 30.0° E. Find the bearing from Vicksburg to Meridian.

Solution
In Figure 5.31, the cities are represented by the first letters in their names. As shown in the figure, the bearing from Laurel to Vicksburg is N 60.0° W. Thus triangle *VLM* is a right triangle with the 90° angle at *L*. To determine the bearing from Vicksburg to Meridian, we need to

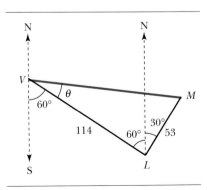

FIGURE 5.31

find the angle θ with vertex at V. We have

$$\tan \theta = \frac{53}{114} = 0.4649,$$

and therefore

$$\theta = 24.9°,$$

to the nearest tenth of a degree. Since $60° + 24.9° = 84.9°$, the direction from Vicksburg to Meridian is S 84.9° E.   ❏

The second method referred to earlier is used to give bearings in air navigation. With this method, a bearing is given by stating the angle measured positive in a clockwise direction from due north. The bearings shown earlier in Figure 5.30 are restated in Figure 5.32 using the air navigation method.

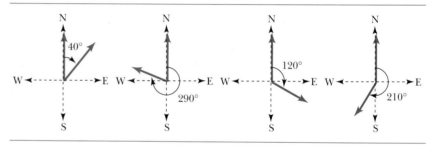

FIGURE 5.32

Example 5 P, Q, and R are towns. A plane leaves P and flies in the direction $220°0'$ at 215 miles per hour for 4 hours and lands at Q. From Q, the plane flies in the direction $310°0'$ to R. If the direction from P to R is $250°10'$, find the distance from Q to R.

Solution
As shown in Figure 5.33(a), angle QPS is $220° - 180° = 40°$. Using the fact that alternate interior angles are equal, we see that the angle PQN is also $40°$. Since angle RQN is $360° - 310° = 50°$, we have a $90°$ angle at Q, and triangle PQR is a right triangle.

To fly from P to Q, it took 4 hours at 215 miles per hour, so the distance from P to Q is 860 miles, as shown in Figure 5.33(b). From this figure, we see that angle QPR is $250°10' - 220° = 30°10'$.

To find the distance p from Q to R, we write

$$\tan 30°10' = \frac{p}{860}$$

$$p = 860 \tan 30°10' = 860(0.5812) = 500.$$

The distance from Q to R is 500 miles.   ❏

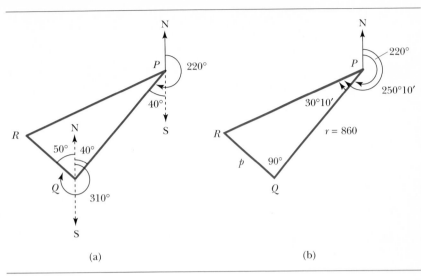

FIGURE 5.33

Exercises 5.6

Solve the following right triangles to the degree of accuracy consistent with the given information. (See Example 1.)

1.
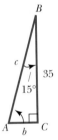

2.
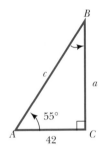

3.
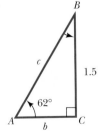

4.
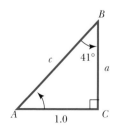

5.
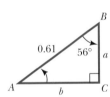

6.
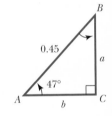

7.
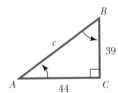

8.
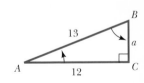

9. $c = 0.19$, $A = 36°$

10. $c = 0.79$, $B = 68°$

11. $a = 3.0$, $c = 5.0$

12. $a = 76$, $b = 51$

13. $a = 172$, $A = 39°40'$

14. $b = 537$, $B = 38°10'$

15. $a = 62.9$, $B = 71.8°$

16. $b = 1.17$, $A = 53.3°$

17. $c = 0.321$, $A = 62°30'$

18. $c = 0.413$, $A = 74°40'$

19. $c = 253$, $B = 42.2°$ **20.** $c = 358$, $A = 37.4°$

21. $a = 4.05$, $b = 12.8$ **22.** $a = 8.13$, $b = 3.91$

23. $a = 0.0625$, $c = 0.144$

24. $b = 0.0502$, $c = 0.209$

25. $a = 143.8$, $B = 63°28'$

26. $a = 342.9$, $b = 512.6$

27. $c = 91.08$, $A = 17.91°$

28. $a = 57.20$, $c = 314.5$

29. $a = 6.013$, $b = 11.54$

30. $b = 0.4772$, $A = 14°34'$

31. $b = 17.24$, $c = 98.48$

32. $c = 117.7$, $B = 66.88°$

(See Example 2.)

33. A ladder 19 feet long leans against a wall and makes an angle of 58° with the ground. How high from the ground is the top end of the ladder?

34. The posts in a fence are 5.0 feet high, with 12 feet of space between posts, as shown in the figure. If a brace is placed from the bottom of one post to the top of the next, what angle will the brace make with the ground?

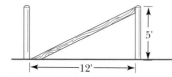

EXERCISE 34

35. A metal brace supports a crossarm on a utility pole. It is attached by bolts in each end through the pole and the centerline of the crossarm. If the holes in the brace are centered 32 inches apart and one end is attached to the pole 14 inches below the centerline of the crossarm, what angle does the brace make with the crossarm? (See the figure below.)

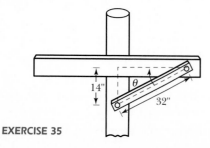

EXERCISE 35

36. A mast for a TV antenna stands on top of a flat hotel roof. A 48-foot guy wire has one end attached to the top of the mast and the other to the roof at a point 26 feet from the base of the mast. How high is the mast?

(See Example 3.)

37. From a point on the ground 200 feet from the base of a flagpole, the angle of elevation of the top of the pole is 15°20'. Find the height of the pole.

38. From the top of a tower that is 300 feet high, the angle of depression of a rock is 35.7°. Find the distance from the base of the tower to the rock.

39. The angle of depression from the top of a pine tree to the base of an oak tree is 66°. If the trees are 30 feet apart, how tall is the pine tree?

40. From atop a vertical cliff 100 feet above the surface of the ocean, the angle of depression to a buoy is 15.0°, as shown in the figure. Find the distance from the buoy to the base of the cliff.

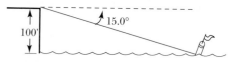

EXERCISE 40

41. When it ran at its maximum height, the giant fountain in Fountain Hills, Arizona, was known as the world's largest fountain. It has since been throttled back so that the angle of elevation of the top of the fountain is 56.0° from a point 236 feet from its base, as shown in the accompanying figure. Find the height h of the fountain since it has been throttled back.

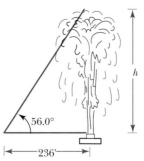

EXERCISE 41

42. The famous fountain in Geneva, Switzerland, is reported to be 400 feet tall. Find the angle of elevation of the top of the fountain from a point 115 feet from its base.

(See Example 4.)

43. A ship sails 128 kilometers from home port on a bearing S 37°20′ W to a point A, then due east to a point B directly south of home port. Find the distance from B to home port.

44. The cruise ship in the figure sails from home port H on a bearing of N 58°10′E for a distance of 158 kilometers to point A. From A, the ship sails 423 kilometers on a bearing S 31°50′ E to point B. Find the bearing of home port from B.

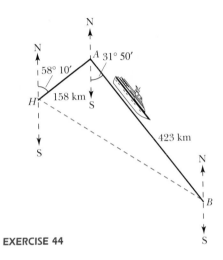

EXERCISE 44

45. A corner lot in the shape of a triangle is bounded on the north and west by streets that intersect at a right angle. The boundary on the third side is a fence that is 322 feet long and makes an angle of 39°40′ with the street along the west side of the lot. Find the length of frontage on the street along the west side of the lot.

46. From one corner of a triangular tract of land, the boundaries run in the directions S 42°40′ E and S 47°20′ W. The boundary opposite this corner has length 892 yards, and the boundary in the direction S 42°40′ E has length 507 yards. Find the angles at the three corners of the triangular tract.

(See Example 5.)

47. Grambling is due north of Quitman at a distance of 13 miles, and Monroe is due east of Grambling at a distance of 35 miles. Find the bearing from Quitman to Monroe.

48. The distance from Lubbock to Abilene is 150 miles, on a bearing of 123.0°. From Abilene, the distance to Wichita Falls is 126 miles, on a bearing of 33.0°. Find the bearing from Lubbock to Wichita Falls.

49. The bearing from Lafayette to Alexandria is 342°, and the bearing from Lafayette to Baton Rouge is 72°. The distance from Lafayette to Alexandria is 74 miles, and the bearing from Alexandria to Baton Rouge is 128°. Find the distance from Alexandria to Baton Rouge.

50. The distance from Kingman to Las Vegas is 87.0 miles, on a bearing of 316.0°. The distance from Kingman to San Diego is 241 miles, on a bearing of 226.0°. Find the bearing from San Diego to Las Vegas.

51. The two hotels shown below face each other across a street that is 97 feet wide. From a window in one hotel, the angle of elevation of the top of the hotel across the street is 41°, and the angle of depression of the bottom at street level is 22°. Find the height of the hotel across the street.

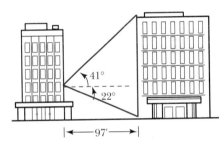

EXERCISE 51

52. A ship travels in a straight line toward a lighthouse that is 300 feet high. A woman on the ship observes that the angle of elevation of the top of the lighthouse is 23.0°. Later she observes that the angle of elevation of the top is 34.0°. How far has the ship traveled between the observations?

53. From the top of a building 152 feet above ground level in the accompanying figure, the angle of depression of the top of a campus street light is 18.0°, and the angle of depression of the base of the light is 32.0°. How high is the light?

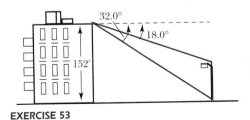

EXERCISE 53

54. From a certain point on the ground in the figure, a sand flea observes that the angle of elevation of the top of a vertical flagpole is 20°. He then advances 30 feet toward the pole and finds the angle of elevation to be 30°. How high is the pole?

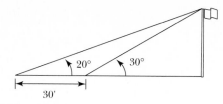

EXERCISE 54

5.7 | Sinusoidal Graphs

We saw in Section 5.3 that the trigonometric functions may be regarded as functions of a real number. When considered in that way, they have graphs just as do the more familiar functions that are studied in algebra.[5] In this chapter we learn to sketch the graphs of trigonometric functions.

Since we plan to draw graphs in the usual xy-plane, we shall write the equations that define functions in a form such as $y = \sin x$ instead of $y = \sin \theta$ or $y = \sin t$.

We start with the sine function, $y = \sin x$, and make a table of values for points with $0 \le x \le 2\pi$, spaced at intervals of $\pi/6$ on the x-axis. This is shown in Table 5.2.

x	0	$\dfrac{\pi}{6}$	$\dfrac{\pi}{3}$	$\dfrac{\pi}{2}$	$\dfrac{2\pi}{3}$	$\dfrac{5\pi}{6}$	π	$\dfrac{7\pi}{6}$	$\dfrac{4\pi}{3}$	$\dfrac{3\pi}{2}$	$\dfrac{5\pi}{3}$	$\dfrac{11\pi}{6}$	2π
y	0	$\dfrac{1}{2}$	$\dfrac{\sqrt{3}}{2}$	1	$\dfrac{\sqrt{3}}{2}$	$\dfrac{1}{2}$	0	$-\dfrac{1}{2}$	$-\dfrac{\sqrt{3}}{2}$	-1	$-\dfrac{\sqrt{3}}{2}$	$-\dfrac{1}{2}$	0

TABLE 5.2 $y = \sin x$

In Section 5.3 values of the sine function were interpreted as ordinates of points on the unit circle. That geometric interpretation, together with the values tabulated in Table 5.2, leads us to draw the smooth curve shown in Figure 5.34. This is the graph of $y = \sin x$ over the interval from 0 to 2π.

We saw in Section 5.3 that $P(x) = P(x + 2\pi)$ for any point $P(x)$ on the unit circle. Since $P(x)$ has coordinates $(\cos x, \sin x)$ and $P(x + 2\pi)$

---

5. Recall that the *graph* of a function f is the set of all points with coordinates of the form $(x, f(x))$.

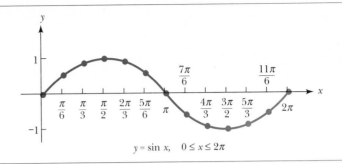

FIGURE 5.34

has coordinates $(\cos (x + 2\pi), \sin (x + 2\pi))$, we see that

$$\sin (x + 2\pi) = \sin x$$

for every real number x. This means that the graph of $y = \sin x$ repeats itself over intervals of width 2π along the x-axis. The complete graph of $y = \sin x$ extends indefinitely both to the left and to the right, as indicated in Figure 5.35.

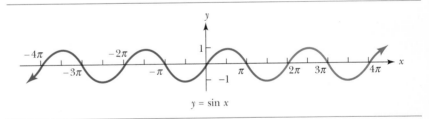

FIGURE 5.35

The graphs of the other trigonometric functions repeat in a manner similar to that of the sine function. Functions that have this repeating property are called *periodic*.

Thus the sine function is periodic with period $P = 2\pi$. Another useful term is given in the following definition.

> ### Definition 5.8
>
> Suppose f is a periodic function that has a maximum value of M and a minimum value of m. The **amplitude** of f, abbreviated **Amp,** is defined to be one-half the difference between M and m:
>
> $$\text{Amp} = \frac{M - m}{2}.$$

The sine function is periodic with a maximum value of $M = 1$ and a minimum value of $m = -1$. According to the definition, its amplitude is given by

$$\text{Amp} = \frac{1 - (-1)}{2} = 1.$$

The usefulness of this number lies in the fact that it tells how high the graph of $y = \sin x$ is at the maximum value and how low it is at the minimum value.

The term **biorhythm** is used to refer to a periodic cycle of change in the functions of an organism. Theoretically, there are three sinusoidal biorhythms, which begin at birth and continue until death, that can be used to predict variations in a person's feelings or abilities. According to this belief, each person has a 23-day physical cycle, a 28-day sensitivity cycle, and a 33-day intellectual cycle. Conditions are supposedly favorable during the first half of each cycle and unfavorable during the last half. Critical days occur on the first day of each cycle and on days when a cycle changes from the favorable phase to the unfavorable phase. A person is thought to be more likely to have bad luck on critical days.

The objective of this section is to learn to sketch quickly the graphs of functions of the type $y = a \sin (bx + c)$ or $y = a \cos (bx + c)$. Such functions are called **sinusoidal**[6] **functions,** and $bx + c$ is referred to as the **argument** of the function.

As an important step toward this objective, we summarize the important features of the graphs in Figures 5.34 and 5.35.

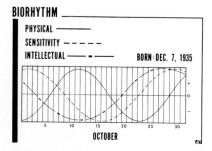

Wide World Photos

---

6. The word *sinusoidal* means "sinelike."

Graph of $y = \sin x$

1. Amplitude: Amp = 1.
2. Period: $P = 2\pi$.
3. The graph crosses the x-axis at the beginning, middle, and end of a period.
4. The maximum value occurs halfway between the beginning and middle of the period.
5. The minimum value occurs halfway between the middle and the end of the period.

The values of the constants a, b, and c in $y = a \sin (bx + c)$ change the graph, but they do not alter its basic shape. The graph of such a function always has **key points** corresponding to the last three items in the preceding list of features for $y = \sin x$. These key points are labeled in Figure 5.36 on a curve that we refer to as the **typical shape for a sine curve.**

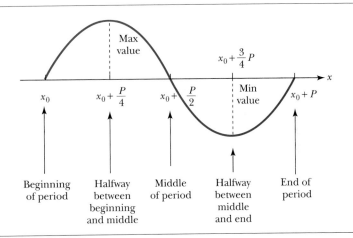

FIGURE 5.36 Typical shape for a sine curve

The following examples illustrate the effect that the constants a, b, and c have on the graph of $y = a \sin (bx + c)$. We consider the effect of a in $y = a \sin x$ in our first two examples.

Example 1 Sketch the graph of the function $y = 3 \sin x$ over the interval $0 \le x \le 2\pi$.

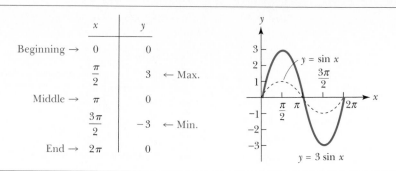

	x	y	
Beginning →	0	0	
	$\dfrac{\pi}{2}$	3	← Max.
Middle →	π	0	
	$\dfrac{3\pi}{2}$	-3	← Min.
End →	2π	0	

FIGURE 5.37

Solution

The multiplier $a = 3$ in $y = 3 \sin x$ simply multiplies all the function values from $y = \sin x$ by 3. The key points on the graph are tabulated in Figure 5.37, and a sketch of the graph is shown there. We note that

1. Amp $= 3$;
2. $P = 2\pi$;
3. The typical shape for a sine curve starts at $x = 0$. ❏

Example 2   Sketch the graph of $y = -2 \sin x$ over the interval $0 \le x \le 2\pi$.

Solution

The multiplier $a = -2$ in $y = -2 \sin x$ multiplies all the function values from $y = \sin x$ by -2. The key points on the graph are tabulated in Figure 5.38, and a sketch of the graph is shown there. We note that

1. Amp $= 2 = |-2|$;
2. $P = 2\pi$;
3. The graph is a reflection through the x-axis of a typical sine curve, starting at $x = 0$.

Generalizing from these two examples, we conclude that the effect of a multiplier a in $y = a \sin x$ is to change the amplitude to the value Amp $= |a|$.

We next consider the effect of b in $y = \sin bx$. Our attention in this chapter is confined to the case in which $b > 0$. The case in which $b < 0$ is treated in Section 6.2. ❏

	x	y	
Beginning →	0	0	
	$\dfrac{\pi}{2}$	-2	← Min.
Middle →	π	0	
	$\dfrac{3\pi}{2}$	2	← Max.
End →	2π	0	

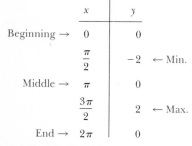
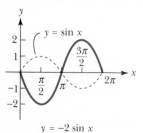

FIGURE 5.38

Example 3   Sketch the graph of $y = \sin 2x$ for $0 \le x \le 2\pi$.

	x	y		x	y
Beginning →	0	0		$\dfrac{5\pi}{4}$	1
	$\dfrac{\pi}{4}$	1 ← Max.		$\dfrac{3\pi}{2}$	0
Middle →	$\dfrac{\pi}{2}$	0		$\dfrac{7\pi}{4}$	-1
	$\dfrac{3\pi}{4}$	-1 ← Min.		2π	0
End →	π	0			

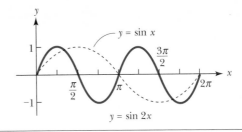

FIGURE 5.39

Solution

Those values that correspond to key points are tabulated in Figure 5.39. From the graph in Figure 5.39, we observe that the multiplier $b = 2$ in $y = \sin 2x$ has changed the period and that

1. Amp $= 1$;

2. $P = \dfrac{2\pi}{2} = \pi$;

3. The typical sine curve shape starts at $x = 0$. ❏

Example 3 illustrates how a positive multiplier b in $y = \sin bx$ changes the period to $2\pi/b$. This happens because an increase in x from 0 to $2\pi/b$ causes the argument bx to increase from 0 to 2π.

The next example illustrates the effect of the constant c in the graph of $y = \sin (x + c)$.

Example 4   Sketch the graph of $y = \sin \left(x + \dfrac{\pi}{4} \right)$ through one complete period locating the key points on the graph.

	x	y
Beginning →	$-\dfrac{\pi}{4}$	0
	$\dfrac{\pi}{4}$	1 ← Max.
Middle →	$\dfrac{3\pi}{4}$	0
	$\dfrac{5\pi}{4}$	-1 ← Min.
End →	$\dfrac{7\pi}{4}$	0

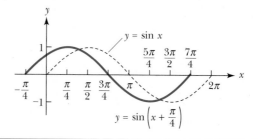

FIGURE 5.40

Solution

The table in Figure 5.40 includes only those values that correspond to key points on the graph. The value $x = -\pi/4$ was chosen to begin with because this locates the point that corresponds to the beginning of a period on the typical sine curve. The corresponding y-value is calculated as

$$y = \sin\left(-\frac{\pi}{4} + \frac{\pi}{4}\right) = \sin 0 = 0.$$

The key points located on the graph in Figure 5.40 indicate that the function has been graphed through one complete period over the interval $-\pi/4 \le x \le 7\pi/4$. The constant $c = \pi/4$ caused the graph to be translated horizontally $\pi/4$ units to the left. A horizontal translation such as this is called a **phase shift.**   ❏

Example 4 indicates that the constant c in $y = a \sin(bx + c)$ causes a phase shift. The phase shift can be found by locating any two corresponding points on the graph. As standard practice, we shall use the beginning of a period on the typical sine curve. The phase shift can also be found by formula, and some people prefer this method. With

$b > 0$, the shift is c/b units to the left if $c > 0$ and $|c/b|$ units to the right if $c < 0$.

Generalizing from our examples, we arrive at the following guide.

Guide for Graphing
$y = a \sin (bx + c),$ **where** $b > 0$

---

1. Amp $= |a|$.

2. $P = \dfrac{2\pi}{b}$.

3. A period begins at the value of x where the argument is 0, that is, where $bx + c = 0$.

4. If $a > 0$, the graph is a typical sine curve. If $a < 0$, the graph is the reflection through the x-axis of a typical sine curve.

The next example illustrates the use of this guide.

Example 5   Sketch the graph of $y = \dfrac{1}{2} \sin \left(3x - \dfrac{\pi}{2} \right)$ through one complete period locating the key points on the graph.

Solution
With $a = \frac{1}{2}$, $b = 3$, and $c = -\pi/2$, we find that the amplitude is $\frac{1}{2}$ and the period is $2\pi/3$. A period begins where

$$3x - \frac{\pi}{2} = 0$$

$$x = \frac{\pi}{6}.$$

To find the end of the period, we add the width of a period to the starting point.

$$x = \frac{\pi}{6} + \frac{2\pi}{3} = \frac{5\pi}{6}$$

To locate the middle of the period, we simply average the x-values at the beginning and end of the period.

$$x = \frac{1}{2}\left(\frac{\pi}{6} + \frac{5\pi}{6} \right) = \frac{\pi}{2}$$

Similarly, we locate the other x-values tabulated in Figure 5.41 and use the properties of the typical sine curve in the following list.

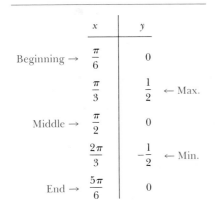

	x	y	
Beginning →	$\dfrac{\pi}{6}$	0	
	$\dfrac{\pi}{3}$	$\dfrac{1}{2}$	← Max.
Middle →	$\dfrac{\pi}{2}$	0	
	$\dfrac{2\pi}{3}$	$-\dfrac{1}{2}$	← Min.
End →	$\dfrac{5\pi}{6}$	0	

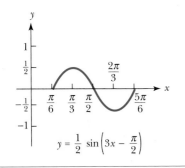

$$y = \frac{1}{2} \sin\left(3x - \frac{\pi}{2}\right)$$

FIGURE 5.41

1. The graph crosses the x-axis at $x = \pi/6$, $x = \pi/2$, and $x = 5\pi/6$.

2. A maximum of $y = |a| = \frac{1}{2}$ occurs at $x = \pi/3$.

3. A minimum of $y = -|a| = -\frac{1}{2}$ occurs at $x = 2\pi/3$.

The graph is drawn in Figure 5.41.   ❏

Functions of the form $y = a \cos (bx + c)$ can be graphed by using a method very similar to the one we have used for sine curves. The procedure is presented here with a minimum of details.

The graph of $y = \cos x$ over the interval $0 \le x \le 2\pi$ is shown in Figure 5.42.

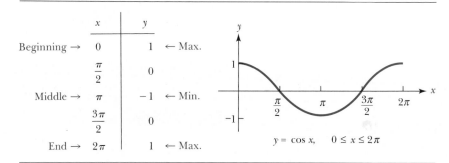

FIGURE 5.42

The unit circle can be used in the same way that it was with the sine function to conclude that the cosine function has period 2π. As it was

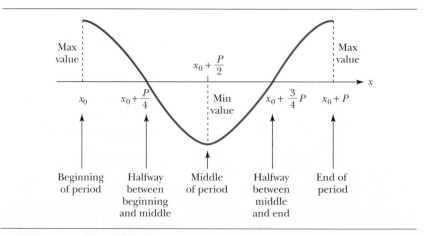

FIGURE 5.43   Typical shape for a cosine curve

with $y = \sin x$, the complete graph of $y = \cos x$ extends indefinitely in both directions along the x-axis.

We take the shape of the graph in Figure 5.42 as a pattern for the **typical shape for a cosine curve** and select as **key points** those points where the graph crosses the x-axis, where it has a maximum value, or where it has a minimum value. These key points are labeled in Figure 5.43.

To sketch the graph of a function of the type $y = a \cos (bx + c)$, we use the typical shape of a cosine curve and the following guide.

Guide for Graphing
$y = a \cos (bx + c)$,   where $b > 0$

---

1. Amp $= |a|$.

2. $P = \dfrac{2\pi}{b}$.

3. A period begins at the value of x where the argument is 0, that is, where $bx + c = 0$.

4. If $a > 0$, the graph is a typical cosine curve. If $a < 0$, it is the reflection through the x-axis of a typical cosine curve.

The following example illustrates the procedure.

Example 6   Sketch the graph of $y = 4 \cos \left(\dfrac{1}{2}x + \dfrac{\pi}{4} \right)$ through one complete period locating the key points on the graph.

Solution
With $a = 4$, $b = \frac{1}{2}$, and $c = \pi/4$, we find that the amplitude is 4 and the period $2\pi/(1/2) = 4\pi$. A period begins where

$$\frac{1}{2}x + \frac{\pi}{4} = 0$$

$$x = -\frac{\pi}{2}.$$

The end of the period is at

$$x = -\frac{\pi}{2} + 4\pi = \frac{7\pi}{2},$$

and the middle of the period is at

$$x = \frac{1}{2}\left(-\frac{\pi}{2} + \frac{7\pi}{2} \right) = \frac{3\pi}{2}.$$

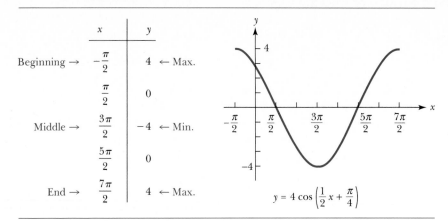

	x	y	
Beginning →	$-\dfrac{\pi}{2}$	4	← Max.
	$\dfrac{\pi}{2}$	0	
Middle →	$\dfrac{3\pi}{2}$	-4	← Min.
	$\dfrac{5\pi}{2}$	0	
End →	$\dfrac{7\pi}{2}$	4	← Max.

$$y = 4\cos\left(\frac{1}{2}x + \frac{\pi}{4}\right)$$

FIGURE 5.44

The key points are tabulated in Figure 5.44. According to the properties of the typical cosine curve,

1. There is a maximum of $y = |a| = 4$ at $x = -\pi/2$ and $x = 7\pi/2$;

2. There is a minimum of $y = -|a| = -4$ at $x = 3\pi/2$;

3. The graph crosses the x-axis at $x = \pi/2$ and $x = 5\pi/2$.

The graph is shown in Figure 5.44.   ❏

The preceding examples have concentrated on sketching the graph of a function over one complete period. It is not difficult to extend such graphs to cover any desired interval.

Example 7   Sketch the graph of $y = 4\cos\left(\dfrac{1}{2}x + \dfrac{\pi}{4}\right)$ over the interval $-5\pi/2 \leq x \leq 9\pi/2$.

Solution

The function $y = 4\cos\left(\dfrac{1}{2}x + \dfrac{\pi}{4}\right)$ is graphed for $-\pi/2 \leq x \leq 7\pi/2$ in Example 6. All we need do is locate key points on the graph until the required interval is covered, and then sketch the graph over that interval. The result is shown in Figure 5.45.   ❏

Sinusoidal graphs are useful in describing many physical quantities. For instance, they are used to describe the motion of a spring under various conditions. We relate one of the simplest situations in the next example.

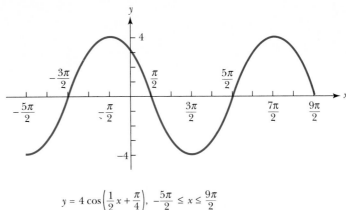

$$y = 4 \cos\left(\frac{1}{2}x + \frac{\pi}{4}\right), \quad -\frac{5\pi}{2} \le x \le \frac{9\pi}{2}$$

FIGURE 5.45

Example 8   A certain spring is such that an object weighing 2 pounds stretches the spring $\frac{1}{2}$ foot and reaches an equilibrium position as shown in Figure 5.46(b). If the spring is stretched an additional $\frac{1}{3}$ foot and released with initial velocity zero, it is possible to describe the position of the object with a cosine function. We assume that the only forces acting on the object are the force of the spring and the weight of the object.

We let x denote the displacement of the object from its equilibrium position t seconds after its release, with x in feet, measured positive downward and negative upward. This is illustrated in Figure 5.46(c).

With these notations, the displacement x is given by the formula

$$x = \frac{1}{3} \cos 8t$$

for $t \ge 0$. The graph of this function is shown in Figure 5.47. Our

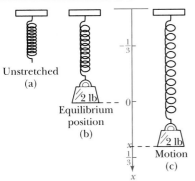

FIGURE 5.46

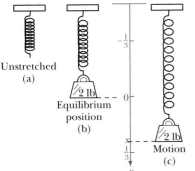

FIGURE 5.47

formula is valid only under the assumption that no friction or other damping force is exerted on the object. Motion of this type is called *free undamped motion*. More complex situations can be described by more complicated sinusoidal functions.   ❏

Exercises 5.7

Find the amplitude, period, and phase shift (if there is one) for the graph of each of the following functions. Do not draw the graphs.

1. $y = 4 \sin 3x$ **2.** $y = 3 \cos 2x$

3. $y = -5 \cos 4x$ **4.** $y = -2 \sin 8x$

5. $y = \frac{1}{2} \cos \left(x - \frac{\pi}{4} \right)$ **6.** $y = \frac{1}{3} \sin \left(x + \frac{\pi}{6} \right)$

7. $y = \frac{1}{3} \sin (3x + \pi)$ **8.** $y = \frac{1}{2} \cos (2x - \pi)$

9. $y = -4 \sin \left(3x - \frac{\pi}{4} \right)$

10. $y = -3 \cos \left(\frac{1}{2}x - \frac{\pi}{2} \right)$

11. $y = -3 \cos \left(2x + \frac{\pi}{4} \right)$

12. $y = -4 \sin \left(2x + \frac{\pi}{3} \right)$

Sketch the graph of the given function through one complete period, locating the key points on each graph. (See Examples 1–6.)

13. $y = 2 \sin x$ **14.** $y = 3 \cos x$

15. $y = -2 \cos x$ **16.** $y = -\sin x$

17. $y = -\frac{1}{3} \sin 3x$ **18.** $y = -\frac{1}{2} \cos 4x$

19. $y = 2 \cos \frac{1}{2}x$ **20.** $y = 6 \sin \frac{1}{3}x$

21. $y = 2 \sin \left(2x - \frac{\pi}{2} \right)$ **22.** $y = 2 \sin (3x + \pi)$

23. $y = 3 \sin \left(x + \frac{\pi}{3} \right)$ **24.** $y = 3 \sin (4x + \pi)$

25. $y = 3 \cos \left(2x - \frac{\pi}{4} \right)$ **26.** $y = 4 \cos \left(2x + \frac{\pi}{4} \right)$

27. $y = 2 \cos \left(3x + \frac{\pi}{4} \right)$ **28.** $y = 4 \cos \left(x + \frac{\pi}{3} \right)$

29. $y = 2 \sin \left(\frac{1}{2}x + \frac{\pi}{4} \right)$ **30.** $y = 3 \sin \left(\frac{1}{2}x - \frac{\pi}{4} \right)$

31. $y = 3 \sin \left(\frac{1}{3}x + \frac{\pi}{3} \right)$

32. $y = 3 \sin \left(\frac{1}{2}x - 2\pi \right)$

33. $y = 3 \cos \left(\frac{1}{2}x + \frac{\pi}{4} \right)$ **34.** $y = 2 \cos \left(\frac{1}{4}x + \frac{\pi}{4} \right)$

35. $y = 2 \cos \left(\frac{1}{4}x - \frac{\pi}{4} \right)$ **36.** $y = 3 \cos \left(\frac{1}{3}x + \frac{\pi}{9} \right)$

In Exercises 37–40, sketch the graph of the given function over the indicated interval. (See Example 7.)

37. $y = \sin \left(2x + \frac{\pi}{2} \right), \quad 0 \le x \le 2\pi$

38. $y = 2 \sin \left(\frac{1}{4}x - \frac{\pi}{2} \right), \quad -2\pi \le x \le 6\pi$

39. $y = 3 \cos \left(\frac{1}{2}x - \frac{\pi}{2} \right), \quad -\pi \le x \le 5\pi$

40. $y = \cos \left(3x + \frac{\pi}{2} \right), \quad -\frac{\pi}{6} \le x \le \frac{5\pi}{6}$

In Exercises 41 and 42, find an equation of the form $y = a \sin (bx + c)$ that has the given graph.

41.

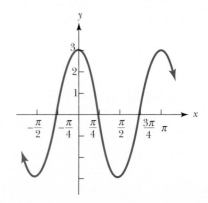

42.

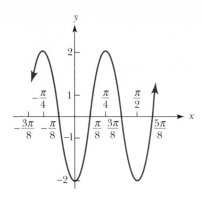

In Exercises 43 and 44, find an equation of the form $y = a \cos (bx + c)$ that has the given graph.

43.

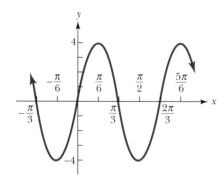

44.

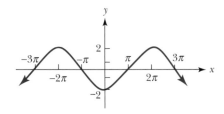

45. The accompanying figure shows a diagram of an electrical circuit that contains an electromotive force $E(t)$ measured in volts, an inductor with inductance L measured in henrys, and a capacitor with capacitance C measured in farads. We assume that $E(0) = 0$ and that the charge on the capacitor is 0 at $t = 0$. If $E(t) = 60$ volts for $t > 0$, $L = 1$ henry, and $C = \frac{1}{16}$ farad, the current $I(t)$ is given by $I(t) = 15 \sin 4t$. Sketch the graph of I through one complete period.

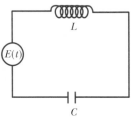

EXERCISE 45

46. We assume that the electrical circuit shown in Exercise 45 has $E(0) = 0$ and that the charge on the capacitor is 0 at $t = 0$. If $E(t) = 36$ volts for $t > 0$, $L = 2$ henrys, and $C = \frac{1}{18}$ farad, the current $I(t)$ is given by $I(t) = 6 \sin 3t$. Sketch the graph of I through one complete period.

5.8 Other Trigonometric Graphs

We consider first the graph of $y = \csc x$. Since $\csc x = 1/\sin x$, values of $\csc x$ can be obtained by using the reciprocals of nonzero values of $\sin x$.

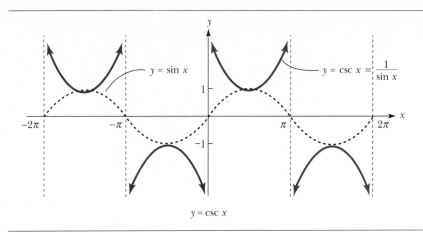

FIGURE 5.48

Since 1 is its own reciprocal, $\csc x = 1$ where $\sin x = 1$, and similarly $\csc x = -1$ where $\sin x = -1$. The value of $\csc x$ is undefined at each x where $\sin x = 0$. As values of $\sin x$ get closer to 0, the values of $\csc x$ increase in absolute value without bound. Thus the graph of $y = \csc x$ has a vertical asymptote at each x where $\sin x = 0$.

In Figure 5.48, the graph of $y = \sin x$ is shown as a dashed curve. By taking reciprocals of the ordinates on the dashed curve, we obtain the solid graph in the figure as the graph of $y = \csc x$.

Since $y = \sin x$ has period 2π and $\csc x = 1/\sin x$, it follows that $y = \csc x$ has period 2π. The amplitude is not defined for the cosecant function since it has neither a maximum nor a minimum.

In the same way that we obtained the graph of $y = \csc x$ from that of $y = \sin x$, the graph of a function of the form $y = a \csc (bx + c)$ can be obtained from the graph of $y = a \sin (bx + c)$.[7] Since $\csc (bx + c) = 1/\sin (bx + c)$, the two functions have the same period, $P = 2\pi/b$. The graph of $y = a \csc (bx + c)$ has vertical asymptotes where $a \sin (bx + c) = 0$. The graphs of $y = a \csc (bx + c)$ and $y = a \sin (bx + c)$ intersect where $\sin (bx + c) = \pm 1$, that is, at maximum and minimum points on the sine curve. Finally, the phase shift for the cosecant function is the same as for the corresponding sine function.

Example 1   Sketch the graph of $y = \dfrac{1}{4} \csc \left(3x + \dfrac{\pi}{2} \right)$ through one complete period.

---

7. We confine our attention here to the case in which $b > 0$, just as we did in Section 5.7.

Solution

We start out as if we were going to graph $y = \frac{1}{4} \sin\left(3x + \frac{\pi}{2}\right)$. The period is $2\pi/3$, and a period begins where

$$3x + \frac{\pi}{2} = 0$$

$$x = -\frac{\pi}{6}.$$

As in the last section, we locate the x-values that correspond to key points on the graph of $y = \frac{1}{4} \sin\left(3x + \frac{\pi}{2}\right)$. These x-values are tabulated in Figure 5.49, and the graph of $y = \frac{1}{4} \sin\left(3x + \frac{\pi}{2}\right)$ is drawn as a dashed curve in the figure. The graph of $y = \frac{1}{4} \csc\left(3x + \frac{\pi}{2}\right)$ is drawn as a solid curve, and it has vertical asymptotes where the sine curve crosses the x-axis. The two function values are equal at maximum and minimum points on the sine curve. We note that the phase shift for the cosecant curve is $\pi/6$ units to the left, the same as it is for $y = \frac{1}{4} \sin\left(3x + \frac{\pi}{2}\right)$.

x	y
$-\dfrac{\pi}{6}$	Undefined
0	$\dfrac{1}{4}$
$\dfrac{\pi}{6}$	Undefined
$\dfrac{\pi}{3}$	$-\dfrac{1}{4}$
$\dfrac{\pi}{2}$	Undefined

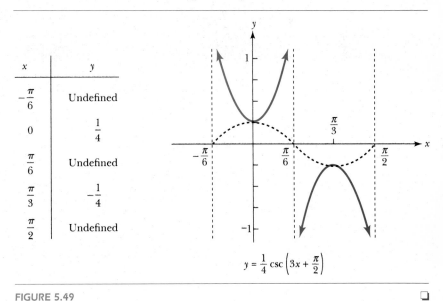

$$y = \frac{1}{4} \csc\left(3x + \frac{\pi}{2}\right)$$

FIGURE 5.49

After a little practice, it is no longer necessary to draw the corresponding sine curve, as we did in Example 1.

We turn our attention now to the graph of the secant function. Since $\sec x = 1/\cos x$, the graph of $y = \sec x$ is related to that of $y = \cos x$ in the same way that the graph of $y = \csc x$ is related to that of $y = \sin x$. This is shown in Figure 5.50, where the graph of $y = \cos x$ appears as a dashed curve.

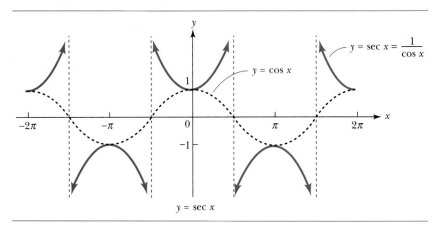

FIGURE 5.50

The graph of $y = a \sec (bx + c)$ can be obtained by using the graph of $y = a \cos (bx + c)$. Since $\sec (bx + c) = 1/\cos (bx + c)$, the two functions have the same period, $P = 2\pi/b$, and the same phase shift. The amplitude is not defined for the secant function, and the graph of $y = a \sec (bx + c)$ has vertical asymptotes at each x where

$$a \cos (bx + c) = 0.$$

The graphs of the two functions intersect where $\cos (bx + c) = \pm 1$, that is, at maximum and minimum points on the cosine curve. All this is illustrated in Example 2.

Example 2   Sketch the graph of $y = 2 \sec \left(\dfrac{1}{2}x - \dfrac{\pi}{6}\right)$ through one complete period.

Solution

We relate this graph to that of $y = 2 \cos \left(\dfrac{1}{2}x - \dfrac{\pi}{6}\right)$. The period is

x	y
$\dfrac{\pi}{3}$	2
$\dfrac{4\pi}{3}$	Undefined
$\dfrac{7\pi}{3}$	-2
$\dfrac{10\pi}{3}$	Undefined
$\dfrac{13\pi}{3}$	2

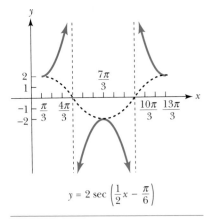

$$y = 2 \sec\left(\frac{1}{2}x - \frac{\pi}{6}\right)$$

FIGURE 5.51

$2\pi/(1/2) = 4\pi$, and a period begins where

$$\frac{1}{2}x - \frac{\pi}{6} = 0$$

$$x = \frac{\pi}{3}.$$

The x-values that correspond to key points on the cosine curve are tabulated in Figure 5.51. The graph of $y = 2\cos\left(\dfrac{1}{2}x - \dfrac{\pi}{6}\right)$ is drawn as a dashed curve, and the graph of $y = 2\sec\left(\dfrac{1}{2}x - \dfrac{\pi}{6}\right)$ is drawn as a solid curve in the figure. Note that both graphs have a phase shift of $\pi/3$ units to the right. ❑

All four of the functions $\sin x$, $\cos x$, $\csc x$, and $\sec x$ have period 2π. Unlike them, $\tan x$ and $\cot x$ have period π. For this reason, we restrict our attention at first to the interval $-\pi/2 < x < \pi/2$ in graphing $y = \tan x$. We start by making a table of values at convenient points, much as we did for the sine function in Table 5.2. These values are displayed in Table 5.3.

x	$-\dfrac{\pi}{3}$	$-\dfrac{\pi}{4}$	$-\dfrac{\pi}{6}$	0	$\dfrac{\pi}{6}$	$\dfrac{\pi}{4}$	$\dfrac{\pi}{3}$
y	$-\sqrt{3}$	-1	$-\dfrac{\sqrt{3}}{3}$	0	$\dfrac{\sqrt{3}}{3}$	1	$\sqrt{3}$

TABLE 5.3 $y = \tan x$

Since $\tan x = \sin x/\cos x$ and $\cos(\pi/2) = 0$, $y = \tan x$ is undefined at $x = \pi/2$. For the moment, we concentrate on values of x between $\pi/3$ and $\pi/2$. As x gets closer to $\pi/2$, $\sin x$ gets closer to 1 and $\cos x$ gets closer to 0. As this happens, $\tan x = \sin x/\cos x$ increases without bound. (This fact can be observed to some extent in the tables or by use of a calculator.) Thus $y = \tan x$ has a vertical asymptote at $x = \pi/2$. A similar situation occurs at $x = -\pi/2$, as shown in Figure 5.52.

We could check on the fact that $y = \tan x$ has period π by plotting more points,[8] but we do not bother with this. Instead, we assume that the graph of $y = \tan x$ given in Figure 5.53 is correct.

---

8. This fact can be proved by using the identities in Chapter 6.

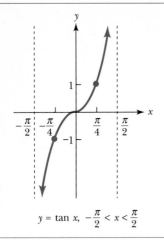

$$y = \tan x, \quad -\frac{\pi}{2} < x < \frac{\pi}{2}$$

FIGURE 5.52

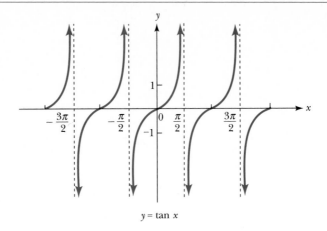

$$y = \tan x$$

FIGURE 5.53

We shall now investigate the effect of the constants a, b, and c on the graph of $y = a \tan (bx + c)$. As we have done before, we restrict ourselves to the case where b is positive.

It is easy to see that the factor a in $y = a \tan x$ simply multiplies the function values in $y = \tan x$ by a. This effect is most conspicuous at $x = \pi/4$ and $x = -\pi/4$, since these are the points where $\tan x = \pm 1$ and $y = \pm a$.

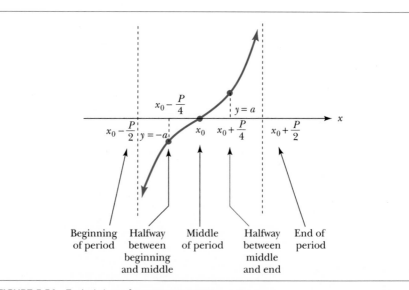

FIGURE 5.54 Typical shape for a tangent curve

The positive multiplier b in $y = a \tan bx$ has the same sort of effect that it did in previous cases. It changes the period from π to π/b. (The difference here is that we are starting with a period of π instead of 2π.)

Finally, the constant c in $y = a \tan (bx + c)$ causes a phase shift in a similar way as it did with the previous trigonometric functions.

We take the shape of the graph in Figure 5.52 as a pattern for the **typical shape for a tangent curve.** This typical shape is shown in Figure 5.54. Those points marked with large dots in the figure are our **key points** on the graph.

To sketch the graph of a function of the type $y = a \tan (bx + c)$ with $b > 0$, we use the typical shape of a tangent curve in Figure 5.54 and the following guide.

Guide for Graphing
$y = a \tan (bx + c),$ where $b > 0$

---

1. $P = \dfrac{\pi}{b}.$

2. The middle of a period is at the value of x where $bx + c = 0$.
3. Vertical asymptotes are located at the beginning and end of the period.
4. If $a > 0$, the graph is a typical tangent curve. If $a < 0$, it is the reflection through the x-axis of a typical tangent curve.

Example 3   Sketch the graph of $y = 2 \tan \left(\dfrac{1}{3}x + \dfrac{\pi}{6} \right)$ through one complete period.

Solution
We have $a = 2$, $b = \frac{1}{3}$, and $c = \pi/6$. The period is $P = \pi/b = \pi/(1/3) = 3\pi$. The middle of a period is where

$$\frac{1}{3}x + \frac{\pi}{6} = 0$$

$$x = -\frac{\pi}{2}.$$

To locate the beginning of the period, we mark off half the length of the period to the left of $x = -\pi/2$. That is, we subtract $\frac{1}{2}(3\pi)$ from $-\pi/2$:

$$x = -\frac{\pi}{2} - \frac{1}{2}(3\pi) = -2\pi.$$

x	y
-2π	Undefined
$-\dfrac{5\pi}{4}$	-2
$-\dfrac{\pi}{2}$	0
$\dfrac{\pi}{2}$	2
π	Undefined

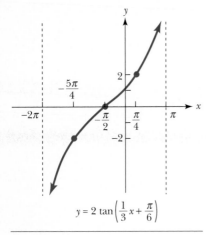

$$y = 2\tan\left(\frac{1}{3}x + \frac{\pi}{6}\right)$$

FIGURE 5.55

Similarly, we add $3\pi/2$ to $-\pi/2$ to locate the end of the period:

$$x = -\frac{\pi}{2} + \frac{1}{2}(3\pi) = \pi.$$

We plot the value $y = a = 2$ halfway between the middle and end of the period, at $x = \pi/4$. Similarly, $y = -a = -2$ at $x = -5\pi/4$. Using the typical shape for a tangent curve, we obtain the graph in Figure 5.55. ❏

Since $\cot x = 1/\tan x$, the function $y = \cot x$ has period π. We can use the graph of $y = \tan x$ to obtain the graph of $y = \cot x$, noting that each of the two functions has a vertical asymptote where the other is zero. To illustrate the relationship between the graphs, the graph of $y = a \cot x$ is drawn as a solid curve and the graph of $y = a \tan x$ is drawn as a dashed curve in Figure 5.56.

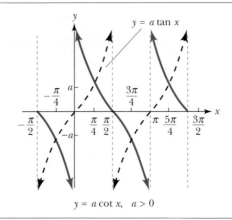

$$y = a \cot x, \quad a > 0$$

FIGURE 5.56

To be consistent with our procedure for graphing tangent functions, we use the portion of the solid graph in Figure 5.56 between $-\pi/2$ and $\pi/2$ as the **typical shape of a cotangent curve.**

Example 4   Sketch the graph of $y = 2\cot\left(\frac{1}{3}x + \frac{\pi}{6}\right)$ through one complete period.

Solution
The period is $P = \pi/b = \pi/(1/3) = 3\pi$. The middle of this period is

x	y
-2π	0
$-\dfrac{5\pi}{4}$	-2
$-\dfrac{\pi}{2}$	Undefined
$\dfrac{\pi}{4}$	2
π	0

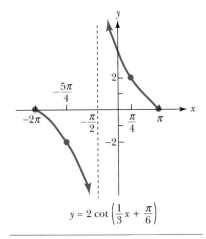

$$y = 2\cot\left(\frac{1}{3}x + \frac{\pi}{6}\right)$$

FIGURE 5.57

where

$$\frac{1}{3}x + \frac{\pi}{6} = 0$$

$$x = -\frac{\pi}{2}.$$

The beginning of the period is at

$$x = -\frac{\pi}{2} - \frac{1}{2}(3\pi) = -2\pi,$$

and the end is at

$$x = -\frac{\pi}{2} + \frac{1}{2}(3\pi) = \pi.$$

We plot $y = a = 2$ at $x = \pi/4$ and $y = -a = -2$ at $x = -5\pi/4$. (The work to this point is exactly the same as for the tangent function in Example 3.) We plot zeros at each end of the period, draw a vertical asymptote in the middle of the period, and sketch the graph as shown in Figure 5.57. ❑

Suppose that a given function $y = f(x)$ can be expressed as the sum $f(x) = g(x) + h(x)$ of two simpler functions, $g(x)$ and $h(x)$. The graph of the sum $y = f(x)$ can be obtained by graphing $y_1 = g(x)$ and $y_2 = h(x)$ separately and then adding their ordinates: $y = y_1 + y_2$ for each value of x. That is, for any value of x, we add the y-value from $g(x)$ to the y-value from $h(x)$ to obtain the y-value for $f(x)$. This method, called **addition of ordinates,** is illustrated in the next two examples.

Example 5 Sketch the graph of $y = x + \cos x$ over the interval $0 \le x \le 2\pi$.

Solution
We first graph $y_1 = x$ and $y_2 = \cos x$ separately. These graphs are shown in Figure 5.58 as dashed curves. We then select some x-values at which it is convenient to add the ordinates y_1 and y_2. The large dots in Figure 5.58 show the results of the selections that were made. Whenever possible, we choose points where the value of either y_1 or y_2 is known without computation. As soon as enough points are plotted to reveal the shape of the graph, we draw a smooth curve through them. ❑

Our next example shows how we handle a situation that involves the *difference* of two functions rather than the sum.

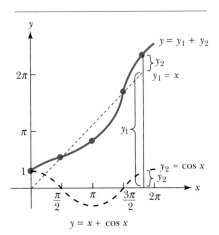

$$y = x + \cos x$$

FIGURE 5.58

Example 6   Sketch the graph of $y = \sin x - \cos 2x$ over the interval $0 \le x \le 2\pi$.

Solution

The graph of $y = \sin x - \cos 2x$ could be obtained by graphing $\sin x$ and $\cos 2x$ separately and then graphically *subtracting* ordinates of $\cos 2x$ from ordinates of $\sin x$. However, it is easier to graph $y_1 = \sin x$ and $y_2 = -\cos 2x$ and then *add* ordinates as we did in Example 5 (see Figure 5.59). ❑

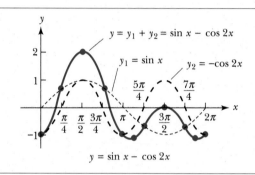

FIGURE 5.59

Exercises 5.8

Find the period, amplitude (if it is defined), and any phase shift or vertical translation for the graph of each of the following functions. Do not draw the graphs.

1. $y = 3 \csc \left(2x - \dfrac{\pi}{4} \right)$ **2.** $y = 2 \csc \left(3x + \dfrac{\pi}{2} \right)$

3. $y = 2 \sec \left(3x + \dfrac{\pi}{3} \right)$

4. $y = 3 \sec \left(4x - \dfrac{2\pi}{3} \right)$

5. $y = -\tan \left(2x + \dfrac{\pi}{4} \right)$ **6.** $y = 4 \tan \left(\dfrac{1}{3}x - \dfrac{\pi}{6} \right)$

7. $y = 4 \cot \left(\dfrac{1}{2}x - \dfrac{\pi}{4} \right)$

8. $y = -2 \cot \left(x + \dfrac{\pi}{4} \right)$

9. $y = 2 \cos x$ **10.** $y = \tfrac{1}{2} \sin 4x$

11. $y = 1 + 2 \sin \left(\dfrac{1}{2}x + \dfrac{\pi}{6} \right)$

12. $y = 2 + 6 \cos \left(\dfrac{1}{3}x - \dfrac{\pi}{6} \right)$

13. $y = 2 + 3 \csc \left(\dfrac{1}{3}x - \dfrac{\pi}{2} \right)$

14. $y = 3 + 4 \sec \left(2x + \dfrac{\pi}{2} \right)$

15. $y = 1 + 5 \tan (4x - \pi)$

16. $y = 4 + 2 \cot (3x - 2\pi)$

Sketch the graph of the given function through one complete period. (See Examples 1–4.)

17. $y = \csc 2x$ **18.** $y = \sec 2x$

19. $y = \tan 2x$ **20.** $y = \cot 2x$

21. $y = \tfrac{1}{3} \sec 2x$ **22.** $y = \tfrac{1}{4} \tan 3x$

23. $y = \frac{1}{2} \cot 3x$

24. $y = 2 \csc \left(\dfrac{x}{3} \right)$

25. $y = 3 \csc \left(\dfrac{1}{2}x - \dfrac{\pi}{2} \right)$

26. $y = \dfrac{1}{4} \csc \left(2x + \dfrac{\pi}{4} \right)$

27. $y = \frac{1}{3} \sec (2x - \pi)$

28. $y = \dfrac{1}{3} \sec \left(2x + \dfrac{\pi}{3} \right)$

29. $y = \tan (2x - \pi)$

30. $y = \tan \left(\dfrac{1}{2}x + \dfrac{\pi}{16} \right)$

31. $y = \dfrac{1}{2} \tan \left(x + \dfrac{\pi}{6} \right)$

32. $y = \dfrac{1}{3} \tan \left(x - \dfrac{\pi}{4} \right)$

33. $y = 3 \tan \left(\dfrac{1}{3}x - \dfrac{\pi}{6} \right)$

34. $y = 2 \tan \left(\dfrac{1}{3}x + \dfrac{\pi}{6} \right)$

35. $y = 2 \cot \left(\dfrac{1}{3}x - \dfrac{\pi}{12} \right)$

36. $y = 3 \cot \left(\dfrac{1}{2}x + \dfrac{\pi}{6} \right)$

Sketch the graph of each of the following functions over the interval $0 \le x \le 2\pi$. (See Examples 5 and 6.)

37. $y = 2 + \sin x$ **38.** $y = -2 + \sin x$

39. $y = x - \cos x$ **40.** $y = x + \sin x$

41. $y = \sin x + \cos 2x$ **42.** $y = \cos x + \sin 2x$

43. $y = 2 \cos x + \sin 2x$ **44.** $y = 3 \sin x + \cos 2x$

45. The figure below shows a diagram of an electrical circuit that contains an electromotive force $E(t) = 60$ volts, an inductance $L = 1$ henry, and a capacitance $C = \frac{1}{16}$ farad. The charge $q(t)$ on the capacitor in the circuit is given by

$$q(t) = \frac{15}{4} - \frac{15}{4} \cos 4t.$$

Sketch the graph of q through one complete period.

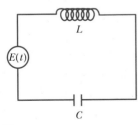

EXERCISE 45

46. For a given location, the number H of hours of daylight in each day of the year can be expressed

as a function of the date. If n is the number of days after January 1, the number $H(n)$ of hours of daylight in Atlanta, Georgia, is approximated by

$$H(n) = \frac{9}{4} \sin \frac{2\pi}{365}(n - 79) + \frac{49}{4}.$$

Sketch the graph of H from $n = 0$ to $n = 365$, and find the values of n that give a maximum value or a minimum value of H.

47. An object weighing 2 pounds stretches a spring $\frac{1}{2}$ foot and reaches an equilibrium position as shown in the figure. The spring is then stretched an additional $\frac{2}{3}$ foot, and the object is released with an upward velocity of $\frac{4}{3}$ foot per second. The only forces acting on the object are its weight and the force of the spring. If x denotes the displacement of the object from its equilibrium position t seconds after its release, then

$$x = \frac{2}{3} \cos 8t - \frac{1}{6} \sin 8t,$$

where x is in feet, measured positive downward and negative upward. Sketch the graph of x through one complete period.

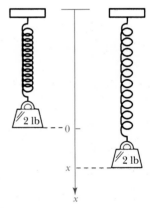

EXERCISE 47

48. a) Sketch the graph of $y = \sin (x + \pi)$ over the interval $[-2\pi, 2\pi]$.
b) Sketch the graph of $y = \cos (x + \pi)$ over the interval $[-2\pi, 2\pi]$.
c) Observe from the graphs in parts (a) and (b) that $\sin (x + \pi) = -\sin x$ and $\cos (x + \pi) = -\cos x$.
d) Use the observations in part (c) to show that $\tan (x + \pi) = \tan x$.

CHAPTER REVIEW

Key Words and Phases

Positive angle	Isosceles triangle	Quotient Identities
Negative angle	Equilateral triangle	Angle of elevation
Revolution	Similar triangles	Angle of depression
Degree	Central angle	Bearing
Minute	Standard position	Period
Second	Coterminal angles	Amplitude
Radian	Trigonometric functions	Sine curve
Right angle	Quadrantal angles	Cosine curve
Straight angle	Related (or reference) angle	Tangent curve
Acute angle	Related Angle Theorem	Cosecant curve
Obtuse angle	Unit circle	Secant curve
Complementary angle	Wrapping function	Cotangent curve
Supplementary angle	Reciprocal Identities	Addition of ordinates
Oblique triangle		

Summary of Important Concepts and Formulas

Conversions in Degree Measure

$$1° = 60', \quad 1' = 60''$$

Basic Equation Relating Radians and Degrees

$$\pi = 180°$$

Conversion Procedures for Radians and Degrees

To change radians to degrees, multiply by $180/\pi$.
To change degrees to radians, multiply by $\pi/180$.

Arc Length

$$s = r\theta$$

Linear and Angular Speed

$$v = r\omega$$

Trigonometric Functions

$$\sin \theta = \frac{y}{r} \quad \cos \theta = \frac{x}{r} \quad \tan \theta = \frac{y}{x}$$

$$\cot \theta = \frac{x}{y} \quad \sec \theta = \frac{r}{x} \quad \csc \theta = \frac{r}{y} \qquad r = \sqrt{x^2 + y^2}$$

Wrapping Function

$$P(t) = (\cos t, \sin t)$$

Trigonometric Functions of a Real Number t

Any trigonometric function of t is equal to the same
function of an angle with measure t radians.

Reciprocal Identities

$$\csc \theta = \frac{1}{\sin \theta} \quad \sec \theta = \frac{1}{\cos \theta} \quad \cot \theta = \frac{1}{\tan \theta}$$

Quotient Identities

$$\tan \theta = \frac{\sin \theta}{\cos \theta} \quad \cot \theta = \frac{\cos \theta}{\sin \theta}$$

Ranges of the Trigonometric Functions

$$-1 \leq \cos \theta \leq 1 \qquad -1 \leq \sin \theta \leq 1$$

$$\sec \theta \leq -1 \text{ or } \sec \theta \geq 1 \quad \csc \theta \leq -1 \text{ or } \csc \theta \geq 1$$

$$\tan \theta: \text{all real numbers} \qquad \cot \theta: \text{all real numbers}$$

Trigonometric Functions in a Right Triangle

$$\sin A = \frac{\text{opp}}{\text{hyp}} \quad \cos A = \frac{\text{adj}}{\text{hyp}} \quad \tan A = \frac{\text{opp}}{\text{adj}}$$

$$\cot A = \frac{\text{adj}}{\text{opp}} \quad \sec A = \frac{\text{hyp}}{\text{adj}} \quad \csc A = \frac{\text{hyp}}{\text{opp}}$$

Guide for Graphing $y = a \sin (bx + c)$, where $b > 0$

a. Amp $= |a|$
b. $P = \dfrac{2\pi}{b}$.
c. A period begins at the value of x where the argument
   is 0, that is, where $bx + c = 0$.
d. If $a > 0$, the graph is a typical sine curve. If $a < 0$,
   the graph is the reflection through the x-axis of a
   typical sine curve.

Guide for Graphing $y = a \cos (bx + c)$, where $b > 0$

a. Amp $= |a|$.
b. $P = \dfrac{2\pi}{b}$.

c. A period begins at the value of x where the argument is 0, that is, where $bx + c = 0$.

d. If $a > 0$, the graph is a typical cosine curve. If $a < 0$, it is the reflection through the x-axis of a typical cosine curve.

Guide for Graphing $y = a \tan (bx + c)$**, where** $b > 0$

a. $P = \dfrac{\pi}{b}$.

b. The middle of a period is at the value of x where $bx + c = 0$.

c. Vertical asymptotes are located at the beginning and end of the period.

d. If $a > 0$, the graph is a typical tangent curve. If $a < 0$, it is the reflection through the x-axis of a typical tangent curve.

Review Problems for Chapter 5

1. Name the angle according to its quadrant. Find four angles (two positive and two negative) that are coterminal to the given angle. Choose your angles with measure as close to 0° as possible.
   a) 123°   b) −44°

2. The brace at the end of a swing set is made from three pipes in the shape of an A. The pipes for the legs are 10 feet long and the pipe for the crossbar is 4 feet long. How far down the legs should the crossbar be bolted so that the legs spread 6 feet apart on the ground?

3. Suppose 7 feet of the 40-foot tilted piling shown below extends 2 feet vertically above the water line. If 12 feet of the tilted piling are embedded in the sand and the water is uniformly deep, how deep is the water?

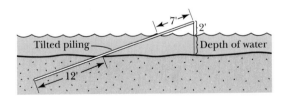

   PROBLEM 3

4. Suppose 9 feet of a tilted piling extends 5 feet vertically above the water line. Assuming the water is uniformly 20 feet deep, and 20 feet of the tilted piling are embedded in the sand, how long is the piling?

5. Determine the exact values of the six trigonometric functions of an angle θ in standard position if the terminal side of θ goes through $(-2\sqrt{10}, 3)$.

6. Determine the exact values of all the other trigonometric functions of the fourth quadrant angle α if $\tan \alpha = -\frac{24}{7}$.

7. Use identities to determine the exact values of the remaining trigonometric functions of the angle β where

$$\csc \beta = \frac{3}{2} \quad \text{and} \quad \cot \beta = \frac{\sqrt{5}}{2}.$$

8. Grand Rapids, Michigan, is almost due north of Gadsden, Alabama. Their latitudes are 43°N and 34°N, respectively. Use 4000 miles as the radius of the earth and find the distance between the cities, to the nearest 10 miles.

9. (See Problem 8.) Walla Walla, Washington, is almost due north of Los Angeles, California; their latitudes are 46°N and 34°N, respectively. Find the distance between the two cities, to the nearest 10 miles.

10. A belt-driven roller with diameter 3.0 feet has an angular speed of 7π radians per minute. Find the speed at which the belt is moving in feet per second.

11. The wheels of a cart are 28 inches in diameter and are turning at the rate of 24 revolutions per minute. Find the speed of the cart in miles per hour.

12. A bicycle with wheels 28 inches in diameter is traveling at a speed of 14 miles per hour. Find the angular speed of a wheel in radians per minute.

13. A wheel on a cart has a radius of length 2 feet and is rolling at a speed of 2 miles per hour. Find the angular speed of the wheel in radians per minute.

14. Convert each of the following radian measures to degree measure.
    a) $-\dfrac{7\pi}{6}$ b) $\dfrac{5\pi}{3}$

15. Convert each of the following to radian measure in terms of π.
    a) 630°   b) 42°11′15″

16. In a circle of diameter 36 centimeters, find the length of the arc intercepted by a central angle measuring 30°.

17. A heavy roller with a radius of 2 feet is being used to pack a roadbed for a highway. If the roller is pulled by a tractor going 10 miles per hour, find the angular speed of the roller in radians per minute.

18. Write out the exact values of all the trigonometric functions of the given angle.
    a) $-270°$ b) $510°$ c) $315°$
    d) $-\dfrac{\pi}{2}$ e) $\dfrac{7\pi}{6}$ f) $\dfrac{4\pi}{3}$

19. Find the rectangular coordinates of each of the following points on the unit circle.
    a) $P\left(-\dfrac{3\pi}{2}\right)$ b) $P\left(\dfrac{2\pi}{3}\right)$

20. Determine whether each of the following statements is possible or impossible.
    a) $\cos \theta = -\frac{3}{2}$
    b) $\tan \alpha = \frac{5}{6}$
    c) $\sin \beta = \frac{5}{4}$ and $\csc \beta = \frac{4}{5}$
    d) $\cot \alpha < 0$ and $\sec \alpha > 0$
    e) $\cot \alpha < 0$ and $\tan \alpha > 0$

21. Find the values of the following trigonometric functions, correct to four digits. Use a calculator or Tables I and II, as instructed by the teacher.
    a) $\sec 13.7°$ b) $\tan 143°10'$
    c) $\csc (-217°20')$ d) $\cos (-125.4°)$
    e) $\sin 438.2°$

22. Find the values of the following trigonometric functions by using either a calculator or Table I. Give answers correct to four significant digits.
    a) $\cos 1.4341$ b) $\tan 2.3969$

23. Suppose θ has the given function value and lies in the given quadrant, with $0° \le \theta < 360°$. According to the instructions of the teacher, use a calculator or Table II to find θ in decimal degrees to the nearest tenth of a degree.
    a) $\sin \theta = 0.4179$, Q I
    b) $\cot \theta = 0.4813$, Q I
    c) $\cos \theta = -0.8829$, Q II
    d) $\tan \theta = 1.1383$, Q III
    e) $\sec \theta = 3.026$, Q IV

24. Suppose θ has the given function value and lies in the given quadrant, with $0 \le \theta < 2\pi$. According to the teacher's instructions, use a calculator or Table II to find θ in radians, rounded to four decimal places.
    a) $\tan \theta = -7.596$, Q II

b) $\sin \theta = -0.2812$, Q III
c) $\cos \theta = -0.3827$, Q II
d) $\cot \theta = -0.6619$, Q IV
e) $\sec \theta = -3.388$, Q III

25. Use Table I or a calculator to find a value of t, $0 \le t \le 1.5708$, that satisfies the given equation. Give answers to four decimal places.
    a) $\tan t = 0.3314$ b) $\sec t = 2.203$

26. With θ in radian measure and in terms of π, find all values of θ for which $0 \le \theta < 2\pi$ and the given equation is true.
    a) $\tan \theta = 1$ b) $\sin \theta = -\frac{1}{2}$

27. Solve the right triangle that has $a = 84$ feet, $A = 15°$.

28. At a point on level ground 87 feet from the base of a vertical flagpole, the angle of elevation of the top of the pole is 37°. Find the height of the pole.

29. From one corner of a triangular field, the boundary fences run in the direction N 37°40′ E for 516 yards and in the direction S 52°20′E. The fence opposite this corner runs in a north–south direction. Find its length.

30. The television broadcasting tower in the figure below is known to be 527 meters in height. From the top of the tower, the angle of depression of an irrigation pump is 29°30′. Find the distance from the base of the tower to the pump.

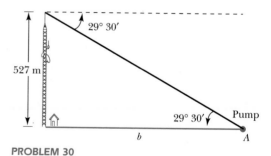

PROBLEM 30

31. Find the period, amplitude (if it is defined), and any phase shift or vertical translation for the graph of each of the following functions. Do not draw the graphs.
    a) $y = 1 + 2\cos(2x - \pi)$
    b) $y = -2 + \frac{1}{2}\sin(4x + \pi)$
    c) $y = 4 + \tan 3x$ d) $y = 3\csc\left(x + \dfrac{\pi}{2}\right)$
    e) $y = 4\sec\left(\dfrac{1}{2}x - \dfrac{\pi}{6}\right)$

In Problems 32–44, sketch the graph of the given function through one complete period, locating key points on each graph.

32. $y = \frac{1}{3} \cos 2x$

33. $y = \tan 3x$

34. $y = 3 \sin \left(x - \frac{\pi}{6} \right)$

35. $y = \frac{1}{4} \csc \left(\frac{3}{2}x + \frac{\pi}{4} \right)$

36. $y = \sec \left(\frac{1}{2}x - \frac{\pi}{6} \right)$

37. $y = \cot \left(x + \frac{\pi}{4} \right)$

38. $y = \frac{3}{2} \sin (2x + \pi)$

39. $y = \frac{2}{5} \sin \left(3x - \frac{\pi}{2} \right)$

40. $y = \frac{1}{2} \cos \left(2x - \frac{\pi}{2} \right)$

41. $y = \cos (3x + \pi)$

42. $y = \tan \left(x + \frac{\pi}{4} \right)$

43. $y = 2 \sin \left(\frac{1}{2}x + \frac{\pi}{2} \right)$

44. $y = 1 + 2 \cos \left(\frac{1}{2}x + \frac{\pi}{4} \right)$

45. Use addition of ordinates to sketch the graph of $y = 2 \cos x + \sin x$ over the interval $0 \le x \le 2\pi$.

In Problems 46 and 47, find an equation of the form $y = a \sin (bx + c)$ that has the given graph.

46.

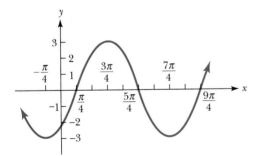

47.

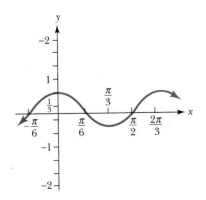

In Problems 48 and 49, find an equation of the form $y = a \cos (bx + c)$ that has the given graph.

48.

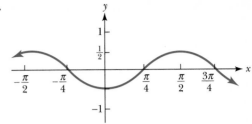

49.

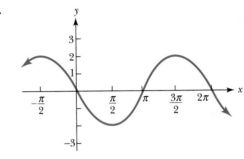

50. If n is the number of days after January 1, the number $H(n)$ of hours of daylight in Tampa, Florida, is approximated by

$$H(n) = \frac{11}{6} \sin \frac{2\pi}{365} (n - 79) + \frac{73}{6}.$$

Sketch the graph of H from $n = 0$ to $n = 365$, and find the values of n that give a maximum value or a minimum value of H.

▲ ▲ CRITICAL THINKING: FIND THE ERRORS ████████

Each of the following nonsolutions has at least one error. Can you find them?

Problem 1 The angle of depression from the top of a 62-foot pine tree to a point on the ground directly behind a log skidder is 38°. How far from the pine tree is the log skidder?

Nonsolution The length x of the side of the right triangle in Figure 5.60 is the required distance. We use the tangent function to find x.

$$\tan 38° = \frac{x}{62}$$

$$x = 62 \tan 38°$$

$$x = 48 \text{ feet}$$

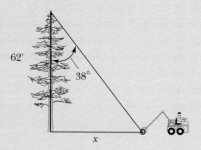

62'

38°

x

FIGURE 5.60

Problem 2 Convert 48°24′ to decimal-degree format.

Nonsolution $$48°24′ = 48.24°$$

Problem 3 Find $\cos \theta$ if $\sin \theta = \frac{4}{5}$ and θ is in quadrant II.

Nonsolution $$\sin \theta = \frac{y}{r} = \frac{4}{5}$$

$$y = 4 \quad \text{and} \quad r = 5$$

$$5^2 = x^2 + 4^2$$

$$9 = x^2$$

$$x = 3$$

$$\cos \theta = \frac{x}{r} = \frac{3}{5}$$

Problem 4   Use special angles to find cos 120°.

Nonsolution   In Figure 5.61, the angle between the terminal side of 120° and the y-axis is 30°. Therefore

$$\cos 120° = -\cos 30° = -\frac{\sqrt{3}}{2}$$

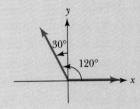

FIGURE 5.61

Problem 5   A wheel on a trailer has a radius of 9 inches. If the wheel is traveling at 36 feet per second, find its angular speed in radians per second.

Nonsolution $$\omega = \frac{v}{r} = \frac{36}{9} = 4 \text{ radians per second}$$

Problem 6   Sketch the graph of $y = 2 \sin\left(\frac{x}{2} - \frac{\pi}{4}\right)$ through one complete period, locating the key points on the graph.

Nonsolution   We have

$$y = 2 \sin\left(\frac{x}{2} - \frac{\pi}{4}\right) = \sin 2\left(\frac{x}{2} - \frac{\pi}{4}\right) = \sin\left(x - \frac{\pi}{2}\right).$$

The amplitude is 1, and the period is 2π. A period starts at $x = \frac{\pi}{2}$. The graph is shown in Figure 5.62.

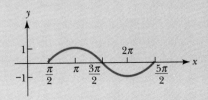

FIGURE 5.62

Problem 7 Sketch the graph of $y = \cos\left(x + \dfrac{\pi}{2}\right)$ through one complete period, locating the key points on the graph.

Nonsolution The amplitude is 1, and the period is 2π. A period starts at $x = -\dfrac{\pi}{2}$. The graph is shown in Figure 5.63.

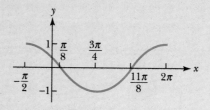

FIGURE 5.63

256

Chapter 6

Analytic Trigonometry

Most of the material in this chapter concerns identities in trigonometric functions. A knowledge of these identities is essential in the calculus, and they play a key role in the analysis of electrical circuits and other physical phenomena. An insight into this role can be obtained by considering the applications in Exercises 6.3 and 6.6.

258 6 ANALYTIC TRIGONOMETRY

6.1 Verifying Identities

In Section 5.4 the Reciprocal Identities

$$\sin \theta \csc \theta = 1, \quad \cos \theta \sec \theta = 1, \quad \tan \theta \cot \theta = 1$$

and the Quotient identities

$$\tan \theta = \frac{\sin \theta}{\cos \theta}, \quad \cot \theta = \frac{\cos \theta}{\sin \theta}$$

are used to find the values of the trigonometric functions of an angle. More important, these identities can be used to rewrite or reduce expressions involving trigonometric functions (called *trigonometric expressions*).

Frequently the algebra involved in working with trigonometric expressions is more troublesome to the student than the trigonometry itself. Fractions involving trigonometric functions can be combined in the same way as fractions involving algebraic expressions. The techniques of factoring extend to trigonometric expressions. Any of the operations on polynomials can be performed when the polynomials contain trigonometric functions as their variables. Once the algebraic operations are mastered, then the trigonometric identities can be used concurrently with the algebra to simplify or reduce trigonometric expressions.

Example 1   Reduce $\cos \theta \tan \theta$ to $\sin \theta$ by using one of the quotient identities.

Solution
We first replace $\tan \theta$ by $\sin \theta / \cos \theta$.

$$\cos \theta \tan \theta = \cos \theta \, \frac{\sin \theta}{\cos \theta} = \sin \theta$$

The last equality follows by dividing out the common factor, $\cos \theta$.   ❑

Recall that the trigonometric functions of θ are defined in terms of $x, y,$ and r, where (x, y) is a point on the terminal side of θ. Three identities can be derived from the Pythagorean relationship existing among $x, y,$ and r:

$$x^2 + y^2 = r^2.$$

Dividing both sides of this equation by r^2 (*note:* $r^2 > 0$) yields

$$\left(\frac{x}{r}\right)^2 + \left(\frac{y}{r}\right)^2 = 1.$$

Since $\cos \theta = x/r$ and $\sin \theta = y/r$, we obtain what is known as one of the **Pythagorean Identities:**

$$\cos^2 \theta + \sin^2 \theta = 1.$$

Note that this equation is true for *all* values of θ.

Dividing both sides of $\cos^2 \theta + \sin^2 \theta = 1$ by $\cos^2 \theta$, for $\cos \theta \neq 0$, yields

$$1 + \left(\frac{\sin \theta}{\cos \theta}\right)^2 = \left(\frac{1}{\cos \theta}\right)^2$$

and then another **Pythagorean Identity,**

$$1 + \tan^2 \theta = \sec^2 \theta.$$

Similarly, dividing both sides of $\cos^2 \theta + \sin^2 \theta = 1$ by $\sin^2 \theta$, for $\sin \theta \neq 0$, yields

$$\left(\frac{\cos \theta}{\sin \theta}\right)^2 + 1 = \left(\frac{1}{\sin \theta}\right)^2,$$

which gives the last **Pythagorean Identity,**

$$\cot^2 \theta + 1 = \csc^2 \theta.$$

The Pythagorean Identities, along with the Reciprocal and Quotient Identities, are known as the eight *Fundamental Identities.* They should be memorized. The student must then be able quickly to obtain or recognize any alternative form.

Fundamental Identities

Reciprocal Identities:	$\sin \theta \csc \theta = 1$
	$\cos \theta \sec \theta = 1$
	$\tan \theta \cot \theta = 1$
Quotient Identities:	$\tan \theta = \dfrac{\sin \theta}{\cos \theta}$
	$\cot \theta = \dfrac{\cos \theta}{\sin \theta}$
Pythagorean Identities:	$\sin^2 \theta + \cos^2 \theta = 1$
	$1 + \tan^2 \theta = \sec^2 \theta$
	$1 + \cot^2 \theta = \csc^2 \theta$

The Fundamental Identities can be used to prove other identities. The phrases "prove the identity" and "verify the identity" mean that we should prove that a given equation is true for all values of the variable such that both members are defined. The typical plan of attack is to reduce the trigonometric expression on one side of the equation to the expression on the other side of the equation.[1]

Usually the proof of an identity is easier if we *start with the more complicated* of the two expressions. It is also helpful to know what trigonometric expression we are trying to obtain. If we keep one eye on the trigonometric expression we are working toward, the path to that result will be easier to find. In other words, *keep the goal in mind and work toward it.*

Sometimes it is advantageous to replace a fraction by a sum (or difference) of two fractions. This technique is illustrated in Example 2.

Example 2   Prove the identity: $2 \cos \theta = \dfrac{\cos \theta \tan \theta + \sin \theta}{\tan \theta}$.

Solution
We reduce the more complicated side

$$\frac{\cos \theta \tan \theta + \sin \theta}{\tan \theta}$$

to the simpler side, $2 \cos \theta$.

$$\frac{\cos \theta \tan \theta + \sin \theta}{\tan \theta} = \frac{\cos \theta \tan \theta}{\tan \theta} + \frac{\sin \theta}{\tan \theta} \qquad \text{Algebra}$$

$$= \cos \theta + \frac{\sin \theta}{\tan \theta} \qquad \text{Algebra}$$

$$= \cos \theta + \sin \theta \cot \theta \qquad \text{Reciprocal Identity}$$

$$= \cos \theta + \sin \theta \frac{\cos \theta}{\sin \theta} \qquad \text{Quotient Identity}$$

$$= \cos \theta + \cos \theta \qquad \text{Algebra}$$

$$\mathbf{= 2 \cos \boldsymbol{\theta}} \qquad \text{Algebra} \qquad \square$$

Note that although we now have another identity,

$$\frac{\cos \theta \tan \theta + \sin \theta}{\tan \theta} = 2 \cos \theta,$$

at our disposal, it is not necessary (or even advisable) to memorize it. We restrict our memory work only to those identities listed on the back endpapers of this text.

---

1. Although this is not the only method of proving identities, it is a method that ensures that each step is reversible.

The conjugate of the binomial $a + b$ is $a - b$. If a binomial appears as the denominator of a fraction, the fraction can often be simplified by multiplying both numerator and denominator by the conjugate of the denominator. This algebraic technique can be applied to fractions containing trigonometric functions, as illustrated in Example 3.

Example 3   Verify the identity: $\dfrac{\sin \beta}{\csc \beta - 1} = \dfrac{1 + \sin \beta}{\cot^2 \beta}$.

Solution
A binomial, $\csc \beta - 1$, appears in the denominator of the left side of the equation. Our first step is to multiply numerator and denominator by the conjugate, $\csc \beta + 1$.

$$\frac{\sin \beta}{\csc \beta - 1} = \frac{\sin \beta(\csc \beta + 1)}{(\csc \beta - 1)(\csc \beta + 1)} \qquad \text{Algebra}$$

$$= \frac{\sin \beta \csc \beta + \sin \beta}{\csc^2 \beta - 1} \qquad \text{Algebra}$$

$$= \frac{1 + \sin \beta}{\cot^2 \beta} \qquad \begin{array}{l}\text{Reciprocal Identity}\\ \text{Pythagorean Identity}\end{array} \quad ❏$$

Sometimes a simplification might involve multiplying or factoring polynomials, as shown in the next example.

Example 4   Verify the identity: $1 + 2 \tan^2 \theta = \sec^4 \theta - \tan^4 \theta$.

Solution
We choose $\sec^4 \theta - \tan^4 \theta$ as the more complicated side, since it contains the higher powers of $\sec \theta$ and $\tan \theta$.

$$\sec^4 \theta - \tan^4 \theta = (\sec^2 \theta - \tan^2 \theta)(\sec^2 \theta + \tan^2 \theta) \qquad \text{Algebra}$$

$$= 1(\sec^2 \theta + \tan^2 \theta) \qquad \begin{array}{l}\text{Pythagorean}\\ \text{Identity}\end{array}$$

$$= 1 + \tan^2 \theta + \tan^2 \theta \qquad \begin{array}{l}\text{Pythagorean}\\ \text{Identity}\end{array}$$

$$= 1 + 2 \tan^2 \theta \qquad \text{Algebra} \quad ❏$$

As a final resort, it might be necessary to rewrite all the trigonometric functions in terms of sines and cosines.

Example 5   Prove the identity: $\dfrac{1}{\tan \alpha + \cot \alpha} = \sin \alpha \cos \alpha$.

Solution

It is tempting to multipy numerator and denominator of the left side by the conjugate, $\tan \alpha - \cot \alpha$, of the denominator. However, doing so would lead us astray. A more direct path to the goal of $\sin \alpha \cos \alpha$ is to rewrite the left side in terms of sines and cosines.

$$\frac{1}{\tan \alpha + \cot \alpha} = \frac{1}{\dfrac{\sin \alpha}{\cos \alpha} + \dfrac{\cos \alpha}{\sin \alpha}} \qquad \text{Quotient Identities}$$

$$= \frac{1}{\dfrac{\sin^2 \alpha + \cos^2 \alpha}{\sin \alpha \cos \alpha}} \qquad \text{Algebra}$$

$$= \frac{1}{\dfrac{1}{\sin \alpha \cos \alpha}} \qquad \text{Pythagorean Identity}$$

$$= \mathbf{\sin \alpha \cos \alpha} \qquad \text{Algebra} \qquad \square$$

The following list summarizes our suggestions for verifying identities. Note that these are only suggestions: There is no rigid step-by-step procedure that guarantees success. Success at verifying identities requires *practice and a thorough knowledge of the Fundamental Identities.*

Verifying Trigonometric Identities

1. Know the Fundamental Identities and be familiar with all the alternative forms.
2. Choose the more complicated side and reduce it to the simpler side.
3. Keep the goal (or alternative forms of the goal) in mind and work toward it. Remember, you cannot score unless you have a goal.
4. If there are fractions, you may need to do one of the following:
   a) Combine two or more fractions into one;
   b) Separate one fraction into two or more;
   c) Multiply numerator and denominator by the conjugate of the denominator (or in some cases, the conjugate of the numerator).
5. If there are trigonometric polynomials, you may need to
   a) Factor or
   b) Multiply.
6. If several trigonometric functions appear, use the Fundamental Identities to eliminate some of them.

7. If all else fails, rewrite all trigonometric functions in terms of sines and cosines and rely on algebraic manipulation of the resulting expression.

In our next example we prove that an equation is not an identity. It is sufficient to find one value of the variable for which both members of the equation are defined but for which the equation is false. Such equations are called *conditional equations.*

Example 6   Show that $\sin \theta + \cos \theta = 1$ is not an identity.

Solution
For $\theta = 180°$, we have

$$\sin 180° + \cos 180° = 0 + (-1) = -1 \neq 1.$$

Thus $\sin \theta + \cos \theta = 1$ is not an identity. In the next section we learn how to find the values of θ that do satisfy this equation.   ❏

In the solution of Example 6, it was necessary only to demonstrate that for one value of the variable both members of the equation were defined but had different values. A demonstration of this type is called a **counterexample.**

In Section 1.7 u-substitutions were made to change algebraic expressions to the form $c(u^2 \pm a^2)$. In the calculus this type of substitution is sometimes used with a Pythagorean Identity to simplify an expression that contains a radical. Our last example in this section illustrates the technique.

Example 7   Simplify the radical $\sqrt{x^2 - a^2}$ by making the substitution $x = a \sec u$, where $a > 0$ and $0 \leq u < \pi/2$.

Solution
With the given substitution, we have

$$\sqrt{x^2 - a^2} = \sqrt{a^2 \sec^2 u - a^2}$$
$$= \sqrt{a^2(\sec^2 u - 1)}$$
$$= \sqrt{a^2 \tan^2 u}.$$

Now $\sqrt{a^2} = a$ since $a > 0$. Similarly, $\sqrt{\tan^2 u} = \tan u$ since $\tan u \geq 0$ for $0 \leq u < \pi/2$. Thus we can write

$$\sqrt{x^2 - a^2} = a \tan u$$

as our final simplification.   ❏

Exercises 6.1

Reduce the first expression to the second. (See Example 1.)

1. $\sin \alpha \cot \alpha;\quad \cos \alpha$ 2. $\sec \beta \cot \beta;\quad \csc \beta$

3. $\dfrac{\cos^2 A + \sin^2 A}{\cos^2 A};\quad \sec^2 A$

4. $\dfrac{\cos^2 A + \sin^2 A}{\sin^2 A};\quad \csc^2 A$

5. $\dfrac{\sec^2 \alpha}{\csc^2 \alpha};\quad \tan^2 \alpha$ 6. $\dfrac{\csc^2 \theta}{\sec^2 \theta};\quad \cot^2 \theta$

7. $\dfrac{1 - \sin^2 \alpha}{1 - \cos^2 \alpha};\quad \cot^2 \alpha$ 8. $\dfrac{1 - \cos^2 \beta}{1 - \sin^2 \beta};\quad \tan^2 \beta$

9. $(1 + \tan^2 \theta)\sin^2 \theta;\quad \tan^2 \theta$

10. $(1 + \cot^2 \theta)\cos^2 \theta;\quad \cot^2 \theta$

11. $\dfrac{\cos \theta + \sin \theta \tan \theta}{\cos \theta};\quad \sec^2 \theta$

12. $\dfrac{\sin \theta + \cos \theta \cot \theta}{\sin \theta};\quad \csc^2 \theta$

13. $\tan \beta + \cot \beta;\quad \csc \beta \sec \beta$

14. $\sec^2 \beta + \csc^2 \beta;\quad \sec^2 \beta \csc^2 \beta$

In Exercises 15–18, write all the trigonometric functions in terms of the given function.

15. $\cos \theta$ 16. $\sin \theta$ 17. $\tan \theta$ 18. $\sec \theta$

In each of the following radicals, replace x by the indicated trigonometric expression and simplify by using the Fundamental Identities. (See Example 7.)

19. $\sqrt{1 + x^2};\quad x = \tan \theta,\quad 0° \le \theta < 90°$

20. $\sqrt{25 - x^2};\quad x = 5 \cos \theta,\quad 0° \le \theta < 90°$

21. $\sqrt{4 - x^2};\quad x = 2 \sin \theta,\quad 0° \le \theta < 90°$

22. $\sqrt{x^2 - 9};\quad x = 3 \sec \theta,\quad 0° \le \theta < 90°$

23. $\dfrac{1}{\sqrt{16 - x^2}};\quad x = 4 \sin u,\quad -\dfrac{\pi}{2} < u < \dfrac{\pi}{2}$

24. $\dfrac{1}{\sqrt{4 + x^2}};\quad x = 2 \tan u,\quad -\dfrac{\pi}{2} < u < \dfrac{\pi}{2}$

25. $(x^2 + 9)^{3/2};\quad x = 3 \tan u,\quad -\dfrac{\pi}{2} < u < \dfrac{\pi}{2}$

26. $(x^2 - 25)^{3/2};\quad x = 5 \sec u,\quad 0 \le u < \dfrac{\pi}{2}$

27. $\dfrac{\sqrt{x^2 + a^2}}{x};\quad x = a \tan t,\ a > 0 \text{ and } 0 < t < \dfrac{\pi}{2}$

28. $\dfrac{x}{\sqrt{a^2 - x^2}};\quad x = a \sin t,\ a > 0 \text{ and } -\dfrac{\pi}{2} < t < \dfrac{\pi}{2}$

29. $\dfrac{1}{x\sqrt{x^2 - a^2}};\quad x = a \sec t,\ a > 0 \text{ and } 0 < t < \dfrac{\pi}{2}$

30. $\dfrac{1}{x^2 + a^2};\quad x = a \tan t,\ a > 0 \text{ and } -\dfrac{\pi}{2} < t < \dfrac{\pi}{2}$

In Exercises 31–100, verify that each equation is an identity. (See Examples 2–5.)

31. $(\sec^2 \theta - 1)(\cot^2 \theta + 1) = \sec^2 \theta$

32. $(1 - \sin^2 \theta)(\sec^2 \theta - 1) = \sin^2 \theta$

33. $\dfrac{1 - \sin^2 \theta}{\cot \theta} = \sin \theta \cos \theta$

34. $\dfrac{1 - \cos^2 \theta}{\tan \theta} = \sin \theta \cos \theta$

35. $\cot^2 \beta - \cos^2 \beta = \cos^2 \beta \cot^2 \beta$

36. $\tan^2 \beta - \sin^2 \beta = \sin^2 \beta \tan^2 \beta$

37. $\sin^2 \alpha - \cos^2 \alpha \tan^2 \alpha = 0$

38. $\cos^2 \alpha - \sin^2 \alpha \cot^2 \alpha = 0$

39. $\sec^2 \alpha = \sin^2 \alpha + \tan^2 \alpha + \cos^2 \alpha$

40. $\cos^2 \alpha = \csc^2 \alpha - \sin^2 \alpha - \cot^2 \alpha$

41. $(1 - \sin \gamma)(1 + \sin \gamma) = \cos^2 \gamma$

42. $(\sec \gamma + 1)(\sec \gamma - 1) = \tan^2 \gamma$

43. $(\csc \gamma - \cot \gamma)(\csc \gamma + \cot \gamma) = 1$

44. $(\sec \gamma + \tan \gamma)(\sec \gamma - \tan \gamma) = 1$

45. $\dfrac{1}{\cos^2 A} = \tan A\,(\cot A + \tan A)$

46. $\dfrac{1}{\sin^2 A} = \cot A\,(\tan A + \cot A)$

47. $\dfrac{\cos A}{\sec A} + \dfrac{\sin A}{\csc A} = 1$

48. $\dfrac{\sec A}{\cos A} - \dfrac{\tan A}{\cot A} = 1$

49. $\dfrac{\csc B - \sin B}{\sin B} = \cot^2 B$

50. $\dfrac{\sec B - \cos B}{\cos B} = \tan^2 B$

51. $\dfrac{\cos B \tan B + \sin B}{1 - \cos^2 B} = 2 \csc B$

52. $\dfrac{\sin B \cot B + \cos B}{1 - \sin^2 B} = 2 \sec B$

53. $\cos t = \dfrac{1}{\sin t \,(\tan t + \cot t)}$

54. $-\sec^2 t = \dfrac{1}{\sin t \,(\sin t - \csc t)}$

55. $(\tan t + 1)^2 = \sec t \,(\sec t + 2 \sin t)$

56. $(\cot t + 1)^2 = \csc t \,(\csc t + 2 \cos t)$

57. $(\cos s + \sin s)^2 + (\cos s - \sin s)^2 = 2$

58. $(\tan s + \sec s)^2 - (\tan s - \sec s)^2 = 4 \sin s \sec^2 s$

59. $(\csc s - \sec s)(\sin s + \cos s) = \cot s - \tan s$

60. $(\cot s - \csc s)(\tan s + \sin s) = \cos s - \sec s$

61. $\dfrac{\cos \theta}{1 - \sin \theta} = \dfrac{1 + \sin \theta}{\cos \theta}$

62. $\dfrac{\sin \theta}{1 - \cos \theta} = \dfrac{1 + \cos \theta}{\sin \theta}$

63. $\dfrac{\tan \theta}{\sec \theta + 1} = \dfrac{\sec \theta - 1}{\tan \theta}$

64. $\dfrac{\cot \theta}{\csc \theta - 1} = \dfrac{\csc \theta + 1}{\cot \theta}$

65. $\dfrac{\sec u - 1}{\tan^2 u} = \dfrac{\cos u}{\cos u + 1}$

66. $\dfrac{\csc u + 1}{\cot^2 u} = \dfrac{\sin u}{1 - \sin u}$

67. $\dfrac{\cos u}{\sec u - \tan u} = 1 + \sin u$

68. $\dfrac{\sin u}{\csc u + \cot u} = 1 - \cos u$

69. $\dfrac{\tan \alpha - \cot \alpha}{\cot \alpha} - \dfrac{\cot \alpha - \tan \alpha}{\tan \alpha} = \sec^2 \alpha - \csc^2 \alpha$

70. $\dfrac{\sec \alpha - \cos \alpha}{\sec \alpha} - \dfrac{\sin \alpha - \csc \alpha}{\csc \alpha} = 1$

71. $\dfrac{1 + \cos \alpha}{1 - \cos \alpha} = \dfrac{\sec \alpha + 1}{\sec \alpha - 1}$

72. $\dfrac{\tan \alpha - 1}{\tan \alpha + 1} = \dfrac{1 - \cot \alpha}{1 + \cot \alpha}$

73. $\dfrac{\cot \gamma - \tan \gamma}{\cot \gamma + \tan \gamma} = \cos^2 \gamma - \sin^2 \gamma$

74. $\dfrac{\tan^2 \gamma - 1}{1 - \cot^2 \gamma} = \tan^2 \gamma$

75. $\dfrac{\csc \gamma + \sin \gamma}{\csc \gamma - \sin \gamma} = 1 + 2 \tan^2 \gamma$

76. $\dfrac{\sec \gamma + \cos \gamma}{\sec \gamma - \cos \gamma} = 1 + 2 \cot^2 \gamma$

77. $\dfrac{\cos x}{1 - \sin x} - \dfrac{\cos x}{1 + \sin x} = 2 \tan x$

78. $\dfrac{\tan x}{1 - \cos x} - \dfrac{\tan x}{1 + \cos x} = 2 \csc x$

79. $\dfrac{\sin x}{\sec x - 1} + \dfrac{\sin x}{\sec x + 1} = 2 \cot x$

80. $\dfrac{\cot x}{\csc x - 1} - \dfrac{\cot x}{\csc x + 1} = 2 \tan x$

81. $\cos^4 t - \sin^4 t = 2 \cos^2 t - 1$

82. $\sec^4 t - \tan^4 t = 2 \sec^2 t - 1$

83. $\dfrac{\cos^3 t - \sin^3 t}{\cos t - \sin t} = 1 + \cos t \sin t$

84. $\dfrac{1 - \tan^3 t}{1 - \tan t} = \sec^2 t + \tan t$

85. $\dfrac{\sin \theta}{1 - \cos \theta} + \dfrac{1 - \cos \theta}{\sin \theta} = 2 \csc \theta$

86. $\dfrac{\sec \theta}{1 - \tan \theta} - \dfrac{1 - \tan \theta}{\sec \theta} = \dfrac{2 \sin \theta}{1 - \tan \theta}$

87. $(\cot \theta - \csc \theta)^2 = \dfrac{1 - \cos \theta}{1 + \cos \theta}$

88. $(\tan \theta + \sec \theta)^2 = \dfrac{\sin \theta + 1}{1 - \sin \theta}$

89. $\dfrac{\sin^2 x}{1 - \cos x} = \dfrac{\sin x}{\csc x - \cot x}$

90. $\dfrac{\tan^2 x}{1 - \sec x} = \dfrac{\tan x}{\cot x - \csc x}$

91. $\dfrac{\tan x + 1}{\sin x + \cos x} = \sec x$

92. $\dfrac{\csc x + 1}{\sin x + 1} = \csc x$

93. $\dfrac{\sin A \cos B + \cos A \sin B}{\cos A \cos B - \sin A \sin B} = \dfrac{\tan A + \tan B}{1 - \tan A \tan B}$

94. $\dfrac{\sin A \cos B - \cos A \sin B}{\cos A \cos B + \sin A \sin B} = \dfrac{\cot B - \cot A}{\cot A \cot B + 1}$

95. $\dfrac{\sin A + \sin B}{\csc A + \csc B} = \sin A \sin B$

96. $\dfrac{\tan A + \tan B}{\cot A + \cot B} = \tan A \tan B$

97. $2 - \cos^2 \theta \csc \theta = \dfrac{\cos \theta}{\sec \theta - \tan \theta} - \dfrac{\cot \theta}{\sec \theta + \tan \theta}$

98. $2 + \sin^2 \theta \sec \theta = \dfrac{\sin \theta}{\csc \theta + \cot \theta} + \dfrac{\tan \theta}{\csc \theta - \cot \theta}$

99. $\dfrac{\cot \theta + \cos \theta}{\cot \theta \cos \theta} = \dfrac{\cot \theta \cos \theta}{\cot \theta - \cos \theta}$

100. $\dfrac{\tan \theta \sin \theta}{\tan \theta + \sin \theta} = \dfrac{\tan \theta - \sin \theta}{\tan \theta \sin \theta}$

Show that each equation in Exercises 101–108 is not an identity by providing a counterexample. (See Example 6.)

101. $(\sin \theta + \cos \theta)^2 = \sin^2 \theta + \cos^2 \theta$

102. $\tan \alpha = \sec \alpha - 1$

103. $\cot^2 \beta = (\csc \beta - 1)^2$

104. $\sin 2\alpha = 2 \sin \alpha$ **105.** $\tan 3A = 3 \tan A$

106. $\cot \frac{1}{2} \beta = \frac{1}{2} \cot \beta$

107. $\cos (\alpha + \beta) = \cos \alpha + \cos \beta$

108. $\sec A + \csc A = \dfrac{1}{\cos A + \sin A}$

6.2	**Cosine of the Sum or Difference of Two Angles**

The Fundamental Identities involve trigonometric functions of a single angle, say θ, α, or β. The sum, $\alpha + \beta$, the difference, $\alpha - \beta$, and multiples, 2θ and $\frac{1}{2}\theta$, of these angles are called **composite angles.** In this chapter we study the interesting and sometimes surprising relationships that exist between the trigonometric functions of composite angles and the trigonometric functions of a single angle. These relationships are known as the **Composite Angle Identities.**

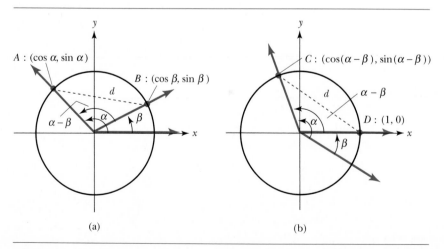

(a)   (b)

FIGURE 6.1

It would be nice if $\sin 2\theta$ were the same as $2 \sin \theta$, for all θ. But unfortunately $\sin 2\theta = 2 \sin \theta$ is true only for integral multiples of $180°$. Similarly, it would be nice if $\cos (\alpha - \beta)$ were equal to $\cos \alpha - \cos \beta$ for all α and β. Again, this is not true. In fact, we now prove the intriguing result that

$$\cos (\alpha - \beta) = \cos \alpha \cos \beta + \sin \alpha \sin \beta.$$

In Figure 6.1(a), angles α and β are drawn in standard position with $\alpha > \beta$. Points A and B are located where the unit circle intersects the terminal sides of α and β, respectively. The angle formed between the terminal sides of α and β is $\alpha - \beta$, and the points A and B have coordinates $(\cos \alpha, \sin \alpha)$ and $(\cos \beta, \sin \beta)$, respectively.

In Figure 6.1(a) the length of the chord subtended by the angle $\alpha - \beta$ is the distance d between the points A and B. The distance formula gives

$$
\begin{aligned}
d^2 &= (\cos \alpha - \cos \beta)^2 + (\sin \alpha - \sin \beta)^2 \\
&= \cos^2 \alpha - 2 \cos \alpha \cos \beta + \cos^2 \beta + \sin^2 \alpha \\
&\quad - 2 \sin \alpha \sin \beta + \sin^2 \beta \\
&= (\cos^2 \alpha + \sin^2 \alpha) + (\cos^2 \beta + \sin^2 \beta) \\
&\quad - 2(\cos \alpha \cos \beta + \sin \alpha \sin \beta) \\
&= 2 - 2(\cos \alpha \cos \beta + \sin \alpha \sin \beta).
\end{aligned}
$$

Next we rotate the angles in Figure 6.1(a) so that $\alpha - \beta$ is in standard position, as in Figure 6.1(b). Then d, the length of the chord subtended by $\alpha - \beta$, is the distance between the points C with coordinates $(\cos (\alpha - \beta), \sin (\alpha - \beta))$ and D with coordinates $(1, 0)$. Then

$$
\begin{aligned}
d^2 &= [\cos (\alpha - \beta) - 1]^2 + [\sin (\alpha - \beta) - 0]^2 \\
&= \cos^2 (\alpha - \beta) - 2 \cos (\alpha - \beta) + 1 + \sin^2 (\alpha - \beta) \\
&= [\cos^2 (\alpha - \beta) + \sin^2 (\alpha - \beta)] + 1 - 2 \cos (\alpha - \beta) \\
&= 2 - 2 \cos (\alpha - \beta).
\end{aligned}
$$

Thus we have

$$2 - 2 \cos (\alpha - \beta) = 2 - 2(\cos \alpha \cos \beta + \sin \alpha \sin \beta),$$

and this leads to the following identity.

Cosine of the Difference of Two Angles

$$\cos (\alpha - \beta) = \cos \alpha \cos \beta + \sin \alpha \sin \beta$$

We illustrate some uses of this identity in the next examples.

Example 1   Without determining α or β, find the exact value of $\cos(\alpha - \beta)$ if $\cos\alpha = -\frac{4}{5}$, $\sin\beta = -\frac{7}{25}$, α is in quadrant III, and β is in quadrant IV.

Solution
Knowing that $\cos\alpha = -\frac{4}{5}$ and α is in quadrant III, we find $\sin\alpha = -\frac{3}{5}$. Also since β is in quadrant IV with $\sin\beta = -\frac{7}{25}$ then $\cos\beta = \frac{24}{25}$. Then

$$\cos(\alpha - \beta) = \cos\alpha\cos\beta + \sin\alpha\sin\beta$$

$$= \left(-\frac{4}{5}\right)\left(\frac{24}{25}\right) + \left(-\frac{3}{5}\right)\left(-\frac{7}{25}\right)$$

$$= \frac{-96 + 21}{125}$$

$$= -\frac{75}{125}$$

$$= -\frac{3}{5}.$$

     In the next example we prove that the equation $\cos(90° - \theta) = \sin\theta$ is an identity. This generalizes to arbitrary angles a cofunction relation that was noted for right triangles in Section 5.6.

Example 2   Verify that the equation $\cos(90° - \theta) = \sin\theta$ is an identity.

Solution
For any θ,

$$\mathbf{\cos(90° - \theta)} = \cos 90°\cos\theta + \sin 90°\sin\theta$$

$$= 0 \cdot \cos\theta + 1 \cdot \sin\theta$$

$$= \mathbf{\sin\theta,}$$

and the identity is proved.

     If we replace θ by $90° - \theta$ in the identity in Example 2, we obtain another identity:

$$\mathbf{\sin(90° - \theta)} = \cos[90° - (90° - \theta)]$$

$$= \mathbf{\cos\theta.}$$

     Proofs are requested in the exercises to show that **any function of a given angle equals the cofunction of the complementary angle** whenever both functions are defined. Stated as equations, these relationships are called the **Cofunction Identities.**

Cofunction Identities

$\cos(90° - \theta) = \sin\theta$	$\sin(90° - \theta) = \cos\theta$
$\cot(90° - \theta) = \tan\theta$	$\tan(90° - \theta) = \cot\theta$
$\csc(90° - \theta) = \sec\theta$	$\sec(90° - \theta) = \csc\theta$

Corresponding statements can be made for radian measure or functions of real numbers: $\cos(\pi/2 - \theta) = \sin\theta$, $\sin(\pi/2 - \theta) = \cos\theta$, and so on.

We write $-\theta = 0 - \theta$ as a first step in expressing the cosine of $-\theta$ in terms of the cosine of θ. The next step is to apply the identity for the cosine of the difference of two angles:

$$\cos(-\theta) = \cos(0 - \theta)$$

$$= \cos 0 \cos\theta + \sin 0 \sin\theta \qquad \text{Cosine of the Difference}$$

$$= 1 \cdot \cos\theta + 0 \cdot \sin\theta \qquad \text{Special Angles}$$

$$= \cos\theta.$$

This identity,

$$\cos(-\theta) = \cos\theta,$$

is one of the **Negative-Angle Identities.**

To derive the Negative-Angle Identity for the sine function, we write

$$\sin(-\theta) = \cos\left[\frac{\pi}{2} - (-\theta)\right] \qquad \text{Cofunction Identity}$$

$$= \cos\left(\frac{\pi}{2} + \theta\right) \qquad \text{Algebra}$$

$$= \cos\left[\theta - \left(-\frac{\pi}{2}\right)\right] \qquad \text{Algebra}$$

$$= \cos\theta\cos\left(-\frac{\pi}{2}\right) + \sin\theta\sin\left(-\frac{\pi}{2}\right) \qquad \text{Cosine of the Difference}$$

$$= \cos\theta \cdot 0 + \sin\theta \cdot (-1) \qquad \text{Special Angles}$$

$$= -\sin\theta.$$

The corresponding Negative-Angle Identities for the remaining trigonometric functions can easily be established by using the Fundamental Identities.

Negative-Angle Identities

$\sin(-\theta) = -\sin\theta$	$\cos(-\theta) = \cos\theta$	$\tan(-\theta) = -\tan\theta$
$\csc(-\theta) = -\csc\theta$	$\sec(-\theta) = \sec\theta$	$\cot(-\theta) = -\cot\theta$

The Negative-Angle Identities are sometimes useful in graphing. This is illustrated in the next example.

Example 3   Sketch the graph of

$$y = 2\sin(-x)$$

through one complete period.

Solution
We use the Negative-Angle Identity to write

$$y = 2\sin(-x)$$
$$= -2\sin x.$$

As shown in Figure 6.2, the graph is a reflection through the x-axis of a typical sine curve with Amp $= 2$, $P = 2\pi$, and no phase shift. ❏

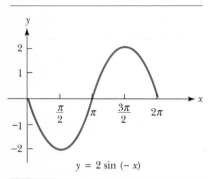

$y = 2\sin(-x)$

FIGURE 6.2

We can use the identities

$$\cos(\alpha - \beta) = \cos\alpha\cos\beta + \sin\alpha\sin\beta$$
$$\cos(-\theta) = \cos\theta$$
$$\sin(-\theta) = -\sin\theta$$

to derive an identity for the cosine of the sum, $\alpha + \beta$. We first write $\alpha + \beta$ as $\alpha - (-\beta)$.

$\cos(\alpha + \beta) = \cos[\alpha - (-\beta)]$	Algebra
$= \cos\alpha\cos(-\beta) + \sin\alpha\sin(-\beta)$	Cosine of the Difference
$= \cos\alpha\cos\beta + \sin\alpha(-\sin\beta)$	Negative-Angle Identities
$= \cos\alpha\cos\beta - \sin\alpha\sin\beta$	Algebra

This establishes the desired identity.

Cosine of the Sum of Two Angles

$$\cos(\alpha + \beta) = \cos\alpha\cos\beta - \sin\alpha\sin\beta$$

The identities for the cosine of the sum and difference of two angles can be used with the function values of the special angles to find the *exact* values of the cosine of certain angles.

Example 4   Find the exact value of cos 285°.

Solution
We can write 285° as the sum of two special angles, 225° and 60°. Thus

$$\cos 285° = \cos(225° + 60°)$$
$$= \cos 225° \cos 60° - \sin 225° \sin 60°$$
$$= \left(-\frac{\sqrt{2}}{2}\right)\left(\frac{1}{2}\right) - \left(-\frac{\sqrt{2}}{2}\right)\left(\frac{\sqrt{3}}{2}\right)$$
$$= \frac{\sqrt{2}}{4}(\sqrt{3} - 1).$$

As a final example, we verify an identity.

Example 5   Verify the identity: $\dfrac{\cos(\alpha + \beta)}{\cos(\alpha - \beta)} = \dfrac{\cot \alpha \cot \beta - 1}{\cot \alpha \cot \beta + 1}$.

Solution

$$\frac{\cos(\alpha + \beta)}{\cos(\alpha - \beta)} = \frac{\cos \alpha \cos \beta - \sin \alpha \sin \beta}{\cos \alpha \cos \beta + \sin \alpha \sin \beta}$$ Cosine of the Sum
Cosine of the Difference

$$= \frac{\dfrac{\cos \alpha \cos \beta}{\sin \alpha \sin \beta} - \dfrac{\sin \alpha \sin \beta}{\sin \alpha \sin \beta}}{\dfrac{\cos \alpha \cos \beta}{\sin \alpha \sin \beta} + \dfrac{\sin \alpha \sin \beta}{\sin \alpha \sin \beta}}$$ Algebra

$$= \frac{\cot \alpha \cot \beta - 1}{\cot \alpha \cot \beta + 1}$$ Quotient Identities

Exercises 6.2

Use an identity to find the exact value of each of the following.

1. cos 127° cos 37° + sin 127° sin 37°
2. cos 200° cos 20° + sin 200° sin 20°
3. cos 1° cos 361° + sin 1° sin 361°
4. cos 257° cos 527° + sin 257° sin 527°
5. cos 42° cos 138° − sin 42° sin 138°
6. cos 23° cos 67° − sin 23° sin 67°

Express each of the following in terms of trigonometric functions of θ.

7. $\cos(\pi + \theta)$

8. $\cos(\pi - \theta)$

9. $\cos\left(\theta - \dfrac{3\pi}{2}\right)$

10. $\cos\left(\theta + \dfrac{\pi}{3}\right)$

11. $\cos\left(\theta - \dfrac{5\pi}{3}\right)$

12. $\cos\left(\theta + \dfrac{7\pi}{6}\right)$

Without determining α or β, find the exact values of $\cos(\alpha - \beta)$ and $\cos(\alpha + \beta)$ if α and β satisfy the given conditions. (See Example 1.)

13. $\cos\alpha = \frac{3}{5}$, $\sin\beta = \frac{3}{5}$; α and β in Q I

14. $\cos\alpha = \frac{7}{25}$, $\cos\beta = \frac{5}{13}$; α and β in Q IV

15. $\cos\alpha = -\frac{4}{5}$, $\sin\beta = -\frac{12}{13}$; α in Q II, β in Q III

16. $\sin\alpha = \frac{15}{17}$, $\sin\beta = \frac{5}{13}$; α in Q II, β in Q I

17. $\cos\alpha = -\frac{24}{25}$, $\sin\beta = -1$; α in Q III

18. $\sin\alpha = 1$, $\sin\beta = -\frac{3}{5}$; β in Q III

19. $\sin\alpha = -\frac{1}{2}$, $\cos\beta = -\frac{5}{13}$; α in Q III, β in Q II

20. $\sin\alpha = \frac{8}{17}$, $\sin\beta = \frac{2}{3}$; α in Q II, β in Q II

21. $\cos\alpha = \dfrac{\sqrt{7}}{3}$, $\sin\beta = \dfrac{\sqrt{5}}{4}$; α in Q IV, β in Q II

22. $\cos\alpha = -\frac{3}{7}$, $\cos\beta = -\dfrac{\sqrt{2}}{3}$; α in Q II, β in Q III

Write each of the following as a single term.

23. $\cos 2x \cos x - \sin 2x \sin x$

24. $\cos 3y \cos 2y + \sin 3y \sin 2y$

25. $\cos\dfrac{x}{4}\cos\dfrac{x}{6} + \sin\dfrac{x}{4}\sin\dfrac{x}{6}$

26. $\cos\dfrac{x}{2}\cos\dfrac{x}{3} - \sin\dfrac{x}{2}\sin\dfrac{x}{3}$

Use the Negative-Angle Identities to simplify and then sketch the graph through one complete period. (See Example 3.)

27. $y = \cos(-x)$

28. $y = \tan(-x)$

29. $y = 3\sin(-x)$

30. $y = 2\cos(-x)$

31. $y = -3\cos(-2x)$

32. $y = -3\sin(-2x)$

Find the exact value of each of the following by using special angles. (See Example 4.)

33. $\cos 75°$

34. $\cos 195°$

35. $\cos 15°$

36. $\cos 165°$

37. $\cos\dfrac{7\pi}{12}$

38. $\cos\dfrac{5\pi}{12}$

39. $\cos\left(-\dfrac{\pi}{12}\right)$

40. $\cos\left(-\dfrac{7\pi}{12}\right)$

Verify the identities. (See Example 5.)

41. $\cos(t+s) + \cos(t-s) = 2\cos t \cos s$

42. $\cos(\alpha+\beta) - \cos(\alpha-\beta) = -2\sin\alpha\sin\beta$

43. $\dfrac{\cos(\alpha+\beta)}{\cos\alpha\cos\beta} = 1 - \tan\alpha\tan\beta$

44. $\dfrac{\cos(x-y)}{\sin x \sin y} = \cot x \cot y + 1$

45. $\cos(\alpha+\beta)\cos(\alpha-\beta) + \sin(\alpha+\beta)\sin(\alpha-\beta) = \cos 2\beta$

46. $\cos(\alpha+\beta)\cos(\alpha-\beta) - \sin(\alpha+\beta)\sin(\alpha-\beta) = \cos 2\alpha$

47. $\dfrac{\cos(\alpha+\beta) + \cos(\alpha-\beta)}{\cos(\alpha+\beta) - \cos(\alpha-\beta)} = -\cot\alpha\cot\beta$

48. $\dfrac{\cos(\alpha-\beta)}{\cos(\alpha+\beta)} = \dfrac{1+\tan\alpha\tan\beta}{1-\tan\alpha\tan\beta}$

49. $\dfrac{\cos\alpha}{\sin\beta} - \dfrac{\sin\alpha}{\cos\beta} = \dfrac{\cos(\alpha+\beta)}{\sin\beta\cos\beta}$

50. $\dfrac{\cos 2\theta}{\sin 3\theta} + \dfrac{\sin 2\theta}{\cos 3\theta} = \dfrac{\cos\theta}{\sin 3\theta\cos 3\theta}$

51. $\cos(\alpha+\beta)\cos(\alpha-\beta) = \cos^2\alpha - \sin^2\beta$

52. $\dfrac{\cos(\theta+h) - \cos\theta}{h} = $
$\cos\theta\left(\dfrac{\cos h - 1}{h}\right) - \sin\theta\left(\dfrac{\sin h}{h}\right)$

Prove the following Cofunction Identities and Negative-Angle Identities.

53. $\tan(90° - \theta) = \cot\theta$ **54.** $\cot(90° - \theta) = \tan\theta$

55. $\sec(90° - \theta) = \csc\theta$ **56.** $\csc(90° - \theta) = \sec\theta$

57. $\tan(-\theta) = -\tan\theta$ **58.** $\cot(-\theta) = -\cot\theta$

59. $\sec(-\theta) = \sec\theta$ **60.** $\csc(-\theta) = -\csc\theta$

Show that each equation is not an identity by providing a counterexample.

61. $\cos (x - y) = \cos x - \cos y$

62. $\cos \left(\dfrac{\pi}{2} - \theta \right) = \cos \theta$

63. $\tan \left(t - \dfrac{\pi}{2} \right) = \cot t$

64. $\sin \left(v - \dfrac{\pi}{2} \right) = \cos v$

6.3 Identities for the Sine and Tangent

The major goal of this section is to extend our list of identities to include identities for the sine and tangent of the sum and difference of two angles.

Sine and Tangent of the Sum and Difference of Two Angles

$$\sin (\alpha + \beta) = \sin \alpha \cos \beta + \cos \alpha \sin \beta$$

$$\sin (\alpha - \beta) = \sin \alpha \cos \beta - \cos \alpha \sin \beta$$

$$\tan (\alpha + \beta) = \frac{\tan \alpha + \tan \beta}{1 - \tan \alpha \tan \beta}$$

$$\tan (\alpha - \beta) = \frac{\tan \alpha - \tan \beta}{1 + \tan \alpha \tan \beta}$$

The first identity is established by using three identities derived in the preceding section.

$\sin (\alpha + \beta) = \cos [90° - (\alpha + \beta)]$	Cofunction Identity
$= \cos [(90° - \alpha) - \beta]$	Algebra
$= \cos (90° - \alpha) \cos \beta$	Cosine of the Difference
$\quad + \sin (90° - \alpha) \sin \beta$	
$= \sin \alpha \cos \beta + \cos \alpha \sin \beta$	Cofunction Identities

The identity for the sine of the difference of two angles follows from the identity for the sine of the sum; we first write $\alpha - \beta$ as $\alpha + (-\beta)$.

$\sin (\alpha - \beta) = \sin [\alpha + (-\beta)]$	Algebra
$= \sin \alpha \cos (-\beta) + \cos \alpha \sin (-\beta)$	Sine of the Sum
$= \sin \alpha \cos \beta - \cos \alpha \sin \beta$	Negative-Angle Identities

Next we derive the identity for the tangent of the sum of two angles by expressing the tangent function as the quotient of the sine and cosine functions.

$$\textbf{tan } (\alpha + \beta) = \frac{\sin (\alpha + \beta)}{\cos (\alpha + \beta)} \qquad \text{Quotient Identity}$$

$$= \frac{\sin \alpha \cos \beta + \cos \alpha \sin \beta}{\cos \alpha \cos \beta - \sin \alpha \sin \beta} \qquad \begin{array}{l}\text{Sine of the Sum}\\ \text{Cosine of the Sum}\end{array}$$

$$= \frac{\dfrac{\sin \alpha \cos \beta}{\cos \alpha \cos \beta} + \dfrac{\cos \alpha \sin \beta}{\cos \alpha \cos \beta}}{\dfrac{\cos \alpha \cos \beta}{\cos \alpha \cos \beta} - \dfrac{\sin \alpha \sin \beta}{\cos \alpha \cos \beta}} \qquad \text{Algebra}$$

$$= \frac{\textbf{tan } \alpha + \textbf{tan } \beta}{1 - \textbf{tan } \alpha \textbf{ tan } \beta} \qquad \text{Quotient Identities}$$

Thus the identity for the tangent of the sum of two angles is established. We leave the proof of the identity for the tangent of the difference of two angles to the student. (See Exercise 39 at the end of this section.)

Some uses of these identities are demonstrated in the next examples.

Example 1 Use identities to find the exact value of each of the following:

a) $\sin 20° \cos 110° - \cos 20° \sin 110°$;

b) $\dfrac{\tan 43° + \tan 137°}{1 - \tan 43° \tan 137°}$;

c) $\tan (\alpha - \beta)$, if the exact values of $\tan \alpha$ and $\tan \beta$ are 1.1 and 3.7, respectively.

Solution
In each part we apply one of the identities listed at the beginning of this section.

a) $\sin 20° \cos 110° - \cos 20° \sin 110° = \sin (20° - 110°) =$
$$\sin (-90°) = -1$$

b) $\dfrac{\tan 43° + \tan 137°}{1 - \tan 43° \tan 137°} = \tan (43° + 137°) = \tan 180° = 0$

c) $\tan (\alpha - \beta) = \dfrac{\tan \alpha - \tan \beta}{1 + \tan \alpha \tan \beta} = \dfrac{1.1 - 3.7}{1 + 1.1(3.7)} = \dfrac{-2.6}{5.07} = -\dfrac{20}{39}$

❏

Example 2 Use the special angles to find the exact value of $\sin 165°$.

Solution

Since we can write 165° as 135° + 30°, we have

$$\sin 165° = \sin (135° + 30°)$$

$$= \sin 135° \cos 30° + \cos 135° \sin 30°$$

$$= \frac{\sqrt{2}}{2} \frac{\sqrt{3}}{2} + \left(-\frac{\sqrt{2}}{2}\right)\left(\frac{1}{2}\right) = \frac{\sqrt{2}}{4}(\sqrt{3} - 1).$$ ❑

Example 3 Verify the identity:

$$\sin (\alpha + \beta) \cos (\alpha - \beta) = \sin \alpha \cos \alpha + \sin \beta \cos \beta.$$

Solution

$$\mathbf{\sin (\alpha + \beta) \cos (\alpha - \beta)} = (\sin \alpha \cos \beta + \cos \alpha \sin \beta)$$

$$\times (\cos \alpha \cos \beta + \sin \alpha \sin \beta)$$

$$= \sin \alpha \cos \alpha \cos^2 \beta + \cos^2 \alpha \sin \beta \cos \beta$$

$$+ \sin^2 \alpha \sin \beta \cos \beta + \cos \alpha \sin \alpha \sin^2 \beta$$

$$= \sin \alpha \cos \alpha (\cos^2 \beta + \sin^2 \beta)$$

$$+ \sin \beta \cos \beta (\cos^2 \alpha + \sin^2 \alpha)$$

$$= \sin \alpha \cos \alpha \cdot 1 + \sin \beta \cos \beta \cdot 1$$

$$\mathbf{= \sin \alpha \cos \alpha + \sin \beta \cos \beta}$$ ❑

In the area of mathematics known as differential equations, expressions of the form $a \sin \theta + b \cos \theta$ are often encountered. The identity for the sine of the sum can be used to rewrite this type of expression in a more convenient form. Given the real numbers a and b, we can draw, as in Figure 6.3, an angle α in standard position whose terminal side passes through the point with coordinates (a, b). Then

$$\cos \alpha = \frac{a}{\sqrt{a^2 + b^2}} \quad \text{and} \quad \sin \alpha = \frac{b}{\sqrt{a^2 + b^2}}.$$

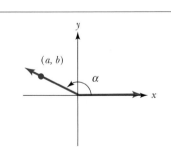

FIGURE 6.3

Now

$$\sqrt{a^2 + b^2} \sin (\theta + \alpha)$$

$$= \sqrt{a^2 + b^2} (\sin \theta \cos \alpha + \cos \theta \sin \alpha) \qquad \text{Sine of the Sum}$$

$$= \sqrt{a^2 + b^2} \left(\sin \theta \cdot \frac{a}{\sqrt{a^2 + b^2}} + \cos \theta \cdot \frac{b}{\sqrt{a^2 + b^2}}\right)$$

$$= a \sin \theta + b \cos \theta.$$

This result is known as the *Reduction Identity*. We use it in this section for graphing and in Section 6.6 to solve certain trigonometric equations.

Reduction Identity

$$a \sin \theta + b \cos \theta = \sqrt{a^2 + b^2} \sin (\theta + \alpha),$$

where α is determined by the equations

$$\cos \alpha = \frac{a}{\sqrt{a^2 + b^2}} \quad \text{and} \quad \sin \alpha = \frac{b}{\sqrt{a^2 + b^2}}.$$

Example 4   Sketch the graph of $y = \sin x - \cos x$ over the interval $0 \le x \le 2\pi$.

Solution

It is possible to sketch the graph of $y = \sin x - \cos x$ by using the method of addition of ordinates, as described in Section 5.8. However, it can be sketched more easily if we rewrite this equation using the Reduction Identity with $a = 1, b = -1$. Then

$$\sqrt{a^2 + b^2} = \sqrt{2} \quad \text{and} \quad \alpha = -\frac{\pi}{4}$$

since

$$\cos \alpha = \frac{1}{\sqrt{2}} \quad \text{and} \quad \sin \alpha = \frac{-1}{\sqrt{2}}.$$

Thus the graph of

$$y = \sin x - \cos x$$

is the same as the graph of

$$y = \sqrt{2} \sin \left(x - \frac{\pi}{4} \right).$$

This is a sine curve with Amp $= \sqrt{2}$, $P = 2\pi$, and phase shift $\pi/4$ units to the right. The curve is sketched in Figure 6.4. ❑

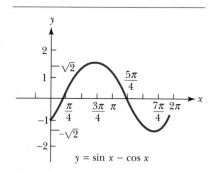

FIGURE 6.4

Exercises 6.3

Use an identity to find the exact value of each of the following. (See Example 1.)

1. $\sin 29° \cos 31° + \cos 29° \sin 31°$

2. $\sin 115° \cos 20° + \cos 115° \sin 20°$

3. $\sin 198° \cos 18° - \cos 198° \sin 18°$

4. $\sin 625° \cos 310° - \cos 625° \sin 310°$

5. $\dfrac{\tan 66° + \tan 54°}{1 - \tan 66° \tan 54°}$

6. $\dfrac{\tan 100° + \tan 80°}{1 - \tan 100° \tan 80°}$

7. $\dfrac{\tan 17° - \tan 47°}{1 + \tan 17° \tan 47°}$ **8.** $\dfrac{\tan 74° - \tan 314°}{1 + \tan 74° \tan 314°}$

9. $\tan (\alpha + \beta)$ if $\tan \alpha = 2.1$ (exact) and $\tan \beta = -0.3$ (exact)

10. $\tan (\alpha - \beta)$ if $\tan \alpha = -1.2$ (exact) and $\tan \beta = 4$ (exact)

11. $\tan (\alpha - \beta)$ if $\cot \alpha = 15$ (exact) and $\cot \beta = 20$ (exact)

12. $\tan (\alpha + \beta)$ if $\cot \alpha = 0.7$ (exact) and $\cot \beta = 1.4$ (exact)

Without determining α or β, find the exact values of (a) $\sin (\alpha + \beta)$, (b) $\sin (\alpha - \beta)$, (c) $\tan (\alpha + \beta)$, and (d) $\tan (\alpha - \beta)$, if α and β satisfy the given conditions.

13. $\cos \alpha = \frac{3}{5}$, $\sin \beta = \frac{3}{5}$; α and β in Q I

14. $\cos \alpha = \frac{7}{25}$, $\cos \beta = \frac{5}{13}$; α and β in Q IV

15. $\cos \alpha = -\frac{4}{5}$, $\sin \beta = -\frac{12}{13}$; α in Q II, β in Q III

16. $\sin \alpha = \frac{15}{17}$, $\sin \beta = \frac{5}{13}$; α in Q II, β in Q I

Find the exact value of each of the following by using special angles. (See Example 2.)

17. $\sin 75°$ **18.** $\sin 105°$ **19.** $\sin \dfrac{\pi}{12}$

20. $\sin \left(-\dfrac{7\pi}{12}\right)$ **21.** $\tan \dfrac{7\pi}{12}$ **22.** $\tan \dfrac{5\pi}{12}$

23. $\tan 165°$ **24.** $\tan 285°$

Express each of the following in terms of trigonometric functions of θ.

25. $\sin (\pi - \theta)$ **26.** $\sin \left(\theta - \dfrac{3\pi}{2}\right)$

27. $\sin \left(\theta + \dfrac{5\pi}{6}\right)$ **28.** $\sin \left(\theta + \dfrac{3\pi}{4}\right)$

29. $\tan (\theta + \pi)$ **30.** $\tan \left(\theta + \dfrac{5\pi}{3}\right)$

31. $\tan \left(\theta - \dfrac{3\pi}{4}\right)$ **32.** $\tan (\theta - 3\pi)$

Sketch the graph over the interval $0 \le x \le 2\pi$. (See Example 4.)

33. $y = \sin x + \cos x$ **34.** $y = -2 \sin x + 2 \cos x$
35. $y = \sqrt{3} \sin x - \cos x$
36. $y = -\sin x - \sqrt{3} \cos x$

Verify the identities. (See Example 3.)

37. $\sin (\alpha + \beta) + \sin (\alpha - \beta) = 2 \sin \alpha \cos \beta$

38. $\sin (\alpha + \beta) - \sin (\alpha - \beta) = 2 \cos \alpha \sin \beta$

39. $\tan (\alpha - \beta) = \dfrac{\tan \alpha - \tan \beta}{1 + \tan \alpha \tan \beta}$

40. $\cot \left(\dfrac{\pi}{2} - \theta\right) = \tan \theta$

41. $\sin (\alpha + \beta) \sin (\alpha - \beta) = \sin^2 \alpha - \sin^2 \beta$

42. $\tan (\alpha - \beta) \tan (\alpha + \beta) = \dfrac{\tan^2 \alpha - \tan^2 \beta}{1 - \tan^2 \alpha \tan^2 \beta}$

43. $\dfrac{\sin (\alpha + \beta)}{\sin \alpha \sin \beta} = \cot \beta + \cot \alpha$

44. $\dfrac{\sin (\alpha + \beta)}{\cos \alpha \cos \beta} = \tan \alpha + \tan \beta$

45. $\dfrac{\sin (\alpha + \beta)}{\sin (\alpha - \beta)} = \dfrac{\tan \alpha + \tan \beta}{\tan \alpha - \tan \beta}$

46. $\dfrac{\cos (\alpha + \beta)}{\sin (\alpha - \beta)} = \dfrac{1 - \tan \alpha \tan \beta}{\tan \alpha - \tan \beta}$

47. $\cot (\alpha + \beta) = \dfrac{\cot \alpha \cot \beta - 1}{\cot \alpha + \cot \beta}$

48. $\cot (\alpha - \beta) = \dfrac{\cot \alpha \cot \beta + 1}{\cot \beta - \cot \alpha}$

49. $\dfrac{\tan (\alpha + \beta) + \tan (\alpha - \beta)}{1 - \tan (\alpha + \beta) \tan (\alpha - \beta)} = \tan 2\alpha$

50. $\tan \left(\theta + \dfrac{\pi}{4}\right) = \dfrac{1 + \tan \theta}{1 - \tan \theta}$

51. $\sin (\alpha + \beta + \gamma) = \sin \alpha \cos \beta \cos \gamma - \sin \alpha \sin \beta \sin \gamma + \cos \alpha \sin \beta \cos \gamma + \cos \alpha \cos \beta \sin \gamma$

52. $\sin (\alpha - \beta - \gamma) = \sin \alpha \cos \beta \cos \gamma - \sin \alpha \sin \beta \sin \gamma - \cos \alpha \sin \beta \cos \gamma - \cos \alpha \cos \beta \sin \gamma$

53. $\cos (\alpha + \beta + \gamma) = \cos \alpha \cos \beta \cos \gamma - \sin \alpha \sin \beta \cos \gamma - \sin \alpha \cos \beta \sin \gamma - \cos \alpha \sin \beta \sin \gamma$

54. $\tan (\alpha + \beta + \gamma)$
$= \dfrac{\tan \alpha + \tan \beta + \tan \gamma - \tan \alpha \tan \beta \tan \gamma}{1 - \tan \alpha \tan \beta - \tan \gamma \tan \alpha - \tan \beta \tan \gamma}$

55. $\dfrac{\sin (x + h) - \sin x}{h}$
$= \sin x \left(\dfrac{\cos h - 1}{h}\right) + \cos x \left(\dfrac{\sin h}{h}\right)$

56. $\dfrac{\tan (x + h) - \tan x}{h}$

$= \left(\dfrac{\sin h}{h}\right) \dfrac{\sec^2 x}{\cos h - \tan x \sin h}$

57. The pendulum shown below has a length of 2 feet.

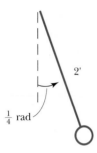

EXERCISE 57

If it is released at $t = 0$ with a displacement of 1/4 radian to the right of the vertical and an angular velocity of $\sqrt{3}$ radians per second, the resulting

motion is described by the equation

$$\theta(t) = \frac{1}{4} \cos 4t + \frac{\sqrt{3}}{4} \sin 4t,$$

where θ is in radians and t is the time in seconds. Use the Reduction Identity to write this equation in the form

$$\theta(t) = A \sin (Bt + C),$$

and then state the amplitude and period for the motion.

58. In Exercise 47 of Exercises 5.8, the displacement $x(t)$ of an object attached to a spring is described by the equation

$$x(t) = \frac{2}{3} \cos 8t - \frac{1}{6} \sin 8t.$$

Use the Reduction Identity to write this equation in the form

$$x(t) = A \sin (Bt + C),$$

and then state the amplitude and period for the motion of the object.

6.4   Double-Angle and Half-Angle Identities

We devote the first part of this section to special cases of the identities for the sum of two angles, called the **Double-Angle Identities.**

Double-Angle Identities

$$\sin 2\theta = 2 \sin \theta \cos \theta$$

$$\cos 2\theta = \cos^2 \theta - \sin^2 \theta$$

$$= 2 \cos^2 \theta - 1$$

$$= 1 - 2 \sin^2 \theta$$

$$\tan 2\theta = \frac{2 \tan \theta}{1 - \tan^2 \theta}$$

Each of these identities follows directly from an identity for the sum of two angles. To determine the sine of twice an angle, we replace both α and β by θ in

$$\sin (\alpha + \beta) = \sin \alpha \cos \beta + \cos \alpha \sin \beta$$

to yield

$$\sin (\theta + \theta) = \sin \theta \cos \theta + \cos \theta \sin \theta.$$

Thus we have the Double-Angle Identity for the sine function

$$\sin 2\theta = 2 \sin \theta \cos \theta.$$

Similarly, we replace both α and β by θ in

$$\cos (\alpha + \beta) = \cos \alpha \cos \beta - \sin \alpha \sin \beta$$

to yield

$$\cos (\theta + \theta) = \cos \theta \cos \theta - \sin \theta \sin \theta.$$

The resulting Double-Angle Identity for cosine is

$$\cos 2\theta = \cos^2 \theta - \sin^2 \theta.$$

A Pythagorean Identity is used to write this identity in two alternative forms. First, we can express $\cos 2\theta$ strictly in terms of $\cos \theta$.

$$
\begin{aligned}
\mathbf{\cos 2\theta} &= \cos^2 \theta - \sin^2 \theta && \text{Double-Angle Identity} \\
&= \cos^2 \theta - (1 - \cos^2 \theta) && \text{Pythagorean Identity} \\
&= \mathbf{2 \cos^2 \theta - 1} && \text{Algebra}
\end{aligned}
$$

Second, we can express $\cos 2\theta$ strictly in terms of $\sin \theta$.

$$
\begin{aligned}
\mathbf{\cos 2\theta} &= \cos^2 \theta - \sin^2 \theta && \text{Double-Angle Identity} \\
&= (1 - \sin^2 \theta) - \sin^2 \theta && \text{Pythagorean Identity} \\
&= \mathbf{1 - 2 \sin^2 \theta} && \text{Algebra}
\end{aligned}
$$

The Double-Angle Identity for the tangent can be derived by replacing both α and β by θ in

$$\tan (\alpha + \beta) = \frac{\tan \alpha + \tan \beta}{1 - \tan \alpha \tan \beta}.$$

Thus

$$\tan (\theta + \theta) = \frac{\tan \theta + \tan \theta}{1 - \tan \theta \tan \theta}$$

gives

$$\mathbf{\tan 2\theta} = \frac{\mathbf{2 \tan \theta}}{\mathbf{1 - \tan^2 \theta}}.$$

The Double-Angle Identities can be used to find the values of the trigonometric functions of 2θ if certain information is known about θ.

Example 1   Suppose $\sin \theta = -(\sqrt{5}/3)$ and $\cos \theta < 0$. Find the exact values of all the trigonometric functions of 2θ.

Solution

The angle θ is in quadrant III since both $\sin \theta$ and $\cos \theta$ are negative. We find

$$\cos \theta = -\frac{2}{3} \quad \text{and} \quad \tan \theta = \frac{\sqrt{5}}{2}$$

by using the Fundamental Identities. Now we turn our attention to the angle 2θ and apply the Double-Angle Identities.

$$\sin 2\theta = 2 \sin \theta \cos \theta = 2\left(-\frac{\sqrt{5}}{3}\right)\left(-\frac{2}{3}\right) = \frac{4\sqrt{5}}{9}$$

$$\cos 2\theta = \cos^2 \theta - \sin^2 \theta = \left(-\frac{2}{3}\right)^2 - \left(-\frac{\sqrt{5}}{3}\right)^2 = \frac{4}{9} - \frac{5}{9} = -\frac{1}{9}$$

$$\tan 2\theta = \frac{2 \tan \theta}{1 - \tan^2 \theta} = \frac{2\left(\frac{\sqrt{5}}{2}\right)}{1 - \left(\frac{\sqrt{5}}{2}\right)^2} = \frac{\sqrt{5}}{1 - \frac{5}{4}} = \frac{\sqrt{5}}{-\frac{1}{4}} = -4\sqrt{5}$$

Finally, we use the Fundamental Identities to evaluate the remaining trigonometric functions.

$$\cot 2\theta = \frac{1}{\tan 2\theta} = \frac{1}{-4\sqrt{5}} = -\frac{\sqrt{5}}{20}$$

$$\sec 2\theta = \frac{1}{\cos 2\theta} = -9$$

$$\csc 2\theta = \frac{1}{\sin 2\theta} = \frac{9}{4\sqrt{5}} = \frac{9\sqrt{5}}{20}$$ ❑

In Example 1 we chose to illustrate the Double-Angle Identities by evaluating all three of $\sin 2\theta$, $\cos 2\theta$, and $\tan 2\theta$. Note that once any one of $\sin 2\theta$, $\cos 2\theta$, or $\tan 2\theta$ is evaluated, then the Fundamental Identities can be used to evaluate the others.

Example 2   Verify the identity: $2 \cos^3 \theta - \cos \theta = \dfrac{\sin 4\theta}{4 \sin \theta}$.

Solution

We use the Double-Angle Identities on the right-hand side.

$$\frac{\sin 4\theta}{4 \sin \theta} = \frac{\sin 2(2\theta)}{4 \sin \theta} \qquad \text{Algebra}$$

$$= \frac{2 \sin 2\theta \cos 2\theta}{4 \sin \theta} \qquad \text{Double-Angle Identity}$$

$$= \frac{2\,(2 \sin \theta \cos \theta)(2 \cos^2 \theta - 1)}{4 \sin \theta} \qquad \text{Double-Angle Identities}$$

$$= \cos \theta\,(2 \cos^2 \theta - 1) \qquad \text{Algebra}$$

$$= 2 \cos^3 \theta - \cos \theta \qquad \text{Algebra} \qquad \square$$

The trigonometric functions of half an angle, $\theta/2$, can also be expressed in terms of trigonometric functions of the angle θ. The resulting identities are called the **Half-Angle Identities,** and we list them next.

Half-Angle Identities

$$\sin \frac{\theta}{2} = \pm \sqrt{\frac{1 - \cos \theta}{2}} \qquad \tan \frac{\theta}{2} = \pm \sqrt{\frac{1 - \cos \theta}{1 + \cos \theta}}$$

$$\cos \frac{\theta}{2} = \pm \sqrt{\frac{1 + \cos \theta}{2}} \qquad\qquad = \frac{\sin \theta}{1 + \cos \theta}$$

$$\qquad\qquad\qquad\qquad\qquad = \frac{1 - \cos \theta}{\sin \theta}$$

We choose either $+$ or $-$ on each radical, depending on the quadrant in which $\theta/2$ terminates.

We establish the Half-Angle Identity for sine by first solving the Double-Angle Identity

$$\cos 2A = 1 - 2 \sin^2 A$$

for $\sin A$.

$$2 \sin^2 A = 1 - \cos 2A$$

$$\sin^2 A = \frac{1 - \cos 2A}{2}$$

$$\sin A = \pm \sqrt{\frac{1 - \cos 2A}{2}}$$

Next, if we replace A by $\theta/2$, we have the desired form:

$$\sin \frac{\theta}{2} = \pm \sqrt{\frac{1 - \cos \theta}{2}}.$$

Since sine is positive in the first and second quadrants, we use + when $\theta/2$ terminates in quadrants I or II and − when $\theta/2$ terminates in quadrants III or IV.

The Half-Angle Identity for cosine can be derived in a similar way from $\cos 2A = 2\cos^2 A - 1$.

The Half-Angle Identity for tangent is derived by using the Quotient Identity for tangent and then the Half-Angle Identities for sine and cosine.

$$\mathbf{\tan \frac{\theta}{2}} = \frac{\sin \dfrac{\theta}{2}}{\cos \dfrac{\theta}{2}} \qquad \text{Quotient Identity}$$

$$= \frac{\pm\sqrt{\dfrac{1 - \cos\theta}{2}}}{\pm\sqrt{\dfrac{1 + \cos\theta}{2}}} \qquad \text{Half-Angle Identities}$$

$$= \pm\sqrt{\frac{1 - \cos\theta}{1 + \cos\theta}} \qquad \text{Algebra}$$

Again, we choose either + or − depending on which quadrant $\theta/2$ terminates in: + in quadrant I or III since tangent is positive there, and − in quadrant II or IV since tangent is negative there.

The alternative forms of the Half-Angle Identity for tangent are sometimes more convenient to use since they do not contain radicals. The derivations of these alternative forms are requested in the exercises for this section.

Example 3   Write the exact values of all the trigonometric functions of $\theta/2$ if $0 \le \theta < 2\pi$, $\cos\theta = \frac{7}{25}$, and $\cot\theta < 0$.

Solution
Since $\cos\theta > 0$ and $\cot\theta < 0$, then θ lies in quadrant IV, and we write

$$\frac{3\pi}{2} < \theta < 2\pi.$$

Dividing all three members of this inequality by 2 locates the angle $\theta/2$ in quadrant II.

$$\frac{3\pi}{4} < \frac{\theta}{2} < \pi$$

We use this information to determine the signs to be used in the Half-Angle Identities.

$$\sin \frac{\theta}{2} = +\sqrt{\frac{1 - \cos \theta}{2}} = \sqrt{\frac{1 - \dfrac{7}{25}}{2}} = \sqrt{\frac{\dfrac{18}{25}}{2}} = \sqrt{\frac{9}{25}} = \frac{3}{5}$$

$$\cos \frac{\theta}{2} = -\sqrt{\frac{1 + \cos \theta}{2}} = -\sqrt{\frac{1 + \dfrac{7}{25}}{2}} = -\sqrt{\frac{\dfrac{32}{25}}{2}} = -\sqrt{\frac{16}{25}} = -\frac{4}{5}$$

$$\tan \frac{\theta}{2} = -\sqrt{\frac{1 - \cos \theta}{1 + \cos \theta}} = -\sqrt{\frac{1 - \dfrac{7}{25}}{1 + \dfrac{7}{25}}} = -\sqrt{\frac{\dfrac{18}{25}}{\dfrac{32}{25}}} = -\sqrt{\frac{9}{16}} = -\frac{3}{4}$$

The Reciprocal Identities then give

$$\csc \frac{\theta}{2} = \frac{5}{3}, \qquad \sec \frac{\theta}{2} = -\frac{5}{4}, \qquad \cot \frac{\theta}{2} = -\frac{4}{3}. \qquad \square$$

Again we note that once one trigonometric function for $\theta/2$ is evaluated, then the Fundamental Identities can be used to evaluate all others.

Exercises 6.4

Express each of the following in terms of a trigonometric function of twice or half the given angle. Assume all variables are such that all expressions are defined.

1. $2 \sin 22° \cos 22°$

2. $8 \sin 17° \cos 17°$

3. $\dfrac{1}{\sin 105° \cos 105°}$

4. $\dfrac{1}{\sin 43° \cos 43°}$

5. $\cos^2 \dfrac{\pi}{18} - \sin^2 \dfrac{\pi}{18}$

6. $1 - 2 \sin^2 \dfrac{\pi}{10}$

7. $\dfrac{1}{2 \cos^2 \gamma - 1}$

8. $\dfrac{1}{\cos^2 2\alpha - \sin^2 2\alpha}$

9. $\dfrac{2 \tan \dfrac{\pi}{7}}{1 - \tan^2 \dfrac{\pi}{7}}$

10. $\dfrac{2 \tan 115°}{1 - \tan^2 115°}$

11. $\dfrac{1 - \tan^2 \alpha}{2 \tan \alpha}$

12. $\dfrac{1 - \tan^2 4\beta}{2 \tan 4\beta}$

13. $\sqrt{\dfrac{1 - \cos 242°}{2}}$

14. $\sqrt{\dfrac{1 - \cos 48°}{2}}$

15. $-\sqrt{\dfrac{1 - \cos 198°}{1 + \cos 198°}}$

16. $-\sqrt{\dfrac{1 - \cos \dfrac{16\pi}{9}}{1 + \cos \dfrac{16\pi}{9}}}$

17. $-\sqrt{\dfrac{1 + \cos 378°}{2}}$

18. $\sqrt{\dfrac{1 + \cos \dfrac{4\pi}{7}}{2}}$

19. $\dfrac{\sin 408°}{1 + \cos 408°}$

20. $\dfrac{1 - \cos 130°}{\sin 130°}$

Determine the exact value of all the trigonometric functions of 2θ if θ satisfies the given conditions. (See Example 1.)

21. $\sin \theta = \frac{4}{5}$; θ in Q II

22. $\cos \theta = \frac{5}{13}$; $\tan \theta < 0$

23. $\sec \theta = -\frac{3}{2}$; $\sin \theta < 0$

24. $\tan \theta = -2\sqrt{2}$; $\sec \theta > 0$

Use the Composite-Angle Identities to express each of the following trigonometric functions of the given multiple angle in terms of the same trigonometric function of θ.

25. $\cos 4\theta$ **26.** $\cos 3\theta$ **27.** $\tan 3\theta$ **28.** $\tan 4\theta$

Determine the exact values of all trigonometric functions of $\theta/2$ if θ satisfies the given conditions and $0 \le \theta < 2\pi$. (See Example 3.)

29. $\sin \theta = \frac{4}{5}$; θ in Q II

30. $\cos \theta = \frac{5}{13}$; $\tan \theta < 0$

31. $\sec \theta = -\frac{3}{2}$; $\sin \theta < 0$

32. $\tan \theta = -2\sqrt{2}$; $\sec \theta > 0$

Determine the exact values of the sine, cosine, and tangent of each of the following.

33. $\dfrac{\pi}{8}$ **34.** $\dfrac{5\pi}{8}$ **35.** $\dfrac{7\pi}{12}$ **36.** $\dfrac{11\pi}{12}$

37. $157.5°$ **38.** $105°$ **39.** $195°$ **40.** $202.5°$

Verify each of the following identities. (See Example 2.)

41. $(\sin \theta - \cos \theta)^2 = 1 - \sin 2\theta$

42. $\cos^4 \theta - \sin^4 \theta = \cos 2\theta$

43. $\dfrac{2}{1 + \cos 2A} = \sec^2 A$

44. $\dfrac{2 + 2 \cos 2A}{\sin^2 2A} = \csc^2 A$

45. $\dfrac{1 - \cos 2\alpha}{\sin 2\alpha} = \tan \alpha$

46. $\dfrac{1 + \cos 2\alpha}{\sin 2\alpha} = \cot \alpha$

47. $\dfrac{\sin 2B \cos B}{1 - \sin^2 B} = 2 \sin B$

48. $\dfrac{\cos B \sin 2B}{1 + \cos 2B} = \sin B$

49. $\dfrac{\cos 2\beta - \sin \beta}{\cos^2 \beta} = \dfrac{1 - 2 \sin \beta}{1 - \sin \beta}$

50. $\dfrac{\sin^2 \beta}{\cos \beta - \cos 2\beta} = \dfrac{1 + \cos \beta}{1 + 2 \cos \beta}$

51. $\dfrac{1 + \sin 2x - \cos 2x}{1 + \cos 2x + \sin 2x} = \tan x$

52. $\dfrac{4 \sin 2x}{(1 + \cos 2x)^2} = 2 \tan x \sec^2 x$

53. $\cot 2\theta = \dfrac{\cot^2 \theta - 1}{2 \cot \theta}$ **54.** $\sec 2\theta = \dfrac{\sec^2 \theta}{2 - \sec^2 \theta}$

55. $\cot 2\theta + \tan \theta = \csc 2\theta$

56. $\tan 2\alpha - \tan \alpha = \tan \alpha \sec 2\alpha$

57. $\csc 2\alpha = \dfrac{1}{2} \csc \alpha \sec \alpha$

58. $\csc 6\alpha \tan 3\alpha = \dfrac{\sec^2 3\alpha}{2}$

59. $\dfrac{\sin 4\alpha}{\cos \alpha} + \dfrac{\cos 4\alpha}{\sin \alpha} = 2 \csc 2\alpha \cos 3\alpha$

60. $\dfrac{\sin 2\alpha}{\sin 3\alpha} + \dfrac{\cos 2\alpha}{\cos 3\alpha} = 2 \sin 5\alpha \csc 6\alpha$

61. $\tan 3y - \tan y = 2 \sin y \sec 3y$

62. $\cot y - \cot 3y = 2 \cos y \csc 3y$

63. $\dfrac{(1 + \tan t)^2}{1 + \tan^2 t} = 1 + \sin 2t$

64. $\dfrac{1 - \tan^2 t}{1 + \tan^2 t} = \cos 2t$

65. $\dfrac{1 + \sin 2\gamma}{\cos 2\gamma} = \dfrac{1 + \tan \gamma}{1 - \tan \gamma}$

66. $1 + \tan \gamma \tan 2\gamma = \sec 2\gamma$

67. $\tan \dfrac{\theta}{2} = \dfrac{1 - \cos \theta}{\sin \theta}$

68. $\tan \dfrac{\theta}{2} = \dfrac{\sin \theta}{1 + \cos \theta}$

69. $\tan \dfrac{\theta}{2} (1 + \sec \theta) = \tan \theta$

70. $\tan \dfrac{\theta}{2} + 2 \sin^2 \dfrac{\theta}{2} \cot \theta = \sin \theta$

71. $\tan \dfrac{\theta}{2} + \cot \dfrac{\theta}{2} = 2 \csc \theta$

72. $\tan^2 \dfrac{\theta}{2} + 1 = 2 \csc \theta \tan \dfrac{\theta}{2}$

6.5 Product and Sum Identities

In some instances it is advantageous to rewrite a product involving sines and/or cosines as a sum of sines and/or cosines, and vice versa. The identities of this section can be applied to those situations.

We begin by first rewriting the four identities (from Sections 6.2 and 6.3) for the sine and cosine of the sum and difference of two angles.

$$\sin (\alpha + \beta) = \sin \alpha \cos \beta + \cos \alpha \sin \beta$$

$$\sin (\alpha - \beta) = \sin \alpha \cos \beta - \cos \alpha \sin \beta$$

$$\cos (\alpha + \beta) = \cos \alpha \cos \beta - \sin \alpha \sin \beta$$

$$\cos (\alpha - \beta) = \cos \alpha \cos \beta + \sin \alpha \sin \beta$$

Adding and subtracting the first two of these equations yield the first two of the following four identities labeled *Product Identities*. Similarly, adding and subtracting the last two of these equations yield the last two of the Product Identities.

Product Identities

$$2 \sin \alpha \cos \beta = \sin (\alpha + \beta) + \sin (\alpha - \beta)$$

$$2 \cos \alpha \sin \beta = \sin (\alpha + \beta) - \sin (\alpha - \beta)$$

$$2 \cos \alpha \cos \beta = \cos (\alpha + \beta) + \cos (\alpha - \beta)$$

$$2 \sin \alpha \sin \beta = \cos (\alpha - \beta) - \cos (\alpha + \beta)$$

Example 1   Write each of the following products as a sum. Use the Negative-Angle Identities if necessary so that no argument contains a minus sign.

a) $2 \sin 3\alpha \cos 2\alpha$

b) $\cos 4\gamma \cos \gamma$

c) $6 \sin 2\theta \sin 6\theta$

Solution

a)   Using the first identity for the product of a sine and a cosine, we have

$$2 \sin 3\alpha \cos 2\alpha = \sin (3\alpha + 2\alpha) + \sin (3\alpha - 2\alpha)$$

$$= \sin 5\alpha + \sin \alpha.$$

b) Since we have a product of two cosines, we use the third of the Product Identities.

$$\cos 4\gamma \cos \gamma = \frac{1}{2} \left[\cos (4\gamma + \gamma) + \cos (4\gamma - \gamma)\right]$$

$$= \frac{1}{2} \cos 5\gamma + \frac{1}{2} \cos 3\gamma$$

c) For the product of two sines we use the last of the Product Identities.

$$6 \sin 2\theta \sin 6\theta = 3[\cos (2\theta - 6\theta) - \cos (2\theta + 6\theta)]$$

$$= 3[\cos (-4\theta) - \cos 8\theta]$$

$$= 3 \cos 4\theta - 3 \cos 8\theta$$

The last equality results from a Negative-Angle Identity.   ❑

Next we derive four additional identities, called the *Sum Identities*. To do this, in each of the four Product Identities we let

$$x = \alpha + \beta \qquad \text{and} \qquad y = \alpha - \beta.$$

The corresponding expressions for α and β are

$$\alpha = \frac{x + y}{2} \qquad \text{and} \qquad \beta = \frac{x - y}{2}.$$

Rewriting the four Product Identities in terms of x and y results in the following Sum Identities.

Sum Identities

$$\sin x + \sin y = 2 \sin \left(\frac{x + y}{2}\right) \cos \left(\frac{x - y}{2}\right)$$

$$\sin x - \sin y = 2 \cos \left(\frac{x + y}{2}\right) \sin \left(\frac{x - y}{2}\right)$$

$$\cos x + \cos y = 2 \cos \left(\frac{x + y}{2}\right) \cos \left(\frac{x - y}{2}\right)$$

$$\cos x - \cos y = -2 \sin \left(\frac{x + y}{2}\right) \sin \left(\frac{x - y}{2}\right)$$

Example 2   Write each of the following sums as a product. Use the Negative-Angle Identities if necessary so that no argument contains a minus sign.

a) $\sin 135° + \sin 45°$ b) $\cos 3\theta - \cos 9\theta$

Solution
In each part we apply the appropriate Sum Identity.

a) $\sin 135° + \sin 45° = 2 \sin \left(\dfrac{135° + 45°}{2} \right) \cos \left(\dfrac{135° - 45°}{2} \right)$

$= 2 \sin 90° \cos 45°$

b) $\cos 3\theta - \cos 9\theta = -2 \sin \left(\dfrac{3\theta + 9\theta}{2} \right) \sin \left(\dfrac{3\theta - 9\theta}{2} \right)$

$= -2 \sin 6\theta \sin (-3\theta)$

$= 2 \sin 6\theta \sin 3\theta$ ❏

Example 3   Verify the identity: $\dfrac{\sin 4\theta + \sin 2\theta}{\cos 4\theta + \cos 2\theta} = \tan 3\theta.$

Solution

$\dfrac{\sin 4\theta + \sin 2\theta}{\cos 4\theta + \cos 2\theta} = \dfrac{2 \sin \left(\dfrac{4\theta + 2\theta}{2} \right) \cos \left(\dfrac{4\theta - 2\theta}{2} \right)}{2 \cos \left(\dfrac{4\theta + 2\theta}{2} \right) \cos \left(\dfrac{4\theta - 2\theta}{2} \right)}$ Sum Identity

$= \dfrac{2 \sin 3\theta \cos \theta}{2 \cos 3\theta \cos \theta}$ Algebra

$= \tan 3\theta$ Quotient Identity
❏

Exercises 6.5

Express each of the following products as a sum. Use the Negative-Angle Identities if necessary so that no argument contains a minus sign. (See Example 1.)

1. $2 \sin 84° \cos 27°$
2. $4 \cos 93° \sin 16°$
3. $8 \cos 12° \cos 13°$
4. $6 \sin 93° \sin 116°$
5. $\cos 3\alpha \sin 8\alpha$
6. $\cos 4\theta \cos 5\theta$
7. $\sin x \sin 2x$
8. $\sin 2y \cos 3y$

Express each of the following sums as a product. Use the Negative-Angle Identities if necessary so that no argument contains a minus sign. (See Example 2.)

9. $\sin 42° + \sin 56°$
10. $\sin 112° - \sin 48°$

11. $\cos 17° - \cos 93°$
12. $\cos 80° + \cos 20°$
13. $\sin 9\alpha - \sin 11\alpha$
14. $\sin 4\gamma + \sin 6\gamma$
15. $\cos x + \cos 3x$
16. $\cos 2t - \cos 4t$

Verify each of the following identities. (See Example 3.)

17. $\dfrac{\sin 3\theta - \sin \theta}{\cos 3\theta + \cos \theta} = \tan \theta$

18. $\dfrac{\cos 5x + \cos 7x}{\sin 5x - \sin 7x} = -\cot x$

19. $\dfrac{\sin 6x - \sin 2x}{\sin 6x + \sin 2x} = \dfrac{\tan 2x}{\tan 4x}$

20. $\dfrac{\cos 3\theta - \cos \theta}{\cos 3\theta + \cos \theta} = \dfrac{2 \tan^2 \theta}{\tan^2 \theta - 1}$

21. $\cos (A + B) \sin (A - B) =$
$$\sin A \cos A - \sin B \cos B$$

22. $\sin (A + B) \sin (A - B) = \cos^2 B - \cos^2 A$

23. $\sin 2x - \sin 4x - \sin 6x = -4 \cos 3x \sin 2x \cos x$

24. $\sin 2x - \sin 4x + \sin 6x = 4 \cos 3x \cos 2x \sin x$

25. $\sin 6t \sin 4t + \cos 8t \cos 2t = \cos 4t \cos 2t$

26. $\sin 12t \sin 4t + \cos 6t \cos 10t = \cos 6t \cos 2t$

27. $\sin 6A \cos 4A - \cos 8A \sin 2A = \sin 4A \cos 2A$

28. $\cos 7A \sin A - \sin 2A \cos 6A = -\cos 5A \sin A$

29. $\sin 2\theta + \sin 4\theta + 2 \cos 3\theta \sin \theta = 2 \sin 4\theta$

30. $\sin 3\theta - \sin \theta + 2 \sin 2\theta \cos \theta = 2 \sin 3\theta$

31. $\cos 4y + \cos 8y - 2 \sin 6y \sin 2y = 2 \cos 8y$

32. $\cos 9y - \cos 7y + 2 \cos 8y \cos y = 2 \cos 9y$

6.6   Trigonometric Equations

A **trigonometric equation** is an equation that involves at least one trigonometric function. To this point we have concentrated mainly on trigonometric equations that were identities. We turn our attention now to conditional trigonometric equations.[2]

The techniques used to solve equations in algebra can be applied to trigonometric equations. We illustrate the parallel procedures in the next three examples.

> **Example 1**   Solve: $4 \sin \theta - \sqrt{3} = 2 \sin \theta$.

Solution
If you can do this .... then you can do this!

$$4x - \sqrt{3} = 2x \dots\dots\dots\dots\dots\dots 4 \sin \theta - \sqrt{3} = 2 \sin \theta$$

$$2x = \sqrt{3} \dots\text{Algebra}\dots\dots\dots\dots 2 \sin \theta = \sqrt{3}$$

$$x = \frac{\sqrt{3}}{2} \dots\text{Algebra}\dots\dots\dots\dots \sin \theta = \frac{\sqrt{3}}{2}$$

$$\text{Related Angle} \dots\dots\dots\dots \theta' = 60°$$

Since $\sin \theta > 0$ only in quadrants I and II, then $\theta = 60°$ and $\theta = 120°$ are solutions, as well as any angle coterminal to $60°$ and $120°$. Thus all solutions are given by

$$\theta = \begin{cases} 60° + n \cdot 360°, & n \text{ any integer;} \\ 120° + n \cdot 360°, & n \text{ any integer.} \end{cases}$$ ❑

---

2. A conditional equation is an equation that is a false statement for at least one value in its domain.

In most cases we restrict our attention to solutions that are non-negative angles measuring less than 1 revolution.

Example 2   Solve for the radian measure of x, where $0 \leq x < 2\pi$.

$$\sin^2 x + 2 \sin x - 3 = 0$$

Solution

If you can do this . . . . then you can do this!

$$y^2 + 2y - 3 = 0 \text{ . . . . } \sin^2 x + 2 \sin x - 3 = 0$$

$$(y + 3)(y - 1) = 0 \text{ . . . . . Algebra . . . . } (\sin x + 3)(\sin x - 1) = 0$$

$$y + 3 = 0 \text{ or } y - 1 = 0 \text{ . . . . . Algebra . . . . } \sin x + 3 = 0 \text{ or}$$
$$\sin x - 1 = 0$$

$$y = -3 \text{ or } \qquad y = 1 \text{ . . . . . Algebra . . . . } \sin x = -3 \text{ or } \qquad \sin x = 1$$

$$\text{Trigonometry . . . . . .no solution} \qquad x = \frac{\pi}{2}$$

$$\text{Solution . . . . } x = \frac{\pi}{2}$$
❏

Example 3   Solve: $2 \sin \theta \cos \theta = \cos \theta$, for $0° \leq \theta < 360°$.

Solution

If you can do this . . . . then you can do this!

$$2xy = y \text{ . . . . } 2 \sin \theta \cos \theta = \cos \theta$$

$$2xy - y = 0 \text{ . . . . . Algebra . . . . } 2 \sin \theta \cos \theta - \cos \theta = 0$$

$$y(2x - 1) = 0 \text{ . . . . . Algebra . . . . } \cos \theta (2 \sin \theta - 1) = 0$$

$$y = 0 \text{ or}$$
$$2x - 1 = 0 \text{ . . . . . Algebra . . . . } \cos \theta = 0 \text{ or } 2 \sin \theta - 1 = 0$$

$$x = \tfrac{1}{2} \text{ . . . . . Algebra . . . . } \cos \theta = 0 \qquad \sin \theta = \tfrac{1}{2}$$

$$\text{Trigonometry . . . Quadrantal} \qquad \theta' = 30°$$

$$\text{Angles: } 90°, 270° \qquad \theta = 30°, 150°$$

$$\text{Solutions . . . . } \theta = 90°, 270°, 30°, 150° \qquad ❏$$

In Example 3 we must resist the temptation to divide out the common factor $\cos \theta$. Doing so would lose the solutions $\theta = 90°, 270°$. Dividing both sides of an equation by 0 ($\cos \theta = 0$ for $\theta = 90°, 270°$) is *not* a valid operation. As a general procedure, we rewrite the equation so that one side is 0 and the other side factors.

The Fundamental Identities are often useful and sometimes necessary in solving trigonometric equations involving more than one trigonometric function.

Example 4   Solve for the radian measure of θ in

$$\sec \theta + 1 = \tan^2 \theta, \qquad 0 \le \theta < 2\pi.$$

Solution

We rewrite the equation $\sec \theta + 1 = \tan^2 \theta$ strictly in terms of $\sec \theta$ by using one of the Pythagorean Identities.

$$\sec \theta + 1 = \sec^2 \theta - 1 \quad \text{Pythagorean Identity}$$

$$\sec^2 \theta - \sec \theta - 2 = 0 \qquad \text{Algebra}$$

$$(\sec \theta - 2)(\sec \theta + 1) = 0 \qquad \text{Algebra}$$

$$\sec \theta - 2 = 0 \quad \text{or} \quad \sec \theta + 1 = 0 \qquad \text{Algebra}$$

$$\sec \theta = 2 \quad \text{or} \quad \sec \theta = -1 \qquad \text{Algebra}$$

$$\text{Related Angle: } \theta' = \frac{\pi}{3} \qquad \text{Quadrantal Angle} \quad \text{Trigonometry}$$

$$\theta = \frac{\pi}{3}, \frac{5\pi}{3} \qquad \theta = \pi \qquad \text{Solutions} \qquad \square$$

The preceding examples illustrate some suggestions for solving trigonometric equations. In the following list, we summarize these suggestions, emphasizing that they are only suggestions and not a rigid step-by-step procedure. Rely on your intuition and do not hesitate to use a trigonometric identity or to perform some algebraic operation that might simplify the equation.

Solving Trigonometric Equations

1. If the equation is linear in one trigonometric function:
   a) Solve for that trigonometric function;
   b) Then solve for the angle by recognizing the function values of the quadrantal or special angles or by using a calculator or the trigonometric tables.
2. If more than one trigonometric function occurs, you might use the Fundamental Identities to rewrite the equation in terms of one trigonometric function.

3. If the equation is not linear, rewrite it so that one side is identically zero and try to factor the other side. If a quadratic expression will not factor, use the quadratic formula.
4. Rely heavily on algebraic techniques for solving equations.

Trigonometric equations involving composite angles can be separated into two categories: those that require simplification by using Composite Angle Identities and those that do not. We begin with the simpler situation in Example 5, keeping in mind the preceding suggestions for solving trigonometric equations.

Example 5   Solve for the values of x in

$$2 \cos 3x = 1, \qquad 0 \le x < 2\pi.$$

Solution
We have

$$\cos 3x = \tfrac{1}{2},$$

so the related angle for $3x$ is $\pi/3$. The problem requires solutions for x to be such that

$$0 \le x < 2\pi.$$

Multiplying all three members of this inequality by 3 gives the required range on the variable $3x$.

$$0 \le 3x < 6\pi$$

Thus $\pi/3$ is the related angle (or number) for nonnegative angles (or numbers) measuring up to 3 revolutions (up to 6π). Hence the solutions for $3x$ in

$$\cos 3x = \tfrac{1}{2}, \qquad 0 \le 3x < 6\pi$$

are

$$3x = \frac{\pi}{3}, \ \frac{5\pi}{3}, \ \frac{7\pi}{3}, \ \frac{11\pi}{3}, \ \frac{13\pi}{3}, \ \text{or} \ \frac{17\pi}{3}.$$

Finally, solving for x, we have

$$x = \frac{\pi}{9}, \ \frac{5\pi}{9}, \ \frac{7\pi}{9}, \ \frac{11\pi}{9}, \ \frac{13\pi}{9}, \ \text{or} \ \frac{17\pi}{9}.$$

These are all the solutions for $0 \le x < 2\pi$.   ❏

The Composite Angle Identities play key roles in the solutions of the remaining examples.

Example 6   Solve: $\cos 2x - \sin x = 0$,   for all values of x such that $0 \leq x < 2\pi$.

Solution

Replacing $\cos 2x$ by $1 - 2 \sin^2 x$ in the equation

$$\cos 2x - \sin x = 0, \qquad 0 \leq x < 2\pi$$

results in a quadratic equation in $\sin x$.

$$1 - 2 \sin^2 x - \sin x = 0, \qquad 0 \leq x < 2\pi$$

We solve this equation by the method of factoring.

$$2 \sin^2 x + \sin x - 1 = 0 \qquad \text{Algebra}$$

$$(2 \sin x - 1)(\sin x + 1) = 0 \qquad \text{Algebra}$$

$$2 \sin x - 1 = 0 \quad \text{or} \quad \sin x + 1 = 0 \qquad \text{Algebra}$$

$$\sin x = \tfrac{1}{2} \qquad \text{or} \quad \sin x = -1 \qquad \text{Algebra}$$

$$x = \frac{\pi}{6}, \frac{5\pi}{6} \qquad\qquad x = \frac{3\pi}{2} \qquad \text{Special Angles}$$

The solutions are $x = \dfrac{\pi}{6}, \dfrac{5\pi}{6}$, or $\dfrac{3\pi}{2}$. ❏

Example 7   Solve for all x in

$$\sin 4x \cos x - \cos 4x \sin x = 1, 0 \leq x < 2\pi.$$

Solution

Recognizing the left side of

$$\sin 4x \cos x - \cos 4x \sin x = 1$$

as part of the identity for the sine of the difference of two angles, we can rewrite the equation in the form

$$\sin (4x - x) = 1, \qquad 0 \leq x < 2\pi$$

or

$$\sin 3x = 1, \qquad 0 \leq 3x < 6\pi.$$

Solving for nonnegative values of $3x$ up to 3 revolutions gives

$$3x = \frac{\pi}{2}, \quad \frac{5\pi}{2}, \quad \text{or} \quad \frac{9\pi}{2}.$$

Finally, solving for x yields the solutions in the range $0 \leq x < 2\pi$:

$$x = \frac{\pi}{6}, \quad \frac{5\pi}{6}, \quad \text{or} \quad \frac{3\pi}{2}. \qquad ❏$$

Example 8 Solve: $\sin x + \cos x = 1$, for all values of x such that $0 \le x < 2\pi$.

Solution

The Reduction Identity can be used to write

$$\sin x + \cos x = \sqrt{2} \sin \left(x + \frac{\pi}{4} \right).$$

Then the equation

$$\sin x + \cos x = 1, \qquad 0 \le x < 2\pi$$

becomes

$$\sqrt{2} \sin \left(x + \frac{\pi}{4} \right) = 1, \qquad \frac{\pi}{4} \le x + \frac{\pi}{4} < 2\pi + \frac{\pi}{4}$$

or

$$\sin \left(x + \frac{\pi}{4} \right) = \frac{1}{\sqrt{2}}.$$

Solving for $x + (\pi/4)$, we have

$$x + \frac{\pi}{4} = \frac{\pi}{4} \quad \text{or} \quad \frac{3\pi}{4}.$$

Subtracting $\pi/4$ finally yields the solutions:

$$x = 0 \quad \text{or} \quad \frac{\pi}{2}. \qquad \qquad \square$$

Exercises 6.6

Solve the equations in Exercises 1–12 for all θ. (See Examples 1 and 2.)

1. $3 \tan \theta = -\sqrt{3}$

2. $\sec \theta + \sqrt{2} = 0$

3. $2 \csc \theta - 2 = \csc \theta$

4. $5 \cot \theta + 3\sqrt{3} = 2\sqrt{3} + 4 \cot \theta$

5. $2 \cos^2 \theta + \cos \theta - 1 = 0$

6. $\sin^2 \theta + 3 \sin \theta + 2 = 0$

7. $2 \cos^2 \theta = 5 \cos \theta + 3$

8. $2 \sin^2 \theta - 3 = \sin \theta$

9. $\sec^2 \theta - 3 \sec \theta = -2$

10. $\tan^2 \theta = 2 \tan \theta - 1$

11. $\cot^2 \theta - 3 = 0$

12. $3 \csc^2 \theta - 4 = 0$

In Exercises 13–20, use the quadratic formula and solve for the degree measure of θ, where $0° \le \theta < 360°$. In Exercises 17–20, express θ to the nearest tenth of a degree.

13. $3 \tan^2 \theta - 2\sqrt{3} \tan \theta + 1 = 0$

14. $4 \sin^2 \theta - 4\sqrt{3} \sin \theta + 3 = 0$

15. $\sec^2 \theta + 2\sqrt{2} \sec \theta + 2 = 0$

16. $2 \cos^2 \theta + 2\sqrt{2} \cos \theta + 1 = 0$

17. $\sin^2 \theta + 3 \sin \theta - 1 = 0$

18. $\cos^2 \theta - 2 \cos \theta - 2 = 0$

19. $\tan^2 \theta - \tan \theta - 3 = 0$

20. $\sin^2 \theta + 4 \sin \theta + 2 = 0$

Solve the equations in Exercises 21–54 for $0 \le x < 2\pi$. (See Examples 3–5.)

21. $\sin x \tan x = \sin x$ **22.** $\cos x \cot x = \cos x$

23. $2 \cos^2 x \sin x = \sin x$

24. $4 \sin^2 x \cos x = \cos x$

25. $3 \sin x = 2 \cos^2 x$

26. $2 \cos^2 x = 3 \sin x + 3$

27. $1 + \cos x = 2 \sin^2 x$

28. $5 \cos x = 1 + 2 \sin^2 x$

29. $\cot^2 x + 3 \csc x + 3 = 0$

30. $\csc^2 x = 2\sqrt{3} \cot x - 2$

31. $\tan^2 x = 3 (\sec x - 1)$

32. $\sec^2 x = 2 \tan x$

33. $\cos x - \sec x = 0$ **34.** $\sin x - \csc x = 0$

35. $\tan x - \cot x = 0$

36. $2 \sin x - 1 - \csc x = 0$

37. $2 \sin x \cos x + 2 \cos x - \sin x - 1 = 0$

38. $4 \sin x \cos x - 2 \cos x + 2 \sin x - 1 = 0$

39. $\sin 2x = 1$ **40.** $\cos 2x = 1$

41. $\tan \dfrac{x}{2} = -1$ **42.** $2 \sin \dfrac{x}{2} = 1$

43. $2 \cos \dfrac{x}{2} = -1$ **44.** $\cot \dfrac{x}{2} = \sqrt{3}$

45. $2 \tan 3x - 2\sqrt{3} = 0$

46. $2 \cot 3x = \sqrt{3} - \cot 3x$

47. $\sin^2 3x - 1 = 0$ **48.** $\cos^2 3x - 1 = 0$

49. $\sin 2x = \cos 2x$ **50.** $4 \cos 2x = \sec 2x$

51. $2 \sin^2 2x + \sin 2x = 1$

52. $2 \cos^2 2x - \cos 2x = 1$

53. $2 \cos^2 3x + 3 \cos 3x - 2 = 0$

54. $2 \sin^2 3x - 5 \sin 3x - 3 = 0$

Use the Composite Angle Identities to solve each of the equations in Exercises 55–68 for all values of x such that $0 \le x < 2\pi$. (See Examples 6 and 7.)

55. $\sin 2x = \sin x$ **56.** $\sin 2x = \cos x$

57. $2 - \cos x = \cos 2x$ **58.** $-2 - \sin x = \cos 2x$

59. $\sin 2x = \cos 2x + 1$ **60.** $\sin 2x = \cos 2x - 1$

61. $2 \cos^2 x + \cos 2x = 0$

62. $\cos 2x - 2 \sin^2 x = 0$

63. $\cos 2x \cos x + \sin 2x \sin x = \dfrac{1}{2}$

64. $\sin 3x \cos x + \cos 3x \sin x = -1$

65. $2 \cos^2 \dfrac{x}{2} - 3 \cos x = 2$

66. $\tan \dfrac{x}{2} + 1 = -\cot x$

67. $\cos 4x - \cos 2x = \sin 3x$

68. $\sin 3x - \sin x = \cos 2x$

Use the Reduction Identity to solve the following equations for all x such that $0 \le x < 2\pi$. (See Example 8.)

69. $\sin x + \cos x = \sqrt{2}$

70. $\sin x - \cos x = -1$

71. $\sqrt{3} \sin x + \cos x = 2$

72. $\sin x + \sqrt{3} \cos x = -1$

73. The current $I(t)$ in an electrical circuit is given by

$$I(t) = 15 \sin 4t$$

volts. Find the nonnegative values of t for which the current is zero.

74. The charge $q(t)$ on a capacitor in an electrical circuit is given by

$$q(t) = \frac{15}{4} - \frac{15}{4} \cos 4t$$

coulombs. Find the positive values of t for which the charge on the capacitor is zero.

75. Exercise 42 of Exercises 5.8 calls for a sketch of the graph of the function

$$f(x) = \cos x + \sin 2x$$

over the interval $[0, 2\pi]$. Find the x-intercepts of the graph in this interval.

76. Exercise 44 of Exercises 5.8 calls for a sketch of the graph of the function

$$f(x) = 3 \sin x + \cos 2x$$

over the interval $[0, 2\pi]$. Find the x-intercepts of the graph in this interval.

77. In Exercise 57 of Exercises 6.3, the angular displacement $\theta(t)$ of a pendulum is given by

$$\theta(t) = \frac{1}{4} \cos 4t + \frac{\sqrt{3}}{4} \sin 4t.$$

Find the positive values of t for which the pendulum is in the equilibrium position.

Photo by Ed Carlin/The Picture Cube

78. In Exercise 47 of Exercises 5.8, the displacement $x(t)$ of an object attached to a spring is given by

$$x(t) = \frac{2}{3} \cos 8t - \frac{1}{6} \sin 8t.$$

Find the smallest positive value of t for which the object is in the equilibrium position.

79. The graph of $y = 2 \sin x - \cos 2x$ over the interval $[0, 2\pi]$ is drawn in the figure. By using the methods of calculus, it can be shown that the turning points A, B, C, D occur at x values where

$$\cos x + \sin 2x = 0.$$

Find the coordinates of these four points.

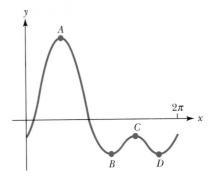

EXERCISE 79

80. The following figure shows a sketch of the graph of $y = \sqrt{3} \sin x + \frac{1}{2} \cos 2x$ over the interval $[0, 2\pi]$. By using calculus, the turning points A, B, C, D can be located at values of x for which

$$\sqrt{3} \cos x - \sin 2x = 0.$$

Find the coordinates of these four points.

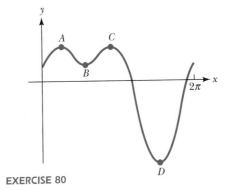

EXERCISE 80

6.7 Inverse Trigonometric Functions

Our main objective in this section is to define an inverse function for each of the six trigonometric functions. We first consider the sine function, $y = \sin x$, where x is a real number.

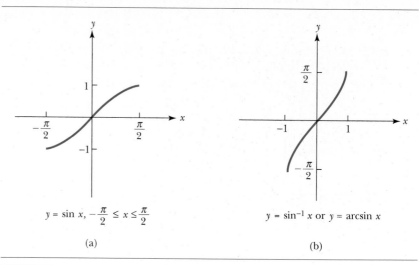

$$y = \sin x, \ -\frac{\pi}{2} \le x \le \frac{\pi}{2}$$

(a)

$$y = \sin^{-1} x \text{ or } y = \arcsin x$$

(b)

FIGURE 6.5

It is clear from the graph of $y = \sin x$ in Figure 5.35 that any horizontal line between $y = -1$ and $y = 1$ intersects the graph of $y = \sin x$ at more than one point, so the sine function is *not* one-to-one. As we learned in Section 2.5, this means that the sine function does not have an inverse function when its domain is the set of all real numbers.

In order to obtain an inverse function, we restrict the domain of the sine function to the interval $-\pi/2 \le x \le \pi/2$. (See Figure 6.5a.) As x varies from $-\pi/2$ to $\pi/2$, $\sin x$ takes on each value from -1 to 1 exactly once.[3] With this restricted domain, an inverse function exists. This inverse function is called the **inverse sine function,** or the **arcsine function.**[4] Both of the following notations are commonly used to denote this function.

Either $y = \sin^{-1} x$ or $y = \arcsin x$

is equivalent to

$$x = \sin y \quad \text{and} \quad -\frac{\pi}{2} \le y \le \frac{\pi}{2}.$$

The graph of the inverse sine function is shown in Figure 6.5(b). Note that it has domain $-1 \le x \le 1$ and range $-\pi/2 \le y \le \pi/2$.

3. The interval $-\pi/2 \le x \le \pi/2$ is sometimes called the **interval of principal values** of the sine function.
4. The term *arcsine* relates to the fact that y is an arclength on the unit circle whose sine is x.

Before considering an example, we note that the variables in the equation $x = \sin y$ have been *real numbers* throughout our discussion. We may think of y as radian measure of an angle if we wish, but we shall avoid any conversions to degrees because of the potential confusion.

Example 1   Find the exact value of arcsin $(-\frac{1}{2})$ without using a calculator or tables.

Solution

Let $y = $ arcsin $(-\frac{1}{2})$. Then $\sin y = -\frac{1}{2}$ and $-\pi/2 \leq y \leq \pi/2$. Because of the restricted range for y and the fact that $\sin y$ is negative, we draw the reference triangle in quadrant IV, as shown in Figure 6.6. From the figure we see that

$$\arcsin\left(-\frac{1}{2}\right) = y = -\frac{\pi}{6}.$$

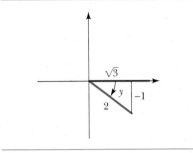

FIGURE 6.6

For each of the other trigonometric functions, we restrict the domain so that the new function is one-to-one and has an inverse function. The resulting **inverse trigonometric functions** have these restricted domains as their ranges. They are listed in Table 6.1 so that their ranges may be easily compared and memorized.

Inverse Trigonometric Functions		
Function	Defining equation	Range
$y = \sin^{-1} x = \arcsin x$	$x = \sin y$	$-\dfrac{\pi}{2} \leq y \leq \dfrac{\pi}{2}$
$y = \csc^{-1} x = \operatorname{arccsc} x$	$x = \csc y$	$-\dfrac{\pi}{2} \leq y \leq \dfrac{\pi}{2}, \quad y \neq 0$
$y = \tan^{-1} x = \arctan x$	$x = \tan y$	$-\dfrac{\pi}{2} < y < \dfrac{\pi}{2}$
$y = \cot^{-1} x = \operatorname{arccot} x$	$x = \cot y$	$0 < y < \pi$
$y = \cos^{-1} x = \arccos x$	$x = \cos y$	$0 \leq y \leq \pi$
$y = \sec^{-1} x = \operatorname{arcsec} x$	$x = \sec y$	$0 \leq y \leq \pi, \quad y \neq \dfrac{\pi}{2}$

TABLE 6.1

The graphs of the inverse trigonometric functions other than the inverse sine are shown in Figure 6.7.

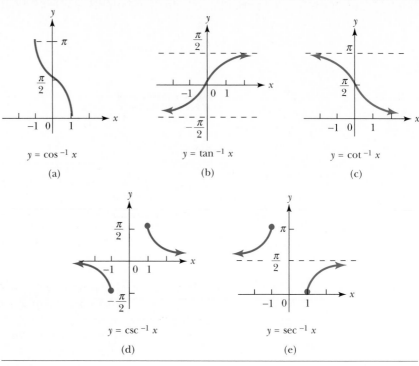

$y = \cos^{-1} x$

(a)

$y = \tan^{-1} x$

(b)

$y = \cot^{-1} x$

(c)

$y = \csc^{-1} x$

(d)

$y = \sec^{-1} x$

(e)

FIGURE 6.7

> **Example 2**   Evaluate without using a calculator or tables.
>
> a) $\cos^{-1}\left(-\frac{1}{2}\right)$ b) $\sin^{-1} 2$

Solution

a) $y = \cos^{-1}\left(-\frac{1}{2}\right)$ is equivalent to $\cos y = -\frac{1}{2}$ and $0 \leq y \leq \pi$. Since $\cos y$ is negative and $0 \leq y \leq \pi$, we find an angle $y = 2\pi/3$ in quadrant II with $\cos y = -\frac{1}{2}$. Thus

$$\cos^{-1}\left(-\frac{1}{2}\right) = y = \frac{2\pi}{3}.$$

It is worth noting that $\cos^{-1}\left(-\frac{1}{2}\right) \neq -4\pi/3$, because $-4\pi/3$ is *not in the range* of the inverse cosine function.

b) $y = \sin^{-1} 2$ is equivalent to $\sin y = 2$ and $-\pi/2 \leq y \leq \pi/2$. This is impossible since $-1 \leq \sin y \leq 1$ for all y. In other words, 2 is not in the domain of $y = \sin^{-1} x$. Thus $\sin^{-1} 2$ is not defined.   ❏

    The equations $f^{-1}(f(x)) = x$ and $f(f^{-1}(x)) = x$ that were obtained in Section 2.5 for inverse functions in general have specific applications here.

$$\sin^{-1}(\sin x) = x \quad \text{or} \quad \arcsin(\sin x) = x \qquad \text{if } -\frac{\pi}{2} \le x \le \frac{\pi}{2}.$$

$$\sin(\sin^{-1} x) = x \quad \text{or} \quad \sin(\arcsin x) = x \qquad \text{if } -1 \le x \le 1.$$

Of course, corresponding equations hold for the other inverse trigonometric functions.

Example 3   Evaluate each of the following without using a calculator or tables.

a) csc (arccsc 3)   b) $\sin(\cos^{-1}(-\tfrac{3}{5}))$

Solution

a) It follows from the discussion in the preceding paragraph that

$$\text{csc (arccsc 3)} = 3.$$

b) Let $y = \cos^{-1}(-\tfrac{3}{5})$. Then $\cos y = -\tfrac{3}{5}$ and $0 \le y \le \pi$. We draw y in quadrant II and complete the reference triangle as shown in Figure 6.8. From the figure,

$$\sin\left(\cos^{-1}\left(-\frac{3}{5}\right)\right) = \sin y = \frac{4}{5}. \qquad \blacksquare$$

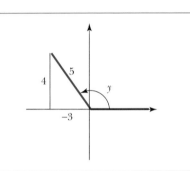

FIGURE 6.8

Some remarks are in order about the use of calculators. The values that a calculator yields when the procedure described in Section 5.5 is followed are values of $\sin^{-1} x$, $\cos^{-1} x$, or $\tan^{-1} x$, given in either degrees or radians, depending on the mode setting of the calculator. In either case, they are values in the ranges specified in Table 6.1. Values of the other inverse trigonometric functions can be obtained by using the following equations.

$$\sec^{-1} x = \cos^{-1}\frac{1}{x} \qquad \text{if } x \le -1 \quad \text{or} \quad x \ge 1.$$

$$\csc^{-1} x = \sin^{-1}\frac{1}{x} \qquad \text{if } x \le -1 \quad \text{or} \quad x \ge 1.$$

$$\cot^{-1} x = \begin{cases} \tan^{-1}\dfrac{1}{x} & \text{if } x > 0; \\[2ex] \dfrac{\pi}{2} & \text{if } x = 0; \\[2ex] \pi + \tan^{-1}\dfrac{1}{x} & \text{if } x < 0. \end{cases}$$

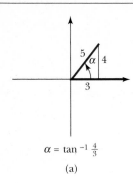

$\alpha = \tan^{-1} \frac{4}{3}$

(a)

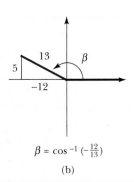

$\beta = \cos^{-1}\left(-\frac{12}{13}\right)$

(b)

FIGURE 6.9

In order to evaluate more complicated expressions involving inverse trigonometric functions, it is frequently necessary to use the identities from the first part of this chapter. The following examples illustrate how substitutions can be used to simplify the work.

Example 4   Evaluate $\sin\left(\tan^{-1}\frac{4}{3} + \cos^{-1}\left(-\frac{12}{13}\right)\right)$ without using a calculator or tables.

Solution
Let $\alpha = \tan^{-1}\left(\frac{4}{3}\right)$ and $\beta = \cos^{-1}\left(-\frac{12}{13}\right)$. We can find $\sin(\alpha + \beta)$ by using the identity $\sin(\alpha + \beta) = \sin\alpha\cos\beta + \cos\alpha\sin\beta$. Both α and β are shown in Figure 6.9. From the figure, we get

$$\sin\alpha = \frac{4}{5}, \quad \cos\alpha = \frac{3}{5}, \quad \sin\beta = \frac{5}{13}, \quad \cos\beta = -\frac{12}{13}.$$

Substitution of these values into the identity yields

$$\sin\left(\tan^{-1}\frac{4}{3} + \cos^{-1}\left(-\frac{12}{13}\right)\right) = \sin(\alpha + \beta)$$

$$= \left(\frac{4}{5}\right)\left(-\frac{12}{13}\right) + \left(\frac{3}{5}\right)\left(\frac{5}{13}\right) = -\frac{33}{65}.$$

Example 5   Assume that x is in the domain of the inverse sine function and find an algebraic expression for $\cos(\sin^{-1}x)$.

Solution
Let $\theta = \sin^{-1}x$. Then $\sin\theta = x$ and θ is one of the angles drawn in Figure 6.10. In either case,

$$\cos(\sin^{-1}x) = \cos\theta = \sqrt{1 - x^2}.$$

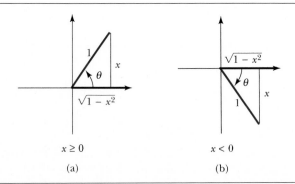

$x \geq 0$

(a)

$x < 0$

(b)

FIGURE 6.10

We can now solve equations in which the variable occurs in an inverse trigonometric function. The basic technique for solving this type of equation is to isolate an inverse trigonometric function on one side of the equation and then use a property of the type $\sin(\sin^{-1}x) = x$. The technique is illustrated in the following example.

Example 6   Solve the equation $\arccos x - \arcsin x = 5\pi/6$.

Solution
We add $\arcsin x$ to both sides and then take the cosine of both sides.

$$\arccos x = \frac{5\pi}{6} + \arcsin x$$

$$\cos(\arccos x) = \cos\left(\frac{5\pi}{6} + \arcsin x\right)$$

$$x = \cos\left(\frac{5\pi}{6} + \arcsin x\right)$$

If we let $\theta = \arcsin x$, then θ is one of the angles shown in Figure 6.10. In both of the cases shown there, $\sin\theta = x$ and $\cos\theta = \sqrt{1-x^2}$. Using the identity for $\cos(\alpha + \beta)$, we have

$$x = \cos\left(\frac{5\pi}{6} + \theta\right)$$

$$x = \cos\frac{5\pi}{6}\cos\theta - \sin\frac{5\pi}{6}\sin\theta$$

$$x = \left(-\frac{\sqrt{3}}{2}\right)\sqrt{1-x^2} - \frac{1}{2}x$$

$$2x = -\sqrt{3}\sqrt{1-x^2} - x$$

$$3x = -\sqrt{3}\sqrt{1-x^2}$$

$$9x^2 = 3 - 3x^2$$

$$12x^2 = 3$$

$$x = \pm\frac{1}{2}.$$

We must check these proposed solutions since we squared both sides of an equation.

Check: $x = \frac{1}{2}$

$$\arccos\frac{1}{2} - \arcsin\frac{1}{2} = \frac{\pi}{3} - \frac{\pi}{6} \neq \frac{5\pi}{6}$$

$x = \frac{1}{2}$ is *not* a solution.

Check: $x = -\frac{1}{2}$

$$\arccos\left(-\frac{1}{2}\right) - \arcsin\left(-\frac{1}{2}\right) = \frac{2\pi}{3} - \left(-\frac{\pi}{6}\right) = \frac{5\pi}{6}$$

$x = -\frac{1}{2}$ is a solution, and the solution set is $\{-\frac{1}{2}\}$.   ❑

Exercises 6.7

Evaluate each of the following. (See Examples 1 and 2.)

1. $\sin^{-1} 1$

2. $\cos^{-1} 1$

3. $\sec^{-1} \sqrt{2}$

4. $\csc^{-1} 2$

5. $\arccos\left(-\frac{1}{\sqrt{2}}\right)$

6. $\arcsec(-2)$

7. $\arccsc(-\sqrt{2})$

8. $\sin^{-1}(-1)$

9. $\arcsin\left(-\frac{\sqrt{3}}{2}\right)$

10. $\cos^{-1}\left(-\frac{\sqrt{3}}{2}\right)$

11. $\cot^{-1}(-\sqrt{3})$

12. $\arctan(-\sqrt{3})$

13. $\cos^{-1} 2$

14. $\sin^{-1} 3$

Find the exact value of each of the following. (See Example 3.)

15. $\tan(\sin^{-1}\frac{1}{2})$

16. $\cos(\tan^{-1}\sqrt{3})$

17. $\csc(\csc^{-1}(-1.3))$

18. $\cot(\cot^{-1}(-3))$

19. $\cos(\sec^{-1}\frac{6}{5})$

20. $\sin(\csc^{-1}\frac{7}{6})$

21. $\cos(\tan^{-1}(-\frac{3}{4}))$

22. $\csc(\arccot(-\frac{5}{12}))$

23. $\cot(\sin^{-1}(-\frac{5}{13}))$

24. $\tan(\cos^{-1}(-\frac{12}{13}))$

25. $\sin^{-1}(\sin 0.99)$

26. $\cos^{-1}(\cos 1.3)$

27. $\cos^{-1}\left(\cos\frac{7\pi}{6}\right)$

28. $\sin^{-1}\left(\sin\frac{4\pi}{3}\right)$

Find the values of the following functions by using either a calculator or Table I, as instructed by the teacher. In either case, give answers to four decimal places.

29. $\arccos 0.9805$

30. $\tan^{-1} 1.829$

31. $\cot^{-1}(-0.7445)$

32. $\sin^{-1}(-0.4514)$

Find the values of the following functions by using either a calculator or Table I, as instructed by the teacher. Give answers to four digits.

33. $\sin(\cos^{-1} 0.8511)$

34. $\cos(\tan^{-1} 1.621)$

35. $\cot(\cos^{-1}(-0.5760))$

36. $\csc(\arctan(-3.108))$

Use identities to find the exact value of each of the following. (See Example 4.)

37. $\cos(2\cos^{-1}\frac{4}{7})$

38. $\cos(2\sin^{-1}\frac{5}{6})$

39. $\sin(2\arccos(-\frac{4}{5}))$

40. $\sin(2\arccot(-\frac{4}{3}))$

41. $\cos(2\arctan(-\frac{5}{12}))$

42. $\cos(2\arccsc(-\frac{13}{12}))$

43. $\sin(\cot^{-1}\frac{5}{12} + \tan^{-1}(-\frac{3}{4}))$

44. $\tan(\sec^{-1}\frac{17}{15} - \cot^{-1}(-\frac{3}{4}))$

45. $\cos(\arcsin(-\frac{4}{5}) - \arctan(-\frac{5}{12}))$

46. $\cos(\arcsin(-\frac{3}{5}) + \arccos(-\frac{5}{13}))$

In Exercises 47–54, assume that x and y are such that each function is defined, and find an algebraic expression for the function value. (See Example 5.)

47. $\tan(\cos^{-1} x)$

48. $\sin(\cos^{-1} x)$

49. $\sec(\cot^{-1} x)$

50. $\csc(\cot^{-1} x)$

51. $\cos(2\tan^{-1} x)$

52. $\cos(2\cot^{-1} x)$

53. $\sin(\sin^{-1} x + \cos^{-1} y)$

54. $\cos(\cos^{-1} x + \sin^{-1} y)$

Solve the following equations. (See Example 6.)

55. $\tan^{-1} x = -\frac{\pi}{3}$

56. $\sin^{-1} x = -\frac{\pi}{6}$

57. $2\cos^{-1} x - \pi = \frac{\pi}{2}$

58. $3\cot^{-1} x - 2\pi = \frac{\pi}{2}$

59. $\arcsin x = \arctan(-1)$

60. $\tan^{-1} x = \sin^{-1}(-\frac{1}{2})$

61. $\sin^{-1} x = \cos^{-1} x$

62. $\tan^{-1} x = \cot^{-1} x$

63. $\arccos x + \arctan\frac{5}{12} = \frac{\pi}{2}$

64. $\cos^{-1} x + \tan^{-1}\frac{3}{4} = \frac{\pi}{2}$

65. $\cos^{-1} x + \sin^{-1} \left(-\frac{3}{5}\right) = \sin^{-1} \frac{4}{5}$

66. $\sin^{-1} x + \sin^{-1} \left(-\frac{5}{13}\right) = \sin^{-1} \frac{12}{13}$

67. $\arccos x + \arcsin (3 - 5x) = \dfrac{\pi}{2}$

68. $\sin^{-1} (-x) + \cos^{-1} (2 + 3x) = \dfrac{\pi}{2}$

69. $\arccos 2x - \arcsin x = \dfrac{7\pi}{6}$

70. $\arccos x - \arcsin x = \dfrac{7\pi}{6}$

71. $\cos^{-1} x + \sin^{-1} x = \dfrac{\pi}{6}$

72. $\cos^{-1} x + \sin^{-1} x = \dfrac{5\pi}{6}$

73. $\arcsin x + \arccos x = \dfrac{\pi}{2}$

74. $\tan^{-1} x + \cot^{-1} x = \dfrac{\pi}{2}$

Sketch the graph of the given equation.

75. $y = \dfrac{\pi}{2} + \sin^{-1} x$ **76.** $y = 2 \sin^{-1} x$

77. $y = \sin^{-1} 2x$ **78.** $y = \sin^{-1} (x - 2)$

79. $y = -\cos^{-1} x$ **80.** $y = \cos^{-1} 2x$

81. $y = \tan^{-1} 2x$ **82.** $y = \tan^{-1} (-x)$

CHAPTER REVIEW

Key Words and Phrases

Fundamental Identities	Reduction Identity	Product Identities
Conditional equation	Sum and Difference Identities	Sum Identities
Cofunction Identities	Double-Angle Identities	Trigonometric equation
Negative-Angle Identities	Half-Angle Identities	Inverse trigonometric functions

Summary of Important Concepts and Formulas

Fundamental Identities

$$\sin \theta \csc \theta = 1 \qquad \sin^2 \theta + \cos^2 \theta = 1$$

$$\cos \theta \sec \theta = 1 \qquad 1 + \tan^2 \theta = \sec^2 \theta$$

$$\tan \theta \cot \theta = 1 \qquad 1 + \cot^2 \theta = \csc^2 \theta$$

$$\tan \theta = \frac{\sin \theta}{\cos \theta} \qquad \cot \theta = \frac{\cos \theta}{\sin \theta}$$

Cofunction Identities

$$\cos \left(\frac{\pi}{2} - \theta\right) = \sin \theta \qquad \sin \left(\frac{\pi}{2} - \theta\right) = \cos \theta$$

$$\cot \left(\frac{\pi}{2} - \theta\right) = \tan \theta \qquad \tan \left(\frac{\pi}{2} - \theta\right) = \cot \theta$$

$$\csc \left(\frac{\pi}{2} - \theta\right) = \sec \theta \qquad \sec \left(\frac{\pi}{2} - \theta\right) = \csc \theta$$

Negative-Angle Identities

$$\sin (-\theta) = -\sin \theta \qquad \csc (-\theta) = -\csc \theta$$

$$\cos (-\theta) = \cos \theta \qquad \sec (-\theta) = \sec \theta$$

$$\tan (-\theta) = -\tan \theta \qquad \cot (-\theta) = -\cot \theta$$

Reduction Identity

$$a \sin \theta + b \cos \theta = \sqrt{a^2 + b^2} \sin (\theta + \alpha),$$

where

$$\cos \alpha = \frac{a}{\sqrt{a^2 + b^2}} \quad \text{and} \quad \sin \alpha = \frac{b}{\sqrt{a^2 + b^2}}$$

Identities for Sum and Difference of Two Angles

$$\cos (\alpha - \beta) = \cos \alpha \cos \beta + \sin \alpha \sin \beta$$

$$\cos (\alpha + \beta) = \cos \alpha \cos \beta - \sin \alpha \sin \beta$$

$$\sin (\alpha + \beta) = \sin \alpha \cos \beta + \cos \alpha \sin \beta$$

$$\sin (\alpha - \beta) = \sin \alpha \cos \beta - \cos \alpha \sin \beta$$

$$\tan (\alpha + \beta) = \frac{\tan \alpha + \tan \beta}{1 - \tan \alpha \tan \beta}$$

$$\tan (\alpha - \beta) = \frac{\tan \alpha - \tan \beta}{1 + \tan \alpha \tan \beta}$$

Double-Angle Identities

$$\sin 2\theta = 2 \sin \theta \cos \theta$$

$$\cos 2\theta = \cos^2 \theta - \sin^2 \theta$$

$$= 2 \cos^2 \theta - 1$$

$$= 1 - 2 \sin^2 \theta$$

$$\tan 2\theta = \frac{2 \tan \theta}{1 - \tan^2 \theta}$$

Half-Angle Identities

$$\sin \frac{\theta}{2} = \pm \sqrt{\frac{1 - \cos \theta}{2}} \qquad \tan \frac{\theta}{2} = \pm \sqrt{\frac{1 - \cos \theta}{1 + \cos \theta}}$$

$$\cos \frac{\theta}{2} = \pm \sqrt{\frac{1 + \cos \theta}{2}} \qquad\qquad = \frac{\sin \theta}{1 + \cos \theta}$$

$$= \frac{1 - \cos \theta}{\sin \theta}$$

Product Identities

$$2 \sin \alpha \cos \beta = \sin (\alpha + \beta) + \sin (\alpha - \beta)$$

$$2 \cos \alpha \sin \beta = \sin (\alpha + \beta) - \sin (\alpha - \beta)$$

$$2 \cos \alpha \cos \beta = \cos (\alpha + \beta) + \cos (\alpha - \beta)$$

$$2 \sin \alpha \sin \beta = \cos (\alpha - \beta) - \cos (\alpha + \beta)$$

Sum Identities

$$\sin x + \sin y = 2 \sin \left(\frac{x + y}{2}\right) \cos \left(\frac{x - y}{2}\right)$$

$$\sin x - \sin y = 2 \cos \left(\frac{x + y}{2}\right) \sin \left(\frac{x - y}{2}\right)$$

$$\cos x + \cos y = 2 \cos \left(\frac{x + y}{2}\right) \cos \left(\frac{x - y}{2}\right)$$

$$\cos x - \cos y = -2 \sin \left(\frac{x + y}{2}\right) \sin \left(\frac{x - y}{2}\right)$$

Inverse Trigonometric Functions

Function	Defining equation	Range
$y = \sin^{-1} x = \arcsin x$	$x = \sin y$	$-\dfrac{\pi}{2} \le y \le \dfrac{\pi}{2}$
$y = \csc^{-1} x = \operatorname{arccsc} x$	$x = \csc y$	$-\dfrac{\pi}{2} \le y \le \dfrac{\pi}{2}, \quad y \ne 0$
$y = \tan^{-1} x = \arctan x$	$x = \tan y$	$-\dfrac{\pi}{2} < y < \dfrac{\pi}{2}$
$y = \cot^{-1} x = \operatorname{arccot} x$	$x = \cot y$	$0 < y < \pi$
$y = \cos^{-1} x = \arccos x$	$x = \cos y$	$0 \le y \le \pi$
$y = \sec^{-1} x = \operatorname{arcsec} x$	$x = \sec y$	$0 \le y \le \pi, \quad y \ne \dfrac{\pi}{2}$

Values of the Inverse Trigonometric Functions

$$\sec^{-1} x = \cos^{-1} \frac{1}{x} \qquad \text{if } x \le -1 \text{ or } x \ge 1$$

$$\csc^{-1} x = \sin^{-1} \frac{1}{x} \qquad \text{if } x \le -1 \text{ or } x \ge 1$$

$$\cot^{-1} x = \begin{cases} \tan^{-1} \dfrac{1}{x} & \text{if } x > 0 \\[2ex] \dfrac{\pi}{2} & \text{if } x = 0 \\[2ex] \pi + \tan^{-1} \dfrac{1}{x} & \text{if } x < 0 \end{cases}$$

Review Problems for Chapter 6

Reduce the first expression to the second.

1. $(\sec^2 \alpha - 1) \cos^2 \alpha$; $\sin^2 \alpha$

2. $\dfrac{1 + \cot^2 \gamma}{1 + \tan^2 \gamma}$; $\cot^2 \gamma$

3. $\dfrac{\sec \theta - \sin \theta \tan \theta}{\cos \theta}$; 1

4. Write all the trigonometric functions in terms of $\csc \theta$.

In Problems 5–10, verify each identity.

5. $(\sin \theta - \csc \theta)^2 = \cot^2 \theta - \cos^2 \theta$

6. $\dfrac{1 + \cot^2 \theta}{\cot^2 \theta} = \sec^2 \theta$

7. $\dfrac{1 - \tan^2 A}{1 + \tan^2 A} = 2 \cos^2 A - 1$

8. $\dfrac{1 - \sin \gamma}{1 + \sin \gamma} = \dfrac{\csc \gamma - 1}{\csc \gamma + 1}$

9. $\dfrac{\cos \alpha - \sec \alpha}{\sec \alpha} + \dfrac{\sin \alpha - \csc \alpha}{\csc \alpha} = -1$

10. $\dfrac{\tan x}{1 - \cos x} = (\sec x + 1) \csc x$

11. Prove that $\tan \theta + \sec \theta = 1$ is not an identity.

12. Use identities to find the exact value of each of the following expressions.

a) $\dfrac{1}{2 \sin 15° \cos 15°}$

b) $\cos 93° \cos 3° + \sin 93° \sin 3°$

c) $\dfrac{2 \tan 22.5°}{1 - \tan^2 22.5°}$

d) $\sin 17° \cos 43° + \cos 17° \sin 43°$

e) $\tan \dfrac{3\pi}{8}$

Write each of the following in terms of trigonometric functions of θ.

13. $\cos\left(\theta + \dfrac{\pi}{4}\right)$ 14. $\sin(\theta + \pi)$ 15. $\tan(\theta - \pi)$

Sketch the graph of each of the following over the interval $0 \le x \le 2\pi$.

16. $y = -2 \sin(-x)$ 17. $y = \sin x + \sqrt{3}\cos x$

18. Without determining α or β, find the exact values of all trigonometric functions of $\alpha - \beta$ if $\cos \alpha = -\frac{5}{13}$, $\sin \beta = \frac{4}{5}$, α is in quadrant III, and β is in quadrant II.

If $0 \le \theta < 2\pi$, $\tan \theta = -\frac{3}{4}$, and $\cos \theta > 0$, find the exact value of each of the following.

19. $\sin 2\theta$ 20. $\cos 2\theta$ 21. $\tan 2\theta$

22. $\sin \dfrac{\theta}{2}$ 23. $\cos \dfrac{\theta}{2}$ 24. $\tan \dfrac{\theta}{2}$

25. Express $\cos 4\theta$ in terms of $\sin \theta$.

In Problems 26–30, verify the identities.

26. $\dfrac{\sin(\alpha - \beta)}{\cos \alpha \cos \beta} = \tan \alpha - \tan \beta$

27. $\dfrac{\cot \beta + \cot \alpha}{\cot \beta \cot \alpha - 1} = \tan(\alpha + \beta)$

28. $\dfrac{\cos 2x + \sin 2x + 1}{\cos 2x - \sin 2x - 1} = -\cot x$

29. $\cos 4\theta + \cos 2\theta + 2 \sin 3\theta \sin \theta = $
    $2 \cos^2 \theta - 2 \sin^2 \theta$

30. $\sin(\alpha + \beta) \cos(\alpha - \beta) = $
    $\sin \alpha \cos \alpha + \sin \beta \cos \beta$

31. Write each of the following products as a sum in which no argument contains a minus sign.
    a) $2 \cos 5\theta \cos 3\theta$
    b) $2 \cos 4x \sin 6x$
    c) $\cos(A + B) \cos(A - B)$

32. Write each of the following sums as a product in which no argument contains a minus sign.
    a) $\sin 4t + \sin 2t$
    b) $\cos 5\alpha + \cos 7\alpha$
    c) $\cos 3t - \cos 9t$

In Problems 33–36, solve for θ, where $0° \le \theta < 360°$.

33. $\sin \theta + 3 = 2 - \sin \theta$

34. $2 \cos^2 \theta = 1 + \cos \theta$

35. $2 \sin^2 \theta - 2\sqrt{2} \sin \theta + 1 = 0$

36. $\sec \theta \sin \theta - \sqrt{2} \sin \theta + \sec \theta - \sqrt{2} = 0$

In Problems 37–42, solve the equations for all x such that $0 \le x < 2\pi$.

37. $\sec x \tan x = \tan x$ 38. $2 \sin x + 3 \csc x = 7$

39. $\cos 2x + \sin x = 0$ 40. $\sin x - \sqrt{3}\cos x = 1$

41. $2 \tan \dfrac{x}{2} + \sqrt{3} = \tan \dfrac{x}{2}$

42. $\cos 3x + \cos x = 0$

Find the exact value of each of the following.

43. $\arcsin 0$

44. $\arccos\left(-\dfrac{1}{\sqrt{2}}\right)$

45. $\tan^{-1}(-1)$

46. $\csc^{-1}\sqrt{2}$

47. $\sin^{-1} 1.2$

48. $\cos(\cos^{-1} 0.2)$

49. $\csc\left(\cos^{-1}\dfrac{1}{\sqrt{2}}\right)$

50. $\tan(\arccos(-1))$

51. $\sin(\operatorname{arccsc} 4)$

52. $\sin(\tan^{-1}(-\tfrac{4}{3}))$

53. $\cos(2\sin^{-1}\tfrac{3}{4})$

54. $\sin(2\arccos(-\tfrac{3}{5}))$

55. $\cos(\sin^{-1}(-\tfrac{4}{5}) + \tan^{-1}(-\tfrac{5}{12}))$

Solve the following equations.

56. $\arctan x = \arcsin \dfrac{1}{\sqrt{1+x^2}}$

57. $\sin^{-1} x - \cos^{-1}\tfrac{12}{13} = \cos^{-1}\tfrac{4}{5}$

58. $\cos^{-1} x + \sin^{-1}\tfrac{4}{5} = \cos^{-1}(-\tfrac{12}{13})$

59. $\sin^{-1}(1 + x) + \cos^{-1} x = \dfrac{5\pi}{6}$

60. $\cos^{-1} 2x + \sin^{-1} x = \dfrac{\pi}{6}$

Sketch the graph of the given equation.

61. $y = \sin^{-1}(-x)$

62. $y = \dfrac{\pi}{2} + \cos^{-1} x$

63. $y = 2\tan^{-1} x$

64. $y = \tan^{-1}(x - 1)$

Prove that the given equation is not an identity by providing a counterexample.

65. $\sin^{-1} x = \dfrac{1}{\sin x}$

66. $\cos^{-1}(-x) = -\cos^{-1} x$

67. $\tan^{-1}(x - 1) = \tan^{-1} x - \tan^{-1} 1$

68. $\cos^{-1}(x + 1) = \cos^{-1} x + \cos^{-1} 1$

Each of the following nonsolutions has at least one error. Can you find them?

Problem 1
Verify the identity: $\tan \theta + \cot \theta = \sec \theta \csc \theta$.

Nonsolution
$$\tan \theta + \cot \theta = \frac{\sin \theta}{\cos \theta} + \frac{\cos \theta}{\sin \theta} = \frac{\sin \theta + \cos \theta}{\cos \theta \sin \theta}$$

$$= \frac{1}{\cos \theta} + \frac{1}{\sin \theta} = \frac{1}{\cos \theta \sin \theta}$$

$$= \frac{1}{\cos \theta} \cdot \frac{1}{\sin \theta} = \sec \theta \csc \theta$$

Problem 2
Solve the following equation for θ, where $0° \leq \theta < 360°$.
$$2 \cos^2 \theta - \cos \theta = 1$$

Nonsolution
$$\cos \theta (2 \cos \theta - 1) = 1$$

$$\cos \theta = 1 \qquad \text{or} \qquad 2 \cos \theta - 1 = 1$$

$$\theta = 0° \qquad\qquad\qquad 2 \cos \theta = 2$$

$$\cos \theta = 1$$

$$\theta = 0°$$

The solution is $\theta = 0°$.

Problem 3
Verify the identity: $\dfrac{\cos 2\theta}{\cos^2 \theta} = 2 - \sec^2 \theta$.

Nonsolution
$$\frac{\cos 2\theta}{\cos^2 \theta} = \frac{\cos^2 \theta + \sin^2 \theta}{\cos^2 \theta}$$

$$= \frac{\cos^2 \theta}{\cos^2 \theta} + \frac{\sin^2 \theta}{\cos^2 \theta} = 1 + \tan^2 \theta$$

$$= 1 + (1 - \sec^2 \theta) = 2 - \sec^2 \theta$$

Problem 4
Verify the identity: $(\cos \theta - \sin\theta)^2 = 1 - \sin 2\theta$.

Nonsolution
$$(\cos \theta - \sin \theta)^2 = \cos^2 \theta - \sin^2 \theta$$

$$= 1 - \sin^2 \theta - \sin^2 \theta = 1 - 2 \sin^2 \theta$$

$$= \cos 2\theta = 1 - \sin 2\theta$$

Problem 5 Solve the following equation for all x such that $0 \le x < 2\pi$.

$$\sin 2x = \cos x$$

Nonsolution

$$2 \sin x \cos x = \cos x$$

$$2 \sin x = 1$$

$$\sin x = \frac{1}{2}$$

$$x = \frac{\pi}{6} \text{ or } x = \frac{5\pi}{6}$$

Problem 6 Find the exact value of arccot $(-\sqrt{3})$.

Nonsolution From Figure 6.11, we get

$$\text{arccot } (-\sqrt{3}) = \arctan \left(-\frac{1}{\sqrt{3}} \right) = -\frac{\pi}{6}.$$

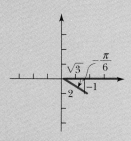

FIGURE 6.11

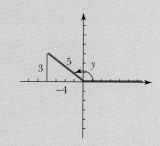

FIGURE 6.12

Problem 7 Find the exact value of $\sin (\tan^{-1} (-\frac{3}{4}))$.

Nonsolution Let $y = \tan^{-1} (-\frac{3}{4})$, as in Figure 6.12. Then $r = \sqrt{(-4)^2 + (3)^2} = 5$, and

$$\sin \left(\tan^{-1} \left(-\frac{3}{4} \right) \right) = \sin y = \frac{3}{5}.$$

Problem 8 Find the exact value of $\sin^{-1} \left(\sin \frac{5\pi}{6} \right)$.

Nonsolution

$$\sin^{-1} \left(\sin \frac{5\pi}{6} \right) = \frac{5\pi}{6}$$

308

Chapter 7

Additional Topics in Trigonometry

The classical applications of trigonometry to surveying and navigation are treated in this chapter. Vectors are introduced and used where applicable. Some modern settings that relate to illegal drug trafficking and offshore oil exploration are included to make these applications more interesting.

7.1   The Law of Sines

We now turn our attention to *oblique triangles* (triangles that are not right triangles). In this and the following section we develop some formulas that are used to solve oblique triangles. The first is called the *Law of Sines*.

Consider the triangles in Figure 7.1, in which the sides and angles are labeled according to the convention described in Section 5.6, with the exception that angle C is not required to be a right angle. In each triangle in the figure, we draw the perpendicular h from the vertex of angle C to side c (extended if necessary). Then we have

$$\sin B = \frac{h}{a} \quad \text{and} \quad \sin A = \frac{h}{b},$$

where we use the fact that $\sin (180° - A) = \sin A$ for the obtuse triangle. Solving for h gives

$$a \sin B = h \quad \text{and} \quad b \sin A = h,$$

and therefore

$$a \sin B = b \sin A.$$

Dividing the equation by $\sin A \sin B$ results in one part of the Law of Sines:

$$\frac{a}{\sin A} = \frac{b}{\sin B}.$$

Similarly, if the perpendicular k is drawn from the vertex of angle A to side a, we can obtain

$$\frac{b}{\sin B} = \frac{c}{\sin C}.$$

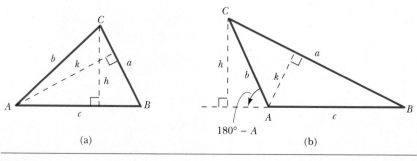

(a)

(b)

FIGURE 7.1

Combining the two parts results in the following general statement.

Law of Sines

$$\frac{a}{\sin A} = \frac{b}{\sin B} = \frac{c}{\sin C}$$

There are actually three parts to the Law of Sines:

$$\frac{a}{\sin A} = \frac{b}{\sin B}, \qquad \frac{a}{\sin A} = \frac{c}{\sin C}, \qquad \frac{b}{\sin B} = \frac{c}{\sin C}$$

The Law of Sines is especially useful in solving triangles in two particular cases when certain information is known about a triangle:

Case 1.   Two angles and one side are known.

Case 2.   Two sides and one angle opposite one of those sides are known.

The next example illustrates the solution of a Case 1 triangle.

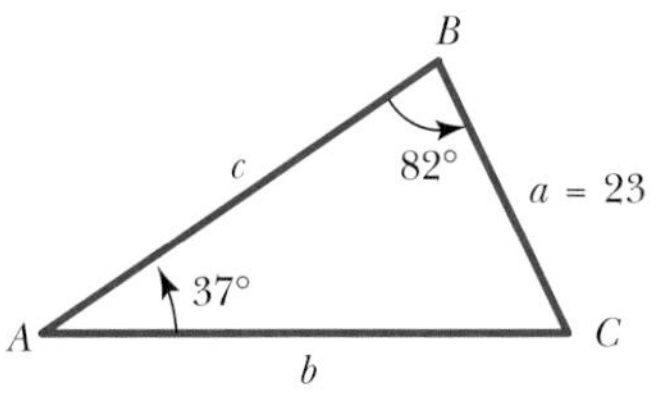

Example 1   Solve the triangle in Figure 7.2, where $A = 37°$, $B = 82°$, and $a = 23$.

Solution
Since the sum of the three angles A, B, and C must be $180°$, we find C to measure $61°$. Sides b and c are found by using two parts of the Law of Sines.

$$\frac{b}{\sin B} = \frac{a}{\sin A} \qquad\qquad \frac{c}{\sin C} = \frac{a}{\sin A}$$

$$\frac{b}{\sin 82°} = \frac{23}{\sin 37°} \qquad\qquad \frac{c}{\sin 61°} = \frac{23}{\sin 37°}$$

$$b = \frac{23 \sin 82°}{\sin 37°} \qquad\qquad c = \frac{23 \sin 61°}{\sin 37°}$$

$$b = 38 \qquad\qquad\qquad c = 33$$

All parts of the triangle are now known:

$$a = 23, \qquad A = 37°,$$
$$\boldsymbol{b = 38,} \qquad B = 82°,$$
$$\boldsymbol{c = 33,} \qquad \boldsymbol{C = 61°.}$$

FIGURE 7.2

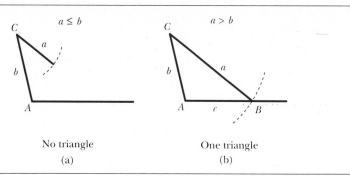

FIGURE 7.3   Obtuse angle A

We now investigate the Case 2 triangles: those in which the known parts are two sides and one angle opposite one of those sides. This case is often called the **ambiguous case** since several possible solutions may occur, depending on the sizes of the known parts.

In a triangle labeled conventionally, suppose angle A and sides a and b are the known parts. We first consider the situation when the known angle is obtuse. By examining Figure 7.3(a), we see that no triangle exists if side a is too short, that is, if $a \leq b$. However, if $a > b$, as in Figure 7.3(b), a unique triangle is determined.

In Case 2 triangles we must first solve for an angle. Therefore it is more convenient to use the Law of Sines in an alternative form:

$$\frac{\sin A}{a} = \frac{\sin B}{b} = \frac{\sin C}{c}.$$

With an obtuse angle A, if the computations result in a value of $\sin B$ between 0 and 1, then there will be a unique triangle satisfying the given conditions. Otherwise, no such triangle exists.

Example 2   Solve the triangle, if one exists, where $A = 123°$, $a = 14$, and $b = 21$.

Solution
We use the Law of Sines to determine angle B.

$$\frac{\sin B}{b} = \frac{\sin A}{a}$$

$$\frac{\sin B}{21} = \frac{\sin 123°}{14}$$

$$\sin B = \frac{21 \sin 123°}{14}$$

$$\sin B = 1.2580 \qquad \text{IMPOSSIBLE}$$

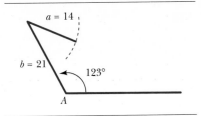

FIGURE 7.4

Since it is true that $-1 \leq \sin B \leq 1$ for any angle B, there is no angle B such that $\sin B = 1.2580 > 1$. Hence NO TRIANGLE satisfies the given conditions. A quick sketch, as in Figure 7.4, where we can see the consequences of the fact that $a \leq b$, would have led us directly to the same conclusion. ❏

Next we consider the Case 2 triangles in which the known angle is acute. Again we choose to use angle A and sides a and b as the known parts. There are four possibilities to consider, as diagrammed in Figure 7.5.

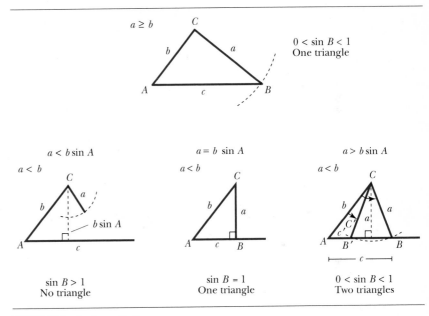

FIGURE 7.5 Acute angle A

Rather than memorizing the conditions stated in the figure, we can again determine which possibility occurs according to the calculations resulting from using the Law of Sines. If it turns out that $\sin B > 1$, then there is no triangle. If $\sin B = 1$, we have a right triangle. If $0 < \sin B < 1$, then there will be either one triangle or two triangles. We use the phrase "solve the triangle" to mean "solve the triangle(s) if one (or two) exists; otherwise show that there is no triangle satisfying the given conditions."

Example 3 Solve the triangle where $A = 48°$, $b = 32$, and $a = 61$.

Solution
Solving for angle B, we have

$$\frac{\sin B}{b} = \frac{\sin A}{a}$$

$$\frac{\sin B}{32} = \frac{\sin 48°}{61}$$

$$\sin B = \frac{32 \sin 48°}{61} = 0.3898.$$

Since the sine function is positive in the first and second quadrants (B can be either acute or obtuse), we consider two possible values for B.

$$B = \begin{cases} 23° \\ 180° - 23° = 157° \qquad \text{\scriptsize IMPOSSIBLE} \end{cases}$$

$$\text{\scriptsize since } A + B = 48° + 157° > 180°$$

Thus there is only one possible triangle. With $B = 23°$, we find C to be

$$C = 180° - (A + B) = 180° - (48° + 23°) = 109°.$$

Then, to determine side c, we use

$$\frac{c}{\sin C} = \frac{a}{\sin A}$$

$$c = \frac{a \sin C}{\sin A} = \frac{61 \sin 109°}{\sin 48°} = 78.$$

All parts of the triangle sketched in Figure 7.6 are known.

$$\begin{array}{ll} a = 61 & A = 48° \\ b = 32 & \boldsymbol{B = 23°} \\ \boldsymbol{c = 78} & \boldsymbol{C = 109°} \end{array} \qquad \square$$

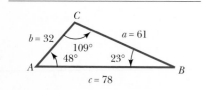

FIGURE 7.6

Finally, the last example illustrates how two triangles satisfying the same given conditions can be determined.

Example 4   Solve the triangle where $A = 40.2°$, $a = 128$, and $b = 179$.

Solution
First we use the Law of Sines to determine angle B.

$$\frac{\sin B}{b} = \frac{\sin A}{a}$$

$$\frac{\sin B}{179} = \frac{\sin 40.2°}{128}$$

$$\sin B = \frac{179 \sin 40.2°}{128} = 0.9026$$

$$B = \begin{cases} 64.5° \\ 180° - 64.5° = 115.5° \end{cases}$$

We have two values, 64.5° and 115.5°, for angle B, since for either value we have $A + B < 180°$. Thus there exist TWO TRIANGLES, one acute and one obtuse. They are sketched in Figure 7.7.

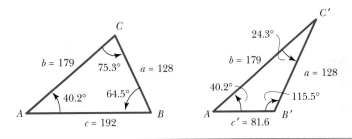

FIGURE 7.7

Acute Angle B	Obtuse Angle B'
$C = 180° - (A + B)$	$C' = 180° - (A + B')$
$= 180° - (40.2° + 64.5°)$	$= 180° - (40.2° + 115.5°)$
$= 75.3°$	$= 24.3°$
$\dfrac{c}{\sin C} = \dfrac{a}{\sin A}$	$\dfrac{c'}{\sin C'} = \dfrac{a}{\sin A}$
$c = \dfrac{128 \sin 75.3°}{\sin 40.2°}$	$c' = \dfrac{128 \sin 24.3°}{\sin 40.2°}$
$c = 192$	$c' = 81.6$

The parts of the acute triangle are:	The parts of the obtuse triangle are:
$a = 128,\quad A = 40.2°,$	$a = 128,\quad A = 40.2°,$
$b = 179,\quad \boldsymbol{B = 64.5°},$	$b = 179,\quad \boldsymbol{B' = 115.5°},$
$\boldsymbol{c = 192},\quad C = 75.3°.$	$c' = 81.6,\quad C' = 24.3°.$ ❑

To be consistent with the labeling in Figure 7.5, the known parts in the triangles of each example have been angle A and sides a and b. However, the method of solution described in each example applies no matter which angle and sides are the known parts of the Case 2 triangles.

Example 5   Point C in Figure 7.8 is on the opposite bank of a river from park headquarters at A. The park director wishes to establish a ranger station at point C and needs to know the distance from A to C. She measures the distance from A along a straight road to a point B where the road ends, and finds this distance is 11 miles. She then measures the angles at A and B, as indicated in the figure. How far is it from A to C?

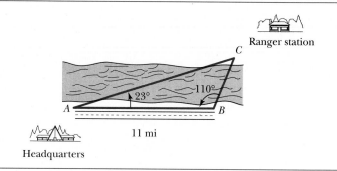

FIGURE 7.8

Solution
Using conventional notation in Figure 7.8, we have $A = 23°$, $B = 110°$, $c = 11$, and we need to find b. Our plan is to find angle C and then use the Law of Sines in the form

$$\frac{b}{\sin B} = \frac{c}{\sin C}.$$

We have

$$C = 180° - (A + B) = 180° - 133° = 47°.$$

Thus

$$\frac{b}{\sin 110°} = \frac{11}{\sin 47°}$$

and

$$b = \frac{11 \sin 110°}{\sin 47°} = 14.$$

The distance from A to C is 14 miles.   ❏

Exercises 7.1

Solve for the indicated part of the triangle to the degree of accuracy consistent with the given information. (See Examples 1–4.)

1. $A = 42°$, $B = 19°$, $a = 1.3$, $c = ?$
2. $B = 39°$, $C = 58°$, $a = 2.5$, $b = ?$
3. $A = 66.6°$, $B = 33.3°$, $c = 392$, $b = ?$
4. $A = 108.2°$, $C = 24.7°$, $c = 423$, $b = ?$

Use a calculator for the following exercises.

5. $A = 42.13°$, $B = 91.46°$, $a = 12.86$, $b = ?$
6. $B = 49.61°$, $C = 88.56°$, $b = 41.39$, $a = ?$

7. $A = 150°$, $a = 40$, $b = 10$, $B = ?$
8. $B = 100°$, $b = 9.0$, $c = 3.0$, $C = ?$
9. $B = 30°$, $a = 14$, $b = 7.0$, $A = ?$
10. $C = 30°$, $b = 2.4$, $c = 1.2$, $B = ?$
11. $B = 60°$, $b = 18$, $c = 15$, $C = ?$
12. $C = 70°$, $a = 12$, $c = 20$, $A = ?$
13. $C = 140°$, $a = 20$, $c = 15$, $A = ?$
14. $B = 72°$, $a = 20$, $b = 17$, $A = ?$
15. $A = 54°$, $a = 80$, $b = 90$, $B = ?$
16. $C = 30°$, $b = 3.4$, $c = 2.0$, $B = ?$

In Exercises 17–32, solve the triangle to the degree of accuracy consistent with the given information. (See Examples 1–4.)

17. $A = 68°$, $B = 62°$, $c = 48$
18. $B = 102°$, $C = 48°$, $a = 52$
19. $B = 79.2°$, $C = 35.1°$, $a = 11.3$
20. $A = 80.8°$, $C = 43.0°$, $c = 423$
21. $A = 120°$, $a = 20$, $b = 40$
22. $C = 121°$, $b = 3.4$, $c = 1.8$
23. $A = 127°$, $a = 40$, $c = 30$
24. $C = 105°$, $b = 22$, $c = 43$
25. $A = 47°$, $a = 80$, $b = 70$
26. $C = 51.3°$, $a = 115$, $c = 127$
27. $A = 17.8°$, $a = 1.21$, $b = 4.89$

28. $A = 68°20'$, $a = 143$, $c = 308$
29. $B = 47°$, $a = 20$, $b = 18$
30. $A = 22.8°$, $a = 1.83$, $b = 4.29$
31. $A = 62.3°$, $a = 178$, $c = 187$
32. $C = 38°40'$, $a = 4.28$, $c = 2.99$

33. Two guy wires are attached 8.0 feet apart on a TV antenna mast and anchored at the same point on the ground, as shown in the figure. The shorter wire makes an angle of 45° with the horizontal, and the longer an angle of 53°. How long is each of the wires?

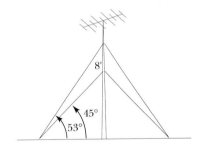

EXERCISE 33

34. A hill in the figure below slopes 12° with the horizontal. The angle of elevation to the top of a pine tree standing 52 feet from the base of the hill is 63°. How tall is the pine tree?

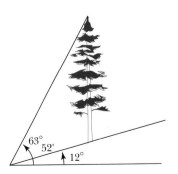

EXERCISE 34

35. The wall of the A-frame cottage in the accompanying figure makes an angle of 68° with the floor. Richard places the top of his 15-foot ladder against the base of the window in which he needs to replace a broken pane. If his ladder forms an angle of 53° with the horizontal, how far from the lower edge of the wall is the base of the ladder?

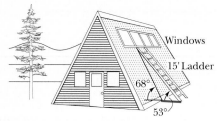

EXERCISE 35

36. The braces on a picnic table are made from 2 × 4's bolted together to form an X, as shown below. Find the length of 2 × 4 needed for each brace.

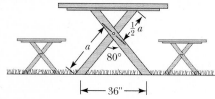

EXERCISE 36

37. A vertical post stands on a hill that is inclined at 16° with the horizontal. The angle of elevation of the top of the post from a point 81 feet down the hill from the base of the post is 26°. Find the height of the post.

38. A flagpole leaning away from the sun forms a 10° angle with the vertical and casts a 22-foot shadow on level ground. From the end of the shadow the angle of elevation to the top of the pole is 47°. Find the length of the pole.

39. The harbor master receives a call for a pilot boat from a freighter in the direction N 58° E from the harbor. The pilot boat is located 60 miles due west of the harbor. The direction from the pilot boat to the freighter is N 75° E. What is the distance between the pilot boat and the freighter?

40. Two angles of a triangular pasture are 36.1° and 45.2°. The side between the angles measures 326 yards. How many yards of fencing are required to enclose the pasture?

41. A vertical tower that is 123 feet high stands on a hill. The angle of elevation of the top of the tower from a point 296 feet down the hill from the base is 26.2°. Find the angle of inclination of the hill with the horizontal.

42. Memphis, Tennessee, is due north of New Orleans, Louisiana, at a distance of 350 miles. If a plane leaves New Orleans and flies in the direction 58.2° until it is 540 miles from Memphis, find the bearing from the plane to Memphis.

7.2 | The Law of Cosines

There are two more cases to consider when certain information is known about a triangle:

Case 3. Two sides and the included angle are known.

Case 4. All three sides are known.

In either of these two cases, no ambiguity arises: If such a triangle exists, it will be unique. One problem does arise, however. If we attempt to use the Law of Sines, we shall discover that it is impossible to find an equation containing only one unknown. This is illustrated in the first example of a Case 3 triangle.

Example 1   Solve the triangle where $A = 52°$, $b = 88$, and $c = 67$.

Solution

In an attempt to solve for angle B by using the Law of Sines, we write one of the two equations:

$$\frac{\sin B}{b} = \frac{\sin A}{a} \qquad \text{or} \qquad \frac{\sin B}{b} = \frac{\sin C}{c}$$

$$\frac{\sin B}{88} = \frac{\sin 52°}{a} \qquad\qquad \frac{\sin B}{88} = \frac{\sin C}{67}.$$

We cannot solve for angle B in either equation, since we have no value for side a in the first and no value of angle C in the second.   ❏

Example 1 illustrates the need for an additional tool for solving triangles. The *Law of Cosines* fills this need. We derive it next.

Suppose angle A and sides b and c are the known parts of a triangle labeled conventionally. We place angle A in standard position with side c lying along the positive x-axis. Angle A may be an acute, obtuse, or right angle. Figure 7.9 illustrates all three possibilities.

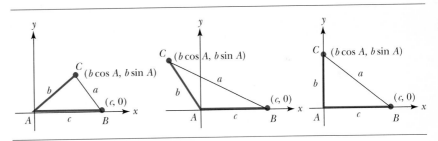

FIGURE 7.9

In each of the triangles in Figure 7.9, the length of side a is the distance between the vertices of angles C and B. The coordinates of the vertex of angle C are $(b \cos A, b \sin A)$ and of angle B are $(c, 0)$. Applying the distance formula, we have

$$\begin{aligned}
\boldsymbol{a^2} &= (b \cos A - c)^2 + (b \sin A - 0)^2 \\
&= b^2 \cos^2 A - 2bc \cos A + c^2 + b^2 \sin^2 A \\
&= b^2 (\cos^2 A + \sin^2 A) + c^2 - 2bc \cos A \\
&= \boldsymbol{b^2 + c^2 - 2bc \cos A},
\end{aligned}$$

which is one of the forms of the *Law of Cosines*.

When angle A is a right angle, side a is the hypotenuse of a right triangle, and the Law of Cosines reduces to the Pythagorean Theorem: $a^2 = b^2 + c^2$.

Two other forms of the Law of Cosines arise when (1) sides a and c and the included angle B are known or (2) sides a and b and the included angle C are known.

Law of Cosines

In any triangle labeled conventionally,

$$a^2 = b^2 + c^2 - 2bc \cos A,$$

$$b^2 = a^2 + c^2 - 2ac \cos B,$$

$$c^2 = a^2 + b^2 - 2ab \cos C.$$

We are now in a position to solve the Case 3 triangle described in Example 1.

Example 2   Solve the triangle where $A = 52°$, $b = 88$, and $c = 67$.

Solution

We use the Law of Cosines to find the length of side a.

$$a^2 = b^2 + c^2 - 2bc \cos A$$

$$= 88^2 + 67^2 - 2(88)(67) \cos 52°$$

$$= 7744 + 4489 - 11792(0.6157)$$

$$= 7744 + 4489 - 7260$$

$$= 4973$$

Taking square roots,

$$a = \sqrt{4973} = 71.$$

We now use the Law of Sines to find angle C. We choose angle C (rather than angle B) since side c is the shortest side. This guarantees that angle C will be acute.

$$\frac{\sin C}{c} = \frac{\sin A}{a}$$

$$\sin C = \frac{c \sin A}{a} = \frac{67 \sin 52°}{71}$$

$$= 0.7436$$

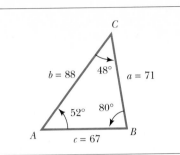

FIGURE 7.10

Since angle C must be acute, we use the first quadrant value for angle C:

$$C = 48°.$$

Then we determine angle B by

$$B = 180° - (A + C) = 180° - (52° + 48°) = 80°.$$

All parts of the triangle sketched in Figure 7.10 are now known.

$$a = 71 \quad A = 52°$$
$$b = 88 \quad B = 80°$$
$$c = 67 \quad C = 48° \qquad ❏$$

The Law of Cosines can also be used to solve the Case 4 triangles, where all three sides are known. For such a triangle to exist, the length of the longest side must be less than the sum of the lengths of the shorter two sides. That is, one of the following three inequalities must be satisfied:

$$a + b > c \quad \text{or} \quad a + c > b \quad \text{or} \quad b + c > a$$

for sides a, b, and c of any triangle. Then each of the three forms of the Law of Cosines can be solved for the cosine of one of the angles A, B, or C in terms of the three sides a, b, and c.

Law of Cosines

In any triangle labeled conventionally,

$$\cos A = \frac{b^2 + c^2 - a^2}{2bc},$$

$$\cos B = \frac{a^2 + c^2 - b^2}{2ac},$$

$$\cos C = \frac{a^2 + b^2 - c^2}{2ab}.$$

Example 3   Solve the triangle where $a = 5.3$, $b = 4.3$, and $c = 7.2$.

Solution
Since $a + b > c$, a triangle exists satisfying the given information. We use the Law of Cosines to solve for angle C, the largest angle in the

triangle. (This assures us that angles A and B will be acute.)

$$\cos C = \frac{a^2 + b^2 - c^2}{2ab}$$

$$= \frac{(5.3)^2 + (4.3)^2 - (7.2)^2}{2(5.3)(4.3)}$$

$$= \frac{28.09 + 18.49 - 51.84}{45.58}$$

$$= -0.1154$$

The cosine is negative in the second and fourth quadrants. But a fourth quadrant angle cannot be an angle in a triangle. Thus C must be the second quadrant angle:

$$C = 97°.$$

Next we use the Law of Sines to find angle B, the smaller of the two remaining angles.

$$\frac{\sin B}{b} = \frac{\sin C}{c}$$

$$\sin B = \frac{b \sin C}{c}$$

$$= \frac{4.3 \sin 97°}{7.2}$$

$$= 0.5928$$

Since B must be acute, we have

$$B = 36°.$$

Then we determine angle A by

$$A = 180° - (B + C) = 180° - (36° + 97°) = 47°.$$

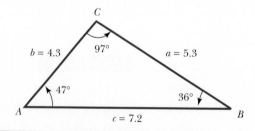

FIGURE 7.11

All parts of the Case 4 triangle shown in Figure 7.11 are now known.

$$a = 5.3 \qquad A = 47°$$
$$b = 4.3 \qquad B = 36°$$
$$c = 7.2 \qquad C = 97°$$ ❏

Example 4   A triangular piece of glass is to be cut to fit in a side window of a small fishing boat. The dimensions of the glass should be 14 inches, 30 inches, and 36 inches. In order to mark off the triangle, the glass cutter needs to know the angle opposite the longest side. Find this angle.

Solution
A sketch of the triangular piece of glass is shown in Figure 7.12.

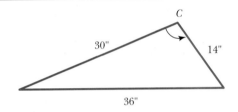

FIGURE 7.12

If we use C to denote the angle opposite the longest side, then $c = 36$ and

$$\cos C = \frac{a^2 + b^2 - c^2}{2ab}$$

$$= \frac{(14)^2 + (30)^2 - (36)^2}{2(14)(30)}$$

$$= -0.2381.$$

Thus $C = \cos^{-1}(-0.2381) = 104°$.   ❏

Exercises 7.2

In Exercises 1–16, solve for the indicated part of the triangle to the degree of accuracy consistent with the given information. (See Examples 2 and 3.)

1. $A = 60°$, $b = 20$, $c = 10$, $a = ?$

2. $B = 60°$, $a = 10$, $c = 40$, $b = ?$

3. $C = 120°$, $a = 4.0$, $b = 6.0$, $c = ?$

4. $A = 120°$, $b = 5.0$, $c = 2.0$, $a = ?$

5. $B = 23°$, $a = 10$, $c = 17$, $b = ?$

6. $C = 48°$, $a = 41$, $b = 36$, $c = ?$

7. $A = 115°$, $b = 3.5$, $c = 2.4$, $a = ?$

8. $B = 108°$, $a = 17$, $c = 22$, $b = ?$

9. $a = 3.0$, $b = 5.0$, $c = 7.0$, $C = ?$

10. $a = 42$, $b = 18$, $c = 30$, $A = ?$

11. $a = 60$, $b = 28$, $c = 45$, $C = ?$

12. $a = 28$, $b = 84$, $c = 70$, $A = ?$

13. $a = 14$, $b = 20$, $c = 80$, $C = ?$

14. $a = 1.8$, $b = 4.5$, $c = 1.2$, $A = ?$

15. $a = 8.23$, $b = 6.81$, $c = 1.01$, $B = ?$

16. $a = 12.4$, $b = 49.8$, $c = 11.3$, $C = ?$

In Exercises 17–28, solve the triangle to the degree of accuracy consistent with the given information. (See Examples 2 and 3.)

17. $A = 56°$, $b = 20$, $c = 30$

18. $C = 35°$, $a = 8.0$, $b = 12$

19. $B = 112°$, $a = 1.5$, $c = 7.6$

20. $A = 128°$, $b = 29$, $c = 42$

21. $a = 3.0$, $b = 4.0$, $c = 6.0$

22. $a = 8.0$, $b = 5.0$, $c = 4.0$

23. $a = 20$, $b = 30$, $c = 11$

24. $a = 18$, $b = 10$, $c = 20$

25. $A = 27°10'$, $b = 308$, $c = 544$

26. $B = 46°50'$, $a = 177$, $c = 293$

27. $B = 122.6°$, $a = 14.3$, $c = 27.9$

28. $C = 128.7°$, $a = 459$, $b = 627$

29. Matt and Beckie are standing on level ground and holding ropes attached to a hot-air balloon. Matt's rope is 125 yards long, and Beckie's is 105 yards long. If the ropes form an angle of 132.1° where they are attached at the balloon, how far apart are Matt and Beckie standing?

30. If one boat left the harbor and traveled N 47.1° E for 25.2 miles, and a second boat left the harbor and traveled S 21.5° E for 18.7 miles, how far apart are the boats?

31. A crewboat leaves harbor H in the following figure and goes 46 miles in the direction S 32° W to deliver the mail to a drilling rig D. From D the boat travels 68 miles in a westerly direction to carry supplies to a pipe barge B that is anchored 97 miles from H. What course should the boat set to go from B back to the harbor?

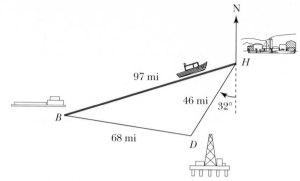

EXERCISE 31

32. A triangular flower bed has sides measuring 3.2 feet, 4.7 feet, and 5.9 feet. Find the measure of the smallest angle.

33. The village of Oak Hill O has built a water reservoir with a pumping station at point A in the accompanying figure and installed a pipeline that is 6.2 miles from A to the village. The reservoir furnishes enough water for Oak Hill and the neighboring village of Bell Camp as well, and plans are under consideration to build a pipeline from A to Bell Camp, as indicated in the figure. Bell Camp is 12 miles directly west of Oak Hill, and the direction from Oak Hill to A is S 55° W. Find the length of the proposed pipeline from A to B.

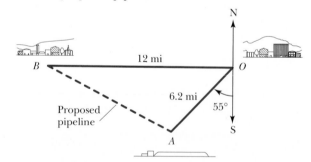

EXERCISE 33

34. Two streets intersect at an angle 137°. A triangular parking lot on the corner has 92-foot frontage on one street and 78-foot frontage on the other. Find the length of the back boundary of the lot.

35. The Coast Guard detects a sunken bargelike vessel and suspects that it was used for smuggling illegal drugs to Florida. With sonar devices they determine that it is located directly in front of the Coast Guard cutter at a distance of 80 feet and an angle of 37° with the horizontal. The Coast Guard then

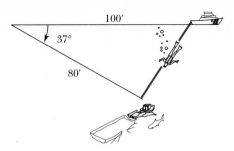

EXERCISE 35

moves forward 100 feet as shown in the figure passing over the bargelike object, and stops. How far must the diver descend in order to reach the object?

36. A 60-inch hose attaches to a 47-inch wand on the vacuum cleaner in the figure below. When the hose is completely outstretched (with no sags), the most comfortable position for use by a person of average height occurs when a 130° angle is formed between the hose and the wand. In this position, how far from the vacuum cleaner is the carpet attachment?

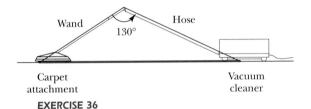

EXERCISE 36

37. Hideaki is at Charleston, West Virginia, and he would like to fly his small plane to Norfolk, Virginia, to see the women's basketball national championship game at Old Dominion University. He knows that the direction from Charleston to Washington, D.C., is 81.0° and the distance is 342 miles. He also knows that the direction from Washington to Norfolk is 166.0° and the distance is 189 miles. Find the distance from Charleston to Norfolk.

38. A plane flies 175 miles in the direction of 150.0° from Toledo Express Airport T to point A, then turns and flies 88.0 miles in the direction of 70.0° before making an emergency landing at point B. (See the figure below.) How far is point B from Toledo Express Airport?

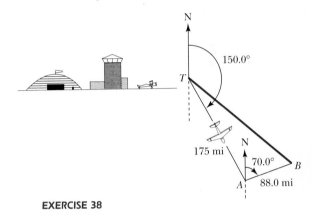

EXERCISE 38

7.3 The Area of a Triangle

Formulas exist for finding the area of a triangle in terms of the known parts for the following three cases.

Case 1. Two angles (and hence all three angles) and one side are known.

Case 3. Two sides and the included angle are known.

Case 4. All three sides are known.

First we derive the area formula for the Case 3 triangle. Since the area of a triangle is one-half the product of the base and the altitude,

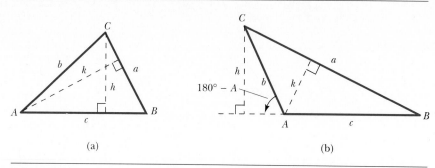

FIGURE 7.13

we see in Figure 7.13 that the area K can be written in three different ways:

1. $K = \frac{1}{2}ch = \frac{1}{2}bc \sin A,$

2. $K = \frac{1}{2}ch = \frac{1}{2}ac \sin B,$

3. $K = \frac{1}{2}ak = \frac{1}{2}ab \sin C.$

Thus we have established the formulas for the area of Case 3 triangles.

Area of a Triangle

In any triangle labeled conventionally, the area K is
$$K = \tfrac{1}{2}bc \sin A = \tfrac{1}{2}ac \sin B = \tfrac{1}{2}ab \sin C.$$

Example 1   Determine the area of the triangle where $A = 42°$, $b = 12$ inches, and $c = 16$ inches.

Solution
Applying the formula for the area, we have

$$K = \tfrac{1}{2}bc \sin A$$

$$= \tfrac{1}{2}(12)(16) \sin 42°$$

$$= 64 \text{ square inches.} \qquad \square$$

We use the formulas just derived and the Law of Sines to obtain formulas for the area of Case 1 triangles.

In the formula for area

$$K = \tfrac{1}{2}bc \sin A,$$

we replace c by

$$c = \frac{b \sin C}{\sin B} \qquad \text{Law of Sines}$$

to yield a formula for area in terms of three angles and one side.

$$K = \frac{b^2 \sin A \sin C}{2 \sin B}$$

Similar replacements yield two additional formulas for the area of Case 1 triangles.

Area of a Triangle

In any triangle labeled conventionally, the area K is

$$K = \frac{a^2 \sin B \sin C}{2 \sin A} = \frac{b^2 \sin A \sin C}{2 \sin B} = \frac{c^2 \sin A \sin B}{2 \sin C}.$$

Example 2   Compute the area of the triangle where $A = 92°$, $B = 28°$, and $b = 10$ feet.

Solution
We first determine angle C by

$$C = 180° - (A + B) = 180° - (92° + 28°) = 60°.$$

Then

$$K = \frac{b^2 \sin A \sin C}{2 \sin B}$$

$$= \frac{10^2 \sin 92° \sin 60°}{2 \sin 28°}$$

$$= 92 \text{ square feet.} \qquad \square$$

Finally, we derive Heron's Formula for the area of Case 4 triangles.

Heron's Formula

The area K of a triangle with sides a, b, and c is

$$K = \sqrt{s(s - a)(s - b)(s - c)},$$

where s is the semiperimeter of the triangle:

$$s = \tfrac{1}{2}(a + b + c).$$

To derive Heron's Formula, we write the square of the area of a triangle as

$$K^2 = \tfrac{1}{4}b^2c^2 \sin^2 A \qquad \text{Area of a Triangle}$$

$$= \frac{b^2c^2}{4}(1 - \cos^2 A) \qquad \text{Pythagorean Identity}$$

$$= \frac{b^2c^2}{4}(1 + \cos A)(1 - \cos A) \qquad \text{Algebra}$$

$$= \frac{b^2c^2}{4}\left(1 + \frac{b^2 + c^2 - a^2}{2bc}\right)\left(1 - \frac{b^2 + c^2 - a^2}{2bc}\right) \qquad \text{Law of Cosines}$$

$$= \frac{b^2c^2}{4}\left[\frac{(b + c)^2 - a^2}{2bc}\right]\left[\frac{a^2 - (b - c)^2}{2bc}\right] \qquad \text{Algebra}$$

$$= \frac{(b + c + a)(b + c - a)(a - b + c)(a + b - c)}{16} \qquad \text{Algebra}$$

$$= \left(\frac{a + b + c}{2}\right)\left(\frac{b + c - a}{2}\right)\left(\frac{a + c - b}{2}\right)\left(\frac{a + b - c}{2}\right) \qquad \text{Algebra}$$

If we let $s = \tfrac{1}{2}(a + b + c)$, then

$$s - a = \tfrac{1}{2}(a + b + c) - a = \frac{b + c - a}{2},$$

$$s - b = \tfrac{1}{2}(a + b + c) - b = \frac{a + c - b}{2},$$

$$s - c = \tfrac{1}{2}(a + b + c) - c = \frac{a + b - c}{2}.$$

Thus

$$K^2 = s(s - a)(s - b)(s - c),$$

and the area K is

$$K = \sqrt{s(s - a)(s - b)(s - c)}.$$

Example 3   Find the area of the triangle with sides measuring $a = 42$ meters, $b = 17$ meters, and $c = 35$ meters.

Solution
We first compute the semiperimeter s.

$$s = \tfrac{1}{2}(a + b + c) = \tfrac{1}{2}(42 + 17 + 35) = 47$$

Then

$$s - a = 47 - 42 = 5,$$

$$s - b = 47 - 17 = 30,$$

$$s - c = 47 - 35 = 12.$$

Finally, the area K is

$$K = \sqrt{s(s - a)(s - b)(s - c)}$$

$$= \sqrt{47(5)(30)(12)}$$

$$= \sqrt{84600}$$

$$= 290 \text{ square meters,}$$

rounded to two significant digits.   ❏

We note that we can also find the area of Case 2 triangles (the ambiguous case). However, we must use the Law of Sines first to determine some additional information about the triangle. Then we can compute the area, remembering that there may be two possible triangles to consider.

Exercises 7.3

Find the area of each of the following triangles. (See Examples 1–3.)

1. $A = 56°$, $b = 20$, $c = 30$
2. $C = 35°$, $a = 8.0$, $b = 12$
3. $B = 112°$, $a = 1.5$, $c = 7.6$
4. $A = 128°$, $b = 29$, $c = 42$
5. $A = 48°$, $C = 92°$, $a = 1.3$
6. $B = 82°$, $C = 59°$, $c = 12$
7. $A = 68°$, $B = 62°$, $c = 48$
8. $B = 102°$, $C = 48°$, $a = 52$
9. $a = 3.0$, $b = 5.0$, $c = 7.0$
10. $a = 42$, $b = 18$, $c = 30$
11. $a = 60$, $b = 28$, $c = 45$
12. $a = 28$, $b = 84$, $c = 70$
13. $A = 47°$, $a = 80$, $b = 70$
14. $B = 22°$, $b = 0.70$, $c = 0.40$

15. $A = 127°$, $a = 40$, $c = 32$
16. $C = 140°$, $a = 5.0$, $c = 7.0$
17. $A = 120°$, $a = 20$, $b = 40$
18. $A = 150°$, $a = 50$, $b = 30$
19. $A = 17.8°$, $a = 1.21$, $b = 4.89$
20. $B = 48.2°$, $b = 204$, $c = 591$
21. $B = 47°$, $a = 20$, $b = 30$
22. $C = 53°$, $a = 80$, $c = 90$
23. $A = 22.8°$, $a = 1.83$, $b = 4.29$
24. $B = 40.2°$, $b = 14.1$, $c = 21.3$

25. How many square feet (to the nearest square foot) of carpet are needed to cover the floor of a triangular conference room with sides of length 20 feet, 22 feet, and 30 feet?

26. How many square feet (to the nearest square foot) of cloth are needed for a triangular hang glider of dimensions 20 feet, 12 feet, and 12 feet?

27. Compute the area of the pasture shown below. (*Hint:* Divide the pasture into two triangles, as indicated in the figure.)

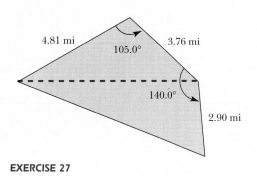

EXERCISE 27

28. Compute the area of the tract of land pictured in the figure. (*Hint:* Divide the land into three triangles, as indicated in the figure.)

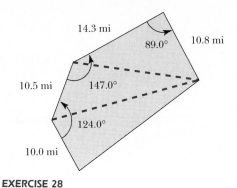

EXERCISE 28

7.4   Vectors

In order to describe many physical quantities, it is necessary to specify both a direction and a magnitude. For example, a statement that the wind is blowing at 30 miles per hour is an incomplete description. For this information to be useful, the direction also needs to be given.

Quantities that possess both magnitude and direction are called **vector quantities.** Displacements, forces, velocities, and accelerations are some of the simpler vector quantities. In contrast, a quantity that can be described by magnitude alone is called a **scalar quantity.** Pressure and temperature are examples of scalar quantities.

A vector quantity can be represented by a directed line segment, which is called a **vector.** That is, a *vector* is a line segment that has been given a direction from one endpoint to the other. An arrowhead is placed at one end to indicate the direction of the segment. The vector from point O to point P may be denoted by $\overrightarrow{OP}$, as indicated in Figure 7.14. The point O is called the **initial point,** or **tail,** of $\overrightarrow{OP}$, and P is called the **terminal point,** or **head,** of $\overrightarrow{OP}$.

It is frequently convenient to use a single letter to name a vector. When this is done, the letter is printed in boldface, such as **V**, or an arrow is written over it, such as $\overrightarrow{V}$. Some illustrations are shown in Figure 7.14.

When a vector quantity is represented by a vector, the **magnitude,** or **length,** of the vector corresponds to the magnitude of the quantity, and the direction of the vector represents the direction of the quantity. The magnitude of the vector **V** is denoted by $|\mathbf{V}|$.

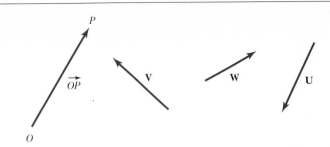

FIGURE 7.14

Vectors **U** and **V** are defined to be **equal** if they have the same direction and equal magnitudes. They need not have the same initial points. (See Figure 7.15.)

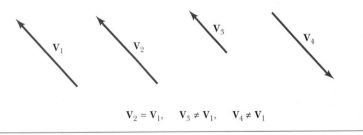

$$\mathbf{V}_2 = \mathbf{V}_1, \quad \mathbf{V}_3 \neq \mathbf{V}_1, \quad \mathbf{V}_4 \neq \mathbf{V}_1$$

FIGURE 7.15

The vector from O to O is called the **zero vector** and is denoted by **0** or $\vec{0}$. The zero vector has length zero and arbitrary direction.

Many applications of vectors involve **vector addition.** When the initial point of **V** is placed at the terminal point of **U**, the vector from the initial point of **U** to the terminal point of **V** is, by definition, the sum **U** + **V**. This is shown in Figure 7.16.

To construct the sum **V** + **U**, the initial point of **U** is placed at the terminal point of **V**, and **V** + **U** is drawn from the initial point of **V** to the terminal point of **U**, as shown in Figure 7.17(a). In Figure 7.17(b), we see that **U** + **V** = **V** + **U** is the diagonal of a parallelogram that has **U** and **V** as sides. Because of this, the rule for vector addition is frequently referred to as the **parallelogram rule.**

The sum **U** + **V** is called the **resultant** of **U** and **V**. One of the reasons vectors are useful is that vector addition provides a faithful representation for many physical results. For example, we learn in physics that when forces are represented by vectors, two forces acting simultaneously on an object combine to produce a resultant force in a manner consistent with addition of vectors.

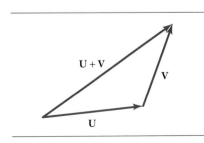

FIGURE 7.16

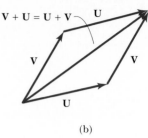

(a)　　　　　　　　　　　(b)

FIGURE 7.17

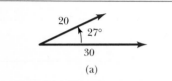

(a)

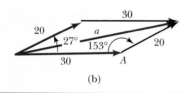

(b)

FIGURE 7.18

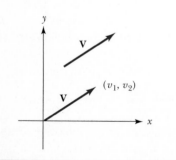

FIGURE 7.19

Example 1　Two forces of magnitudes 20 pounds and 30 pounds act on a point in the plane. If the angle between the directions of the two forces is 27°, find the magnitude of the resultant force.

Solution

The vectors representing the two forces are shown in Figure 7.18(a). The magnitude of the resultant force is the length a of the diagonal in the parallelogram in Figure 7.18(b). We see that angle A measures 153°.

We use the Law of Cosines to find a.

$$a^2 = 30^2 + 20^2 - 2(30)(20) \cos 153° = 2369$$

Thus the magnitude of the resultant force is

$$a = \sqrt{2369} = 49 \text{ pounds.} \qquad \square$$

As the preceding example suggests, the representation of vector quantities by directed line segments has many useful applications. When the calculus is applied to vector quantities, however, another representation is required. To obtain this representation, we restrict our attention to the set of all vectors in an xy coordinate plane.

Since vectors are equal if they have the same direction and magnitude, any vector $\mathbf{V}$ in the plane is equal to a vector with its initial point at the origin and its terminal point at (v_1, v_2), as shown in Figure 7.19. This fact associates each vector $\mathbf{V}$ in the xy plane with an ordered pair (v_1, v_2), and the association is actually a one-to-one correspondence since either the vector or the ordered pair uniquely determines the other. We use this one-to-one correspondence to represent $\mathbf{V}$ by the ordered pair $\langle v_1, v_2 \rangle$. The corner brackets $\langle$ and $\rangle$ are used instead of parentheses to avoid confusion with the coordinates of points. We write

$$\mathbf{V} = \langle v_1, v_2 \rangle,$$

and we call v_1 and v_2 the **x-component** and the **y-component** of $\mathbf{V}$, respectively. It is clear from Figure 7.19 that

$$|\mathbf{V}| = \sqrt{v_1^2 + v_2^2}$$

and

$$\langle u_1, u_2 \rangle = \langle v_1, v_2 \rangle \quad \text{if and only if} \quad u_1 = v_1 \text{ and } u_2 = v_2.$$

The parallelogram rule for vectors in component form appears as

$$\langle u_1, u_2 \rangle + \langle v_1, v_2 \rangle = \langle u_1 + v_1, u_2 + v_2 \rangle.$$

This rule is diagrammed in Figure 7.20.

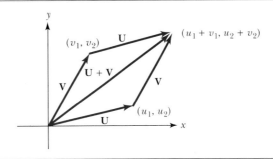

FIGURE 7.20

The product of a number a and a vector $\mathbf{V} = \langle v_1, v_2 \rangle$ is given by

$$a\mathbf{V} = \langle av_1, av_2 \rangle.$$

Direct computation can be used to show that

$$|a\mathbf{V}| = |a| \cdot |\mathbf{V}|.$$

In connection with the product $a\mathbf{V}$, a real number a is called a **scalar,** and the operation of computing $a\mathbf{V}$ is called **scalar multiplication.**

As directed line segments, $a\mathbf{V}$ has the same direction as $\mathbf{V}$ if a is positive, and the opposite direction to $\mathbf{V}$ if a is negative. Thus $(-1)\mathbf{V}$ is the same as the vector $-\mathbf{V} = \langle -v_1, -v_2 \rangle$, which has the property that

$$\mathbf{V} + (-\mathbf{V}) = (-\mathbf{V}) + \mathbf{V} = \mathbf{0}.$$

Subtraction of vectors is defined by $\mathbf{U} - \mathbf{V} = \mathbf{U} + (-\mathbf{V})$, or

$$\langle u_1, u_2 \rangle - \langle v_1, v_2 \rangle = \langle u_1 - v_1, u_2 - v_2 \rangle.$$

Example 2   Let $\mathbf{U} = \langle 2, -1 \rangle$ and $\mathbf{V} = \langle 3, 4 \rangle$, as shown in Figure 7.21. Find

a) $\mathbf{U} + \mathbf{V}$,   b) $2\mathbf{V}$,   c) $-\mathbf{U}$,   d) $5\mathbf{U} - 4\mathbf{V}$,   e) $|\mathbf{V}|$ and $|2\mathbf{V}|$.

Solution

a) $\mathbf{U} + \mathbf{V} = \langle 2, -1 \rangle + \langle 3, 4 \rangle = \langle 2 + 3, -1 + 4 \rangle = \langle 5, 3 \rangle$

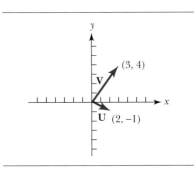

FIGURE 7.21

b) $2\mathbf{V} = 2\langle 3, 4\rangle = \langle (2)(3), (2)(4)\rangle = \langle 6, 8\rangle$

c) $-\mathbf{U} = (-1)\langle 2, -1\rangle = \langle -2, 1\rangle$

d) $5\mathbf{U} - 4\mathbf{V} = 5\langle 2, -1\rangle - 4\langle 3, 4\rangle = \langle 10, -5\rangle - \langle 12, 16\rangle = \langle -2, -21\rangle$

e)   We have

$$|\mathbf{V}| = \sqrt{3^2 + 4^2} = \sqrt{25} = 5$$

and

$$|2\mathbf{V}| = \sqrt{6^2 + 8^2} = \sqrt{100} = 10,$$

illustrating the fact that

$$|2\mathbf{V}| = 2|\mathbf{V}|. \qquad \square$$

The following list presents the basic properties of vector addition and scalar multiplication.

Basic Properties of Vector Operations

1. $\mathbf{U} + \mathbf{V} = \mathbf{V} + \mathbf{U}$
2. $\mathbf{U} + (\mathbf{V} + \mathbf{W}) = (\mathbf{U} + \mathbf{V}) + \mathbf{W}$
3. $\mathbf{V} + \mathbf{0} = \mathbf{V}$
4. $\mathbf{V} + (-\mathbf{V}) = \mathbf{0}$
5. $a(b\mathbf{V}) = (ab)\mathbf{V}$
6. $a(\mathbf{U} + \mathbf{V}) = a\mathbf{U} + a\mathbf{V}$
7. $(a + b)\mathbf{V} = a\mathbf{V} + b\mathbf{V}$
8. $1\,\mathbf{V} = \mathbf{V}$
9. $|a\mathbf{V}| = |a||\mathbf{V}|$

The basic properties of vector operations are consequences of familiar properties of operations with real numbers, such as the commutative and associative properties of addition. We illustrate these with the proof of property 7 and leave the proofs of the remaining properties as exercises. For arbitrary $\mathbf{V} = \langle v_1, v_2\rangle$ and scalars a, b, we have the following:

$(a + b)\mathbf{V} = (a + b)\langle v_1, v_2\rangle$

$= \langle (a + b)v_1, (a + b)v_2\rangle$ Scalar Multiplication

$= \langle av_1 + bv_1, av_2 + bv_2\rangle$ Distributive Property of Real Numbers

$= \langle av_1, av_2\rangle + \langle bv_1, bv_2\rangle$ Vector Addition

$= a\mathbf{V} + b\mathbf{V}$ Scalar Multiplication

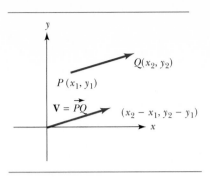

FIGURE 7.22

We have seen in Figure 7.19 that any directed line segment in the xy plane is equal to a vector **V** with initial point at the origin and terminal point (v_1, v_2). A vector with its initial point at the origin is called a **position vector.** If the original directed line segment extended from $P(x_1, y_1)$ to $Q(x_2, y_2)$, the position vector $\mathbf{V} = \overrightarrow{PQ}$ extends from $(0, 0)$ to $(x_2 - x_1, y_2 - y_1)$, as shown in Figure 7.22.

Example 3   The vector $\overrightarrow{PQ}$ from $P(-4, 7)$ to $Q(2, 4)$ is given by

$$\overrightarrow{PQ} = \langle 2 - (-4), 4 - 7 \rangle$$

$$= \langle 6, -3 \rangle.$$

It is worth noting that $\overrightarrow{QP} = \langle -6, 3 \rangle$, and that $\overrightarrow{QP} = -\overrightarrow{PQ}$ is always true.   ❑

The simple equation

$$\langle v_1, v_2 \rangle = v_1 \langle 1, 0 \rangle + v_2 \langle 0, 1 \rangle$$

shows that every vector in the plane can be written as the sum of a multiple of $\langle 1, 0 \rangle$ and a multiple of $\langle 0, 1 \rangle$. This fact leads to an alternative notation for vectors in the plane. If we let $\mathbf{i} = \langle 1, 0 \rangle$ and $\mathbf{j} = \langle 0, 1 \rangle$, we have

$$\mathbf{V} = \langle v_1, v_2 \rangle = v_1 \mathbf{i} + v_2 \mathbf{j}$$

for arbitrary **V**. This notation is used in calculus and is very common in many areas of application. The vectors **i** and **j** are called **base vectors.**

In future courses it will be helpful to be familiar with the **i, j** notation. We have

$$(u_1 \mathbf{i} + u_2 \mathbf{j}) + (v_1 \mathbf{i} + v_2 \mathbf{j}) = (u_1 + v_1)\mathbf{i} + (u_2 + v_2)\mathbf{j},$$

$$a(v_1 \mathbf{i} + v_2 \mathbf{j}) = (av_1)\mathbf{i} + (av_2)\mathbf{j},$$

and

$$|v_1 \mathbf{i} + v_2 \mathbf{j}| = \sqrt{v_1^2 + v_2^2}.$$

Example 4   The use of the formulas in the preceding paragraph is illustrated by the following computations.

a) $(3\mathbf{i} + 7\mathbf{j}) + (2\mathbf{i} + 5\mathbf{j}) = (3 + 2)\mathbf{i} + (-7 + 5)\mathbf{j}$

$$= 5\mathbf{i} - 2\mathbf{j}$$

b) $2(4\mathbf{i} - 6\mathbf{j}) = (2)(4)\mathbf{i} + (2)(-6)\mathbf{j}$

$$= 8\mathbf{i} - 12\mathbf{j}$$

c) $|5\mathbf{i} - 2\mathbf{j}| = \sqrt{5^2 + (-2)^2}$

$$= \sqrt{29}$$ ❑

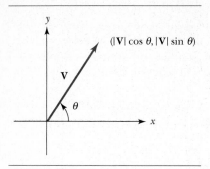

FIGURE 7.23

We note that each of $\mathbf{i} = \langle 1, 0 \rangle$ and $\mathbf{j} = \langle 0, 1 \rangle$ has magnitude 1. Any vector with magnitude 1 is called a **unit vector.**

There is yet another formulation that can be made for vectors. For an arbitrary nonzero position vector $\mathbf{V} = \langle v_1, v_2 \rangle = v_1\mathbf{i} + v_2\mathbf{j}$, let θ be an angle in standard position with $\mathbf{V}$ as its terminal side. (See Figure 7.23.) The angle θ is called the **direction angle** for $\mathbf{V}$. It follows at once from the definitions of $\sin\theta$ and $\cos\theta$ that $\cos\theta = v_1/|\mathbf{V}|$ and $\sin\theta = v_2/|\mathbf{V}|$. Thus $v_1 = |\mathbf{V}| \cos\theta$, $v_2 = |\mathbf{V}| \sin\theta$, and

$$\mathbf{V} = \langle |\mathbf{V}| \cos\theta, |\mathbf{V}| \sin\theta \rangle$$

$$= |\mathbf{V}| \cos\theta\,\mathbf{i} + |\mathbf{V}| \sin\theta\,\mathbf{j},$$

where $|\mathbf{V}| = \sqrt{v_1^2 + v_2^2}$ and $\tan\theta = v_2/v_1$. This formula expresses $\mathbf{V}$ in terms of its magnitude $|\mathbf{V}|$ and its direction angle θ.

Example 5

a) Find the component form of a vector that has magnitude 6 and direction angle $5\pi/6$.

b) Find the magnitude and direction angle for the vector $\mathbf{V} = -\mathbf{i} + \sqrt{3}\mathbf{j}$.

Solution

a) Since we have $|\mathbf{V}| = 6$ and $\theta = 5\pi/6$, the vector $\mathbf{V}$ is given by

$$\mathbf{V} = |\mathbf{V}| \cos\theta\,\mathbf{i} + |\mathbf{V}| \sin\theta\,\mathbf{j}$$

$$= 6 \cos\frac{5\pi}{6}\,\mathbf{i} + 6 \sin\frac{5\pi}{6}\,\mathbf{j}$$

$$= 6\left(-\frac{\sqrt{3}}{2}\right)\mathbf{i} + 6\left(\frac{1}{2}\right)\mathbf{j}$$

$$= -3\sqrt{3}\,\mathbf{i} + 3\mathbf{j}.$$

b) For $\mathbf{V} = -\mathbf{i} + \sqrt{3}\mathbf{j}$, we have

$$|\mathbf{V}| = \sqrt{(-1)^2 + (\sqrt{3})^2} = \sqrt{4} = 2.$$

We also have $\tan\theta = \sqrt{3}/(-1) = -\sqrt{3}$, but $\theta \neq \tan^{-1}(-\sqrt{3})$ since θ is a second quadrant angle and $\tan^{-1}(-\sqrt{3}) = -\pi/3$. We must choose the second quadrant angle

$$\theta = \pi - \frac{\pi}{3} = \frac{2\pi}{3}$$

as the direction angle for $-\mathbf{i} + \sqrt{3}\mathbf{j}$.   ❏

In applications involving navigation, the words *course* and *heading* have technical meanings. The **course** of a ship or a plane is the direction of the path in which the craft is moving, and the **heading** is the direction in which the craft is pointed. The word **speed** is used when speaking of magnitude only. The **airspeed** of a plane is its speed in still air, whereas the **ground speed** of a plane is its speed relative to the ground. The word **velocity** is used when referring to both speed and direction.

Example 6   Find the airspeed and heading of an airplane if a wind of 30.0 miles per hour from the direction of 343.0° results in the airplane flying a course of 43.0° with a ground speed of 250 miles per hour.

Solution
In Figure 7.24, the length of side x represents the airspeed, and the angle β represents the heading of the airplane. We first find x by using the Law of Cosines.

$$x^2 = (30.0)^2 + (250)^2 - 2(30.0)(250) \cos 120.0°$$

$$= 900 + 62500 - 15000(-0.5)$$

$$= 900 + 62500 + 7500$$

$$= 70900$$

Thus the airspeed is

$$x = \sqrt{70900}$$

$$= 266 \text{ miles per hour,}$$

rounded to the nearest mile per hour.
    To find β, we first determine the angle labeled α in Figure 7.24.

$$\frac{\sin \alpha}{30.0} = \frac{\sin 120.0°}{266}$$

$$\sin \alpha = \frac{30.0 \sin 120.0°}{266}$$

$$\sin \alpha = 0.0977$$

Since α is an acute angle, we have $\alpha = 5.6°$. Then the angle β is $43.0° - 5.6° = 37.4°$, and the heading of the airplane is $37.4°$.   ❑

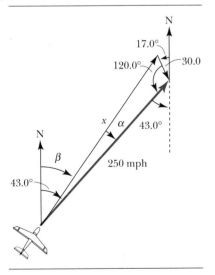

FIGURE 7.24

    Example 6 illustrates the fact that velocities acting simultaneously on an object combine according to the parallelogram rule for addition of vectors.

Example 7   A barrel of oil weighing 350 pounds lies on its side on a loading ramp that makes an angle of 8.4° with the horizontal. Neglecting friction, what force parallel to the ramp is required to keep the barrel from rolling down the ramp?

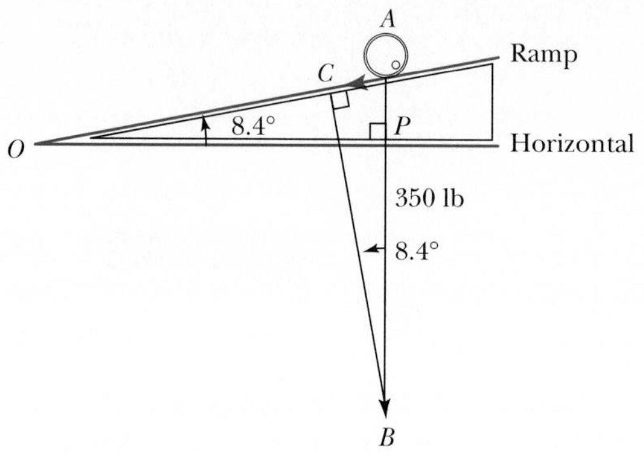

FIGURE 7.25

Solution

As shown in Figure 7.25, the weight of the barrel is represented by a vector $\overrightarrow{AB}$ with length 350 units and direction vertically downward. This vector can be written as the sum of a vector $\overrightarrow{AC}$ parallel to the ramp and a vector $\overrightarrow{CB}$ perpendicular to the ramp.

The triangles ABC and AOP are similar, and the angle at vertex B is 8.4°. To find the length of $\overrightarrow{AC}$, we note that

$$\sin 8.4° = \frac{|\overrightarrow{AC}|}{350}$$

and

$$|\overrightarrow{AC}| = 350 \sin 8.4° = 51.1,$$

where the result is rounded to three digits to agree with the given data. A force of 51.1 pounds parallel to the ramp is required to hold the barrel in place. ❏

Exercises 7.4

In Exercises 1–4, sketch the vectors **U, V, U + V,** and **U − V** on the same coordinate system.

1. $\mathbf{U} = \langle 1, 2 \rangle$, $\mathbf{V} = \langle 3, -4 \rangle$

2. $\mathbf{U} = \langle -3, 2 \rangle$, $\mathbf{V} = \langle -2, -4 \rangle$

3. $\mathbf{U} = -2\mathbf{i} + 3\mathbf{j}$, $\mathbf{V} = 3\mathbf{i} - \mathbf{j}$

4. $\mathbf{U} = 2\mathbf{i} - \mathbf{j}$, $\mathbf{V} = -\mathbf{i} + 3\mathbf{j}$

In Exercises 5–8, find $2\mathbf{U} - 3\mathbf{V}$ and write your answer in both the ordered pair notation and the **i, j** notation. (See Examples 2 and 4.)

5. $\mathbf{U} = \langle 1, -2 \rangle$, $\mathbf{V} = \langle -3, 5 \rangle$

6. $\mathbf{U} = \langle 0, -6 \rangle$, $\mathbf{V} = \langle 2, 7 \rangle$

7. $\mathbf{U} = 7\mathbf{i} + \mathbf{j}$, $\mathbf{V} = -5\mathbf{i} + 3\mathbf{j}$

8. $\mathbf{U} = -8\mathbf{i} + 5\mathbf{j}$, $\mathbf{V} = 7\mathbf{j}$

Find $\overrightarrow{PQ}$ and $|\overrightarrow{PQ}|$ for each of the following pairs P, Q. (See Examples 2 and 3.)

9. $P(-4, -1), Q(-5, 3)$ **10.** $P(6, -3), Q(1, 4)$

11. $P(-3, 0), Q(-2, 4)$ **12.** $P(-8, 5), Q(0, 7)$

In Exercises 13–20, find the magnitude and direction angle for the given vector. (See Example 5.)

13. $-4\mathbf{j}$ **14.** $5\mathbf{j}$

15. $8\mathbf{i}$ **16.** $-9\mathbf{i}$

17. $\mathbf{i} - \mathbf{j}$ **18.** $\sqrt{3}\mathbf{i} - \mathbf{j}$

19. $-2\sqrt{2}\,\mathbf{i} + 2\sqrt{2}\,\mathbf{j}$ **20.** $-3\mathbf{i} - 3\mathbf{j}$

21. Two forces of 25 pounds and 45 pounds act on a point in the plane. If the angle between the directions of the two forces is 120°, find the magnitude of the resultant force.

22. Two forces of 37 pounds and 13 pounds act on a point in the plane. If the angle between the directions of the two forces is 78°, find the magnitude of the resultant force.

23. A 13-pound force and a 12-pound force produce a 20-pound resultant force. Find the angle between the directions of the 13-pound force and the resultant force in the figure below.

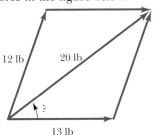

12 lb   20 lb

?

13 lb

EXERCISE 23

24. The angle between the directions of two forces is 113.2°. One of the forces has magnitude 219 pounds, and the resultant has magnitude 242 pounds. Find the angle between the 219-pound force and the resultant force.

25. Two forces act on a point in the plane, producing a resultant force with magnitude 487 pounds. The magnitude of one of the forces is 232 pounds. Find the magnitude of the other force if the angle between the 232-pound force and the resultant force is 27.5°.

26. Two forces act on a point in the plane, producing a resultant force with magnitude 31.3 pounds. The magnitude of one of the forces is 77.1 pounds. Find the magnitude of the other force if the angle between the 77.1-pound force and the resultant force is 63.2°.

(See Examples 6 and 7.)

27. Alex Jones's canoe is headed due west across a stream that is flowing south at the rate of 2.1 miles per hour. If Alex is rowing at the rate of 3.4 miles per hour in still water, find the speed and course at which the canoe is traveling. (See the figure.)

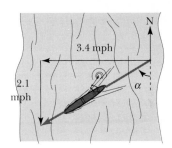

N

3.4 mph

2.1 mph

α

EXERCISE 27

28. A boat that travels at 16 miles per hour in still water wishes to travel due east across a stream that is flowing south at the rate of 2.8 miles per hour. On what heading should the boat be set?

29. A train is traveling in the direction N 41° E at 95 miles per hour, and a ball is thrown from the train at 52 miles per hour in the direction N 49° W. Find the speed and direction of the path of the ball.

30. A duck heads S 20°40′ E flying at an airspeed of 63.1 miles per hour. If the wind is blowing from the north at 18.2 miles per hour, find the ground speed of the duck. (See the following figure.)

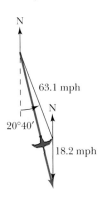

N

63.1 mph

20°40′

N

18.2 mph

EXERCISE 30

31. A duck heads in the direction 159°50′ with an airspeed of 63.2 miles per hour. If the wind is blowing from the direction 49°50′ at 18.0 miles per hour, find the ground speed of the duck and the course it is traveling.

32. A plane is flying with an airspeed of 300 miles per hour and a heading of 285.0°. If the wind is blowing from the direction 230.0° at 45.0 miles per hour, find the ground speed and the course of the plane.

33. A plane is headed in the direction 128.0° with an airspeed of 350 miles per hour, and the wind is blowing from the direction 18.0°. If the course of the plane is in the direction 133.0°, find the speed of the wind.

34. The airplane in the figure is headed due east with an airspeed of 324 miles per hour in a wind blowing from the direction 200.0°. If the ground speed of the plane is 331 miles per hour, find the course of the flight and the speed of the wind.

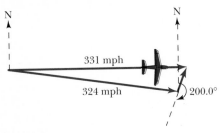

EXERCISE 34

35. An airplane flies with a heading of 221.0° and an airspeed of 267 miles per hour. The ground speed is 275 miles per hour, and the wind speed is 29.2 miles per hour. If the course of the plane is north of the heading, find the angle between the heading and the path of the flight.

36. An airplane flies with a heading of 167.0° and an airspeed of 318 miles per hour. The ground speed is 288 miles per hour, and the wind speed is 42.1 miles per hour. If the course of the plane is north of the heading, find the angle between the heading and the path of the flight.

37. A boat is headed N 42° E across a current flowing due south at the rate of 11 miles per hour. If the motor is driving the boat at the rate of 15 miles per hour, find the speed of the boat and the direction of its course.

38. A plane is flying with an airspeed of 312 miles per hour and a heading of 287.2°. The wind is blowing

from the west at 57.3 miles per hour. Find the ground speed and the course of the flight.

39. In the accompanying figure, a car weighing 1.4 tons sits on a driveway that makes an angle of 16° with the horizontal. If friction is neglected, what force parallel to the driveway is required to hold the car in place?

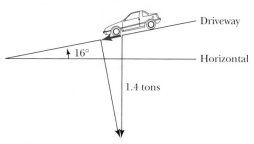

EXERCISE 39

40. In Exercise 39, find the magnitude of the force that the car's weight exerts perpendicular to the driveway.

41. Prove the following properties of vector addition.
a) $\mathbf{U} + \mathbf{V} = \mathbf{V} + \mathbf{U}$
b) $\mathbf{U} + (\mathbf{V} + \mathbf{W}) = (\mathbf{U} + \mathbf{V}) + \mathbf{W}$
c) $\mathbf{V} + \mathbf{0} = \mathbf{V}$
d) $\mathbf{V} + (-\mathbf{V}) = \mathbf{0}$

42. Prove the following properties of scalar multiplication.
a) $a(b\mathbf{V}) = (ab)\mathbf{V}$
b) $a(\mathbf{U} + \mathbf{V}) = a\mathbf{U} + a\mathbf{V}$
c) $1\,\mathbf{V} = \mathbf{V}$
d) $|a\mathbf{V}| = |a||\mathbf{V}|$

43. a) Prove that $0 \cdot \mathbf{V} = \mathbf{0}$ for any $\mathbf{V}$.
b) Prove that $a \cdot \mathbf{0} = \mathbf{0}$ for any a.

44. a) Prove that if $a\mathbf{V} = \mathbf{0}$ and $\mathbf{V} \neq \mathbf{0}$, then $a = 0$.
b) Prove that if $a\mathbf{V} = \mathbf{0}$ and $a \neq \mathbf{0}$, then $\mathbf{V} = \mathbf{0}$.

45. The **dot product** or **inner product** $\mathbf{U} \cdot \mathbf{V}$ of two vectors $\mathbf{U} = u_1\mathbf{i} + u_2\mathbf{j}$ and $\mathbf{V} = v_1\mathbf{i} + v_2\mathbf{j}$ is defined by $\mathbf{U} \cdot \mathbf{V} = u_1 v_1 + u_2 v_2$. Prove the following properties of the dot product.
a) $\mathbf{U} \cdot \mathbf{V} = \mathbf{V} \cdot \mathbf{U}$
b) $\mathbf{V} \cdot \mathbf{V} = |\mathbf{V}|^2$

46. (See Problem 45.) Use the Law of Cosines to prove that

$$\mathbf{U} \cdot \mathbf{V} = |\mathbf{U}||\mathbf{V}| \cos \theta,$$

where θ is the angle between the position vectors $\mathbf{U}$ and $\mathbf{V}$.

7.5 Trigonometric Form of Complex Numbers

We have seen that real numbers may be represented geometrically by the points in the real number line. It is also possible to represent complex numbers geometrically. To do this, we begin with a conventional Cartesian coordinate system in the plane. With each complex number $a + bi$ in standard form, we associate the point that has coordinates (a, b). We label this point $a + bi$ to emphasize that the point (a, b) corresponds to the complex number $a + bi$. Several complex numbers are located in Figure 7.26.

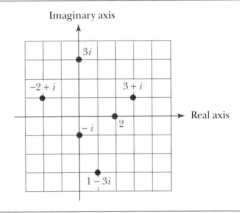

FIGURE 7.26

Points on the horizontal axis correspond to the real numbers $a + 0i$, and consequently the horizontal axis is referred to as the **real axis.** Points on the vertical axis correspond to the imaginary numbers $0 + bi$, so the vertical axis is called the **imaginary axis.**

Complex numbers are sometimes represented geometrically by vectors. In this approach, the complex number $a + bi$ is represented by the vector with initial point at the origin and terminal point with coordinates (a, b), or by any other vector with the same length and direction. This is shown in Figure 7.27.

Note that, in this interpretation, addition of complex numbers corresponds to the usual *parallelogram rule* for adding vectors. This is illustrated in Example 1.

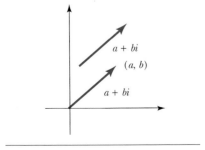

FIGURE 7.27

Example 1   Illustrate the parallelogram rule with the complex numbers $1 + 3i$ and $-2 + 2i$.

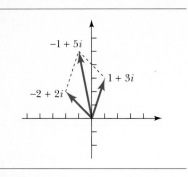

FIGURE 7.28

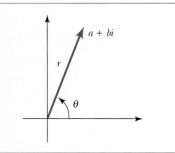

FIGURE 7.29

Solution

In Figure 7.28, the diagonal of the parallelogram is the vector representing the sum $-1 + 5i$ of $1 + 3i$ and $-2 + 2i$.

Any vector with initial point at the origin can be described by designating its length r and its direction θ, where we do *not* restrict θ to be between 0° and 360°. Figure 7.29 shows r and θ for a complex number $a + bi$ in standard form.

From Figure 7.29 we see that r and θ are related to a and b by the equations

$$a = r \cos \theta, \quad b = r \sin \theta, \quad r = \sqrt{a^2 + b^2}.$$

The complex number $a + bi$ can thus be written in the form

$$a + bi = r(\cos \theta + i \sin \theta).$$

Definition 7.1

The **trigonometric form** (or **polar form**) of the complex number $a + bi$ is[1]

$$r(\cos \theta + i \sin \theta),$$

where r and θ are determined by the equations

$$r = \sqrt{a^2 + b^2}, \quad a = r \cos \theta, \quad b = r \sin \theta.$$

The number r is called the **absolute value** (or **modulus**) of $a + bi$, and the angle θ is called the **argument** (or **amplitude**) of $a + bi$.

The usual absolute value notation is used for the absolute value of a complex number:

$$|a + bi| = r = \sqrt{a^2 + b^2}.$$

The absolute value, r, is unique, but the angle θ is not, since there are many angles in standard position that determine the same vector. We usually choose to use the smallest positive value for θ. We note that any equation of the form

$$r_1(\cos \theta_1 + i \sin \theta_1) = r_2(\cos \theta_2 + i \sin \theta_2)$$

requires that $r_1 = r_2$ and that θ_1 and θ_2 be coterminal. Hence[2]

$$\theta_2 = \theta_1 + k(360°)$$

for some integer k.

---

1. The expression $\cos \theta + i \sin \theta$ is sometimes abbreviated as cis θ.
2. We use degree measure for the angle. Radian measure can be used just as well.

Example 2   Express each complex number in trigonometric form.

a) $-1 - i$ b) $\sqrt{3} - i$ c) $-3i$

Solution
In each part of Figure 7.30, we sketch the vector representing the corresponding complex number. We then determine the values of r and θ by recognizing the special angles.

a) $-1 - i = \sqrt{2}(\cos 225° + i \sin 225°)$

b) $\sqrt{3} - i = 2(\cos 330° + i \sin 330°)$

c) $-3i = 3(\cos 270° + i \sin 270°)$ ❏

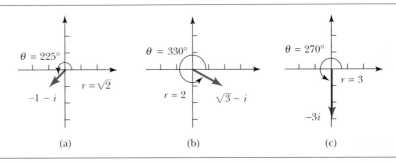

(a)   (b)   (c)

FIGURE 7.30

Example 3   Convert each of the following complex numbers from trigonometric form to standard form. Also, draw the vector representing each complex number.

a) $4(\cos 210° + i \sin 210°)$

b) $6[\cos(-45°) + i \sin (-45°)]$

c) $3(\cos \pi + i \sin \pi)$

Solution
In each part we simply evaluate the cosine and sine of the given angle and then simplify. The vectors are drawn in Figure 7.31.

a) $4(\cos 210° + i \sin 210°) = 4\left[-\dfrac{\sqrt{3}}{2} + i\left(-\dfrac{1}{2} \right) \right] = -2\sqrt{3} - 2i$

b) $6[\cos (-45°) + i \sin (-45°)] = 6\left[\dfrac{\sqrt{2}}{2} + i\left(-\dfrac{\sqrt{2}}{2} \right) \right] =$

$3\sqrt{2} - 3i\sqrt{2}$

c) $3(\cos \pi + i \sin \pi) = 3[(-1) + i(0)] = -3$ ❏

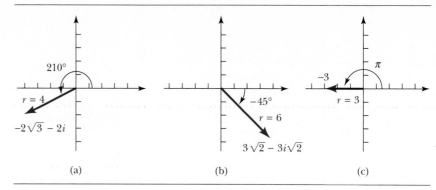

FIGURE 7.31

The trigonometric form for complex numbers can be used to compute products and quotients of complex numbers. Suppose z_1 and z_2 are complex numbers with trigonometric forms $r_1 (\cos \theta_1 + i \sin \theta_1)$ and $r_2 (\cos \theta_2 + i \sin \theta_2)$, respectively. First, we consider the product.

$$z_1 z_2 = r_1(\cos \theta_1 + i \sin \theta_1) \cdot r_2(\cos \theta_2 + i \sin \theta_2)$$

$$= r_1 r_2 [(\cos \theta_1 \cos \theta_2 - \sin \theta_1 \sin \theta_2) + i(\sin \theta_1 \cos \theta_2 + \cos \theta_1 \sin \theta_2)]$$

$$= r_1 r_2 [\cos(\theta_1 + \theta_2) + i \sin (\theta_1 + \theta_2)]$$

The last equality follows from the identities for the sum of two angles.

Next we consider the quotient of z_1 and z_2, where $z_2 \neq 0$. This time we use the identities for the difference of two angles.

$$\frac{z_1}{z_2} = \frac{r_1(\cos \theta_1 + i \sin \theta_1)}{r_2(\cos \theta_2 + i \sin \theta_2)}$$

$$= \left(\frac{r_1}{r_2}\right) \frac{(\cos \theta_1 + i \sin \theta_1)(\cos \theta_2 - i \sin \theta_2)}{(\cos \theta_2 + i \sin \theta_2)(\cos \theta_2 - i \sin \theta_2)}$$

$$= \left(\frac{r_1}{r_2}\right) \frac{(\cos \theta_1 \cos \theta_2 + \sin \theta_1 \sin \theta_2) + i(\sin \theta_1 \cos \theta_2 - \cos \theta_1 \sin \theta_2)}{\cos^2 \theta_2 - i^2 \sin^2 \theta_2}$$

$$= \left(\frac{r_1}{r_2}\right) \frac{\cos (\theta_1 - \theta_2) + i \sin (\theta_1 - \theta_2)}{\cos^2 \theta_2 + \sin^2 \theta_2}$$

$$= \left(\frac{r_1}{r_2}\right) [\cos (\theta_1 - \theta_2) + i \sin (\theta_1 - \theta_2)]$$

We summarize these results as follows.

Product and Quotient Rules

If two complex numbers z_1 and z_2 have trigonometric forms

$$z_1 = r_1(\cos \theta_1 + i \sin \theta_1) \quad \text{and} \quad z_2 = r_2(\cos \theta_2 + i \sin \theta_2),$$

then the product is given by

$$z_1 z_2 = r_1 r_2 [\cos (\theta_1 + \theta_2) + i \sin (\theta_1 + \theta_2)].$$

That is, to multiply two complex numbers, multiply their absolute values and add their arguments.

     Also, if $z_2 \neq 0$, the quotient is given by

$$\frac{z_1}{z_2} = \frac{r_1}{r_2} [\cos (\theta_1 - \theta_2) + i \sin (\theta_1 - \theta_2)].$$

That is, to divide two complex numbers, divide their absolute values and subtract their arguments.

Example 4   Suppose $z_1 = 12(\cos 135° + i \sin 135°)$ and $z_2 = 8(\cos 15° + i \sin 15°)$. Write the product $z_1 z_2$ and the quotient z_1/z_2 in both trigonometric form and standard form.

Solution

To multiply, we multiply the absolute values and add the arguments.

$$z_1 z_2 = 12 \cdot 8[\cos (135° + 15°) + i \sin (135° + 15°)]$$

$$= 96(\cos 150° + i \sin 150°) \qquad \text{Trigonometric Form}$$

$$= 96\left(-\frac{\sqrt{3}}{2} + \frac{1}{2}i\right)$$

$$= -48\sqrt{3} + 48i \qquad \text{Standard Form}$$

To divide, we divide absolute values and subtract the arguments.

$$\frac{z_1}{z_2} = \frac{12}{8}[\cos (135° - 15°) + i \sin (135° - 15°)]$$

$$= \frac{3}{2}(\cos 120° + i \sin 120°) \qquad \text{Trigonometric Form}$$

$$= \frac{3}{2}\left(-\frac{1}{2} + \frac{\sqrt{3}}{2}i\right)$$

$$= -\frac{3}{4} + \frac{3\sqrt{3}}{4}i \qquad \text{Standard Form}$$

We have stated each trigonometric form using the smallest positive value for the argument. ❏

Exercises 7.5

Illustrate the parallelogram rule with each pair of complex numbers. (See Example 1.)

1. $2 + i$, $\ -4 + 3i$ **2.** $3 - 2i$, $\ 1 + 6i$

3. $-1 + 2i$, $\ 3 - i$ **4.** $2 - 3i$, $\ -4 - 2i$

5. -2, $\ 1 + 3i$ **6.** $4i$, $\ -3 + i$

7. $-3i$, $\ 5$ **8.** -4, $\ 2i$

Express each complex number in trigonometric form. In Exercises 25–28, take $r = 1$ and round the angles to the nearest tenth of a degree. (See Example 2.)

9. $-1 + i$ **10.** $1 + i\sqrt{3}$

11. $3 - 3i$ **12.** $2\sqrt{2} - 2i\sqrt{2}$

13. -4 **14.** $-5i$

15. $7i$ **16.** 3

17. $-4\sqrt{3} - 4i$ **18.** $-\dfrac{\sqrt{3}}{2} + \dfrac{1}{2}i$

19. $\frac{1}{4} + \frac{1}{4}i$ **20.** $0.8 - 0.8i$

21. $1 - i\sqrt{3}$ **22.** $-\sqrt{3} - i$

23. $-3\sqrt{3} + 3i$ **24.** $-10 + 10i$

25. $0.9033 + 0.4289i$ **26.** $0.6743 + 0.7385i$

27. $-0.7206 + 0.6934i$ **28.** $-0.6018 - 0.7986i$

Express each complex number in standard form. In Exercises 37–40, round to four decimal places. (See Example 3.)

29. $4(\cos 45° + i \sin 45°)$

30. $2(\cos 60° + i \sin 60°)$

31. $3[\cos (-120°) + i \sin (-120°)]$

32. $12(\cos 315° + i \sin 315°)$

33. $9\left(\cos \dfrac{3\pi}{2} + i \sin \dfrac{3\pi}{2} \right)$

34. $2\left[\cos \left(-\dfrac{\pi}{2} \right) + i \sin \left(-\dfrac{\pi}{2} \right) \right]$

35. $\sqrt{2}\left(\cos \dfrac{5\pi}{4} + i \sin \dfrac{5\pi}{4} \right)$

36. $\sqrt{3}\left(\cos \dfrac{11\pi}{6} + i \sin \dfrac{11\pi}{6} \right)$

37. $3(\cos 27° + i \sin 27°)$

38. $5(\cos 114° + i \sin 114°)$

39. $7(\cos 312° + i \sin 312°)$

40. $2(\cos 200° + i \sin 200°)$

Write the exact results of the indicated operations in standard form. (See Example 4.)

41. $12(\cos 93° + i \sin 93°) \cdot 2(\cos 27° + i \sin 27°)$

42. $2(\cos 47° + i \sin 47°) \cdot 8(\cos 43° + i \sin 43°)$

43. $(\cos 142° + i \sin 142°) \cdot (\cos 38° + i \sin 38°)$

44. $(\cos 182° + i \sin 182°) \cdot (\cos 133° + i \sin 133°)$

45. $3(\cos 324° + i \sin 324°) \cdot 2(\cos 396° + i \sin 396°)$

46. $10(\cos 421° + i \sin 421°) \cdot 8(\cos 119° + i \sin 119°)$

47. $8[\cos (-18°) + i \sin (-18°)] \cdot$
$\qquad\qquad\qquad 2(\cos 138° + i \sin 138°)$

48. $(\cos 93° + i \sin 93°) \cdot$
$\qquad\qquad\qquad 4[\cos (-153°) + i \sin (-153°)]$

49. $\dfrac{12(\cos 47° + i \sin 47°)}{6(\cos 2° + i \sin 2°)}$

50. $\dfrac{3(\cos 278° + i \sin 278°)}{2(\cos 38° + i \sin 38°)}$

51. $\dfrac{8(\cos 192° + i \sin 192°)}{4(\cos 102° + i \sin 102°)}$

52. $\dfrac{5(\cos 560° + i \sin 560°)}{4(\cos 260° + i \sin 260°)}$

53. $\dfrac{\cos 62° + i \sin 62°}{\cos 107° + i \sin 107°}$

54. $\dfrac{\cos 48° + i \sin 48°}{\cos 138° + i \sin 138°}$

55. $\dfrac{\cos (-20°) + i \sin (-20°)}{\cos 130° + i \sin 130°}$

56. $\dfrac{\cos (-142°) + i \sin (-142°)}{\cos 218° + i \sin 218°}$

Use the trigonometric forms to find the exact values of z_1z_2 and z_1/z_2. Write the results in both trigonometric form and standard form. (See Example 4.)

57. $z_1 = 4\left(\cos \dfrac{\pi}{8} + i \sin \dfrac{\pi}{8} \right)$,

$\qquad z_2 = \cos \dfrac{5\pi}{8} + i \sin \dfrac{5\pi}{8}$

58. $z_1 = 3\left(\cos\dfrac{7\pi}{6} + i\sin\dfrac{7\pi}{6}\right),$

$z_2 = \cos\dfrac{2\pi}{3} + i\sin\dfrac{2\pi}{3}$

59. $z_1 = 2\left(\cos\dfrac{5\pi}{6} + i\sin\dfrac{5\pi}{6}\right),$

$z_2 = 3\left[\cos\left(-\dfrac{\pi}{6}\right) + i\sin\left(-\dfrac{\pi}{6}\right)\right]$

60. $z_1 = 6\left(\cos\dfrac{5\pi}{3} + i\sin\dfrac{5\pi}{3}\right),$

$z_2 = 5\left[\cos\left(-\dfrac{\pi}{3}\right) + i\sin\left(-\dfrac{\pi}{3}\right)\right]$

61. $z_1 = \sqrt{3} + i,\quad z_2 = -1 + i\sqrt{3}$

62. $z_1 = 1 - i\sqrt{3},\quad z_2 = -2\sqrt{3} + 2i$

63. $z_1 = -1 - i,\quad z_2 = 2 - 2i$

64. $z_1 = -2 + 2i,\quad z_2 = 4 + 4i$

65. $z_1 = 2,\quad z_2 = 5 + 5i\sqrt{3}$

66. $z_1 = 5i,\quad z_2 = -\sqrt{3} + i$

67. $z_1 = -3i,\quad z_2 = -2 - 2i$

68. $z_1 = -8,\quad z_2 = 4 - 4i$

Use the trigonometric form to prove each of the following statements, where z, z_1, and z_2 are arbitrary complex numbers.

69. $|\bar{z}| = |z|$ **70.** $z\bar{z} = |z|^2$

71. $|z_1 z_2| = |z_1||z_2|$ **72.** If $z_2 \neq 0$, $\left|\dfrac{z_1}{z_2}\right| = \dfrac{|z_1|}{|z_2|}$

7.6 Powers and Roots of Complex Numbers

Any positive integral power of a complex number z written in trigonometric form can be determined by repeated application of the product rule. If

$$z = r(\cos\theta + i\sin\theta),$$

we have

$$z^2 = r(\cos\theta + i\sin\theta) \cdot r(\cos\theta + i\sin\theta)$$
$$= r^2(\cos 2\theta + i\sin 2\theta).$$

Also,

$$z^3 = z^2 \cdot z$$
$$= r^2(\cos 2\theta + i\sin 2\theta) \cdot r(\cos\theta + i\sin\theta)$$
$$= r^3(\cos 3\theta + i\sin 3\theta).$$

Similarly,

$$z^4 = r^4(\cos 4\theta + i\sin 4\theta)$$
$$z^5 = r^5(\cos 5\theta + i\sin 5\theta),$$

and so on. In general, we have the next result, which begins to reveal the true usefulness of the trigonometric form.

DeMoivre's Theorem

If the complex number z has the trigonometric form

$$z = r(\cos \theta + i \sin \theta)$$

and n is any integer, then

$$z^n = r^n(\cos n\theta + i \sin n\theta).$$

Example 1   Use DeMoivre's Theorem to write $(1 - i)^{10}$ in standard form.

Solution
We first write $1 - i$ in trigonometric form and then apply DeMoivre's Theorem.

$$\begin{aligned}
(1 - i)^{10} &= [\sqrt{2}(\cos 315° + i \sin 315°)]^{10} \\
&= (\sqrt{2})^{10}[\cos (10 \cdot 315°) + i \sin (10 \cdot 315°)] \\
&= 32(\cos 3150° + i \sin 3150°) \\
&= 32[\cos (270° + 8 \cdot 360°) + i \sin (270° + 8 \cdot 360°)] \\
&= 32(\cos 270° + i \sin 270°) \qquad \text{Trigonometric Form} \\
&= 32(0 - i) \\
&= -32i \qquad\qquad\qquad\qquad \text{Standard Form} \qquad \square
\end{aligned}$$

If n is a positive integer greater than 1 and if $u^n = z$ for complex numbers u and z, then u is called an nth root of z. There are, in fact, exactly n nth roots of any nonzero complex number z.

Suppose that the trigonometric forms of z and u are

$$z = r(\cos \theta + i \sin \theta)$$

and

$$u = s(\cos \omega + i \sin \omega).$$

Whenever $u^n = z$, we have

$$[s(\cos \omega + i \sin \omega)]^n = r(\cos \theta + i \sin \theta).$$

Applying DeMoivre's Theorem to the left-hand side yields

$$s^n(\cos n\omega + i \sin n\omega) = r(\cos \theta + i \sin \theta).$$

This statement of equality of complex numbers requires that the absolute values be equal and that the arguments be coterminal. That is,

$$s^n = r \qquad \text{and} \qquad n\omega = \theta + k \cdot 360°$$

for some integer k. Hence u can be written as

$$u = \sqrt[n]{r}\left[\cos\left(\frac{\theta + k \cdot 360°}{n}\right) + i \sin\left(\frac{\theta + k \cdot 360°}{n}\right)\right].$$

If we use any n consecutive values of k, we obtain n distinct nth roots of z. For convenience we use $k = 0, 1, 2, \ldots, n - 1$.

Roots of Complex Numbers

If z is a nonzero complex number with trigonometric form

$$z = r(\cos \theta + i \sin \theta),$$

then the nth roots of z are given by

$$\sqrt[n]{r}\left[\cos\left(\frac{\theta + k \cdot 360°}{n}\right) + i \sin\left(\frac{\theta + k \cdot 360°}{n}\right)\right]$$

for $k = 0, 1, 2, \ldots, n - 1$.

When the n angles

$$\frac{\theta}{n}, \quad \frac{\theta + 360°}{n}, \quad \frac{\theta + 2 \cdot 360°}{n}, \ldots, \quad \frac{\theta + (n - 1)360°}{n}$$

are placed in standard position, their terminal sides are equally spaced $360°/n$ apart around a circle. Each nth root of z has absolute value $\sqrt[n]{r}$. Therefore the n vectors representing the n nth roots of z divide a circle of radius $\sqrt[n]{r}$ into n equal parts. This is illustrated in the next example.

Example 2 Find the three cube roots of $64i$ and draw the vectors representing them.

Solution
We first write $64i$ in trigonometric form.

$$64i = 64(\cos 90° + i \sin 90°)$$

According to our formula, the three cube roots of $64i$ are

$$\sqrt[3]{64}\left[\cos\left(\frac{90° + k \cdot 360°}{3}\right) + i \sin\left(\frac{90° + k \cdot 360°}{3}\right)\right]$$

$$= 4[\cos(30° + k \cdot 120°) + i \sin(30° + k \cdot 120°)], \quad k = 0, 1, 2.$$

Using the three values of k, we obtain the following three cube roots of

64i, each written in both trigonometric form and standard form.

	Trigonometric Form		Standard Form
$k = 0,$	$4(\cos 30° + i \sin 30°)$	$=$	$2\sqrt{3} + 2i$
$k = 1,$	$4(\cos 150° + i \sin 150°)$	$=$	$-2\sqrt{3} + 2i$
$k = 2,$	$4(\cos 270° + i \sin 270°)$	$=$	$-4i$

In Figure 7.32, the vectors representing the three cube roots of 64i are spaced 120° apart in the circle of radius 4 units.   ❑

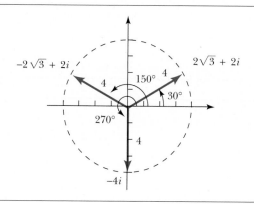

FIGURE 7.32

Exercises 7.6

Use DeMoivre's Theorem to find the exact value of each of the following. Leave your answers in trigonometric form. (See Example 1.)

1. $(\cos 42° + i \sin 42°)^3$

2. $(\cos 68° + i \sin 68°)^4$

3. $[2(\cos 12° + i \sin 12°)]^9$

4. $[\sqrt{2}(\cos 3° + i \sin 3°)]^{16}$

5. $\left(\cos \dfrac{\pi}{8} + i \sin \dfrac{\pi}{8}\right)^{15}$

6. $\left(\cos \dfrac{\pi}{10} + i \sin \dfrac{\pi}{10}\right)^{16}$

7. $\left[\sqrt{3}\left(\cos \dfrac{2\pi}{5} + i \sin \dfrac{2\pi}{5}\right)\right]^4$

8. $\left[\sqrt{7}\left(\cos \dfrac{2\pi}{7} + i \sin \dfrac{2\pi}{7}\right)\right]^6$

Use DeMoivre's Theorem to evaluate and write the result in standard form. (See Example 1.)

9. $\left(\dfrac{\sqrt{3}}{2} + \dfrac{1}{2}i\right)^7$

10. $\left(\dfrac{1}{2} + \dfrac{\sqrt{3}}{2}i\right)^5$

11. $\left(-\dfrac{1}{2} + \dfrac{\sqrt{3}}{2}i\right)^{18}$

12. $\left(\dfrac{\sqrt{3}}{2} - \dfrac{1}{2}i\right)^{21}$

13. $\left(-\dfrac{\sqrt{2}}{2} - \dfrac{\sqrt{2}}{2}i\right)^6$

14. $\left(\dfrac{\sqrt{2}}{2} - \dfrac{\sqrt{2}}{2}i\right)^7$

15. $(1 - i\sqrt{3})^8$

16. $(-\sqrt{3} + i)^9$

17. $(\sqrt{2} + i\sqrt{2})^{10}$

18. $(-\sqrt{2} + i\sqrt{2})^{12}$

19. $(2 - 2i\sqrt{3})^5$

20. $(-2\sqrt{3} - 2i)^4$

21. $(-1 + i)^8$

22. $(\sqrt{3} - i)^6$

23. $(5 + 5i)^3$

24. $(3 - 3i)^4$

Find the indicated roots and write the results in standard form. Also, draw the vectors representing them. (See Example 2.)

25. Cube roots of 1

26. Fourth roots of 1

27. Fourth roots of -1

28. Sixth roots of -1

29. Cube roots of $-i$

30. Square roots of i

31. Eighth roots of 256

32. Cube roots of -8

33. Square roots of $-64i$

34. Sixth roots of 64

35. Fourth roots of $-8 + 8i\sqrt{3}$

36. Fourth roots of $-8 - 8i\sqrt{3}$

Find the indicated roots and write the results in trigonometric form. (See Example 2.)

37. Sixth roots of $-64i$

38. Fifth roots of $-32i$

39. Cube roots of $\dfrac{\sqrt{3}}{2} + \dfrac{1}{2}i$

40. Cube roots of $-\dfrac{\sqrt{3}}{2} + \dfrac{1}{2}i$

41. Fourth roots of $\dfrac{1}{2} - \dfrac{\sqrt{3}}{2}i$

42. Sixth roots of $-\dfrac{1}{2} - \dfrac{\sqrt{3}}{2}i$

43. Fifth roots of $-\dfrac{\sqrt{2}}{2} - \dfrac{\sqrt{2}}{2}i$

44. Cube roots of $-\dfrac{\sqrt{2}}{2} + \dfrac{\sqrt{2}}{2}i$

45. Fifth roots of $-16\sqrt{2} - 16i\sqrt{2}$

46. Sixth roots of $32\sqrt{3} - 32i$

47. Cube roots of $4\sqrt{2} - 4i\sqrt{2}$

48. Fifth roots of $16\sqrt{2} - 16i\sqrt{2}$

Solve each equation and write the solutions in standard form.

49. $x^3 + 27 = 0$

50. $x^8 - 1 = 0$

51. $x^3 - i = 0$

52. $x^3 + 8i = 0$

53. $x^6 + 64 = 0$

54. $x^3 + 64i = 0$

55. $x^4 + \dfrac{1}{2} - \dfrac{\sqrt{3}}{2}i = 0$

56. $x^4 + \dfrac{1}{2} + \dfrac{\sqrt{3}}{2}i = 0$

Solve the equations and write the solutions in trigonometric form.

57. $x^5 - 32i = 0$

58. $x^5 - i = 0$

59. $x^3 - \left(\dfrac{1}{2} + \dfrac{\sqrt{3}}{2}i \right) = 0$

60. $x^4 + \dfrac{\sqrt{3}}{2} + \dfrac{1}{2}i = 0$

CHAPTER REVIEW

Key Words and Phrases

Ambiguous case
Vector quantity
Scalar quantity
Vector
Initial point
Terminal point
Magnitude
Length
Equal vectors
Zero vector
Parallelogram rule

Resultant
x-component
y-component
Base vectors
Unit vector
Direction angle
Course
Heading
Speed
Airspeed

Ground speed
Velocity
Real axis
Imaginary axis
Trigonometric form
Polar form
Absolute value
Modulus
Argument
Amplitude

Summary of Important Concepts and Formulas

Law of Sines

$$\frac{a}{\sin A} = \frac{b}{\sin B} = \frac{c}{\sin C}$$

Law of Cosines

$$\cos A = \frac{b^2 + c^2 - a^2}{2bc}$$

$$a^2 = b^2 + c^2 - 2bc \cos A$$

$$b^2 = a^2 + c^2 - 2ac \cos B \qquad \cos B = \frac{a^2 + c^2 - b^2}{2ac}$$

$$c^2 = a^2 + b^2 - 2ab \cos C$$

$$\cos C = \frac{a^2 + b^2 - c^2}{2ab}$$

Area of a Triangle

$$K = \tfrac{1}{2}bc \sin A = \tfrac{1}{2}ac \sin B = \tfrac{1}{2}ab \sin C$$

$$K = \frac{a^2 \sin B \sin C}{2 \sin A} = \frac{b^2 \sin A \sin C}{2 \sin B}$$

$$= \frac{c^2 \sin A \sin B}{2 \sin C}$$

$$K = \sqrt{s(s - a)(s - b)(s - c)},$$

where $s = \tfrac{1}{2}(a + b + c)$

Vectors

Vector addition:

$$(u_1\mathbf{i} + u_2\mathbf{j}) + (v_1\mathbf{i} + v_2\mathbf{j}) = (u_1 + v_1)\mathbf{i} + (u_2 + v_2)\mathbf{j}$$

Scalar multiplication: $a(v_1\mathbf{i} + v_2\mathbf{j}) = (av_1)\mathbf{i} + (av_2)\mathbf{j}$

Magnitude: $|v_1\mathbf{i} + v_2\mathbf{j}| = \sqrt{v_1^2 + v_2^2}$

Product and Quotient Rules

If z_1 and z_2 have trigonometric forms

$$z_1 = r_1(\cos \theta_1 + i \sin \theta_1)$$

and

$$z_2 = r_2(\cos \theta_2 + i \sin \theta_2),$$

then

$$z_1 z_2 = r_1 r_2[\cos (\theta_1 + \theta_2) + i \sin (\theta_1 + \theta_2)]$$

$$\frac{z_1}{z_2} = \frac{r_1}{r_2}[\cos (\theta_1 - \theta_2) + i \sin (\theta_1 - \theta_2)].$$

De Moivre's Theorem

If z has the trigonometric form $z = r(\cos \theta + i \sin \theta)$, then

$$z^n = r^n(\cos n\theta + i \sin n\theta).$$

Roots of Complex Numbers

If z is a nonzero complex number with trigonometric form

$$z = r(\cos \theta + i \sin \theta),$$

then the n nth roots of z are given by

$$\sqrt[n]{r}\left[\cos \left(\frac{\theta + k \cdot 360°}{n}\right) + i \sin \left(\frac{\theta + k \cdot 360°}{n}\right)\right]$$

for $k = 0, 1, 2, \ldots, n - 1$.

Review Problems for Chapter 7

In Problems 1–7, solve the triangles to the degree of accuracy consistent with the given information.

1. $B = 87°$, $C = 43°$, $a = 17$
2. $A = 138°$, $a = 2.1$, $b = 8.8$
3. $C = 37.2°$, $b = 10.1$, $c = 14.2$
4. $B = 37.2°$, $c = 14.2$, $b = 10.1$
5. $a = 47.2$, $b = 41.3$, $c = 16.1$
6. $A = 120°$, $b = 14$, $c = 22$
7. $C = 108°$, $a = 83$, $b = 32$
8. Find the area of the triangle in Problem 1.

9. Find the area of the triangle in Problem 5.
10. Find the area of the triangle in Problem 6.

Write each complex number in standard form.

11. $2\left(\cos \dfrac{2\pi}{3} + i \sin \dfrac{2\pi}{3}\right)$
12. $4[\cos (-150°) + i \sin (-150°)]$

Write each complex number in trigonometric form.

13. $-3i$
14. $2 - 2i$

Write the exact results of the indicated operations in standard form.

15. $3(\cos 17° + i \sin 17°) \cdot 4(\cos 253° + i \sin 253°)$

16. $\dfrac{4(\cos 198° + i \sin 198°)}{3(\cos 48° + i \sin 48°)}$

17. Use the trigonometric forms to find the exact values of $z_1 z_2$ and z_1/z_2, where $z_1 = 4 - 4i\sqrt{3}$ and $z_2 = -\sqrt{3} - i$. Write the results in both trigonometric form and standard form.

Use DeMoivre's Theorem to evaluate and write the results in standard form.

18. $\left[\sqrt{3}\left(\cos \dfrac{3\pi}{8} + i \sin \dfrac{3\pi}{8} \right) \right]^4$

19. $(1 - i)^{14}$ **20.** $(-\sqrt{3} + i)^6$

21. Find the four fourth roots of $-\dfrac{1}{2} - \dfrac{\sqrt{3}}{2}i$ and write the results in standard form.

22. Find the six sixth roots of $64i$ and write the results in trigonometric form.

23. Solve the equation $x^6 - 64 = 0$ and write the solutions in standard form.

24. Matt and Beckie are standing on a level beach where Matt is flying a kite on a 63-foot string. The angle of elevation from Matt to the kite is $47°$, and the angle of elevation from Beckie to the kite is $48°$. How far apart are Matt and Beckie if the kite is in a vertical plane between them?

25. Suppose 3 steel pipes of lengths 6.0 feet, 14 feet, and 18 feet are to be welded together to form the truss for the roof of the produce stand shown in the figure. Find the measure of the angle α of inclination of the roof to the horizontal.

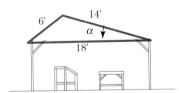

PROBLEM 25

26. In a billiards game a ball, struck by the cue ball, travels 12 inches, hitting the end bumper at an angle of $58°$. It then bounces off the end and travels 42 inches, landing in a side pocket. What is the shortest distance from the ball's original position in the figure below to the point where it dropped off into the side pocket?

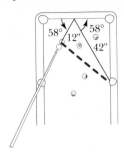

PROBLEM 26

27. An airplane is headed due north with an airspeed of 252 miles per hour, and the wind is blowing from the east at 31.8 miles per hour. Find the ground speed and the course of the plane.

28. If the airspeed of a small plane is 210 miles per hour, and the wind is blowing from the north at 18.0 miles per hour, what heading should the pilot give his plane in order to fly on a course due west from Alba to Whitehall?

29. A freighter is running at the rate of 19 miles per hour in still water and is headed in the direction N 53° E. If the ship is traveling in an ocean current flowing at 4.8 miles per hour in the direction S 37° E, what is its course? At what speed is it traveling?

30. Two forces of magnitudes 12.8 pounds and 21.2 pounds act on a point in the plane. If the angle between the directions of the two forces is $108.2°$, find the magnitude of the resultant force.

31. A 28-pound force and a 42-pound force produce a 30-pound resultant force. Find the angle between the directions of the 42-pound force and the resultant force.

32. The angle between the directions of two forces is $54.0°$. One of the forces has magnitude 197 pounds, and the resultant has magnitude 402 pounds. Find the angle between the 197-pound force and the resultant force.

33. A steel ball weighing 500 pounds is to be rolled up a ramp that is inclined at an angle of $7.6°$ with the horizontal. What force must the ramp withstand in the direction perpendicular to the ramp?

34. In Problem 33, what force parallel to the ramp is required to hold the ball in place on the ramp, if friction is neglected?

35. The airplane in the accompanying figure is flying with a heading of 317.0°, an airspeed of 280 miles per hour, and a ground speed of 312 miles per hour. If the wind is blowing from the south, find its speed.

36. A plane is flying with a heading of 285.0°, an airspeed of 300 miles per hour, and a ground speed of 247 miles per hour. If the wind is blowing from the west and the course of the plane is in a northwesterly direction, find the speed of the wind.

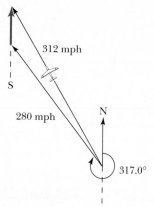

PROBLEM 35

▲ ▲ CRITICAL THINKING: FIND THE ERRORS ▬▬▬▬

Each of the following nonsolutions has at least one error. Can you find them?

Problem 1

Find side c in a triangle that has $a = 7.0$, $b = 3.0$, $C = 60°$.

Nonsolution

$$c^2 = a^2 + b^2 - 2ab \cos C$$
$$= (7.0)^2 + (3.0)^2 - 2(7.0)(3.0) \cos 60°$$
$$= 49 + 9 - 42(\tfrac{1}{2})$$
$$= 16(\tfrac{1}{2})$$
$$= 8$$
$$c = \sqrt{8} = 2\sqrt{2}$$

Problem 2

A plane is flying with a heading of 120.0° and an airspeed of 380 miles per hour. If the wind is blowing from the south at 40.0 miles per hour, find the ground speed of the plane.

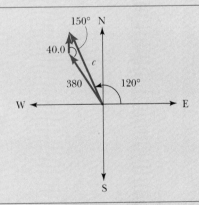

FIGURE 7.33

Nonsolution

The ground speed is represented by c in Figure 7.33. We use the Law of Cosines to find c.

$$c^2 = (380)^2 + (40)^2 - 2(380)(40) \cos 150°$$
$$= 144,400 + 1600 - 30,400(-0.8660)$$
$$= 172,326$$
$$c = 415 \text{ miles per hour}$$

Express $-\sqrt{3} - i$ in trigonometric form.

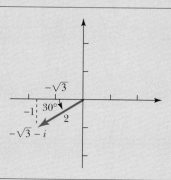

FIGURE 7.34

Nonsolution   The vector representing $-\sqrt{3} - i$ is shown in Figure 7.34. From the figure we get

$$-\sqrt{3} - i = 2(\cos 30° + i \sin 30°).$$

Problem 4   Use DeMoivre's Theorem to evaluate $(1 + i\sqrt{3})^5$.

Nonsolution

$$
\begin{aligned}
(1 + i\sqrt{3})^5 &= 2(\cos 60° + i \sin 60°)^5 \\
&= 2(\cos 300° + i \sin 300°) \\
&= 2\left(\frac{1}{2} - \frac{\sqrt{3}}{2}i\right) \\
&= 1 - i\sqrt{3}
\end{aligned}
$$

Chapter 8

Systems of Equations and Inequalities

In the last half of the twentieth century, the use of computers with the matrix methods presented in this chapter has made a revolutionary change in the kinds of applications that are possible for linear systems. Solutions of systems are routinely accomplished in business and industry today that were completely impossible at the middle of this century. As the material in this chapter becomes more familiar, the power of the methods will become clearer. The last section, "Linear Programming," is especially revealing.

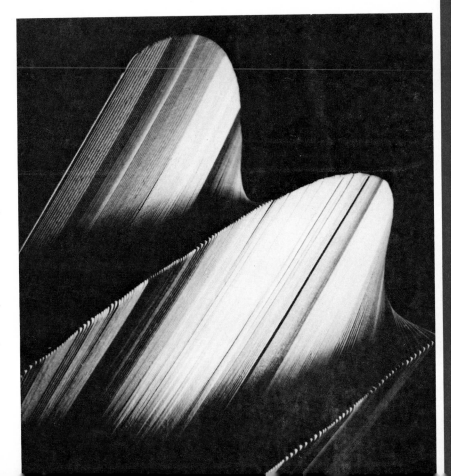

8.1 | Systems of Equations

The material in this chapter is unusually rich in applications. Business and industry utilize systems of equations and inequalities in making plans for efficient operation, growth, and future profits. In connection with the use of calculus, engineers and scientists also frequently apply the knowledge and methods presented in this chapter.

To begin our study, we consider an important problem that frequently arises with respect to the graphs of equations. The problem is to find the coordinates (x, y) of the points that are common to the graphs of two equations in x and y, that is, the **points of intersection** of the two graphs. When we are working with this type of problem, the pair of equations involved is referred to as a **system of equations.** A set of values for x and y that satisfies both equations is called a **simultaneous solution** of the system, or simply a **solution** of the system.

In this section we consider two methods for solving this problem. Each method is based on the idea that a point which is on both graphs must have coordinates that satisfy both equations. The simpler of these two methods is known as the **substitution method.** With this method the idea is to use one of the given equations, solve it for one of the variables in terms of the other, then *substitute* this value in the other equation, thereby obtaining an equation that involves only one variable. This procedure, which sounds more complicated than it really is, is illustrated in the following example.

Example 1   Find the point of intersection of the lines $3y + x = 7$ and $3x - 2y + 12 = 0$. That is, solve the following system of equations.

$$3y + x = 7$$

$$3x - 2y + 12 = 0$$

Solution
As indicated in the preceding discussion, we shall employ the substitution method. The first step is to solve for one of the variables in one of the given equations. The simplest possibility here is to solve for x in the first equation, obtaining

$$x = 7 - 3y.$$

When this value for x is substituted into the other equation, we have

$$3(7 - 3y) - 2y + 12 = 0.$$

This simplifies to

$$33 - 11y = 0$$

$$33 = 11y$$

and

$$y = 3.$$

Substituting this value for y in $x = 7 - 3y$, we get $x = -2$. Thus the point of intersection is $(-2, 3)$, and it is easy to check that the coordinates of this point do indeed satisfy both of the original equations. The situation is shown geometrically in Figure 8.1.   ❏

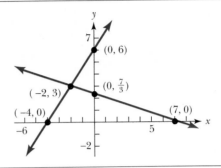

FIGURE 8.1

Another illustration of solution by the substitution method is provided in Example 2.

Example 2   Solve the following system. Graph the two equations on the same coordinate system and label the points of intersection.

$$x^2 - 5x - 4y - 28 = 0$$

$$x - 4y = 1$$

Solution

It is convenient here to solve for x in the second equation, obtaining

$$x = 4y + 1.$$

Substituting into the first equation, we have

$$(4y + 1)^2 - 5(4y + 1) - 4y - 28 = 0,$$

which simplifies to

$$16y^2 - 16y - 32 = 0$$

and

$$y^2 - y - 2 = 0.$$

This factors as

$$(y - 2)(y + 1) = 0.$$

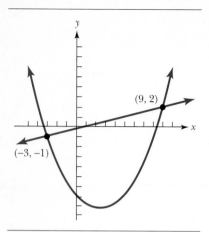

FIGURE 8.2

Thus the y-coordinates of the points of intersection are $y = 2$ and $y = -1$. When these values are substituted into $x = 4y + 1$, we find that the points of intersection are given by $(9, 2)$ and $(-3, -1)$.

To draw the graph of the first equation, we solve for y, obtaining

$$y = \frac{1}{4}x^2 - \frac{5}{4}x - 7.$$

We recognize this as the equation of a parabola that opens upward with vertex at $x = -b/(2a) = \frac{5}{2}$, $y = -\frac{137}{16}$. The graph of $x - 4y = 1$ is a straight line with slope $\frac{1}{4}$ and y-intercept $-\frac{1}{4}$. The graphs and the points of intersection are shown in Figure 8.2.

One more point should be made before leaving this example. It is important to notice that when the values $y = 2$ and $y = -1$ were obtained, they were substituted into the equation for the line, and not into the equation for the parabola. If they had been used in the equation for the parabola, *two* values of x would have been obtained for each y, and only one of these values would have satisfied the linear equation. More specifically, when $y = -1$ is substituted into $x^2 - 5x - 4y - 28 = 0$, we obtain $x = -3$ and $x = 8$. The value $x = 8$ is an extraneous solution, since $(8, -1)$ is not on the line. Similarly, $y = 2$ yields $x = 9$ and $x = -4$, with $x = -4$ an extraneous solution.   ❏

Graphs that picture the solutions of the system have been provided in both of the preceding examples of this section. These are valuable for understanding, and they provide a rough check on the solutions. They are not indispensable, however, and we omit them in the remainder of the section.

The other method of solution referred to earlier is the **elimination method.** This method derives its name from the fact that one of the variables is eliminated by an appropriate combination of the two equations in a system. Suppose that the system of equations is represented symbolically by

$$R(x, y) = c_1$$

$$S(x, y) = c_2,$$

where $R(x, y)$ and $S(x, y)$ denote algebraic expressions in the variables x and y, and c_1 and c_2 denote constants (real numbers). We first observe that if m_1 and m_2 are nonzero real numbers, then any solution of the given system is also a solution to the system

$$m_1 R(x, y) = m_1 c_1$$

$$m_2 S(x, y) = m_2 c_2,$$

and is therefore a solution to the sum of these equations:

$$m_1 R(x, y) + m_2 S(x, y) = m_1 c_1 + m_2 c_2.$$

The key to success in the elimination method is to choose the multipliers m_1 and m_2 so as to eliminate one of the variables when the sum is formed. This method is used in the next example.

Example 3   Solve the following system of equations by the elimination method.

$$9x^2 + y^2 = 36$$
$$x^2 + 3y^2 = 56$$

Solution

If we multiply the first equation by 3, the second equation by -1, and then add, we have

$$
\begin{aligned}
27x^2 + 3y^2 &= 108 \\
-x^2 - 3y^2 &= -56 \\
\hline
26x^2 &= 52.
\end{aligned}
$$

Thus y is eliminated, and this equation simplifies to

$$x^2 = 2$$

and

$$x = \pm\sqrt{2}.$$

To complete the solution, we substitute $x^2 = 2$ into the first equation and obtain

$$9(2) + y^2 = 36$$
$$y^2 = 18$$

and

$$y = \pm 3\sqrt{2}.$$

This gives the four pairs

$$(\sqrt{2}, 3\sqrt{2}), \quad (\sqrt{2}, -3\sqrt{2}), \quad (-\sqrt{2}, 3\sqrt{2}), \quad (-\sqrt{2}, -3\sqrt{2}),$$

and each of these checks in the original system. It is good practice *always to check solutions in each of the original equations.* This is especially true when extraction of roots has been employed in obtaining the solutions. ❏

The substitution method and the elimination method are not adequate for the solution of all types of systems that might be encountered. The examples presented here were chosen because they lend themselves to solution by these methods. Some more difficult problems can be solved by a combination of the two methods. This is illustrated in the next example.

Example 4   Solve the following system.

$$x^2 + 2xy + 2y^2 = 5$$
$$x^2 + xy + 2y^2 = 4$$

Solution

It is impossible to employ the elimination method directly. Further, the substitution method is not an attractive prospect since the quadratic formula would have to be used to obtain one variable in terms of the other. However, all terms involving a square can be eliminated by subtracting the second equation from the first. When this is done, we have

$$\begin{array}{r} x^2 + 2xy + 2y^2 = 5 \\ x^2 + xy + 2y^2 = 4 \\ \hline xy \qquad\quad = 1 \end{array}$$

and

$$y = \frac{1}{x}.$$

This gives us an expression for y that can be substituted into either of the original equations. Substitution into the first equation yields

$$x^2 + 2x\left(\frac{1}{x}\right) + 2\left(\frac{1}{x}\right)^2 = 5,$$

which simplifies to

$$x^2 + \frac{2}{x^2} = 3.$$

Clearing the equation of fractions, we have

$$x^4 + 2 = 3x^2$$

and

$$x^4 - 3x^2 + 2 = 0.$$

This factors as

$$(x^2 - 2)(x^2 - 1) = 0.$$

Setting $x^2 - 2 = 0$, we obtain $x = \pm\sqrt{2}$, and $x^2 - 1 = 0$ yields $x = \pm 1$. Using each of these values for x in the equation $y = 1/x$, we obtain the solution set

$$\{(\sqrt{2}, \sqrt{2}/2), \quad (-\sqrt{2}, -\sqrt{2}/2), \quad (1, 1), \quad (-1, -1)\}.$$

It is easy to confirm that all these solutions check in the original system. The interested reader may wish to verify that substitution of the values for x in either of the original equations yields extraneous solutions for y. ❏

It may well happen that a given system of equations has no solution. This is illustrated below.

Example 5   Solve the following system.

$$x^2 + y^2 = 1$$
$$x^2 - y = 4$$

Solution
Subtracting the second equation from the first, we have

$$
\begin{aligned}
x^2 + y^2 &= 1 \\
x^2 - y &= 4 \\
\hline
y^2 + y &= -3,
\end{aligned}
$$

or

$$y^2 + y + 3 = 0.$$

The value of the discriminant for this quadratic equation is $b^2 - 4ac = (1)^2 - 4(1)(3) = -11$, so there is no real solution for y. This in turn indicates that the original system has no real solution. That is, the circle ($x^2 + y^2 = 1$) and the parabola ($x^2 - y = 4$) do not intersect. ❏

Example 6   Suppose an aluminum can with volume 64π cubic inches is to be constructed so that the amount of aluminum used to form the two ends is the same as the amount used in the side of the can. Find the dimensions of such a can if it is in the shape of a right circular cylinder.

Solution
We draw the right circular cylinder in Figure 8.3(a) with height h and radius of the ends r. A flat sheet of aluminum as seen in Figure 8.3(b) with width h and length $2\pi r$ (the circumference of the circle) is used to form the side of the can. Its area is $2\pi rh$. The sum of the area of the ends is $2\pi r^2$. We obtain a nonlinear system of equations by setting the area of the ends equal to the area of the side and by using the formula for the volume of the cylinder.

$$\text{Areas:} \quad 2\pi r^2 = 2\pi rh$$

$$\text{Volume:} \quad \pi r^2 h = 64\pi$$

Solving for h in terms of r in the second equation yields

$$h = \frac{64}{r^2}.$$

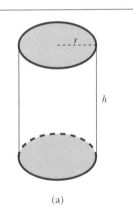

(a)

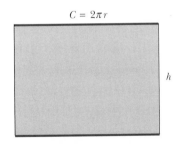

$C = 2\pi r$

h

(b)

FIGURE 8.3

Then dividing both sides of the first equation by 2π and substituting $h = 64/r^2$ yields an equation in r.

$$r^2 = rh \qquad \text{Dividing by } 2\pi$$

$$r^2 = r\frac{64}{r^2} \qquad \text{Substituting}$$

$$r^3 = 64 \qquad \text{Simplifying}$$

The only real solution to this equation is $r = 4$. Then $h = 64/r^2 = \frac{64}{16} = 4$. Thus the can will have ends of radius 4 inches and will be 4 inches tall. ❑

Exercises 8.1

Solve each of the following systems. Sketch the graphs of the two equations on the same coordinate system, and label the points of intersection.

1. $x + 2y = 4$
   $3x - 2y = -12$

2. $2y = x + 4$
   $3x + 2y + 12 = 0$

3. $x - 2y = 7$
   $2x = 4y - 14$

4. $2x = 3y - 5$
   $9y = 6x - 15$

5. $y = 2x + 6$
   $x^2 = 2y$

6. $x^2 - y = 0$
   $2x - y + 3 = 0$

7. $y = 3x^2 + 12x$
   $2x - y = 16$

8. $y = 3 - 2x - x^2$
   $4x + y = 5$

9. $x^2 + y^2 = 4$
   $2x - y = 2$

10. $x^2 + y^2 = 25$
    $y - x = 7$

11. $x^2 + y^2 = 10x$
    $4y = 3x - 8$

12. $x^2 + y^2 = 25$
    $x + 3y = 5$

Solve the following systems of equations. It is not necessary to graph the equations.

13. $2x - 5y = 11$
    $5x + y = 14$

14. $7x + 3y = 5$
    $5x - y = 13$

15. $3y = 2x + 6$
    $6y + 5x = 39$

16. $3y = 20 - 5x$
    $5x = 10 - 2y$

17. $2x + 9y = 3$
    $5x + 7y = -8$

18. $3x + 5y = 9$
    $5x + 7y = 13$

19. $x^2 - 2y^2 = -1$
    $2x - y = -1$

20. $3x^2 + 2y^2 = 5$
    $x - y = -2$

21. $4x^2 + 25y^2 = 100$
    $x + 2y = 8$

22. $36x^2 + 16y^2 = 25$
    $x - 2y = 12$

23. $9x^2 + 16y^2 = 144$
    $3x^2 + 4y^2 = 36$

24. $4x^2 + 25y^2 = 83$
    $9x^2 + 16y^2 = 66$

25. $x^2 + y^2 = 25$
    $(3x - 4y)(3x + 4y) = 0$

26. $4x^2 + 36y^2 = 100$
    $(x - 4y)(x + 4y) = 0$

27. $x^2 + y^2 = 9$
    $y^2 + 2x = 10$

28. $x^2 + y^2 = 9$
    $x^2 + 2y = 6$

29. $x^2 + y^2 = 4$
    $x^2 - 2y = 1$

30. $x^2 + y^2 = 4$
    $3x + y^2 = 0$

31. $2x + 7y = 15$
    $xy = 1$

32. $6x + y + 7 = 0$
    $xy = 2$

33. $9y^2 - 4x^2 = 7$
    $xy = -2$

34. $2x^2 + y^2 = 19$
    $xy = 3$

35. $2x^2 + 3xy + y^2 = 12$
    $2x^2 - xy + y^2 = 4$

36. $2x^2 - 3xy - 3y^2 = 11$
    $2x^2 - xy - 3y^2 = 7$

37. $5x^2 - 4xy - 3y^2 = -8$
    $5x^2 + 5xy - 3y^2 = 28$

38. $8x^2 + 6xy - 9y^2 = 8$
    $8x^2 - 3xy - 9y^2 = 17$

39. $x^2 - 3y^2 = -2$
    $xy + 2y^2 = 3$

40. $3x^2 - 4xy = 25$
    $2x^2 - 4y^2 = 9$

41. Suppose the cost $C(x)$ and the revenue $R(x)$ of producing and selling x items are given by $C(x) = 2000 - 50x$ and $R(x) = 50x$, where $0 \le x \le 30$. Find the break-even point, that is, the point where cost and revenue are equal.

42. The sum of the digits of a two-digit number is 8. If the digits are reversed, the new number is 71 less than twice the original number. What is the original number?

43. A woman has $15,000 invested, part in stocks paying a simple interest rate of 9%, and the rest in bonds paying a simple interest rate of 5%. If her total annual interest income is $1130, find the amount invested at each rate.

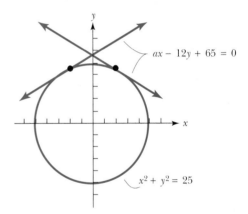

Photo by Susan Van Etten

44. Suppose it takes 1 hour 15 minutes for a plane to fly from Natchez, Mississippi, to Cross Anchor, South Carolina, an air distance of 600 miles, and 1 hour 30 minutes for the return flight. If the wind is blowing at a constant rate and direction, as indicated in the figure, find the rate of the wind and the rate of the plane in still air.

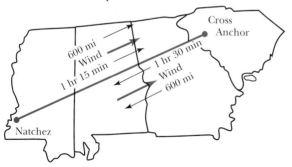

EXERCISE 44

45. A dietician is doing research in which she needs 28 grams of a substance that is 30% protein. How many grams of each of two ingredients, one that is 50% protein and the other 25% protein, should she mix together?

46. A hostess needs 10 gallons of punch for her party. How much pineapple juice should be mixed with a drink that is 10% fruit juice to obtain a punch that is 50% fruit juice?

47. Find two numbers whose difference is 15 and whose product is 76.

48. Find two numbers whose sum is 29 if the difference of their squares is 377.

49. Find two numbers such that the sum of their squares is 65 and the difference of their squares is 33.

50. Find the dimensions of a rectangle if its perimeter is 42 feet and its area is 108 square feet.

51. Find a value of the real number m so that the line $y = mx - 5$ is tangent to the circle $x^2 + y^2 = 9$.

52. Find a value of the real number a so that the straight line $ax - 12y + 65 = 0$ is tangent to the circle $x^2 + y^2 = 25$ in the following figure.

[figure: coordinate axes with circle $x^2 + y^2 = 25$ and line $ax - 12y + 65 = 0$]

EXERCISE 52

53. Suppose it has been determined that the cost $C(x)$ and the revenue $R(x)$ (both in thousands of dollars) of producing and selling x hundred items are given by $C(x) = 24 + 20x$ and $R(x) = x(10 + x)$, $x \geq 0$. Find the break-even point, that is, the point where cost and revenue are equal.

54. The city council of Midcity, U.S.A., plans to fence off a 1200-square-meter area next to an existing building for a playground. No fence is needed for the side against the building. Fencing material for the side parallel to the building costs $4 per meter, whereas fencing for the other two sides costs $3 per meter. Find the dimensions of the playground that can be so constructed if the council spends $340 on fencing.

55. A wooden freight carrier with a volume of 24 cubic feet is to be specially constructed with ends in the shape of equilateral triangles, as shown. Find the dimensions s and ℓ if three times more wood is used for the three sides than for the two ends.

56. Find the dimensions of a 12-cubic-foot box with square bottom and top that can be constructed from 32 square feet of cardboard (assuming no waste in construction).

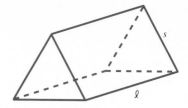

EXERCISE 55

8.2 | Matrix Algebra

While watching a football game on television, a football fan often sees flashed on the screen a tabulation similar to the following:

	Falcons	Rams
Number of offensive plays	8	14
Yards rushing	63	84
Yards passing	102	83
Total yards	165	167
Time of possession	5:20	9:40
Number of turnovers	2	0

Wide World Photos

If the fan is interested in the number of "yards passing" by the Rams, he looks at the entry in the third row and second column, and finds 83.

Suppose we consider only the array of numbers in this tabulation, omitting the row and column headings and placing square brackets around the array. We have

$$
\begin{bmatrix}
8 & 14 \\
63 & 84 \\
102 & 83 \\
165 & 167 \\
5\!:\!20 & 9\!:\!40 \\
2 & 0
\end{bmatrix}.
$$

This array of entries is called a **matrix** (plural: **matrices**), and each number in the array is called an **element** of the matrix. The formal definition follows.

Definition 8.1

An **m × n matrix** is a rectangular array of elements arranged in m rows and n columns. Such a matrix can be written in the form

$$
A = \begin{bmatrix}
a_{11} & a_{12} & \cdots & a_{1n} \\
a_{21} & a_{22} & \cdots & a_{2n} \\
a_{31} & a_{32} & \cdots & a_{3n} \\
\vdots & \vdots & & \vdots \\
a_{m1} & a_{m2} & \cdots & a_{mn}
\end{bmatrix},
$$

where a_{ij} denotes the element in row i and column j of the matrix A. The matrix A is referred to as a matrix of **dimension m × n** (read: m by n).

It is customary to denote matrices by capital letters. An $m \times n$ matrix A can be denoted compactly by

$$
A = [a_{ij}]_{(m,n)} \qquad \text{or} \qquad A_{(m,n)}.
$$

The elements of the form $a_{11}, a_{22}, a_{33}, \ldots$ are called the **main diagonal elements.**

A matrix that has n rows and n columns is called a **square matrix of order n.** A matrix that has only one column is called a **column matrix,** and a matrix that has only one row is called a **row matrix.** A square matrix that has zeros for all its elements except the main diagonal elements is called a **diagonal matrix.**

For example, the matrix

$$\begin{bmatrix} 2 & 1 \\ -3 & 4 \\ -2 & 0 \end{bmatrix}$$

has dimension 3×2; the matrix $[-1 \quad 0 \quad 4 \quad 7]$ is a row matrix of dimension 1×4; and the matrix

$$\begin{bmatrix} 1 & 0 \\ 0 & -1 \end{bmatrix}$$

is a diagonal matrix (and hence a square matrix) of order 2.

Definition 8.2

Two matrices are **equal** if they have the same dimensions and if the elements in corresponding positions are equal.

Example 1   The meaning of the definition of equality is illustrated by the following pairs of matrices.

a) $\begin{bmatrix} 1 \\ 2 \\ -5 \end{bmatrix}$ and $[1 \quad 2 \quad -5]$ are not equal, since they have different dimensions.

b) $\begin{bmatrix} x & 2y \\ 1 & z \\ 0 & 2 \end{bmatrix}$ and $\begin{bmatrix} -3 & 4 \\ 1 & z \\ 0 & \frac{4}{2} \end{bmatrix}$ are equal only if $x = -3$ and $y = 2$.

c) $[1 \quad 3]$ and $[3 \quad 1]$ are not equal, since elements in corresponding positions are not equal.   ❑

Addition of matrices is defined in a natural manner.

Definition 8.3

The **sum** of two $m \times n$ matrices A and B is the $m \times n$ matrix $A + B$ formed by adding the elements in corresponding positions of A and B.

Notice that the sum of two matrices of different dimensions is *not* defined.

Example 2   Addition of matrices is illustrated below.

a) $\begin{bmatrix} 2 & -2 & 1 \\ 1 & 0 & -3 \end{bmatrix} + \begin{bmatrix} 1 & -5 & 7 \\ 2 & 1 & 0 \end{bmatrix} = \begin{bmatrix} 3 & -7 & 8 \\ 3 & 1 & -3 \end{bmatrix}$

b) $\begin{bmatrix} 1 & 4 & 0 \\ -1 & 2 & 1 \end{bmatrix} + \begin{bmatrix} 0 & 1 \\ 2 & -2 \end{bmatrix}$ is not defined, since the dimensions of

the two matrices are not equal.   ❏

It can be shown that addition of matrices is associative and commutative. That is:

Let A, B, and C be $m \times n$ matrices. Then

$$A + (B + C) = (A + B) + C \qquad \text{and} \qquad A + B = B + A.$$

The **additive identity** for the set of all $m \times n$ matrices is an $m \times n$ matrix with all elements equal to 0. This additive identity matrix is also called the **zero matrix** of dimension $m \times n$. Thus

$$\begin{bmatrix} 0 & 0 \\ 0 & 0 \\ 0 & 0 \end{bmatrix}$$

is the 3×2 zero matrix.

The **additive inverse** of the $m \times n$ matrix A is the $m \times n$ matrix $-A$, the matrix obtained by multiplying each element of A by -1. For example, the additive inverse of

$$A = \begin{bmatrix} -2 & 4 & 1 & -1 \\ -1 & 0 & 3 & 2 \\ 0 & -2 & 0 & 1 \end{bmatrix} \qquad \text{is} \qquad -A = \begin{bmatrix} 2 & -4 & -1 & 1 \\ 1 & 0 & -3 & -2 \\ 0 & 2 & 0 & -1 \end{bmatrix}.$$

Subtraction of matrices is defined by using additive inverses:

$$A - B = A + (-B).$$

Example 3   Compute $A - B$ if $A = \begin{bmatrix} -2 & 1 \\ 3 & -2 \end{bmatrix}$ and $B = \begin{bmatrix} 1 & 7 \\ 2 & -4 \end{bmatrix}.$

Solution

$$A - B = A + (-B) = \begin{bmatrix} -2 & 1 \\ 3 & -2 \end{bmatrix} + \begin{bmatrix} -1 & -7 \\ -2 & 4 \end{bmatrix} = \begin{bmatrix} -3 & -6 \\ 1 & 2 \end{bmatrix}$$ ❏

There are two types of multiplication to be considered when working with matrices. The simplest is the product of a real number and a matrix. In this sort of product the real number is often called a **scalar.**

Definition 8.4

To multiply a matrix A by a scalar c, multiply each element of A by c. This **product** is denoted by cA.

Example 4   Compute $3A$, where $A = \begin{bmatrix} 2 & -1 & -2 \\ 1 & 4 & 0 \end{bmatrix}$.

Solution

$$3A = 3\begin{bmatrix} 2 & -1 & -2 \\ 1 & 4 & 0 \end{bmatrix} = \begin{bmatrix} 3(2) & 3(-1) & 3(-2) \\ 3(1) & 3(4) & 3(0) \end{bmatrix} = \begin{bmatrix} 6 & -3 & -6 \\ 3 & 12 & 0 \end{bmatrix}$$

The multiplication of two matrices is much more involved. We begin with a formal statement of the definition and then illustrate the definition with some examples.

Definition 8.5

The **product** of an $m \times n$ matrix A and an $n \times p$ matrix B is an $m \times p$ matrix $C = AB$, where the element c_{ij} in row i and column j of AB is found by using the elements in row i of A and the elements in column j of B in the following manner:

$$
\underset{\substack{\text{Row } i \\ \text{of } A}}{}
\begin{bmatrix}
\vdots & \vdots & \vdots & & \vdots \\
a_{i1} & a_{i2} & a_{i3} & \cdots & a_{in} \\
\vdots & \vdots & \vdots & & \vdots
\end{bmatrix}
\cdot
\begin{bmatrix}
\cdots & b_{1j} & \cdots \\
\cdots & b_{2j} & \cdots \\
\cdots & b_{3j} & \cdots \\
& \vdots & \\
\cdots & b_{nj} & \cdots
\end{bmatrix}
=
\begin{bmatrix}
& \vdots & \\
\cdots & c_{ij} & \cdots \\
& \vdots &
\end{bmatrix}
\underset{\substack{\text{Row } i \\ \text{of } C}}{}
$$

where

$$c_{ij} = a_{i1}b_{1j} + a_{i2}b_{2j} + a_{i3}b_{3j} + \cdots + a_{in}b_{nj}.$$

That is, the element c_{ij} in row i and column j of AB is found by adding the products formed from corresponding elements of row i in A and column j in B (first times first, second times second, and so on).

Notice that the number of columns in *A must* equal the number of rows in *B* in order to form the product *AB*. If this is the case, then *A* and *B* are said to be *conformable for multiplication*. A simple diagram illustrates this fact.

$$A_{(m,n)} \cdot B_{(n,p)} \qquad = \qquad C_{(m,p)}$$

Must be
equal

Dimension of product matrix

Some examples are helpful in understanding the definition of matrix multiplication.

Example 5   Form the products

a) *AB*,   b) *BA*,   c) *AC*, and   d) *CA*,
if possible, where

$$A = \begin{bmatrix} 3 & 1 \\ -2 & -1 \\ 4 & -2 \end{bmatrix}, \quad B = \begin{bmatrix} 2 & 1 & 0 \\ -1 & 3 & 5 \end{bmatrix}, \quad C = \begin{bmatrix} 2 & 7 \\ 0 & -4 \end{bmatrix}.$$

Solution

a)   The product *AB* exists since *A* has two columns and *B* has two rows.

$$A_{(3,2)} \qquad \cdot \qquad B_{(2,3)}$$

Equal

Dimension of
product matrix

Performing the multiplication yields the following result. (The shading indicates how the element in row 1, column 2, is computed.)

$$\begin{bmatrix} 3 & 1 \\ -2 & -1 \\ 4 & -2 \end{bmatrix}\begin{bmatrix} 2 & 1 & 0 \\ -1 & 3 & 5 \end{bmatrix}$$

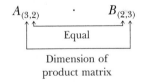

$$= \begin{bmatrix} 3(2) + 1(-1) & 3(1) + 1(3) & 3(0) + 1(5) \\ -2(2) + (-1)(-1) & -2(1) + (-1)(3) & -2(0) + (-1)(5) \\ 4(2) + (-2)(-1) & 4(1) + (-2)(3) & 4(0) + (-2)(5) \end{bmatrix}$$

and

$$AB = \begin{bmatrix} 5 & 6 & 5 \\ -3 & -5 & -5 \\ 10 & -2 & -10 \end{bmatrix}$$

b) The product BA exists since B has three columns and A has three rows.

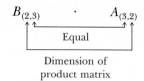

Performing the multiplication, we have the following result.

$$\begin{bmatrix} 2 & 1 & 0 \\ -1 & 3 & 5 \end{bmatrix}\begin{bmatrix} 3 & 1 \\ -2 & -1 \\ 4 & -2 \end{bmatrix}$$

$$= \begin{bmatrix} 2(3) + 1(-2) + 0(4) & 2(1) + 1(-1) + 0(-2) \\ -1(3) + 3(-2) + 5(4) & -1(1) + 3(-1) + 5(-2) \end{bmatrix}$$

and

$$BA = \begin{bmatrix} 4 & 1 \\ 11 & -14 \end{bmatrix}$$

Notice that the dimensions of the products AB and BA are not the same. Hence we have $AB \neq BA$. (The commutative law of multiplication does *not* hold for matrices.)

c) The product AC exists since A has two columns and C has two rows. The multiplication is given by

$$AC = \begin{bmatrix} 3 & 1 \\ -2 & -1 \\ 4 & -2 \end{bmatrix}\begin{bmatrix} 2 & 7 \\ 0 & -4 \end{bmatrix} = \begin{bmatrix} 6 & 17 \\ -4 & -10 \\ 8 & 36 \end{bmatrix}.$$

d) The product CA does not exist, since C has two columns and A has three rows:

$$C_{(2,2)} \cdot A_{(3,2)} \text{ does not exist.}$$
$$\underbrace{\phantom{C_{(2,2)} \cdot A_{(3,2)}}}_{\text{Not equal}}$$

❑

Even when the two products have the same dimension, AB and BA are usually different. Although multiplication of matrices does not have the commutative property, it can be shown that matrix multiplication is associative, provided that the products involved are defined.

Let A be an $m \times n$ matrix, B be an $n \times p$ matrix, and C be a $p \times q$ matrix. Then

$$A(BC) = (AB)C.$$

It can also be shown that the following distributive properties hold.

> Let A be an $m \times n$ matrix, B and C be $n \times p$ matrices, and D be a $p \times q$ matrix. Let a and b be real numbers. Then
>
> a) $A(B + C) = AB + AC$,
> b) $(B + C)D = BD + CD$,
> c) $a(B + C) = aB + aC$,
> d) $(a + b)A = aA + bA$.

One type of product is worthy of special note. For every square matrix A of order n, there is a square matrix I_n of order n such that $AI_n = I_nA = A$. The matrix I_n is called the **identity matrix of order n.** It is a diagonal matrix with 1's on the main diagonal.

$$I_n = \begin{bmatrix} 1 & 0 & 0 & \cdots & 0 \\ 0 & 1 & 0 & \cdots & 0 \\ 0 & 0 & 1 & \cdots & 0 \\ \vdots & \vdots & \vdots & & \vdots \\ 0 & 0 & 0 & \cdots & 1 \end{bmatrix}$$

If the order is understood from the context, then we write I for I_n.

Example 6

$$I_2 \begin{bmatrix} 1 & 4 \\ 2 & 1 \end{bmatrix} = \begin{bmatrix} 1 & 0 \\ 0 & 1 \end{bmatrix}\begin{bmatrix} 1 & 4 \\ 2 & 1 \end{bmatrix} = \begin{bmatrix} 1 & 4 \\ 2 & 1 \end{bmatrix}$$

$$\begin{bmatrix} 1 & 4 \\ 2 & 1 \end{bmatrix} I_2 = \begin{bmatrix} 1 & 4 \\ 2 & 1 \end{bmatrix}\begin{bmatrix} 1 & 0 \\ 0 & 1 \end{bmatrix} = \begin{bmatrix} 1 & 4 \\ 2 & 1 \end{bmatrix}$$ ❏

Many practical problems can be formulated in terms of matrices so that matrix multiplication is instrumental in the solution of the problems.

Example 7

Consider the following problem of determining the cost of 10 pounds of four chicken feed mixtures. The number of pounds of oats, barley, and corn contained in 10 pounds of each of the four mixtures A, B, C, and D are given by the following matrix.

$$\begin{array}{cc} & \textit{Ingredient} \\ & \begin{array}{ccc} \text{Oats} & \text{Barley} & \text{Corn} \end{array} \\ \textit{Mixture} \begin{array}{c} \text{A} \\ \text{B} \\ \text{C} \\ \text{D} \end{array} & \begin{bmatrix} 2 & 4 & 4 \\ 3 & 2 & 5 \\ 2 & 2 & 6 \\ 5 & 1 & 4 \end{bmatrix} \end{array}$$

For example, there are 2 pounds of oats, 2 pounds of barley, and 6 pounds of corn in 10 pounds of mixture C.

There are two suppliers of ingredients, and the cost per pound of each ingredient from each supplier is given in the following matrix.

Supplier

		1	2
	Oats	$0.17	$0.18
Ingredient	Barley	$0.14	$0.16
	Corn	$0.15	$0.13

The product of these two matrices,

$$\begin{bmatrix} 2 & 4 & 4 \\ 3 & 2 & 5 \\ 2 & 2 & 6 \\ 5 & 1 & 4 \end{bmatrix} \begin{bmatrix} \$0.17 & \$0.18 \\ \$0.14 & \$0.16 \\ \$0.15 & \$0.13 \end{bmatrix},$$

will yield the matrix whose elements represent the cost of 10 pounds of each mixture with ingredients supplied by each supplier.

Supplier

		1	2
	A	$1.50	$1.52
	B	$1.54	$1.51
Mixture	C	$1.52	$1.46
	D	$1.59	$1.58

For example, 10 pounds of mixture B costs $1.54 when ingredients are furnished by supplier 1 and $1.51 when furnished by supplier 2. ❏

Exercises 8.2

Determine whether or not the following pairs of matrices are equal. (See Example 1.)

1. $\begin{bmatrix} 1 & 2 \\ -1 & 3 \\ 1 & 7 \end{bmatrix}, \begin{bmatrix} 2 & 1 \\ 3 & -1 \\ 7 & 1 \end{bmatrix}$

2. $\begin{bmatrix} 0 & 0 \\ 0 & 0 \end{bmatrix}, \begin{bmatrix} 0 & 0 & 0 \\ 0 & 0 & 0 \\ 0 & 0 & 0 \end{bmatrix}$

3. $\begin{bmatrix} 1 & -7 \\ 5 & 3.2 \end{bmatrix}, \begin{bmatrix} 3-2 & -7 \\ 5 & \frac{32}{10} \end{bmatrix}$

4. $[1 \quad 3 \quad -1], \begin{bmatrix} 1 \\ 3 \\ -1 \end{bmatrix}$

5. $\begin{bmatrix} x & 1 \\ 0 & -y \end{bmatrix}, \begin{bmatrix} 2 & 1 \\ 0 & -3 \end{bmatrix}$

6. $\begin{bmatrix} x^2 \\ -y \end{bmatrix}, \begin{bmatrix} 4 \\ 2 \end{bmatrix}$

7. $[2x-5], [-1]$

8. $[2x \quad x], [6 \quad 3]$

Perform the indicated operations, if possible. (See Examples 2–4.)

9. $\begin{bmatrix} 3 & -1 & 1 \\ 2 & 7 & -4 \end{bmatrix} + \begin{bmatrix} 2 & 1 & 0 \\ 1 & 3 & -1 \end{bmatrix}$

10. $\begin{bmatrix} 1 & 0 \\ 0 & 1 \end{bmatrix} + \begin{bmatrix} 2 & 1 \\ -3 & 2 \end{bmatrix}$

11. $\begin{bmatrix} -2 \\ 3 \\ -6 \end{bmatrix} + [1 \quad 1 \quad -2]$

12. $\begin{bmatrix} -3 & 1 \\ 0 & -2 \\ 1 & -1 \end{bmatrix} + \begin{bmatrix} -3 & 1 & 0 \\ -2 & 1 & -1 \end{bmatrix}$

13. $\begin{bmatrix} 4 & -2 \\ 1 & 7 \end{bmatrix} - \begin{bmatrix} 2 & 6 \\ -3 & 0 \end{bmatrix}$

14. $\begin{bmatrix} 3 & 7 & 11 \\ -2 & 4 & -1 \end{bmatrix} - \begin{bmatrix} 1 & -1 & 2 \\ 1 & -3 & 1 \end{bmatrix}$

15. $3\begin{bmatrix} -1 & 3 \\ 2 & 0 \\ -1 & 1 \end{bmatrix} - 2\begin{bmatrix} 4 & -2 \\ -5 & 4 \\ 0 & -3 \end{bmatrix}$

16. $-6\begin{bmatrix} 2 & 0 \\ -5 & 4 \end{bmatrix} - \begin{bmatrix} -1 & -3 \\ 1 & 7 \end{bmatrix}$

17. $2\begin{bmatrix} -11 & 2 \\ 0 & -3 \end{bmatrix} + 3\begin{bmatrix} -11 \\ 0 \end{bmatrix}$

18. $2\begin{bmatrix} -11 & 0 \\ 0 & 0 \end{bmatrix} - 3\begin{bmatrix} -11 \\ 0 \end{bmatrix}$

19. $-3\begin{bmatrix} -1 \\ 6 \end{bmatrix} + 2\begin{bmatrix} -2 \\ 5 \end{bmatrix}$

20. $9[-1 \quad 4] - 3[-6 \quad 0]$

21. $7\begin{bmatrix} 2 & 1 \\ 3 & 5 \\ 0 & 2 \end{bmatrix} - \begin{bmatrix} 6 & 1 \\ 2 & 3 \\ -2 & 1 \end{bmatrix}$

22. $-\begin{bmatrix} 1 & 1 \\ 3 & 8 \\ 0 & -2 \end{bmatrix} - 2\begin{bmatrix} 9 & 2 \\ 1 & 1 \\ 4 & 3 \end{bmatrix}$

23. $2\begin{bmatrix} 4 & -2 & 5 & 1 \\ 7 & 1 & 8 & 2 \end{bmatrix} + 3\begin{bmatrix} 6 & 2 & 1 & -3 \\ 8 & -1 & 4 & -2 \end{bmatrix} -$

$\qquad 4\begin{bmatrix} -6 & -1 & 0 & 4 \\ 1 & 5 & 1 & 3 \end{bmatrix}$

24. $5\begin{bmatrix} 1 & 5 & -3 \\ 2 & 1 & 1 \\ 4 & 0 & -2 \end{bmatrix} - 2\begin{bmatrix} -1 & 3 & 5 \\ 4 & 8 & 6 \\ 0 & 1 & 3 \end{bmatrix} -$

$\qquad 3\begin{bmatrix} 0 & 8 & 3 \\ 1 & 0 & 5 \\ -3 & 4 & 0 \end{bmatrix}$

Perform the following matrix multiplications, if possible (See Examples 5 and 6.)

25. $\begin{bmatrix} 1 & 3 & 0 \\ -2 & 4 & 1 \end{bmatrix}\begin{bmatrix} 1 & 2 \\ 2 & -1 \\ 1 & 3 \end{bmatrix}$

26. $\begin{bmatrix} 1 & 2 \\ 2 & -1 \\ 1 & 3 \end{bmatrix}\begin{bmatrix} 1 & 3 & 0 \\ -2 & 4 & 1 \end{bmatrix}$

27. $\begin{bmatrix} -2 & 1 & 3 \\ 4 & 7 & 0 \\ -1 & 4 & -2 \end{bmatrix}\begin{bmatrix} 3 & 5 \\ -1 & 0 \\ 2 & 4 \end{bmatrix}$

28. $\begin{bmatrix} 3 & 5 \\ -1 & 0 \\ 2 & 4 \end{bmatrix}\begin{bmatrix} -2 & 1 & 3 \\ 4 & 7 & 0 \\ -1 & 4 & -2 \end{bmatrix}$

29. $\begin{bmatrix} 1 & 0 \\ 5 & -1 \end{bmatrix}I_3$ **30.** $I_2\begin{bmatrix} 1 & 7 & 0 \\ 0 & 0 & 1 \end{bmatrix}$

31. $[-3 \quad -2 \quad 0 \quad 1][1 \quad 0 \quad 3 \quad -1]$

32. $\begin{bmatrix} 1 \\ 5 \\ 9 \end{bmatrix}\begin{bmatrix} -9 \\ -5 \\ -1 \end{bmatrix}$

33. $[0 \quad -3 \quad 1]\begin{bmatrix} 10 \\ -4 \\ 10 \end{bmatrix}$ **34.** $\begin{bmatrix} 10 \\ -4 \\ 10 \end{bmatrix}[0 \quad -3 \quad 1]$

35. $\begin{bmatrix} -2 \\ 1 \\ 1 \end{bmatrix}[4 \quad 11 \quad -2]$ **36.** $[4 \quad 11 \quad -2]\begin{bmatrix} -2 \\ 1 \\ 1 \end{bmatrix}$

37. $\begin{bmatrix} 1 & 0 \\ 0 & 1 \\ 0 & 0 \end{bmatrix}\begin{bmatrix} 1 & 3 \\ 2 & -2 \\ 1 & 0 \end{bmatrix}$ **38.** $\begin{bmatrix} 1 & 3 \\ 2 & -2 \\ 1 & 0 \end{bmatrix}\begin{bmatrix} 1 & 0 \\ 0 & 1 \\ 0 & 0 \end{bmatrix}$

39. $\begin{bmatrix} 11 & 3 \\ 4 & 1 \end{bmatrix}\begin{bmatrix} -1 & 3 \\ 4 & -11 \end{bmatrix}$

40. $\begin{bmatrix} 1 & 1 & 2 \\ 1 & 1 & 0 \\ 0 & -1 & 1 \end{bmatrix}\begin{bmatrix} -\frac{1}{2} & \frac{3}{2} & 1 \\ \frac{1}{2} & -\frac{1}{2} & -1 \\ \frac{1}{2} & -\frac{1}{2} & 0 \end{bmatrix}$

41. $\begin{bmatrix} 10 & 1 & -4 \\ -8 & 1 & 5 \\ -1 & -1 & 4 \end{bmatrix}\begin{bmatrix} \frac{1}{9} & 0 & \frac{1}{9} \\ \frac{3}{9} & \frac{4}{9} & -\frac{2}{9} \\ \frac{1}{9} & \frac{1}{9} & \frac{2}{9} \end{bmatrix}$

42. $\begin{bmatrix} -1 & 0 \\ 0 & -1 \end{bmatrix}\begin{bmatrix} -1 & 0 \\ 0 & -1 \end{bmatrix}$

Write the matrix whose elements are defined by each equation.

43. $a_{ij} = 2i + j;$ $i = 1, 2,$ $j = 1, 2, 3, 4$

44. $a_{ij} = i \cdot j$; $i = 1, 2, 3$, $j = 1, 2, 3, 4$

45. Write a 4×3 matrix such that $a_{ij} = 2i + 3j$ if $i < j$, and $a_{ij} = 1$ if $i \geq j$.

46. Write a 4×2 matrix such that $a_{ij} = 1$ if $i \neq j$, and $a_{ij} = -1$ if $i = j$.

47. Find two matrices A and B such that $AB \neq BA$.

48. Find two square matrices A and B, of order 2, such that $AB \neq BA$.

49. Find two nonzero square matrices A and B, of order 2, such that $AB = BA$.

50. Find two nonzero matrices A and B such that $AB = O$, where O is a zero matrix.

51. Evaluate $AB + AC$ and $A(B + C)$, and compare the results for

$$A = \begin{bmatrix} -1 & 3 \\ -2 & 0 \\ -1 & 1 \end{bmatrix}, \quad B = \begin{bmatrix} 2 & 1 \\ -3 & 0 \end{bmatrix}, \quad C = \begin{bmatrix} 5 & 3 \\ 0 & -1 \end{bmatrix}.$$

52. Evaluate $A(BC)$ and $(AB)C$, and compare the results for

$$A = \begin{bmatrix} 4 & 1 \\ 0 & -1 \\ -1 & 1 \end{bmatrix}, \quad B = \begin{bmatrix} 1 & 6 & 0 & 1 \\ 5 & -3 & 4 & 0 \end{bmatrix},$$

$$C = \begin{bmatrix} 2 \\ 3 \\ -1 \\ 1 \end{bmatrix}.$$

53. Evaluate $(A - B)(A + B)$ and $A^2 - B^2$, and compare the results for

$$A = \begin{bmatrix} -6 & 4 \\ 1 & 3 \end{bmatrix}, \quad B = \begin{bmatrix} 0 & 1 \\ 1 & 2 \end{bmatrix}.$$

(*Note:* $A^2 = A \cdot A$.)

54. For the same matrices as in Exercise 53, evaluate $(A + B)^2$ and $A^2 + 2AB + B^2$, and compare the results.

(See Example 7.)

55. Adrienne found that she needed to purchase 2 cans of peaches, 1 sack of flour, $\frac{1}{2}$ dozen eggs, and 1 sack of sugar in order to have the ingredients for a new recipe. From advertisements in the newspaper, she found the prices of each item at three supermarkets:

		Item			
		1 can peaches	1 sack flour	1 dozen eggs	1 sack sugar
Store	A	$0.65	$0.69	$0.84	$1.03
	B	0.63	0.89	0.62	0.78
	C	0.72	0.90	0.78	0.82

Use matrix multiplication to determine the grocery bill at each of the three stores.

56. Suppose that a nut vendor wants to determine the cost per pound of three mixtures, each containing peanuts, cashews, and pecans, as given here:

		Nut		
		Peanuts	Cashews	Pecans
Mixture	A	$\frac{1}{2}$	$\frac{1}{6}$	$\frac{1}{3}$
	B	$\frac{1}{3}$	$\frac{1}{3}$	$\frac{1}{3}$
	C	$\frac{3}{10}$	$\frac{3}{10}$	$\frac{2}{5}$

If peanuts cost $1.90 per pound, cashews $2.20 per pound, and pecans $3.80 per pound, use matrix multiplication to determine the cost per pound of each mixture.

8.3 Solution of Linear Systems by Matrix Methods

Matrices are useful tools in solving systems of linear equations. In this section we consider two of the best-known matrix methods, the *Gaussian elimination method* and the *Gauss–Jordan elimination method*. Some new terminology is needed to describe these methods.

Although the methods can be presented in a more general setting, we restrict our attention to systems of n linear equations in n unknowns. A system of n linear equations in n unknowns $x_1, x_2, \ldots, x_n$ is a system of the following form.

$$a_{11}x_1 + a_{12}x_2 + a_{13}x_3 + \cdots + a_{1n}x_n = b_1$$

$$a_{21}x_1 + a_{22}x_2 + a_{23}x_3 + \cdots + a_{2n}x_n = b_2$$

$$\vdots \qquad\qquad \vdots$$

$$a_{n1}x_1 + a_{n2}x_2 + a_{n3}x_3 + \cdots + a_{nn}x_n = b_n$$

For this system, the matrices

$$A = \begin{bmatrix} a_{11} & a_{12} & a_{13} & \cdots & a_{1n} \\ a_{21} & a_{22} & a_{23} & \cdots & a_{2n} \\ \vdots & \vdots & \vdots & & \vdots \\ a_{n1} & a_{n2} & a_{n3} & \cdots & a_{nn} \end{bmatrix} \quad \text{and} \quad B = \begin{bmatrix} b_1 \\ b_2 \\ \vdots \\ b_n \end{bmatrix}$$

are called the **coefficient matrix** and the **constant matrix,** respectively. The matrix

$$[A \mid B] = \begin{bmatrix} a_{11} & a_{12} & a_{13} & \cdots & a_{1n} & b_1 \\ a_{21} & a_{22} & a_{23} & \cdots & a_{2n} & b_2 \\ \vdots & \vdots & \vdots & & \vdots & \vdots \\ a_{n1} & a_{n2} & a_{n3} & \cdots & a_{nn} & b_n \end{bmatrix}$$

is called the **augmented matrix** for the system.

Example 1   The augmented matrix for

$$\begin{aligned} x - y + 2z &= 9 \\ -2x + 3y - z &= -11 \\ 3x + y + z &= 4 \end{aligned} \quad \text{is the matrix} \quad \begin{bmatrix} 1 & -1 & 2 & 9 \\ -2 & 3 & -1 & -11 \\ 3 & 1 & 1 & 4 \end{bmatrix}.$$

Conversely, a system associated with the augmented matrix

$$\begin{bmatrix} 1 & 3 & 0 & 4 & 1 \\ 0 & 1 & -1 & 2 & -1 \\ 1 & -1 & 5 & 3 & 1 \\ 2 & 1 & 1 & 0 & -2 \end{bmatrix} \quad \text{is} \quad \begin{aligned} x_1 + 3x_2 + 4x_4 &= 1, \\ x_2 - x_3 + 2x_4 &= -1, \\ x_1 - x_2 + 5x_3 + 3x_4 &= 1, \\ 2x_1 + x_2 + x_3 &= -2. \end{aligned}$$

A system associated with the augmented matrix

$$\begin{bmatrix} 1 & 0 & 0 & a \\ 0 & 1 & 0 & b \\ 0 & 0 & 1 & c \end{bmatrix} \quad \text{is} \quad \begin{aligned} x &= a, \\ y &= b, \\ z &= c, \end{aligned}$$

which has the obvious solution $x = a, y = b, z = c$.   ❏

The elimination method used in Section 8.1 can be modified to obtain the method known as *Gaussian elimination*. This modification is motivated by the following example.

Example 2   Suppose that we wish to solve the system

$$2x - y + z = 6$$
$$y + 2z = 2 \quad \text{with augmented matrix} \quad \begin{bmatrix} 2 & -1 & 1 & \vdots & 6 \\ 0 & 1 & 2 & \vdots & 2 \\ 1 & 1 & 1 & \vdots & 1 \end{bmatrix}.$$
$$x + y + z = 1$$

We can interchange the first and third equations and obtain the equivalent system

$$x + y + z = 1$$
$$y + 2z = 2 \quad \text{with augmented matrix} \quad \begin{bmatrix} 1 & 1 & 1 & \vdots & 1 \\ 0 & 1 & 2 & \vdots & 2 \\ 2 & -1 & 1 & \vdots & 6 \end{bmatrix}.$$
$$2x - y + z = 6$$

Adding -2 times the first equation to the third equation yields the equivalent system

$$x + y + z = 1$$
$$y + 2z = 2 \quad \text{with augmented matrix} \quad \begin{bmatrix} 1 & 1 & 1 & \vdots & 1 \\ 0 & 1 & 2 & \vdots & 2 \\ 0 & -3 & -1 & \vdots & 4 \end{bmatrix}.$$
$$-3y - z = 4$$

Adding 3 times the second equation to the third equation yields the equivalent system

$$x + y + z = 1$$
$$y + 2z = 2 \quad \text{with augmented matrix} \quad \begin{bmatrix} 1 & 1 & 1 & \vdots & 1 \\ 0 & 1 & 2 & \vdots & 2 \\ 0 & 0 & 5 & \vdots & 10 \end{bmatrix}.$$
$$5z = 10$$

Multiplying the third equation by $\frac{1}{5}$ yields the equivalent system

$$x + y + z = 1$$
$$y + 2z = 2 \quad \text{with augmented matrix} \quad \begin{bmatrix} 1 & 1 & 1 & \vdots & 1 \\ 0 & 1 & 2 & \vdots & 2 \\ 0 & 0 & 1 & \vdots & 2 \end{bmatrix}.$$
$$z = 2$$

We can now find the solution by these substitutions:

Substitute $z = 2$ in $y + 2z = 2$ to obtain $y = -2$

and then

substitute $y = -2, z = 2$ in $x + y + z = 1$ to obtain $x = 1$.

This procedure of substitution into prior equations is called **back substitution.** ❑

Focusing our attention on the matrices in Example 2, we can solve the system by making the following changes in the augmented matrix and then using back substitution.

$$\begin{bmatrix} 2 & -1 & 1 & \vdots & 6 \\ 0 & 1 & 2 & \vdots & 2 \\ 1 & 1 & 1 & \vdots & 1 \end{bmatrix} \xrightarrow{R_1 \leftrightarrow R_3} \begin{bmatrix} 1 & 1 & 1 & \vdots & 1 \\ 0 & 1 & 2 & \vdots & 2 \\ 2 & -1 & 1 & \vdots & 6 \end{bmatrix}$$

$$\xrightarrow{-2R_1 + R_3} \begin{bmatrix} 1 & 1 & 1 & \vdots & 1 \\ 0 & 1 & 2 & \vdots & 2 \\ 0 & -3 & -1 & \vdots & 4 \end{bmatrix} \xrightarrow{3R_2 + R_3} \begin{bmatrix} 1 & 1 & 1 & \vdots & 1 \\ 0 & 1 & 2 & \vdots & 2 \\ 0 & 0 & 5 & \vdots & 10 \end{bmatrix}$$

$$\xrightarrow{\frac{1}{5}R_3} \begin{bmatrix} 1 & 1 & 1 & \vdots & 1 \\ 0 & 1 & 2 & \vdots & 2 \\ 0 & 0 & 1 & \vdots & 2 \end{bmatrix}$$

The notation $R_1 \leftrightarrow R_3$ indicates that rows 1 and 3 are interchanged; $-2R_1 + R_3$ indicates that row 3 is replaced by -2 times row 1 plus row 3; and $\frac{1}{5}R_3$ indicates that row 3 was multiplied by $\frac{1}{5}$.

In the ordinary elimination method that was used in Section 8.1, equivalent systems are obtained by using the following operations.

1. Interchange two equations.
2. Multiply (or divide) both members of an equation by the same nonzero number.
3. Add (or subtract) a multiple of one equation to (from) another equation.

Example 2 shows how these operations on systems correspond to performing the following row operations on the augmented matrix.

Row Operations

1. Interchange two rows.
2. Multiply (or divide) every element in a row by the same nonzero number.
3. Add (or subtract) a multiple of one row to (from) another row.

The **Gaussian elimination** method uses row operations on the augmented matrix to solve a system of linear equations. The procedure in this method follows.

Gaussian Elimination Method

1. Write the augmented matrix $[A \vdots B]$ for the system, where A is the coefficient matrix and B is the constant matrix.

> **2.** Use any of the row operations (1), (2), (3) to change $[A \mid B]$ to a matrix in which all elements on the main diagonal are 1's and all elements below the main diagonal are 0's.
>
> **3.** Use back substitution on the system that has the augmented matrix obtained in step 2.

This method is illustrated in the following example.

Example 3   Use Gaussian elimination to solve the following system.

$$3x + 2y - 3z = -1$$
$$x + 3y + 2z = 1$$
$$x + y - 2z = -3$$

Solution

A straightforward approach is to change one column of the augmented matrix at a time into the desired form, working from left to right. In the first column we use appropriate row operations to obtain a 1 in the first row. We then use the third type of row operation to obtain 0's below that 1.

$$\begin{bmatrix} 3 & 2 & -3 & \vdots & -1 \\ 1 & 3 & 2 & \vdots & 1 \\ 1 & 1 & -2 & \vdots & -3 \end{bmatrix} \xrightarrow{R_1 \leftrightarrow R_3} \begin{bmatrix} 1 & 1 & -2 & \vdots & -3 \\ 1 & 3 & 2 & \vdots & 1 \\ 3 & 2 & -3 & \vdots & -1 \end{bmatrix}$$

$$\xrightarrow{-R_1 + R_2} \begin{bmatrix} 1 & 1 & -2 & \vdots & -3 \\ 0 & 2 & 4 & \vdots & 4 \\ 3 & 2 & -3 & \vdots & -1 \end{bmatrix} \xrightarrow{-3R_1 + R_2} \begin{bmatrix} 1 & 1 & -2 & \vdots & -3 \\ 0 & 2 & 4 & \vdots & 4 \\ 0 & -1 & 3 & \vdots & 8 \end{bmatrix}$$

Moving now to the second column, we obtain a 1 in the second position on the main diagonal and then use row operations of the third type to obtain 0's below that 1.

$$\xrightarrow{\frac{1}{2}R_2} \begin{bmatrix} 1 & 1 & -2 & \vdots & -3 \\ 0 & 1 & 2 & \vdots & 2 \\ 0 & -1 & 3 & \vdots & 8 \end{bmatrix} \xrightarrow{R_2 + R_3} \begin{bmatrix} 1 & 1 & -2 & \vdots & -3 \\ 0 & 1 & 2 & \vdots & 2 \\ 0 & 0 & 5 & \vdots & 10 \end{bmatrix}$$

As a final step, we get a 1 in the third position on the main diagonal.

$$\xrightarrow{\frac{1}{5}R_3} \begin{bmatrix} 1 & 1 & -2 & \vdots & -3 \\ 0 & 1 & 2 & \vdots & 2 \\ 0 & 0 & 1 & \vdots & 2 \end{bmatrix}$$

This augmented matrix corresponds to the following system.

$$x + y - 2z = -3$$
$$y + 2z = 2$$
$$z = 2$$

Back substitution yields the solution $x = 3, y = -2, z = 2$. ❏

The back substitution step in Gaussian elimination can be avoided by performing additional row operations on the augmented matrix. This procedure is described in the following method, known as **Gauss–Jordan elimination.**

Gauss–Jordan Elimination

---

1. Write the augmented matrix $[A \mid B]$ for the system.
2. Use any of the row operations (1), (2), (3) to change $[A \mid B]$ to $[I \mid C]$, where I is an identity matrix of the same size as A, and C is a column matrix.
3. Read the solutions from the column matrix C.

This method is illustrated in the following example.

Example 4   Use Gauss–Jordan elimination to solve the following system.

$$x + y + 2z = 3$$
$$2x - y + z = 6$$
$$-x + 3y = -5$$

Solution
The augmented matrix is

$$[A \mid B] = \begin{bmatrix} 1 & 1 & 2 & 3 \\ 2 & -1 & 1 & 6 \\ -1 & 3 & 0 & -5 \end{bmatrix}.$$

We shall use row operations to transform $[A \mid B]$ into

$$[I \mid C] = \begin{bmatrix} 1 & 0 & 0 & a \\ 0 & 1 & 0 & b \\ 0 & 0 & 1 & c \end{bmatrix},$$

which is the augmented matrix for the system whose solution is

$$x = a, \quad y = b, \quad z = c.$$

The most straightforward approach to use to transform $[A \mid B]$ into $[I \mid C]$ is to change one column of A at a time into a column of the identity matrix I, working from left to right. In the first column we use appropriate row operations to obtain a 1 in the first row. Then, with that 1, we use the third type of row operation to obtain 0's in the remaining positions of column 1.

$$\begin{bmatrix} 1 & 1 & 2 & 3 \\ 2 & -1 & 1 & 6 \\ -1 & 3 & 0 & -5 \end{bmatrix} \xrightarrow{-2R_1 + R_2} \begin{bmatrix} 1 & 1 & 2 & 3 \\ 0 & -3 & -3 & 0 \\ -1 & 3 & 0 & -5 \end{bmatrix}$$

$$\xrightarrow{R_1 + R_3} \begin{bmatrix} 1 & 1 & 2 & 3 \\ 0 & -3 & -3 & 0 \\ 0 & 4 & 2 & -2 \end{bmatrix}$$

Next, we use row operations to obtain a 1 in the second-row, second-column, position. Then, with that 1, we use row operations of the third type to obtain 0's in the remaining positions of column 2.

$$\xrightarrow{-\frac{1}{3}R_2} \begin{bmatrix} 1 & 1 & 2 & 3 \\ 0 & 1 & 1 & 0 \\ 0 & 4 & 2 & -2 \end{bmatrix} \xrightarrow{-R_2 + R_1} \begin{bmatrix} 1 & 0 & 1 & 3 \\ 0 & 1 & 1 & 0 \\ 0 & 4 & 2 & -2 \end{bmatrix}$$

$$\xrightarrow{-4R_2 + R_3} \begin{bmatrix} 1 & 0 & 1 & 3 \\ 0 & 1 & 1 & 0 \\ 0 & 0 & -2 & -2 \end{bmatrix}$$

Proceeding to column 3, we use row operations to obtain a 1 in the third row. Then with that 1 we use row operations of type 3 to obtain 0's in the remaining positions of column 3.

$$\xrightarrow{-\frac{1}{2}R_3} \begin{bmatrix} 1 & 0 & 1 & 3 \\ 0 & 1 & 1 & 0 \\ 0 & 0 & 1 & 1 \end{bmatrix} \xrightarrow{-R_3 + R_2} \begin{bmatrix} 1 & 0 & 1 & 3 \\ 0 & 1 & 0 & -1 \\ 0 & 0 & 1 & 1 \end{bmatrix}$$

$$\xrightarrow{-R_3 + R_1} \begin{bmatrix} 1 & 0 & 0 & 2 \\ 0 & 1 & 0 & -1 \\ 0 & 0 & 1 & 1 \end{bmatrix}$$

Thus the solution is

$$x = 2, \quad y = -1, \quad z = 1. \qquad \square$$

Sometimes it is impossible to transform the augmented matrix $[A \mid B]$ into $[I \mid C]$ by row operations. If any of the row operations on $[A \mid B]$ yields a row with all zero elements except possibly the last element, then either there is no solution to the system or the solution is not unique. This is illustrated in the next two examples.

Example 5 Solve the following system.

$$x - 2y - 2z = -1$$
$$x + y + z = 2$$
$$x + 2y + 2z = 1$$

Solution

Using the procedure described in Examples 3 or 4 yields the following result.

$$\begin{bmatrix} 1 & -2 & -2 & \vdots & -1 \\ 1 & 1 & 1 & \vdots & 2 \\ 1 & 2 & 2 & \vdots & 1 \end{bmatrix} \xrightarrow{-R_1 + R_2} \begin{bmatrix} 1 & -2 & -2 & \vdots & -1 \\ 0 & 3 & 3 & \vdots & 3 \\ 1 & 2 & 2 & \vdots & 1 \end{bmatrix}$$

$$\xrightarrow{-R_1 + R_3} \begin{bmatrix} 1 & -2 & -2 & \vdots & -1 \\ 0 & 3 & 3 & \vdots & 3 \\ 0 & 4 & 4 & \vdots & 2 \end{bmatrix} \xrightarrow{\frac{1}{3}R_2} \begin{bmatrix} 1 & -2 & -2 & \vdots & -1 \\ 0 & 1 & 1 & \vdots & 1 \\ 0 & 4 & 4 & \vdots & 2 \end{bmatrix}$$

$$\xrightarrow{-4R_2 + R_3} \begin{bmatrix} 1 & -2 & -2 & \vdots & -1 \\ 0 & 1 & 1 & \vdots & 1 \\ 0 & 0 & 0 & \vdots & -2 \end{bmatrix}$$

Since the third row in the last matrix contains all zero elements except the last element, it is impossible to obtain $[I \vdots C]$ by row operations. The last matrix obtained is the augmented matrix for a system with a last equation that reads $0 = -2$. This indicates there is no solution to the original system. ❏

Example 6 Solve the following system.

$$x + 2y = 1$$
$$x + 3y + z = 4$$
$$2y + 2z = 6$$

Solution

$$\begin{bmatrix} 1 & 2 & 0 & \vdots & 1 \\ 1 & 3 & 1 & \vdots & 4 \\ 0 & 2 & 2 & \vdots & 6 \end{bmatrix} \xrightarrow{-R_1 + R_2} \begin{bmatrix} 1 & 2 & 0 & \vdots & 1 \\ 0 & 1 & 1 & \vdots & 3 \\ 0 & 2 & 2 & \vdots & 6 \end{bmatrix} \xrightarrow{-2R_2 + R_3} \begin{bmatrix} 1 & 2 & 0 & \vdots & 1 \\ 0 & 1 & 1 & \vdots & 3 \\ 0 & 0 & 0 & \vdots & 0 \end{bmatrix}$$

Since the last row contains only zeros, it is impossible to obtain $[I \vdots C]$ by row operations. The following system corresponds to the last matrix we obtained.

$$x + 2y = 1$$
$$y + z = 3$$
$$0 = 0$$

This system has many solutions. To display them, we can let r represent an arbitrary real number, set $z = r$, and then solve for x and y in terms of r. We get

$$x = 2r - 5, \qquad y = 3 - r, \qquad z = r. \qquad \square$$

Exercises 8.3

Write the augmented matrix for each of the following systems. (See Example 1.)

1. $\begin{aligned} 3x - y &= 0 \\ -x + y &= 1 \end{aligned}$

2. $\begin{aligned} r + s &= 10 \\ r - s &= -4 \end{aligned}$

3. $\begin{aligned} 3x - 2y + 5z &= 0 \\ 4x + 7y - z &= 0 \\ x \qquad + z &= 2 \end{aligned}$

4. $\begin{aligned} x_1 + x_3 &= 4 \\ x_2 + x_3 &= 0 \\ x_1 + x_2 &= 2 \end{aligned}$

5. $\begin{aligned} p - q &= 0 \\ r + s &= 0 \\ 3p + 2s &= 0 \\ 5q - r &= 0 \end{aligned}$

6. $\begin{aligned} a + 3b + c + d &= 1 \\ a - 3b - c + 2d &= 5 \\ 2a \qquad + c + 3d &= 7 \\ b + c \qquad &= 0 \end{aligned}$

Write a system of linear equations that has the given augmented matrix. (See Example 1.)

7. $\begin{bmatrix} 1 & 2 & \vdots & 5 \\ 3 & 4 & \vdots & 6 \end{bmatrix}$

8. $\begin{bmatrix} -2 & 1 & \vdots & -1 \\ 4 & 3 & \vdots & 0 \end{bmatrix}$

9. $\begin{bmatrix} 1 & 0 & 0 & \vdots & a \\ 0 & 0 & 1 & \vdots & b \\ 0 & 1 & 0 & \vdots & c \end{bmatrix}$

10. $\begin{bmatrix} 1 & 1 & 0 & \vdots & 0 \\ 3 & 0 & -3 & \vdots & 0 \\ -1 & 1 & -1 & \vdots & -6 \end{bmatrix}$

11. $\begin{bmatrix} 1 & 0 & 1 & 0 & \vdots & 0 \\ 0 & 2 & 1 & 3 & \vdots & 7 \\ 3 & 1 & 1 & 1 & \vdots & 1 \\ -3 & 1 & -1 & 2 & \vdots & 5 \end{bmatrix}$

12. $\begin{bmatrix} 0 & 0 & 0 & 1 & \vdots & 1 \\ 0 & -1 & 0 & 0 & \vdots & 2 \\ 1 & 0 & 0 & 0 & \vdots & 3 \\ 0 & 0 & -1 & 0 & \vdots & 4 \end{bmatrix}$

Solve the following systems. (See Examples 3–6.)

13. $\begin{aligned} x + y &= 1 \\ 2x + 3y &= -2 \end{aligned}$

14. $\begin{aligned} 3x + 2y &= 1 \\ x + 2y &= 7 \end{aligned}$

15. $\begin{aligned} 4a + 5b &= -22 \\ 3a - 4b &= -1 \end{aligned}$

16. $\begin{aligned} 7x_1 - x_2 &= -2 \\ -3x_1 + x_2 &= 6 \end{aligned}$

17. $\begin{aligned} x - 2y &= 0 \\ 2x - 4y &= 1 \end{aligned}$

18. $\begin{aligned} -3g + h &= 2 \\ 6g - 2h &= 0 \end{aligned}$

19. $\begin{aligned} 2x - 3y &= -1 \\ -5x + 8y &= 0 \end{aligned}$

20. $\begin{aligned} 9x - y &= 0 \\ 3x + 3y &= -10 \end{aligned}$

21. $\begin{aligned} 2x + 7y &= 0 \\ x - 2y &= 0 \end{aligned}$

22. $\begin{aligned} 3x &= 0 \\ x + 5y &= -15 \end{aligned}$

23. $\begin{aligned} x - 2y \qquad &= 1 \\ y + z &= 0 \\ 2x \qquad + 3z &= 3 \end{aligned}$

24. $\begin{aligned} x - 2y + 3z &= 0 \\ 3y - 2z &= 0 \\ -3x + 4y - z &= 0 \end{aligned}$

25. $\begin{aligned} x - y - 4z &= -4 \\ -3x + 4y + 2z &= 6 \\ -x + 3y + 2z &= 10 \end{aligned}$

26. $\begin{aligned} 2a + 2b - c &= -2 \\ -b + 4c &= 11 \\ a - b \qquad &= 6 \end{aligned}$

27. $\begin{aligned} -2y - 2z &= -2 \\ 2x - y + z &= -3 \\ x + y + 3z &= -2 \end{aligned}$

28. $\begin{aligned} 3x_1 + 2x_2 + 5x_3 &= 2 \\ 2x_1 \qquad + 4x_3 &= 2 \\ x_1 + 3x_2 - x_3 &= -2 \end{aligned}$

29. $\begin{aligned} r - t &= 1 \\ 3r + s &= 6 \\ 5s + 6t &= -12 \end{aligned}$

30. $\begin{aligned} -3x_2 + 4x_3 &= -3 \\ 3x_1 - x_3 &= 6 \\ -x_1 + 3x_2 &= 1 \end{aligned}$

31. $\begin{aligned} 2x - 2y - 2z &= -2 \\ x - 4y + z &= -2 \\ 5x - 8y - 3z &= -2 \end{aligned}$

32. $\begin{aligned} x + 4y - z &= 0 \\ 3x - 5y + z &= 1 \\ 5x - 14y + 3z &= 1 \end{aligned}$

33.
$$\begin{aligned}
x + y - t &= 2 \\
y + z + 2t &= -3 \\
-2x + t &= -4 \\
x + y + z + t &= 0
\end{aligned}$$

34.
$$\begin{aligned}
2x_1 - 4x_2 + x_3 + 10x_4 &= -2 \\
3x_1 - x_2 + 3x_3 &= -3 \\
x_1 + 2x_3 &= 0 \\
x_2 - x_3 + x_4 &= 4
\end{aligned}$$

35.
$$\begin{aligned}
-a + b - c &= 5 \\
b - c + 3d &= 10 \\
a - c - d &= 5 \\
a + 3d &= 5
\end{aligned}$$

36.
$$\begin{aligned}
x + y - 2t &= -1 \\
2y - z &= -7 \\
z - 2t &= 1 \\
x + y + z - 4t &= 0
\end{aligned}$$

Some of the coefficients and constant terms in the following systems involve an unspecified constant c. For each system, find the values of c for which the system has (**a**) no solution, (**b**) exactly one solution, (**c**) many solutions.

37.
$$\begin{aligned}
x + 2y &= 1 \\
2x + c^2 y &= c
\end{aligned}$$

38.
$$\begin{aligned}
x + y &= 1 \\
x + cy &= c^2
\end{aligned}$$

39.
$$\begin{aligned}
x + 2y - z &= 1 \\
x + y &= 0 \\
x + 2y + (c^2 - 1)z &= c + 1
\end{aligned}$$

40.
$$\begin{aligned}
x + 2y - z &= 1 \\
2x + 4y - 2z &= 0 \\
x + 2y + (c^2 - 1)z &= c + 1
\end{aligned}$$

8.4   Inverses of Matrices

In Section 8.2 we saw that

1. A matrix with all elements zero acts as an additive identity for matrix addition;

2. Every matrix has an additive inverse, $-A$;

3. An identity matrix I_n acts as a multiplicative identity for matrix multiplication.

A question that arises in connection with (3) is this: Does every nonzero matrix A have a multiplicative inverse? In other words, for any nonzero matrix A, is there a matrix B such that $AB = BA = I_n$? The answer to this question might be somewhat disappointing, since *not all nonzero matrices have multiplicative inverses.*

We first notice that multiplicative inverses can exist only for square matrices, since the only way that AB and BA can be equal is for A and B to be square and of the same order. But we shall see shortly that some nonzero square matrices do not have multiplicative inverses.

From now on, we shall use the term *inverse* to mean "multiplicative inverse." If the matrix A has an inverse, we shall denote it by A^{-1}.

For square matrices of order 2, we have the following theorem, whose proof is requested in Exercises 40 and 41 at the end of this section.

Theorem 8.6

For a given 2×2 matrix

$$A = \begin{bmatrix} a & b \\ c & d \end{bmatrix},$$

let $\delta(A)$ denote the number $\delta(A) = ad - bc$. The inverse of A exists if $\delta(A) \neq 0$ and is given by

$$A^{-1} = \frac{1}{\delta(A)} \begin{bmatrix} d & -b \\ -c & a \end{bmatrix}.$$

If $\delta(A) = 0$, then A does not have an inverse.

Formulas for A^{-1} that are similar to the one in Theorem 8.6 exist for matrices of order greater than 2, but they are much more complicated, and they more properly belong in a course in linear algebra. Another method for finding A^{-1} is presented after the following example. This method is easier to use than the formulas for matrices of order greater than 2.

Example 1   Use Theorem 8.6 to find the inverse of each of the following matrices, if it exists.

a) $\begin{bmatrix} 1 & 4 \\ 3 & -2 \end{bmatrix}$ b) $\begin{bmatrix} 1 & -3 \\ -2 & 6 \end{bmatrix}$

Solution

a) For $A = \begin{bmatrix} 1 & 4 \\ 3 & -2 \end{bmatrix}$, we have $\delta(A) = (1)(-2) - (3)(4) = -14$.

Thus A^{-1} exists and is given by

$$A^{-1} = \frac{1}{-14} \begin{bmatrix} -2 & -4 \\ -3 & 1 \end{bmatrix} = \begin{bmatrix} \frac{1}{7} & \frac{2}{7} \\ \frac{3}{14} & -\frac{1}{14} \end{bmatrix}.$$

b) For $A = \begin{bmatrix} 1 & -3 \\ -2 & 6 \end{bmatrix}$, we have $\delta(A) = (1)(6) - (-2)(-3) = 0$.

Thus A^{-1} does not exist for this matrix A.   ❏

The easier method for finding A^{-1} is the **Gauss–Jordan elimination method.** It can be used to calculate the inverse of a square matrix A of any order n, provided that A^{-1} exists. The procedure to be followed is given by the three steps below.

1. Augment A with an identity matrix of the same size: $[A \mid I_n]$.
2. Use row operations to transform $[A \mid I_n]$ into the form $[I_n \mid B]$. If this is not possible, then A^{-1} does not exist.
3. If $[A \mid I_n]$ is transformed into $[I_n \mid B]$ by row operations, the inverse can be read from the last n columns of $[I_n \mid B]$. That is, $B = A^{-1}$.

Example 2   Use the Gauss–Jordan elimination method to find the inverse of each matrix below, if it exists.

a) $\begin{bmatrix} 1 & 0 & 2 \\ 0 & -1 & 3 \\ 2 & 1 & 3 \end{bmatrix}$

b) $\begin{bmatrix} 1 & -2 \\ -2 & 4 \end{bmatrix}$

Solution

a) We use row operations to transform $[A \mid I]$ into $[I \mid A^{-1}]$.

$$\begin{bmatrix} 1 & 0 & 2 & 1 & 0 & 0 \\ 0 & -1 & 3 & 0 & 1 & 0 \\ 2 & 1 & 3 & 0 & 0 & 1 \end{bmatrix} \xrightarrow{-2R_1 + R_3} \begin{bmatrix} 1 & 0 & 2 & 1 & 0 & 0 \\ 0 & -1 & 3 & 0 & 1 & 0 \\ 0 & 1 & -1 & -2 & 0 & 1 \end{bmatrix}$$

$$\xrightarrow{-R_2} \begin{bmatrix} 1 & 0 & 2 & 1 & 0 & 0 \\ 0 & 1 & -3 & 0 & -1 & 0 \\ 0 & 1 & -1 & -2 & 0 & 1 \end{bmatrix} \xrightarrow{-R_2 + R_3} \begin{bmatrix} 1 & 0 & 2 & 1 & 0 & 0 \\ 0 & 1 & -3 & 0 & -1 & 0 \\ 0 & 0 & 2 & -2 & 1 & 1 \end{bmatrix}$$

$$\xrightarrow{\frac{1}{2}R_3} \begin{bmatrix} 1 & 0 & 2 & 1 & 0 & 0 \\ 0 & 1 & -3 & 0 & -1 & 0 \\ 0 & 0 & 1 & -1 & \frac{1}{2} & \frac{1}{2} \end{bmatrix} \xrightarrow{3R_3 + R_2} \begin{bmatrix} 1 & 0 & 2 & 1 & 0 & 0 \\ 0 & 1 & 0 & -3 & \frac{1}{2} & \frac{3}{2} \\ 0 & 0 & 1 & -1 & \frac{1}{2} & \frac{1}{2} \end{bmatrix}$$

$$\xrightarrow{-2R_3 + R_1} \begin{bmatrix} 1 & 0 & 0 & 3 & -1 & -1 \\ 0 & 1 & 0 & -3 & \frac{1}{2} & \frac{3}{2} \\ 0 & 0 & 1 & -1 & \frac{1}{2} & \frac{1}{2} \end{bmatrix}$$

This gives $A^{-1} = \begin{bmatrix} 3 & -1 & -1 \\ -3 & \frac{1}{2} & \frac{3}{2} \\ -1 & \frac{1}{2} & \frac{1}{2} \end{bmatrix}$.

b) We augment A with I and find

$$\begin{bmatrix} 1 & -2 & 1 & 0 \\ -2 & 4 & 0 & 1 \end{bmatrix} \xrightarrow{2R_1 + R_2} \begin{bmatrix} 1 & -2 & 1 & 0 \\ 0 & 0 & 2 & 1 \end{bmatrix}.$$

The zeros in the shaded portion of the last matrix indicate that it is impossible to transform $[A \mid I]$ into $[I \mid B]$ by row operations, and A^{-1} does not exist. ❑

Inverses of matrices can be used in solving certain types of systems of n linear equations in n unknowns. The use of inverses depends on the fact that any system of linear equations can be represented by a single matrix equation. For example, the system

$$ax + by = e \qquad\qquad (1)$$
$$cx + dy = f$$

is equivalent to the single matrix equation

$$\begin{bmatrix} ax + by \\ cx + dy \end{bmatrix} = \begin{bmatrix} e \\ f \end{bmatrix},$$

and this equation can be written in factored form as

$$\begin{bmatrix} a & b \\ c & d \end{bmatrix} \begin{bmatrix} x \\ y \end{bmatrix} = \begin{bmatrix} e \\ f \end{bmatrix}.$$

Thus the system of linear equations given in Eq. (1) is equivalent to the matrix equation $AX = B$, where

$$A = \begin{bmatrix} a & b \\ c & d \end{bmatrix}, \quad X = \begin{bmatrix} x \\ y \end{bmatrix}, \quad B = \begin{bmatrix} e \\ f \end{bmatrix}.$$

The matrix X is called the **unknown matrix.** As in Section 8.3, the matrix A is the **coefficient matrix,** and B is the **constant matrix.**

To find the values of x and y that satisfy the system of Eq. (1), we need only solve for X in the matrix equation $AX = B$. This can always be done when A^{-1} exists.

$$AX = B$$
$$A^{-1}(AX) = A^{-1}B \qquad \text{Multiplying by } A^{-1}$$
$$(A^{-1}A)X = A^{-1}B \qquad \text{Associative Property}$$
$$IX = A^{-1}B \qquad \text{Since } A^{-1}A = I$$
$$X = A^{-1}B \qquad \text{Since } IX = X$$

To check that $X = A^{-1}B$ does indeed satisfy the equation, we can substitute $X = A^{-1}B$ into the left member of $AX = B$.

$$AX = A(A^{-1}B) \qquad \text{Substitution}$$
$$= (AA^{-1})B \qquad \text{Associative Property}$$
$$= IB \qquad \text{Since } AA^{-1} = I$$
$$= B \qquad \text{Since } IB = B$$

These results extend easily to larger systems and are recorded in the following theorem.

Theorem 8.7

If A^{-1} exists, then $X = A^{-1}B$ satisfies the equation $AX = B$, and this is the only value of X that satisfies the equation.

Example 3 Use the inverse of the coefficient matrix to solve

$$x \qquad + 2z = 1,$$
$$-y + 3z = -3,$$
$$2x + y + 3z = 3.$$

Solution

The matrices involved in the matrix form are given by

$$A = \begin{bmatrix} 1 & 0 & 2 \\ 0 & -1 & 3 \\ 2 & 1 & 3 \end{bmatrix}, \quad X = \begin{bmatrix} x \\ y \\ z \end{bmatrix}, \quad B = \begin{bmatrix} 1 \\ -3 \\ 3 \end{bmatrix}.$$

The inverse of the coefficient matrix was found to be

$$A^{-1} = \begin{bmatrix} 3 & -1 & -1 \\ -3 & \frac{1}{2} & \frac{3}{2} \\ -1 & \frac{1}{2} & \frac{1}{2} \end{bmatrix}$$

in Example 2(a). Thus

$$X = A^{-1}B = \begin{bmatrix} 3 & -1 & -1 \\ -3 & \frac{1}{2} & \frac{3}{2} \\ -1 & \frac{1}{2} & \frac{1}{2} \end{bmatrix} \begin{bmatrix} 1 \\ -3 \\ 3 \end{bmatrix} = \begin{bmatrix} 3 \\ 0 \\ -1 \end{bmatrix},$$

and the solution to the system is

$$x = 3, \quad y = 0, \quad z = -1.$$

Exercises 8.4

Find the inverse of each matrix, if it exists. (See Examples 1 and 2.)

1. $\begin{bmatrix} 0 & 3 \\ -2 & 4 \end{bmatrix}$

2. $\begin{bmatrix} 2 & 1 \\ 1 & 1 \end{bmatrix}$

3. $\begin{bmatrix} -1 & 3 \\ 2 & 2 \end{bmatrix}$

4. $\begin{bmatrix} 4 & -2 \\ 1 & -1 \end{bmatrix}$

5. $\begin{bmatrix} -3 & 5 \\ 12 & -20 \end{bmatrix}$

6. $\begin{bmatrix} 5 & -15 \\ -1 & 3 \end{bmatrix}$

7. $\begin{bmatrix} -2 & 3 \\ 5 & 7 \end{bmatrix}$

8. $\begin{bmatrix} -4 & 6 \\ 2 & 1 \end{bmatrix}$

9. $\begin{bmatrix} 0 & 1 & 0 \\ 1 & 4 & 1 \\ 0 & 3 & -1 \end{bmatrix}$

10. $\begin{bmatrix} 1 & 3 & -2 \\ 2 & 5 & -7 \\ 1 & 4 & 0 \end{bmatrix}$

11. $\begin{bmatrix} 1 & -4 & 2 \\ 2 & -9 & 5 \\ 1 & -5 & 4 \end{bmatrix}$

12. $\begin{bmatrix} 2 & -3 & 3 \\ -3 & 1 & 0 \\ 1 & -1 & 1 \end{bmatrix}$

13. $\begin{bmatrix} 1 & 0 & 1 \\ -1 & 2 & 1 \\ 0 & 1 & 3 \end{bmatrix}$ **14.** $\begin{bmatrix} 0 & -1 & -2 \\ 2 & 4 & 8 \\ -1 & 1 & 0 \end{bmatrix}$

15. $\begin{bmatrix} 5 & 3 & -2 \\ -1 & 2 & 5 \\ 11 & 4 & -9 \end{bmatrix}$ **16.** $\begin{bmatrix} -5 & -3 & 1 \\ 6 & 3 & 0 \\ 1 & 2 & -3 \end{bmatrix}$

17. $\begin{bmatrix} 1 & 0 & -1 & 2 \\ 3 & -1 & -1 & 6 \\ 2 & 0 & -3 & 8 \\ 1 & 2 & -2 & -9 \end{bmatrix}$ **18.** $\begin{bmatrix} 1 & -2 & 1 & 0 \\ 0 & 1 & 0 & -2 \\ 1 & -1 & 0 & 1 \\ 2 & -4 & 1 & 5 \end{bmatrix}$

Solve each system by using the inverse of the coefficient matrix. (See Example 3.)

19. $\begin{aligned} 2x + 3y &= 1 \\ 2x + y &= 7 \end{aligned}$ **20.** $\begin{aligned} 8a - 5b &= 2 \\ -3a + 2b &= 1 \end{aligned}$

21. $\begin{aligned} x_1 - 3x_2 &= -6 \\ 6x_1 - 3x_2 &= 9 \end{aligned}$ **22.** $\begin{aligned} x + y &= 1 \\ x + 2y &= -3 \end{aligned}$

23. $\begin{aligned} 3x + 2y &= 22 \\ x + 2y &= 10 \end{aligned}$ **24.** $\begin{aligned} 2x + 8y &= 3 \\ 3x - 2y &= 1 \end{aligned}$

25. $\begin{aligned} x + y &= 1 \\ 7x + 3y &= 0 \end{aligned}$ **26.** $\begin{aligned} 10x + 5y &= 3 \\ 4x - y &= 0 \end{aligned}$

27. $\begin{aligned} x_1 + 5x_2 &= -4 \\ 2x_1 - 2x_3 &= 0 \\ 4x_2 - x_3 &= 4 \end{aligned}$ **28.** $\begin{aligned} x + 3y - 2z &= 5 \\ x + 2y - 5z &= -2 \\ x + 4y &= 5 \end{aligned}$

29. $\begin{aligned} x - y + z &= -2 \\ y - 3z &= 10 \\ 3x - 3y + 2z &= -3 \end{aligned}$ **30.** $\begin{aligned} a + b + 3c &= -3 \\ 2b + c &= -6 \\ a - b &= -1 \end{aligned}$

31. $\begin{aligned} -y + 3z &= 0 \\ -x + 5z &= 0 \\ x - y - z &= 1 \end{aligned}$

32. $\begin{aligned} x - 3z &= -2 \\ 2y - 3z &= -16 \\ -x - 2y + 7z &= 20 \end{aligned}$

33. $\begin{aligned} x + 2y + z &= 4 \\ y + z &= 0 \\ x + y + z &= 3 \end{aligned}$

34. $\begin{aligned} x + 2y + z &= 0 \\ x + y - 2z &= -1 \\ 2x + 4y + 3z &= \tfrac{1}{2} \end{aligned}$

35. $\begin{aligned} x + y + 2z - w &= 0 \\ -2x - y - 2z + 2w &= -1 \\ 4x - 2y + z &= -4 \\ y + z - w &= 1 \end{aligned}$

36. $\begin{aligned} a - 2b + c &= -1 \\ b - 2d &= 0 \\ b - c + d &= -2 \\ 2a - 4b + c + 5d &= -4 \end{aligned}$

37. Find D^{-1} if $D = \begin{bmatrix} a & 0 & 0 \\ 0 & b & 0 \\ 0 & 0 & c \end{bmatrix}$,

where a, b, and c are nonzero real numbers.

38. Let A and B be $n \times n$ matrices such that A^{-1} and B^{-1} exist. Prove that $(AB)^{-1}$ exists and that $(AB)^{-1} = B^{-1}A^{-1}$.

39. Let A, B, and C be $n \times n$ matrices such that A^{-1}, B^{-1}, and C^{-1} exist. Prove that $(ABC)^{-1}$ exists and that $(ABC)^{-1} = C^{-1}B^{-1}A^{-1}$.

40. Let $A = \begin{bmatrix} a & b \\ c & d \end{bmatrix}$.

Assume that $\delta(A) = ad - bc \neq 0$, and prove that

$$A^{-1} = \frac{1}{\delta(A)} \begin{bmatrix} d & -b \\ -c & a \end{bmatrix}.$$

41. Let $A = \begin{bmatrix} a & b \\ c & d \end{bmatrix}$.

Prove that if $ad - bc = 0$, then A does not have an inverse.

42. Assuming that A^{-1} exists, solve for X in the matrix equation $XA = B$. (X and B are not column matrices here.)

43. Given that A^{-1} and C^{-1} exist, solve for X in the matrix equation $AXC = B$.

8.5   Partial Fractions

In Section 1.2 we studied the procedure for combining (adding or subtracting) rational expressions (fractions). Often in the calculus it is necessary to perform the reverse process, that is, separating a fraction into a sum or difference of simpler, or **partial, fractions.** We call this type of separation the **partial fraction decomposition.** Suppose that $P(x)/Q(x)$ is a rational expression, reduced to lowest terms, where the degree of $P(x)$ is less than the degree of $Q(x)$.[1] We first factor $Q(x)$ into its linear and irreducible quadratic factors over the real numbers.

Linear Factors of $Q(x)$

For each factor of $Q(x)$ of the form $(ax + b)^n$, where $n \geq 1$, in the partial fraction decomposition, there corresponds a sum of n partial fractions of the form

$$\frac{A_1}{ax + b} + \frac{A_2}{(ax + b)^2} + \frac{A_3}{(ax + b)^3} + \cdots + \frac{A_n}{(ax + b)^n},$$

where $A_1, A_2, \ldots, A_n$ are constants.

Example 1   Resolve $\dfrac{10x - 2}{x^3 - x}$ into partial fractions.

Solution
The irreducible factors of the denominator $Q(x)$ are $x, x - 1$, and $x + 1$, since $Q(x) = x^3 - x = x(x - 1)(x + 1)$. Thus we can write

$$\frac{10x - 2}{x^3 - x} = \frac{A}{x} + \frac{B}{x - 1} + \frac{C}{x + 1}. \tag{1}$$

We must determine constants A, B, and C so the sum of the three fractions on the right will be the fraction on the left. Multiplying both sides of Eq. (1) by the least common denominator, $x(x - 1)(x + 1)$, yields

$$10x - 2 \equiv A(x - 1)(x + 1) + B(x)(x + 1) + C(x)(x - 1). \tag{2}$$

The notation $\equiv$ (read "identically equals") means that the equation is true for all values of x. There are two methods for determining the constants A, B, and C.

Method 1
Since Eq. (2) is an identity, it must be true for all values of x. Each time a value is assigned to x, there results an equation in A, B, and C. The

---

1. If the degree of $P(x)$ is not less than the degree of $Q(x)$, we divide $P(x)$ by $Q(x)$ and write

$$\frac{P(x)}{Q(x)} = D(x) + \frac{R(x)}{Q(x)},$$

where the degree of $R(x)$ is less than the degree of $Q(x)$. Then we consider the rational expression $R(x)/Q(x)$.

best choices for values of x are those that make one of the linear factors zero, because these give simple equations in A, B, and C. In Eq. (2) the best choices are $x = 0$, $x = 1$, and $x = -1$.

If $x = 0$, then $-2 = A(-1)(1)$, and $A = 2$.

If $x = 1$, then $8 = B(1)(2)$, and $B = 4$.

If $x = -1$, then $-12 = C(-1)(-2)$, and $C = -6$.

Therefore the partial fraction decomposition is

$$\frac{10x - 2}{x^3 - x} = \frac{2}{x} + \frac{4}{x - 1} - \frac{6}{x + 1}.$$

This may be checked by adding the three fractions on the right.

Method 2

Combining like terms in Eq. (2) yields

$$10x - 2 \equiv (A + B + C)x^2 + (B - C)x + (-A).$$

Since this equation is an identity, the coefficients of like powers of x on each side of the equation must be equal. Equating coefficients of like powers of x yields a system of three linear equations.

$$A + B + C = 0$$
$$B - C = 10$$
$$-A = -2$$

The solution to this system is $A = 2$, $B = 4$, $C = -6$, and this gives the partial fraction decomposition that was obtained by Method 1. ❏

Example 2   Use Method 1 to decompose $\dfrac{x^3 + 1}{x^2(x - 1)^2}$ into partial fractions.

Solution

Since both linear factors of the denominator are repeated, the partial fraction decomposition will have the form

$$\frac{x^3 + 1}{x^2(x - 1)^2} = \frac{A}{x} + \frac{B}{x^2} + \frac{C}{x - 1} + \frac{D}{(x - 1)^2}.$$

Mutiplying by $x^2(x - 1)^2$ gives

$$x^3 + 1 \equiv Ax(x - 1)^2 + B(x - 1)^2 + Cx^2(x - 1) + Dx^2. \qquad (3)$$

Since Eq. (3) is true for all values of x, we can assign values to x and obtain equations in the unknown constants.

If $x = 0$, then $1 = B(-1)^2$, and $B = 1$.

If $x = 1$, then $2 = D(1)^2$, and $D = 2$.

If $x = 2$, then $9 = A(2)(1)^2 + B(1)^2 + C(2)^2(1) + D(2)^2$,
and $0 = A + 2C$.

If $x = -1$, then $0 = A(-1)(-2)^2 + B(-2)^2$
$+ C(-1)^2(-2) + D(-1)^2$, and $3 = 2A + C$.

Solving the system

$$A + 2C = 0,$$

$$2A + C = 3,$$

we obtain $A = 2$ and $C = -1$. Therefore the partial fraction decomposition is

$$\frac{x^3 + 1}{x^2(x - 1)^2} = \frac{2}{x} + \frac{1}{x^2} - \frac{1}{x - 1} + \frac{2}{(x - 1)^2}.$$ ❏

Quadratic Factors of $Q(x)$

For each factor of $Q(x)$ of the form $(ax^2 + bx + c)^n$, where $n \geq 1$ and $ax^2 + bx + c$ is irreducible, there corresponds a sum of n partial fractions of the form

$$\frac{A_1 x + B_1}{ax^2 + bx + c} + \frac{A_2 x + B_2}{(ax^2 + bx + c)^2} + \cdots + \frac{A_n x + B_n}{(ax^2 + bx + c)^n},$$

where all A_i and B_i are constants.

Example 3   Write $\dfrac{x^4 + x^3 + 3x^2 + 2x + 1}{x^3 + x}$ as a sum of partial fractions.

Solution
Since the degree of the numerator is greater than the degree of the denominator, we divide to obtain

$$\frac{x^4 + x^3 + 3x^2 + 2x + 1}{x^3 + x} = 1 + x + \frac{2x^2 + x + 1}{x^3 + x}$$

and consider the partial fraction decomposition of

$$\frac{2x^2 + x + 1}{x^3 + x}.$$

The irreducible factors of the denominator are x and $x^2 + 1$. Thus the partial fraction decomposition is of the form

$$\frac{2x^2 + x + 1}{x^3 + x} = \frac{A}{x} + \frac{Bx + C}{x^2 + 1}.$$

Multiplying by $x(x^2 + 1)$ gives the identity

$$2x^2 + x + 1 \equiv A(x^2 + 1) + (Bx + C)x. \qquad (4)$$

It is possible to use either method to determine the constants, A, B, and C, but we shall use Method 2. Collecting like terms in Eq. (4) yields

$$2x^2 + x + 1 \equiv (A + B)x^2 + Cx + A.$$

Equating coefficients, we have the system

$$A + B \quad = 2,$$
$$C = 1,$$
$$A \qquad = 1,$$

and the solution is easily seen to be $A = 1$, $B = 1$, $C = 1$. Thus

$$\frac{x^4 + x^3 + 3x^2 + 2x + 1}{x^3 + x} = 1 + x + \frac{1}{x} + \frac{x + 1}{x^2 + 1}. \qquad \square$$

Example 4 Resolve $\dfrac{5x^2 - 2x}{(x^2 + 1)^2}$ into partial fractions.

Solution

Since the irreducible quadratic factor is repeated, the partial fraction decomposition has the form

$$\frac{5x^2 - 2x}{(x^2 + 1)^2} = \frac{Ax + B}{x^2 + 1} + \frac{Cx + D}{(x^2 + 1)^2}.$$

Multiplying by $(x^2 + 1)^2$ yields

$$5x^2 - 2x \equiv (Ax + B)(x^2 + 1) + Cx + D.$$

Using Method 2 to determine the constants A, B, C, and D, we collect like terms and obtain

$$5x^2 - 2x \equiv Ax^3 + Bx^2 + (A + C)x + (B + D).$$

Equating coefficients of like powers of x yields the following system:

$$A \qquad\qquad = 0$$
$$B \qquad\qquad = 5$$
$$A \quad + C \qquad = -2$$
$$B \qquad + D = 0$$

The solution to this system is

$$A = 0, \quad B = 5, \quad C = -2, \quad D = -5,$$

and the partial fraction decomposition is

$$\frac{5x^2 - 2x}{(x^2 + 1)^2} = \frac{5}{x^2 + 1} - \frac{2x + 5}{(x^2 + 1)^2}.$$ ❑

Three important points are worth remembering in decomposing rational expressions into partial fractions.

1. The number of constants to be found is always equal to the degree of the denominator of the original fraction.
2. A decomposition can be checked by adding the partial fractions to see if the sum is equal to the original fraction.
3. The decomposition of a given rational expression into partial fractions is unique.

Exercises 8.5

Write each of the following as a sum of partial fractions.

1. $\dfrac{3x + 2}{x^2 - 5x + 6}$

2. $\dfrac{5x - 1}{x^2 - 1}$

3. $\dfrac{3x - 2}{x^2 + 2x}$

4. $\dfrac{2x}{x^2 - 4}$

5. $\dfrac{3x - 4}{x^3 - 3x^2 + 2x}$

6. $\dfrac{3x^2 - 4x - 5}{2x^3 - x^2 - x}$

7. $\dfrac{x - 1}{x^3 - x^2 - 2x}$

8. $\dfrac{1 - 4x^2}{(x - 2)(x - 1)(x + 3)}$

9. $\dfrac{4x^2 + 10x + 8}{(3x + 2)(x + 1)(x + 2)}$

10. $\dfrac{1 + 3x - 8x^2}{(1 - 4x)(1 - x^2)}$

11. $\dfrac{x - 3}{(x - 2)^2}$

12. $\dfrac{x}{(x + 3)^2}$

13. $\dfrac{x^2 + 1}{(x - 1)^3}$

14. $\dfrac{x^2 + x + 1}{(x + 2)^4}$

15. $\dfrac{x^2 + 1}{(x + 1)(x - 1)(x + 1)}$

16. $\dfrac{3x^2 - x + 6}{(x + 2)(1 - 2x)(x + 2)}$

17. $\dfrac{1}{x^4 - 4x^2}$

18. $\dfrac{x - 1}{x^4 + x^3}$

19. $\dfrac{3x - 2}{x^2(x - 1)^2}$

20. $\dfrac{12x^3 - 13x^2 + 7x - 1}{x^2(1 - 2x)^2}$

21. $\dfrac{8x^2 + 5x + 3}{(x - 2)^2(3x - 1)}$

22. $\dfrac{3x^2 + 1}{x(x - 1)^3}$

23. $\dfrac{-4}{(x - 1)(x^2 + x + 2)}$

24. $\dfrac{3x - 2}{x^3 + 2x^2 + 2x}$

25. $\dfrac{5x^2 - 1}{(x - 3)(x^2 + x - 1)}$

26. $\dfrac{3x^3 - 7x^2 + 7x - 1}{(x - 1)^2(x^2 + 1)}$

27. $\dfrac{5x^3 + 3x^2 - 3x + 3}{(x^2 - 1)(x^2 + 1)}$

28. $\dfrac{4x^3 - 17x^2 - 60}{(x^2 - 4)(x^2 + 4)}$

29. $\dfrac{x^3 - 1}{(1 + x^2)^2}$

30. $\dfrac{3x^2 + 1}{(1 + x + x^2)^2}$

31. $\dfrac{4x^3 + x}{(2x^2 - x + 1)^2}$ **32.** $\dfrac{6x^2 + 4x - 5}{(3x^2 + x - 5)^2}$ **35.** $\dfrac{2x^4 - 5x^3 + x^2 + 6x - 3}{x(x - 1)^2}$

33. $\dfrac{x^3 - 2x^2 + 5x}{x^2 - 2x - 3}$ **36.** $\dfrac{x^5 - 2x^4 + 6x^2 - 12x + 10}{(x - 2)(x^2 + 1)}$

34. $\dfrac{10 + 5x - 17x^2 - 6x^3}{1 + x - 6x^2}$

8.6   The Definition of a Determinant

In Theorem 8.6, $\delta(A)$ denoted the number $ad - bc$, which is associated with the 2×2 matrix

$$A = \begin{bmatrix} a & b \\ c & d \end{bmatrix}.$$

This number, $\delta(A)$, is called the determinant of the 2×2 matrix A. Every square matrix A, of any order, has such a number associated with it. This number is called the **determinant** of A. The **order** of the determinant of A is the same as the order of the matrix A. The determinant of A is denoted by $\delta(A)$, or by $|A|$, or by simply replacing the brackets around the elements of A by straight lines. Thus if

$$A = \begin{bmatrix} a & b \\ c & d \end{bmatrix},$$

then

$$\delta(A) = |A| = \begin{vmatrix} a & b \\ c & d \end{vmatrix} = ad - bc.$$

The material in this section leads to a method for evaluating the determinant of any square matrix. It is important to note that the determinant of a matrix is a number.

Example 1   Evaluate the determinant of each of the following matrices.

a) $A = \begin{bmatrix} -1 & 7 \\ 2 & -3 \end{bmatrix}$ b) $B = \begin{bmatrix} 1 & 0 \\ 0 & 1 \end{bmatrix}$ c) $C = \begin{bmatrix} 2 & 0 \\ -1 & 0 \end{bmatrix}$

Solution

a) $\delta(A) = (-1)(-3) - (7)(2) = -11,$ or $\begin{vmatrix} -1 & 7 \\ 2 & -3 \end{vmatrix} = -11$

b) $\delta(B) = |B| = (1)(1) - (0)(0) = 1,$ or $\begin{vmatrix} 1 & 0 \\ 0 & 1 \end{vmatrix} = 1$

c) $\delta(C) = |C| = (2)(0) - (0)(-1) = 0,$ or $\begin{vmatrix} 2 & 0 \\ -1 & 0 \end{vmatrix} = 0$ ❑

The determinant of a square matrix of order 3 is given in the next definition.

Definition 8.8

For a given 3×3 matrix

$$A = \begin{bmatrix} a_{11} & a_{12} & a_{13} \\ a_{21} & a_{22} & a_{23} \\ a_{31} & a_{32} & a_{33} \end{bmatrix},$$

the **determinant** of A is the number

$$|A| = a_{11}a_{22}a_{33} + a_{12}a_{23}a_{31} + a_{13}a_{21}a_{32}$$
$$- a_{11}a_{23}a_{32} - a_{12}a_{21}a_{33} - a_{13}a_{22}a_{31}.$$

Example 2   Evaluate $|A|$ for $A = \begin{bmatrix} 3 & 2 & 1 \\ -4 & -1 & 5 \\ -3 & -4 & -2 \end{bmatrix}.$

Solution
Using Definition 8.8, we have

$$|A| = (3)(-1)(-2) + (2)(5)(-3) + (1)(-4)(-4)$$
$$- (3)(5)(-4) - (2)(-4)(-2) - (1)(-1)(-3)$$
$$= 6 - 30 + 16 + 60 - 16 - 3 = 33.$$ ❑

Similar and even more unwieldy definitions can be made for determinants of higher-order matrices. Although we could use these definitions to evaluate a determinant, they are very inefficient. A much more efficient way to proceed is to evaluate determinants by a method called the *cofactor expansion*. The cofactor expansion applies to determinants of any order. In order to describe the method, some preliminary definitions are needed.

Definition 8.9

Let A be a square matrix of order $n \geq 2$, and let a_{ij} be the element in row i and column j of A. Then the **minor**, denoted by M_{ij}, of the element a_{ij} is the determinant of the matrix formed by deleting row i and column j from the matrix A. The **cofactor**, denoted by C_{ij}, of the element a_{ij} is the product of $(-1)^{i+j}$ and the minor M_{ij}. That is,

$$C_{ij} = (-1)^{i+j}M_{ij}.$$

Note that if $i + j$ is even, then $C_{ij} = M_{ij}$, and if $i + j$ is odd, then $C_{ij} = -M_{ij}$.

Example 3   Consider the 3×3 matrix

$$A = \begin{bmatrix} 6 & -3 & 2 \\ 0 & 1 & -2 \\ 2 & 4 & 5 \end{bmatrix}.$$

a) The element in row 2 and column 1 is 0; that is, $a_{21} = 0$.

b) The minor of a_{21} is $\begin{vmatrix} -3 & 2 \\ 4 & 5 \end{vmatrix}$; hence $M_{21} = -23$.

c) The cofactor of a_{21} is $(-1)^{2+1}\begin{vmatrix} -3 & 2 \\ 4 & 5 \end{vmatrix}$; hence $C_{21} = 23$.   ❏

With the definitions of minor and cofactor in mind, we consider again the definition of a determinant of order 3 (Definition 8.8). In the equation

$$|A| = a_{11}a_{22}a_{33} + a_{12}a_{23}a_{31} + a_{13}a_{21}a_{32}$$
$$- a_{11}a_{23}a_{32} - a_{12}a_{21}a_{33} - a_{13}a_{22}a_{31},$$

there are two terms involving a_{11}, two terms involving a_{12}, and two terms involving a_{13}. Regrouping the terms and factoring, we have

$$|A| = a_{11}(a_{22}a_{33} - a_{23}a_{32}) - a_{12}(a_{21}a_{33} - a_{23}a_{31}) + a_{13}(a_{21}a_{32} - a_{22}a_{31}).$$

Since each expression in parentheses can be expressed as a second-order determinant, we can write

$$|A| = a_{11}\begin{vmatrix} a_{22} & a_{23} \\ a_{32} & a_{33} \end{vmatrix} - a_{12}\begin{vmatrix} a_{21} & a_{23} \\ a_{31} & a_{33} \end{vmatrix} + a_{13}\begin{vmatrix} a_{21} & a_{22} \\ a_{31} & a_{32} \end{vmatrix}.$$

Expressing each second-order determinant as a minor yields

$$|A| = a_{11}M_{11} - a_{12}M_{12} + a_{13}M_{13}$$

or

$$|A| = a_{11}C_{11} + a_{12}C_{12} + a_{13}C_{13}.$$

The last expression for $|A|$ is called the **cofactor expansion** of the determinant of A about the first row. The cofactor expansion of $|A|$ about the first row is a sum of terms formed by multiplying each element in the first row by its cofactor.

The terms in the equation defining $|A|$ in Definition 8.8 could also have been regrouped and factored in such a way as to have a cofactor expansion of $|A|$ about any certain row or any certain column. In general, the **cofactor expansion** of $|A|$ about row i ($i = 1, 2,$ or 3) is

$$|A| = a_{i1}C_{i1} + a_{i2}C_{i2} + a_{i3}C_{i3},$$

and the cofactor expansion of $|A|$ about column j ($j = 1, 2,$ or 3) is

$$|A| = a_{1j}C_{1j} + a_{2j}C_{2j} + a_{3j}C_{3j}.$$

Example 4   Evaluate $|A|$ by expanding about the second column.

$$|A| = \begin{vmatrix} 3 & 2 & 1 \\ -4 & -1 & 5 \\ -3 & -4 & -2 \end{vmatrix}$$

Solution
Using

$$|A| = a_{12}C_{12} + a_{22}C_{22} + a_{32}C_{32},$$

we have

$$|A| = (2)(-1)^3 \begin{vmatrix} -4 & 5 \\ -3 & -2 \end{vmatrix} + (-1)(-1)^4 \begin{vmatrix} 3 & 1 \\ -3 & -2 \end{vmatrix} + (-4)(-1)^5 \begin{vmatrix} 3 & 1 \\ -4 & 5 \end{vmatrix}$$

$$= (-2)[8 - (-15)] + (-1)[-6 - (-3)] + (4)[15 - (-4)]$$

$$= (-2)(23) + (-1)(-3) + (4)(19)$$

$$= 33.$$

This agrees with the value obtained in Example 2 for the same determinant.   ❑

Determinants of order $n > 3$ can also be evaluated by the method of cofactor expansion. The cofactor expansion of an nth-order determinant is given in the following theorem.

Theorem 8.10

Let $A = [a_{ij}]_{(n,n)}$ be a square matrix of order $n \geq 2$, and let C_{ij} be the cofactor of element a_{ij}, for $i = 1, 2, \ldots, n$ and $j = 1, 2, \ldots, n$. Then the cofactor expansion of $|A|$ about row i is

$$|A| = a_{i1}C_{i1} + a_{i2}C_{i2} + \cdots + a_{in}C_{in},$$

and the cofactor expansion of $|A|$ about column j is

$$|A| = a_{1j}C_{1j} + a_{2j}C_{2j} + \cdots + a_{nj}C_{nj}.$$

Theorem 8.10 is proved in the study of linear algebra. That is, it is shown that $|A|$ does not depend on the choice of the row number i or on the choice of the column number j. This proof is very complicated and does not properly belong in this text.

Example 5   Evaluate the following determinant.

$$|A| = \begin{vmatrix} -2 & 1 & 0 & 1 \\ 0 & -2 & 1 & 4 \\ -3 & 1 & 0 & -2 \\ 1 & 0 & -2 & 1 \end{vmatrix}$$

Solution

We expand $|A|$ about the third column since it contains more zeros than any other column or row, and consequently this cofactor expansion will have the fewest nonzero terms. We obtain

$$|A| = a_{13}C_{13} + a_{23}C_{23} + a_{33}C_{33} + a_{43}C_{43}$$

$$= 0 \cdot C_{13} + (1)(-1)^5 \begin{vmatrix} -2 & 1 & 1 \\ -3 & 1 & -2 \\ 1 & 0 & 1 \end{vmatrix} + 0 \cdot C_{33}$$

$$+ (-2)(-1)^7 \begin{vmatrix} -2 & 1 & 1 \\ 0 & -2 & 4 \\ -3 & 1 & -2 \end{vmatrix}$$

$$= - \begin{vmatrix} -2 & 1 & 1 \\ -3 & 1 & -2 \\ 1 & 0 & 1 \end{vmatrix} + 2 \begin{vmatrix} -2 & 1 & 1 \\ 0 & -2 & 4 \\ -3 & 1 & -2 \end{vmatrix}.$$

Expanding the first third-order determinant about its third row and the second third-order determinant about its first column, we have

$$|A| = -\left[(1)(-1)^4\begin{vmatrix} 1 & 1 \\ 1 & -2 \end{vmatrix} + 0 + (1)(-1)^6\begin{vmatrix} -2 & 1 \\ -3 & 1 \end{vmatrix}\right]$$

$$+ 2\left[(-2)(-1)^2\begin{vmatrix} -2 & 4 \\ 1 & -2 \end{vmatrix} + 0 + (-3)(-1)^4\begin{vmatrix} 1 & 1 \\ -2 & 4 \end{vmatrix}\right]$$

$$= -[(1)(-2 - 1) + 0 + (1)(-2 - (-3))]$$
$$+ 2[(-2)(4 - 4) + 0 + (-3)(4 - (-2))]$$

$$= -[-3 + 0 + 1] + 2[0 + 0 - 18]$$
$$= 2 - 36$$
$$= -34. \qquad \square$$

Theorem 8.10 leads to the following theorem.

Theorem 8.11

If every element in one row (or one column) of a square matrix A is zero, then $|A| = 0$.

Exercises 8.6

Evaluate the following determinants. Letters represent real numbers. (See Example 1.)

1. $\begin{vmatrix} -3 & 2 \\ 1 & 1 \end{vmatrix}$ **2.** $\begin{vmatrix} 0 & -3 \\ -1 & 3 \end{vmatrix}$

3. $\begin{vmatrix} 5 & 8 \\ 2 & 3 \end{vmatrix}$ **4.** $\begin{vmatrix} 4 & -3 \\ 3 & -2 \end{vmatrix}$

5. $\begin{vmatrix} 7 & 0 \\ 0 & -3 \end{vmatrix}$ **6.** $\begin{vmatrix} 0 & 0 \\ -1 & 1 \end{vmatrix}$

7. $\begin{vmatrix} -1 & -1 \\ 1 & 1 \end{vmatrix}$ **8.** $\begin{vmatrix} 2 & 9 \\ -4 & 10 \end{vmatrix}$

9. $\begin{vmatrix} -\frac{1}{2} & \frac{1}{3} \\ \frac{1}{6} & \frac{1}{9} \end{vmatrix}$ **10.** $\begin{vmatrix} \frac{1}{4} & -1 \\ 0 & -4 \end{vmatrix}$

11. $\begin{vmatrix} x & -x \\ 2 & 3 \end{vmatrix}$ **12.** $\begin{vmatrix} x & y \\ 2x & 3y \end{vmatrix}$

Use a cofactor expansion to evaluate the following determinants. (See Examples 4 and 5.)

13. $\begin{vmatrix} 2 & -1 & 2 \\ 0 & 2 & 0 \\ -3 & 1 & 0 \end{vmatrix}$ **14.** $\begin{vmatrix} -2 & 6 & 1 \\ 0 & 0 & 4 \\ 1 & -4 & 0 \end{vmatrix}$

15. $\begin{vmatrix} 2 & 0 & -1 \\ -1 & 2 & 0 \\ 0 & -1 & 2 \end{vmatrix}$ **16.** $\begin{vmatrix} 2 & 0 & 1 \\ 0 & 1 & 1 \\ 4 & -1 & 1 \end{vmatrix}$

17. $\begin{vmatrix} 1 & 4 & 0 \\ -1 & 2 & 0 \\ 0 & 3 & 0 \end{vmatrix}$ **18.** $\begin{vmatrix} 1 & 0 & 1 \\ 0 & 0 & 0 \\ -1 & 2 & 1 \end{vmatrix}$

19. $\begin{vmatrix} 1 & 3 & -2 \\ 1 & 4 & 0 \\ 2 & 5 & -7 \end{vmatrix}$ **20.** $\begin{vmatrix} 1 & 0 & 3 \\ -1 & 1 & -3 \\ 1 & -3 & 2 \end{vmatrix}$

21. $\begin{vmatrix} -1 & 1 & 1 \\ 1 & -1 & 1 \\ 1 & 1 & 1 \end{vmatrix}$ **22.** $\begin{vmatrix} -2 & 3 & 3 \\ 3 & 1 & -2 \\ 1 & 1 & -1 \end{vmatrix}$

23. $\begin{vmatrix} 1 & -2 & 3 \\ 2 & 1 & 2 \\ 3 & -3 & 6 \end{vmatrix}$
    24. $\begin{vmatrix} 2 & 1 & 1 \\ 9 & 4 & 3 \\ 6 & 3 & 4 \end{vmatrix}$

25. $\begin{vmatrix} 4 & -5 & 1 \\ 5 & -9 & 2 \\ 2 & -4 & 1 \end{vmatrix}$
    26. $\begin{vmatrix} 4 & -3 & -8 \\ 1 & 3 & -2 \\ 2 & 1 & -3 \end{vmatrix}$

27. $\begin{vmatrix} 2 & 4 & 3 \\ 5 & -9 & -2 \\ -1 & 11 & 5 \end{vmatrix}$
    28. $\begin{vmatrix} -3 & -1 & 2 \\ 1 & 6 & -3 \\ -5 & 4 & 1 \end{vmatrix}$

29. $\begin{vmatrix} 1 & 0 & 1 & 2 \\ -2 & 1 & -1 & -4 \\ 1 & 0 & 0 & 1 \\ 0 & -2 & 1 & 5 \end{vmatrix}$

30. $\begin{vmatrix} 6 & -1 & -1 & 3 \\ 8 & 0 & -3 & 2 \\ -9 & 2 & -2 & 1 \\ 2 & 0 & -1 & 1 \end{vmatrix}$

31. $\begin{vmatrix} 0 & 4 & -2 & 1 \\ -1 & -2 & -1 & 1 \\ -2 & -1 & -3 & 3 \\ 1 & 0 & 2 & -1 \end{vmatrix}$

32. $\begin{vmatrix} 1 & 3 & -1 & 1 \\ 2 & 9 & -2 & 3 \\ 2 & 10 & -3 & 2 \\ 3 & -8 & 0 & 1 \end{vmatrix}$

Solve for x in each of the following equations.

33. $\begin{vmatrix} 3 & x \\ -2 & 1 \end{vmatrix} = 1$
    34. $\begin{vmatrix} -x & 3 \\ -4 & -1 \end{vmatrix} = -3$

35. $\begin{vmatrix} 2 & 0 & 1 \\ 4 & x & 5 \\ 1 & 4 & -3 \end{vmatrix} = -10$
    36. $\begin{vmatrix} -2 & 0 & 1 \\ 1 & -3 & -2 \\ x & 1 & 1 \end{vmatrix} = 0$

The values of x that satisfy the equation $|A - xI| = 0$ are called the **eigenvalues,** or the **characteristic values,** of the matrix A. Find the eigenvalues of the following matrices.

37. $\begin{bmatrix} 2 & 0 \\ 0 & -3 \end{bmatrix}$
    38. $\begin{bmatrix} 5 & 6 \\ 4 & 0 \end{bmatrix}$

39. $\begin{bmatrix} 1 & -2 \\ 3 & -4 \end{bmatrix}$
    40. $\begin{bmatrix} 1 & 3 \\ 9 & 7 \end{bmatrix}$

41. $\begin{bmatrix} 5 & 1 & -1 \\ 0 & -3 & 2 \\ 0 & 0 & 2 \end{bmatrix}$
    42. $\begin{bmatrix} 1 & 3 & 15 \\ -2 & 0 & -2 \\ 1 & 0 & 1 \end{bmatrix}$

43. $\begin{bmatrix} 1 & -2 & 0 \\ 0 & 0 & -3 \\ 2 & -4 & 0 \end{bmatrix}$
    44. $\begin{bmatrix} 1 & -1 & 1 \\ 1 & -1 & -1 \\ -1 & 1 & -3 \end{bmatrix}$

45. Use Definition 8.8 to show that a determinant $|A|$ of order 3 can be evaluated by any one of the following equations.
   a) $|A| = a_{11}C_{11} + a_{21}C_{21} + a_{31}C_{31}$
   b) $|A| = a_{21}C_{21} + a_{22}C_{22} + a_{23}C_{23}$
   c) $|A| = a_{12}C_{12} + a_{22}C_{22} + a_{32}C_{32}$
   d) $|A| = a_{31}C_{31} + a_{32}C_{32} + a_{33}C_{33}$
   e) $|A| = a_{13}C_{13} + a_{23}C_{23} + a_{33}C_{33}$

46. Prove Theorem 8.11.

8.7   Evaluation of Determinants

In Section 8.3 we used three types of row operations on augmented matrices to solve linear systems by Gaussian elimination. These operations are listed below.

Row Operations

1. Interchange any two rows.
2. Multiply (or divide) every element in a row by the same nonzero number.
3. Add (or subtract) a multiple of one row to (from) another row.

Corresponding to each of these row operations is a similar *column operation*. These column operations are described below.

Column Operations

---

1. Interchange any two columns.
2. Multiply (or divide) every element in a column by the same non-zero number.
3. Add (or subtract) a multiple of one column to (from) another column.

When a row or column operation is performed on a square matrix A, the value of $|A|$ is sometimes (but not always) changed. The effect of performing an operation of each type is described in the next three theorems. Each theorem can be stated in terms of rows or in terms of columns. We indicate this by stating each theorem in terms of rows and then inserting the word *column* in parentheses. A dual theorem is obtained simply by replacing the word *row* by the word *column*. As with Theorem 8.10, the proofs of these theorems for the general case are more appropriate in a linear algebra course and are not presented here.

Theorem 8.12

If the matrix B is obtained from a matrix A by interchanging any two rows (columns) of A, then $|B| = -|A|$.

Example 1   Every time two rows or two columns are interchanged, the sign of the determinant changes. Thus

$$\begin{vmatrix} 1 & -2 & 3 \\ 4 & -1 & 0 \\ -1 & 1 & 2 \end{vmatrix} = - \begin{vmatrix} -1 & 1 & 2 \\ 4 & -1 & 0 \\ 1 & -2 & 3 \end{vmatrix}, \qquad \text{since rows 1 and 3 have been interchanged;}$$

$$= \begin{vmatrix} -1 & 2 & 1 \\ 4 & 0 & -1 \\ 1 & 3 & -2 \end{vmatrix}, \qquad \text{since columns 2 and 3 have been interchanged.} \qquad ❏$$

Theorem 8.13

If the matrix B is obtained from a matrix A by multiplying each element of a row (column) of A by the same number c, then $|B| = c|A|$.

Example 2   By applying Theorem 8.13 twice, we have

$$\begin{vmatrix} 12 & -4 \\ 9 & 2 \end{vmatrix} = 4 \begin{vmatrix} 3 & -1 \\ 9 & 2 \end{vmatrix} = (4)(3) \begin{vmatrix} 1 & -1 \\ 3 & 2 \end{vmatrix} = 12 \begin{vmatrix} 1 & -1 \\ 3 & 2 \end{vmatrix}.$$ ❏

Multiplication of a determinant by a number c has the same effect as multiplying one column or one row of the determinant by c. By contrast, in multiplication of a matrix by the number c, every element in the matrix is multiplied by c. This difference between multiplication of a matrix by a number and multiplication of a determinant by a number is illustrated in the next example.

Example 3   If $A = \begin{bmatrix} -7 & -4 \\ 2 & 5 \end{bmatrix}$, then $|A| = -27$. Also $3A = \begin{bmatrix} -21 & -12 \\ 6 & 15 \end{bmatrix}$, and $|3A| = -243$. Thus $|3A| \neq 3|A|$.   ❏

Theorem 8.14

If the matrix B is obtained from a matrix A by adding a multiple of one row (column) to another row (column), then $|B| = |A|$.

Example 4   Adding 2 times the first row to the third row in

$$\begin{vmatrix} 1 & 3 & -1 \\ 0 & 1 & 5 \\ -2 & 1 & 4 \end{vmatrix} \qquad \text{yields} \qquad \begin{vmatrix} 1 & 3 & -1 \\ 0 & 1 & 5 \\ 0 & 7 & 2 \end{vmatrix},$$

and these determinants are both equal to -33.   ❏

Theorem 8.14 proves to be most useful in evaluating determinants. By introducing zeros into a row (or column), the cofactor expansion can be reduced to only one term. This is illustrated in Example 5. In the example, we use a notation for row and column operations that is similar to that used in Section 8.3:

$R_i \leftrightarrow R_j$ indicates that row i and row j are interchanged.
$C_i \leftrightarrow C_j$ indicates that column i and column j are interchanged.
$cR_i + R_j$ indicates that row j is replaced by the sum of row j and c times row i.
$cC_i + C_j$ indicates that column j is replaced by the sum of column j and c times column i.

Example 5 Evaluate the following determinant by introducing zeros into column 2.

$$|A| = \begin{vmatrix} 3 & 2 & -2 \\ -1 & -1 & 4 \\ 2 & 4 & -1 \end{vmatrix}$$

Solution

We use row operations to introduce two zeros into column 2 as follows:

$$|A| = \begin{vmatrix} 3 & 2 & -2 \\ -1 & -1 & 4 \\ 2 & 4 & -1 \end{vmatrix} \overset{2R_2 + R_1}{=} \begin{vmatrix} 1 & 0 & 6 \\ -1 & -1 & 4 \\ 2 & 4 & -1 \end{vmatrix}$$

$$\overset{4R_2 + R_3}{=} \begin{vmatrix} 1 & 0 & 6 \\ -1 & -1 & 4 \\ -2 & 0 & 15 \end{vmatrix}.$$

The cofactor expansion of this determinant about the second column reduces to one term.

$$|A| = (0) \cdot C_{12} + (-1)C_{22} + (0) \cdot C_{32}$$

$$= (-1)\begin{vmatrix} 1 & 6 \\ -2 & 15 \end{vmatrix} = (-1)(15 + 12) = -27 \qquad \square$$

The use of column operations is illustrated in the next example.

Example 6 Evaluate the following determinant by introducing zeros before expanding about a row or column.

$$|A| = \begin{vmatrix} 3 & 1 & -5 & 0 \\ 2 & 1 & 4 & 1 \\ -1 & 2 & -4 & 1 \\ 0 & -3 & 1 & 5 \end{vmatrix}$$

Solution

Since there is already one zero in the first row, we choose to introduce two more zeros in this row by using column operations.

$$|A| = \begin{vmatrix} 3 & 1 & -5 & 0 \\ 2 & 1 & 4 & 1 \\ -1 & 2 & -4 & 1 \\ 0 & -3 & 1 & 5 \end{vmatrix} \overset{-3C_2 + C_1}{=} \begin{vmatrix} 0 & 1 & -5 & 0 \\ -1 & 1 & 4 & 1 \\ -7 & 2 & -4 & 1 \\ 9 & -3 & 1 & 5 \end{vmatrix}$$

$$\overset{5C_2 + C_3}{=} \begin{vmatrix} 0 & 1 & 0 & 0 \\ -1 & 1 & 9 & 1 \\ -7 & 2 & 6 & 1 \\ 9 & -3 & -14 & 5 \end{vmatrix} \overset{\text{Expanding about } R_1}{=} (1)(-1)\begin{vmatrix} -1 & 9 & 1 \\ -7 & 6 & 1 \\ 9 & -14 & 5 \end{vmatrix}$$

Next we choose to introduce zeros into the third column. (Any column or row can be used.) Using row operations, we have

$$|A| = - \begin{vmatrix} -1 & 9 & 1 \\ -7 & 6 & 1 \\ 9 & -14 & 5 \end{vmatrix} \overset{-R_1 + R_2}{=} - \begin{vmatrix} -1 & 9 & 1 \\ -6 & -3 & 0 \\ 9 & -14 & 5 \end{vmatrix}$$

$$\overset{-5R_1 + R_3}{=} - \begin{vmatrix} -1 & 9 & 1 \\ -6 & -3 & 0 \\ 14 & -59 & 0 \end{vmatrix} \overset{\substack{\text{Expanding} \\ \text{about } C_3}}{=} -(1) \begin{vmatrix} -6 & -3 \\ 14 & -59 \end{vmatrix}$$

$$= -396. \qquad \square$$

Exercises 8.7

Without evaluating the determinants, determine the value of the variable that makes each of the following statements true. (See Examples 1, 2, and 4.)

1. $\begin{vmatrix} 3 & -4 \\ 1 & 5 \end{vmatrix} = - \begin{vmatrix} 1 & 5 \\ 3 & x \end{vmatrix}$

2. $\begin{vmatrix} 11 & -2 \\ 7 & 3 \end{vmatrix} = - \begin{vmatrix} -2 & x \\ 3 & 7 \end{vmatrix}$

3. $\begin{vmatrix} 5 & -1 \\ 2 & -3 \end{vmatrix} = a \begin{vmatrix} 5 & 1 \\ 2 & 3 \end{vmatrix}$

4. $\begin{vmatrix} 3 & -2 \\ 7 & 0 \end{vmatrix} = y \begin{vmatrix} -3 & 2 \\ -7 & 0 \end{vmatrix}$

5. $\begin{vmatrix} 1 & 0 & 2 \\ -2 & 1 & 4 \\ 0 & 1 & 5 \end{vmatrix} = x \begin{vmatrix} 0 & 2 & 1 \\ 1 & 4 & -2 \\ 1 & 5 & 0 \end{vmatrix}$

6. $\begin{vmatrix} 2 & 1 & 11 \\ 3 & 1 & 4 \\ 0 & 1 & 0 \end{vmatrix} = t \begin{vmatrix} 3 & 1 & 4 \\ 0 & 1 & 0 \\ 2 & 1 & 11 \end{vmatrix}$

7. $\begin{vmatrix} -3 & 4 \\ 9 & -2 \end{vmatrix} = x \begin{vmatrix} -1 & -2 \\ 3 & 1 \end{vmatrix}$

8. $\begin{vmatrix} 1 & 1 & 3 \\ 4 & 20 & 12 \\ 5 & 25 & 30 \end{vmatrix} = y \begin{vmatrix} 1 & 1 & 1 \\ 1 & 5 & 1 \\ 1 & 5 & 2 \end{vmatrix}$

9. $\begin{vmatrix} 3 & 2 & 1 \\ 5 & -4 & -3 \\ 1 & 1 & 2 \end{vmatrix} = \begin{vmatrix} 3 & 2 & 1 \\ 14 & x & 0 \\ 1 & 1 & 2 \end{vmatrix}$

10. $\begin{vmatrix} -1 & -3 & 1 \\ 1 & -2 & 1 \\ 4 & -2 & 2 \end{vmatrix} = \begin{vmatrix} x & -5 & 3 \\ 1 & -2 & 1 \\ 4 & -2 & 2 \end{vmatrix}$

11. $\begin{vmatrix} -1 & 11 & -1 & 4 \\ 0 & 3 & 0 & 1 \\ 1 & 1 & -2 & 1 \\ 4 & 1 & 5 & -2 \end{vmatrix} = \begin{vmatrix} -1 & -1 & -1 & 4 \\ 0 & 0 & 0 & 1 \\ 1 & x & -2 & 1 \\ 4 & 7 & 5 & -2 \end{vmatrix}$

12. $\begin{vmatrix} 1 & -1 & 2 & 3 \\ -2 & 1 & 1 & 4 \\ 1 & 5 & 0 & 2 \\ 1 & 1 & -2 & 3 \end{vmatrix} = \begin{vmatrix} 1 & -1 & 2 & 3 \\ -1 & 0 & x & 7 \\ 1 & 5 & 0 & 2 \\ 1 & 1 & -2 & 3 \end{vmatrix}$

Evaluate the following determinants by introducing zeros before expanding about a row or column. (See Examples 5 and 6.)

13. $\begin{vmatrix} 2 & -1 & 3 \\ 1 & 2 & -1 \\ 0 & 2 & 1 \end{vmatrix}$

14. $\begin{vmatrix} 1 & 3 & -2 \\ 0 & 1 & 3 \\ 2 & 0 & 5 \end{vmatrix}$

15. $\begin{vmatrix} 7 & 4 & -2 \\ 1 & 2 & -1 \\ 4 & 2 & 0 \end{vmatrix}$

16. $\begin{vmatrix} 4 & -2 & 1 \\ -2 & 2 & -1 \\ 5 & 0 & 2 \end{vmatrix}$

17. $\begin{vmatrix} 1 & 3 & -1 \\ 3 & -2 & 4 \\ 2 & 1 & 3 \end{vmatrix}$

18. $\begin{vmatrix} 13 & 3 & 1 \\ 11 & 4 & -2 \\ 4 & -1 & 3 \end{vmatrix}$

19. $\begin{vmatrix} 5 & 0 & 0 & 3 \\ 2 & 4 & -3 & 1 \\ 0 & 1 & 0 & -1 \\ 1 & -1 & 2 & -1 \end{vmatrix}$

20. $\begin{vmatrix} 4 & -3 & 1 & 0 \\ 2 & 1 & -1 & 0 \\ -1 & 0 & 1 & 2 \\ 2 & 1 & 1 & 1 \end{vmatrix}$

21. $\begin{vmatrix} 5 & -2 & 2 & 3 \\ 0 & 1 & 1 & 3 \\ 1 & 0 & 1 & 1 \\ -2 & -1 & 0 & 6 \end{vmatrix}$ **22.** $\begin{vmatrix} 24 & 10 & 4 & 1 \\ 21 & 3 & 0 & -2 \\ 3 & 2 & 3 & 0 \\ 1 & 1 & 1 & -4 \end{vmatrix}$

23. $\begin{vmatrix} 2 & 2 & -1 & 3 \\ 1 & 1 & -1 & 1 \\ 1 & -1 & 1 & 1 \\ 4 & 2 & 1 & 5 \end{vmatrix}$

24. $\begin{vmatrix} 3 & -2 & -1 & 2 \\ 4 & 1 & 2 & -3 \\ -9 & -5 & 7 & -8 \\ 1 & 5 & 3 & -2 \end{vmatrix}$

25. $\begin{vmatrix} 4 & 1 & 0 & -1 & 3 \\ 0 & -1 & 1 & 0 & 1 \\ 1 & 1 & -2 & 1 & -1 \\ -2 & 0 & 0 & 1 & 1 \\ -1 & 0 & 0 & 1 & 4 \end{vmatrix}$

26. $\begin{vmatrix} 1 & -1 & 2 & 0 & 1 \\ -1 & 4 & -1 & 0 & 1 \\ 1 & 1 & -1 & 2 & 0 \\ 0 & 0 & 1 & 0 & 2 \\ 1 & 0 & 1 & 1 & -1 \end{vmatrix}$

27. Show that

$$\begin{vmatrix} 1 & 1 & 1 \\ x & y & z \\ x^2 & y^2 & z^2 \end{vmatrix} = (x - y)(y - z)(z - x).$$

28. Show that

$$\begin{vmatrix} -x & 1 & 0 & 0 \\ 0 & -x & 1 & 0 \\ 0 & 0 & -x & 1 \\ -c_0 & -c_1 & -c_2 & -c_3 - x \end{vmatrix}$$

$$= x^4 + c_3 x^3 + c_2 x^2 + c_1 x + c_0.$$

29. Show that if $A = \begin{bmatrix} a & b \\ c & d \end{bmatrix}$ and $B = \begin{bmatrix} e & f \\ g & h \end{bmatrix}$, then

$$|AB| = |A| \cdot |B|.$$

30. Show that

$$\begin{vmatrix} a & b & c \\ d & e & f \\ h & i & j \end{vmatrix} + \begin{vmatrix} a & b & c \\ k & m & n \\ h & i & j \end{vmatrix} = \begin{vmatrix} a & b & c \\ d+k & e+m & f+n \\ h & i & j \end{vmatrix}.$$

31. If the rows of a 3×3 matrix A are the columns of B in the same order, prove that $|A| = |B|$.

32. Show that if A is a square matrix of order 3 and c is a number, then $|cA| = c^3|A|$.

8.8 | Cramer's Rule

Determinants can be used to solve linear systems. The method that uses determinants is called **Cramer's Rule.** We present the method in this section.

We begin with a linear system of two equations in two variables.

$$a_1 x + b_1 y = c_1$$

$$a_2 x + b_2 y = c_2$$

Suppose first that $a_1 b_2 - a_2 b_1 \neq 0$. We shall use the elimination method to solve for x and then formulate the solution in terms of determinants. To eliminate y, we multiply the first equation by b_2, the second equation by b_1, and then subtract.

$$a_1 b_2 x + b_1 b_2 y = c_1 b_2$$
$$\underline{a_2 b_1 x + b_1 b_2 y = c_2 b_1}$$
$$(a_1 b_2 - a_2 b_1)x \qquad = c_1 b_2 - c_2 b_1$$

Since $a_1b_2 - a_2b_1 \neq 0$, this gives

$$x = \frac{c_1b_2 - c_2b_1}{a_1b_2 - a_2b_1} = \frac{\begin{vmatrix} c_1 & b_1 \\ c_2 & b_2 \end{vmatrix}}{\begin{vmatrix} a_1 & b_1 \\ a_2 & b_2 \end{vmatrix}}.$$

Similarly, x can be eliminated by multiplying the first equation by a_2 and the second equation by a_1. The solution for y is found to be

$$y = \frac{a_1c_2 - a_2c_1}{a_1b_2 - a_2b_1} = \frac{\begin{vmatrix} a_1 & c_1 \\ a_2 & c_2 \end{vmatrix}}{\begin{vmatrix} a_1 & b_1 \\ a_2 & b_2 \end{vmatrix}}.$$

The two fractions for x and y have the same denominator, which is the determinant of the coefficients. We denote this determinant by D:

$$D = \begin{vmatrix} a_1 & b_1 \\ a_2 & b_2 \end{vmatrix}.$$

If the first column of D is replaced by the column of constants, the resulting determinant is the numerator in the expression for x (the first variable in the system). If the second column of D is replaced by the column of constants, the resulting determinant is the numerator in the expression for y (the second variable in the system). The standard notations for these determinants are

$$D_x = \begin{vmatrix} c_1 & b_1 \\ c_2 & b_2 \end{vmatrix} \qquad \text{and} \qquad D_y = \begin{vmatrix} a_1 & c_1 \\ a_2 & c_2 \end{vmatrix}.$$

With this notation, the solutions are given by

$$x = \frac{D_x}{D}, \qquad y = \frac{D_y}{D}.$$

The expressions that we have obtained for x and y constitute the first part of Cramer's Rule. The second and third parts describe the possibilities when $D = 0$, and we omit the proofs of these statements.

Cramer's Rule

1. If $D \neq 0$, the solution is given by

$$x = \frac{D_x}{D}, \qquad y = \frac{D_y}{D}.$$

2. If $D = 0$ and either D_x or D_y is not zero, there is no solution. In this case the system is called *inconsistent*.

3. If $D = 0$ and both D_x and D_y are zero, there are many solutions. In this case the system is called *dependent*.

Example 1   Use Cramer's Rule to solve the following system.

$$3x + y = -1$$
$$4x - y = -13$$

Solution
We first calculate D:

$$D = \begin{vmatrix} 3 & 1 \\ 4 & -1 \end{vmatrix} = -3 - 4 = -7.$$

Since $D \neq 0$, the solution is given by the formulas in part (1) of Cramer's Rule. The determinants D_x and D_y are given by

$$D_x = \begin{vmatrix} -1 & 1 \\ -13 & -1 \end{vmatrix} = 1 + 13 = 14,$$

$$D_y = \begin{vmatrix} 3 & -1 \\ 4 & -13 \end{vmatrix} = -39 + 4 = -35.$$

Thus

$$x = \frac{D_x}{D} = \frac{14}{-7} = -2 \quad \text{and} \quad y = \frac{D_y}{D} = \frac{-35}{-7} = 5.$$

These values can easily be checked in the original system.   ❏

Example 2   Apply Cramer's Rule to the following system.

$$x - 4y = 2$$
$$2x - 8y = 2$$

Solution
Since

$$D = \begin{vmatrix} 1 & -4 \\ 2 & -8 \end{vmatrix} = -8 + 8 = 0$$

and

$$D_y = \begin{vmatrix} 1 & 2 \\ 2 & 2 \end{vmatrix} = 2 - 4 \neq 0,$$

the system has no solution. In other words, the system is inconsistent.   ❏

Let us consider now a general system of n linear equations in n unknowns.

$$a_{11}x_1 + a_{12}x_2 + \cdots + a_{1n}x_n = b_1$$

$$a_{21}x_1 + a_{22}x_2 + \cdots + a_{2n}x_n = b_2$$

$$\vdots \qquad\qquad \vdots$$

$$a_{n1}x_1 + a_{n2}x_2 + \cdots + a_{nn}x_n = b_n$$

We let D denote the determinant of the coefficient matrix, and we let D_{x_i} denote the determinant of the matrix formed by replacing the elements in the ith column of the coefficient matrix by the column of constants. Using these notations, we state Cramer's Rule as follows:

Cramer's Rule

1. If $D \neq 0$, the solution is given by

$$x_1 = \frac{D_{x_1}}{D}, \quad x_2 = \frac{D_{x_2}}{D}, \quad \ldots, \quad x_n = \frac{D_{x_n}}{D}.$$

2. If $D = 0$ and one or more D_{x_i} is not zero, there is no solution. In this case the system is called *inconsistent*.
3. If $D = 0$ and all $D_{x_i} = 0$, the system is *dependent*, and there are many solutions.

Example 3 Use Cramer's Rule to solve the following system.

$$2x + 3y - z = 4$$
$$-x + y + 2z = -2$$
$$3x - y - 2z = 1$$

Solution
Evaluation of the four determinants involved yields

$$D = \begin{vmatrix} 2 & 3 & -1 \\ -1 & 1 & 2 \\ 3 & -1 & -2 \end{vmatrix} = 14, \quad D_x = \begin{vmatrix} 4 & 3 & -1 \\ -2 & 1 & 2 \\ 1 & -1 & -2 \end{vmatrix} = -7,$$

$$D_y = \begin{vmatrix} 2 & 4 & -1 \\ -1 & -2 & 2 \\ 3 & 1 & -2 \end{vmatrix} = 15, \quad D_z = \begin{vmatrix} 2 & 3 & 4 \\ -1 & 1 & -2 \\ 3 & -1 & 1 \end{vmatrix} = -25.$$

Thus the solution is given by

$$x = \frac{D_x}{D} = \frac{-7}{14} = -\frac{1}{2}, \quad y = \frac{D_y}{D} = \frac{15}{14}, \quad z = \frac{D_z}{D} = -\frac{25}{14}. \qquad \square$$

Exercises 8.8

Use Cramer's Rule to find the solution if the determinant of the coefficients is not zero. If the system is dependent or inconsistent, state so.

1. $2x + 9y = 3$
$3x - 2y = -11$

2. $x + 2y = 4$
$3x - 2y = -12$

3. $7x - y = 9$
$3x + 5y = -7$

4. $4x + 3y = -6$
$5x - y = 27$

5. $4a + 3b = 1$
$2a - 5b = -19$

6. $a + 2b = 4$
$3a - 2b = -12$

7. $3x - 12y = 9$
$-x + 4y = -3$

8. $5x + 2y = 8$
$15x + 6y = 24$

9. $7x - y = 1$
$-14x + 2y = -1$

10. $-5x + 4y = 0$
$10x - 8y = 2$

11. $x + y = 3$
$x + 2y - z = 5$
$2y + z = 1$

12. $4x + z = -6$
$y + 2z = -3$
$2x + 5y + z = 1$

13. $3x + 2y + z = -1$
$x - 2z = -3$
$y + 2z = -2$

14. $-2x - y + 3z = 1$
$2x + y - z = -2$
$-x + 3y + 2z = 4$

15. $3x + y + 2z = 3$
$x - 5y - z = 0$
$2x + 3y + 2z = 0$

16. $x + 3y - z = 4$
$2x + y + 3z = 11$
$3x - 2y + 4z = 11$

17. $a + 3b - c = 4$
$3a - 2b + 4c = 11$
$2a + b + 3c = 13$

18. $2r - s + 2t = 2$
$2r - s + t = -5$
$r - 2s + 3t = 4$

19. $2x + y - z = 5$
$x - y + 2z = -3$
$-3y + 5z = -11$

20. $4x - y + 2z = -7$
$x + y + z = 4$
$5x - 5y + z = -26$

21. $2x_1 - x_2 + 3x_3 = 17$
$5x_1 - 2x_2 + 4x_3 = 28$
$3x_1 + 3x_2 - x_3 = 1$

22. $5x_1 + 3x_2 - x_3 = 4$
$2x_1 - 7x_2 + 3x_3 = -36$
$3x_1 - x_2 - 2x_3 = -13$

23. $x - y - 2z = 2$
$2x + 3y - z = 4$
$3x - y - 2z = 1$

24. $x + y + z = 1$
$x - 3y - 5z = -1$
$4x - 5y + 2z = -35$

25. $x + 2y + 3z = 0$
$y + z = 0$
$x + y + 3z = 1$

26. $3x - y + 2z = 4$
$x + 2y - z = -2$
$x + 2y - 3z = -2$

27. $x - y - z - w = 2$
$x + y - w = 0$
$3y + 2z + w = -2$
$y + w = 0$

28. $x - 2y + 3z = -3$
$x + 2z + w = 0$
$-2x + 4y + 5w = 10$
$2x - 4z - 3w = -2$

29. $x + 3y - 2z + w = 2$
$2x + 4y + 2z - 3w = -1$
$x - y - z - 2w = -1$
$2x + 5y - z - w = 1$

30. $4x - 2y - 5z - 6w = 1$
$2x + y + 3w = 3$
$7x + 2y - 3z + 2w = 4$
$4x - y + 7z + 5w = 2$

31. Find m and b so that the straight line with equation $y = mx + b$ passes through the points $(1, -2)$ and $(3, 2)$.

32. Find c and d so that the straight line with equation $cx + dy = 2$ passes through the points $(-3, -4)$ and $(2, 6)$.

33. Find a, b, and c so that the parabola with equation $y = ax^2 + bx + c$ passes through the points $(2, 10)$, $(-1, -5)$, and $(-3, 5)$.

34. Find a, b, and c so that the circle with equation $x^2 + ax + y^2 + by = c$ passes through the points $(1, 1)$, $(3, 3)$, and $(5, 1)$.

8.9 | Systems of Inequalities

We have seen that a solution to a given equation in x and y is a set of values for x and y that satisfies the equation, and that the graph of the equation consists of the points with coordinates (x, y) that correspond to solutions of the equation. The situation is much the same for inequalities involving x and y. A **solution** to a given inequality in x and y is a set of values for x and y that makes the given inequality a true statement. The **graph** of the inequality consists of the points with coordinates that correspond to solutions of the inequality.

The simplest type of inequality in x and y is one that is obtained from a linear equality $ax + by = c$ by replacing the equality sign with one of the inequality symbols $>$, $<$, $\geq$, or $\leq$. Such inequalities are called **linear inequalities.**

Suppose that a certain line in the plane has an equation given by $ax + by = c$. The line separates the points of the plane into three distinct subsets:

1. The points (x, y), where $ax + by = c$;

2. The points (x, y), where $ax + by > c$;

3. The points (x, y), where $ax + by < c$.

Geometrically the points described in (1) are the points on the line, the points described in (2) are the points on one side of the line, and the points described in (3) are the points on the other side of the line. Each of the regions described in (2) and (3) is called a **half-plane,** and the line is the **boundary** of the half-planes.

> **Example 1**　Sketch the graph of $3x + 4y > 24$.

Solution

We first locate the points on the line $3x + 4y = 24$. It is easy to see that the x-intercept of the line is 8 and the y-intercept is 6. The line is drawn in Figure 8.4 as a dashed line to indicate that the points on the boundary are not part of the solution set. To see how the solutions to $3x + 4y > 24$ consist of all points on one side of the line, let us start with a particular point (x_0, y_0) on the line. If the x-coordinate increases to a new value $x > x_0$, the value of $3x$ increases, and

$$3x + 4y_0 > 3x_0 + 4y_0 = 24.$$

That is, $3x + 4y_0 > 24$ for a point (x, y_0) to the right of (x_0, y_0) on the line. Any point on the right side of the line is located to the right of some point on the line, so we see that all points on the right side of the line satisfy $3x + 4y > 24$. Similar reasoning shows that the inequality

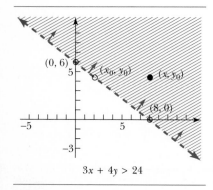

$3x + 4y > 24$

FIGURE 8.4

$3x + 4y < 24$ holds for all points on the left side of the line. The solution set to $3x + 4y > 24$ is shaded in Figure 8.4, and the side of the line that yields solutions is indicated by arrows based on the line.   ❑

The **solution set** to a system of inequalities is the intersection of the solution sets to the individual inequalities in the system. An example of this is furnished below.

Example 2   Graph the solution set of the following system.

$$3x + 4y > 24$$

$$x - 2y \geq -2$$

Solution
We use the solution of the first inequality from Example 1. This solution set is indicated in Figure 8.5 by shading with grey lines that slope upward to the right and by arrows based on the boundary.

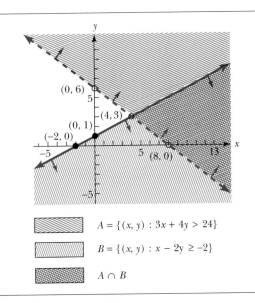

▨	$A = \{(x, y) : 3x + 4y > 24\}$
▨	$B = \{(x, y) : x - 2y \geq -2\}$
▨	$A \cap B$

FIGURE 8.5

To graph the solution set of the second inequality, we first sketch the line $x - 2y = -2$. The x-intercept is -2, and the y-intercept is 1. To determine which half-plane is the solution to $x - 2y > -2$, we simply test a point on one side of the line. The origin $(0, 0)$ is the simplest choice to use, and $0 > -2$ indicates that the solutions to $x - 2y > -2$

are on the same side of the line as the origin. This region is indicated in Figure 8.5 by shading with colored lines that slope upward to the right and by arrows based on the boundary. The boundary is included in the solution set since equality is included in $x - 2y \geq -2$.

The shading done leads to a crosshatching of the intersection of the individual solution sets, and the region shaded by crosshatching is the solution set of the system.   ❑

The description that was given for the graphs of linear inequalities generalizes to other types of inequalities in x and y. For a given inequality, a corresponding equation may be obtained by replacing the inequality symbol by an equality symbol. The graph of the equation separates the plane into regions, and it forms the boundary of these regions. Each region either consists entirely of solutions or it contains no solutions at all. The solution set can be determined by simply testing one point from each region. The boundary is included or omitted, according to whether or not equality is permitted in the original statement of inequality.

Example 3   Graph the solution set for the following system.

$$x^2 + y^2 \leq \frac{25}{9}$$

$$16y - 9x^2 > 0$$

Solution
We begin by graphing the two equations that correspond to the given inequalities.

The equation

$$x^2 + y^2 = \frac{25}{9}$$

is an equation of a circle with center $(0, 0)$ and radius $\frac{5}{3}$. If we solve for y in $16y - 9x^2 = 0$, we have

$$y = \frac{9}{16}x^2,$$

which is an equation of a parabola that has vertex at the origin and opens upward. To find the points of intersection, we can substitute $x^2 = \frac{16}{9}y$ into the equation of the circle. This gives

$$\frac{16}{9}y + y^2 = \frac{25}{9},$$

or

$$9y^2 + 16y - 25 = 0.$$

This can be factored as

$$(9y + 25)(y - 1) = 0,$$

so we obtain $y = -\frac{25}{9}$ and $y = 1$ as the solutions to this equation. Substituting $y = -\frac{25}{9}$ into $x^2 = \frac{16}{9}y$ yields

$$x^2 = \frac{16}{9}\left(-\frac{25}{9}\right)$$

$$= -\frac{400}{81},$$

which is impossible for a real number x. This indicates that there is no point of intersection at $y = -\frac{25}{9}$. Substituting $y = 1$ in $x^2 = \frac{16}{9}y$ yields

$$x^2 = \frac{16}{9}$$

and

$$x = \pm\frac{4}{3}.$$

The coordinates $(\frac{4}{3}, 1)$ and $(-\frac{4}{3}, 1)$ check in both original equations and give the points of intersection.

It is clear that the solutions to $x^2 + y^2 \leq \frac{25}{9}$ are the points interior to, or on, the circle $x^2 + y^2 = \frac{25}{9}$. To determine which points satisfy $16y - 9x^2 > 0$, we choose $(0, 3)$ as a test point. Since

$$16(3) - 9(0)^2 > 0,$$

$(0, 3)$ is a solution. This indicates that the solution set to $16y - 9x^2 > 0$ is the region above the parabola $16y - 9x^2 = 0$. The solution set of the system, then, is the set of points that lie above the parabola and on, or interior to, the circle. This is shown in Figure 8.6. ❏

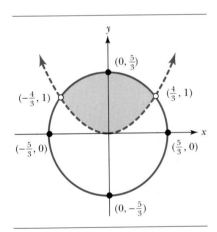

FIGURE 8.6

Exercises 8.9

Graph the solution set of the given inequality.

1. $3x + 2y \leq 12$ **2.** $3x - 2y \geq 6$

3. $5x + y > 3x + 2y - 6$

4. $3x - y < x + 2y + 6$

5. $y > (x - 1)^2$ **6.** $y < x^2 - 1$

7. $y \leq 4 - x^2$ **8.** $y \geq x^2 - 4x$

9. $x^2 + y^2 \leq 16$

10. $(x - 1)^2 + (y + 2)^2 \leq 9$

In Exercises 11–18, graph the solution set of the given system of inequalities, showing all points of intersection of the boundaries. These exercises correspond to those numbered 5–12 in Exercises 8.1.

11. $\begin{aligned} y &\leq 2x + 6 \\ x^2 &\leq 2y \end{aligned}$ **12.** $\begin{aligned} x^2 - y &\leq 0 \\ 2x - y + 3 &\geq 0 \end{aligned}$

13. $\begin{aligned} y &> 3x^2 + 12x \\ 2x - y &\geq 16 \end{aligned}$ **14.** $\begin{aligned} y &< 3 - 2x - x^2 \\ 4x + y &\geq 5 \end{aligned}$

15. $\begin{aligned} x^2 + y^2 &\leq 4 \\ 2x - y &< 2 \end{aligned}$ **16.** $\begin{aligned} x^2 + y^2 &\leq 25 \\ y - x &< 7 \end{aligned}$

17. $x^2 + y^2 < 10x$
$4y > 3x - 8$

18. $x^2 + y^2 < 25$
$x + 3y > 5$

25. $y > x^2 + 1$
$y - x < 3$

26. $y > x^2 - 4x + 3$
$y - x < 3$

In Exercises 19–32, graph the solution set of the given system. Label all points of intersection of the boundaries.

27. $x^2 + y^2 \geq 9$
$x^2 + 2y \leq 10$

28. $x^2 + y^2 \leq 9$
$x^2 + 2y \leq 6$

19. $x + y = 3$
$3x - y > 6$

20. $x + y = 1$
$2x + 3y < 4$

29. $x^2 + y^2 \leq 4$
$x^2 - 2y \leq 1$

30. $x^2 + y^2 \leq 4$
$3y + x^2 \geq 0$

21. $x - y > 3$
$x + 2y > 4$

22. $3x + 2y < 6$
$x - y > 2$

31. $y \geq 2^x$
$y \geq 2^{-x}$

32. $y < \dfrac{1}{x}$

23. $y \leq 2x + 2$
$y + x + 1 \geq 0$
$2y + 5x \leq 13$

24. $y + 6 > 3x$
$x + y < 6$
$x \geq 1$

$y < x$
$y \geq 0$
$y \geq 0$

In Exercises 33–40, write a system of inequalities that has the shaded region for its solution set.

33.

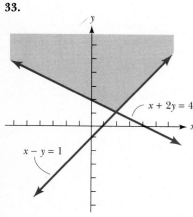

$x + 2y = 4$

$x - y = 1$

34.

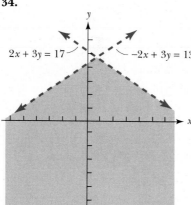

$2x + 3y = 17$ $-2x + 3y = 13$

35.

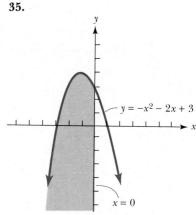

$y = -x^2 - 2x + 3$

$x = 0$

36.

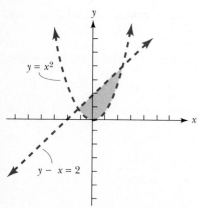

$y = x^2$

$y - x = 2$

37.

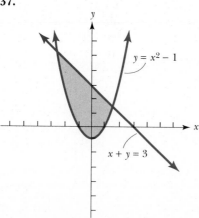

$y = x^2 - 1$

$x + y = 3$

38.

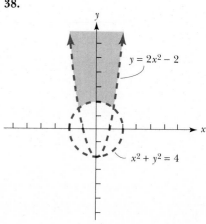

$y = 2x^2 - 2$

$x^2 + y^2 = 4$

39.

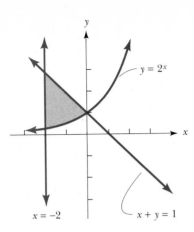

40.

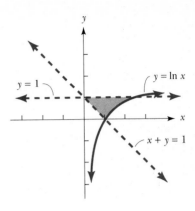

8.10 | Linear Programming

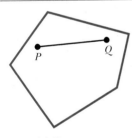

(a) Convex

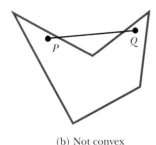

(b) Not convex

FIGURE 8.7

If the values of a variable f are given by an equation of the form

$$f = ax + by + c,$$

where a, b, and c are constants, then f is said to be a **linear function** of the two variables x and y.

In this section we consider problems that call for a maximum (largest) or a minimum (smallest) value of a linear function of two variables with a certain type of domain. Many problems in business, science, and engineering have a mathematical formulation of this type. These problems involve such things as profit and cost, diet and nutrition, or production scheduling, and they usually involve a linear function of several variables. We restrict our attention here to the two-variable case.

The linear functions that we consider have domains that are *convex regions* in the plane. A region in the plane is **convex** if, for every pair of points P and Q of the region, the line segment from P to Q lies entirely in the region. This is illustrated in Figure 8.7.

A **linear programming problem** is a problem in which it is desired to find the maximum (or minimum) value of a linear function that has its domain restricted to a convex solution set of a system of linear inequalities. The linear inequalities are called **constraints,** and the solution set of the system is called the **region of feasible solutions.** A point of the region where two boundaries intersect is called a **vertex.** The following example is typical of the problems that we shall consider.

Example 1   Find the maximum value of $f = 2x + y$, subject to the constraint that (x, y) must be a solution of the following system of inequalities.

$$x + 2y \geq 8$$

$$x - 4y \geq -10$$

$$x - y \leq 2$$

Solution

The solution set of the system of inequalities is shown in Figure 8.8. For each fixed value of d, the points (x, y) satisfying $f = d$ lie on the straight line $2x + y = d$. Different values of d give different straight lines with slope -2 and y-intercept d. Geometrically the problem is to find the largest possible value of the y-intercept for a line that intersects the solution set. The dashed lines corresponding to $d = 2, 7, 12$, and 16 are shown in the figure. From these graphs, it is easily seen that the maximum d is for the line through the vertex $C(6, 4)$. Substituting the values $x = 6$ and $y = 4$ yields the maximum value 16 for f on the solution set.

❏

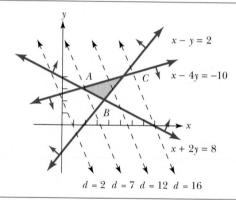

FIGURE 8.8

The following theorem assures us that it was no accident that the maximum value of f in Example 1 was attained at a vertex. The theorem states the facts that are needed in our work with linear programming problems, and we accept it without proof.

Theorem 8.15

Suppose the linear function $f = ax + by + c$ has its domain restricted to points (x, y) in a convex solution set of a system of linear inequalities.

> **1.** If f has a maximum or minimum value on this domain, it will occur at a vertex of the region.
> **2.** If the domain is a convex polygon together with its interior, then f will have a maximum and a minimum value on this domain.

The words **maximize** and **minimize** are frequently used in stating linear programming problems. To maximize f is to find its maximum value, and to minimize f is to find its minimum value.

Example 2   Minimize $f = 5x - 4y$, subject to the following constraints.

$$x - y \le 1$$
$$x + 2y \ge 4$$
$$2x + 3y \le 17$$
$$2x - 3y \ge -13$$

Solution
We first graph the system of inequalities and locate the vertices by the methods of Section 8.9. This is shown in Figure 8.9. Since the minimum value of f occurs at a vertex, all we need do is calculate the values of f at each vertex and select the smallest of them. The table in Figure 8.9 lists the value of f at each vertex. From the table, we see that f has a minimum value of -22 at the vertex $(-2, 3)$.

Vertex	$f = 5x - 4y$
$(2, 1)$	6
$(4, 3)$	8
$(1, 5)$	-15
$(-2, 3)$	-22

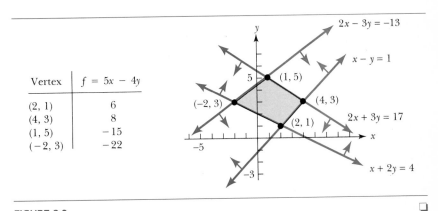

FIGURE 8.9

Our next example provides some insight into the situations that lead to linear programming problems.

Example 3   A fuel refinery has two processes for manufacturing three grades of gasoline: G_1, G_2, and G_3. The first process will produce 1 tank-car load of G_1, 3 loads of G_2, and 2 loads of G_3 in 10 hours. The second process will produce 3 loads of G_1, 4 loads of G_2, and 1 load of G_3 in 8 hours. The refinery has received an order for 8 loads of G_1, 19 loads of G_2, and 6 loads of G_3 that must be filled within 80 hours. How many times should each process be employed to fill the order as soon as possible, and what is the minimum time necessary?

Solution

If the first process is employed x times and the second process is employed y times, the time required is given by

$$T = 10x + 8y.$$

We need to minimize T, subject to the production requirements.

Employing the first process x times will produce x loads of G_1, $3x$ loads of G_2, and $2x$ loads of G_3. Employing the second process y times will produce $3y$ loads of G_1, $4y$ loads of G_2, and y loads of G_3. Filling the order imposes the following constraints.

$$x + 3y \geq 8$$

$$3x + 4y \geq 19$$

$$2x + y \geq 6$$

$$0 \leq x \leq 8$$

$$0 \leq y \leq 10$$

Vertex	$T = 10x + 8y$
(0, 10)	80
(0, 6)	48
(1, 4)	42
(5, 1)	58
(8, 0)	80
(8, 10)	160

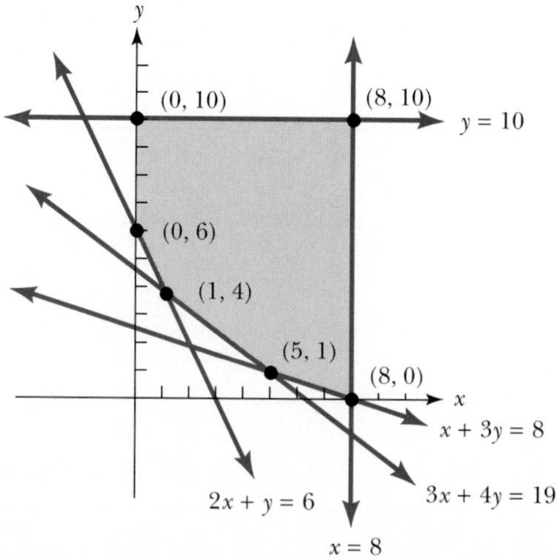

FIGURE 8.10

The last two constraints reflect the facts that the number of times a process is employed cannot be negative and that the order must be filled within 80 hours.

We graph the system of inequalities and tabulate the values of T at the vertices, as shown in Figure 8.10. From the table, we see that T has a **minimum** value of 42 at the vertex $(1, 4)$. The first process should be employed one time and the second should be employed four times, for a minimum time of 42 hours. ❏

Exercises 8.10

1. Maximize $f = 2x + y$, subject to:
$$2x - y \le 10$$
$$x + y \le 8$$
$$x \ge 0$$
$$y \ge 0.$$

2. Maximize $f = 5x + 2y$, subject to:
$$x + y \le 7$$
$$x + 2y \le 10$$
$$x \ge 0$$
$$y \ge 0.$$

3. Maximize $f = x + 4y + 1$, subject to:
$$x + y \le 2$$
$$x - y \le 1$$
$$x \ge 0$$
$$y \ge 0.$$

4. Maximize $f = 2x + 5y - 3$, subject to:
$$x + y \le 2$$
$$x - 2y \ge -1$$
$$x \ge 0$$
$$y \ge 0.$$

5. Minimize $f = 3x + y + 4$, subject to:
$$x - y \le 3$$
$$x + 5y \le 21$$
$$x + y \ge 5.$$

6. Minimize $f = 2x + y - 1$, subject to:
$$x + 2y \ge 4$$
$$2x - y \le 4$$
$$3x - 2y \ge -6.$$

7. Maximize $f = 2x + y$, subject to:
$$2x - y \le 10$$
$$x + y \le 8$$
$$x - y \ge -5$$
$$x \ge 0$$
$$y \ge 0.$$

8. Maximize $f = 14x + 3y$, subject to:
$$x + y \ge 2$$
$$x - y \le 2$$
$$x + y \le 4$$
$$x \ge 0.$$

9. Minimize $f = y - x + 3$, subject to:
$$2x + y \le 9$$
$$x - 2y \le 2$$
$$2x - 3y \ge 3$$
$$y \ge 0.$$

10. Minimize $f = y - 3x + 5$, subject to:
$$x - y \le 2$$
$$x + y \le 6$$
$$x + 2y \le 10$$
$$x \ge 0$$
$$y \ge 0.$$

11. Maximize $f = x + y - 2$, subject to:
$$x - 2y \le 2$$
$$2x + y \le 9$$
$$x - 2y \ge -3$$
$$x \ge 0$$
$$y \ge 0.$$

12. Maximize $f = 4x + y - 1$, subject to:
$$2x - 5y \le 10$$
$$3x - 5y \le 20$$
$$x - 5y \ge -30$$
$$x \ge 0$$
$$y \ge 0.$$

13. Matt supplies to a wholesaler two mixtures of nuts packed in 8-pound cans. The first mixture contains 6 pounds of peanuts and 2 pounds of cashews in each can. The second mixture contains 5 pounds of peanuts and 3 pounds of cashews. The profit on the first mixture is $4 per can, and the profit on the second is $5 per can. From a supply of 240 pounds of peanuts and 96 pounds of cashews, how many cans of each mixture should be made for a maximum profit?

14. A company owns two factories that produce barrels and pressure tanks. During each day of operation, factory A produces 3000 barrels and 1000 pressure tanks; factory B produces 2000 barrels and 2000 pressure tanks. The cost at factory A is $10 per barrel and $20 per pressure tank, whereas the cost at factory B is $20 per barrel and $10 per pressure tank. The company has an order for 16,000 barrels and 8000 pressure tanks. How many days should each factory operate if the cost of filling the order is to be minimized?

15. A furniture company manufactures two types of desk using oak and mahogany lumber. The first type requires 10 board feet of oak and 5 board feet of mahogany, whereas the second requires 6 board feet of oak and 4 board feet of mahogany. A profit of $45 is made on each desk of the first type and $30 on each desk of the second type. From a supply of 1000 board feet of oak and 600 board feet of mahogany, how many desks of each type should be made in order to yield a maximum profit?

16. A certain company owns two small factories that produce iceboxes, ski belts, and minnow buckets. During each day of operation, factory A produces 4000 iceboxes, 1000 ski belts, and 2000 minnow buckets. Factory B produces 1000 iceboxes, 1000 ski belts, and 7000 minnow buckets each day. The company has an order to supply 8000 iceboxes, 5000 ski belts, and 20,000 minnow buckets. It costs $6000 per day to run factory A and $2000 per day to run factory B. Find the number of days each factory must operate in order for the cost of filling the order to be a minimum.

17. A small manufacturing company can sell all the chairs and bar stools that it can produce. Each chair requires $1\frac{3}{5}$ hours in the assembly room and $\frac{2}{3}$ hour in the finishing room. Each bar stool requires $\frac{4}{5}$ hour in the assembly room and $\frac{4}{3}$ hours in the finishing room. Both the assembly room and the finishing room operate 24 hours a day, and each of them can work on only one item at a time. If the company makes a profit of $30 on each chair and $20 on each stool, find the number of chairs and stools that should be manufactured to make a maximum profit.

18. During the noon hour each day at the county fair in Topsfield, Massachusetts, Joe sells hero sandwiches and hot dogs from his trailer. He sells only

Photo by Ulrike Welsch/Stock Boston

these two items, and he makes a profit of 30¢ on each sandwich and 20¢ on each hot dog. Experience has shown him that the total number of items sold never exceeds 150 and that the number of hot dogs is always at least twice the number of sandwiches. Find the amount of sales of each item that yields a maximum profit.

19. A supermarket prepares fresh daily two types of deli trays, a cold cut tray and a seafood tray. The meat department can supply no more than 6 hours and 5 minutes of labor daily to prepare the meat, and the delicatessen can supply no more than 5 hours of labor daily to assemble the trays. Each cold cut tray requires 10 minutes in the meat department and 15 minutes in the delicatessen, whereas each seafood tray requires 25 minutes in the meat department and 10 minutes in the delicatessen. Because of demand, the supermarket wants to be sure to supply at least 4 cold cut trays daily. If there is a $2 profit on each cold cut tray sold and $3.50 profit on each seafood tray sold, how many trays of each type should the supermarket produce daily to maximize profit, assuming that all can be sold.

20. A small clock company produces two types of clock, a grandfather clock and a brass mantle clock. Each grandfather clock sold nets a profit of $150 and each mantle clock $30. In order to stay in business, the company must make a profit of at least $4500 per month. Each grandfather clock requires 1 pound of brass, and each mantle clock requires 3 pounds of brass. Each month the company has available at most 100 pounds of brass. Because of space limitations in the factory, no more than 79 grandfather clocks can be produced per month. Shipping costs are $75 per grandfather clock and $20 per mantle clock. Determine the number of each type of clock to be produced monthly that will minimize the shipping cost.

CHAPTER REVIEW

Key Words and Phrases

Points of intersection
System of equations
Substitution method
Elimination method
Matrix
Dimension of a matrix
Main diagonal elements
Square matrix
Column matrix
Row matrix
Diagonal matrix
Equal matrices
Sum of two matrices
Additive identity
Additive inverse

Scalar
Scalar product
Product of two matrices
Identity matrix
Coefficient matrix
Constant matrix
Augmented matrix
Row operations on a matrix
Gaussian elimination method
Gauss–Jordan elimination
Inverse of a matrix
Unknown matrix
Partial fraction decomposition
Determinant

Order of a determinant
Minor
Cofactor
Cofactor expansion
Row and column operations on a
   determinant
Cramer's Rule
System of inequalities
Convex region
Linear programming problem
Constraints
Feasible solutions
Maximize
Minimize

Summary of Important Concepts and Formulas

Properties of Matrix Addition and Multiplication

$$A + (B + C) = (A + B) + C$$

$$A + B = B + A$$

$$A(BC) = (AB)C$$

$$A(B + C) = AB + AC$$

$$(B + C)A = BA + CA$$

$$AI = A = IA$$

$$AA^{-1} = I = A^{-1}A$$

Row Operations on a Matrix

a. Interchange two rows.
b. Multiply (or divide) every element in a row by the same nonzero number.
c. Add (or subtract) a multiple of one row to (from) another row.

Gaussian Elimination Method

a. Write the augmented matrix $[A \vdots B]$ for the system, where A is the coefficient matrix and B is the constant matrix.

b. Use any of the row operations to change $[A \vdots B]$ to a matrix in which all elements on the main diagonal are 1's and all elements below the main diagonal are 0's.
c. Use back substitution on the system that has the augmented matrix obtained in step (b).

Gauss–Jordan Elimination Method for Solving a Linear System

a. Write the augmented matrix $[A \vdots B]$ for the system.
b. Use any of the row operations to change $[A \vdots B]$ to $[I \vdots C]$, where I is an identity matrix of the same size as A, and C is a column matrix.
c. Read the solutions from the column matrix C.

Gauss–Jordan Elimination Method for Finding A^{-1}

a. Augment A with an identity matrix of the same size: $[A \vdots I_n]$.
b. Use row operations to transform $[A \vdots I_n]$ into the form $[I_n \vdots B]$. If this is not possible, then A^{-1} does not exist.
c. If $[A \vdots I_n]$ is transformed into $[I_n \vdots B]$ by row operations, the inverse can be read from the last n columns of $[I_n \vdots B]$. That is, $B = A^{-1}$.

Solution of Linear Systems by Matrix Inverses
If A^{-1} exists, $X = A^{-1}B$ is the unique solution to $AX = B$.

Partial Fraction Decomposition of $P(x)/Q(x)$
a. Linear Factors of $Q(x)$

For each factor of $Q(x)$ of the form $(ax + b)^n$, where $n \geq 1$, in the partial fraction decomposition, there corresponds a sum of n partial fractions of the form

$$\frac{A_1}{ax + b} + \frac{A_2}{(ax + b)^2} + \frac{A_3}{(ax + b)^3} + \cdots + \frac{A_n}{(ax + b)^n},$$

where $A_1, A_2, \ldots, A_n$ are constants.

b. Quadratic Factors of $Q(x)$

For each factor of $Q(x)$ of the form $(ax^2 + bx + c)^n$, where $n \geq 1$ and $ax^2 + bx + c$ is irreducible, there corresponds a sum of n partial fractions of the form

$$\frac{A_1 x + B_1}{ax^2 + bx + c} + \frac{A_2 x + B_2}{(ax^2 + bx + c)^2}$$
$$+ \cdots + \frac{A_n x + B_n}{(ax^2 + bx + c)^n},$$

where all A_i and B_i are constants.

Cofactor Expansion of a Determinant

$$|A| = a_{i1}C_{i1} + a_{i2}C_{i2} + \cdots + a_{in}C_{in}$$

$$|A| = a_{1j}C_{1j} + a_{2j}C_{2j} + \cdots a_{nj}C_{nj}$$

Row Operations on a Determinant
a. Interchange any two rows.
b. Multiply (or divide) every element in a row by the same nonzero number.
c. Add (or subtract) a multiple of one row to (from) another row.

Column Operations on a Determinant
a. Interchange any two columns.
b. Multiply (or divide) every element in a column by the same nonzero number.
c. Add (or subtract) a multiple of one column to (from) another column.

Evaluation of Determinants
Theorem 8.12 If the matrix B is obtained from a matrix A by interchanging any two rows (columns) of A, then $|B| = -|A|$.

Theorem 8.13 If the matrix B is obtained from a matrix A by multiplying each element of a row (column) of A by the same number c, then $|B| = c|A|$.

Theorem 8.14 If the matrix B is obtained from a matrix A by adding a multiple of one row (column) to another row (column), then $|B| = |A|$.

Cramer's Rule
a. If $D \neq 0$, the solution is given by

$$x_1 = \frac{D_{x_1}}{D}, \quad x_2 = \frac{D_{x_2}}{D}, \quad \ldots, \quad x_n = \frac{D_{x_n}}{D}.$$

b. If $D = 0$ and one or more D_{x_i} is not zero, there is no solution. In this case the system is called *inconsistent*.
c. If $D = 0$ and all $D_{x_i} = 0$, the system is *dependent*, and there are many solutions.

Review Problems for Chapter 8

In Problems 1–6, solve the given system of equations. It is not necessary to graph the equations.

1. $3x + 2y = 12$
    $3x - 2y = 6$

2. $2x - y = 10$
    $x + y = 8$

3. $x = y^2 + 1$
    $x - y = 3$

4. $x = y^2 - 4y + 3$
    $x - y = 3$

5. $x^2 - 3y = 1$
    $2x - y = 3$

6. $x^2 + 3xy - 6y^2 = 8$
    $x^2 - xy - 6y^2 = 4$

Perform the indicated operations, if possible.

7. $\begin{bmatrix} 2 & -3 \\ -4 & -5 \\ 0 & 2 \end{bmatrix} - 2\begin{bmatrix} -1 & 0 \\ 3 & -11 \\ 1 & 0 \end{bmatrix}$

8. $[1 \quad 3 \quad 0] + [2 \quad -7]$

9. $2\begin{bmatrix} -2 & 1 & 3 \\ 1 & 5 & 1 \end{bmatrix} - 3\begin{bmatrix} 1 & 7 & -2 \\ 0 & 1 & -5 \end{bmatrix}$

Perform the matrix multiplications, if possible.

10. $\begin{bmatrix} 1 & 4 & 2 \\ -2 & 1 & 0 \end{bmatrix}\begin{bmatrix} -3 & 1 & 2 \\ 0 & 5 & -2 \end{bmatrix}$

11. $\begin{bmatrix} 1 & -2 & 0 \\ 0 & 3 & 2 \\ 5 & 0 & 1 \end{bmatrix}\begin{bmatrix} -1 & 0 \\ 3 & 4 \\ 0 & -1 \end{bmatrix}$

12. $\begin{bmatrix} 2 & 1 & 0 \\ -1 & 3 & 5 \end{bmatrix}\begin{bmatrix} 2 & 7 \\ 0 & -4 \end{bmatrix}$

13. $\begin{bmatrix} 2 & 7 \\ 0 & -4 \end{bmatrix} \begin{bmatrix} 2 & 1 & 0 \\ -1 & 3 & 5 \end{bmatrix}$

Use Gaussian or Gauss–Jordan elimination to find a solution, if one exists.

14. $\begin{aligned} 2x - y &= 2 \\ -x + 3y &= 14 \end{aligned}$

15. $\begin{aligned} x - 2y - 4z &= -1 \\ 3x \quad\;\; - z &= 4 \\ x + 4y + 7z &= 2 \end{aligned}$

16. $\begin{aligned} -x + 2y + z &= -1 \\ 3x + y - z &= 7 \\ y + z &= -3 \end{aligned}$

17. $\begin{aligned} x + 2y + z &= 1 \\ 2x + 5y + 3z &= 2 \\ x \quad\quad - 2z &= 3 \end{aligned}$

18. $\begin{aligned} x + 2y + z &= 1 \\ x + y - z &= 1 \\ y + 3z &= 1 \end{aligned}$

19. $\begin{aligned} x \quad\;\; + z &= 3 \\ x + y + z &= 1 \\ 2x - y + 3z &= 12 \end{aligned}$

Find the inverse of the given matrix, if it exists.

20. $\begin{bmatrix} 2 & -4 \\ -1 & 3 \end{bmatrix}$

21. $\begin{bmatrix} 8 & 12 \\ 6 & 9 \end{bmatrix}$

22. $\begin{bmatrix} 1 & 0 & 0 \\ -1 & 1 & 1 \\ 0 & 1 & 2 \end{bmatrix}$

23. $\begin{bmatrix} 1 & 0 & 2 \\ 2 & 1 & 5 \\ 0 & 1 & 2 \end{bmatrix}$

24. $\begin{bmatrix} 1 & 0 & 2 \\ 2 & -2 & 5 \\ 0 & 1 & -1 \end{bmatrix}$

25. $\begin{bmatrix} 1 & 1 & -3 \\ 1 & 0 & 3 \\ -2 & 1 & -12 \end{bmatrix}$

Solve each system by using the inverse of the coefficient matrix.

26. $\begin{aligned} -2r + 3s &= 3 \\ r + 2s &= -19 \end{aligned}$

27. $\begin{aligned} 4x - y &= 2 \\ 7x - 3y &= 1 \end{aligned}$

28. $\begin{aligned} 2x + y - z &= 1 \\ y - 2z &= 9 \\ x + 3y &= 1 \end{aligned}$

29. $\begin{aligned} 4x - 5y - 3z &= 8 \\ 3x - 3y - 2z &= 7 \\ x - y - z &= 1 \end{aligned}$

Write each fraction as a sum of partial fractions.

30. $\dfrac{x - 8}{x^2 - x - 6}$

31. $\dfrac{5x^2 + 5x + 6}{(x + 1)^2(x - 2)}$

32. $\dfrac{3x^2 + 8x - 8}{(x - 4)(x^2 + 2)}$

33. $\dfrac{2x^2 + 5x + 1}{(x^2 + x + 1)^2}$

34. Evaluate the determinant: $\begin{vmatrix} 4 & -2 \\ -1 & -3 \end{vmatrix}$.

Use a cofactor expansion to evaluate the determinant.

35. $\begin{vmatrix} -1 & 2 & 3 \\ 4 & -2 & 1 \\ 1 & 5 & 0 \end{vmatrix}$

36. $\begin{vmatrix} 2 & 1 & -2 \\ 1 & 5 & 0 \\ 1 & -1 & 3 \end{vmatrix}$

37. $\begin{vmatrix} 3 & -2 & 1 \\ 4 & 0 & 1 \\ 2 & 5 & -2 \end{vmatrix}$

38. Solve for x in the equation $\begin{vmatrix} -2 & x & -2 \\ 3 & 1 & 0 \\ 1 & 2x & 1 \end{vmatrix} = 0$.

Without evaluating the determinants, determine the value of the variable that makes each statement true.

39. $\begin{vmatrix} 2 & 1 & 1 \\ -3 & 5 & -3 \\ 4 & 2 & 1 \end{vmatrix} = \begin{vmatrix} 2 & 1 & 1 \\ -3 & 5 & -3 \\ 0 & 0 & x \end{vmatrix}$

40. $\begin{vmatrix} 1 & -2 & -1 \\ 3 & 0 & 2 \\ 2 & 1 & 1 \end{vmatrix} = x \begin{vmatrix} -1 & -2 & 1 \\ 2 & 0 & 3 \\ 1 & 1 & 2 \end{vmatrix}$

41. $\begin{vmatrix} -2 & 4 & 8 \\ -4 & 2 & -6 \\ 2 & 2 & -4 \end{vmatrix} = x \begin{vmatrix} -1 & 2 & 4 \\ -2 & 1 & -3 \\ 1 & 1 & -2 \end{vmatrix}$

42. Evaluate the determinant by introducing zeros before expanding about a row or column.

$$\begin{vmatrix} -1 & 0 & 2 & 0 \\ 2 & 3 & 1 & 2 \\ 1 & 1 & 1 & 0 \\ -1 & 2 & -1 & 3 \end{vmatrix}$$

Use Cramer's Rule to find the solution if the determinant of the coefficients is not zero. If the system is dependent or inconsistent, state so.

43. $\begin{aligned} 3x - y &= -2 \\ 6x - 2y &= -4 \end{aligned}$

44. $\begin{aligned} 5x + 2y &= 4 \\ -x + 2y &= -8 \end{aligned}$

45. $\begin{aligned} x + y + 2z &= 3 \\ -x + 2y + z &= 9 \\ 3x + y + 3z &= 0 \end{aligned}$

46. $\begin{aligned} 2x - y &= 3 \\ x + y - z &= 1 \\ 3x - 3y + z &= 5 \end{aligned}$

In Problems 47–50, graph the solution set of the given system of inequalities, showing all points of intersection of the boundaries.

47. $\begin{aligned} 3x + y &> 7 \\ x - y &\le 1 \\ y &< 4 \end{aligned}$

48. $\begin{aligned} 2x + y &< -2 \\ 4x - 3y &\ge -24 \\ y &\ge -4 \end{aligned}$

49. $\begin{aligned} x^2 + 2x + 3 &\le y \\ 3x + y + 1 &\le 0 \end{aligned}$

50. $\begin{aligned} 2x^2 - y &\le 2 \\ 2x - y &\ge -2 \end{aligned}$

51. Maximize $f = 2x + 7y$, subject to:
$$\begin{aligned} 3x + 4y &\le 24 \\ x - 2y &\ge -2 \\ x &\ge 0 \\ y &\ge 0. \end{aligned}$$

52. Maximize $f = 15x - 4y$, subject to:
$$3x + 2y \leq 12$$
$$3x - 2y \leq 6$$
$$x - 2y \geq -4$$
$$x \geq 0$$
$$y \geq 0.$$

53. A nut company has 400 pounds of peanuts and 200 pounds of cashews to sell as two different mixes. One mix will contain half peanuts and half cashews and will sell for \$6 per pound. The other mix will contain $\frac{3}{4}$ peanuts and $\frac{1}{4}$ cashews and will sell for \$5 per pound. How many pounds of each mix should the company prepare for maximum revenue?

▲ ▲ CRITICAL THINKING: FIND THE ERRORS ▬▬▬

Each of the following nonsolutions has at least one error. Can you find them?

Problem 1   Perform the indicated operations, if possible.

a) $[2 \quad -5 \quad 0] + [1 \quad 3]$ b) $\begin{bmatrix} 0 & 1 \\ 2 & 3 \end{bmatrix} \begin{bmatrix} 4 & 5 \\ -1 & -2 \end{bmatrix}$

Nonsolution   a) $[2 \quad -5 \quad 0] + [1 \quad 3] = [3 \quad -2 \quad 0]$

b) $\begin{bmatrix} 0 & 1 \\ 2 & 3 \end{bmatrix} \begin{bmatrix} 4 & 5 \\ -1 & -2 \end{bmatrix} = \begin{bmatrix} 0 & 5 \\ -2 & -6 \end{bmatrix}$

Problem 2   Use Gauss–Jordan elimination to find a solution, if one exists.

$$x + y + 2z = 2$$
$$y + 3z = 1$$
$$x \qquad - 2z = 1$$

Nonsolution $\begin{bmatrix} 1 & 1 & 2 & | & 2 \\ 0 & 1 & 3 & | & 1 \\ 1 & 0 & -2 & | & 1 \end{bmatrix} \longrightarrow \begin{bmatrix} 1 & 1 & 2 & | & 2 \\ 0 & 1 & 3 & | & 1 \\ 0 & -1 & -4 & | & 1 \end{bmatrix} \longrightarrow \begin{bmatrix} 1 & 1 & 2 & | & 2 \\ 0 & 1 & 3 & | & 1 \\ 0 & 0 & -1 & | & 1 \end{bmatrix}$

$\longrightarrow \begin{bmatrix} 1 & 0 & -1 & | & 2 \\ 0 & 1 & 3 & | & 1 \\ 0 & 0 & 1 & | & 1 \end{bmatrix} \longrightarrow \begin{bmatrix} 1 & 0 & 0 & | & 2 \\ 0 & 1 & 0 & | & 1 \\ 0 & 0 & 1 & | & 1 \end{bmatrix}$

The solution is $x = 2, y = 1, z = 1$.

Problem 3   Write $\dfrac{2x - 4}{(x - 1)^2(x + 1)}$ as a sum of partial fractions.

Nonsolution   Since the denominator contains linear factors, we write

$$\frac{2x - 4}{(x - 1)^2(x + 1)} = \frac{A}{x - 1} + \frac{B}{x + 1},$$

or

$$2x - 4 = A(x + 1) + B(x - 1).$$
$$\text{If } x = 1, \text{ then } -2 = A(2), \text{ and } A = -1.$$
$$\text{If } x = -1, \text{ then } -6 = B(-2), \text{ and } B = 3.$$

Hence the partial fraction decomposition is

$$\frac{2x - 4}{(x - 1)^2(x + 1)} = \frac{-1}{x - 1} + \frac{3}{x + 1}.$$

427

Evaluate the following determinant by expanding about the first column.

$$\begin{vmatrix} 3 & 1 & -1 \\ -1 & 1 & 2 \\ 1 & 2 & 1 \end{vmatrix}$$

Nonsolution

$$\begin{vmatrix} 3 & 1 & -1 \\ -1 & 1 & 2 \\ 1 & 2 & 1 \end{vmatrix} = 3\begin{vmatrix} 1 & 2 \\ 2 & 1 \end{vmatrix} - 1\begin{vmatrix} 1 & -1 \\ 2 & 1 \end{vmatrix} + 1\begin{vmatrix} 1 & -1 \\ 1 & 2 \end{vmatrix}$$

$$= 3(1-4) - 1(1+2) + 1(2+1)$$

$$= 3(-3) - 1(3) + 1(3)$$

$$= -9$$

Problem 5 Evaluate the following determinant by introducing zeros before expanding.

$$\begin{vmatrix} 0 & 1 & 4 \\ -1 & 2 & 3 \\ 1 & -2 & 0 \end{vmatrix}$$

Nonsolution

$$\begin{vmatrix} 0 & 1 & 4 \\ -1 & 2 & 3 \\ 1 & -2 & 0 \end{vmatrix} \overset{R_2 + 1}{=} \begin{vmatrix} 0 & 1 & 4 \\ 0 & 3 & 4 \\ 1 & -2 & 0 \end{vmatrix}$$

$$= 1\begin{vmatrix} 1 & 4 \\ 3 & 4 \end{vmatrix}$$

$$= 4 - 12$$

$$= -8$$

Problem 6 Use Cramer's Rule to determine the value of y in the solution of the system

$$2x - 4y + 2z = -8$$

$$3y + 2z = 4$$

$$3x \quad + z = 2.$$

Nonsolution

$$D_y = \begin{vmatrix} 2 & -8 & 2 \\ 0 & 4 & 2 \\ 3 & 2 & 1 \end{vmatrix} \overset{\frac{1}{2}R_2}{=} \begin{vmatrix} 2 & -8 & 2 \\ 0 & 2 & 1 \\ 3 & 2 & 1 \end{vmatrix} \overset{-2C_3 + C_2}{=} \begin{vmatrix} 2 & -12 & 2 \\ 0 & 0 & 1 \\ 3 & 0 & 1 \end{vmatrix}$$

$$= -1\begin{vmatrix} 2 & -12 \\ 3 & 0 \end{vmatrix} = -36$$

Thus $y = -36$.

Chapter 9

Further Topics in Algebra

The topics presented in this chapter are truly "precalculus" in nature. All of them are contained in the typical calculus sequence, and most of them are somewhat difficult. They are included in this book so that the student will have some familiarity with them when encountering them again in the calculus.

9.1   Sequences and Series

Higher-level mathematics makes extensive use of certain types of functions called sequences. Informally, a sequence is an ordered list in which there is a first element, a second element, a third element, and so on. Some familiar sequences are the sequence of positive integers

$$1, 2, 3, 4, \ldots$$

and the sequence of even positive integers

$$2, 4, 6, 8, \ldots.$$

These are examples of *infinite sequences*. Sequences that have a last, or terminal, element are called *finite sequences*. An example of a finite sequence is the ordered listing

$$2, 5, 8, 11, 14, 17.$$

These ideas are formalized in Definition 9.1.

Definition 9.1

A **finite sequence** is a function that has a domain of the form $\{1, 2, 3, \ldots, k\}$, where k is a fixed integer. An **infinite sequence** is a function that has the set of all positive integers as its domain.

According to this definition, a finite sequence has function values of the form

$$f(1), f(2), f(3), \ldots, f(k),$$

where f is the function that has the set $\{1, 2, 3, \ldots, k\}$ as its domain. Similarly, an infinite sequence with f as its defining function has function values of the form

$$f(1), f(2), f(3), \ldots,$$

where the dots at the end indicate that there is no terminal element. In working with sequences, it is traditional to use a subscripted letter, such as a_n, instead of $f(n)$ to indicate the function value at the positive integer n. A finite sequence is written as

$$a_1, a_2, a_3, \ldots, a_k,$$

and an infinite sequence is written as

$$a_1, a_2, a_3, \ldots.$$

With this notation, $\boldsymbol{a_n}$ stands for the **nth term,** or the **general term,** in the sequence. Thus a_1 denotes the first term, a_2 denotes the second term, a_{18} denotes the eighteenth term, and so on.

Example 1   Write the first four terms of the sequence that has the given general term.

a) $a_n = \dfrac{1}{n}$ b) $a_n = 2n - 1$ c) $b_n = (-1)^n \dfrac{n}{n+1}$

Solution
To find the first four terms of each sequence, we assign the values 1, 2, 3, 4 to n in succession. This gives

a) $a_1 = 1$, $a_2 = \frac{1}{2}$, $a_3 = \frac{1}{3}$, $a_4 = \frac{1}{4}$;

b) $a_1 = 1$, $a_2 = 3$, $a_3 = 5$, $a_4 = 7$;

c) $b_1 = -\frac{1}{2}$, $b_2 = \frac{2}{3}$, $b_3 = -\frac{3}{4}$, $b_4 = \frac{4}{5}$.   ❏

Occasionally a sequence is defined by specifying the first few terms and a rule for finding the nth term from the preceding terms. Consider the following example.

Example 2   Find the first six terms of the following sequence, known as the Fibonacci sequence.

$$a_1 = 1, \quad a_2 = 1, \quad a_{n+1} = a_n + a_{n-1}$$

Solution
We have $a_1 = 1$ and $a_2 = 1$. Thus

$$a_3 = a_2 + a_1 = 1 + 1 = 2,$$

$$a_4 = a_3 + a_2 = 2 + 1 = .3,$$

$$a_5 = a_4 + a_3 = 3 + 2 = 5,$$

$$a_6 = a_5 + a_4 = 5 + 3 = 8,$$

and the first six terms are 1, 1, 2, 3, 5, 8.   ❏

Definition 9.2

A **series** is the sum of the terms of a sequence.

In this section we consider only series that are formed from a finite sequence. Such a series has the form

$$a_1 + a_2 + a_3 + \cdots + a_k.$$

It is possible in some cases for the sum of an infinite series to have meaning. Some series of this type are considered in Section 9.3.

If a formula for the general term is known, it is common practice to write a series in a compact form called the **sigma notation.** Using this notation, we write the series

$$a_1 + a_2 + a_3 + \cdots + a_k$$

as $\displaystyle\sum_{n=1}^{k} a_n$. That is,

$$\sum_{n=1}^{k} a_n = a_1 + a_2 + a_3 + \cdots + a_k.$$

The capital Greek letter Σ (sigma) is used to indicate a *sum*, and the notations at the bottom and top of the sigma give the initial and terminal values of n. The letter n is called the **index of summation.**

Example 3   Write the following series in expanded form and find the value of the sum.

$$\sum_{n=1}^{5} (2n - 1)$$

Solution

The given expression represents the series which has terms that are obtained by substituting, in succession, the values 1, 2, 3, 4, and 5 for n in $(2n - 1)$. Thus

$$\sum_{n=1}^{5} (2n - 1) = 1 + 3 + 5 + 7 + 9$$

$$= 25. \qquad \square$$

Two points need to be made in connection with the sigma notation. First, the index of summation is an arbitrary symbol, or a **dummy variable.** That is,

$$\sum_{n=1}^{k} a_n = \sum_{i=1}^{k} a_i = \sum_{j=1}^{k} a_j,$$

since each of these notations represents the sum

$$a_1 + a_2 + a_3 + \cdots + a_k.$$

The second point is that the initial value of the index is not necessarily 1. For example,

$$\sum_{j=3}^{5} \frac{1}{2^j} = \frac{1}{2^3} + \frac{1}{2^4} + \frac{1}{2^5} = \frac{1}{8} + \frac{1}{16} + \frac{1}{32} = \frac{7}{32}.$$

Note that the value $j = 3$ does not give the third term.

Exercises 9.1

In Exercises 1–18, write the first five terms of the sequence that has the given general term. (See Example 1.)

1. $a_n = \dfrac{1}{n+1}$

2. $a_n = 2n^2 - 1$

3. $a_n = \dfrac{n-1}{n}$

4. $a_n = \dfrac{n^2 - 1}{2n}$

5. $a_j = (-2)^j$

6. $a_j = (-1)^j$

7. $a_n = 2$

8. $a_n = 3$

9. $a_i = \dfrac{(-1)^i}{i^2}$

10. $a_i = (-1)^i i^2$

11. $a_j = \dfrac{j}{2j-1}$

12. $a_j = \dfrac{j}{j+1}$

13. $a_i = 3i - 2$

14. $a_i = 5i + 3$

15. $a_n = (x-1)^n$

16. $a_n = x^{n-1}, \quad x \neq 0$

17. $a_j = (-1)^j x^{2j}$

18. $a_j = (-1)^j x^{2j-1}$

Write the first six terms of each sequence. (See Example 2.)

19. $a_1 = 1, \quad a_{n+1} = 2a_n$

20. $a_1 = 2, \quad a_{n+1} = a_n + 3$

21. $a_1 = -5, \quad a_{n+1} = (-1)^n a_n$

22. $a_1 = 1, \quad a_{n+1} = x \cdot a_n$

23. $a_1 = 2, \quad a_2 = 3, \quad a_{n+1} = 2a_n + a_{n-1}$

24. $a_1 = -1, \quad a_2 = 1, \quad a_{n+1} = 2a_n - a_{n-1}$

Write each series in expanded form and find the value of the sum. (See Example 3.)

25. $\displaystyle\sum_{n=1}^{5} 2^n$

26. $\displaystyle\sum_{n=1}^{7} (-2)^n$

27. $\displaystyle\sum_{n=3}^{9} \dfrac{n-2}{n+1}$

28. $\displaystyle\sum_{n=2}^{9} \dfrac{n-1}{n}$

29. $\displaystyle\sum_{j=1}^{7} \dfrac{j^2 - j}{2}$

30. $\displaystyle\sum_{j=1}^{11} \dfrac{j+1}{2}$

31. $\displaystyle\sum_{i=1}^{7} (-1)^i$

32. $\displaystyle\sum_{i=4}^{10} 10^i$

33. $\displaystyle\sum_{n=2}^{7} 2$

34. $\displaystyle\sum_{n=2}^{6} 3$

35. $\displaystyle\sum_{j=2}^{7} j^2(-1)^j$

36. $\displaystyle\sum_{j=-1}^{5} j$

37. $\displaystyle\sum_{i=-1}^{5} i^3$

38. $\displaystyle\sum_{i=2}^{5} 6i(-1)^i$

39. $\displaystyle\sum_{n=1}^{6} \left(\dfrac{1}{n} - \dfrac{1}{n+1}\right)$

40. $\displaystyle\sum_{n=1}^{5} \left(\dfrac{1}{2^n} - \dfrac{1}{2^{n+1}}\right)$

41. $\displaystyle\sum_{n=2}^{5} (-1)^n \cdot 2^{-n}$

42. $\displaystyle\sum_{n=2}^{7} 3^{-n}$

Write each of the following sums in sigma notation.

43. $1 + 2 + 3 + \cdots + 17$

44. $1^2 + 2^2 + 3^2 + \cdots + 9^2$

45. $2 + 4 + 8 + \cdots + 128$

46. $3 + 9 + 27 + \cdots + 729$

47. $1 - 3 + 5 - 7 + 9 - 11 + 13$

48. $2 - 4 + 6 - 8 + 10 - 12 + 14$

49. $\dfrac{1}{2} + \dfrac{2}{3} + \dfrac{3}{4} + \dfrac{4}{5} + \dfrac{5}{6}$

50. $\dfrac{1}{2} + \dfrac{3}{4} + \dfrac{5}{8} + \dfrac{7}{16} + \dfrac{9}{32}$

51. $\dfrac{1}{2} - \dfrac{3}{4} + \dfrac{5}{8} - \dfrac{7}{16} + \dfrac{9}{32} - \dfrac{11}{64}$

52. $\dfrac{1}{3} + \dfrac{3}{5} + \dfrac{5}{7} + \dfrac{7}{9} + \dfrac{9}{11} + \dfrac{11}{13}$

53. $\dfrac{1}{1 \cdot 3} + \dfrac{1}{3 \cdot 5} + \dfrac{1}{5 \cdot 7} + \cdots + \dfrac{1}{85 \cdot 87}$

54. $\dfrac{1}{1 \cdot 2} + \dfrac{1}{2 \cdot 3} + \dfrac{1}{3 \cdot 4} + \cdots + \dfrac{1}{49 \cdot 50}$

55. $1 + \dfrac{x}{2} + \dfrac{x^2}{3} + \cdots + \dfrac{x^n}{n+1}$

56. $\dfrac{x}{2} + \dfrac{x^2}{4} + \dfrac{x^3}{8} + \cdots + \dfrac{x^n}{2^n}$

57. One of the important uses of calculus is to find the area of regions that have curved boundaries. The basic idea is to approximate such areas by sums of areas of rectangles. In part (a) of the figure below, the shaded region is bounded by the x-axis, the graph of $f(x) = x^2$, and the line $x = 2$. Approximate the area of this region by forming the sum of the areas of the four rectangles with bases on the subintervals

$$[0, \tfrac{1}{2}], \quad [\tfrac{1}{2}, 1], \quad [1, \tfrac{3}{2}], \quad \text{and} \quad [\tfrac{3}{2}, 2]$$

as shown in part (b).

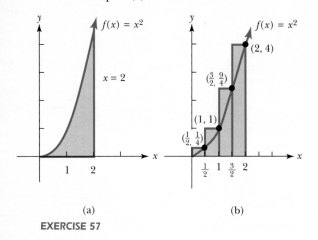

(a)   (b)

EXERCISE 57

58. (See Exercise 57.) Approximate the area of the shaded region in part (a) of the preceding figure by forming the sum of the areas of eight rectangles of the type used in Exercise 57 with bases $\tfrac{1}{4}$ unit wide on the subintervals

$$[0, \tfrac{1}{4}], \quad [\tfrac{1}{4}, \tfrac{1}{2}], \quad \cdots \quad , [\tfrac{3}{2}, \tfrac{7}{4}], \quad [\tfrac{7}{4}, 2].$$

59. Express the sum of the areas of the four rectangles in Exercise 57 in sigma notation where each term has the form

$$(\text{length})(\text{width}).$$

60. Express the sum of the areas of the eight rectangles in Exercise 58 in sigma notation where each term has the form

$$(\text{length})(\text{width}).$$

9.2 Arithmetic Sequences

An arithmetic sequence is a special type of sequence, defined as follows.

Definition 9.3

An **arithmetic sequence**[1] is a sequence in which each term after the first is obtained by adding the same number, d, to the preceding term. The constant d is called the **common difference.**

Example 1   Find the first five terms of the arithmetic sequence that has the first term and common difference as specified below.

$$a_1 = -2, \quad d = 3$$

Solution
The first five terms are computed as follows:

$$a_1 = -2,$$

$$a_2 = a_1 + d = -2 + 3 = 1,$$

---

1. An arithmetic sequence is also called an **arithmetic progression.**

$$a_3 = a_2 + d = 1 + 3 = 4,$$

$$a_4 = a_3 + d = 4 + 3 = 7,$$

$$a_5 = a_4 + d = 7 + 3 = 10. \qquad \square$$

It is easy to arrive at a formula for the nth term in an arithmetic sequence. Observing the pattern in

$$a_1 = a_1,$$

$$a_2 = a_1 + d,$$

$$a_3 = a_2 + d = a_1 + d + d = a_1 + 2d,$$

$$a_4 = a_3 + d = a_1 + 2d + d = a_1 + 3d,$$

$$a_5 = a_4 + d = a_1 + 3d + d = a_1 + 4d,$$

we see that the coefficient of d is always one less than the number of the term, and

$$a_n = a_1 + (n - 1)d. \qquad (1)$$

Example 2 Find the nineteenth term and a formula for the nth term of the following arithmetic sequence.

$$-1, 2, 5, 8, \ldots$$

Solution
The nineteenth term of the sequence is $a_{19} = a_1 + 18d$. To find the common difference, we need only subtract one of the terms from the succeeding term. Using the first two terms, we get

$$d = 2 - (-1) = 3.$$

Since $a_1 = -1$, we have

$$a_{19} = -1 + (18)(3) = 53$$

and

$$a_n = -1 + (n - 1)(3) = 3n - 4. \qquad \square$$

Example 3 A certain arithmetic sequence has $a_5 = 5$ and $a_{13} = 21$. Find the first term a_1, the common difference d, and a_{25}.

Solution
Using the formula $a_n = a_1 + (n - 1)d$ with $a_5 = 5$ and $a_{13} = 21$, we have

$$a_1 + 4d = 5 \qquad \text{Fifth term,}$$

$$a_1 + 12d = 21 \qquad \text{Thirteenth term.}$$

Subtracting the top equation from the bottom one, we get

$$8d = 16$$
$$d = 2.$$

Using $d = 2$ in $a_1 + 4d = 5$, we have

$$a_1 + 8 = 5$$
$$a_1 = -3.$$

Then

$$a_{25} = -3 + (24)(2) = 45,$$

and this completes the solution.　❏

The series formed by adding the first n terms of a general arithmetic sequence is denoted by S_n:

$$S_n = a_1 + a_2 + a_3 + \cdots + a_n.$$

A formula for S_n in terms of a_1 and a_n may be obtained as follows. First we write the sum S_n in the form

$$S_n = a_1 + (a_1 + d) + (a_1 + 2d) + \cdots + [a_1 + (n-1)d]. \quad (2)$$

We then write the same sum in reverse order, subtracting d from each term to get the next one.

$$S_n = a_n + (a_n - d) + (a_n - 2d) + \cdots + [a_n - (n-1)d] \quad (3)$$

Adding Eqs. (2) and (3), we have

$$2S_n = (a_1 + a_n) + (a_1 + a_n) + (a_1 + a_n) + \cdots + (a_1 + a_n). \quad (4)$$

The right member of Eq. (4) has n of the terms $a_1 + a_n$, so

$$2S_n = n(a_1 + a_n)$$

and

$$S_n = \frac{n}{2}(a_1 + a_n). \quad (5)$$

We can obtain another formula for S_n in terms of a_1 and d by substituting $a_n = a_1 + (n-1)d$ into this formula. This substitution gives

$$S_n = \frac{n}{2}[a_1 + a_1 + (n-1)d],$$

or

$$S_n = \frac{n}{2}[2a_1 + (n-1)d]. \quad (6)$$

For convenient reference, we record the formulas relating to arithmetic sequences in the following theorem.

Theorem 9.4

In an arithmetic sequence with first term a_1 and common difference d,

a) The **nth term** is given by

$$a_n = a_1 + (n - 1)d;$$

b) The **sum of the first n terms** is given by

$$S_n = \frac{n}{2}(a_1 + a_n),$$

or

$$S_n = \frac{n}{2}[2a_1 + (n - 1)d].$$

Example 4 Use the information given for each arithmetic sequence to find the twentieth term and the sum of the first 20 terms.

a) $-3, -6, -9, -12, \ldots$ b) $a_1 = 5, \quad a_{19} = 59$
c) $S_{10} = 145, \quad a_7 = 19$

Solution

a) It is clear that $a_1 = -3$ and $d = -3$. By the formula in Theorem 9.4(a), we have

$$a_{20} = a_1 + 19d$$

$$= -3 + (19)(-3) = -60.$$

By the first formula in Theorem 9.4(b),

$$S_{20} = \frac{20}{2}(a_1 + a_{20})$$

$$= 10(-3 - 60) = -630.$$

b) Substituting $a_1 = 5$ and $a_{19} = 59$ in the formula from Theorem 9.4(a), we have

$$59 = 5 + 18d$$

and

$$d = 3.$$

Thus

$$a_{20} = a_{19} + d = 59 + 3 = 62$$

and

$$S_{20} = \frac{20}{2}(a_1 + a_{20})$$

$$= 10(5 + 62)$$

$$= 670.$$

c)   Using $S_{10} = 145$ in the second formula from Theorem 9.4(b), we have

$$145 = \frac{10}{2}(2a_1 + 9d),$$

and this gives

$$145 = 10a_1 + 45d. \tag{7}$$

Using $a_7 = 19$ in Theorem 9.4(a), we have

$$19 = a_1 + 6d. \tag{8}$$

To find a_1 and d, we solve Eqs. (7) and (8) simultaneously. Multiplying both sides of Eq. (8) by 10 and rewriting Eq. (7), we have

$$190 = 10a_1 + 60d,$$

$$145 = 10a_1 + 45d.$$

Subtraction and division yield

$$d = 3.$$

Substituting $d = 3$ in Eq. (8) yields

$$a_1 = 1.$$

Thus

$$a_{20} = a_1 + 19d$$

$$= 1 + (19)(3)$$

$$= 58$$

and

$$S_{20} = \frac{20}{2}(a_1 + a_{20})$$

$$= 10(1 + 58)$$

$$= 590.$$

Note that this way of computing S_{20} is a lot shorter than actually adding $1 + 4 + 7 + 10 + \cdots + 58$ and is less prone to errors in arithmetic.   ❏

Exercises 9.2

Use the information given to find the first six terms of an arithmetic sequence that satisfies the stated conditions. (See Example 1.)

1. $a_1 = 1, \quad d = 1$

2. $a_1 = -3, \quad d = -2$

3. $a_4 = 7, \quad d = -2$

4. $a_4 = -7, \quad d = -5$

5. $a_3 = -5, \quad a_6 = -17$

6. $a_3 = 6, \quad a_4 = 5.9$

7. $a_3 - a_2 = 4, \quad a_4 = 14$

8. $a_5 = -7, \quad d = 2$

Determine whether or not each sequence is an arithmetic sequence. If it is, find d and a_n.

9. $-17, -12, -7, -2, \ldots$

10. $5, 2, -1, -4, \ldots$

11. $-6, -9, -12, -15, \ldots$

12. $\frac{5}{2}, 2, \frac{3}{2}, 1, \ldots$

13. $8, 4, 2, 1, \ldots$

14. $2, 4, 8, 16, \ldots$

15. $7, -7, -21, -35, \ldots$

16. $3, 3.04, 3.08, 3.12, \ldots$

17. $x, x + y, x + 2y, x + 3y, \ldots$

18. $-k, -k + x, -k + 2x, -k + 3x, \ldots$

19. $x, x^2, x^3, x^4, \ldots$

20. $x, x/2, x/4, x/8, \ldots$

21. $x, 2x, 3x, 4x, \ldots$

22. $2x, 4x, 6x, 8x, \ldots$

Use the information given to find a_{11} and S_{11} for an arithmetic sequence that satisfies the stated conditions. (See Example 4.)

23. $a_1 = -5, \quad d = -2$

24. $a_1 = 7, \quad a_5 = 15$

25. $a_7 = -5, \quad d = 3$

26. $a_5 = -3, \quad d = -3$

27. $a_5 = 5, \quad a_9 = -7$

28. $a_3 = 10, \quad a_7 = -2$

29. $a_1 = m, \quad d = -x$

30. $a_4 = m, \quad d = -x$

31. $a_5 = x + 2k, \quad a_7 = x - 2k$

32. $a_5 = 3 - 4p, \quad d = p$

Use the information given to find the value of the requested quantities in an arithmetic sequence. (See Examples 2–4.)

33. $a_1 = -7, \quad d = 3; \quad$ find a_{21}.

34. $a_1 = 17, \quad d = -5; \quad$ find a_{17}.

35. $a_3 = -7, \quad d = 16; \quad$ find S_4.

36. $a_{10} = -\frac{5}{2}, \quad d = \frac{1}{2}; \quad$ find S_{17}.

37. $-4, -1, 2, 5, \ldots; \quad$ find a_{10}.

38. $\frac{5}{2}, \frac{3}{2}, \frac{1}{2}, -\frac{1}{2}, \ldots; \quad$ find a_{15}.

39. $a_1 = 16, \quad a_{17} = -3; \quad$ find S_{57}.

40. $a_1 = -3, \quad a_{11} = 16; \quad$ find S_{200}.

41. $S_5 = -30, \quad a_7 = -4; \quad$ find a_1 and d.

42. $S_5 = \frac{5}{2}, \quad a_7 = -\frac{7}{2}; \quad$ find a_1 and d.

43. $S_7 = 14, \quad a_{10} = -28; \quad$ find a_1 and d.

44. $S_8 = 52, \quad a_4 = 5; \quad$ find a_1 and d.

Find the sum of each of the following series, using the fact that each is formed from an arithmetic sequence. (See Exercises 53 and 54 below.)

45. $\displaystyle\sum_{i=5}^{11} (2i - 3)$

46. $\displaystyle\sum_{i=3}^{16} (3i - 2)$

47. $\displaystyle\sum_{k=0}^{10} \left(5 - \frac{k}{3}\right)$

48. $\displaystyle\sum_{k=0}^{20} \left(-3 - \frac{k}{2}\right)$

49. $\displaystyle\sum_{n=1}^{50} 2n$

50. $\displaystyle\sum_{n=1}^{100} n$

51. $\displaystyle\sum_{n=1}^{13} (5 - 2n)$

52. $\displaystyle\sum_{n=1}^{19} (2n - \frac{11}{2})$

53. Prove that if $a_1, a_2, a_3, \ldots$ is a sequence that has a defining function f that is linear [that is, $a_n = f(n) = an + b$], then the sequence is an arithmetic sequence.

54. Prove that every arithmetic sequence has a defining function that is linear (that is, $a_n = an + b$ for some fixed a and b).

55. Insert numbers in the blanks below so that the sequence forms an arithmetic sequence.

$$-15, \underline{\quad}, \underline{\quad}, \underline{\quad}, \underline{\quad}, -40$$

56. Insert seven numbers between -5 and -26 so that the sequence thus formed is an arithmetic sequence.

57. Find the sum of the multiples of 5 between 7 and 213.

58. Find the sum of the integers from -13 to -71, inclusive.

59. Find the number n of positive integers between 17 and 199 that are multiples of 5.

60. Find the number n of positive integers between 17 and 199 that are multiples of 3.

61. If a child puts 1 penny in her piggy bank one week, 3 pennies the second week, 5 pennies the third week, 7 pennies the fourth week, and so on, how many pennies will she put in the bank on the twentieth week?

62. In Exercise 61, how much money will she have in her bank after she puts her money in on the twentieth week?

63. A stock boy is to display cans of pork and beans in a triangular-shaped stack like that in the accompanying photograph, with 1 can in the top row, 2 cans in the second row, 3 cans in the third row, and so on. Since he has to begin stacking from the bottom, how many cans should he place in the bottom row if he must display a total of 45 cans?

Photo by Susan Van Etten

64. In Exercise 63, how many cans should the stockboy place in the bottom row if he must display a total of 66 cans?

9.3 Geometric Sequences

In this section we consider another special type of sequence, called geometric sequences.

> **Definition 9.5**
>
> A **geometric sequence**[2] is a sequence in which each term after the first is obtained by multiplying the preceding term by a fixed nonzero number r. The constant r is called the **common ratio.**

A geometric sequence with first term a_1 and common ratio r is given by

$$a_1, a_1r, a_1r^2, a_1r^3, \ldots.$$

---

2. A geometric sequence is also called a **geometric progression.**

Example 1   Find the first four terms of the geometric sequence that has the first term and common ratio specified below.

a) $a_1 = -3$, $r = -2$ b) $a_1 = 16$, $r = \frac{1}{4}$

Solution

a) The first four terms of the geometric sequence are computed as follows:

$$a_1 = -3,$$

$$a_2 = a_1 r = (-3)(-2) = 6,$$

$$a_3 = a_1 r^2 = (-3)(-2)^2 = -12,$$

$$a_4 = a_1 r^3 = (-3)(-2)^3 = 24.$$

Thus the first four terms of the geometric sequence are -3, 6, -12, 24.

b) This time we illustrate another method of computation. We obtain the terms after the first one by multiplying by r.

$$a_1 = 16$$

$$a_2 = a_1 r = (16)\left(\frac{1}{4}\right) = 4$$

$$a_3 = a_2 r = (4)\left(\frac{1}{4}\right) = 1$$

$$a_4 = a_3 r = (1)\left(\frac{1}{4}\right) = \frac{1}{4}$$

Thus the first four terms are given by $16, 4, 1, \frac{1}{4}$.   ❑

From the pattern in the terms

$$a_1, a_1 r, a_1 r^2, a_1 r^3, a_1 r^4, \ldots$$

of a geometric sequence, it is easy to see that the nth term in a geometric sequence is given by the formula

$$a_n = a_1 r^{n-1}. \tag{1}$$

Example 2   Find the sixth term and a formula for the nth term of the following geometric sequence.

$$1, 2, 4, 8, \ldots$$

Solution
To find r, we need only choose one of the terms, other than the first, and divide it by the preceding term. Using the third and second terms,

we obtain $r = \frac{4}{2} = 2$. Since $a_1 = 1$, we have

$$a_6 = a_1 r^5 = (1)(2)^5 = 32$$

and

$$a_n = a_1 r^{n-1} = (1)(2)^{n-1} = 2^{n-1}. \qquad \square$$

Example 3   A geometric sequence has $a_4 = -3$ and $a_7 = -24$. Find the first term a_1 and the common ratio r.

Solution
Using the formula $a_n = a_1 r^{n-1}$ with $a_4 = -3$ and $a_7 = -24$, we have

$$-3 = a_1 r^3 \qquad \text{Fourth term,}$$

$$-24 = a_1 r^6 \qquad \text{Seventh term.}$$

We can solve for r by forming quotients of corresponding sides in these two equations:

$$\frac{a_1 r^6}{a_1 r^3} = \frac{-24}{-3}.$$

This gives

$$r^3 = 8$$

and

$$r = 2.$$

Substituting $r = 2$ and $a_4 = -3$ in $a_4 = a_1 r^3$, we have

$$-3 = a_1 (2)^3$$

and

$$a_1 = -\frac{3}{8}. \qquad \square$$

As was the case with arithmetic sequences, we can find a formula for the series formed by adding the first n terms of a general geometric sequence. To obtain the formula, let

$$S_n = a_1 + a_1 r + a_1 r^2 + \cdots + a_1 r^{n-3} + a_1 r^{n-2} + a_1 r^{n-1}. \qquad (2)$$

When both sides of this equation are multiplied by r, we have

$$rS_n = a_1 r + a_1 r^2 + a_1 r^3 + \cdots + a_1 r^{n-2} + a_1 r^{n-1} + a_1 r^n. \qquad (3)$$

Subtracting Eq. (3) from Eq. (2) yields

$$S_n - rS_n = a_1 - a_1 r^n,$$

or

$$(1 - r)S_n = a_1(1 - r^n).$$

If $r \neq 1$, both sides of this equation can be divided by $1 - r$ to obtain

$$S_n = a_1 \frac{1 - r^n}{1 - r} \qquad \text{if } r \neq 1. \tag{4}$$

This formula cannot be used if $r = 1$, but Eq. (2) easily gives $S_n = na_1$ for $r = 1$.

For easy reference, the formulas in Eqs. (1) and (4) are stated in the following theorem.

Theorem 9.6

In a geometric sequence with first term a_1 and common ratio r,

a)   The **nth term** is

$$a_n = a_1 r^{n-1};$$

b)   The **sum of the first n terms** is

$$S_n = a_1 \frac{1 - r^n}{1 - r} \qquad \text{if } r \neq 1.$$

Example 4   Find the sum of the first six terms in the geometric sequence

$$-\frac{1}{3}, \ -\frac{1}{9}, \ -\frac{1}{27}, \ -\frac{1}{81}, \ \dots \dots$$

Solution
Using $a_1 = -\frac{1}{3}$ and $r = \frac{1}{3}$ in the formula from Theorem 9.6(b), we have

$$S_6 = -\frac{1}{3} \cdot \frac{1 - \left(\dfrac{1}{3}\right)^6}{1 - \dfrac{1}{3}}$$

$$= -\frac{1}{3} \cdot \frac{1 - \dfrac{1}{729}}{1 - \dfrac{1}{3}}$$

$$= -\frac{1}{3} \cdot \frac{\dfrac{728}{729}}{\dfrac{2}{3}}$$

$$= -\frac{364}{729}.$$

Example 5   Use Theorem 9.6 to find the value of $\displaystyle\sum_{i=1}^{5} 3 \cdot 2^i$.

Solution

Expanding the sum, we have

$$\sum_{i=1}^{5} 3 \cdot 2^i = 3 \cdot 2 + 3 \cdot 2^2 + 3 \cdot 2^3 + 3 \cdot 2^4 + 3 \cdot 2^5.$$

From this expansion, it is clear that the terms in the sum form a geometric sequence with $a_1 = 3 \cdot 2 = 6$ and $r = 2$. By Theorem 9.6(b),

$$\sum_{i=1}^{5} 3 \cdot 2^i = 6 \cdot \frac{1 - 2^5}{1 - 2}$$

$$= 6 \cdot \frac{-31}{-1}$$

$$= 186.$$

Using the formula in Theorem 9.6 is a lot shorter than adding the terms in the sum. ❏

Up to this point, we have considered only series that are formed from a finite sequence. We now examine some cases where the sum of an infinite series has meaning.

Consider an infinite geometric sequence with first term a_1 and common ratio r:

$$a_1, a_1r, a_1r^2, a_1r^3, \ldots .$$

We have seen that the sum S_n of the first n terms of this sequence is given by

$$S_n = a_1 + a_1r + a_1r^2 + \cdots + a_1r^{n-1}$$

$$= a_1 \frac{1 - r^n}{1 - r} \qquad \text{if } r \neq 1.$$

If $|r| < 1$, that is, if $-1 < r < 1$, the term r^n steadily decreases in absolute value as n increases, getting closer and closer to 0. The fact that r^n gets nearer and nearer to 0 as n takes on larger and larger values suggests that the sums S_n should themselves be getting closer and closer to some certain value S. This is actually what happens, and we write

$$\lim_{n \to \infty} S_n = S$$

to indicate that the sums S_n get closer and closer to S as n increases without bound.

To illustrate how the sums S_n behave when $|r| < 1$, consider the particular case where $a_1 = 1$ and $r = \frac{1}{2}$. The geometric sequence is given by

$$1, \frac{1}{2}, \frac{1}{4}, \frac{1}{8}, \frac{1}{16}, \ldots,$$

and the sum of the first n terms is

$$S_n = 1 + \frac{1}{2} + \frac{1}{4} + \cdots + \frac{1}{2^{n-1}}$$

$$= 1 \cdot \frac{1 - \left(\frac{1}{2}\right)^n}{1 - \frac{1}{2}}$$

$$= 2\left[1 - \left(\frac{1}{2}\right)^n\right].$$

As n takes on the values 1, 2, 3, 4, 5, 6, $\ldots$ in succession, the corresponding values of S_n are given by

$$S_1 = 1 = 2\left(1 - \frac{1}{2}\right) = 2 - 1,$$

$$S_2 = 1 + \frac{1}{2} = 2\left(1 - \frac{1}{4}\right) = 2 - \frac{1}{2},$$

$$S_3 = 1 + \frac{1}{2} + \frac{1}{4} = 2\left(1 - \frac{1}{8}\right) = 2 - \frac{1}{4},$$

$$S_4 = 1 + \frac{1}{2} + \frac{1}{4} + \frac{1}{8} = 2\left(1 - \frac{1}{16}\right) = 2 - \frac{1}{8},$$

$$S_5 = 1 + \frac{1}{2} + \frac{1}{4} + \frac{1}{8} + \frac{1}{16} = 2\left(1 - \frac{1}{32}\right) = 2 - \frac{1}{16},$$

$$S_6 = 1 + \frac{1}{2} + \frac{1}{4} + \frac{1}{8} + \frac{1}{16} + \frac{1}{32} = 2\left(1 - \frac{1}{64}\right) = 2 - \frac{1}{32}.$$

From these computations it can be seen that, as n increases without bound, the sums S_n get closer and closer to the value 2. Thus

$$\lim_{n \to \infty} S_n = 2$$

for this geometric sequence. We say that 2 is the **sum** of the infinite geometric sequence, and we have

$$1 + \frac{1}{2} + \frac{1}{4} + \frac{1}{8} + \cdots = 2,$$

even though it is impossible to find the sum by actually performing the indicated additions. Another notation that is commonly used is

$$\sum_{n=1}^{\infty} \frac{1}{2^{n-1}} = 2.$$

The development in the particular case above can be carried out for any geometric sequence with $|r| < 1$. Since r^n gets closer and closer to 0 as n increases without bound, we have

$$\lim_{n\to\infty} r^n = 0.$$

With this fact in mind, it is clear that

$$\lim_{n\to\infty} S_n = \lim_{n\to\infty} a_1 \frac{1-r^n}{1-r}$$

$$= \frac{a_1}{1-r}.$$

If $|r| \geq 1$ and $a_1 \neq 0$, it can be shown that $\lim_{n\to\infty} S_n$ does not exist. This discussion is summarized in the following theorem.

Theorem 9.7

For an infinite geometric sequence $a_1, a_1r, a_1r^2, \ldots$, let
$$S_n = a_1 + a_1r + a_1r^2 + \cdots + a_1r^{n-1}.$$

a) If $|r| < 1$, the sum $a_1 + a_1r + a_1r^2 + \cdots$ of the infinite geometric sequence is given by

$$S = \lim_{n\to\infty} S_n = \frac{a_1}{1-r}.$$

b) If $|r| \geq 1$ and $a_1 \neq 0$, the sum of the series $\sum_{n=1}^{\infty} a_1r^{n-1}$ does not exist.

In the case where $|r| < 1$, we also write

$$a_1 + a_1r + a_1r^2 + \cdots = \frac{a_1}{1-r},$$

or

$$\sum_{n=1}^{\infty} a_1r^{n-1} = \frac{a_1}{1-r}.$$

Example 6 For each of the following geometric sequences, determine whether or not the sum exists. If the sum exists, find its value.

a) $\frac{3}{5}, \frac{1}{5}, \frac{1}{15}, \frac{1}{45}, \ldots$ b) $\frac{2}{3}, 1, \frac{3}{2}, \frac{9}{4}, \ldots$

Solution

a) The common ratio is given by

$$r = \frac{\dfrac{1}{5}}{\dfrac{3}{5}} = \frac{1}{3}.$$

Since $|r| < 1$, the sum exists, by Theorem 9.7(a), and

$$\frac{3}{5} + \frac{1}{5} + \frac{1}{15} + \frac{1}{45} + \cdots = \frac{a_1}{1 - r}$$

$$= \frac{\dfrac{3}{5}}{1 - \dfrac{1}{3}}$$

$$= \frac{9}{10}.$$

b) It is easy to see that $r = \frac{3}{2}$. Since $|r| > 1$, the sum does not exist, by Theorem 9.7(b). ❏

The formula in Theorem 9.7(a) can be used to write a repeating decimal as a quotient of integers.

Example 7 Write the rational number $3.4173173173\ldots$ as a quotient of integers.

Solution
We first write the number as follows:

$$3.4173173173\ldots = 3.4 + 0.0173 + 0.0000173 + \cdots.$$

The terms on the right after 3.4 are from the geometric sequence with $a_1 = 0.0173$ and $r = 0.001$. The sum of this sequence is

$$\frac{0.0173}{1 - 0.001} = \frac{0.0173}{0.999}$$

$$= \frac{173}{9990}.$$

This means that the rational number 3.4173173173 . . . is equal to

$$3.4 + \frac{173}{9990} = \frac{34}{10} + \frac{173}{9990} = \frac{34,139}{9990}.$$

This can be checked by division. ❏

Exercises 9.3

For the given value of n, write out the first n terms of the geometric sequence that satisfies the stated conditions. (See Example 1.)

1. $a_1 = -1$, $r = -2$, $n = 5$
2. $a_1 = 3$, $r = \frac{1}{2}$, $n = 4$
3. $a_1 = 4$, $r = \frac{1}{4}$, $n = 4$
4. $a_1 = -6$, $r = 2$, $n = 5$
5. $a_1 = \frac{3}{4}$, $r = 4$, $n = 4$
6. $a_1 = -\frac{1}{8}$, $r = -2$, $n = 5$
7. $a_2 = -\frac{1}{2}$, $a_3 = 1$, $n = 5$
8. $a_3 = \frac{2}{3}$, $a_4 = 1$, $n = 4$
9. $a_1 = -3$, $a_3 = -12$, $n = 3$
10. $a_1 = 4$, $a_3 = \frac{1}{16}$, $n = 4$

Find the fifth term, the nth term, and the sum of the first five terms of a geometric sequence that satisfies the stated conditions. (See Examples 2 and 4.)

11. $\frac{1}{4}, \frac{1}{2}, 1, 2, \ldots$
12. $-1, -\sqrt{3}, -3, -3\sqrt{3}, \ldots$
13. $a_1 = \frac{1}{3}$, $r = -3$
14. $a_1 = 4$, $r = \frac{1}{4}$
15. $a_3 = -2$, $a_4 = 4$
16. $a_3 = -3$, $a_4 = \frac{3}{2}$

Determine whether or not each of the following sequences is a geometric sequence. If it is, find r and a_n.

17. $7, \frac{7}{2}, \frac{7}{4}, \frac{7}{8}, \ldots$
18. $\sqrt{3}, 3, 3\sqrt{3}, 9, \ldots$
19. $4, 2\sqrt{2}, 2, \sqrt{2}, \ldots$
20. $\frac{5}{6}, \frac{5}{3}, \frac{10}{3}, \frac{20}{3}, \ldots$
21. $2, -4, 6, -8, \ldots$
22. $1, 3, 7, 15, \ldots$
23. $1, 4, 8, 12, \ldots$
24. $3, 6, 18, 108, \ldots$
25. $-4, 2, -1, \frac{1}{2}, \ldots$
26. $-5, \frac{5}{3}, -\frac{5}{9}, \frac{5}{27}, \ldots$
27. $343, -49, 7, -1, \ldots$
28. $\frac{1}{9}, -\frac{1}{3}, 1, -3, \ldots$

Find the sums of each of the following series, using the fact that each is from a geometric sequence. (See Example 5.)

29. $\sum_{i=1}^{5} 2^{i-1}$
30. $\sum_{n=0}^{5} 4^n$
31. $\sum_{i=1}^{5} (\frac{3}{5})^i$
32. $\sum_{n=4}^{9} \frac{1}{3^n}$
33. $\sum_{n=0}^{4} 5(-\frac{2}{3})^n$
34. $\sum_{j=1}^{6} 128(-\frac{3}{2})^j$
35. $\sum_{j=0}^{4} 64(\frac{5}{4})^j$
36. $\sum_{n=2}^{7} 9(\frac{5}{3})^{n-2}$

Find r and a_1 for the geometric sequence that satisfies the given conditions. (See Example 3.)

37. $a_7 = -9$, $a_{11} = -81$
38. $a_4 = -\frac{2}{3}$, $a_7 = -\frac{9}{4}$
39. $a_6 = 4(1.01)^4$, $a_8 = 4(1.01)^6$
40. $a_4 = 2$, $a_8 = \frac{2}{81}$

For each of the following geometric sequences, determine whether or not the sum exists. If the sum exists, find its value. (See Example 6.)

41. $-3, \frac{3}{2}, -\frac{3}{4}, \frac{3}{8}, \ldots$
42. $-4, \frac{4}{3}, -\frac{4}{9}, \frac{4}{27}, \ldots$
43. $\frac{9}{2}, -\frac{3}{2}, \frac{1}{2}, -\frac{1}{6}, \ldots$
44. $\frac{5}{9}, \frac{1}{9}, \frac{1}{45}, \frac{1}{225}, \ldots$
45. $-\frac{4}{3}, 4, -12, 36, \ldots$
46. $8, -4, 2, -1, \ldots$
47. $1, 1.01, (1.01)^2, (1.01)^3, \ldots$
48. $1, 1.2, 1.44, 1.728, \ldots$
49. $2, \sqrt{2}, 1, 1/\sqrt{2}, \ldots$
50. $-0.9, 0.81, -0.729, 0.6561, \ldots$
51. $5, 0.5, 0.05, 0.005, \ldots$
52. $5, 2.5, 1.25, 0.625, \ldots$

Find the value of each of the following sums if it exists. If it does not exist, give a reason. (See Example 6.)

53. $15 + \frac{15}{2} + \frac{15}{4} + \frac{15}{8} + \cdots$
54. $52 + 4 + \frac{4}{13} + \frac{4}{169} + \cdots$

55. $4 - \frac{4}{3} + \frac{4}{9} - \frac{4}{27} + \cdots$

56. $-17 + \frac{17}{3} - \frac{17}{9} + \frac{17}{27} - \cdots$

57. $\frac{1}{36} + \frac{1}{6} + 1 + 6 + \cdots$

58. $\frac{1}{16} + \frac{1}{4} + 1 + 4 + \cdots$

59. $\displaystyle\sum_{i=0}^{\infty} 5\left(\frac{1}{2}\right)^i$

60. $\displaystyle\sum_{i=2}^{\infty} -15\left(\frac{2}{3}\right)^i$

61. $\displaystyle\sum_{i=3}^{\infty} 17(-3)^i$

62. $\displaystyle\sum_{n=4}^{\infty} 7(2)^n$

63. $\displaystyle\sum_{i=1}^{\infty} \frac{13}{4^i}$

64. $\displaystyle\sum_{n=1}^{\infty} \frac{8}{5^n}$

Express each rational number as a quotient of integers. (See Example 7.)

65. $0.9999\ldots$

66. $0.1111\ldots$

67. $0.010101\ldots$

68. $0.313131\ldots$

69. $3.111111\ldots$

70. $3.8787878\ldots$

71. $-2.2917917\ldots$

72. $-9.01727272\ldots$

73. Adrienne decided to save money to buy a new television set. She saved 1 dollar the first day, and on each day following she saved twice as much as she had the day before. To her surprise, she had saved exactly enough to buy the set at the end of the eighth day. How much did the television set cost?

74. It is known that the population of a country in 1920 was 10,000,000 people. If the population doubles every 20 years, what will the population be in the year 2000?

75. At a certain rate of inflation, the cost of living doubles every 6 years. If a person earns $24,000 now,

how much must she earn 36 years from now to keep her salary in pace with inflation?

76. The number of bacteria in a culture is observed to triple every hour. If initially there are 1000 bacteria, approximately how long will it take for 1,000,000 bacteria to be present?

77. Mary has a ball that will bounce back to $\frac{3}{7}$ of the height from which it is dropped. If she drops the ball from a height of 4 feet, find the approximate distance it will travel before coming to rest. (See the figure.)

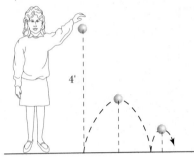

EXERCISE 77

78. Work Exercise 77 for a ball that bounces back each time to $\frac{1}{3}$ of the height from which it falls.

79. Suppose that a culture of bacteria doubles every 20 minutes. If the culture initially contains 4000 bacteria, how many are present after 2 hours?

80. Sharky, the local lender, computes interest by doubling the amount of interest owed every week. If his favorite customer owes him $50 in interest today, how much interest will he owe at the end of 10 weeks?

9.4 Mathematical Induction

Mathematical induction is a method of proof used mainly to prove theorems which assert that a certain statement holds true for all positive integers.

The method of proof by mathematical induction is based on the following property of the positive integers: If T is a set such that

a) 1 is in T,

b) $k \in T$ always implies $k + 1 \in T$,

then T contains all the positive integers.

To get some intuitive feeling for the method, assume that T is a set that satisfies the two conditions above, and let us apply the conditions a few times. We have

$1 \in T$, by condition (a);

$1 \in T$ implies $1 + 1 = 2 \in T$, by condition (b);

$2 \in T$ implies $2 + 1 = 3 \in T$, by condition (b);

$3 \in T$ implies $3 + 1 = 4 \in T$, by condition (b);

$4 \in T$ implies $4 + 1 = 5 \in T$, by condition (b);

and condition (b) can be applied repeatedly, as long as we wish. By applying condition (b) enough times, we can arrive at the statement that any given positive integer is in T. Thus T must contain all the positive integers.

Mathematical induction is usually employed in connection with a certain statement $S(n)$ about the positive integer n. As an illustration, $S(n)$ might be the statement that

$$1 + 2 + 3 + \cdots + n = \frac{n(n + 1)}{2}$$

for an arbitrary positive integer n. As another example, $S(n)$ might be the statement that

$$1 + 2n \leq 3^n.$$

Such statements as these can be proved by using the Principle of Mathematical Induction, stated in the following theorem.

Theorem 9.8 **Principle of Mathematical Induction**

For each positive integer n, suppose that $S(n)$ represents a statement about n that is either true or false. If

a) $S(1)$ is true,

and if

b) the truth of $S(k)$ always implies the truth of $S(k + 1)$,

then $S(n)$ is true for all positive integers n.

A proof based on the Principle of Mathematical Induction consists of three parts:

1. The statement is verified for $n = 1$.

2. The statement is *assumed*[3] true for $n = k$, and with this assumption made,

3. The statement is then proved to be true for $n = k + 1$.

It then follows that the statement is true for all positive integers n.
    This type of proof is illustrated in the following examples.

Example 1   Use mathematical induction to prove that

$$1 + 2 + 3 + \cdots + n = \frac{n(n + 1)}{2} \tag{1}$$

for any positive integer n.

Solution
Let $S(n)$ be the statement that Eq. (1) holds true.

1. We first verify that $S(1)$ is true. (In a formula such as this, it is understood that when $n = 1$, there is only one term on the left side, and no addition is performed in this case.) The value of the left side is 1 when $n = 1$, and the value of the right side is

$$\frac{1(1 + 1)}{2} = \frac{(1)(2)}{2} = 1.$$

Thus $S(1)$ is true.

2. Assume that $S(k)$ is true. That is, assume that

$$1 + 2 + 3 + \cdots + k = \frac{k(k + 1)}{2}. \tag{2}$$

3. We must now prove that $S(k + 1)$ is true. By adding $k + 1$ to both sides of Eq. (2), we obtain

$$1 + 2 + 3 + \cdots + k + (k + 1) = \frac{k(k + 1)}{2} + k + 1$$

$$= \frac{k(k + 1) + 2(k + 1)}{2}$$

$$= \frac{(k + 1)(k + 2)}{2}$$

$$= \frac{(k + 1)[(k + 1) + 1]}{2}.$$

---

3. This assumption is frequently referred to as the induction hypothesis.

The last expression exactly matches the right member of Eq. (1) with n replaced by $k + 1$. We have shown that the truth of $S(k)$ implies the truth of $S(k + 1)$. Therefore the statement $S(n)$ is true for all positive integers n. ❏

Example 2   Use mathematical induction to prove that

$$\frac{1}{1 \cdot 2} + \frac{1}{2 \cdot 3} + \frac{1}{3 \cdot 4} + \cdots + \frac{1}{n(n + 1)} = \frac{n}{n + 1} \tag{3}$$

for any positive integer n.

Solution

1. For $n = 1$, the left member of Eq. (3) is

$$\frac{1}{1 \cdot 2} = \frac{1}{2},$$

and the right member is

$$\frac{1}{1 + 1} = \frac{1}{2}.$$

Thus the statement is true for $n = 1$.

2. Assume that

$$\frac{1}{1 \cdot 2} + \frac{1}{2 \cdot 3} + \frac{1}{3 \cdot 4} + \cdots + \frac{1}{k(k + 1)} = \frac{k}{k + 1}. \tag{4}$$

3. To change the left member of Eq. (4) to the left member of Eq. (3) when $n = k + 1$, we need to add

$$\frac{1}{(k + 1)[(k + 1) + 1]} = \frac{1}{(k + 1)(k + 2)}$$

to both sides of Eq. (4). Doing this, we have

$$\frac{1}{1 \cdot 2} + \frac{1}{2 \cdot 3} + \frac{1}{3 \cdot 4} + \cdots + \frac{1}{k(k + 1)} + \frac{1}{(k + 1)(k + 2)}$$

$$= \frac{k}{k + 1} + \frac{1}{(k + 1)(k + 2)}$$

$$= \frac{k(k + 2) + 1}{(k + 1)(k + 2)}$$

$$= \frac{k^2 + 2k + 1}{(k + 1)(k + 2)}$$

$$= \frac{(k + 1)^2}{(k + 1)(k + 2)}$$

$$= \frac{k + 1}{k + 2}$$

$$= \frac{k + 1}{(k + 1) + 1}.$$

The last expression matches the right member of Eq. (3), with n replaced by $k + 1$. Thus the truth of the statement for $n = k$ implies the truth of the statement for $n = k + 1$. By the Principle of Mathematical Induction, Eq. (3) holds for all positive integers n. ❑

In the calculus, statements $S(n)$ are sometimes found that are false for a few values of the positive integer n but are true for all positive integers n that are sufficiently large. Statements of this form can be proved by a modified form of mathematical induction. Let m be a positive integer. To prove that $S(n)$ is true for all $n \geq m$, we alter our previous steps so that we

1. Verify that the statement is true for $n = m$;

2. Assume that the statement is true for $n = k$, where $k \geq m$;

3. Prove the statement is true for $n = k + 1$.

The next example demonstrates this method of proof.

Example 3   Prove that

$$1 + 2n < 2^n \tag{5}$$

for every positive integer $n \geq 3$.

Solution
For $n = 3$,

$$1 + 2n = 1 + 6 = 7 \quad \text{and} \quad 2^n = 2^3 = 8.$$

Since $7 < 8$, the statement is true for $n = 3$.
    Assume now that the statement is true for k, where $k \geq 3$:

$$1 + 2k < 2^k.$$

When $n = k + 1$, the left member of Eq. (5) is $1 + 2(k + 1)$, and

$$1 + 2(k + 1) = 1 + 2k + 2$$

$$< 2^k + 2, \qquad \text{since } 1 + 2k < 2^k;$$

$$< 2^k + 2^k, \qquad \text{since } 2 < 2^k \text{ for } k \geq 3.$$

But $2^k + 2^k = 2^k(1 + 1) = 2^k(2) = 2^{k+1}$ is the right member of Eq. (5) when $n = k + 1$. Thus we have proved that

$$1 + 2n < 2^n$$

is true when $n = k + 1$. Therefore the statement is true for all positive integers $n \geq 3$. ❑

Exercises 9.4

Use mathematical induction to prove that each of the following statements is true for all positive integers n.

1. $1 + 3 + 5 + \cdots + (2n - 1) = n^2$

2. $2 + 4 + 6 + \cdots + (2n) = n(n + 1)$

3. $1^2 + 2^2 + 3^2 + \cdots + n^2 = \dfrac{n(n + 1)(2n + 1)}{6}$

4. $1 \cdot 2 + 2 \cdot 3 + 3 \cdot 4 + \cdots + n(n + 1) = \dfrac{n(n + 1)(n + 2)}{3}$

5. $1 \cdot 2 + 2 \cdot 2^2 + 3 \cdot 2^3 + \cdots + n \cdot 2^n = (n - 1)2^{n+1} + 2$

6. $2 + 4 + 8 + \cdots + 2^n = 2^{n+1} - 2$

7. $4 + 4^2 + 4^3 + \cdots + 4^n = \dfrac{4(4^n - 1)}{3}$

8. $4 + 8 + 12 + 16 + \cdots + 4n = 2n(n + 1)$

9. $3 + 6 + 9 + 12 + \cdots + 3n = \dfrac{3n(n + 1)}{2}$

10. $\dfrac{1}{1 \cdot 4} + \dfrac{1}{4 \cdot 7} + \dfrac{1}{7 \cdot 10} + \cdots +$
$\dfrac{1}{(3n - 2)(3n + 1)} = \dfrac{n}{3n + 1}$

11. $\dfrac{1}{1 \cdot 2 \cdot 3} + \dfrac{1}{2 \cdot 3 \cdot 4} + \dfrac{1}{3 \cdot 4 \cdot 5} + \cdots +$
$\dfrac{1}{n(n + 1)(n + 2)} = \dfrac{n(n + 3)}{4(n + 1)(n + 2)}$

12. $1^3 + 2^3 + 3^3 + \cdots + n^3 = \dfrac{n^2(n + 1)^2}{4}$

13. $\dfrac{1}{3} + \dfrac{1}{3^2} + \dfrac{1}{3^3} + \cdots + \dfrac{1}{3^n} = \dfrac{1}{2}\left[1 - \left(\dfrac{1}{3}\right)^n\right]$

14. $a + (a + d) + (a + 2d) + \cdots + [a + (n - 1)d] = \dfrac{n}{2}[2a + (n - 1)d]$

15. $a + ar + ar^2 + \cdots + ar^{n-1} = a\dfrac{1 - r^n}{1 - r}$ if $r \neq 1$

16. 3 is a factor of $4^n - 1$.

17. 3 is a factor of $n^3 + 2n$.

18. $1 - x$ is a factor of $1 - x^n$.

19. $a - b$ is a factor of $a^n - b^n$.
[Hint: $a^{k+1} - b^{k+1} = a^k(a - b) + (a^k - b^k)b$]

20. $a + b$ is a factor of $a^{2n} - b^{2n}$.

21. $x^{2n} > 0$ if $x \neq 0$.

22. $\left(\dfrac{a}{b}\right)^n < 1$ if $0 < a < b$.

23. $1 + 2n \leq 3^n$

24. Use mathematical induction and the result in Example 3 to prove that $n^2 < 2^n$ if $n \geq 5$.

25. Assume that $|a + b| \leq |a| + |b|$ for all real numbers a and b. Then prove that $|a_1 + a_2 + \cdots + a_n| \leq |a_1| + |a_2| + \cdots + |a_n|$ for any real numbers $a_1, a_2, \ldots, a_n$.

26. Use mathematical induction to prove that $\overline{z^n} = (\bar{z})^n$ for any complex number z and any positive integer n.

27. Show that if the statement

$$1 + 2 + 3 + \cdots + n = \dfrac{n(n + 1)}{2} + 2$$

is assumed to be true for $n = k$, the same equation can be proved to hold for $n = k + 1$. Is the statement true for all positive integers?

28. Show that $n^2 - n + 5$ is a prime integer when n is equal to 1, 2, 3, or 4, but that it is not true that $n^2 - n + 5$ is always a prime integer. Investigate the same set of statements for the polynomial $n^2 - n + 11$.

In Exercises 29–31, use mathematical induction to prove the stated property of the sigma notation.

29. $\displaystyle\sum_{i=1}^{n} ca_i = c\sum_{i=1}^{n} a_i$

30. $\displaystyle\sum_{i=1}^{n} (a_i + b_i) = \sum_{i=1}^{n} a_i + \sum_{i=1}^{n} b_i$

31. $\displaystyle\sum_{i=1}^{n} (a_i - b_i) = \sum_{i=1}^{n} a_i - \sum_{i=1}^{n} b_i$

9.5 The Binomial Theorem

In this section a formula is presented for expanding a positive integral power $(a + b)^n$ of a binomial $a + b$ into a sum of terms. The statement of the formula is called the *Binomial Theorem*.

The *factorial* notation described in the next definition is needed in the statement of the Binomial Theorem.

Definition 9.9

If n is a positive integer, then **n factorial**, denoted by $n!$, is the product of the n integers $n, n - 1, n - 2, \ldots, 2, 1$; that is,

$$n! = n \cdot (n - 1) \cdot (n - 2) \cdots 2 \cdot 1.$$

Also,

$$0! = 1.$$

The reason for defining $0! = 1$ will be clear after Theorem 9.10.

Example 1 Some illustrations of the use of the factorial notation are given below.

a) $3! = 3 \cdot 2 \cdot 1 = 6$

b) $7! = 7 \cdot 6 \cdot 5 \cdot 4 \cdot 3 \cdot 2 \cdot 1 = 5040$

c) $1! = 1$

d) $(n - 1)! = (n - 1)(n - 2)(n - 3) \cdots (2)(1)$

e) $n(n - 1)! = n(n - 1)(n - 2)(n - 3) \cdots (2)(1) = n!$ ❏

When n is a positive integer, $(a + b)^n$ represents a product of n factors, with each factor being $(a + b)$. By direct multiplication, we have the following:

$$(a + b)^1 = a + b,$$
$$(a + b)^2 = a^2 + 2ab + b^2,$$

$$(a + b)^3 = a^3 + 3a^2b + 3ab^2 + b^3,$$
$$(a + b)^4 = a^4 + 4a^3b + 6a^2b^2 + 4ab^3 + b^4,$$
$$(a + b)^5 = a^5 + 5a^4b + 10a^3b^2 + 10a^2b^3 + 5ab^4 + b^5.$$

By examining these expansions, we can make the following observations about $(a + b)^n$.

1. There are $n + 1$ terms.
2. The highest power of a and b is n.
3. In each term, the sum of the powers of a and b is n.
4. As the power of a decreases in each successive term, the power of b increases.

The preceding expansions illustrate the fact that the terms in the expansion of $(a + b)^n$ are terms involving products of the form

$$a^n, a^{n-1}b, a^{n-2}b^2, \ldots, a^2b^{n-2}, ab^{n-1}, b^n.$$

In general, the coefficient of $a^{n-r}b^r$ in the expansion of $(a + b)^n$ is given by $\dbinom{n}{r}$, where

$$\binom{n}{r} = \frac{n!}{(n - r)!r!}.$$

The numbers $\dbinom{n}{r}$, $r = 0, 1, 2, \ldots, n$, are called the **binomial coefficients.** We can now state the Binomial Theorem.

Theorem 9.10   **The Binomial Theorem**

Let n be a positive integer, and let a and b be real or complex numbers. Then $(a + b)^n$ is a sum of $n + 1$ terms of the form

$$\binom{n}{r}a^{n-r}b^r, \qquad r = 0, 1, 2, \ldots, n.$$

More specifically,

$$(a + b)^n = a^n + \binom{n}{1}a^{n-1}b + \binom{n}{2}a^{n-2}b^2 + \cdots + \binom{n}{r}a^{n-r}b^r + \cdots$$

$$+ \binom{n}{n - 2}a^2b^{n-2} + \binom{n}{n - 1}ab^{n-1} + b^n.$$

Mathematical induction can be used to prove the Binomial Theorem, but the details are somewhat involved. We omit the proof here and concentrate on some examples that illustrate its use.

We note that with the definition $0! = 1$, the formula $\binom{n}{r}a^{n-r}b^r$ works for *all* the terms, including those with $r = 0$ and $r = n$.

Example 2   Expand $(x^2 - 2)^6$.

Solution
Using the Binomial Theorem, we first write $(x^2 - 2)^6$ as $[x^2 + (-2)]^6$ and

$$[x^2 + (-2)]^6 = (x^2)^6 + \binom{6}{1}(x^2)^5(-2) + \binom{6}{2}(x^2)^4(-2)^2$$

$$+ \binom{6}{3}(x^2)^3(-2)^3 + \binom{6}{4}(x^2)^2(-2)^4$$

$$+ \binom{6}{5}(x^2)(-2)^5 + (-2)^6$$

$$= x^{12} + 6 \cdot x^{10} \cdot (-2) + 15 \cdot x^8 \cdot (4) + 20 \cdot x^6 \cdot (-8)$$

$$+ 15 \cdot x^4 \cdot (16) + 6 \cdot x^2 \cdot (-32) + 64.$$

Thus

$$(x^2 - 2)^6 = x^{12} - 12x^{10} + 60x^8 - 160x^6 + 240x^4 - 192x^2 + 64.$$

❏

The Binomial Theorem can also be used to determine the coefficient of a particular term in the expansion of $(a + b)^n$.

Example 3   Find the coefficient of $t^8 s^9$ in the expansion of $(t^2 - s)^{13}$.

Solution
The terms in the expansion of $(t^2 - s)^{13}$ are of the form

$$\binom{n}{r}(t^2)^{n-r}(-s)^r.$$

The coefficient required is in the term with $r = 9$ and $n - r = 13 - 9 = 4$. Therefore the term in the expansion is

$$\binom{13}{9}(t^2)^4(-s)^9,$$

and the required coefficient is

$$-\binom{13}{9} = -\frac{13!}{4!9!} = -715.$$

❏

Notice in the binomial expansion that the power of b is always one less than the term number. For example, b^1 occurs in term number 2, b^2 occurs in term number 3, . . . , b^n occurs in term number $n + 1$. Thus b^{r-1} occurs in the rth term. Since the sum of the powers of a and b must be equal to n, then the corresponding power of a is $n - (r - 1)$. Knowing the power of b leads to the correct selection of the binomial coefficient, $\binom{n}{r-1}$. Thus we have the following result.

---

rth Term in the Binomial Expansion

The rth term $(1 \le r \le n + 1)$ in the expansion of $(a + b)^n$ is

$$\binom{n}{r-1}a^{n-r+1}b^{r-1}.$$

Example 4   Find the twelfth term in the expansion of $(3x^2 - y^3)^{15}$.

Solution

The value of n is 15, and $r - 1$ is 11 in the twelfth term. Thus the twelfth term is given by

$$\binom{15}{11}(3x^2)^4(-y^3)^{11} = -\frac{15!}{4!11!}3^4x^8y^{33}$$

$$= -\frac{15 \cdot \overset{7}{\cancel{14}} \cdot 13 \cdot \cancel{12} \cdot \cancel{11!}}{\cancel{4} \cdot \cancel{3} \cdot \cancel{2} \cdot 1 \cdot \cancel{11!}} \cdot 81x^8y^{33}$$

$$= -110{,}565x^8y^{33}.$$

❏

It is shown in calculus that the Binomial Theorem is valid under certain conditions when n is any positive or negative rational number. Before we formulate these conditions, we make the following observation. The quantity $\binom{n}{r}$ can be rewritten as

$$\binom{n}{r} = \frac{n!}{(n-r)!r!}$$

$$= \frac{n(n-1)(n-2)\cdots(n-r+1)(n-r)!}{(n-r)!r!}$$

$$= \frac{n(n-1)(n-2)\cdots(n-r+1)}{r!}.$$

The binomial expansion can now be restated as

$$(a+b)^n = a^n + na^{n-1}b + \frac{n(n-1)}{2!}a^{n-2}b^2 + \frac{n(n-1)(n-2)}{3!}a^{n-3}b^3$$

$$+ \cdots + \frac{n(n-1)(n-2)\cdots(n-r+1)}{r!}a^{n-r}b^r + \cdots.$$

If n is not a positive integer, the binomial expansion never terminates, but continues indefinitely, since $n - r$ is never zero. In the calculus, it is shown that if $a = 1$ and $b = x$, then the binomial expansion for

$$(a+b)^n = (1+x)^n$$

is valid for all x such that $|x| < 1$.

Example 5   Find the first five terms of the expansion of $(1 + x)^{1/2}$.

Solution
The expansion is valid for all x such that $|x| < 1$. We have

$$(1+x)^{1/2} = 1 + \frac{1}{2}\cdot x + \frac{\left(\frac{1}{2}\right)\left(\frac{1}{2}-1\right)}{2!}x^2 + \frac{\frac{1}{2}\left(\frac{1}{2}-1\right)\left(\frac{1}{2}-2\right)}{3!}x^3$$

$$+ \frac{\left(\frac{1}{2}\right)\left(\frac{1}{2}-1\right)\left(\frac{1}{2}-2\right)\left(\frac{1}{2}-3\right)}{4!}x^4 + \cdots.$$

$$= 1 + \frac{1}{2}x - \frac{1}{8}x^2 + \frac{1}{16}x^3 - \frac{5}{128}x^4 + \cdots. \qquad \square$$

We close this section with an interesting pattern that emerges when only the coefficients of $(a + b)^n$ for $n = 0, 1, 2, \ldots$ are examined.

$(a + b)^n$	Coefficients of $(a + b)^n$
$(a + b)^0$	1
$(a + b)^1$	1 1
$(a + b)^2$	1 2 1
$(a + b)^3$	1 3 3 1
$(a + b)^4$	1 4 6 4 1
$(a + b)^5$	1 5 10 10 5 1

Blaise Pascal
Courtesy of the Bettmann Archive

This triangular pattern is called **Pascal's Triangle**; it is named for the French mathematician Blaise Pascal (1623–1662). Notice that once the pattern is begun, it can be continued by adding two consecutive coefficients together in one row to obtain a coefficient in the next row.

Example 6   Use Pascal's Triangle to write out the expansion of $(x - 2y)^4$.

Solution

The coefficients for the expansion of $(x - 2y)^4$ come from row 5 of Pascal's Triangle.

$$(x - 2y)^4 = \mathbf{1}x^4 + \mathbf{4}x^3(-2y)^1 + \mathbf{6}x^2(-2y)^2 + \mathbf{4}x(-2y)^3 + \mathbf{1}(-2y)^4$$

$$= x^4 + 4(-2)x^3y + 6(4)x^2y^2 + 4(-8)xy^3 + 1(16)y^4$$

$$= x^4 - 8x^3y + 24x^2y^2 - 32xy^3 + 16y^4 \qquad \square$$

Exercises 9.5

Expand by using the Binomial Theorem. (See Examples 2 and 6.)

1. $(x + y)^7$

2. $(a + b)^8$

3. $(2x + y)^4$

4. $(x + 3y)^6$

5. $(x^2 + 2y)^5$

6. $(3x + y^2)^3$

7. $(x^2 - y^2)^4$

8. $(a^3 - b^2)^5$

9. $\left(2x - \dfrac{1}{y}\right)^4$

10. $\left(x + \dfrac{1}{x}\right)^3$

11. $\left(3x + \dfrac{y}{3}\right)^4$

12. $\left(\dfrac{x}{4} - 2y\right)^6$

13. $\left(x^3 - \dfrac{x^2}{2}\right)^6$

14. $(3x^2 - x)^4$

15. $(\tfrac{1}{2}a - 3b^2)^5$

16. $(\tfrac{1}{3}a^2 + 4b^3)^4$

Find the coefficient of the indicated term in the given expansion. (See Example 3.)

17. x^8y^2 in $(x - y)^{10}$

18. a^3b^6 in $(a - b)^9$

19. x^4y^9 in $(x^2 + 2y^3)^5$

20. a^2b^{10} in $(3a^2 + b^2)^6$

Find the indicated term in the given expansion. (See Example 4.)

21. Tenth term of $(2x + y)^{12}$

22. Ninth term of $\left(\dfrac{x}{2} + y\right)^{13}$

23. Fourth term of $(x - 4y^2)^{15}$

24. Fifth term of $(x^2 - 2)^9$

25. Ninth term of $(p^2 + q^3)^{14}$

26. Eighth term of $(x^3 + z)^{10}$

27. Fourteenth term of $(x + 2y)^{13}$

28. Twenty-first term of $\left(\dfrac{x}{3} - y^2\right)^{20}$

Determine the first five terms in the expansions of the following. Also determine the values of the variable for which each expansion is valid. (See Example 5.)

29. $(1 + x)^{1/4}$

30. $(1 + x)^{1/3}$

31. $(1 - y)^{2/3}$

32. $(1 - y)^{3/4}$

33. $(1 + 2x)^{-1/2}$

34. $(1 + 3x)^{-1/4}$

35. $(1 + x^2)^{-2}$

36. $(1 + x^2)^{-3}$

37. $\left(1 - \dfrac{a}{2}\right)^{-2}$

38. $(1 - 3a)^{-3}$

Expand by using the Binomial Theorem.

39. $(x + y + 1)^4$

40. $(x - 1 - z)^3$

41. $(x - y + z - w)^3$

42. $(x + y + z + w)^4$

Approximate the following by using the first four terms in the binomial expansion.

43. $(1.01)^6$ [*Hint:* $(1.01)^6 = (1 + 0.01)^6$]

44. $(1.03)^4$

45. $(0.99)^8$ [*Hint:* $(0.99)^8 = (1 - 0.01)^8$]

46. $(2.99)^3$

47. $(3.01)^3$

48. $(1.03)^8$

49. $\sqrt{1.02}$

50. $\sqrt[3]{1.1}$

CHAPTER REVIEW

Key Words and Phrases

Finite sequence	Index of summation	Mathematical induction
Infinite sequence	Arithmetic sequence	n factorial
nth term	Common difference	Binomial coefficients
Series	Geometric sequence	Binomial Theorem
Sigma notation	Common ratio	Pascal's Triangle

Summary of Important Concepts and Formulas

Formulas for Arithmetic Sequences

$$a_n = a_1 + (n - 1)d$$

$$S_n = \frac{n}{2}(a_1 + a_n)$$

$$S_n = \frac{n}{2}[2a_1 + (n - 1)d]$$

Formulas for Geometric Sequences

$$a_n = a_1 r^{n-1}$$

$$S_n = a_1 \frac{1 - r^n}{1 - r} \quad \text{if} \quad r \neq 1.$$

If $|r| < 1$, $\displaystyle\sum_{n=1}^{\infty} a_1 r^{n-1} = \frac{a_1}{1 - r}$.

If $|r| \geq 1$ and $a_1 \neq 0$, $\displaystyle\sum_{n=1}^{\infty} a_1 r^{n-1}$ does not exist.

Binomial Coefficients

$$\binom{n}{r} = \frac{n!}{(n - r)! \, r!}$$

Principle of Mathematical Induction

Let $S(n)$ represent a statement about each positive integer n that is either true or false. If

$$S(1) \text{ is true,}$$

and if

the truth of $S(k)$ always implies the truth of $S(k + 1)$,

then $S(n)$ is true for all positive integers n.

n Factorial

$$n! = n \cdot (n - 1) \cdot (n - 2) \cdots 2 \cdot 1$$
if n is a positive integer.

$$0! = 1$$

rth Term in the Expansion of $(a + b)^n$

$$\binom{n}{r - 1} a^{n-r+1} b^{r-1}$$

Binomial Theorem

$$(a + b)^n = a^n + \binom{n}{1}a^{n-1}b + \binom{n}{2}a^{n-2}b^2 + \cdots + \binom{n}{r}a^{n-r}b^r + \cdots + \binom{n}{n - 2}a^2 b^{n-2} + \binom{n}{n - 1}ab^{n-1} + b^n$$

Review Problems for Chapter 9

Write the first five terms of the sequence that has the given general term.

1. $a_j = \dfrac{j-1}{j+1}$ **2.** $a_n = \dfrac{(-2)^n}{n!}$

Write the first six terms of each sequence.

3. $a_1 = 1$, $a_n = a_{n-1} + 2$

4. $a_1 = 2$, $a_n = 2a_{n-1} - 3$

Write the given series in expanded form and find the value of the sum.

5. $\displaystyle\sum_{n=1}^{5} (-2)^n$ **6.** $\displaystyle\sum_{n=1}^{6} \dfrac{1}{2^{n-1}}$

7. $\displaystyle\sum_{n=1}^{7} 5$ **8.** $\displaystyle\sum_{n=1}^{4} \dfrac{(2n)!}{2^n}$

9. $\displaystyle\sum_{n=1}^{5} \dfrac{n!}{(-1)^n}$ **10.** $\displaystyle\sum_{n=1}^{4} \dfrac{(2n)!}{n!}$

Write each of the following sums in sigma notation.

11. $2 + 4 + 6 + 8 + 10 + 12 + 14$

12. $(1)(2) + (2)(4) + (4)(6) + (8)(8) + (16)(10)$

13. $1 - 3 + 5 - 7 + 9 - 11$

14. $\dfrac{1}{2} + \dfrac{4}{3} + \dfrac{9}{4} + \dfrac{16}{5} + \dfrac{25}{6} + \dfrac{36}{7} + \dfrac{49}{8}$

15. $\dfrac{1}{3} + \dfrac{2}{4} + \dfrac{3}{5} + \dfrac{4}{6} + \dfrac{5}{7} + \dfrac{6}{8}$

16. $1 + \dfrac{1 \cdot 2}{2} + \dfrac{1 \cdot 2 \cdot 3}{2^2} + \cdots + \dfrac{1 \cdot 2 \cdot 3 \cdot 4 \cdot 5 \cdot 6}{2^5}$

In Problems 17 and 18, use the information given to find the value of the requested quantities of an arithmetic sequence.

17. $a_4 = 6$, $a_9 = -4$;   find a_{11} and S_{11}.

18. $a_6 = 8$, $S_8 = 40$;   find a_1 and d.

19. Find the first five terms of the arithmetic sequence that has $a_1 = 5$ and $d = -2$.

20. Find the nineteenth term and a formula for the nth term of the following arithmetic sequence.

$$3, 1, -1, -3, \ldots$$

21. Find a_5 and S_5 for a geometric sequence that has $a_3 = -2$, $a_4 = 3$.

22. Determine whether or not each sequence is a geometric sequence. If it is, find r and a_n.
a) $24, -36, 54, -81, \ldots$ b) $3, 4, 12, 20, \ldots$

23. Find r and a_1 for the geometric sequence that has $a_3 = 81$ and $a_6 = -24$.

24. Find the sixth term and a formula for the nth term of the geometric sequence with $a_1 = 4$, $r = -\frac{1}{2}$.

25. For each of the following infinite geometric sequences, determine whether or not the sum exists. If the sum exists, find its value.
a) $\frac{9}{16}, \frac{3}{4}, 1, \frac{4}{3}, \ldots$ b) $9, -6, 4, -\frac{8}{3}, \ldots$

26. Find the value of $\displaystyle\sum_{n=1}^{\infty} 25\left(-\dfrac{2}{3}\right)^n$ if it exists. If it does not exist, give a reason.

27. Express $3.212121\ldots$ as a quotient of integers.

28. Expand $(x - 2y)^6$ by using the Binomial Theorem.

29. Find the sixth term of $(2x - y^2)^9$.

30. Approximate $(1.02)^7$ by using the first four terms in a binomial expansion.

Use mathematical induction to prove that the following statements are true for all positive integers n.

31. $\dfrac{1}{1 \cdot 3} + \dfrac{1}{3 \cdot 5} + \dfrac{1}{5 \cdot 7} + \cdots +$

$\dfrac{1}{(2n - 1)(2n + 1)} = \dfrac{n}{2n + 1}$

32. $\dfrac{1}{1 \cdot 5} + \dfrac{1}{5 \cdot 9} + \dfrac{1}{9 \cdot 13} + \cdots +$

$\dfrac{1}{(4n - 3)(4n + 1)} = \dfrac{n}{4n + 1}$

33. $3 + 6 + 11 + \cdots + (n^2 + 2) =$

$\dfrac{n(2n^2 + 3n + 13)}{6}$

34. $n! > n^2$ if $n \geq 4$.

35. $n! > n^3$ if $n \geq 7$.

36. $a - b$ is a factor of $a^{2n} - b^{2n}$.

37. (**DeMoivre's Theorem**)   If $z = r(\cos \theta + i \sin \theta)$ is a complex number in trigonometric form, then $z^n = r^n(\cos n\theta + i \sin n\theta)$.

Each of the following nonsolutions has at least one error. Can you find them?

Problem 1 Write $\displaystyle\sum_{n=2}^{7} (2n - 1)$ in expanded form and find the value of the sum.

Nonsolution

$$\sum_{n=2}^{7} (2n - 1) = 1 + 3 + \dot{5} + 7$$

$$= 16$$

Problem 2 Find the seventh term in the arithmetic sequence whose first three terms are $2, -1, -4$.

Nonsolution

$$a_7 = a_1 r^6 = 2\left(-\frac{1}{2}\right)^6 = 2\left(\frac{1}{64}\right) = \frac{1}{32}$$

Problem 3 If the following sum exists, find its value.

$$\frac{2}{3} - 1 + \frac{3}{2} - \frac{9}{4} + \cdots$$

Nonsolution Since $a = \frac{2}{3}$ and $r = -\frac{3}{2}$, the value of the sum is

$$\frac{a}{1 - r} = \frac{\dfrac{2}{3}}{1 - \left(-\dfrac{3}{2}\right)} = \frac{\dfrac{2}{3}}{\dfrac{5}{2}} = \frac{4}{15}.$$

Problem 4 Determine the third term in the expansion of $(x - 2y)^9$.

Nonsolution

$$\binom{9}{3} x^6 (2y)^3 = \frac{9 \cdot 8 \cdot 7 \cdot 6!}{3 \cdot 2 \cdot 1 \cdot 6!} x^6 8y^3 = 672 x^6 y^3$$

Chapter 10

Analytic Geometry

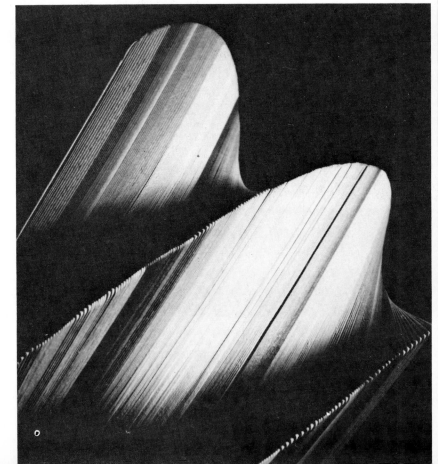

Much of the content of this chapter is repeated in a briefer form in most calculus sequences. Even so, a prior exposure is advantageous and adds depth to understanding. A sufficient number of applications are included to show that this material is important in its own right, as well as being useful in the calculus.

In this chapter our work is restricted to real variables. We saw in Section 2.4 that the graph of a quadratic relation of the form $y = ax^2 + bx + c$ is a parabola. The parabola is one of three types of curves known as the **conic sections.** The term *conic sections* comes from the fact that each type of curve can be obtained by intersecting a plane with a right circular cone.[1]

If the intersecting plane has a suitable inclination to the axis of the cone, the curve of intersection is a **parabola,** as shown in Figure 10.1(a). With a greater inclination to the axis of the cone, the curve of intersection may be an **ellipse,** which is an oval-shaped curve, as shown in Figure 10.1(b). The **circle** is a special case of the ellipse. With a lesser inclination, the curve of intersection may be a **hyperbola,** which is a curve having two branches, as shown in Figure 10.1(c).

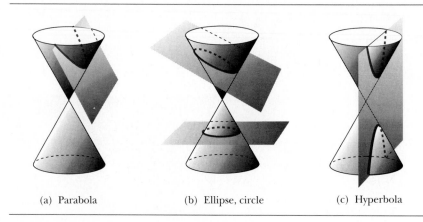

(a) Parabola   (b) Ellipse, circle   (c) Hyperbola

FIGURE 10.1

10.1   The Parabola

Just as circles can be described as the set of all points equally distant from a given point, parabolas can be described as a set of points satisfying a certain property involving distances.

---

1. By a *cone,* we mean a complete cone, with two nappes, as shown in Figure 10.1.

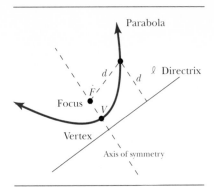

FIGURE 10.2

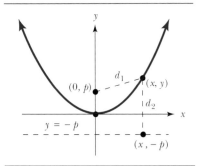

FIGURE 10.3

> ### Definition 10.1
>
> A **parabola** is the set of all points equidistant from a given point F, called the **focus**, and a given line ℓ, called the **directrix**.

With this definition of a parabola in mind, we see in Figure 10.2 that the vertex of a parabola is the point on the parabola closest to both the focus and the directrix. Also, the **axis of symmetry** is the line that goes through the focus and the vertex, and runs perpendicular to the directrix.

Definition 10.1 and the Distance Formula can be used to derive an equation of a parabola. We first consider the parabola in Figure 10.3 with vertex at the origin and focus at $(0, p)$. Then $y = -p$ is the equation of the directrix. If (x, y) is any point on the parabola, then $(x, -p)$ is the point on the directrix that is closest to (x, y). The distance d_1 from (x, y) to the focus $(0, p)$ must be equal to the distance d_2 from (x, y) to $(x, -p)$. Setting $d_1 = d_2$, we have

$$d_1 = d_2 \qquad \text{Definition 10.1}$$

$$\sqrt{(x - 0)^2 + (y - p)^2} = \sqrt{(x - x)^2 + [y - (-p)]^2} \quad \text{Distance Formula}$$

$$x^2 + (y - p)^2 = (y + p)^2 \qquad \text{Squaring both sides}$$

$$x^2 + y^2 - 2py + p^2 = y^2 + 2py + p^2 \qquad \text{Removing parentheses}$$

$$x^2 = 4py \qquad \text{Simplifying}$$

Thus the equation of a parabola that has vertex at the origin and the y-axis as its axis of symmetry is

$$x^2 = 4py.$$

We note that this equation can be put in the form

$$y = a(x - h)^2 + k$$

studied in Section 2.4, with $a = \dfrac{1}{4p}$, $h = 0$, and $k = 0$:

$$y = \frac{1}{4p}(x - 0)^2 + 0.$$

The parabola opens upward if $p > 0$ and downward if $p < 0$.

Suppose now we translate the parabola $x^2 = 4py$ in Figure 10.4(a) away from the origin. If we make a horizontal translation of h units and a vertical translation of k units as in Figure 10.4(b), the resulting parabola is described by the equation

$$(x - h)^2 = 4p(y - k).$$

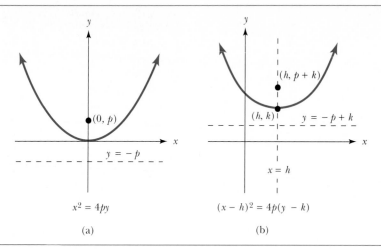

FIGURE 10.4

Finally, if we interchange the roles of x and y, we can obtain dual results concerning parabolas opening to the left or to the right. These results are summarized in Table 10.1.

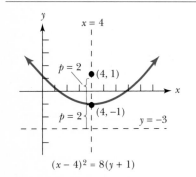

FIGURE 10.5

Example 1   Write the equation and sketch the graph of the parabola with directrix $y = -3$ and focus at $(4, 1)$.

Solution
Since the directrix is horizontal, the axis of symmetry is the vertical line going through the focus. Hence the equation of the axis is $x = 4$ and also $h = 4$. Since the focus is $2p$ units from the directrix, we have $2p = 4$ and $p = 2$. Thus the vertex is at $(4, -3 + 2) = (4, -1)$. Thus $k = -1$. We now have enough information to write out the equation

$$(x - 4)^2 = 4(2)(y + 1)$$
$$(x - 4)^2 = 8(y + 1)$$

and sketch the graph in Figure 10.5.   ❏

Example 2   Find the equation of the parabola opening to the left with vertex at $(2, -3)$ and passing through the point $(-2, 1)$. Sketch the graph locating the focus and the directrix.

Solution
A parabola opening to the left with vertex at $(2, -3)$ has an equation of the form

$$[y - (-3)]^2 = 4p(x - 2),$$

The Parabola		
Equation: $p \neq 0$	$(x - h)^2 = 4p(y - k)$	$(y - k)^2 = 4p(x - h)$
Vertex: V	(h, k)	(h, k)
Focus: F	$(h, k + p)$	$(h + p, k)$
Directrix: ℓ	$y = k - p$	$x = h - p$
Axis of symmetry	$x = h$	$y = k$
Direction of opening	Upward if $p > 0$; downward if $p < 0$	To the right if $p > 0$; to the left if $p < 0$
Graph: $p > 0$	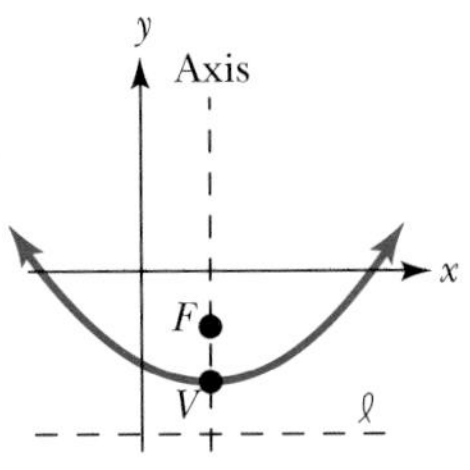	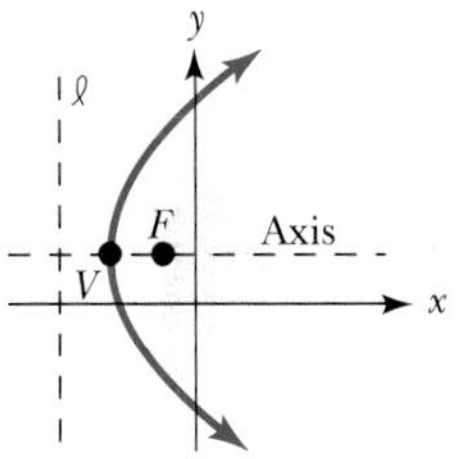
Graph: $p < 0$	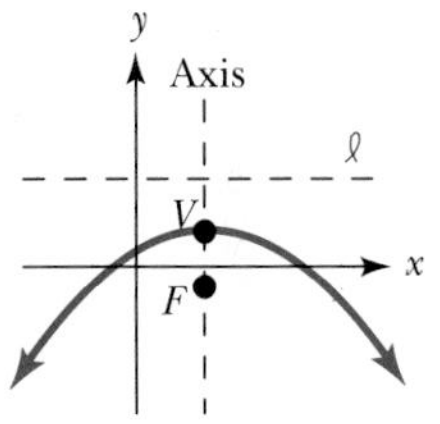	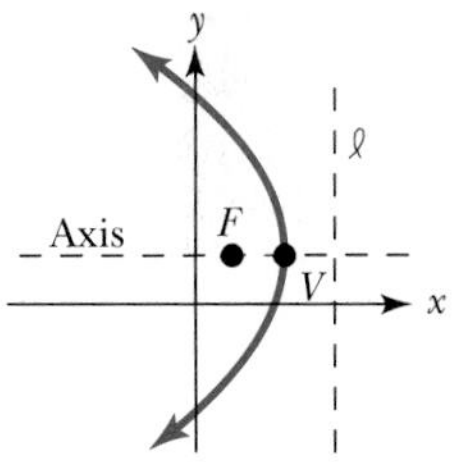

TABLE 10.1

with $p < 0$. Since $(-2, 1)$ must satisfy this equation, we have

$$(1 + 3)^2 = 4p(-2 - 2)$$
$$4^2 = 4p(-4)$$
$$-1 = p.$$

Therefore the equation of the parabola is

$$(y + 3)^2 = -4(x - 2).$$

Since the directrix is a vertical line located $|p| = |-1| = 1$ unit to the

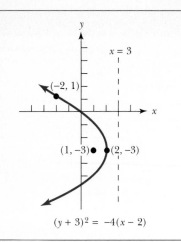

$$(y + 3)^2 = -4(x - 2)$$

FIGURE 10.6

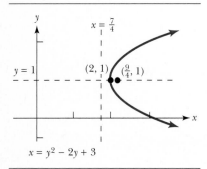

$$x = y^2 - 2y + 3$$

FIGURE 10.7

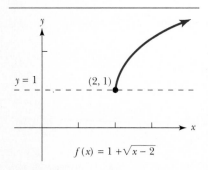

$$f(x) = 1 + \sqrt{x - 2}$$

FIGURE 10.8

right of the vertex, then the equation of the directrix is $x = 2 + 1 = 3$. Similarly, the focus is located $|p| = |-1| = 1$ unit to the left of the vertex. The focus is $(2 - 1, -3) = (1, -3)$. The graph is given in Figure 10.6. ❏

Example 3   Sketch the graph of the equation $x = y^2 - 2y + 3$.

Solution
Since this equation is quadratic in y and linear in x, it can be rewritten in the form of a parabola opening to the right or to the left. This can be accomplished by completing the square on y.

$$x = y^2 - 2y + 3$$

$$x - 3 = y^2 - 2y \qquad \text{Isolating the } y\text{-terms}$$

$$x - 3 + 1 = y^2 - 2y + 1 \qquad \text{Completing the square}$$

$$x - 2 = (y - 1)^2 \qquad \text{Rewriting}$$

This is the equation of a parabola with $p = \frac{1}{4}$, $h = 2$, and $k = 1$. Thus the parabola opens to the right with vertex at $(2, 1)$, focus at $(2 + \frac{1}{4}, 1) = (\frac{9}{4}, 1)$, and directrix $x = 2 - \frac{1}{4} = \frac{7}{4}$, as shown in Figure 10.7. ❏

When relations of the form $x = ay^2 + by + c$ are encountered in the calculus, they usually appear in a form similar to that in the following example.

Example 4   Sketch the graph of f, where $f(x) = 1 + \sqrt{x - 2}$.

Solution
The graph is not familiar in the given form, but it can be recognized after we set $y = f(x)$ and square both sides to eliminate the radical. We have

$$y - 1 = \sqrt{x - 2}$$

and

$$(y - 1)^2 = x - 2.$$

We recognize this as the equation of the parabola in Example 3 as shown in Figure 10.7. However, this parabola is not the graph of the original equation, because $y = 1 + \sqrt{x - 2}$ requires that $y \geq 1$. (Recall that $\sqrt{x - 2}$ indicates the nonnegative square root of $x - 2$.) The graph of the original equation consists only of the top half of the parabola and is shown in Figure 10.8. This top half is the graph of the function f, where $f(x) = 1 + \sqrt{x - 2}$. ❏

Exercises 10.1

In Exercises 1–16, write an equation of the parabola satisfying the given conditions. (See Examples 1 and 2.)

1. Directrix: $y = 3$
Focus: $(-1, 5)$

2. Directrix: $y = -2$
Focus: $(2, 2)$

3. Directrix: $y = 1$
Focus: $(1, -1)$

4. Directrix: $y = 4$
Focus: $(-2, 1)$

5. Directrix: $x = -4$
Focus: $(2, -2)$

6. Directrix: $x = 1$
Focus: $(2, 3)$

7. Directrix: $x = -3$
Focus: $(-7, 3)$

8. Directrix: $x = 0$
Focus: $(-5, -5)$

9. Vertex $(1, 1)$
Focus: $(-3, 1)$

10. Vertex: $(-1, 1)$
Focus: $(-1, 0)$

11. Vertex: $(-1, 3)$
Directrix: $x = -3$

12. Vertex: $(-1, -2)$
Directrix: $x = -6$

13. Vertex: $(1, 2)$
Passes through $(-2, 5)$
Opens upward

14. Vertex: $(0, -2)$
Passes through $(2, 4)$
Opens to the right

15. Vertex: $(-1, -2)$
Passes through $(2, -4)$
Axis parallel to x-axis

16. Vertex: $(2, 6)$
Passes through $(0, 3)$
Axis parallel to y-axis

In Exercises 17–32, sketch the graph of the parabola, locating the vertex, focus, and directrix. (See Example 3.)

17. $x^2 = 8(y - 1)$

18. $(x - 2)^2 = 2y$

19. $(x + 1)^2 + 4y = 0$

20. $(x + 1)^2 - 3y = 0$

21. $(y - 2)^2 = x + 1$

22. $(y + 2)^2 = -4x - 4$

23. $(y + 1)^2 + 8x - 16 = 0$

24. $y^2 - 3x = 6$

25. $x = 12 - 3(y + 2)^2$

26. $x = 4 - 2(y - 1)^2$

27. $x = y^2 - 10y + 25$

28. $x = y^2 - 4y + 4$

29. $x = 2y^2 - 8y + 5$

30. $x = 3y^2 + 6y - 20$

31. $x = -2y^2 + 8y - 11$

32. $x = -2y^2 + 4y - 3$

In Exercises 33–40, sketch the graph of the given equation. (See Example 4.)

33. $y = \sqrt{x - 3}$

34. $y = 1 - \sqrt{x - 2}$

35. $x = \sqrt{y - 2}$

36. $x = 2 - 2\sqrt{y - 1}$

37. $y = -\sqrt{1 - x^2}$

38. $2x = -\sqrt{1 - 4y^2}$

39. $y + 1 = \sqrt{16 - x^2}$

40. $x + 2 = -\sqrt{25 - y^2}$

Graph the solution set of the given system of inequalities. Label all points of intersection of the boundaries.

41. $x > y^2 + 1$
$x - y \le 3$

42. $x > y^2 - 4y + 3$
$x - y \le 3$

In Exercises 43–46, write an equation for the given parabola.

43.

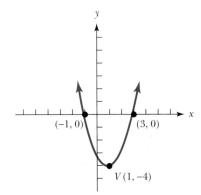

44.

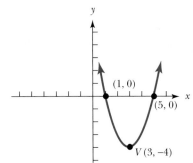

45.

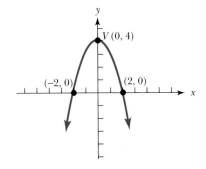

46.

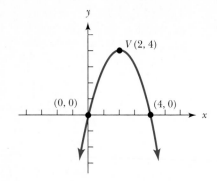

47. An arch under the bridge in the following figure spans an opening in the shape of a parabola. The arch is 40 meters wide at the base, and the highest point on the arch is at the center, 12 meters above the base. Find the height of the arch at a point 10 meters from the center of the base.

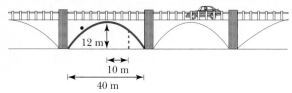

EXERCISE 47

48. The satellite dish in the accompanying figure has a cross section in the shape of a parabola 8 feet in diameter across the opening and 3 feet deep. If the receiver is located at the focus, find the distance from the vertex of the dish to the receiver.

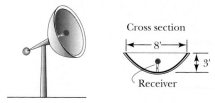

EXERCISE 48

49. A spotlight reflector has a cross section in the shape of a parabola, with the light source at the focus. If the reflector is 60 centimeters in diameter at the opening and 20 centimeters deep, find the distance from the vertex of the reflector to the light source.

50. If a ball is thrown straight up from 6 feet above ground level with an initial velocity of 88 feet per second, its height $h(t)$ in feet above ground after t seconds is given by $h(t) = 6 + 88t - 16t^2$. Find the maximum height that the ball reaches.

10.2   The Ellipse

One of the most familiar applications of ellipses is their use in describing the orbits of earth satellites. They have many other important applications, a few of which can be found in the exercises for this section.

There are standard forms for the equation of an ellipse that are somewhat analogous to the standard equation of a circle. In this section we obtain these standard equations of an ellipse by using the following definition.

> **Definition 10.2**
>
> Let F_1 and F_2 denote two fixed points in a plane. An **ellipse** is the set of all points P in the plane such that the sum of the distance from P to F_1 and the distance from P to F_2 is a constant. Each of the fixed points is called a **focus** (plural: **foci**) of the ellipse.

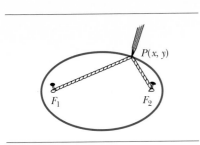

FIGURE 10.9

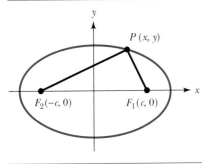

FIGURE 10.10

Figure 10.9 shows a method of drawing an ellipse that illustrates the definition very well. With this method, we fasten the two ends of a string with two thumbtacks, leaving slack in the string. When we pull the string taut, but not stretched, with a pencil and move the pencil to all possible positions, the pencil tip traces out an ellipse with foci at the thumbtacks.

To obtain a standard equation for an ellipse, we choose a coordinate system in the plane such that the foci F_1 and F_2 are located at $(c, 0)$ and $(-c, 0)$, where $c > 0$. This is indicated in Figure 10.10. In order to have a form of the equation that is as simple as possible, we let $2a$ denote the constant, which is the sum of the distances from P to F_1 and from P to F_2. This requires that $a > c$, since $2c$ is the distance between F_1 and F_2. A point P with coordinates (x, y) is then on the ellipse if and only if

$$\sqrt{(x - c)^2 + (y - 0)^2} + \sqrt{(x + c)^2 + (y - 0)^2} = 2a.$$

If we isolate the first radical on the left and square both sides, we have

$$\left[\sqrt{(x - c)^2 + y^2}\right]^2 = \left[2a - \sqrt{(x + c)^2 + y^2}\right]^2,$$

or

$$x^2 - 2cx + c^2 + y^2 = 4a^2 - 4a\sqrt{(x + c)^2 + y^2} + x^2 + 2cx + c^2 + y^2.$$

This simplifies to

$$-2cx = 4a^2 - 4a\sqrt{(x + c)^2 + y^2} + 2cx.$$

In order to remove the remaining radical, we isolate it on the left and obtain

$$4a\sqrt{(x + c)^2 + y^2} = 4a^2 + 4cx.$$

Dividing both sides by 4, we have

$$a\sqrt{(x + c)^2 + y^2} = a^2 + cx.$$

Squaring both sides now yields

$$a^2(x^2 + 2cx + c^2 + y^2) = a^4 + 2a^2cx + c^2x^2,$$

or

$$a^2x^2 + 2a^2cx + a^2c^2 + a^2y^2 = a^4 + 2a^2cx + c^2x^2.$$

This simplifies to

$$(a^2 - c^2)x^2 + a^2y^2 = a^2(a^2 - c^2).$$

We divide both sides by $a^2(a^2 - c^2)$ and obtain

$$\frac{x^2}{a^2} + \frac{y^2}{a^2 - c^2} = 1.$$

As one last simplification, we let

$$b = \sqrt{a^2 - c^2}.$$

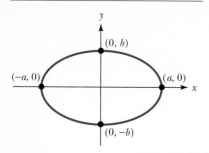

$$\frac{x^2}{a^2} + \frac{y^2}{b^2} = 1$$

$$a > b > 0$$

$$b^2 = a^2 - c^2$$

(a)

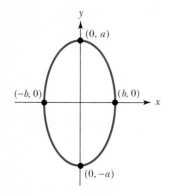

$$\frac{y^2}{a^2} + \frac{x^2}{b^2} = 1$$

$$a > b > 0$$

$$b^2 = a^2 - c^2$$

(b)

FIGURE 10.11

This is possible, since $a > c$. This substitution yields

$$\frac{x^2}{a^2} + \frac{y^2}{b^2} = 1. \tag{1}$$

Up to this point, our derivation shows that if a point (x, y) is on the ellipse, then the coordinates must satisfy Eq. (1). It is also true that any point (x, y) with coordinates that satisfy Eq. (1) must be on the ellipse. We leave this verification as an exercise.

A typical graph corresponding to Eq. (1) is shown in Figure 10.11(a). The graph is symmetric with respect to the x-axis, the y-axis, and the origin. The two line segments, one joining the x-intercepts and the other joining the y-intercepts, are called the *axes* of the ellipse. The longer axis is the **major axis**. It has length $2a$ and always contains the foci. The endpoints of the major axis are called the **vertices** of the ellipse. The shorter axis is designated as the **minor axis** and has length $2b$. The **center** of the ellipse is the point where the two axes cross.

The preceding discussion is summarized in the first statement of the following theorem. An interchange of the roles of x and y yields the statement concerning Eq. (2). A typical graph for Eq. (2) is shown in Figure 10.11(b).

Theorem 10.3

An ellipse with foci $(\pm c, 0)$ and vertices $(\pm a, 0)$ has a **standard equation** of the form

$$\frac{x^2}{a^2} + \frac{y^2}{b^2} = 1, \tag{1}$$

where $a > b > 0$ and $b^2 = a^2 - c^2$. Similarly, an ellipse with foci $(0, \pm c)$ and vertices $(0, \pm a)$ has a **standard equation** of the form

$$\frac{x^2}{b^2} + \frac{y^2}{a^2} = 1, \tag{2}$$

where $a > b > 0$ and $b^2 = a^2 - c^2$.

It is worth noting that the condition $a > b > 0$ is required in both types of equation in Theorem 10.3. The larger square occurs under the variable that corresponds to the location of the foci.

We also note that if one puts $a = b$ in an equation of either type (1) or type (2), the resulting equation appears as

$$\frac{x^2}{a^2} + \frac{y^2}{a^2} = 1,$$

where $a > 0$. This is equivalent to

$$x^2 + y^2 = a^2,$$

which we recognize as the equation of a circle. In this sense, then, a circle can be regarded as a special case of an ellipse.

As with the circle, a degenerate conic section is obtained when 1 is replaced by 0 in Eq. (1) or Eq. (2). In these cases the resulting equation is satisfied only by the coordinates of the origin.

Example 1 Sketch the graph of the ellipse $4x^2 + 9y^2 = 36$.

Solution
Since the given equation is not in either of the forms given in Theorem 10.3, it may not be immediately clear that this is an equation of an ellipse. In order to obtain one of the standard forms, we divide both members by 36. This gives

$$\frac{x^2}{9} + \frac{y^2}{4} = 1.$$

This fits the form of Eq. (1) in Theorem 10.3, with $a = 3$ and $b = 2$, so we have an ellipse with foci on the x-axis. It is easy to see that the x-intercepts are ± 3, and the y-intercepts are ± 2. With this information, we can draw the graph as in Figure 10.12. ❏

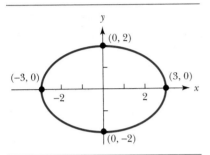

FIGURE 10.12

Example 1 illustrates how easy it is to sketch the graph of an ellipse by using the intercepts. The intercepts can actually be read by inspection from either of the equations in Theorem 10.3. They are always given by $\pm a$ and $\pm b$, *with the square of the x-intercepts under x^2, and the square of the y-intercepts under y^2.* It is usually easier to find the intercepts directly from the original than it is to obtain one of the standard forms.

Example 2 Sketch the graph of the ellipse $16x^2 + 9y^2 = 144$.

Solution
With $y = 0$, we get $16x^2 = 144$ and $x = \pm 3$. With $x = 0$, we get $9y^2 = 144$ and $y = \pm 4$. The graph is shown in Figure 10.13. ❏

If an ellipse with center at the origin is translated h units horizontally and k units vertically, the result is ellipse with center at (h, k). Such an ellipse has an equation and graph of one of the forms given in Table 10.2.

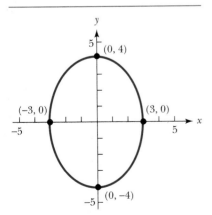

FIGURE 10.13

Ellipse with Center at (h, k), $a > b > 0$		
Equation	$\dfrac{(x - h)^2}{a^2} + \dfrac{(y - k)^2}{b^2} = 1$	$\dfrac{(x - h)^2}{b^2} + \dfrac{(y - k)^2}{a^2} = 1$
Vertices	$(h \pm a, k)$	$(h, k \pm a)$
Foci	$(h \pm c, k), \quad c = \sqrt{a^2 - b^2}$	$(h, k \pm c), \quad c = \sqrt{a^2 - b^2}$
Graph		

TABLE 10.2

Example 3　　Sketch the graph of $x^2 + 4x + 4y^2 - 8y = 8$ and determine the center, vertices, and foci.

Solution

We must complete the square on x and y to locate the center of the ellipse.

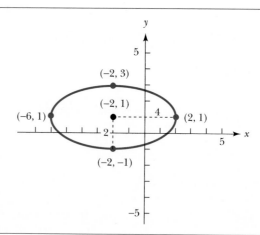

FIGURE 10.14

$$x^2 + 4x + 4 + 4(y^2 - 2y + 1) = 8 + 4 + 4$$

$$(x + 2)^2 + 4(y - 1)^2 = 16$$

$$\frac{(x + 2)^2}{4^2} + \frac{(y - 1)^2}{2^2} = 1$$

Thus the ellipse has center at $(-2, 1)$. We use $a = 4$ and $b = 2$ to plot the vertices at $(2, 1)$ and $(-6, 1)$ and the endpoints of the minor axis at $(-2, 3)$ and $(-2, -1)$ in Figure 10.14. We have $c^2 = 4^2 - 2^2 = 12$, so $c = 2\sqrt{3}$ and the foci are at $(-2 \pm 2\sqrt{3}, 1)$.   ❏

Exercises 10.2

Sketch the graphs of the following equations. If the graph is an ellipse, locate the vertices and foci. (See Examples 1–3.)

1. $\dfrac{x^2}{4} + \dfrac{y^2}{9} = 1$

2. $\dfrac{x^2}{25} + \dfrac{y^2}{16} = 1$

3. $25x^2 + 4y^2 = 100$

4. $9x^2 + 16y^2 = 144$

5. $x^2 + 4y^2 = 16$

6. $y^2 = 36 - 9x^2$

7. $4x^2 + 25y^2 = 25$

8. $9x^2 = 9 - 16y^2$

9. $\dfrac{4x^2}{9} + \dfrac{9y^2}{16} = 1$

10. $\dfrac{9x^2}{4} + \dfrac{16y^2}{25} = 1$

11. $x^2 + y^2 = 25$

12. $9x^2 + 9y^2 = 25$

13. $64x^2 + 36y^2 = 100$

14. $36x^2 + 16y^2 = 25$

15. $2x^2 + y^2 = 8$

16. $3x^2 + y^2 = 12$

17. $2x^2 + 3y^2 = 24$

18. $3x^2 + 5y^2 = 45$

19. $x^2 + 4y^2 = 0$

20. $4x^2 + 9y^2 = 0$

21. $\dfrac{(x - 2)^2}{4} + \dfrac{(y - 1)^2}{9} = 1$

22. $\dfrac{(x + 2)^2}{25} + \dfrac{(y - 1)^2}{16} = 1$

23. $\dfrac{(x + 3)^2}{16} + \dfrac{(y - 1)^2}{4} = 1$

24. $\dfrac{(x - 1)^2}{9} + \dfrac{(y + 1)^2}{16} = 1$

25. $25(x + 1)^2 + 4(y - 3)^2 = 100$

26. $9x^2 + 16(y + 2)^2 = 144$

27. $9(x + 2)^2 + (y - 3)^2 = 81$

28. $4(x - 1)^2 + 25(y + 3)^2 = 100$

29. $4x^2 - 8x + y^2 + 6y = 3$

30. $9x^2 + 4y^2 + 36x - 8y + 4 = 0$

31. $x^2 + 9y^2 + 6x = 27$

32. $4x^2 + 25y^2 - 32x - 50y = 11$

In Exercises 33–36, sketch the graph of the given equation.

33. $2y = \sqrt{16 - x^2}$

34. $y = -\sqrt{16 - 9x^2}$

35. $5x = -\sqrt{9 - 4y^2}$

36. $x = \dfrac{\sqrt{25 - 16y^2}}{2}$

In Exercises 37–48, write the standard equation of the ellipse satisfying the given conditions.

37. Center: $(0, 0)$
x-intercepts: ± 3
y-intercepts: ± 2

38. Center: $(0, 0)$
x-intercepts: ± 6
y-intercepts: ± 3

39. Center: $(0, 0)$
x-intercepts: ± 1
y-intercepts: ± 4

40. Center: $(0, 0)$
x-intercepts: ± 2
y-intercepts: ± 7

41. Vertices: $(3, 0), (-1, 0)$
Endpoints of minor axis: $(1, 1), (1, -1)$

42. Vertices: $(-4, 2), (4, 2)$
Endpoints of minor axis: $(0, 5), (0, -1)$

43. Vertices: $(2, 6), (2, 0)$
Length of minor axis: 4

44. Vertices: $(1, 2), (1, -6)$
Length of minor axis: 6

45. Vertices: $(-7, 1), (3, 1)$
Foci: $(-6, 1), (2, 1)$

46. Vertices: $(3, -6)$, $(3, 4)$
Foci: $(3, -4)$, $(3, 2)$

47. Vertices: $(\pm 3, 0)$
Through $(2, \frac{5}{3})$

48. Vertices: $(0, \pm 5)$
Through $(\frac{12}{5}, 3)$

Graph the solution set of the given system of inequalities. Label all points of intersection of the boundaries.

49. $4x^2 + 25y^2 \le 100$
$x + 2y > 8$

50. $36x^2 + 16y^2 \le 25$
$x - 2y < 12$

51. $x^2 + y^2 > 4$
$4x^2 + 9y^2 \le 36$

52. $x^2 + y^2 \ge 9$
$9x^2 + 16y^2 < 144$

In Exercises 53–56, write an equation for the given ellipse.

53.

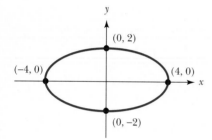

54.

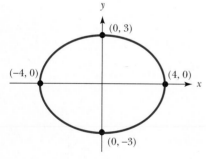

55.

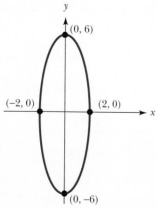

56.

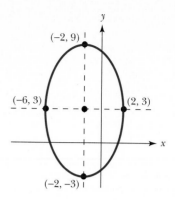

57. The arch over the gateway in the accompanying figure is in the shape of half an ellipse, with the major axis vertical. The arch is 12 feet across at the bottom and 9 feet high at the center. How high is the arch at a point 3 feet from the center of the gateway?

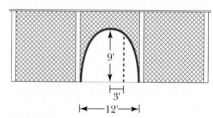

EXERCISE 57

58. The chimney cap in the figure below forms an arch in the shape of half an ellipse. The base of the arch is 16 inches across, and the highest point on the arch is 12 inches above the base. Find the height of the arch above the base at a point 2 inches from the center of the base.

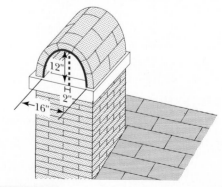

EXERCISE 58

59. A metal culvert under a street has a cross section that is an ellipse with a horizontal major axis of length 6 feet and a vertical minor axis of length 4 feet. Find the horizontal distance across the culvert 1 foot above the center of the ellipse.

60. A cross section of a roadway tunnel has the shape of half an ellipse with the minor axis vertical. The road in the tunnel is 24 feet wide, and the tunnel is 9 feet high at a point 6 feet from the center of the road. How high is the tunnel at its highest point?

61. The **eccentricity** e of an ellipse is defined by $e = c/a$. Ellipses with eccentricity close to 0 are nearly circular, and ellipses with eccentricity close to 1 are extremely elongated. Show that $0 < e < 1$ for any ellipse by using the facts that $0 < a^2 - b^2 < a^2$.

62. (See Exercise 61.) The planets in our solar system revolve about the sun in elliptical orbits with the sun at one focus. The orbit of the earth is an ellipse with eccentricity $e \approx 0.0161$ and a major axis of approximate length 185,200,000 miles. Find the shortest possible distance between the earth and the sun.

63. Complete the derivation of the first statement in Theorem 10.3 by showing that any point with coordinates (x, y) that satisfy Equation (1) must be on the ellipse.

64. Suppose that an ellipse has foci at $(0, \pm c)$ on the y-axis and that the sum of the distances from a point on the ellipse to the foci is $2a$, where $a > c > 0$. Show that the ellipse has an equation of the form

$$\frac{x^2}{b^2} + \frac{y^2}{a^2} = 1,$$

where $b^2 = a^2 - c^2$.

10.3 The Hyperbola

In this section we formulate some standard forms for the equation of a hyperbola. A hyperbola can be defined in a manner analogous to the way in which we defined an ellipse. The formal definition is as follows.

Definition 10.4

Let F_1 and F_2 denote two fixed points in a plane. A **hyperbola** is the set of all points P in the plane such that the absolute value of the difference of the distances from P to the fixed points is a constant. Each of the fixed points is called a **focus** (plural: **foci**) of the hyperbola.

A standard equation for a hyperbola can be derived in a manner very similar to that used in Section 10.2 for an ellipse. We once again choose a coordinate system in the plane such that the foci F_1 and F_2 are located at $(c, 0)$ and $(-c, 0)$, where $c > 0$. This is illustrated in Figure 10.15. In order to obtain a simple form of the equation, we let $2a$ denote the constant that equals the absolute value of the difference of the distances from P to the foci. We have

$$\left| \sqrt{(x - c)^2 + (y - 0)^2} - \sqrt{(x + c)^2 + (y - 0)^2} \right| = 2a,$$

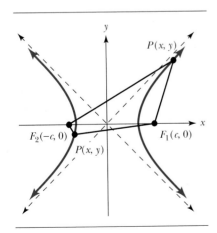

FIGURE 10.15

which is equivalent to

$$\sqrt{(x - c)^2 + (y - 0)^2} - \sqrt{(x + c)^2 + (y - 0)^2} = \pm 2a.$$

In writing these equations, we have assumed that $0 < a < c$. This condition is necessary in order to have points on the hyperbola that are not on the x-axis. (Notice that the distance between F_1 and F_2 is $2c$.)

From this point on, the radicals can be eliminated by the same type of procedure we employed in the derivation of the equation of an ellipse. We leave as an exercise the details of this procedure, which concludes with the equation

$$\frac{x^2}{a^2} - \frac{y^2}{c^2 - a^2} = 1.$$

Since $c > a$ here, we can make the substitution

$$b = \sqrt{c^2 - a^2},$$

and this yields

$$\frac{x^2}{a^2} - \frac{y^2}{b^2} = 1. \qquad (1)$$

This is the standard form for the equation of a hyperbola with foci on the x-axis. The derivation that we have outlined shows that any point on the hyperbola has coordinates that satisfy Eq. (1). It can also be demonstrated that any point with coordinates that satisfy Eq. (1) is on the hyperbola. This is left as one of the exercises for this section.

The numbers a and b, which appear in the standard equation, are closely related to the graph of the hyperbola. This is exhibited in Figure 10.16. The points on the branches of the curve nearest the other branch are called **vertices** and are labeled $V_1(a, 0)$ and $V_2(-a, 0)$ in the figure. The line segment from V_1 to V_2 is called the **transverse axis** and has length $2a$. The line segment with endpoints at $(0, \pm b)$ is called the **conjugate axis** and has length $2b$. The dashed lines and line segments are *not* part of the graph, of course, but serve as guides in drawing the hyperbola. The dashed lines passing through the origin and the vertices of the rectangle have equations $y = \pm(b/a)x$. As the distance between x and 0 increases from a, the points (x, y) on the hyperbola are nearer these dashed lines. These dashed lines are the **asymptotes** of the hyperbola, and their point of intersection is called the **center**. The rectangle is drawn to serve as a guide in sketching the asymptotes, and it emphasizes that the slopes of the asymptotes are $\pm b/a$.

It is an interesting fact that if the number 1 in the right member of Eq. (1) is replaced by 0, we obtain an equation of the asymptotes. This is easy to confirm, for the equation

$$\frac{x^2}{a^2} - \frac{y^2}{b^2} = 0$$

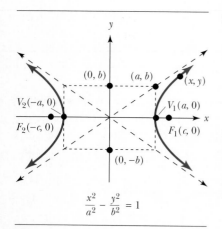

$$\frac{x^2}{a^2} - \frac{y^2}{b^2} = 1$$

FIGURE 10.16

is equivalent to

$$b^2x^2 - a^2y^2 = 0,$$

or

$$(bx - ay)(bx + ay) = 0.$$

This product is zero if and only if one of the factors is zero, that is, if and only if

$$y = \pm\frac{b}{a}x.$$

The main results of this discussion are summarized in the following theorem.

Theorem 10.5

A hyperbola with foci $(\pm c, 0)$ and vertices $(\pm a, 0)$ has a **standard equation** of the form

$$\frac{x^2}{a^2} - \frac{y^2}{b^2} = 1, \tag{1}$$

where $c > a > 0$ and $b = \sqrt{c^2 - a^2}$.

The equations of the asymptotes are given by

$$y = \pm\frac{b}{a}x,$$

which can be obtained from Eq. (1) by replacing the 1 with 0.

We note that the condition $a > b$, which was required in the case of the ellipse, is *not present* in the case of the hyperbola.

Example 1 Sketch the graph of the following hyperbola. Locate the vertices, foci, and asymptotes, and write the equations of the asymptotes.

$$9x^2 - 4y^2 = 36$$

Solution
We first divide both members by 36 to obtain the standard form of the equation. This gives

$$\frac{x^2}{4} - \frac{y^2}{9} = 1,$$

or

$$\frac{x^2}{(2)^2} - \frac{y^2}{(3)^2} = 1.$$

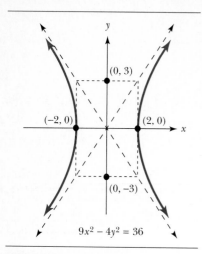

$9x^2 - 4y^2 = 36$

FIGURE 10.17

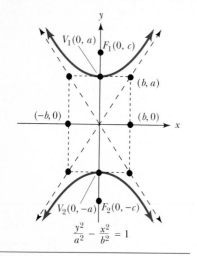

$$\frac{y^2}{a^2} - \frac{x^2}{b^2} = 1$$

FIGURE 10.18

Using $a = 2$ and $b = 3$ in the same manner as in Figure 10.16, we obtain the graph in Figure 10.17. The vertices are located at $(\pm 2, 0)$, and the asymptotes are given by $y = \pm\frac{3}{2}x$. Since $b^2 = c^2 - a^2$, we have

$$c = \sqrt{a^2 + b^2} = \sqrt{4 + 9} = \sqrt{13},$$

and the foci are at $(\pm\sqrt{13}, 0)$.   ❏

As we would expect, Theorem 10.5 has a counterpart for the situation in which the foci are on the y-axis. This is stated in Theorem 10.6 and illustrated in Figure 10.18.

Theorem 10.6

A hyperbola with foci $(0, \pm c)$ and vertices $(0, \pm a)$ has a **standard equation** of the form

$$\frac{y^2}{a^2} - \frac{x^2}{b^2} = 1, \tag{2}$$

where $c > a > 0$ and $b = \sqrt{c^2 - a^2}$.

The equations of the asymptotes are given by

$$y = \pm\frac{a}{b}x,$$

which can be obtained from Eq. (2) by replacing the 1 with 0.

It is easy to see that Theorem 10.6 can be obtained from Theorem 10.5 simply by interchanging x and y.

Example 2   Sketch the graph of the following hyperbola. Locate the vertices, foci, and asymptotes, and write the equations of the asymptotes.

$$\frac{y^2}{16} - \frac{x^2}{4} = 1$$

Solution
We have $a = 4$ and $b = 2$ from the equation $y^2/16 - x^2/4 = 1$. Using Figure 10.18 as a guide, we draw the graph as shown in Figure 10.19. The vertices are at $(0, \pm 4)$, and the asymptotes have equations $y = \pm 2x$. We have

$$c = \sqrt{a^2 + b^2} = \sqrt{20} = 2\sqrt{5},$$

and the foci are at $(0, \pm 2\sqrt{5})$.   ❏

If a hyperbola with center at the origin is translated h units horizontally and k units vertically, the result is a hyperbola with center at (h, k). The corresponding equations and graphs are given in Table 10.3.

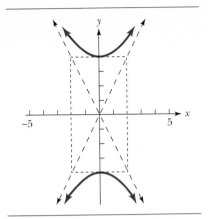

FIGURE 10.19

Hyperbola with Center at (h, k)		
Equation	$\dfrac{(x-h)^2}{a^2} - \dfrac{(y-k)^2}{b^2} = 1$	$\dfrac{(y-k)^2}{a^2} - \dfrac{(x-h)^2}{b^2} = 1$
Vertices	$(h \pm a, k)$	$(h, k \pm a)$
Foci	$(h \pm c, k), \quad c = \sqrt{a^2 + b^2}$	$(h, k \pm c), \quad c = \sqrt{a^2 + b^2}$
Asymptotes	$y - k = \pm \dfrac{b}{a}(x - h)$	$y - k = \pm \dfrac{a}{b}(x - h)$
Graph		

TABLE 10.3

Example 3 Sketch the graph of the following hyperbola. Find the center, vertices, and foci, and write equations for the asymptotes.

$$\frac{(x+1)^2}{9} - \frac{(y-1)^2}{4} = 1$$

Solution
This is a hyperbola with center at $(-1, 1)$, $a = 3$, and $b = 2$. Since $a = 3$, we move three units to the right and left of the center to obtain the vertices of the hyperbola at $(-4, 1)$ and $(2, 1)$. Since $b = 2$, we move two units up and down from the center to obtain the endpoints of the conjugate axis. With these five points we can first sketch the rectangle, then the asymptotes, and finally the hyperbola in Figure 10.20. The

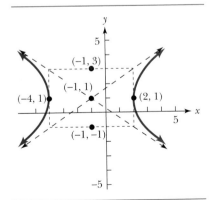

FIGURE 10.20

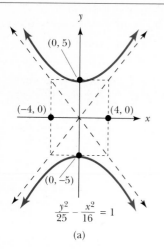

$$\frac{y^2}{25} - \frac{x^2}{16} = 1$$

(a)

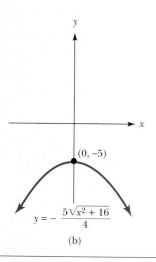

$$y = -\frac{5\sqrt{x^2 + 16}}{4}$$

(b)

FIGURE 10.21

asymptotes have equations $y - 1 = \pm\frac{2}{3}(x + 1)$. Since $c = \sqrt{9 + 4} = \sqrt{13}$, the foci are at $(-1 \pm \sqrt{13}, 1)$. ❏

There are variations in the equations that correspond to only a portion of a hyperbola.

Example 4   Sketch the graph of $y = -\dfrac{5\sqrt{x^2 + 16}}{4}$.

Solution

If we multiply both sides of the given equation by 4 and then square both sides, we have

$$(4y)^2 = 25(x^2 + 16),$$

or

$$16y^2 = 25x^2 + 400.$$

In order to obtain a standard equation, we subtract $25x^2$ from both sides and then divide both sides by 400. This gives

$$\frac{y^2}{25} - \frac{x^2}{16} = 1,$$

or

$$\frac{y^2}{(5)^2} - \frac{x^2}{(4)^2} = 1.$$

We recognize this as the standard equation of a hyperbola with $a = 5$, $b = 4$, and foci on the y-axis. The graph of this hyperbola is given in Figure 10.21(a). But the original equation $y = (-5\sqrt{x^2 + 16})/4$ requires that y be nonpositive. Thus we must take only the bottom half of the hyperbola to have the graph of the original equation. This is shown in Figure 10.21(b). ❏

Exercises 10.3

Sketch the graphs of the following equations. Locate the center, vertices, and foci of all hyperbolas and write the equations of the asymptotes. (See Examples 1–3.)

1. $\dfrac{x^2}{9} - \dfrac{y^2}{16} = 1$ **2.** $\dfrac{x^2}{9} - \dfrac{y^2}{4} = 1$

3. $\dfrac{y^2}{4} - \dfrac{x^2}{25} = 1$ **4.** $\dfrac{y^2}{36} - \dfrac{x^2}{25} = 1$

5. $y^2 - x^2 = 9$ **6.** $y^2 - x^2 = 16$

7. $4x^2 - 9y^2 = 36$ **8.** $36x^2 - 4y^2 = 144$

9. $9x^2 + 4y^2 = 0$ **10.** $x^2 + y^2 = 0$

11. $4x^2 - 25y^2 = 0$

12. $16x^2 - 9y^2 = 0$

13. $x^2 - 4y^2 = 4$

14. $9x^2 - y^2 = 9$

15. $4x^2 + 25y^2 = 100$

16. $16x^2 + 9y^2 = 144$

17. $\dfrac{16y^2}{25} - \dfrac{9x^2}{4} = 1$

18. $\dfrac{25y^2}{16} - \dfrac{4x^2}{9} = 1$

19. $x^2 + y^2 = 16$

20. $x^2 + y^2 = 9$

21. $2x^2 - y^2 = 8$

22. $y^2 - 3x^2 = 75$

23. $\dfrac{(x-1)^2}{9} - \dfrac{(y-2)^2}{25} = 1$

24. $\dfrac{(x+2)^2}{4} - \dfrac{(y-1)^2}{9} = 1$

25. $\dfrac{(y+3)^2}{4} - (x-2)^2 = 1$

26. $\dfrac{(y-1)^2}{16} - \dfrac{(x+4)^2}{4} = 1$

27. $4(x+2)^2 - 9(y+2)^2 = 36$

28. $4(x-3)^2 - y^2 = 4$

29. $9(y-1)^2 - x^2 = 9$

30. $36(y+2)^2 - 4(x+4)^2 = 144$

31. $4x^2 - y^2 + 2y = 17$

32. $y^2 - x^2 - 4y - 2x + 2 = 0$

33. $4y^2 - 9x^2 + 8y - 36x = 68$

34. $x^2 - 25y^2 + 4x + 150y = 246$

In Exercises 35–38, sketch the graph of the given equation. (See Example 4.)

35. $y = -\sqrt{9 + x^2}$

36. $x = -\sqrt{16 + y^2}$

37. $5y = \sqrt{9 + 4x^2}$

38. $2x = \sqrt{16 + y^2}$

In Exercises 39–47, write the standard equation of the hyperbola satisfying the given conditions.

39. Foci: $(\pm 5, 0)$
Vertices: $(\pm 4, 0)$

40. Foci: $(\pm 3, 0)$
Vertices: $(\pm 2, 0)$

41. Foci: $(0, \pm 4)$
Vertices: $(0, \pm 2)$

42. Foci: $(0, \pm 3)$
Vertices: $(0, \pm 1)$

43. Vertices: $(\pm 4, 0)$
Through $(8, 3)$

44. Vertices: $(0, \pm 3)$
Through $(9, 6)$

45. Vertices: $(\pm 8, 0)$
Asymptotes: $y = \pm\dfrac{3}{4}x$

46. Vertices: $(0, \pm 1)$
Asymptotes: $y = \pm 2x$

Graph the solution set of the given set of inequalities. Label all points of intersection of the boundaries.

47. $\begin{aligned} x^2 - y^2 &> 1 \\ 4x^2 + 9y^2 &< 36 \end{aligned}$

48. $\begin{aligned} x^2 + 4y^2 &< 100 \\ 4x^2 - 9y^2 &> 36 \end{aligned}$

49. $\begin{aligned} x^2 + 2y^2 &\geq 64 \\ x^2 - y^2 &\leq 16 \end{aligned}$

50. $\begin{aligned} x^2 - y^2 &\geq -1 \\ 4x^2 + 9y^2 &\geq 36 \end{aligned}$

In Exercises 51–54, write an equation for the given hyperbola.

51.

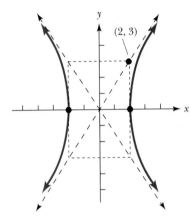

52.

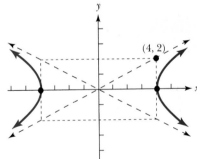

53.

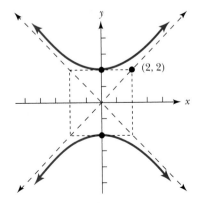

54.

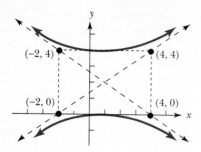

55. Show that

$$\sqrt{(x - c)^2 + (y - 0)^2} - \sqrt{(x + c)^2 + (y - 0)^2} = \pm 2a$$

is equivalent to

$$\frac{x^2}{a^2} - \frac{y^2}{b^2} = 1,$$

where a, b, and c are positive, $c > a$, and $b^2 = c^2 - a^2$.

56. Consider a hyperbola with foci at $(0, \pm c)$. Let $2a$ denote the constant absolute value of the difference of the distances from a point on the hyperbola to the foci. Assuming that $0 < a < c$, show that the hyperbola has an equation of the form

$$\frac{y^2}{a^2} - \frac{x^2}{b^2} = 1,$$

with $b^2 = c^2 - a^2$.

57. Show that Eq. (1) in Theorem 10.5 is equivalent to

$$y = \pm \frac{b}{a} x \sqrt{1 - \frac{a^2}{x^2}}.$$

From this equation, it can be seen that, as $|x|$ increases from a, the points (x, y) on the hyperbola in Exercise 55 move nearer the lines $y = \pm (b/a)x$.

58. Show that if $x \neq 0$, Eq. (2) in Theorem 10.6 is equivalent to

$$y = \pm \frac{a}{b} x \sqrt{1 + \frac{b^2}{x^2}}.$$

From this equation, it can be seen that, as $|x|$ increases, the points (x, y) on the hyperbola in Exercise 56 move nearer the lines $y = \pm (a/b)x$.

10.4 Rotation of Axes

In the last three sections and in some earlier material, we worked with the conic sections from two approaches. Given an equation, we sketched the graph by locating special points such as the intercepts, vertices, center, or foci, and by making use of special lines such as the directrix or the asymptotes. Given the location of some of the special points or lines, we also learned how to write the equation of the conic in one of the following standard forms.

Circle: $(x - h)^2 + (y - k)^2 = r^2$

Parabola: $(x - h)^2 = 4p(y - k)$ or $(y - k)^2 = 4p(x - h)$

Ellipse: $\dfrac{(x - h)^2}{a^2} + \dfrac{(y - k)^2}{b^2} = 1$ or $\dfrac{(x - h)^2}{b^2} + \dfrac{(y - k)^2}{a^2} = 1$

Hyperbola: $\dfrac{(x - h)^2}{a^2} - \dfrac{(y - k)^2}{b^2} = 1$ or $\dfrac{(y - k)^2}{a^2} - \dfrac{(x - h)^2}{b^2} = 1$

Any conic section has an equation that can be written in what is called a **general form**:

$$Ax^2 + Cy^2 + Dx + Ey + F = 0, \quad \text{not both } A \text{ and } C \text{ zero.}$$

One of the desirable skills for the calculus is to be able to recognize the type of conic section we are dealing with by examining the coefficients A, C, D, E and F. Table 10.4 lists conditions the coefficients must meet in order to have a particular type of conic section (or a degenerate case) as the graph of the equation. All these conics have their axes parallel to either the x-axis or the y-axis.

The General Quadratic Equation: $Ax^2 + Cy^2 + Dx + Ey + F = 0$		
Conic section	**Conditions**	**Example**
Parabola (opens up/down)	$A \neq 0, \quad C = 0, \quad E \neq 0$	$x^2 + 2x - y + 1 = 0$
Parabola (opens left/right)	$A = 0, \quad C \neq 0, \quad D \neq 0$	$-y^2 + 2x + 4 = 0$
Circle	$A = C \neq 0$	$2x^2 + 2y^2 + 3x - 2y = 0$
Ellipse	A and C same sign	$4x^2 + y^2 + x - 1 = 0$
Hyperbola	A and C opposite sign	$-x^2 + 2y^2 + 4y + 10 = 0$

TABLE 10.4

We next study the conics whose axes are not parallel to one of the coordinate axes. The general form for the equation will contain an xy-term:

$$Ax^2 + Bxy + Cy^2 + Dx + Ey + F = 0, \quad B \neq 0$$

The **discriminant** of this equation is defined as the number $B^2 - 4AC$. It can be shown that the sign of the discriminant can be used to determine the conic section in the following way.

If $B^2 - 4AC = 0$, the graph of the equation is a parabola.

If $B^2 - 4AC > 0$, the graph of the equation is a hyperbola.

If $B^2 - 4AC < 0$, the graph of the equation is an ellipse.

However, the most efficient way to sketch the graph is to first eliminate the xy-term, that is, rewrite the equation so that B is zero. This can be accomplished by a **rotation of axes,** which we describe next.

In Figure 10.22, we rotate the x- and y-coordinate axes through an angle θ and label the rotated coordinate axes as the x'-axis and the y'-axis. By doing this, we actually have two coordinate systems, the xy-system and the $x'y'$-system, one superimposed on the other, sharing a common origin O.

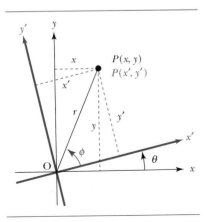

FIGURE 10.22

Any point P in the coordinate plane can be expressed in terms of its xy-coordinates or its $x'y'$-coordinates. In either system, the distance r between P and the origin is the same. Let ϕ be the angle formed by the line segment from the origin to P and the x'-axis. Then, using the trigonometric functions, we can write

$$x = r \cos (\phi + \theta), \qquad x' = r \cos \phi,$$
$$y = r \sin (\phi + \theta), \qquad y' = r \sin \phi.$$

The identities for the cosine and sine of the sum allow us to express x and y in terms of x', y', and θ as follows:

$$x = r(\cos \phi \cos \theta - \sin \phi \sin \theta) \qquad y = r(\cos \phi \sin \theta + \sin \phi \cos \theta)$$
$$= (r \cos \phi) \cos \theta - (r \sin \phi) \sin \theta \qquad = (r \cos \phi) \sin \theta + (r \sin \phi) \cos \theta$$
$$= x' \cos \theta - y' \sin \theta; \qquad = x' \sin \theta + y' \cos \theta.$$

Solving these two equations for x' and y' provides formulas for x' and y' in terms of x, y, and θ.

$$x' = x \cos \theta + y \sin \theta, \qquad y' = -x \sin \theta + y \cos \theta$$

We call these two pairs of equations the **rotation formulas.**

Rotation Formulas

If the xy-coordinate axes are rotated about the origin through an angle θ to form a new $x'y'$-system, then the xy-coordinates of a point P are related to the $x'y'$-coordinates of P by

$$x = x' \cos \theta - y' \sin \theta, \qquad y = x' \sin \theta + y' \cos \theta.$$

Similarly, the $x'y'$-coordinates of P are related to the xy-coordinates of P by

$$x' = x \cos \theta + y \sin \theta, \qquad y' = -x \sin \theta + y \cos \theta.$$

We illustrate how the rotation formulas can be used to eliminate the xy-term in Example 1.

Example 1   Use a 45° rotation to eliminate the xy-term in

$$x^2 - xy + y^2 = 2.$$

Sketch the graph of the resulting equation, showing both the xy-system and the $x'y'$-system.

Solution
With $\theta = 45°$, the rotation formulas become

$$x = x' \cos 45° - y' \sin 45° = \frac{1}{\sqrt{2}}(x' - y'),$$

$$y = x' \sin 45° + y' \cos 45° = \frac{1}{\sqrt{2}}(x' + y').$$

Substituting these expressions for x and y into $x^2 - xy + y^2 = 2$ will eliminate the xy-term.

$$\left[\frac{1}{\sqrt{2}}(x' - y')\right]^2 - \frac{1}{\sqrt{2}}(x' - y')\frac{1}{\sqrt{2}}(x' + y')$$

$$+ \left[\frac{1}{\sqrt{2}}(x' + y')\right]^2 = 2 \qquad \text{Substitution}$$

$$\frac{1}{2}[(x')^2 - 2x'y' + (y')^2] - \frac{1}{2}[(x')^2 - (y')^2]$$

$$+ \frac{1}{2}[(x')^2 + 2x'y' + (y')^2] = 2 \qquad \text{Expanding}$$

$$\frac{1}{2}(x')^2 + \frac{3}{2}(y')^2 = 2 \qquad \text{Collecting like terms}$$

$$\frac{(x')^2}{4} + \frac{(y')^2}{\frac{4}{3}} = 1 \qquad \begin{array}{l}\text{Writing in}\\ \text{standard form}\end{array}$$

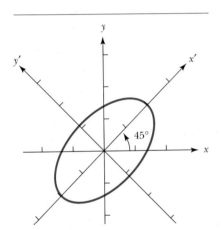

FIGURE 10.23

This is the equation of an ellipse in the $x'y'$-system with intercepts at $x' = \pm 2$ and $y' = \pm\frac{2}{\sqrt{3}}$. In Figure 10.23, we rotate the xy-axes through 45° and sketch the ellipse in the $x'y'$-system. ❏

In Example 1 the angle of rotation was given as 45°. The question that naturally arises is: Will any angle θ work? If not, how is an appropriate angle θ determined? We answer these questions next by considering the general equation containing an xy-term:

$$Ax^2 + Bxy + Cy^2 + Dx + Ey + F = 0, \qquad B \neq 0.$$

Substitution of

$$x = x' \cos \theta - y' \sin \theta$$

$$y = x' \sin \theta + y' \cos \theta,$$

followed by many algebraic simplifications, yields this equation in the variables x' and y':

$$(A \cos^2 \theta + B \sin \theta \cos \theta + C \sin^2 \theta)(x')^2$$
$$+ (-2A \cos \theta \sin \theta - B \sin^2 \theta + B \cos^2 \theta + 2C \sin \theta \cos \theta) \, x'y'$$
$$+ (A \sin^2 \theta - B \sin \theta \cos \theta + C \cos^2 \theta)(y')^2$$
$$+ (D \cos \theta + E \sin \theta)x' + (-D \sin \theta + E \cos \theta)y' + F = 0.$$

Recalling that our goal is to eliminate the $x'y'$-term, we set its coefficient equal to 0.

$$-2A \cos \theta \sin \theta + B(\cos^2 \theta - \sin^2 \theta) + 2C \sin \theta \cos \theta = 0$$
$$-A \sin 2\theta + B \cos 2\theta + C \sin 2\theta = 0 \qquad \text{Trigonometric Identities}$$

$$B \cos 2\theta = (A - C) \sin 2\theta \qquad \text{Algebra}$$

$$\cot 2\theta = \frac{A - C}{B} \qquad \text{Solving for } \cot 2\theta$$

Thus the angle of rotation θ can be determined from the coefficients A, B, and C.

Angle of Rotation

The xy-term in the general quadratic equation
$$Ax^2 + Bxy + Cy^2 + Dx + Ey + F = 0$$
will be eliminated if the rotation formulas are used to rotate the xy-coordinate system through an angle θ to form a new $x'y'$-system, where θ is determined by

$$\cot 2\theta = \frac{A - C}{B}.$$

We determine the angle of rotation θ in the next example.

Example 2   Use a rotation to eliminate the xy-term in
$$5x^2 - 6\sqrt{3}\,xy - y^2 + 16 = 0.$$

Sketch the graph of the resulting equation, showing both the xy-system and the $x'y'$-system.

Solution
With $A = 5$, $B = -6\sqrt{3}$, and $C = -1$, we have

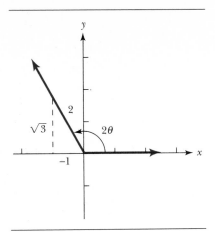

FIGURE 10.24

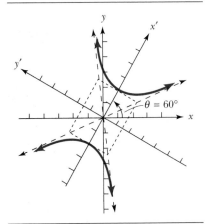

FIGURE 10.25

$$\cot 2\theta = \frac{A - C}{B} = \frac{5 - (-1)}{-6\sqrt{3}} = -\frac{1}{\sqrt{3}}.$$

We must determine θ subject to the condition that the cotangent of 2θ is $-1/\sqrt{3}$. To assure that θ is a positive acute angle, we draw 2θ as a second quadrant angle and construct the reference triangle for 2θ as shown in Figure 10.24. We recognize 2θ as one of the special angles. That is, $2\theta = 120°$ and $\theta = 60°$. Then the rotation formulas become

$$x = x' \cos 60° - y' \sin 60° = \frac{1}{2}x' - \frac{\sqrt{3}}{2}y',$$

$$y = x' \sin 60° + y' \cos 60° = \frac{\sqrt{3}}{2}x' + \frac{1}{2}y'.$$

We substitute these expressions into the given equation.

$$5\left(\frac{1}{2}x' - \frac{\sqrt{3}}{2}y'\right)^2 - 6\sqrt{3}\left(\frac{1}{2}x' - \frac{\sqrt{3}}{2}y'\right)\left(\frac{\sqrt{3}}{2}x' + \frac{1}{2}y'\right)$$
$$- \left(\frac{\sqrt{3}}{2}x' + \frac{1}{2}y'\right)^2 + 16 = 0$$

This simplifies to an equation of a hyperbola in the $x'y'$-system.

$$\frac{(x')^2}{4} - \frac{(y')^2}{2} = 1$$

To graph this hyperbola, we first rotate the xy-axes $60°$ and then sketch the hyperbola in the new $x'y'$-system. Both systems and the hyperbola are shown in Figure 10.25.   ❏

Our final example illustrates the situation in which the angle of rotation is not one of the special angles.

Example 3 Sketch the graph of

$$x^2 + 4xy + 4y^2 + 5\sqrt{5}\,y + 5 = 0.$$

Solution
Since $A = 1$, $B = 4$, and $C = 4$, we have

$$\cot 2\theta = \frac{1 - 4}{4} = -\frac{3}{4}.$$

If we draw 2θ as a second quadrant angle, then θ will be a positive acute angle. In Figure 10.26, we construct the reference triangle for 2θ. Thus

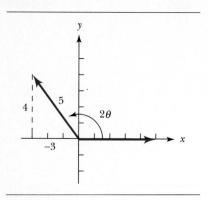

FIGURE 10.26

$\cos 2\theta = -\frac{3}{5}$, which we will use in the Half-Angle Identities to determine the values of $\sin \theta$ and $\cos \theta$.

$$\sin \theta = \sqrt{\frac{1 - \cos 2\theta}{2}} = \sqrt{\frac{1 - \left(-\frac{3}{5}\right)}{2}} = \sqrt{\frac{5 + 3}{10}} = \sqrt{\frac{4}{5}} = \frac{2}{\sqrt{5}}$$

$$\cos \theta = \sqrt{\frac{1 + \cos 2\theta}{2}} = \sqrt{\frac{1 + \left(-\frac{3}{5}\right)}{2}} = \sqrt{\frac{5 - 3}{10}} = \sqrt{\frac{1}{5}} = \frac{1}{\sqrt{5}}$$

The rotation formulas become

$$x = x' \cos \theta - y' \sin \theta = \frac{1}{\sqrt{5}}x' - \frac{2}{\sqrt{5}}y',$$

$$y = x' \sin \theta + y' \cos \theta = \frac{2}{\sqrt{5}}x' + \frac{1}{\sqrt{5}}y'.$$

We substitute these expressions into the given equation.

$$\left(\frac{1}{\sqrt{5}}x' - \frac{2}{\sqrt{5}}y'\right)^2 + 4\left(\frac{1}{\sqrt{5}}x' - \frac{2}{\sqrt{5}}y'\right)\left(\frac{2}{\sqrt{5}}x' + \frac{1}{\sqrt{5}}y'\right)$$

$$+ 4\left(\frac{2}{\sqrt{5}}x' + \frac{1}{\sqrt{5}}y'\right)^2 + 5\sqrt{5}\left(\frac{2}{\sqrt{5}}x' + \frac{1}{\sqrt{5}}y'\right) + 5 = 0,$$

which simplifies as a quadratic equation with no $x'y'$-term:

$$(x')^2 + 2x' + y' + 1 = 0.$$

Rewriting this equation as

$$y' = -(x' + 1)^2,$$

we then sketch the parabola in Figure 10.27 opening downward with vertex at $(-1, 0)$ in the $x'y'$-system. We note that the angle of rotation is

$$\theta = \sin^{-1}\frac{2}{\sqrt{5}} \approx 63°.$$

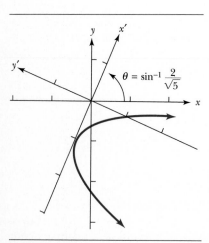

FIGURE 10.27

Exercises 10.4

The graph of each of the following equations is a non-degenerate conic section. Identify the type of conic section, but do not graph the equation. (See the examples in Table 10.4.)

1. $x^2 + y^2 - 16 = 0$ **2.** $x^2 + 4y^2 - 12 = 0$

3. $2x^2 + y^2 + 2x - y - 43 = 0$

4. $9x^2 - 9y^2 + x - 3y - 30 = 0$

5. $y^2 + 5x - y + 12 = 0$

6. $x^2 + 3x - 4y + 11 = 0$

7. $-x^2 + 4y^2 - 6x + 2y - 15 = 0$

8. $-x^2 - y^2 + 4x - 3y + 3 = 0$

9. $x^2 - y^2 = 49$ **10.** $4x^2 + 3y^2 = 12$

11. $43x^2 + 43y^2 = 77$ **12.** $-7x^2 + 9y^2 = 63$

In Exercises 13–20, use the discriminant to identify the graph of the equation. Do not sketch the graph.

13. $2xy - 2\sqrt{2}x + 2\sqrt{2}y - 5 = 0$

14. $3x^2 + xy + 3y^2 - 4 = 0$

15. $x^2 - 2xy + y^2 + 2x - 2y + 4 = 0$

16. $x^2 + 2\sqrt{3}xy - y^2 - 2 = 0$

17. $30x^2 - 120xy - 5y^2 + 150 = 0$

18. $9x^2 + 24xy + 16y^2 + 50x + 25y - 75 = 0$

19. $73x^2 + 72xy + 52y^2 + 160x + 120y = 0$

20. $34x^2 - 24xy + 41y^2 - 20x + 10y + 1 = 0$

Use a rotation to eliminate the xy-term in each of the following equations. Sketch the graph of the resulting equation, showing both the xy-system and the $x'y'$-system. (See Examples 1–3.)

21. $xy = 2$ **22.** $3x^2 - 2xy + 3y^2 = 8$

23. $3x^2 + 2\sqrt{3}xy + y^2 + x - \sqrt{3}y = 0$

24. $9x^2 + 2\sqrt{3}xy + 7y^2 = 150$

25. $11x^2 - 6\sqrt{3}xy + 5y^2 = 14$

26. $x^2 - 2\sqrt{3}xy - y^2 = 2$

27. $8x^2 + 4xy + 5y^2 = 36$

28. $4x^2 - 12xy + 9y^2 - 12\sqrt{13}x - 8\sqrt{13}y = 0$

29. $9x^2 + 24xy + 16y^2 - 20x + 15y = 0$

30. $73x^2 - 72xy + 52y^2 = 100$

31. $5x^2 - 24xy - 5y^2 + 13 = 0$

32. $18x^2 - 12xy + 2y^2 + \sqrt{10}x + 3\sqrt{10}y = 0$

33. $xy - \sqrt{2}x - 2 = 0$

34. $xy + \sqrt{2}y + 2 = 0$

35. $3x^2 + 2\sqrt{3}xy + y^2 + x - \sqrt{3}y + 2 = 0$

36. $3x^2 + 2\sqrt{3}xy + 5y^2 - 2\sqrt{3}x + 2y - 4 = 0$

37. $91x^2 - 24xy + 84y^2 + 340x + 120y + 100 = 0$

38. $4x^2 + 4xy + y^2 + 3\sqrt{5}x + 4\sqrt{5}y + 5 = 0$

10.5   Polar Coordinates

We use the Cartesian (rectangular) coordinate system to locate points that correspond to ordered pairs of the form (x, y). However, another coordinate system is often more convenient to use. It is called the **polar coordinate system**. Points in this system are located by specifying a distance r from a fixed point, called the **pole,** and the amount of rotation θ from a fixed ray, called the **polar axis.** These two quantities are specified in an ordered-pair format (r, θ), where r is the distance from the pole and θ is the amount of rotation from the polar axis. The ordered pair (r, θ) is called the **polar coordinates** of the point. We note that θ may be specified in either degree or radian measure.

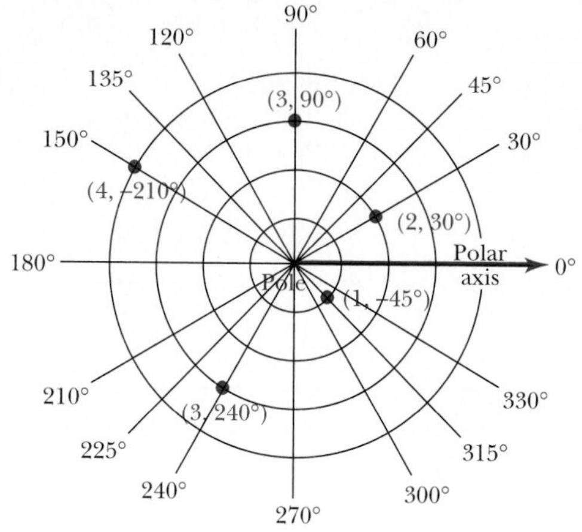

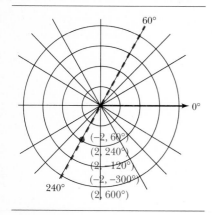

FIGURE 10.28

If a Cartesian coordinate system were superimposed on a polar co-ordinate system, the positive x-axis would coincide with the polar axis. Hence θ can be thought of as an angle in standard position. As before, positive angles are rotated counterclockwise and negative angles clockwise. In the polar coordinate system in Figure 10.28, we locate several points using polar coordinates.

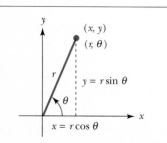

FIGURE 10.29

We also allow the possibility that r might be negative. For example, to plot a point with polar coordinates $(-2, 60°)$, we rotate $60°$ from the polar axis, extend the $60°$ ray through the pole, and fix the point two units along the extension. This is illustrated in Figure 10.29. We observe that the point with polar coordinates $(-2, 60°)$ can also be described by the polar coordinates $(2, 240°)$, $(2, -120°)$, $(-2, -300°)$, or $(2, 600°)$, as well as by many others. Hence a given point may correspond to many pairs of polar coordinates; however, a given pair of polar coordinates determines one and only one point.

The relationship between the polar coordinates (r, θ) and the rectangular coordinates (x, y) of a point agrees with the relationship between the r, θ of the trigonometric form and the a, b of the standard form of a complex number. This is obvious from Figure 10.30.

FIGURE 10.30

Relationship between Rectangular and Polar Coordinates

If (r, θ) are the polar coordinates of a point with rectangular coordinates (x, y), then

$$r^2 = x^2 + y^2, \qquad x = r \cos \theta, \qquad y = r \sin \theta.$$

Example 1   Determine the rectangular coordinates for the point with polar coordinates (3, 150°).

Solution
We have $r = 3$ and $\theta = 150°$. Thus x and y are given by

$$x = 3 \cos 150° \quad \text{and} \quad y = 3 \sin 150°$$

$$= 3\left(-\frac{\sqrt{3}}{2}\right) \qquad\qquad = 3\left(\frac{1}{2}\right)$$

$$= \frac{-3\sqrt{3}}{2}, \qquad\qquad = \frac{3}{2}.$$

The rectangular coordinates are $\left(-\dfrac{3\sqrt{3}}{2}, \dfrac{3}{2}\right)$.   ❑

Example 2   Suppose a point has $(-2, -2)$ as its rectangular coordinates. Determine the corresponding polar coordinates, where $0° \le \theta < 360°$.

Solution
We are given $x = -2$ and $y = -2$. Then $r = \sqrt{(-2)^2 + (-2)^2} = \sqrt{8} = 2\sqrt{2}$. The angle θ satisfies the equations

$$-2 = 2\sqrt{2} \cos \theta \quad \text{and} \quad -2 = 2\sqrt{2} \sin \theta$$

$$-\frac{1}{\sqrt{2}} = \cos \theta, \qquad\qquad -\frac{1}{\sqrt{2}} = \sin \theta.$$

Thus $\theta = 225°$, and the polar coordinates are $(2\sqrt{2}, 225°)$.   ❑

An equation expressed in terms of the rectangular coordinates (x, y) can be transformed into an equation in terms of the polar coordinates (r, θ), and vice versa.

Example 3   Write the equation $x^2 + y^2 = 4y$ in terms of polar coordinates (r, θ).

Solution
Since $x^2 + y^2 = r^2$ and $y = r \sin \theta$, the equation

$$x^2 + y^2 = 4y$$

becomes

$$r^2 = 4r \sin \theta.$$

The factor r can be divided out and still have an equivalent equation, because the pole is on the graph of

$$r = 4 \sin \theta.$$ ❏

Suppose we consider the graphs of the equation $x^2 + y^2 = 4$ and the corresponding polar coordinate equation $r^2 = 4$. The equation $x^2 + y^2 = 4$ represents the circle in Figure 10.31(a), with center $(0, 0)$ and radius 2. The equation $r^2 = 4$ is equivalent to $r = \pm 2$. The graph of $r = \pm 2$ is the set of all points $(\pm 2, \theta)$, where θ is arbitrary. These points lie on the circle in Figure 10.31(b), with center at the pole and radius 2. Hence we obtain the same graph whether we use rectangular or polar coordinates.

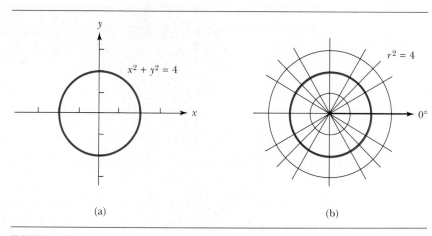

(a)   (b)

FIGURE 10.31

Example 4   Convert $r \cos \theta + r \sin \theta = 2$ to an equation in rectangular coordinates.

Solution
Since $x = r \cos \theta$ and $y = r \sin \theta$, the equation

$$r \cos \theta + r \sin \theta = 2$$

becomes

$$x + y = 2.$$ ❏

A **polar equation** is an equation expressed in terms of the polar coordinates r and θ. Many polar equations have special types of curves as their graphs. We examine a few of these next.

Example 5 Sketch the graph of the equation $r = 2 + 2 \cos \theta$.

Solution
The points with coordinates $(r, \theta) = (2 + 2 \cos \theta, \theta)$ lie on the graph of the equation. We can locate some of these points by choosing values of θ and computing corresponding values of $r = 2 + 2 \cos \theta$. Such a tabulation is shown in Figure 10.32 for several special angles between $0°$ and $180°$. These points all lie on the upper portion of the graph. Since $\cos (-\theta) = \cos \theta$, whenever (r, θ) lies on the graph, so does $(r, -\theta)$. Using $-\theta$ in place of θ yields points on the lower portion of the graph. The resulting curve is called a **cardioid.**

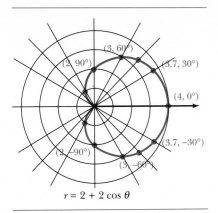

$r = 2 + 2 \cos \theta$

FIGURE 10.32

θ	0°	30°	45°	60°	90°	120°	135°	150°	180°
$2 + 2 \cos \theta$	4	3.7	3.4	3	2	1	0.6	0.3	0
(r, θ)	$(4, 0°)$	$(3.7, 30°)$	$(3.4, 45°)$	$(3, 60°)$	$(2, 90°)$	$(1, 120°)$	$(0.6, 135°)$	$(0.3, 150°)$	$(0, 180°)$

Alternate Solution
Graphs of this type might be analyzed more easily by examining the graphs of their rectangular counterparts; in this case $y = 2 + 2 \cos x$. A typical analysis might resemble the following.

First we sketch the graph of $y = 2 + 2 \cos x$ and then, as in Figure 10.33(a), we rename the x-axis as the θ-axis[2] and the y-axis as the r-axis. We separate this graph into four parts determined by the key points of the cosine curve.

In Part 1 we observe that as θ ranges from 0 to $\pi/2$, the corresponding r-values decrease from 4 to 2. We transfer this information to Part 1 of the polar graph in Figure 10.33(b). Similarly, in Part 2, as θ ranges from $\pi/2$ to π, the r-values decrease from 2 to 0. This information is transferred to Part 2 of the polar graph. Continuing in Part 3, as θ ranges from π to $3\pi/2$, r increases from 0 to 2. This provides sufficient information to sketch Part 3 of the polar graph. In Part 4, r increases from 2 to 4 as θ moves from $3\pi/2$ to 2π. This completes the graph in Figure 10.33(b) of the cardioid with equation $r = 2 + 2 \cos \theta$.

---

2. We are switching to radian measure of θ, since in Chapter 5 the x-coordinates represented real numbers.

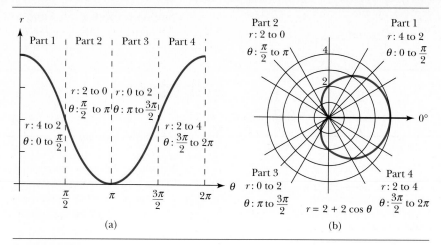

FIGURE 10.33

Example 6   Sketch the graph of the equation $r = 4 \sin 2\theta$.

Solution

We analyze the graph of $r = 4 \sin 2\theta$ by examining its rectangular counterpart $y = 4 \sin 2x$. The sketch of $y = 4 \sin 2x$ is shown in Figure 10.34(a), where the x- and y-axes are renamed the θ- and r-axes, respectively. The graph is divided into eight parts determined by the key points of the sine curve. The table in Figure 10.34 describes how the points "act" in each of the eight parts. Notice that Parts 3, 4, 7, and 8 result from points with negative r-values. The corresponding polar graph is

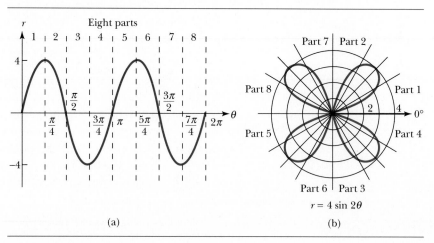

FIGURE 10.34

sketched in Figure 10.34(b). This type of curve is called a **four-leaved rose.**

	Part 1	Part 2	Part 3	Part 4	Part 5	Part 6	Part 7	Part 8
θ	0 to $\dfrac{\pi}{4}$	$\dfrac{\pi}{4}$ to $\dfrac{\pi}{2}$	$\dfrac{\pi}{2}$ to $\dfrac{3\pi}{4}$	$\dfrac{3\pi}{4}$ to π	π to $\dfrac{5\pi}{4}$	$\dfrac{5\pi}{4}$ to $\dfrac{3\pi}{2}$	$\dfrac{3\pi}{2}$ to $\dfrac{7\pi}{4}$	$\dfrac{7\pi}{4}$ to 2π
r	0 to 4	4 to 0	0 to -4	-4 to 0	0 to 4	4 to 0	0 to -4	-4 to 0

❑

Exercises 10.5

Find the rectangular coordinates of the points whose polar coordinates are given. (See Example 1.)

1. $(4, 30°)$

2. $(3, 60°)$

3. $(1, 135°)$

4. $(2, 330°)$

5. $\left(2, \dfrac{2\pi}{3}\right)$

6. $\left(5, \dfrac{\pi}{2}\right)$

7. $\left(3, \dfrac{7\pi}{6}\right)$

8. $\left(1, \dfrac{3\pi}{4}\right)$

9. $(-2, 45°)$

10. $(-8, 150°)$

11. $(-5, 270°)$

12. $(-1, 300°)$

13. $\left(-2, -\dfrac{\pi}{3}\right)$

14. $\left(-4, -\dfrac{5\pi}{4}\right)$

15. $(-3, -\pi)$

16. $(-5, 0)$

Find the polar coordinates for the points whose rectangular coordinates are given. Choose θ such that $0 \le \theta < 360°$ and $r > 0$. (See Example 2.)

17. $(-\sqrt{2}, \sqrt{2})$

18. $\left(\dfrac{1}{2}, \dfrac{\sqrt{3}}{2}\right)$

19. $(-\sqrt{3}, -1)$

20. $(-1, 0)$

21. $(0, -1)$

22. $(3, 0)$

23. $(4, -4)$

24. $(3, 3)$

Convert each of the following equations into a polar equation. (See Example 3.)

25. $x = 3$

26. $y = -2$

27. $x = y$

28. $2x + 3y = 1$

29. $x^2 + y^2 = 1$

30. $x^2 - y^2 = 4$

31. $x^2 + y^2 - 4x = 0$

32. $x^2 + y^2 - 6y = 0$

Convert each of the following polar equations into an equation in the rectangular coordinates x and y. (See Example 4.)

33. $r \sin \theta = 2$

34. $r \cos \theta = -3$

35. $r = \sin \theta - \cos \theta$

36. $r = 2 \sin \theta + 2 \cos \theta$

37. $r = \dfrac{1}{\sin \theta + \cos \theta}$

38. $r = \dfrac{-3}{\sin \theta + \cos \theta}$

39. $r = \dfrac{1}{r - \cos \theta}$

40. $r = \dfrac{1}{r + 4 \sin \theta}$

41. $r = \dfrac{2}{1 + \sin \theta}$

42. $r = \dfrac{6}{2 + \sin \theta}$

43. $r = \dfrac{6}{2 - \cos \theta}$

44. $r = \dfrac{4}{2 + 3 \cos \theta}$

Sketch the graph of each of the following equations in a polar coordinate system. (See Examples 5 and 6.)

45. $r = 2$

46. $r = 3$

47. $\theta = \dfrac{\pi}{4}$

48. $\theta = \dfrac{2\pi}{3}$

49. $r = 2 \sin \theta$

50. $r = 3 \cos \theta$

51. $r = -2 \cos \theta$

52. $r = -\sin \theta$

53. $r = 1 + \cos \theta$ (cardioid)

54. $r = 2 + 2 \sin \theta$ (cardioid)

55. $r = 1 - \sin \theta$ (cardioid)

56. $r = 2 - 2 \cos \theta$ (cardioid)

57. $r = 2 + \cos \theta$ (This type of curve is called a limacon.)

58. $r = 3 + \sin \theta$ (limacon)

59. $r = 1 + 3 \cos \theta$ (limacon)

60. $r = 1 + 3 \sin \theta$ (limacon)

61. $r = 3 \sin 2\theta$ (four-leaved rose)

62. $r = 3 \cos 2\theta$ (four-leaved rose)

63. $r = 4 \cos 3\theta$ (three-leaved rose)

64. $r = 2 \sin 3\theta$ (three-leaved rose)

65. $r = 2\theta$, $0 \le \theta \le 2\pi$ (spiral)

66. $r = 3\theta$, $0 \le \theta \le 2\pi$ (spiral)

67. $r^2 = \cos 2\theta$ (This type of curve is called a lemniscate.)

68. $r^2 = \sin 2\theta$ (lemniscate)

10.6 Polar Equations of Conics

In Definition 10.1, a parabola is defined as the set of all points equidistant from a given point F and a given line ℓ. The point F is the *focus* and the line ℓ is the *directrix* of the parabola. It is an interesting fact that hyperbolas and noncircular ellipses can be given similar formulations in terms of a focus and a directrix. This fact is a unifying property for the parabola, the ellipse, and the hyperbola, and it is sometimes referred to as the *focus–directrix* property of conic sections. This property is stated formally in the following theorem.

> ### Theorem 10.7
>
> Suppose that the distance $|\overrightarrow{PF}|$ from the point P to a given point F is a constant multiple of the distance $|\overrightarrow{PD}|$ from P to a given line ℓ, so that
>
> $$|\overrightarrow{PF}| = e \cdot |\overrightarrow{PD}|$$
>
> for a positive constant e. The set of all such points P is a conic section that forms
>
> a) A parabola if $e = 1$,
>
> b) An ellipse if $0 < e < 1$,
>
> c) A hyperbola if $e > 1$.
>
> The point F is the **focus,** the line ℓ is the **directrix,** and the number e is the **eccentricity** of the conic section.

To show that the theorem is true, we choose a coordinate system with the focus F at the origin and the directrix ℓ with equation $x = k > 0$, as shown in Figure 10.35.

From the figure, we see that $|\overrightarrow{PF}| = \sqrt{x^2 + y^2}$ and $|\overrightarrow{PD}| = k - x$, so the equation $|\overrightarrow{PF}| = e \cdot |\overrightarrow{PD}|$ appears as

$$\sqrt{x^2 + y^2} = e(k - x). \tag{1}$$

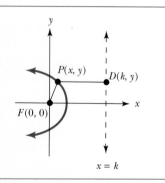

FIGURE 10.35

Our goal now is to get an equivalent equation that can be used to obtain the statements in the theorem.

Squaring both sides of Eq. (1), we have

$$x^2 + y^2 = e^2(k^2 - 2kx + x^2). \tag{2}$$

This is equivalent to Eq. (1), since both $\sqrt{x^2 + y^2}$ and $e(k - x)$ are positive numbers. Collecting like terms on the left side gives

$$(1 - e^2)x^2 + y^2 + 2ke^2x - e^2k^2 = 0. \tag{3}$$

Equation (3) fits the form given in Table 10.4 with $A = 1 - e^2$, $C = 1$, $D = 2ke^2$, $F = -e^2k^2$, and the following conclusions (a), (b), and (c) can be drawn from this fact.

a) If $e = 1$, we have $A = 0$, $C \neq 0$, and $D = 2k \neq 0$ in Table 10.4, so the set of points forms a parabola.

b) If $0 < e < 1$, then $0 < e^2 < e < 1$ and $1 - e^2 > 0$. Thus we have $A = 1 - e^2 > 0$, $C = 1$, and $F = -e^2k^2 < 0$. Since $AC > 0$ and $AF < 0$, we have an ellipse in this case.

c) If $e > 1$, then $A = 1 - e^2 < 0$ and $AC < 0$, so that we have a hyperbola in this case.

It is easy to see that the use of the terms *focus* and *directrix* in Theorem 10.7 agrees with their use in Definition 10.1 for the parabola. It can be shown that the use of the term *eccentricity* in the theorem also agrees with its use in the definition

$$e = \frac{c}{a}$$

that is given for an ellipse in Exercise 61 of Exercises 10.2, where $c = \sqrt{a^2 - b^2}$. With the notation of Section 10.3, it can also be shown that the same formula $e = c/a$ works for a hyperbola, but $c = \sqrt{a^2 + b^2}$ in the case of a hyperbola.

We do not go into any of the details here, but a statement that is the converse of Theorem 10.7 is true. That is, every nondegenerate conic section that is not a circle has a focus F and a directrix ℓ such that the points P on the conic section satisfy the conditions stated in Theorem 10.7.

The focus–directrix property of conic sections leads to some simple polar equations for conic sections. Consider first the situation in Theorem 10.7 in which the focus F is at the origin and the directrix is located at $x = k > 0$. With $r = \sqrt{x^2 + y^2}$ and $x = r \cos \theta$, Eq. (1) appears in polar coordinates as

$$r = e(k - r \cos \theta).$$

Solving this equation for r yields

$$r = \frac{ek}{1 + e \cos \theta}. \tag{4}$$

This is one of the standard forms for polar equations of conics with a focus at the origin. There are three other forms, corresponding to different locations of the directrix. With directrix $x = -k$ and $k > 0$, for example, Eq. (1) is replaced by

$$\sqrt{x^2 + y^2} = e(k + x),$$

or

$$r = e(k + r \cos \theta).$$

When this equation is solved for r, it yields

$$r = \frac{ek}{1 - e \cos \theta}. \tag{5}$$

Comparing Eqs. (4) and (5), we note that the $+$ sign applies when the directrix is to the right of the pole, and the $-$ sign applies when the directrix is to the left of the pole.

Similar derivations can be made for a directrix $y = k$ or $y = -k$, where k is positive. The results are summarized in Table 10.5.

Polar Equations for Conic Sections with Focus $(0, 0)$, Eccentricity $e > 0$, and Distance $k > 0$ between the Focus and Directrix	
Polar equation of conic	**Cartesian equation of directrix**
$r = \dfrac{ek}{1 + e \cos \theta}$	$x = k$
$r = \dfrac{ek}{1 - e \cos \theta}$	$x = -k$
$r = \dfrac{ek}{1 + e \sin \theta}$	$y = k$
$r = \dfrac{ek}{1 - e \sin \theta}$	$y = -k$

TABLE 10.5

As the following example shows, important information about a conic can be found by matching its equation to one of the forms in Table 10.5.

Example 1   Find the eccentricity of the conic that has the following equation, identify the type of conic, and find a polar equation of the directrix.

$$r = \frac{6}{2 + \cos \theta}$$

Solution
To obtain an equation that matches one of those in Table 10.5, we need to change the constant in the denominator to 1. We get

$$r = \frac{\frac{1}{2}(6)}{\frac{1}{2}(2 + \cos \theta)} = \frac{3}{1 + \frac{1}{2}\cos \theta},$$

and this matches

$$r = \frac{ek}{1 + e \cos \theta},$$

with $e = \frac{1}{2}$ and $k = 6$. Thus the eccentricity is $\frac{1}{2}$, and the conic is an ellipse. According to Table 10.5, the directrix has Cartesian equation $x = 6$, and this transforms to the polar equations

$$r \cos \theta = 6, \quad \text{or} \quad r = 6 \sec \theta. \qquad \qquad \Box$$

Example 2   Sketch the graph of the conic with equation

$$r = \frac{6}{2 + \cos \theta}.$$

Solution

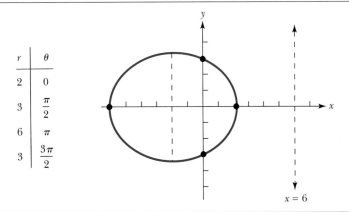

r	θ
2	0
3	$\dfrac{\pi}{2}$
6	π
3	$\dfrac{3\pi}{2}$

FIGURE 10.36

As in Example 1, the given equation is equivalent to

$$r = \frac{3}{1 + \frac{1}{2}\cos\theta}.$$

We have $e = \frac{1}{2}$, $k = 6$, and the conic is an ellipse with directrix at $x = 6$. To obtain the graph, we need only plot points corresponding to $\theta = 0$, $\pi/2$, π, and $3\pi/2$ and use these to sketch the ellipse. This is shown in Figure 10.36. ❏

As Example 2 illustrates, we can identify a conic by its eccentricity and use points corresponding to quadrantal angles to sketch its graph. The accuracy obtained by this method is ordinarily good enough, but more detail can be obtained by using some knowledge from Sections 10.1–10.3.

For more detail in Example 2, we first note that the polar coordinates of the vertices of the ellipse are $(2, 0)$ and $(6, \pi)$. With the notation from Section 10.2, this means that $2a = 2 + 6$, so $a = 4$ and the center is at $(2, \pi)$. We can then use $e = c/a$ to obtain

$$c = ae = (4)\left(\frac{1}{2}\right) = 2.$$

The formula $c = \sqrt{a^2 - b^2}$ from Section 10.2 then yields $b^2 = a^2 - c^2 = 12$ and $b = 2\sqrt{3}$ as half the length of the minor axis. The ends of the minor axes can now be located as the ends of the dashed segment in Figure 10.36.

If the asymptotes of a hyperbola are needed as part of the sketch, the values of a and b in the standard equations from Section 10.3 must be found. The technique for this is demonstrated in the next example.

Example 3   Sketch the graph of the conic with equation

$$r = \frac{3}{1 + 2\sin\theta}.$$

Solution
We recognize at once that $e = 2$ and $ek = 3$, so $k = \frac{3}{2}$ and the conic is a hyperbola with one focus at the pole and directrix with equation $y = \frac{3}{2}$. Using quadrantal angles for θ, we obtain the points

$$(3, 0), (1, \pi/2), (3, \pi), (-3, 3\pi/2).$$

With these points plotted and the focus located at $(0, 0)$, we can draw the sketch as shown in Figure 10.37(a).

If a sketch that includes the asymptotes is desired, we proceed as follows: The vertices are at $(1, \pi/2)$ and $(-3, 3\pi/2)$, so $2a = 2$, $a = 1$,

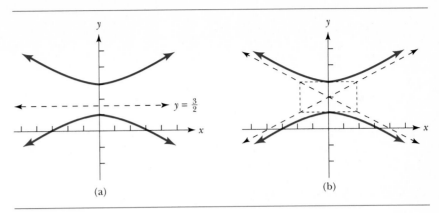

(a)   (b)

FIGURE 10.37

and the center is at $(2, \pi/2)$. Using $e = 2$ and $a = 1$ in $e = c/a$ gives

$$c = ae = (1)(2) = 2.$$

From the formula $c = \sqrt{a^2 + b^2}$ of Section 10.3, we get

$$b^2 = c^2 - a^2 = 4 - 1 = 3$$

and $b = \sqrt{3}$. We now use $a = 1$ and $b = \sqrt{3}$ to draw the dashed rectangle and asymptotes as we did in Section 10.3. This is shown in Figure 10.37(b).   ❏

Example 4   Find a polar equation of the conic with eccentricity $e = \frac{2}{3}$ and focus at the pole, if the equation of a directrix is $r \sin \theta = -2$.

Solution
In Cartesian coordinates, the directrix equation is $y = -2$. From Table 10.5, we recognize that $k = 2$ and the form of the equation is

$$r = \frac{ek}{1 - e \sin \theta}.$$

The values $e = \frac{2}{3}$ and $k = 2$ give

$$r = \frac{\dfrac{4}{3}}{1 - \dfrac{2}{3} \sin \theta},$$

or

$$r = \frac{4}{3 - 2 \sin \theta}. \qquad ❏$$

Exercises 10.6

In Exercises 1–12, find the eccentricity of the conic section that has the given equation, identify the type of conic, and find a polar equation of the directrix. (See Example 1.)

1. $r = \dfrac{2}{1 + \sin \theta}$ **2.** $r = \dfrac{3}{1 - \cos \theta}$

3. $r = \dfrac{3}{2 - 2 \cos \theta}$ **4.** $r = \dfrac{2}{3 + 3 \sin \theta}$

5. $r = \dfrac{6}{2 + \sin \theta}$ **6.** $r = \dfrac{6}{2 - \sin \theta}$

7. $r = \dfrac{4}{2 + 3 \cos \theta}$ **8.** $r = \dfrac{2}{2 - 3 \cos \theta}$

9. $2r = 3r \cos \theta + 6$ **10.** $3r = 6 - 2r \sin \theta$

11. $r\sqrt{2} = 2 + r \sin \theta$ **12.** $r\sqrt{2} = 2 - r \cos \theta$

In Exercises 13–24, identify the type of conic section represented by the given equation, state the eccentricity, and sketch the graph of the equation. (See Examples 2 and 3.)

13. $r = \dfrac{3}{4 + 2 \cos \theta}$ **14.** $r = \dfrac{3}{4 - 2 \cos \theta}$

15. $r = \dfrac{3}{4 + 2 \sin \theta}$ **16.** $r = \dfrac{3}{4 - 2 \sin \theta}$

17. $r = \dfrac{3}{2 - 2 \sin \theta}$ **18.** $r = \dfrac{3}{2 + 2 \cos \theta}$

19. $r = 3 + 2r \cos \theta$ **20.** $r = 3 - 2r \cos \theta$

21. $r = \dfrac{6 \sec \theta}{2 \sec \theta - 1}$ **22.** $r = \dfrac{3 \csc \theta}{4 \csc \theta + 2}$

23. $r = \dfrac{2}{\cos \theta - 1}$ **24.** $r = \dfrac{2}{\sin \theta - 1}$

In Exercises 25–36, find a polar equation of the conic section that has a focus at the pole and satisfies the given conditions. (See Example 4.)

25. Eccentricity 3; equation of a directrix, $r \cos \theta = -4$

26. Eccentricity 2; equation of a directrix, $r \sin \theta = 6$

27. Eccentricity 1; equation of a directrix, $r = 2 \csc \theta$

28. Eccentricity 1; equation of a directrix, $r = -3 \sec \theta$

29. Ellipse with vertices at $(1, 0)$ and $(5, \pi)$ in polar coordinates

30. Ellipse with vertices at $(2, \pi/2)$ and $(4, 3\pi/2)$ in polar coordinates

31. Parabola with vertex at $(3, 0)$ in polar coordinates

32. Parabola with vertex at $(2, \pi/2)$ in polar coordinates

33. Hyperbola with vertices at $(1, 0)$ and $(-3, \pi)$ in polar coordinates

34. Hyperbola with vertices at $(2, 0)$ and $(-8, \pi)$ in polar coordinates

35. Eccentricity $\frac{2}{3}$, vertex at $(6, \pi/2)$ in polar coordinates

36. Eccentricity $\frac{4}{3}$, vertex at $(12, 0)$ in polar coordinates

10.7 Parametric Equations

In many sections of this book, we have seen graphs of functions and graphs of equations that involve two coordinate variables. All these graphs are examples of *plane curves*. The graphs and their related equations give useful information about various situations. Some situations, however, can be described in a more natural way by using a third variable. This is especially true when the location of a moving object is being given. Our first example presents a simple case of this type.

Example 1   The Cartesian coordinates of a point in the xy-plane are given in centimeters. The point starts moving at the origin, with its *x*-coordinate increasing at 2 centimeters per second and its *y*-coordinate increasing at 6 centimeters per second. If the motion lasts for 7 seconds, find the point's final location and describe its path.

Solution

Let *t* be the number of seconds that have elapsed since the point started moving. Since the *x*-coordinate increases at 2 centimeters per second, $x = 2$ when $t = 1$, $x = 4$ when $t = 2$, and $x = 2t$ for *t* in the interval [0, 7]. Similarly, $y = 6t$ for *t* in [0, 7]. Since the motion stops at $t = 7$, the final location is where $x = 2(7) = 14$, $y = 6(7) = 42$. That is, the point stops at (14, 42).

To obtain a familiar representation of the path, we eliminate *t* from the equations

$$x = 2t, \qquad y = 6t, \quad \text{for} \quad t \in [0, 7].$$

Solving for *t* in the first equation, we have $t = x/2$. Substitution of this value in the second equation yields

$$y = 6\left(\frac{x}{2}\right), \quad \text{or} \quad y = 3x.$$

Thus the moving point always lies on the straight line $y = 3x$. However, its path is not the entire line but only the segment from (0, 0) to (14, 42), including the endpoints. This is shown in Figure 10.38. ❏

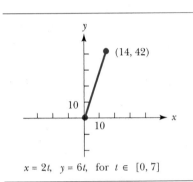

$x = 2t, \ y = 6t, \ \text{for} \ t \in \ [0, 7]$

FIGURE 10.38

Example 1 illustrates the terms in the following definition.

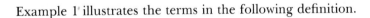

> **Definition 10.8**
>
> A **plane curve** is a set *C* of points with coordinates (*x*, *y*) that are determined by equations
>
> $$x = f(t), \qquad y = g(t),$$
>
> where *f* and *g* are functions of a third real variable *t*. The domain *I* of *t* is either $\mathcal{R}$ or an interval of one of the types defined in Section 1.1. The equations $x = f(t), y = g(t)$ are called **parametric equations,** and *t* is called the **parameter.**

A plane curve is frequently referred to simply as a **curve,** and some variable other than *t* may be used for the parameter.

The graph of any function *f* is a plane curve. To obtain parametric equations from $y = f(x)$, we need only write

$$x = t, \qquad y = f(t)$$

and restrict *t* to the domain of *f*. To see that these parametric equations are not unique, consider the function *f* defined by $y = x^2 - 1$. The graph of f is determined by

$$x = t, \qquad y = t^2 - 1, \quad \text{for} \quad t \in \mathcal{R}.$$

This same graph is also determined by the equations

$$x = v + 1, \qquad y = v^2 + 2v, \quad \text{for} \quad v \in \mathcal{R},$$

since $(v + 1)^2 - 1 = v^2 + 2v$. Thus a given curve may be determined by many different sets of parametric equations.

Two important points illustrated in Example 1 need emphasis. The first point is that frequently the type of curve under consideration can be recognized by eliminating the parameter to obtain a Cartesian equation. The second point is that the curve that has the given parametric equations *may be* only *part* of the graph of the Cartesian equation. In Example 1 the parametric equations determined a line segment that was only *part* of the straight line with equation $y = 3x$. We will see in Example 3, however, that the parametric equations do sometimes determine the whole graph of the Cartesian equation.

Example 2 Graph the curve C that has parametric equations

$$x = 2t^2, \qquad y = 8t^4 + 1, \quad \text{for} \quad t \in \mathcal{R}.$$

Solution

To eliminate the parameter, we can solve for t^2 in the first equation to get $t^2 = x/2$ and then substitute for $t^4 = (t^2)^2$ in the second equation. This yields

$$y = 8\left(\frac{x}{2}\right)^2 + 1, \quad \text{or} \quad y = 2x^2 + 1.$$

We recognize this as an equation of a parabola that opens upward with vertex at $(0, 1)$. However, the equation $x = 2t^2$ requires that $x \geq 0$, since t^2 is never negative. Also, $y \geq 1$, since $t^4 \geq 0$ for all t in $\mathcal{R}$. Thus the curve C consists of the right half of the parabola, as drawn in Figure 10.39. A table of values is included in the figure to illustrate how the points on C are traced out as t varies. As t increases through the value 0, the corresponding points move down the parabola to the vertex at $t = 0$ and then reverse direction to move back up the parabola. ❑

t	-2	-1	0	1	2
x	8	2	0	2	8
y	129	9	1	9	129

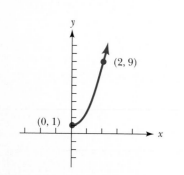

FIGURE 10.39

In Examples 1 and 2 the parameter was eliminated by substitution. Sometimes the parameter may be eliminated by other methods, such as addition or subtraction of equations. In some cases it is impossible to eliminate the parameter. Then we simply plot as many points as are necessary to sketch the desired curve.

If the parametric equations for C involve trigonometric functions, the parameter may be eliminated by use of an identity. This important technique is illustrated in our next example. In this same example, we have a situation where the parametric equations determine the whole graph of the Cartesian equation.

Example 3   A curve C is determined by the parametric equations

$$x = 3 \cos \theta, \qquad y = 2 \sin \theta, \quad \text{for} \quad \theta \in [0, 2\pi].$$

Find a Cartesian equation for a graph that contains C and sketch the curve C.

Solution
To eliminate the parameter θ, we substitute for $\cos \theta$ and $\sin \theta$ in the identity $\cos^2 \theta + \sin^2 \theta = 1$. From the parametric equations, we have

$$\cos \theta = \frac{x}{3} \quad \text{and} \quad \sin \theta = \frac{y}{2}.$$

Substituting in the identity, we have

$$\left(\frac{x}{3}\right)^2 + \left(\frac{y}{2}\right)^2 = 1, \quad \text{or} \quad \frac{x^2}{9} + \frac{y^2}{4} = 1.$$

This is the equation of the ellipse shown in Figure 10.40. By assigning values of t at intervals of $\pi/2$, we obtain the table of values in that figure. As t increases from 0 to $\pi/2$, the corresponding point on C moves along the ellipse in the first quadrant from $(3, 0)$ to $(0, 2)$. As t continues to increase from $\pi/2$ to 2π, the point on C traces out the remaining portion of the ellipse.

As a final remark, we note that the domain $[0, 2\pi]$ for θ was crucial in obtaining the entire ellipse as the curve C. If the domain for θ had been $[0, \pi/2]$, for instance, the curve C would have consisted of only the first quadrant portion of the ellipse, together with $(3, 0)$ and $(0, 2)$.   ❏

Example 4   Find a Cartesian equation for a graph that contains C and sketch the graph of the curve C that is given parametrically by

$$x = \cos 2t, \qquad y = \cos t, \quad \text{for} \quad t \in \mathcal{R}.$$

Solution
We use the identity $\cos 2t = 2 \cos^2 t - 1$ to eliminate the parameter, getting

$$x = 2y^2 - 1.$$

This is an equation of a parabola that opens to the right with vertex at $(-1, 0)$. However, the parametric equations

$$x = \cos 2t, \qquad y = \cos t, \quad t \in \mathcal{R}$$

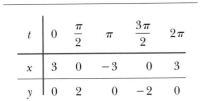

t	0	$\dfrac{\pi}{2}$	π	$\dfrac{3\pi}{2}$	2π
x	3	0	-3	0	3
y	0	2	0	-2	0

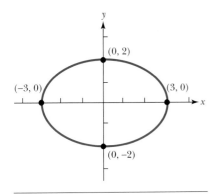

FIGURE 10.40

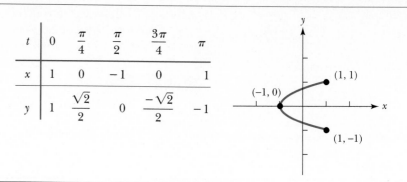

t	0	$\dfrac{\pi}{4}$	$\dfrac{\pi}{2}$	$\dfrac{3\pi}{4}$	π
x	1	0	-1	0	1
y	1	$\dfrac{\sqrt{2}}{2}$	0	$\dfrac{-\sqrt{2}}{2}$	-1

FIGURE 10.41

force $-1 \le x \le 1$ and $-1 \le y \le 1$. In the table of Figure 10.41, enough values have been assigned to t to determine that the curve C consists of the portion of the parabola that is shown in the figure.   ❑

Example 5   The curve traced out by a point P on a circle of radius a as the circle rolls along a line without slipping is called a **cycloid.** We assume that the point P starts at the origin O and the circle rolls along the x-axis in the positive direction, as shown in Figure 10.42.

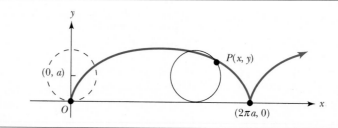

FIGURE 10.42

Let Q be the center of the circle and let θ be the positive angle of rotation in radians through which the radius QP has turned, as shown in Figure 10.43. Using θ as the parameter, find a set of parametric equations for the cycloid.

Solution
In Figure 10.43, the angle θ' is the related angle for θ, so that $\theta' = \pi - \theta$ and

$$\sin \theta' = \sin (\pi - \theta) = \sin \theta,$$

$$\cos \theta' = \cos (\pi - \theta) = -\cos \theta.$$

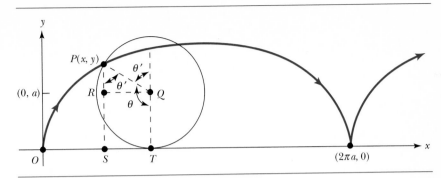

FIGURE 10.43

We have $|\overrightarrow{PQ}| = a$ in the right triangle PQR, so that

$$|\overrightarrow{RQ}| = a \sin \theta' = a \sin \theta,$$

$$|\overrightarrow{RP}| = a \cos \theta' = -a \cos \theta.$$

Since the circle rolled without slipping, the distance $|\overrightarrow{OT}|$ is the same as the length of the arc from P to T. The arc length formula $s = r\theta$ of Section 5.1 gives $|\overrightarrow{OT}| = a\theta$. Thus

$$x = |\overrightarrow{OT}| - |\overrightarrow{ST}| = |\overrightarrow{OT}| - |\overrightarrow{RQ}|$$
$$= a\theta - a \sin \theta = a(\theta - \sin \theta).$$

Similarly,

$$y = |\overrightarrow{SR}| + |\overrightarrow{RP}| = |\overrightarrow{TQ}| + |\overrightarrow{RP}|$$
$$= a - a \cos \theta = a(1 - \cos \theta).$$

Hence we have the parametric equations

$$x = a(\theta - \sin \theta), \qquad y = a(1 - \cos \theta).$$

These equations are neat and effective. They also illustrate a case where it is very difficult to eliminate the parameter, and it surely is not worth the effort. ❏

Exercises 10.7

Parametric equations for a curve C are given in Exercises 1–34. In each exercise, (a) find a Cartesian equation for a graph that contains C and (b) sketch the graph of C. (See Examples 1–5.)

1. $x = t + 1$, $y = t - 5$, $t \in [-1, 3]$
2. $x = 2t$, $y = t - 1$, $t \in [-1, 3]$
3. $x = -2t$, $y = t + 1$, $t \in [0, \infty)$

4. $x = -t$, $y = 2t + 1$, $t \in [0, \infty)$

5. $x = t^2 - 1$, $y = 2t^4 + 1$, $t \in \mathcal{R}$

6. $x = 6t^2$, $y = 2t^2 - 8$, $t \in \mathcal{R}$

7. $x = t^4$, $y = t^2 + 1$, $t \in \mathcal{R}$

8. $x = 4t^4 + 1$, $y = 2t^2$, $t \in \mathcal{R}$

9. $x = t - 1$, $y = t^2 + 2$, $t \in \mathcal{R}$

10. $x = 2t + 1$, $y = 4t^2$, $t \in \mathcal{R}$

11. $x = e^t$, $y = e^{2t}$, $t \in \mathcal{R}$

12. $x = e^t$, $y = e^{-t}$, $t \in \mathcal{R}$

13. $x = t^2$, $y = 6 - 2t^2$, $t \in [0, 2]$

14. $x = t^4$, $y = 1 + 3t^4$, $t \in [0, 1]$

15. $x = \sqrt{t}$, $y = t$, $t \in [0, \infty)$

16. $x = \sqrt{t}$, $y = \sqrt{t}$, $t \in [0, \infty)$

17. $x = t$, $y = \sqrt{4 - t}$, $t \in [-2, 0]$

18. $x = \sqrt{t}$, $y = -\sqrt{1 - t}$, $t \in [0, 1]$

19. $x = t$, $y = \dfrac{1}{t}$, $t \in \mathcal{R}$ and $t \neq 0$

20. $x = t$, $y = \dfrac{1}{t^2}$, $t \in \mathcal{R}$ and $t \neq 0$

21. $x = \cos \theta$, $y = \sec \theta$, $\theta \in [0, \pi/2)$

22. $x = \csc \theta$, $y = \sin \theta$, $\theta \in (0, \pi)$

23. $x = \cos \theta$, $y = \sin \theta$, $\theta \in [\pi, 2\pi]$

24. $x = 3 \sin \theta$, $y = 2 \cos \theta$, $\theta \in [0, \pi/2]$

25. $x = \tan \theta$, $y = \sec^2 \theta$, $\theta \in [0, \pi/4]$

26. $x = \cot \theta$, $y = \csc^2 \theta$, $\theta \in [\pi/4, \pi/2]$

27. $x = \sec^2 \theta$, $y = \tan \theta$, $\theta \in (-\pi/2, \pi/2)$

28. $x = \csc \theta$, $y = \cot^2 \theta$, $\theta \in (0, \pi)$

29. $x = \csc \theta$, $y = \cot \theta$, $\theta \in (0, \pi)$

30. $x = \sec \theta$, $y = \tan \theta$, $\theta \in (-\pi/2, \pi/2)$

31. $x = \sin \theta + 2$, $y = \cos \theta - 1$, $\theta \in \mathcal{R}$

32. $x = \cos \theta + 1$, $y = \sin \theta - 2$, $\theta \in \mathcal{R}$

33. $x = \sin \theta$, $y = \cos 2\theta$, $\theta \in \mathcal{R}$

34. $x = \cos 2\theta$, $y = \sin \theta$, $\theta \in \mathcal{R}$

35. The projectile in the figure is launched from ground level at the origin with initial velocity v_0 at an angle θ with the positive x-axis. If $v_0 = 800$ feet per second and $\theta = 30°$, the projectile is located at time t by

$$x = 400t\sqrt{3}, \quad y = 400t - 16t^2.$$

Find the distance from the origin to where the projectile strikes the ground. (This distance is the **range** of the projectile.)

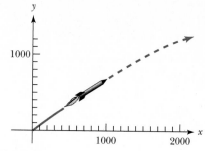

EXERCISE 35

36. The projectile in Exercise 35 has maximum range when $\theta = 45°$. With $v_0 = 800$ feet per second and $\theta = 45°$, the parametric equations are

$$x = 400t\sqrt{2}, \quad y = 400t\sqrt{2} - 16t^2.$$

Find the range of the projectile in this case.

37. Assume the projectile in Exercise 35 stops when it hits the ground and sketch its path.

38. Assume the projectile in Exercise 36 stops when it hits the ground and sketch its path.

39. If a projectile is launched as described in Exercise 35 with $0 < \theta < \pi/2$ and air resistance is neglected, the parametric equations for the path are

$$x = v_0 t \cos \theta, \quad y = v_0 t \sin \theta - 16t^2.$$

Eliminate the parameter and find a Cartesian equation for the path.

40. Show that the range R of the projectile in Exercise 39 is given by the formula $R = (v_0^2 \sin 2\theta)/32$ feet.

Launch of Apollo 15
Courtesy of Bettmann Newsphotos

<table>
<tr><td>**10.8**</td><td>## Three-Dimensional Rectangular Coordinates</td></tr>
</table>

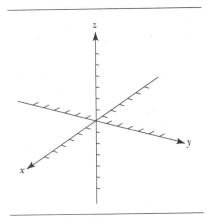

FIGURE 10.44

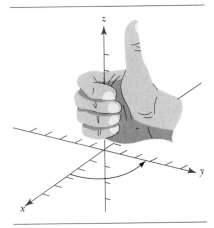

FIGURE 10.45

Ordered triples of real numbers can be used as rectangular coordinates to locate points in space. We begin with an xy-coordinate plane in a horizontal position, as shown in Figure 10.44. A third axis, the z-axis, is introduced as a vertical number line through the origin of the xy-plane. The z-axis is placed so that zero on the z-axis is located at the xy-origin. The three **coordinate axes** are then mutually perpendicular and intersect at the **origin** of the three-dimensional system.

The most common convention is to choose the positive direction upward on the z-axis and the negative direction downward. This orientation of axes is shown in Figure 10.45 and is known as a **right-handed system**. To understand the reason for this term, consider a rotation of 90° about the z-axis that moves the positive x-axis to the former position of the positive y-axis. If the z-axis is grasped with the fingers of the right hand pointing in the direction of rotation, the extended thumb points in the direction of the positive z-axis. If the positive direction on the z-axis were chosen downward in Figure 10.45, a **left-handed system** would result.

As a matter of choice, our coordinate systems will be right-handed. Ordinarily we draw the yz-plane in the plane of the page with the positive y-axis to the right, the positive z-axis upward, and the positive x-axis perpendicular to the page, pointing toward the viewer. See Drawing Tip 1.

The pairs of coordinate axes determine three **coordinate planes:** the xy-plane, the xz-plane, and the yz-plane. We set up a correspondence from points P in space to ordered triples (x, y, z) of real numbers, where x is the directed distance from the yz-plane to P, y is the directed distance from the xz-plane to P, and z is the directed distance from the xy-plane to P. This correspondence is one-to-one, and the ordered triple (x, y, z)

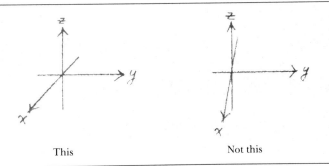

This   Not this

DRAWING TIP 1 Make the angle between the positive x-axis and the positive y-axis large enough.

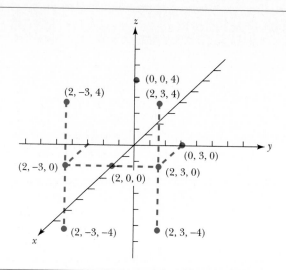

FIGURE 10.46

is referred to as the **rectangular coordinates,** or **Cartesian coordinates,** of the point P. To locate the point P with coordinates (x_0, y_0, z_0), we use x_0 and y_0 as usual in the xy-plane and then move up or down, depending on whether z_0 is positive or negative. Several points are plotted in Figure 10.46.

The three coordinate planes partition space into eight regions that have the coordinate planes as their boundaries. Each of these regions is called an **octant,** and the octant where all coordinates are positive is called the **first octant.**

The formula for the distance between two points in space is a natural extension of the familiar formula in the xy-plane. Let $P(x_1, y_1, z_1)$ and $Q(x_2, y_2, z_2)$ be two points as shown in Figure 10.47, and let $R(x_2, y_2, z_1)$ be the point directly below Q with the same z-coordinate as P. Then $\overrightarrow{PR}$ is parallel to the xy-plane and has the same length as the segment joining $(x_1, y_1, 0)$ and $(x_2, y_2, 0)$ in the xy-plane. Thus

$$|\overrightarrow{PR}| = \sqrt{(x_2 - x_1)^2 + (y_2 - y_1)^2}.$$

It is clear from the figure that

$$|\overrightarrow{RQ}| = |z_2 - z_1| = \sqrt{(z_2 - z_1)^2}.$$

Since $\triangle PRQ$ is a right triangle, we have

$$|\overrightarrow{PQ}|^2 = |\overrightarrow{PR}|^2 + |\overrightarrow{RQ}|^2$$
$$= (x_2 - x_1)^2 + (y_2 - y_1)^2 + (z_2 - z_1)^2$$

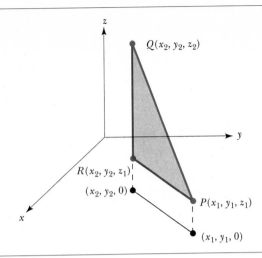

FIGURE 10.47

and

$$\overrightarrow{|PQ|} = \sqrt{(x_2 - x_1)^2 + (y_2 - y_1)^2 + (z_2 - z_1)^2}.$$

The same formula holds for all points P and Q in space, no matter how they are located relative to each other. (See Drawing Tip 2.)

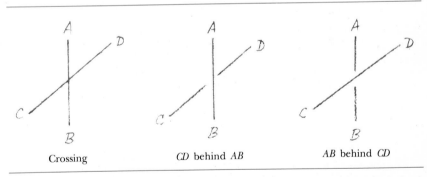

Crossing CD behind AB AB behind CD

DRAWING TIP 2 Break lines. When one line passes behind another, break it to show that it doesn't touch and that part of it is hidden.

As a special case of the distance formula, the distance from the origin O to the point $P(x, y, z)$ is

$$\overrightarrow{|PQ|} = \sqrt{x^2 + y^2 + z^2}.$$

Example 1　Find the distance between the points $(-2, 1, 4)$ and $(1, -1, -2)$.

Solution

With $P(-2, 1, 4)$ and $Q(1, -1, -2)$ in the distance formula, we find that

$$|\overrightarrow{PQ}| = \sqrt{(1 + 2)^2 + (-1 - 1)^2 + (-2 - 4)^2}$$
$$= \sqrt{9 + 4 + 36} = \sqrt{49} = 7. \qquad \square$$

A hint is given in Exercise 56 of Exercises 10.8 that suggests a way to prove the following theorem.

Theorem 10.9　**The Midpoint Formula**

The midpoint of the line segment joining $P(x_1, y_1, z_1)$ and $Q(x_2, y_2, z_2)$ has coordinates

$$\left(\frac{x_1 + x_2}{2}, \frac{y_1 + y_2}{2}, \frac{z_1 + z_2}{2} \right).$$

Example 2　The midpoint of the segment joining $(-2, 1, 4)$ and $(1, -1, -2)$ has coordinates

$$\left(\frac{-2 + 1}{2}, \frac{1 - 1}{2}, \frac{4 - 2}{2} \right) = (-\tfrac{1}{2}, 0, 1). \qquad \square$$

A **sphere** is the set of all points in space equally distant from a given point, called the **center** of the sphere (see Figure 10.48). A line segment between the center of the sphere and any point on the sphere is called a **radius** of the sphere. A line segment that joins two points of a sphere and has the center of the sphere as its midpoint is called a **diameter** of the sphere.

The **graph** of a given equation in x, y, and z is the set of all points in space with coordinates that are solutions of the given equation. Using the distance formula, it is easy to derive an equation of a sphere with center (x_0, y_0, z_0) and radius of length r. A point $P(x, y, z)$ is on the sphere if and only if

$$\sqrt{(x - x_0)^2 + (y - y_0)^2 + (z - z_0)^2} = r.$$

Since both sides of this equation are nonnegative, we can square both sides and obtain the following result.

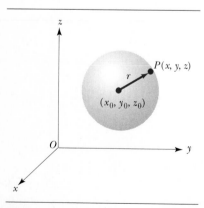

FIGURE 10.48

> ### Theorem 10.10
>
> A sphere with center (x_0, y_0, z_0) and radius of length r has the equation
> $$(x - x_0)^2 + (y - y_0)^2 + (z - z_0)^2 = r^2.$$

Example 3 Find an equation of the sphere that has endpoints of a diameter at $(3, -1, 5)$ and $(-1, -5, 7)$.

Solution

The center (x_0, y_0, z_0) of the sphere is at the midpoint of the segment joining $(3, -1, 5)$ and $(-1, -5, 7)$, and this midpoint is given by

$$\left(\frac{3 - 1}{2}, \frac{-1 - 5}{2}, \frac{5 + 7}{2} \right), \quad \text{or} \quad (1, -3, 6).$$

The radius r is the distance between the center $(1, -3, 6)$ and either of the diameter endpoints. Using $(-1, -5, 7)$, we get

$$r = \sqrt{(-1 - 1)^2 + (-5 + 3)^2 + (7 - 6)^2}$$

$$= \sqrt{4 + 4 + 1}$$

$$= 3.$$

Thus an equation of the sphere is

$$(x - 1)^2 + (y + 3)^2 + (z - 6)^2 = 9. \qquad \square$$

The graph in space of a given equation in x, y, and z is usually referred to as a **surface.** The sphere in Example 3 is our first concrete example of a surface. See Drawing Tip 3.

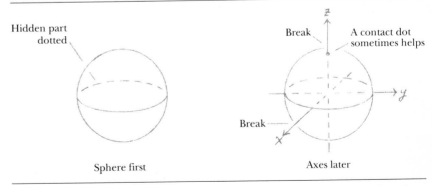

Hidden part dotted

Sphere first

Break

A contact dot sometimes helps

Break

Axes later

DRAWING TIP 3 Spheres: Draw the sphere first (outline and equator); draw axes, if any, later. Use line breaks and dotted lines.

If the binomials in the equation of Theorem 10.10 are expanded, the resulting equation may be rewritten in the form

$$x^2 + y^2 + z^2 + ax + by + cz + d = 0.$$

For example, the equation obtained in Example 3 can be rewritten as

$$x^2 + y^2 + z^2 - 2x + 6y - 12z + 37 = 0.$$

Example 4   Determine whether or not the graph of

$$x^2 + y^2 + z^2 + 4x - 2y - 8z = -12$$

is an equation of a sphere. If it is, find the center and radius.

Solution

Applying the same technique that was used with circles in the plane, we complete the square on all three variables to get the following equivalent equation.

$$x^2 + 4x + 4 + y^2 - 2y + 1 + z^2 - 8z + 16 = -12 + 4 + 1 + 16$$

This can be rewritten as

$$(x + 2)^2 + (y - 1)^2 + (z - 4)^2 = 3^2.$$

We recognize this as the equation of a sphere with center $(-2, 1, 4)$ and radius 3.   ❑

The simplest examples of surfaces are planes and, as might be expected, planes have equations of a very simple form. The exact form of these equations is stated in the next theorem, which we accept without proof.

Theorem 10.11

The graph of any equation of the form

$$ax + by + cz = d$$

is a plane if a, b, c and d are real numbers and at least one of a, b, and c is not zero. Conversely, every plane is the graph of an equation of this form.

From one point of view, Theorem 10.11 might seem surprising. All coordinate variables appearing in the equation have first degree, and our past experience might lead us to expect the graph to be a straight line instead of a plane. Some simple examples can serve to set our intuition straight.

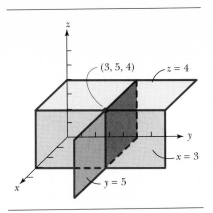

FIGURE 10.49

Example 5   Sketch the graph of each of the planes

a) $x = 3$,   b) $y = 5$,   c) $z = 4$.

Solution
We graph all three planes in the same coordinate system in Figure 10.49. We usually sketch a plane by drawing a parallelogram or a triangle that lies in the plane. Frequently, as in the situation here, it is sufficient to draw only the positive axes and sketch the first octant part of the plane.

The drawing in Figure 10.49 shows that the graph of $x = 3$ consists of all points located three units in front of the yz-plane. This is the same as the set of all points with coordinates of the form $(3, y, z)$.

Similarly, the plane $y = 5$ consists of all points $(x, 5, z)$ that lie five units to the right of the xz-plane, and the plane $z = 4$ consists of all points $(x, y, 4)$ that are four units above the xy-plane. As indicated in the figure, these three planes intersect at the point $(3, 5, 4)$.   ❑

Planes can easily be sketched by following Drawing Tips 4, 5, and 6.

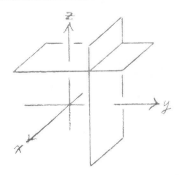

DRAWING TIP 4   Draw planes parallel to the coordinate planes as if they were rectangles with sides parallel to the coordinate axes.

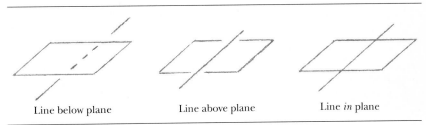

Line below plane   Line above plane   Line *in* plane

DRAWING TIP 5   Dot hidden portions of lines. Don't let the line touch the boundary of the parallelogram that represents the plane, unless the line lies in the plane.

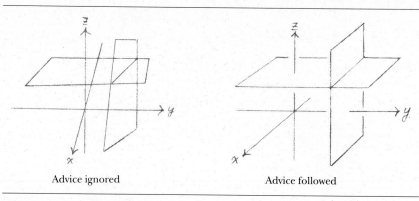

Advice ignored   Advice followed

DRAWING TIP 6 A general rule for perspective: Draw the object as if it lies some distance away, below, and to the left.

Example 6   Sketch the graph of the plane $3x + 4y = 12$.

Solution

In the xy-plane, the graph of $3x + 4y = 12$ is a straight line with x-intercept 4 and y-intercept 3. Hence the straight line through $(4, 0, 0)$ and $(0, 3, 0)$ is part of the graph of $3x + 4y = 12$ in the xyz-coordinate system. This line is called the **trace**[3] of the plane $3x + 4y = 12$ in the xy-plane, and the first octant part of the line is drawn in Figure 10.50.

Since z does not appear in the equation $3x + 4y = 12$, any point directly above or directly below the trace in the xy-plane is also on the plane $3x + 4y = 12$. In particular, all points with coordinates $(4, 0, z)$ or $(0, 3, z)$ are in the plane, as indicated in Figure 10.50. We sketch the plane by drawing the parallelogram shown in Figure 10.50.   ❏

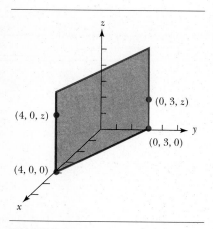

FIGURE 10.50

Generalizing from Example 6, we see that the absence of a coordinate variable in an equation is a signal that there is no intercept for that variable, and consequently the surface is parallel to that coordinate axis.

Example 7   Sketch the graph of the surface

$$3x + 4y + 2z = 12.$$

Solution

The equation fits the form given in Theorem 10.11, so we recognize the surface as a plane. We begin by finding the intercepts on the coordinate axes.

---

3. In general, the intersection of a surface with a coordinate plane is called the *trace* of the surface in that plane.

When $y = 0$ and $z = 0$, then $3x = 12$ and $x = 4$;

When $x = 0$ and $z = 0$, then $4y = 12$ and $y = 3$;

When $x = 0$ and $y = 0$, then $2z = 12$ and $z = 6$.

Thus the points $(4, 0, 0)$, $(0, 3, 0)$, and $(0, 0, 6)$ are on the plane. The lines connecting these points are the *traces* in the coordinate planes. As shown in Figure 10.51, the first octant parts of the traces form a triangle in the first octant. This triangle is our sketch of the plane. ❏

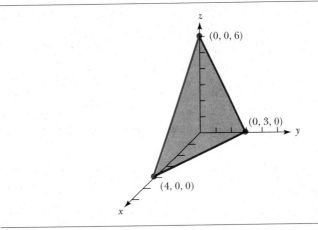

FIGURE 10.51

Drawing Tip 7 illustrates the technique used to sketch the surface in Example 7.

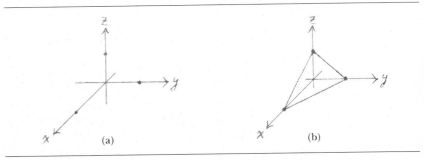

DRAWING TIP 7 To draw a plane that intersects all three coordinate axes, follow the steps shown here: (a) Sketch the axes and mark the intercepts. (b) Connect the intercepts to form a triangle.

Exercises 10.8

Plot the following points in a rectangular coordinate system. (See Figure 10.46.)

1. $(2, 4, 0)$
2. $(0, 4, 1)$
3. $(2, 0, 3)$
4. $(2, 1, 0)$
5. $(3, 2, 1)$
6. $(4, 1, 3)$
7. $(-2, 3, 1)$
8. $(2, 0, -3)$
9. $(-2, 0, 4)$
10. $(-1, -4, 2)$
11. $(4, -3, 2)$
12. $(3, 0, 5)$

Find the length and midpoint of the line segment joining the given points. (See Examples 1 and 2.)

13. $(0, 2, 1), (1, 0, -1)$
14. $(1, 0, 2), (3, -2, 3)$
15. $(1, 0, 2), (3, -1, 4)$
16. $(3, -2, 1), (1, 4, 5)$
17. $(-2, 1, 6), (-6, 5, 0)$
18. $(3, -2, 2), (7, 10, 2)$
19. $(3, -1, 2), (-4, 1, 5)$
20. $(3, 7, -1), (4, 5, 1)$

Find an equation of the sphere that has the given center C and radius r. (See Example 3.)

21. $C(2, -3, 1), r = 1$
22. $C(5, -3, 2), r = 3$
23. $C(1, -2, 2), r = 4$
24. $C(3, 0, -1), r = 2$
25. $C(-2, 4, 0), r = \sqrt{11}$
26. $C(-1, 3, -2), r = \sqrt{15}$

Determine whether or not the graph of the given equation is a sphere. If it is, find the center and radius. (See Example 4.)

27. $x^2 + y^2 + z^2 + 8x - 4y + 6z + 13 = 0$
28. $x^2 + y^2 + z^2 - 2x + 4y + 6z - 11 = 0$
29. $x^2 + y^2 + z^2 - 8x + 2y - 4z - 4 = 0$
30. $x^2 + y^2 + z^2 + 4x - 6y + 2z - 2 = 0$
31. $x^2 + y^2 + z^2 + 6x + 4y - 4z + 20 = 0$

32. $x^2 + y^2 + z^2 + 2x + 8y - 6z + 30 = 0$

Sketch the graph of the given equation. (See Examples 5–7.)

33. $x = 2$
34. $z = 3$
35. $y = 4$
36. $z = 5$
37. $x + y = 2$
38. $2x + z = 4$
39. $2y + 3z = 6$
40. $4x + 3y = 12$
41. $y = x$
42. $y = 2x$
43. $x + y + z = 3$
44. $2x + 3y + 6z = 6$
45. $6x + 4y + 3z = 12$
46. $2x + 3y + 4z = 24$
47. $x^2 + y^2 + z^2 = 16$
48. $x^2 + y^2 + z^2 = 9$
49. $(x - 2)^2 + y^2 + z^2 = 4$
50. $x^2 + y^2 + (z - 3)^2 = 9$

51. Find an equation of the sphere that has center at the origin and passes through $(-1, 4, -3)$.

52. Find an equation of the sphere that has center at $(2, -1, 4)$ and passes through the point $(5, 0, -1)$.

53. Find an equation of the sphere that has endpoints of a diameter at $(0, 0, 0)$ and $(-4, -6, 2)$.

54. Find an equation of the sphere that has endpoints of a diameter at $(-5, -1, 3)$ and $(1, 3, -5)$.

55. Describe the graph of $x^2 + y^2 + z^2 - 2x + 10y - 4z + 30 = 0$.

56. Show that the midpoint R of the line segment joining $P(x_1, y_1, z_1)$ and $Q(x_2, y_2, z_2)$ has coordinates

$$\left(\frac{x_1 + x_2}{2}, \frac{y_1 + y_2}{2}, \frac{z_1 + z_2}{2} \right).$$

(*Hint:* Show that $|\overrightarrow{PR}| = |\overrightarrow{RQ}| = \frac{1}{2}|\overrightarrow{PQ}|$ and use the fact that P, R, and Q lie on a straight line if $|\overrightarrow{PR}| + |\overrightarrow{RQ}| = |\overrightarrow{PQ}|$.)

CHAPTER REVIEW

Key Words and Phrases

Parabola
Focus (foci)
Vertex (vertices)
Directrix
Axis of symmetry
Ellipse

Major axis
Minor axis
Hyperbola
Transverse axis
Conjugate axis
Asymptotes

General quadratic equation
Discriminant
Rotation of axes
Polar coordinate system
Pole
Polar axis

Polar coordinates of a point
Polar equation
Cardioid
Four-leaved rose
Polar equations of conics
Focus
Directrix
Eccentricity
Parametric equations

Plane curves
Parameter
Cycloid
Three-dimensional rectangular
  coordinates
Right-handed system
Coordinate planes
Octants

First octant
Sphere
Center of a sphere
Radius of a sphere
Diameter of a sphere
Surface
Plane
Trace

Summary of Important Concepts and Formulas

The Parabola

Equation: $p \neq 0$	$(x - h)^2 = 4p(y - k)$	$(y - k)^2 = 4p(x - h)$
Vertex: V	(h, k)	(h, k)
Focus: F	$(h, k + p)$	$(h + p, k)$
Directrix: ℓ	$y = k - p$	$x = h - p$
Axis of symmetry	$x = h$	$y = k$
Direction of opening	Upward if $p > 0$; downward if $p < 0$	To the right if $p > 0$; to the left if $p < 0$

Ellipse with Center at (h, k), $a > b > 0$

Equation	$\dfrac{(x - h)^2}{a^2} + \dfrac{(y - k)^2}{b^2} = 1$	$\dfrac{(x - h)^2}{b^2} + \dfrac{(y - k)^2}{a^2} = 1$
Vertices	$(h \pm a, k)$	$(h, k \pm a)$
Foci	$(h \pm c, k), c = \sqrt{a^2 - b^2}$	$(h, k \pm c), c = \sqrt{a^2 - b^2}$

Hyperbola with Center at (h, k)

Equation	$\dfrac{(x - h)^2}{a^2} - \dfrac{(y - k)^2}{b^2} = 1$	$\dfrac{(y - k)^2}{a^2} - \dfrac{(x - h)^2}{b^2} = 1$
Vertices	$(h \pm a, k)$	$(h, k \pm a)$
Foci	$(h \pm c, k), c = \sqrt{a^2 + b^2}$	$(h, k \pm c), c = \sqrt{a^2 + b^2}$
Asymptotes	$y - k = \pm\dfrac{b}{a}(x - h)$	$y - k = \pm\dfrac{a}{b}(x - h)$

Discriminant of $Ax^2 + Bxy + Cy^2 + Dx + Ey + F = 0$

If $B^2 - 4AC$ $\begin{cases} > 0, \text{ the graph is a hyperbola;} \\ = 0, \text{ the graph is a parabola;} \\ < 0, \text{ the graph is an ellipse.} \end{cases}$

Rotation Formulas

$x = x' \cos \theta - y' \sin \theta \qquad x' = x \cos \theta + y \sin \theta$

$y = x' \sin \theta + y' \cos \theta \qquad y' = -x \sin \theta + y \cos \theta$

Angle of Rotation

$$\cot 2\theta = \frac{A - C}{B}$$

Relationship between Rectangular (x, y) and Polar (r, θ) Coordinates

$$r^2 = x^2 + y^2, \quad x = r \cos \theta, \quad y = r \sin \theta$$

Polar Equations for the Conic Sections

Equation	$r = \dfrac{ek}{1 + e \cos \theta}$	$r = \dfrac{ek}{1 - e \cos \theta}$	$r = \dfrac{ek}{1 + e \sin \theta}$	$r = \dfrac{ek}{1 - e \sin \theta}$
Directrix	$x = k$	$x = -k$	$y = k$	$y = -k$

Eccentricity: $e = \dfrac{c}{a} > 0$

If e $\begin{cases} > 1, \text{ the conic is a hyperbola;} \\ = 1, \text{ the conic is a parabola;} \\ < 1, \text{ the conic is an ellipse.} \end{cases}$

Distance between $P(x_1, y_1, z_1)$ and $Q(x_2, y_2, z_2)$

$$|\overrightarrow{PQ}| = \sqrt{(x_2 - x_1)^2 + (y_2 - y_1)^2 + (z_2 - z_1)^2}$$

Midpoint between (x_1, y_1, z_1) and (x_2, y_2, z_2)

$$\left(\frac{x_1 + x_2}{2}, \frac{y_1 + y_2}{2}, \frac{z_1 + z_2}{2} \right)$$

Sphere with Center (x_0, y_0, z_0) and Radius r

$$(x - x_0)^2 + (y - y_0)^2 + (z - z_0)^2 = r^2$$

Equation of a Plane

$$ax + by + cz = d$$

Review Problems for Chapter 10

In Problems 1 and 2, write an equation of the parabola that satisfies the given conditions.

1. Directrix: $y = -2$
   Focus: $(-2, 4)$

2. Vertex: $(4, -1)$
   Focus: $(1, -1)$

In Problems 3–7, sketch the graph of the given equation and state any of the following that are appropriate: center, vertices, foci, asymptotes.

3. $x = 2(y + 1)^2 - 6$

4. $x = -2y^2 + 8y - 6$

5. $4(x - 2)^2 + 9(y + 1)^2 = 36$

6. $25y^2 - 4x^2 = 100$

7. $(x + 3)^2 - 4(y - 2)^2 = 16$

Sketch the graphs of the following equations.

8. $2y = \sqrt{1 - 4x^2}$

9. $2x = -\sqrt{16 - y^2}$

10. $y = -1 + \sqrt{x - 2}$

11. $y = 3\sqrt{x^2 + 1}$

In Problems 12 and 13, write an equation of the ellipse satisfying the given conditions.

12. Vertices: $(-3, 2)$, $(5, 2)$
    Length of minor axis: 6

13. Vertices: $(1, -4)$, $(1, 6)$
    Foci: $(1, -3)$, $(1, 5)$

In Problems 14 and 15, write an equation of the hyperbola that satisfies the given conditions.

14. Vertices: $(0, \pm 6)$
    Asymptotes: $y = \pm 3x$

15. Foci: $(\pm 5, 0)$
    Vertices: $(\pm 3, 0)$

Identify the graphs of the following equations. Do not draw the graphs.

16. $9x^2 + 4y^2 - 54x + 16y + 61 = 0$

17. $9x^2 + 16y^2 - 18x + 64y - 71 = 0$

18. $y^2 - 6y - 2x + 7 = 0$

19. $4x^2 - 24x - 9y^2 + 18y + 9 = 0$

Use the discriminant to identify the graph of the equation. Do not draw the graphs.

20. $x^2 - xy + y^2 = 2$

21. $y^2 + \sqrt{3}\,xy = 3$

22. $2x^2 - 24xy + 9y^2 = 75$

23. $y^2 + 24xy - 6x^2 = 30$

Use a rotation to eliminate the xy-term in each of the following equations. Sketch the graph of the resulting equation, showing both the xy- and $x'y'$-systems.

24. $3x^2 - 4\sqrt{3}\,xy + 7y^2 = 9$

25. $52x^2 - 72xy + 73y^2 = 100$

26. $x^2 + 2\sqrt{3}\,xy + 3y^2 - \sqrt{3}\,x + y = 0$

27. $xy - \sqrt{2}\,x + \sqrt{2}\,y - \dfrac{5}{2} = 0$

Convert each of the following equations into a polar equation.

28. $x^2 + y^2 = 2x$ **29.** $3x + 4y = 12$

Convert each of the following into an equation in the rectangular coordinates x and y.

30. $r(\sin\theta - 2\cos\theta) = 4$

31. $r^2 \sin\theta \cos\theta = 1$

Sketch the graph of each of the following equations in a polar coordinate system.

32. $r = 2 - \sin\theta$ (limacon)

33. $r = 1 + 2\cos\theta$ (limacon)

34. $r = 1 + \sin\theta$ (cardioid)

35. $r = 1 - \cos\theta$ (cardioid)

Identify the type of conic section represented by the given equation, state the eccentricity, and sketch the graph of the equation.

36. $r = 3 + r\sin\theta$ **37.** $r = 2 - r\cos\theta$

38. $r = \dfrac{4\sec\theta}{2\sec\theta + 1}$ **39.** $r = \dfrac{6\csc\theta}{2\csc\theta - 1}$

Find a polar equation of the conic section that has a focus at the pole and satisfies the given conditions.

40. Eccentricity 1; equation of a directrix, $r = 2\sec\theta$

41. Eccentricity $\dfrac{2}{3}$; equation of a directrix, $r = 4\csc\theta$

42. Hyperbola with vertices at $(2, 0)$ and $(-4, \pi)$

43. Ellipse with vertices at $(2, 0)$ and $(6, \pi)$

Parametric equations for a curve C are given in Problems 44–49. In each problem, (a) find a Cartesian equation for a graph that contains C and (b) sketch the graph of C.

44. $x = e^t$, $y = e^{-2t}$, $t \in \mathcal{R}$

45. $x = 4t^2$, $y = 2t - 5$, $t \in \mathcal{R}$

46. $x = 2\sin\theta$, $y = 3\cos\theta$, $\theta \in [0, \pi]$

47. $x = 3\sin\theta$, $y = 4\cos\theta$, $\theta \in [0, \pi/2]$

48. $x = \cos\theta$, $y = 1 + \cos 2\theta$, $\theta \in [0, \pi]$

49. $x = 2\sin\theta$, $y = \cos^2\theta$, $\theta \in [0, 2\pi]$

Find the length and midpoint of the line segment joining the given points.

50. $(1, 5, 4)$, $(-1, 3, 10)$ **51.** $(-4, 1, 0)$, $(2, -3, 6)$

52. $(3, 0, 1)$, $(2, -2, 3)$ **53.** $(3, -4, 4)$, $(1, 0, -3)$

Find an equation of the sphere that has the given center C and radius r.

54. $C(2, -3, 1)$, $r = 4$ **55.** $C(-3, 0, 4)$, $r = 1$

56. $C(-2, 3, 5)$, $r = 2$ **57.** $C(-3, 6, -1)$, $r = 3$

Determine whether or not the graph of the given equation is a sphere. If it is, find the center and radius.

58. $x^2 + y^2 + z^2 = 12x - 4y + 6z$

59. $x^2 + y^2 + z^2 = 8y - 6z$

60. $x^2 + y^2 + z^2 - 2y + 8z + 36 = 0$

61. $x^2 + y^2 + z^2 + 4x - 12z + 64 = 0$

Sketch the graph of the given equation in a three-dimensional coordinate system.

62. $2y + 3z = 12$ **63.** $6x + 2y + 3z = 12$

64. $x^2 + y^2 + z^2 = 4$ **65.** $y = 3x$

▲ ▲ CRITICAL THINKING: FIND THE ERRORS ▰▰▰▰▰

Each of the following nonsolutions has at least one error. Can you find them?

Problem 1 Find equations for x and y in a change of variables by rotation of axes that will remove the xy-term in

$$5x^2 + 2\sqrt{3}\,xy + 3y^2 = 24.$$

Nonsolution We have

$$\cot 2\theta = \frac{5 - 3}{2\sqrt{3}}$$

$$= \frac{1}{\sqrt{3}},$$

so $2\theta = 60°$ and

$$x = x' \cos 60° - y' \sin 60° = \frac{1}{2}(x' - \sqrt{3}\,y')$$

$$y = x' \sin 60° + y' \cos 60° = \frac{1}{2}(\sqrt{3}\,x' + y')$$

Problem 2 Find the midpoint of the line segment joining $(1, 3, 4)$ and $(5, 5, 10)$.

Nonsolution The midpoint is given by

$$\left(\frac{5 - 1}{2}, \frac{5 - 3}{2}, \frac{10 - 4}{2}\right) = (2, 1, 3).$$

Problem 3 If the graph of the following equation is a sphere, find the center and the radius.

$$x^2 + y^2 + z^2 = 8y - 6x + 2z + 25$$

Nonsolution

$$x^2 + 6x + y^2 - 8y + z^2 - 2z = 25$$

$$x^2 + 6x + 9 + y^2 - 8y + 16 + z^2 - 2z + 1 = 25$$

$$(x + 3)^2 + (y - 4)^2 + (z - 1)^2 = 5^2$$

The center is at $(-3, 4, 1)$, and the radius is 5.

Problem 4	Sketch the graph of the equation $3y + 5z = 15$ in a three-dimensional coordinate system.
Nonsolution	The graph is a line in the yz-plane with y-intercept 5 and z-intercept 3, as shown in Figure 10.52.

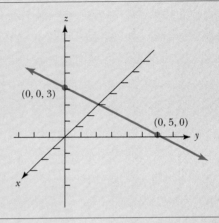

FIGURE 10.52

Appendix

A.1 Table Evaluation of Logarithms

In this section we see how values of logarithms to base 10 can be found by using Table III in the back of this book. As a starting point, we recall from Exercises 1.2 that any positive number N can be written as the product of a power of 10 and a number between 1 and 10: $N = a \times 10^n$, where $1 \le a < 10$. For example,

$$5520 = 5.52 \times 10^3$$

and

$$0.0436 = 4.36 \times 10^{-2}.$$

The expression $a \times 10^n$, where $1 \le a < 10$, is called the **scientific notation** for the number N. The equality $N = a \times 10^n$ means that $\log N$ differs from $\log a$ by an integer, since

$$\log N = \log (a \times 10^n)$$
$$= \log a + \log 10^n$$
$$= \log a + n$$
$$= n + \log a.$$

With the same numbers as above, we would have

$$\log 5520 = \log (5.52 \times 10^3)$$
$$= \log 5.52 + \log 10^3$$
$$= 3 + \log 5.52$$

and

$$\log 0.0436 = -2 + \log 4.36.$$

The significance of this is that the logarithms of numbers between 1 and 10 can be used to find the logarithms of all positive numbers. The Table of Common Logarithms gives the values of logarithms of numbers from 1 to 9.99, at intervals of 0.01. A portion of this table is reproduced in Table A.1. The logarithm of a three-digit number[1] between 1 and 10 can be found in the table by locating the first two digits of the number in the column under N and the last digit in the column headings at the top of the table. The logarithm of the number is located in the row of the table that starts with the first two digits and in the column that has the last digit at the top.

---

1. By a three-digit number, we mean that there are three digits when the number is written in scientific notation. Zeros used only to place the decimal are not counted.

N	0	1	2	3	4	5	6	7	8	9
5.5	.7404	.7412	.7419	.7427	.7435	.7443	.7451	.7459	.7466	.7474
5.6	.7482	.7490	.7497	.7505	.7513	.7520	.7528	.7536	.7543	.7551
5.7	.7559	.7566	.7574	.7582	.7589	.7597	.7604	.7612	.7619	.7627
5.8	.7634	.7642	.7649	.7657	.7664	.7672	.7679	.7686	.7694	.7701
5.9	.7709	.7716	.7723	.7731	.7738	.7745	.7752	.7760	.7767	.7774

TABLE A.1

Example 1   Use the log table to find the value of log 5.84.

Solution

In the log table, or in the portion reproduced in Table A.1, log 5.84 is located by matching up the row that has 5.8 under N and the column that has 4 at the top. The appropriate row and column are shaded in Table A.1. The value of log 5.84 is found at their intersection:

$$\log 5.84 = 0.7664.$$

Actually, the value 0.7664 is an approximation to four decimal places of log 5.84, as are most of the values in the log table. It would be more precise to write $\log 5.84 \approx 0.7664$, but we choose to write $=$ instead of $\approx$ as a matter of convenience.   ❏

We have noted before that any positive number N can be written in scientific notation as $N = a \times 10^n$, where $1 \le a < 10$, and consequently

$$\log N = n + \log a,$$

where $0 \le \log a < 1$. Thus log N can be expressed as the sum of an integer and a nonnegative decimal fraction less than 1. The integral part, n, of the logarithm is called the **characteristic,** and the fractional part, $\log a$, is called the **mantissa.** The integral part can be found by writing the number in scientific notation, and the mantissa for three-digit numbers can be read from the log table, with accuracy to four decimal places.

Example 2   Use the log table to find the value of log 584.

Solution

Writing 584 in scientific notation, we have

$$584 = 5.84 \times 10^2,$$

so the characteristic is 2, and the mantissa is

$$\log 5.84 = 0.7664$$

from the log table. Thus

$$\log 584 = 2 + 0.7664 = 2.7664.$$ ❏

It may happen, of course, that the characteristic is a negative number.

Example 3   Use the log table to find the value of log 0.00584.

Solution

Writing the number in scientific notation, we have

$$0.00584 = 5.84 \times 10^{-3},$$

and the characteristic is -3. As in Example 1, the mantissa is 0.7664, so

$$\log 0.00584 = -3 + 0.7664.$$

The convention in a situation like this is to avoid combining the negative integer and the positive fraction, and to write the logarithm in the form

$$\log 0.00584 = 7.7664 - 10. \qquad ❑$$

As was done here, a negative characteristic is usually written as a positive integer minus a multiple of 10. We shall see in Section A.2 that this is very convenient in numerical computations.

In order to perform calculations by use of logarithms, we must be able to use the tables to find a number N when log N is known.

Example 4   Find the number N, given that log $N = 8.7657 - 10$.

Solution

The digits in N are determined by the mantissa 0.7657, and the position of the decimal in N is determined by the characteristic, which is $8 - 10 = -2$. To find the digits in N, we search through the body of the log table until we find the entry 0.7657 (note that the mantissa increases as N increases in the table). This entry is found in the row that starts with 5.8 and in the column headed by 3. Thus

$$N = 5.83 \times 10^{-2} = 0.0583. \qquad ❑$$

When a number N is found by using log N, the number N is referred to as the *antilogarithm* of log N (abbreviated *antilog*). Thus in Example 4 we would say that

$$\text{antilog } (8.7657 - 10) = 0.0583.$$

As mentioned earlier, the log table furnishes logarithms only for numbers with three digits. Tables exist that give logarithms for numbers with more than three digits. However, the log table can be used to approximate the logarithm of a four-digit number with very good accuracy, employing a procedure called **linear interpolation.**

Example 5   To illustrate the procedure of linear interpolation, let us consider the problem of finding the value of log 10.37. The two numbers nearest 10.37 that have logarithms in the log table are 10.30 and 10.40. Their logarithms are given by

$$\log 10.30 = 1.0128$$

and

$$\log 10.40 = 1.0170.$$

Thinking geometrically, this means that the points $P(10.30, 1.0128)$ and $Q(10.40, 1.0170)$ are on the graph of $y = \log x$. A portion of the graph that contains these points is shown in Figure A.1. The coordinates at R on the curve are $(10.37, \log 10.37)$. The idea behind linear interpolation is this: Use the straight-line segment PQ as an approximation to the curve $y = \log x$, and use the ordinate at point S as an approximation to log 10.37. The ordinate at S can be found by adding the difference d to the ordinate at P. Since 10.37 is $\frac{7}{10}$ of the distance from 10.30 to 10.40, the difference d is $\frac{7}{10}$ of the difference between the ordinates at P and Q. These differences can be set up as follows:

$$10\left[7\left[\begin{array}{l}\log 10.30 = 1.0128 \\ \log 10.37 = \underline{\quad\quad} \\ \log 10.40 = 1.0170\end{array}\right]d\right]0.0042.$$

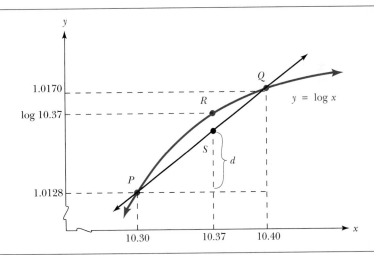

FIGURE A.1

To find d, we use the proportion

$$\frac{d}{0.0042} = \frac{7}{10}$$

and obtain

$$d = \left(\frac{7}{10}\right)(0.0042)$$

$$= 0.00294$$

$$= 0.0029,$$

where d is rounded to the number of decimal places in the table. Adding d to the ordinate at P, we have

$$\log 10.37 = 1.0128 + 0.0029 = 1.0157.$$

(A calculator gives $\log 10.37 = 1.0157788$.) ❏

The next example shows how linear interpolation can be used in finding antilogarithms.

Example 6   If $\log N = 8.4089 - 10$, use interpolation to approximate N to four digits.

Solution
The mantissa 0.4089 is located between 0.4082 and 0.4099 in the tables, and these mantissas correspond to the digits 256 and 257. Thus we have the following arrangement:

$$10\left[x\left[\begin{array}{l} \log 0.02560 = 8.4082 - 10 \\ \log N \quad\;\; = 8.4089 - 10 \end{array} \right]0.0007 \atop \log 0.02570 = 8.4099 - 10 \right]0.0017$$

$$\frac{x}{10} = \frac{0.0007}{0.0017}$$

$$x = \frac{70}{17} = 4.$$

The number x represents the last digit in N and is rounded to the nearest whole number. Thus we have

$$N = 0.02564.$$

(A calculator gives $N = 0.02563894$.) ❏

Exercises A.1

Find the common logarithms of each number, using the log table as necessary. (See Examples 1–3.)

1. 4.16×10^{-9}

2. 1.73×10^3

3. 30.7

4. 17.3

5. 4.51

6. 3.01

7. 10.0

8. 10^{-4}

9. 1,070,000   **10.** 10,800   **21.** 10.11   **22.** 243.6

11. 0.00107   **12.** 0.000132   **23.** 4.171   **24.** 1.017

Find N in each equation, using the log table as necessary. (See Example 4.)

25. 417,800   **26.** 7,103

27. 0.007717   **28.** 0.08354

13. $\log N = 0.8561$ **14.** $\log N = 0.9186$

15. $\log N = 8.4518$ **16.** $\log N = 4.6385$

17. $\log N = 7.6776 - 10$ **18.** $\log N = 6.9943 - 10$

19. $\log N = -3$ **20.** $\log N = 4$

Use the log table and linear interpolation to find N in each equation. Check the accuracy with a calculator if one is available. (See Example 6.)

29. $\log N = 0.1113$ **30.** $\log N = 0.7549$

31. $\log N = 4.9433$ **32.** $\log N = 5.4686$

Use the log table and linear interpolation to find the logarithm of each number. Check the accuracy with a calculator if one is available. (See Example 5.)

33. $\log N = 7.6950 - 10$ **34.** $\log N = 8.9963 - 10$

35. $\log N = 9.3881 - 10$ **36.** $\log N = 9.5240 - 10$

A.2	**Computations with Logarithms**

The properties of logarithms stated in Section 4.3 are the basis for their usefulness in computational work. For easy reference, we restate these properties for logarithms with base 10 in the following theorem.

Theorem A.1

If u, v and r are real numbers with u and v positive, then

a) $\log (u \cdot v) = \log u + \log v$,

b) $\log \dfrac{u}{v} = \log u - \log v$,

c) $\log u^r = r \log u$.

In our first example, we illustrate the use of part (a) of the theorem with a simple computation.

Example 1   Use logarithms to compute the value of $(3.85)(705)$ as a three-digit number.

Solution
As a notational convenience, let

$$N = (3.85)(705).$$

Then

$$\log N = \log 3.85 + \log 705,$$

Using the log table, we find that

$$\log 3.85 = 0.5855,$$

$$\log 705 = 2.8482.$$

Adding these logarithms, we have

$$\log N = 3.4337.$$

The mantissa 0.4337 is not found in the log table, but we do not interpolate, since we need accuracy only to three digits. The mantissa nearest 0.4337 in the table is 0.4330, which occurs with the digits 271. Thus we write

$$N = 2.71 \times 10^3 = 2710.$$

The zero in N is not a significant digit, since it is used only to place the decimal. (The actual product is 2714.25.)   ❏

Example 2   Use the log table and linear interpolation to compute the value of

$$\frac{\sqrt[3]{37.12}}{(2.931)^4}$$

with four-digit accuracy.

Solution

Let

$$N = \frac{\sqrt[3]{37.12}}{(2.931)^4}.$$

By parts (b) and (c) of Theorem A.1, we have

$$\log N = \log \sqrt[3]{37.12} - \log (2.931)^4$$

$$= \frac{1}{3} \log 37.12 - 4 \log 2.931.$$

The interpolation to find $\log 37.12$ is as follows:

$$10\left[2\left[\begin{array}{l} \log 37.10 = 1.5694 \\ \log 37.12 = \underline{\hspace{1cm}} \\ \log 37.20 = 1.5705 \end{array}\right]d\right]0.0011$$

$$\frac{d}{0.0011} = \frac{2}{10}$$

$$d = \left(\frac{2}{10}\right)(0.0011)$$

$$= 0.0002$$

$$\log 37.12 = 1.5694 + 0.0002$$
$$= 1.5696.$$

Similarly, we find that

$$\log 2.931 = 0.4670.$$

Thus

$$\log N = \frac{1}{3}(1.5696) - 4(0.4670)$$

$$= 0.5232 - 1.8680.$$

At this point, we *do not* perform the subtraction as it is indicated, because this yields a number with a *negative* decimal fraction, and the log table requires *positive* decimal fractions as mantissas. To retain a logarithm with a positive decimal part, we proceed as follows:

$$\log N = 0.5232 - 1.8680$$

$$= (10.5232 - 10) - 1.8680$$

$$= (10.5232 - 1.8680) - 10$$

$$= 8.6552 - 10.$$

To find $N = $ antilog $(8.6552 - 10)$, we use interpolation. (With a little practice, interpolation can be performed mentally.)

$$10\left[x\left[\begin{array}{l} \log 0.04520 = 8.6551 - 10 \\ \log N \quad\ \ = 8.6552 - 10 \end{array} \right]0.0001 \right]0.0010 \\ \log 0.04530 = 8.6561 - 10 $$

$$x = 1$$

$$N = 0.04521$$

(A calculator gives the answer 0.0452001.)   ❑

The work in Example 2 illustrates one of the "fine points" in using logarithms for computation; it shows how negative decimal fractions are avoided in the values of logarithms. Another fine point is brought out in Example 3.

Example 3   Use logarithms to compute

$$N = \frac{(4.16)\sqrt[3]{0.0235}}{(0.0787)^{3/2}}$$

as a three-digit number.

Solution

Using all three parts of Theorem A.1, we have

$$\log N = \log [(4.16)\sqrt[3]{0.0235}] - \log (0.0787)^{3/2}$$

$$= \log 4.16 + \log\sqrt[3]{0.0235} - \log (0.0787)^{3/2}$$

$$= \log 4.16 + \frac{1}{3}\log 0.0235 - \frac{3}{2}\log 0.0787$$

$$= 0.6191 + \frac{1}{3}(8.3711 - 10) - \frac{3}{2}(8.8960 - 10).$$

The fine point comes up first in the multiplication

$$\frac{1}{3}(8.3711 - 10).$$

If each term in parentheses is multiplied by $\frac{1}{3}$, the resulting difference involves two decimal fractions instead of one. This trouble is avoided by adding and subtracting a multiple of 10, which is chosen so as to make the last term evenly divisible by 3. We write

$$\frac{1}{3}(8.3711 - 10) = \frac{1}{3}(28.3711 - 30)$$

$$= 9.4570 - 10.$$

Similarly,

$$\frac{3}{2}(8.8960 - 10) = \frac{3}{2}(18.8960 - 20)$$

$$= 3(9.4480 - 10)$$

$$= 28.3440 - 30$$

$$= 8.3440 - 10.$$

Substituting these values in the equation for $\log N$, we have

$$\log N = 0.6191 + (9.4570 - 10) - (8.3440 - 10)$$

$$= (10.0761 - 10) - (8.3440 - 10)$$

$$= 1.7321.$$

To three digits, this gives

$$N = \text{antilog} (1.7321)$$

$$= 54.0.$$

(A calculator gives 53.970179.)

Exercises A.2

Use logarithms to compute a three-digit value for each of the following quantities. (See Examples 1 and 3.)

1. $(2.17)(30.1)$

2. $(11.7)(4.91)$

3. $\dfrac{6.01}{31.7}$

4. $\dfrac{23.1}{3.08}$

5. $(4.63)^5$

6. $(40.5)^4$

7. $\sqrt{112}$

8. $\sqrt[5]{17.1}$

9. $\dfrac{13.1}{\sqrt{41.1}}$

10. $\dfrac{\sqrt[3]{3.92}}{2.46}$

11. $\dfrac{(8.12)\sqrt[3]{0.147}}{(39.1)^2}$

12. $\dfrac{(0.0417)^2}{(0.374)\sqrt[5]{0.195}}$

Use logarithms and linear interpolation to compute a four-digit value for each quantity. Check the accuracy with a calculator if one is available. (See Example 2.)

13. $(0.006213)^3(429.5)^2$

14. $(0.1735)(66.17)^2$

15. $\sqrt[6]{0.3704}$

16. $\sqrt[4]{0.6713}$

17. $\dfrac{(27.11)^{1/3}}{(14.13)^{1/2}}$

18. $\dfrac{\sqrt[5]{171.3}}{\sqrt[3]{317.1}}$

19. $\dfrac{(4.813)^2(17.13)}{(0.3612)^{1/3}}$

20. $\dfrac{(0.3179)^{1/3} \cdot (41.95)^2}{(29.71)^4}$

21. Use logarithms to approximate 2^{32} with four significant digits.

22. Use logarithms to approximate $3^{\sqrt{3}}$ with four significant digits. (Use $\sqrt{3} = 1.732$.)

In Exercises 23 and 24, compute a four-digit value of the reciprocal of each number.

23. 2.173

24. 31.19

25. Use logarithms to compute a three-digit value of $57.3/\log 696$.

26. Use logarithms to evaluate

$$\frac{(-3.17) \cdot (6.13)}{(-4.19) \cdot (-3.11)}$$

as a three-digit number.

27. The volume V of a right circular cone with radius r and altitude h is given by $V = \frac{1}{3}\pi r^2 h$. A certain conical funnel has radius 12.8 centimeters and altitude 48.2 centimeters. Compute the volume of the funnel with three-digit accuracy.

28. The area K of a triangle that has sides of lengths a, b, and c is given by $K = \sqrt{s(s-a)(s-b)(s-c)}$, where s is one-half of the perimeter. A triangular-shaped plot of land has sides of length 112 meters, 121 meters, and 157 meters. Find the area of the plot of land.

A.3 | Table Evaluation of Trigonometric Functions

In this section we see how values of trigonometric functions can be found by the use of tables. There are tables of trigonometric functions in the back of this book.

Table I gives values of all six trigonometric functions for angles from 0° to 90°, measured in degrees and minutes, at intervals of 10 minutes. A portion of this table is reproduced in Table A.2. The use of the columns labeled *Radians* in Table I is described later in this section. Angles from 0° to 45° are listed in the leftmost column of the table, and function values for these angles are read by using the function labels at the *top* of the columns. Angles from 45° to 90° are listed in the rightmost column, and function values for these angles are read by using the function labels at the *bottom* of the column. Some illustrations of the use of Table I are given in Example 1.

Angle θ									
Degrees	Radians	sin θ	csc θ	tan θ	cot θ	sec θ	cos θ		
27° 00′	.4712	.4540	2.203	.5095	1.963	1.122	.8910	1.0996	63° 00′
10	.4741	.4566	2.190	.5132	1.949	1.124	.8897	1.0966	50
20	.4771	.4592	2.178	.5169	1.000	1.126	.8884		40
30	.4801	.4626	2.166			.927			
40									
50			2.751	.6959		.8	.8208		10
35° 00′	.6109	.5736	1.743	.7002	1.428	1.221	.8192	.9599	55° 00′
10	.6138	.5760	1.736	.7046	1.419	1.223	.8175	.9570	50
20	.6167	.5783	1.729	.7089	1.411	1.226	.8158	.9541	40
30	.6196	.5807	1.722	.7133	1.402	1.228	.8141	.9512	30
40	.6225	.5831	1.715	.7177	1.393	1.231	.8124	.9483	20
50	.6254	.5854	1.708	.7221	1.385	1.233	.8107	.9454	10
36° 00′	.6283	.5878	1.701	.7265	1.376	1.236	.8090	.9425	54° 00′
		cos θ	sec θ	cot θ	tan θ	csc θ	sin θ	Radians	Degrees
								Angle θ	

TABLE A.2

Example 1 Use Table I to find the following function values.

a) tan 35°40′   b) sec 54°50′   c) csc 35°10′

Solution

a) tan 35°40′ is found in the row of the table with 35°40′ at the left end and in the column with **tan** at the top:

$$\tan 35°40′ = 0.7177.$$

b) sec 54°50′ is in the row that has 54°50′ at the right end and in the column that has **sec** at the bottom:

$$\sec 54°50′ = 1.736.$$

c) csc 35°10′ is in the row with 35°10′ at the left end and in the column with **csc** at the top:

$$\csc 35°10′ = 1.736.$$ ❏

The fact that sec 54°50′ = csc 35°10′ in Example 1 is no accident. It illustrates some general facts about Table I that are worth noting. First of all, we observe that the angles at opposite ends of each row are *complementary:* Their sum is 90°. Next we see that the functions paired together at the top and bottom of each column have related names:

sin θ and cos θ,   tan θ and cot θ,   sec θ and csc θ.

Each function in one of these pairs is called the **complementary function,** or **cofunction,** of the other. The structure that we have observed in the table reflects the fact that **any function of a given angle equals the cofunction of the complementary angle.** The identities in Chapter 6 show why this is true.

For angles outside the range $0 \leq \theta \leq 90°$, the related angle is used as it was in Section 6.4. Some illustrations are given in Example 2.

Example 2 Use Table I to find the following function values.

a) cos 217°40' b) sin 476°10'

Solution

a) As shown in Figure A.2, the related angle is

$$217°40' - 180° = 37°40'.$$

Since 217°40' is a third quadrant angle, its cosine is a negative number. Reading cos 37°40' from Table I, we obtain

$$\cos 217°40' = -\cos 37°40' = -0.7916.$$

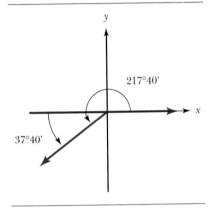

FIGURE A.2

b) We first find an angle between 0° and 360° that is coterminal with 476°10':

$$476°10' - 360° = 116°10'.$$

This is shown in Figure A.3, along with the related angle, which is

$$180° - 116°10' = 63°50'.$$

Since sin θ is positive for a second quadrant angle,

$$\sin 476°10' = \sin 116°10'$$
$$= \sin 63°50'$$
$$= 0.8975. \qquad \square$$

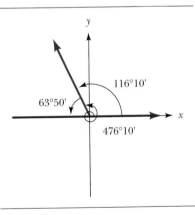

FIGURE A.3

If a trigonometric function value of θ is known, then Table I can be used to find the value of θ, correct to the nearest 10 minutes.[2] If the given function value does not appear in the table, we choose the angle whose function value is nearest the given value.

Example 3 Use Table I to find a value of θ, correct to the nearest 10 minutes, that satisfies the given equation.

a) cos θ = 0.5398 b) tan θ = 0.4597

---

2. More accuracy can be obtained by use of interpolation.

Solution

a) The value 0.5398 is found in Table I in the column with **cos** at the bottom and in the row with 57°20′ at the right end. Thus

$$\cos \theta = 0.5398$$

has the solution

$$\theta = 57°20'.$$

b) The value $\tan \theta = 0.4597$ does not occur in Table I. The two tangent values nearest 0.4597 are

$$\tan 24°40' = 0.4592,$$

$$\tan 24°50' = 0.4628.$$

Since 0.4597 is nearer 0.4592 than 0.4628, we take the value of θ to the nearest 10 minutes to be

$$\theta = 24°40'.$$ ❑

Table II is read in much the same way as Table I. The essential differences are that angles are given in decimal degrees at intervals of 0.1°, and the only functions included in the table are sine, cosine, tangent, and cotangent.

Table I gives values of all six trigonometric functions for angles in radian measure at increments of 0.0029 from 0 to 1.5708 radians. For angles outside the range $0 \leq \theta \leq 1.5708$, we use the radian measure of the related angle.[3] This is illustrated in the next example.

Example 4   Use Table I to find cos 2.5598.

Solution
We think of $t = 2.5598$ as radian measure of an angle. Since

$$\frac{\pi}{2} < 2.5598 < \pi,$$

the angle terminates in quadrant II and has a related angle with radian measure

$$\pi - 2.5598 \approx 0.5818.$$

Since the cosine function is negative in quadrant II,

$$\cos 2.5598 = -\cos 0.5818 = -0.8355.$$

---

3. Radian measure of the related angle is sometimes called the *related number*.

The value cos 0.5818 = 0.8355 is read from Table I.   ❏

As this example shows, there is little that is new in finding values of trigonometric functions of numbers.

The problems in Exercises 5.5 may be worked at this point.

Tables

I. Trigonometric Functions—Degrees and Minutes or Radians

II. Trigonometric Functions—Decimal Degrees

III. Common Logarithms

TABLES

TABLE I Trigonometric Functions—Degrees and Minutes or Radians

Angle θ

Degrees	Radians	sin θ	csc θ	tan θ	cot θ	sec θ	cos θ		
0° 00′	.0000	.0000	No value	.0000	No value	1.000	1.0000	1.5708	90° 00′
10	.0029	.0029	343.8	.0029	343.8	1.000	1.0000	1.5679	50
20	.0058	.0058	171.9	.0058	171.9	1.000	1.0000	1.5650	40
30	.0087	.0087	114.6	.0087	114.6	1.000	1.0000	1.5621	30
40	.0116	.0116	85.95	.0116	85.94	1.000	.9999	1.5592	20
50	.0145	.0145	68.76	.0145	68.75	1.000	.9999	1.5563	10
1° 00′	.0175	.0175	57.30	.0175	57.29	1.000	.9998	1.5533	89° 00′
10	.0204	.0204	49.11	.0204	49.10	1.000	.9998	1.5504	50
20	.0233	.0233	42.98	.0233	42.96	1.000	.9997	1.5475	40
30	.0262	.0262	38.20	.0262	38.19	1.000	.9997	1.5446	30
40	.0291	.0291	34.38	.0291	34.37	1.000	.9996	1.5417	20
50	.0320	.0320	31.26	.0320	31.24	1.001	.9995	1.5388	10
2° 00′	.0349	.0349	28.65	.0349	28.64	1.001	.9994	1.5359	88° 00′
10	.0378	.0378	26.45	.0378	26.43	1.001	.9993	1.5330	50
20	.0407	.0407	24.56	.0407	24.54	1.001	.9992	1.5301	40
30	.0436	.0436	22.93	.0437	22.90	1.001	.9990	1.5272	30
40	.0465	.0465	21.49	.0466	21.47	1.001	.9989	1.5243	20
50	.0495	.0494	20.23	.0495	20.21	1.001	.9988	1.5213	10
3° 00′	.0524	.0523	19.11	.0524	19.08	1.001	.9986	1.5184	87° 00′
10	.0553	.0552	18.10	.0553	18.07	1.002	.9985	1.5155	50
20	.0582	.0581	17.20	.0582	17.17	1.002	.9983	1.5126	40
30	.0611	.0610	16.38	.0612	16.35	1.002	.9981	1.5097	30
40	.0640	.0640	15.64	.0641	15.60	1.002	.9980	1.5068	20
50	.0669	.0669	14.96	.0670	14.92	1.002	.9978	1.5039	10
4° 00′	.0698	.0698	14.34	.0699	14.30	1.002	.9976	1.5010	86° 00′
10	.0727	.0727	13.76	.0729	13.73	1.003	.9974	1.5981	50
20	.0756	.0756	13.23	.0758	13.20	1.003	.9971	1.5952	40
30	.0785	.0785	12.75	.0787	12.71	1.003	.9969	1.5923	30
40	.0814	.0814	12.29	.0816	12.25	1.003	.9967	1.5893	20
50	.0844	.0843	11.87	.0846	11.83	1.004	.9964	1.5864	10
5° 00′	.0873	.0872	11.47	.0875	11.43	1.004	.9962	1.4835	85° 00′
10	.0902	.0901	11.10	.0904	11.06	1.004	.9959	1.4806	50
20	.0931	.0929	10.76	.0934	10.71	1.004	.9957	1.4777	40
30	.0960	.0958	10.43	.0963	10.39	1.005	.9954	1.4748	30
40	.0989	.0987	10.13	.0992	10.08	1.005	.9951	1.4719	20
50	.1018	.1016	9.839	.1022	9.788	1.005	.9948	1.4690	10
6° 00′	.1047	.1045	9.567	.1051	9.514	1.006	.9945	1.4661	84° 00′
10	.1076	.1074	9.309	.1080	9.255	1.006	.9942	1.4632	50
20	.1105	.1103	9.065	.1110	9.010	1.006	.9939	1.4603	40
30	.1134	.1132	8.834	.1139	8.777	1.006	.9936	1.4573	30
40	.1164	.1161	8.614	.1169	8.556	1.007	.9932	1.4544	20
50	.1193	.1190	8.405	.1198	8.345	1.007	.9929	1.4515	10
7° 00′	.1222	.1219	8.206	.1228	8.144	1.008	.9925	1.4486	83° 00′
10	.1251	.1248	8.016	.1257	7.953	1.008	.9922	1.4457	50
20	.1280	.1276	7.834	.1287	7.770	1.008	.9918	1.4428	40
30	.1309	.1305	7.661	.1317	7.596	1.009	.9914	1.4399	30
40	.1338	.1334	7.496	.1346	7.429	1.009	.9911	1.4370	20
50	.1367	.1363	7.337	.1376	7.269	1.009	.9907	1.4341	10
8° 00′	.1396	.1392	7.185	.1405	7.115	1.010	.9903	1.4312	82° 00′
10	.1425	.1421	7.040	.1435	6.968	1.010	.9899	1.4283	50
20	.1454	.1449	6.900	.1465	6.827	1.011	.9894	1.4254	40
30	.1484	.1478	6.765	.1495	6.691	1.011	.9890	1.4224	30
40	.1513	.1507	6.636	.1524	6.561	1.012	.9886	1.4195	20
50	.1542	.1536	6.512	.1554	6.435	1.012	.9881	1.4166	10
9° 00′	.1571	.1564	6.392	.1584	6.314	1.012	.9877	1.4137	81° 00′
		cos θ	sec θ	cot θ	tan θ	csc θ	sin θ	Radians	Degrees

Angle θ

TABLE I Trigonometric Functions—Degrees and Minutes or Radians (continued)

Angle θ Degrees	Angle θ Radians	sin θ	csc θ	tan θ	cot θ	sec θ	cos θ		
9° 00′	.1571	.1564	6.392	.1584	6.314	1.012	.9877	1.4137	81° 00′
10	.1600	.1593	6.277	.1614	6.197	1.013	.9872	1.4108	50
20	.1629	.1622	6.166	.1644	6.084	1.013	.9868	1.4079	40
30	.1658	.1650	6.059	.1673	5.976	1.014	.9863	1.4050	30
40	.1687	.1679	5.955	.1703	5.871	1.014	.9858	1.4021	20
50	.1716	.1708	5.855	.1733	5.769	1.015	.9853	1.3992	10
10° 00′	.1745	.1736	5.759	.1763	5.671	1.015	.9848	1.3963	80° 00′
10	.1774	.1765	5.665	.1793	5.576	1.016	.9843	1.3934	50
20	.1804	.1794	5.575	.1823	5.485	1.016	.9838	1.3904	40
30	.1833	.1822	5.487	.1853	5.396	1.017	.9833	1.3875	30
40	.1862	.1851	5.403	.1883	5.309	1.018	.9827	1.3846	20
50	.1891	.1880	5.320	.1914	5.226	1.018	.9822	1.3817	10
11° 00′	.1920	.1908	5.241	.1944	5.145	1.019	.9816	1.3788	79° 00′
10	.1949	.1937	5.164	.1974	5.066	1.019	.9811	1.3759	50
20	.1978	.1965	5.089	.2004	4.989	1.020	.9805	1.3730	40
30	.2007	.1994	5.016	.2035	4.915	1.020	.9799	1.3701	30
40	.2036	.2022	4.945	.2065	4.843	1.021	.9793	1.3672	20
50	.2065	.2051	4.876	.2095	4.773	1.022	.9787	1.3643	10
12° 00′	.2094	.2079	4.810	.2126	4.705	1.022	.9781	1.3614	78° 00′
10	.2123	.2108	4.745	.2156	4.638	1.023	.9775	1.3584	50
20	.2153	.2136	4.682	.2186	4.574	1.024	.9769	1.3555	40
30	.2182	.2164	4.620	.2217	4.511	1.024	.9763	1.3526	30
40	.2211	.2193	4.560	.2247	4.449	1.025	.9757	1.3497	20
50	.2240	.2221	4.502	.2278	4.390	1.026	.9750	1.3468	10
13° 00′	.2269	.2250	4.445	.2309	4.331	1.026	.9744	1.3439	77° 00′
10	.2298	.2278	4.390	.2339	4.275	1.027	.9737	1.3410	50
20	.2327	.2306	4.336	.2370	4.219	1.028	.9730	1.3381	40
30	.2356	.2334	4.284	.2401	4.165	1.028	.9724	1.3352	30
40	.2385	.2363	4.232	.2432	4.113	1.029	.9717	1.3323	20
50	.2414	.2391	4.182	.2462	4.061	1.030	.9710	1.3294	10
14° 00′	.2443	.2419	4.134	.2493	4.011	1.031	.9703	1.3265	76° 00′
10	.2473	.2447	4.086	.2524	3.962	1.031	.9696	1.3235	50
20	.2502	.2476	4.039	.2555	3.914	1.032	.9689	1.3206	40
30	.2531	.2504	3.994	.2586	3.867	1.033	.9681	1.3177	30
40	.2560	.2532	3.950	.2617	3.821	1.034	.9674	1.3148	20
50	.2589	.2560	3.906	.2648	3.776	1.034	.9667	1.3119	10
15° 00′	.2618	.2588	3.864	.2679	3.732	1.035	.9659	1.3090	75° 00′
10	.2647	.2616	3.822	.2711	3.689	1.036	.9652	1.3061	50
20	.2676	.2644	3.782	.2742	3.647	1.037	.9644	1.3032	40
30	.2705	.2672	3.742	.2773	3.606	1.038	.9636	1.3003	30
40	.2734	.2700	3.703	.2805	3.566	1.039	.9628	1.2974	20
50	.2763	.2728	3.665	.2836	3.526	1.039	.9621	1.2945	10
16° 00′	.2793	.2756	3.628	.2867	3.487	1.040	.9613	1.2915	74° 00′
10	.2822	.2784	3.592	.2899	3.450	1.041	.9605	1.2886	50
20	.2851	.2812	3.556	.2931	3.412	1.042	.9596	1.2857	40
30	.2880	.2840	3.521	.2962	3.376	1.043	.9588	1.2828	30
40	.2909	.2868	3.487	.2944	3.340	1.044	.9580	1.2799	20
50	.2938	.2896	3.453	.3026	3.305	1.045	.9572	1.2770	10
17° 00′	.2967	.2924	3.420	.3057	3.271	1.046	.9563	1.2741	73° 00′
10	.2996	.2952	3.388	.3089	3.237	1.047	.9555	1.2712	50
20	.3025	.2979	3.357	.3121	3.204	1.048	.9546	1.2683	40
30	.3054	.3007	3.326	.3153	3.172	1.048	.9537	1.2654	30
40	.3083	.3035	3.295	.3185	3.140	1.049	.9528	1.2625	20
50	.3113	.3062	3.265	.3217	3.108	1.050	.9520	1.2595	10
18° 00′	.3142	.3090	3.236	.3249	3.078	1.051	.9511	1.2566	72° 00′
		cos θ	sec θ	cot θ	tan θ	csc θ	sin θ	Radians	Degrees
								Angle θ	

TABLE I   Trigonometric Functions—Degrees and Minutes or Radians (continued)

Angle θ									
Degrees	Radians	sin θ	csc θ	tan θ	cot θ	sec θ	cos θ		
18° 00′	.3142	.3090	3.236	.3249	3.078	1.051	.9511	1.2566	72° 00′
10	.3171	.3118	3.207	.3281	3.047	1.052	.9502	1.2537	50
20	.3200	.3145	3.179	.3314	3.018	1.053	.9492	1.2508	40
30	.3229	.3173	3.152	.3346	2.989	1.054	.9483	1.2479	30
40	.3258	.3201	3.124	.3378	2.960	1.056	.9474	1.2450	20
50	.3287	.3228	3.098	.3411	2.932	1.057	.9465	1.2421	10
19° 00′	.3316	.3256	3.072	.3443	2.904	1.058	.9455	1.2392	71° 00′
10	.3345	.3283	3.046	.3476	2.877	1.059	.9446	1.2363	50
20	.3374	.3311	3.021	.3508	2.850	1.060	.9436	1.2334	40
30	.3403	.3338	2.996	.3541	2.824	1.061	.9426	1.2305	30
40	.3432	.3365	2.971	.3574	2.798	1.062	.9417	1.2275	20
50	.3462	.3393	2.947	.3607	2.773	1.063	.9407	1.2246	10
20° 00′	.3491	.3420	2.924	.3640	2.747	1.064	.9397	1.2217	70° 00′
10	.3520	.3448	2.901	.3673	2.723	1.065	.9387	1.2188	50
20	.3549	.3475	2.878	.3706	2.699	1.066	.9377	1.2159	40
30	.3578	.3502	2.855	.3739	2.675	1.068	.9367	1.2130	30
40	.3607	.3529	2.833	.3772	2.651	1.069	.9356	1.2101	20
50	.3636	.3557	2.812	.3805	2.628	1.070	.9346	1.2072	10
21° 00′	.3665	.3584	2.790	.3839	2.605	1.071	.9336	1.2043	69° 00′
10	.3694	.3611	2.769	.3872	2.583	1.072	.9325	1.2014	50
20	.3723	.3638	2.749	.3906	2.560	1.074	.9315	1.1985	40
30	.3752	.3665	2.729	.3939	2.539	1.075	.9304	1.1956	30
40	.3782	.3692	2.709	.3973	2.517	1.076	.9293	1.1926	20
50	.3811	.3719	2.689	.4006	2.496	1.077	.9283	1.1897	10
22° 00′	.3840	.3746	2.669	.4040	2.475	1.079	.9272	1.1868	68° 00′
10	.3869	.3773	2.650	.4074	2.455	1.080	.9261	1.1839	50
20	.3898	.3800	2.632	.4108	2.434	1.081	.9250	1.1810	40
30	.3927	.3827	2.613	.4142	2.414	1.082	.9239	1.1781	30
40	.3956	.3854	2.595	.4176	2.394	1.084	.9228	1.1752	20
50	.3985	.3881	2.577	.4210	2.375	1.085	.9216	1.1723	10
23° 00′	.4014	.3907	2.559	.4245	2.356	1.086	.9205	1.1694	67° 00′
10	.4043	.3934	2.542	.4279	2.337	1.088	.9194	1.1665	50
20	.4072	.3961	2.525	.4314	2.318	1.089	.9182	1.1636	40
30	.4102	.3987	2.508	.4348	2.300	1.090	.9171	1.1606	30
40	.4131	.4014	2.491	.4383	2.282	1.092	.9159	1.1577	20
50	.4160	.4041	2.475	.4417	2.264	1.093	.9147	1.1548	10
24° 00′	.4189	.4067	2.459	.4452	2.246	1.095	.9135	1.1519	66° 00′
10	.4218	.4094	2.443	.4487	2.229	1.096	.9124	1.1490	50
20	.4247	.4120	2.427	.4522	2.211	1.097	.9112	1.1461	40
30	.4276	.4147	2.411	.4557	2.194	1.099	.9100	1.1432	30
40	.4305	.4173	2.396	.4592	2.177	1.100	.9088	1.1403	20
50	.4334	.4200	2.381	.4628	2.161	1.102	.9075	1.1374	10
25° 00′	.4363	.4226	2.366	.4663	2.145	1.103	.9063	1.1345	65° 00′
10	.4392	.4253	2.352	.4699	2.128	1.105	.9051	1.1316	50
20	.4422	.4279	2.337	.4734	2.112	1.106	.9038	1.1286	40
30	.4451	.4305	2.323	.4770	2.097	1.108	.9026	1.1257	30
40	.4480	.4331	2.309	.4806	2.081	1.109	.9013	1.1228	20
50	.4509	.4358	2.295	.4841	2.066	1.111	.9001	1.1199	10
26° 00′	.4538	.4384	2.281	.4877	2.050	1.113	.8988	1.1170	64° 00′
10	.4567	.4410	2.268	.4913	2.035	1.114	.8975	1.1141	50
20	.4596	.4436	2.254	.4950	2.020	1.116	.8962	1.1112	40
30	.4625	.4462	2.241	.4986	2.006	1.117	.8949	1.1083	30
40	.4654	.4488	2.228	.5022	1.991	1.119	.8936	1.1054	20
50	.4683	.4514	2.215	.5059	1.977	1.121	.8923	1.1025	10
27° 00′	.4712	.4540	2.203	.5095	1.963	1.122	.8910	1.0996	63° 00′
		cos θ	sec θ	cot θ	tan θ	csc θ	sin θ	Radians	Degrees
								Angle θ	

TABLE I Trigonometric Functions—Degrees and Minutes or Radians (continued)

Angle θ Degrees	Angle θ Radians	$\sin\theta$	$\csc\theta$	$\tan\theta$	$\cot\theta$	$\sec\theta$	$\cos\theta$		
27° 00′	.4712	.4540	2.203	.5095	1.963	1.122	.8910	1.0996	63° 00′
10	.4741	.4566	2.190	.5132	1.949	1.124	.8897	1.0966	50
20	.4771	.4592	2.178	.5169	1.935	1.126	.8884	1.0937	40
30	.4800	.4617	2.166	.5206	1.921	1.127	.8870	1.0908	30
40	.4829	.4643	2.154	.5243	1.907	1.129	.8857	1.0879	20
50	.4858	.4669	2.142	.5280	1.894	1.131	.8843	1.0850	10
28° 00′	.4887	.4695	2.130	.5317	1.881	1.133	.8829	1.0821	62° 00′
10	.4916	.4720	2.118	.5354	1.868	1.134	.8816	1.0792	50
20	.4945	.4746	2.107	.5392	1.855	1.136	.8802	1.0763	40
30	.4974	.4772	2.096	.5430	1.842	1.138	.8788	1.0734	30
40	.5003	.4797	2.085	.5467	1.829	1.140	.8774	1.0705	20
50	.5032	.4823	2.074	.5505	1.816	1.142	.8760	1.0676	10
29° 00′	.5061	.4848	2.063	.5543	1.804	1.143	.8746	1.0647	61° 00′
10	.5091	.4874	2.052	.5581	1.792	1.145	.8732	1.0617	50
20	.5120	.4899	2.041	.5619	1.780	1.147	.8718	1.0588	40
30	.5149	.4924	2.031	.5658	1.767	1.149	.8704	1.0559	30
40	.5178	.4950	2.020	.5696	1.756	1.151	.8689	1.0530	20
50	.5207	.4975	2.010	.5735	1.744	1.153	.8675	1.0501	10
30° 00′	.5236	.5000	2.000	.5774	1.732	1.155	.8660	1.0472	60° 00′
10	.5265	.5025	1.990	.5812	1.720	1.157	.8646	1.0443	50
20	.5294	.5050	1.980	.5851	1.709	1.159	.8631	1.0414	40
30	.5323	.5075	1.970	.5890	1.698	1.161	.8616	1.0385	30
40	.5352	.5100	1.961	.5930	1.686	1.163	.8601	1.0356	20
50	.5381	.5125	1.951	.5969	1.675	1.165	.8587	1.0327	10
31° 00′	.5411	.5150	1.942	.6009	1.664	1.167	.8572	1.0297	59° 00′
10	.5440	.5175	1.932	.6048	1.653	1.169	.8557	1.0268	50
20	.5469	.5200	1.923	.6088	1.643	1.171	.8542	1.0239	40
30	.5498	.5225	1.914	.6128	1.632	1.173	.8526	1.0210	30
40	.5527	.5250	1.905	.6168	1.621	1.175	.8511	1.0181	20
50	.5556	.5275	1.896	.6208	1.611	1.177	.8496	1.0152	10
32° 00′	.5585	.5299	1.887	.6249	1.600	1.179	.8480	1.0123	58° 00′
10	.5614	.5324	1.878	.6289	1.590	1.181	.8465	1.0094	50
20	.5643	.5348	1.870	.6330	1.580	1.184	.8450	1.0065	40
30	.5672	.5373	1.861	.6371	1.570	1.186	.8434	1.0036	30
40	.5701	.5398	1.853	.6412	1.560	1.188	.8418	1.0007	20
50	.5730	.5422	1.844	.6453	1.550	1.190	.8403	.9977	10
33° 00′	.5760	.5446	1.836	.6494	1.540	1.192	.8387	.9948	57° 00′
10	.5789	.5471	1.828	.6536	1.530	1.195	.8371	.9919	50
20	.5818	.5495	1.820	.6577	1.520	1.197	.8355	.9890	40
30	.5847	.5519	1.812	.6619	1.511	1.199	.8339	.9861	30
40	.5876	.5544	1.804	.6661	1.501	1.202	.8323	.9832	20
50	.5905	.5568	1.796	.6703	1.492	1.204	.8307	.9803	10
34° 00′	.5934	.5592	1.788	.6745	1.483	1.206	.8290	.9774	56° 00′
10	.5963	.5616	1.781	.6787	1.473	1.209	.8274	.9745	50
20	.5992	.5640	1.773	.6830	1.464	1.211	.8258	.9716	40
30	.6021	.5664	1.766	.6873	1.455	1.213	.8241	.9687	30
40	.6050	.5688	1.758	.6916	1.446	1.216	.8225	.9657	20
50	.6080	.5712	1.751	.6959	1.437	1.218	.8208	.9628	10
35° 00′	.6109	.5736	1.743	.7002	1.428	1.221	.8192	.9599	55° 00′
10	.6138	.5760	1.736	.7046	1.419	1.223	.8175	.9570	50
20	.6167	.5783	1.729	.7089	1.411	1.226	.8158	.9541	40
30	.6196	.5807	1.722	.7133	1.402	1.228	.8141	.9512	30
40	.6225	.5831	1.715	.7177	1.393	1.231	.8124	.9483	20
50	.6254	.5854	1.708	.7221	1.385	1.233	.8107	.9454	10
36° 00′	.6283	.5878	1.701	.7265	1.376	1.236	.8090	.9425	54° 00′
		$\cos\theta$	$\sec\theta$	$\cot\theta$	$\tan\theta$	$\csc\theta$	$\sin\theta$	Radians	Degrees
								Angle θ	

TABLE I Trigonometric Functions—Degrees and Minutes or Radians (continued)

Degrees	Radians	sin θ	csc θ	tan θ	cot θ	sec θ	cos θ		
36° 00′	.6283	.5878	1.701	.7265	1.376	1.236	.8090	.9425	54° 00′
10	.6312	.5901	1.695	.7310	1.368	1.239	.8073	.9396	50
20	.6341	.5925	1.688	.7355	1.360	1.241	.8056	.9367	40
30	.6370	.5948	1.681	.7400	1.351	1.244	.8039	.9338	30
40	.6400	.5972	1.675	.7445	1.343	1.247	.8021	.9308	20
50	.6429	.5995	1.668	.7490	1.335	1.249	.8004	.9279	10
37° 00′	.6458	.6018	1.662	.7536	1.327	1.252	.7986	.9250	53° 00′
10	.6487	.6041	1.655	.7581	1.319	1.255	.7969	.9221	50
20	.6516	.6065	1.649	.7627	1.311	1.258	.7951	.9192	40
30	.6545	.6088	1.643	.7673	1.303	1.260	.7934	.9163	30
40	.6574	.6111	1.636	.7720	1.295	1.263	.7916	.9134	20
50	.6603	.6134	1.630	.7766	1.288	1.266	.7898	.9105	10
38° 00′	.6632	.6157	1.624	.7813	1.280	1.269	.7880	.9076	52° 00′
10	.6661	.6180	1.618	.7860	1.272	1.272	.7862	.9047	50
20	.6690	.6202	1.612	.7907	1.265	1.275	.7844	.9018	40
30	.6720	.6225	1.606	.7954	1.257	1.278	.7826	.8988	30
40	.6749	.6248	1.601	.8002	1.250	1.281	.7808	.8959	20
50	.6778	.6271	1.595	.8050	1.242	1.284	.7790	.8930	10
39° 00′	.6807	.6293	1.589	.8098	1.235	1.287	.7771	.8901	51° 00′
10	.6836	.6316	1.583	.8146	1.228	1.290	.7753	.8872	50
20	.6865	.6338	1.578	.8195	1.220	1.293	.7735	.8843	40
30	.6894	.6361	1.572	.8243	1.213	1.296	.7716	.8814	30
40	.6923	.6383	1.567	.8292	1.206	1.299	.7698	.8785	20
50	.6952	.6406	1.561	.8342	1.199	1.302	.7679	.8756	10
40° 00′	.6981	.6428	1.556	.8391	1.192	1.305	.7660	.8727	50° 00′
10	.7010	.6450	1.550	.8441	1.185	1.309	.7642	.8698	50
20	.7039	.6472	1.545	.8491	1.178	1.312	.7623	.8668	40
30	.7069	.6494	1.540	.8541	1.171	1.315	.7604	.8639	30
40	.7098	.6517	1.535	.8591	1.164	1.318	.7585	.8610	20
50	.7127	.6539	1.529	.8642	1.157	1.322	.7566	.8581	10
41° 00′	.7156	.6561	1.524	.8693	1.150	1.325	.7547	.8552	49° 00′
10	.7185	.6583	1.519	.8744	1.144	1.328	.7528	.8523	50
20	.7214	.6604	1.514	.8796	1.137	1.332	.7509	.8494	40
30	.7243	.6626	1.509	.8847	1.130	1.335	.7490	.8465	30
40	.7272	.6648	1.504	.8899	1.124	1.339	.7470	.8436	20
50	.7301	.6670	1.499	.8952	1.117	1.342	.7451	.8407	10
42° 00′	.7330	.6691	1.494	.9004	1.111	1.346	.7431	.8378	48° 00′
10	.7359	.6713	1.490	.9057	1.104	1.349	.7412	.8348	50
20	.7389	.6734	1.485	.9110	1.098	1.353	.7392	.8319	40
30	.7418	.6756	1.480	.9163	1.091	1.356	.7373	.8290	30
40	.7447	.6777	1.476	.9217	1.085	1.360	.7353	.8261	20
50	.7476	.6799	1.471	.9271	1.079	1.364	.7333	.8232	10
43° 00′	.7505	.6820	1.466	.9325	1.072	1.367	.7314	.8203	47° 00′
10	.7534	.6841	1.462	.9380	1.066	1.371	.7294	.8174	50
20	.7563	.6862	1.457	.9435	1.060	1.375	.7274	.8145	40
30	.7592	.6884	1.453	.9490	1.054	1.379	.7254	.8116	30
40	.7621	.6905	1.448	.9545	1.048	1.382	.7234	.8087	20
50	.7650	.6926	1.444	.9601	1.042	1.386	.7214	.8058	10
44° 00′	.7679	.6947	1.440	.9657	1.036	1.390	.7193	.8029	46° 00′
10	.7709	.6967	1.435	.9713	1.030	1.394	.7173	.7999	50
20	.7738	.6988	1.431	.9770	1.024	1.398	.7153	.7970	40
30	.7767	.7009	1.427	.9827	1.018	1.402	.7133	.7941	30
40	.7796	.7030	1.423	.9884	1.012	1.406	.7112	.7912	20
50	.7825	.7050	1.418	.9942	1.006	1.410	.7092	.7883	10
45° 00′	.7854	.7071	1.414	1.000	1.000	1.414	.7071	.7854	45° 00′
		cos θ	sec θ	cot θ	tan θ	csc θ	sin θ	Radians	Degrees
								Angle θ	

TABLE II Trigonometric Functions—Decimal Degrees
(csc $\theta = 1/\sin \theta$; sec $\theta = 1/\cos \theta$)

↱	sin	cos	tan	cot	
0.0	0.00000	1.0000	0.00000	∞	90.0
.1	.00175	1.0000	.00175	573.0	89.9
.2	.00349	1.0000	.00349	286.5	.8
.3	.00524	1.0000	.00524	191.0	.7
.4	.00698	1.0000	.00698	143.24	.6
.5	.00873	1.0000	.00873	114.59	.5
.6	.01047	0.9999	.01047	95.49	.4
.7	.01222	.9999	.01222	81.85	.3
.8	.01396	.9999	.01396	71.62	.2
.9	.01571	.9999	.01571	63.66	89.1
1.0	0.01745	0.9998	0.01746	57.29	89.0
.1	.01920	.9998	.01920	52.08	88.9
.2	.02094	.9998	.02095	47.74	.8
.3	.02269	.9997	.02269	44.07	.7
.4	.02443	.9997	.02444	40.92	.6
.5	.02618	.9997	.02619	38.19	.5
.6	.02792	.9996	.02793	35.80	.4
.7	.02967	.9996	.02968	33.69	.3
.8	.03141	.9995	.03143	31.82	.2
.9	.03316	.9995	.03317	30.14	88.1
2.0	0.03490	0.9994	0.03492	28.64	88.0
.1	.03664	.9993	.03667	27.27	87.9
.2	.03839	.9993	.03842	26.03	.8
.3	.04013	.9992	.04016	24.90	.7
.4	.04188	.9991	.04191	23.86	.6
.5	.04362	.9990	.04366	22.90	.5
.6	.04536	.9990	.04541	22.02	.4
.7	.04711	.9989	.04716	21.20	.3
.8	.04885	.9988	.04891	20.45	.2
.9	.05059	.9987	.05066	19.74	87.1
3.0	0.05234	0.9986	0.05241	19.081	87.0
.1	.05408 ·	.9985	.05416	18.464	86.9
.2	.05582	.9984	.05591	17.886	.8
.3	.05756	.9983	.05766	17.343	.7
.4	.05931	.9982	.05941	16.832	.6
.5	.06105	.9981	.06116	16.350	.5
.6	.06279	.9980	.06291	15.895	.4
.7	.06453	.9979	.06467	15.464	.3
.8	.06627	.9978	.06642	15.056	.2
.9	.06802	.9977	.06817	14.669	86.1
4.0	0.06976	0.9976	0.06993	14.301	86.0
	cos	sin	cot	tan	↰

TABLE II Trigonometric Functions—Decimal Degrees
(csc $\theta = 1/\sin \theta$; sec $\theta = 1/\cos \theta$) (continued)

↓ →	sin	cos	tan	cot	
4.0	0.06976	0.9976	0.06993	14.301	86.0
.1	.07150	.9974	.07168	13.951	85.9
.2	.07324	.9973	.07344	13.617	.8
.3	.07498	.9972	.07519	13.300	.7
.4	.07672	.9971	.07695	12.996	.6
.5	.07846	.9969	.07870	12.706	.5
.6	.08020	.9968	.08046	12.429	.4
.7	.08194	.9966	.08221	12.163	.3
.8	.08368	.9965	.08397	11.909	.2
.9	.08542	.9963	.08573	11.664	85.1
5.0	0.08716	0.9962	0.08749	11.430	85.0
.1	.08889	.9960	.08925	11.205	84.9
.2	.09063	.9959	.09101	10.988	.8
.3	.09237	.9957	.09277	10.780	.7
.4	.09411	.9956	.09453	10.579	.6
.5	.09585	.9954	.09629	10.385	.5
.6	.09758	.9952	.09805	10.199	.4
.7	.09932	.9951	.09981	10.019	.3
.8	.10106	.9949	.10158	9.845	.2
.9	.10279	.9947	.10334	9.677	84.1
6.0	0.10453	0.9945	0.10510	9.514	84.0
.1	.10626	.9943	.10687	9.357	83.9
.2	.10800	.9942	.10863	9.205	.8
.3	.10973	.9940	.11040	9.058	.7
.4	.11147	.9938	.11217	8.915	.6
.5	.11320	.9936	.11394	8.777	.5
.6	.11494	.9934	.11570	8.643	.4
.7	.11667	.9932	.11747	8.513	.3
.8	.11840	.9930	.11924	8.386	.2
.9	.12014	.9928	.12101	8.264	83.1
7.0	0.12187	0.9925	0.12278	8.144	83.0
.1	.12360·	.9923	.12456	8.028	82.9
.2·	.12533	.9921	.12633	7.916	.8
.3	.12706	.9919	.12810	7.806	.7
.4	.12880	.9917	.12988	7.700	.6
.5	.13053	.9914	.13165	7.596	.5
.6	.13226	.9912	.13343	7.495	.4
.7	.13399	.9910	.13521	7.396	.3
.8	.13572	.9907	.13698	7.300	.2
.9	.13744	.9905	.13876	7.207	82.1
8.0	0.13917	0.9903	0.14054	7.115	82.0
	cos	sin	cot	tan	↑←↲

TABLE II (continued)

↓→	sin	cos	tan	cot	
8.0	0.13917	0.9903	0.14054	7.115	82.0
.1	.14090	.9900	.14232	7.026	81.9
.2	.14263	.9898	.14410	6.940	.8
.3	.14436	.9895	.14588	6.855	.7
.4	.14608	.9893	.14767	6.772	.6
.5	.14781	.9890	.14945	6.691	.5
.6	.14954	.9888	.15124	6.612	.4
.7	.15126	.9885	.15302	6.535	.3
.8	.15299	.9882	.15481	6.460	.2
.9	.15471	.9880	.15660	6.386	81.1
9.0	0.15643	0.9877	0.15838	6.314	81.0
.1	.15816	.9874	.16017	6.243	80.9
.2	.15988	.9871	.16196	6.174	.8
.3	.16160	.9869	.16376	6.107	.7
.4	.16333	.9866	.16555	6.041	.6
.5	.16505	.9863	.16734	5.976	.5
.6	.16677	.9860	.16914	5.912	.4
.7	.16849	.9857	.17093	5.850	.3
.8	.17021	.9854	.17273	5.789	.2
.9	.17193	.9851	.17453	5.730	80.1
10.0	0.1736	0.9848	0.1763	5.671	80.0
.1	.1754	.9845	.1781	5.614	79.9
.2	.1771	.9842	.1799	5.558	.8
.3	.1788	.9839	.1817	5.503	.7
.4	.1805	.9836	.1835	5.449	.6
.5	.1822	.9833	.1853	5.396	.5
.6	.1840	.9829	.1871	5.343	.4
.7	.1857	.9826	.1890	5.292	.3
.8	.1874	.9823	.1908	5.242	.2
.9	.1891	.9820	.1926	5.193	79.1
11.0	0.1908	0.9816	0.1944	5.145	79.0
.1	.1925	.9813	.1962	5.097	78.9
.2	.1942	.9810	.1980	5.050	.8
.3	.1959	.9806	.1998	5.005	.7
.4	.1977	.9803	.2016	4.959	.6
.5	.1994	.9799	.2035	4.915	.5
.6	.2011	.9796	.2053	4.872	.4
.7	.2028	.9792	.2071	4.829	.3
.8	.2045	.9789	.2089	4.787	.2
.9	.2062	.9785	.2107	4.745	78.1
12.0	0.2079	0.9781	0.2126	4.705	78.0
	cos	sin	cot	tan	←↑

TABLE II Trigonometric Functions—Decimal Degrees
($\csc \theta = 1/\sin \theta$; $\sec \theta = 1/\cos \theta$) (continued)

	sin	cos	tan	cot	
12.0	0.2079	0.9781	0.2126	4.705	78.0
.1	.2096	.9778	.2144	4.665	77.9
.2	.2113	.9774	.2162	4.625	.8
.3	.2130	.9770	.2180	4.586	.7
.4	.2147	.9767	.2199	4.548	.6
.5	.2164	.9763	.2217	4.511	.5
.6	.2181	.9759	.2235	4.474	.4
.7	.2198	.9755	.2254	4.437	.3
.8	.2215	.9751	.2272	4.402	.2
.9	.2233	.9748	.2290	4.366	77.1
13.0	0.2250	0.9744	0.2309	4.331	77.0
.1	.2267	.9740	.2327	4.297	76.9
.2	.2284	.9736	.2345	4.264	.8
.3	.2300	.9732	.2364	4.230	.7
.4	.2317	.9728	.2382	4.198	.6
.5	.2334	.9724	.2401	4.165	.5
.6	.2351	.9720	.2419	4.134	.4
.7	.2368	.9715	.2438	4.102	.3
.8	.2385	.9711	.2456	4.071	.2
.9	.2402	.9707	.2475	4.041	76.1
14.0	0.2419	0.9703	0.2493	4.011	76.0
.1	.2436	.9699	.2512	3.981	75.9
.2	.2453	.9694	.2530	3.952	.8
.3	.2470	.9690	.2549	3.923	.7
.4	.2487	.9686	.2568	3.895	.6
.5	.2504	.9681	.2586	3.867	.5
.6	.2521	.9677	.2605	3.839	.4
.7	.2538	.9673	.2623	3.812	.3
.8	.2554	.9668	.2642	3.785	.2
.9	.2571	.9664	.2661	3.758	75.1
15.0	0.2588	0.9659	0.2679	3.732	75.0
.1	.2605	.9655	.2698	3.706	74.9
.2	.2622	.9650	.2717	3.681	.8
.3	.2639	.9646	.2736	3.655	.7
.4	.2656	.9641	.2754	3.630	.6
.5	.2672	.9636	.2773	3.606	.5
.6	.2689	.9632	.2792	3.582	.4
.7	.2706	.9627	.2811	3.558	.3
.8	.2723	.9622	.2830	3.534	.2
.9	.2740	.9617	.2849	3.511	74.1
16.0	0.2756	0.9613	0.2867	3.487	74.0
	cos	sin	cot	tan	

TABLE II (continued)

↱	sin	cos	tan	cot	
16.0	0.2756	0.9613	0.2867	3.487	74.0
.1	.2773	.9608	.2886	3.465	73.9
.2	.2790	.9603	.2905	3.442	.8
.3	.2807	.9598	.2924	3.420	.7
.4	.2823	.9593	.2943	3.398	.6
.5	.2840	.9588	.2962	3.376	.5
.6	.2857	.9583	.2981	3.354	.4
.7	.2874	.9578	.3000	3.333	.3
.8	.2890	.9573	.3019	3.312	.2
.9	.2907	.9568	.3038	3.291	73.1
17.0	0.2924	0.9563	0.3057	3.271	73.0
.1	.2940	.9558	.3076	3.251	72.9
.2	.2957	.9553	.3096	3.230	.8
.3	.2974	.9548	.3115	3.211	.7
.4	.2990	.9542	.3134	3.191	.6
.5	.3007	.9537	.3153	3.172	.5
.6	.3024	.9532	.3172	3.152	.4
.7	.3040	.9527	.3191	3.133	.3
.8	.3057	.9521	.3211	3.115	.2
.9	.3074	.9516	.3230	3.096	72.1
18.0	0.3090	0.9511	0.3249	3.078	72.0
.1	.3107	.9505	.3269	3.060	71.9
.2	.3123	.9500	.3288	3.042	.8
.3	.3140	.9494	.3307	3.024	.7
.4	.3156	.9489	.3327	3.006	.6
.5	.3173	.9483	.3346	2.989	.5
.6	.3190	.9478	.3365	2.971	.4
.7	.3206	.9472	.3385	2.954	.3
.8	.3223	.9466	.3404	2.937	.2
.9	.3239	.9461	.3424	2.921	71.1
19.0	0.3256	0.9455	0.3443	2.904	71.0
.1	.3272	.9449	.3463	2.888	70.9
.2	.3289	.9444	.3482	2.872	.8
.3	.3305	.9438	.3502	2.856	.7
.4	.3322	.9432	.3522	2.840	.6
.5	.3338	.9426	.3541	2.824	.5
.6	.3355	.9421	.3561	2.808	.4
.7	.3371	.9415	.3581	2.793	.3
.8	.3387	.9409	.3600	2.778	.2
.9	.3404	.9403	.3620	2.762	70.1
20.0	0.3420	0.9397	0.3640	2.747	70.0
	cos	sin	cot	tan	↲

TABLE II Trigonometric Functions—Decimal Degrees
$(\csc \theta = 1/\sin \theta;\ \sec \theta = 1/\cos \theta)$ (continued)

	sin	cos	tan	cot	
20.0	0.3420	0.9397	0.3640	2.747	70.0
.1	.3437	.9391	.3659	2.733	69.9
.2	.3453	.9385	.3679	2.718	.8
.3	.3469	.9379	.3699	2.703	.7
.4	.3486	.9373	.3719	2.689	.6
.5	.3502	.9367	.3739	2.675	.5
.6	.3518	.9361	.3759	2.660	.4
.7	.3535	.9354	.3779	2.646	.3
.8	.3551	.9348	.3799	2.633	.2
.9	.3567	.9342	.3819	2.619	69.1
21.0	0.3584	0.9336	0.3839	2.605	69.0
.1	.3600	.9330	.3859	2.592	68.9
.2	.3616	.9323	.3879	2.578	.8
.3	.3633	.9317	.3899	2.565	.7
.4	.3649	.9311	.3919	2.552	.6
.5	.3665	.9304	.3939	2.539	.5
.6	.3681	.9298	.3959	2.526	.4
.7	.3697	.9291	.3979	2.513	.3
.8	.3714	.9285	.4000	2.500	.2
.9	.3730	.9278	.4020	2.488	68.1
22.0	0.3746	0.9272	0.4040	2.475	68.0
.1	.3762	.9265	.4061	2.463	67.9
.2	.3778	.9259	.4081	2.450	.8
.3	.3795	.9252	.4101	2.438	.7
.4	.3811	.9245	.4122	2.426	.6
.5	.3827	.9239	.4142	2.414	.5
.6	.3843	.9232	.4163	2.402	.4
.7	.3859	.9225	.4183	2.391	.3
.8	.3875	.9219	.4204	2.379	.2
.9	.3891	.9212	.4224	2.367	67.1
23.0	0.3907	0.9205	0.4245	2.356	67.0
.1	.3923	.9198	.4265	2.344	66.9
.2	.3939	.9191	.4286	2.333	.8
.3	.3955	.9184	.4307	2.322	.7
.4	.3971	.9178	.4327	2.311	.6
.5	.3987	.9171	.4348	2.300	.5
.6	.4003	.9164	.4369	2.289	.4
.7	.4019	.9157	.4390	2.278	.3
.8	.4035	.9150	.4411	2.267	.2
.9	.4051	.9143	.4431	2.257	66.1
24.0	0.4067	0.9135	0.4452	2.246	66.0
	cos	sin	cot	tan	

TABLE II   (continued)

	sin	cos	tan	cot	
24.0	0.4067	0.9135	0.4452	2.246	66.0
.1	.4083	.9128	.4473	2.236	65.9
.2	.4099	.9121	.4494	2.225	.8
.3	.4115	.9114	.4515	2.215	.7
.4	.4131	.9107	.4536	2.204	.6
.5	.4147	.9100	.4557	2.194	.5
.6	.4163	.9092	.4578	2.184	.4
.7	.4179	.9085	.4599	2.174	.3
.8	.4195	.9078	.4621	2.164	.2
.9	.4210	.9070	.4642	2.154	65.1
25.0	0.4226	0.9063	0.4663	2.145	65.0
.1	.4242	.9056	.4684	2.135	64.9
.2	.4258	.9048	.4706	2.125	.8
.3	.4274	.9041	.4727	2.116	.7
.4	.4289	.9033	.4748	2.106	.6
.5	.4305	.9026	.4770	2.097	.5
.6	.4321	.9018	.4791	2.087	.4
.7	.4337	.9011	.4813	2.078	.3
.8	.4352	.9003	.4834	2.069	.2
.9	.4368	.8996	.4856	2.059	64.1
26.0	0.4384	0.8988	0.4877	2.050	64.0
.1	.4399	.8980	.4899	2.041	63.9
.2	.4415	.8973	.4921	2.032	.8
.3	.4431	.8965	.4942	2.023	.7
.4	.4446	.8957	.4964	2.014	.6
.5	.4462	.8949	.4986	2.006	.5
.6	.4478	.8942	.5008	1.997	.4
.7	.4493	.8934	.5029	1.988	.3
.8	.4509	.8926	.5051	1.980	.2
.9	.4524	.8918	.5073	1.971	63.1
27.0	0.4540	0.8910	0.5095	1.963	63.0
.1	.4555	.8902	.5117	1.954	62.9
.2	.4571	.8894	.5139	1.946	.8
.3	.4586	.8886	.5161	1.937	.7
.4	.4602	.8878	.5184	1.929	.6
.5	.4617	.8870	.5206	1.921	.5
.6	.4633	.8862	.5228	1.913	.4
.7	.4648	.8854	.5250	1.905	.3
.8	.4664	.8846	.5272	1.897	.2
.9	.4679	.8838	.5295	1.889	62.1
28.0	0.4695	0.8829	0.5317	1.881	62.0
	cos	sin	cot	tan	

TABLE II Trigonometric Functions—Decimal Degrees (csc θ = 1/sin θ; sec θ = 1/cos θ) (continued)

↓→	sin	cos	tan	cot	
28.0	0.4695	0.8829	0.5317	1.881	62.0
.1	.4710	.8821	.5340	1.873	61.9
.2	.4726	.8813	.5362	1.865	.8
.3	.4741	.8805	.5384	1.857	.7
.4	.4756	.8796	.5407	1.849	.6
.5	.4772	.8788	.5430	1.842	.5
.6	.4787	.8780	.5452	1.834	.4
.7	.4802	.8771	.5475	1.827	.3
.8	.4818	.8763	5498	1.819	.2
.9	.4833	.8755	.5520	1.811	61.1.
29.0	0.4848	0.8746	0.5543	1.804	61.0
.1	.4863	.8738	.5566	1.797	60.9
.2	.4879	.8729	.5589	1.789	.8
.3	.4894	.8721	.5612	1.782	.7
.4	.4909	.8712	.5635	1.775	.6
.5	.4924	.8704	.5658	1.767	.5
.6	.4939	.8695	.5681	1.760	.4
.7	.4955	.8686	.5704	1.753	.3
.8	.4970	.8678	.5727	1.746	.2
.9	.4985	.8669	.5750	1.739	60.1
30.0	0.5000	0.8660	0.5774	1.7321	60.0
.1	.5015	.8652	.5797	1.7251	59.9
.2	.5030	.8643	.5820	1.7182	.8
.3	.5045	.8634	.5844	1.7113	.7
.4	.5060	.8625	.5867	1.7045	.6
.5	.5075	.8616	.5890	1.6977	.5
.6	.5090	.8607	.5914	1.6909	.4
.7	.5105	.8599	.5938	1.6842	.3
.8	.5120	.8590	.5961	1.6775	.2
.9	.5135	.8581	.5985	1.6709	59.1
31.0	0.5150	0.8572	0.6009	1.6643	59.0
.1	.5165	.8563	.6032	1.6577	58.9
.2	.5180	.8554	.6056	1.6512	.8
.3	.5195	.8545	.6080	1.6447	.7
.4	.5210	.8536	.6104	1.6383	.6
.5	.5225	.8526	.6128	1.6319	.5
.6	.5240	.8517	.6152	1.6255	.4
.7	.5255	.8508	.6176	1.6191	.3
.8	.5270	.8499	.6200	1.6128	.2
.9	.5284	.8490	.6224	1.6066	58.1
32.0	0.5299	0.8480	0.6249	1.6003	58.0
	cos	sin	cot	tan	←↑

TABLE II (continued)

↓→	sin	cos	tan	cot	
32.0	0.5299	0.8480	0.6249	1.6003	58.0
.1	.5314	.8471	.6273	1.5941	57.9
.2	.5329	.8462	.6297	1.5880	.8
.3	.5344	.8453	.6322	1.5818	.7
.4	.5358	.8443	.6346	1.5757	.6
.5	.5373	.8434	.6371	1.5697	.5
.6	.5388	.8425	.6395	1.5637	.4
.7	.5402	.8415	.6420	1.5577	.3
.8	.5417	.8406	.6445	1.5517	.2
.9	.5432	.8396	.6469	1.5458	57.1
33.0	0.5446	0.8387	0.6494	1.5399	57.0
.1	.5461	.8377	.6519	1.5340	56.9
.2	.5476	.8368	.6544	1.5282	.8
.3	.5490	.8358	.6569	1.5224	.7
.4	.5505	.8348	.6594	1.5166	.6
.5	.5519	.8339	.6619	1.5108	.5
.6	.5534	.8329	.6644	1.5051	.4
.7	.5548	.8320	.6669	1.4994	.3
.8	.5563	.8310	.6694	1.4938	.2
.9	.5577	.8300	.6720	1.4882	56.1
34.0	0.5592	0.8290	0.6745	1.4826	56.0
.1	.5606	.8281	.6771	1.4770	55.9
.2	.5621	.8271	.6796	1.4715	.8
.3	.5635	.8261	.6822	1.4659	.7
.4	.5650	.8251	.6847	1.4605	.6
.5	.5664	.8241	.6873	1.4550	.5
.6	.5678	.8231	.6899	1.4496	.4
.7	.5693	.8221	.6924	1.4442	.3
.8	.5707	.8211	.6950	1.4388	.2
.9	.5721	.8202	.6976	1.4335	55.1
35.0	0.5736	0.8192	0.7002	1.4281	55.0
.1	.5750	.8181	.7028	1.4229	54.9
.2	.5764	.8171	.7054	1.4176	.8
.3	.5779	.8161	.7080	1.4124	.7
.4	.5793	.8151	.7107	1.4071	.6
.5	.5807	.8141	.7133	1.4019	.5
.6	.5821	.8131	.7159	1.3968	.4
.7	.5835	.8121	.7186	1.3916	.3
.8	.5850	.8111	.7212	1.3865	.2
.9	.5864	.8100	.7239	1.3814	54.1
36.0	0.5878	0.8090	0.7265	1.3764	54.0
	cos	sin	cot	tan	←↑

TABLE II Trigonometric Functions—Decimal Degrees
(csc θ = 1/sin θ; sec θ = 1/cos θ) (continued)

↳→	sin	cos	tan	cot	
36.0	0.5878	0.8090	0.7265	1.3764	54.0
.1	.5892	.8080	.7292	1.3713	53.9
.2	.5906	.8070	.7319	1.3663	.8
.3	.5920	.8059	.7346	1.3613	.7
.4	.5934	.8049	.7373	1.3564	.6
.5	.5948	.8039	.7400	1.3514	.5
.6	.5962	.8028	.7427	1.3465	.4
.7	.5976	.8018	.7454	1.3416	.3
.8	.5990	.8007	.7481	1.3367	.2
.9	.6004	.7997	.7508	1.3319	53.1
37.0	0.6018	0.7986	0.7536	1.3270	53.0
.1	.6032	.7976	.7563	1.3222	52.9
.2	.6046	.7965	.7590	1.3175	.8
.3	.6060	.7955	.7618	1.3127	.7
.4	.6074	.7944	.7646	1.3079	.6
.5	.6088	.7934	.7673	1.3032	.5
.6	.6101	.7923	.7701	1.2985	.4
.7	.6115	.7912	.7729	1.2938	.3
.8	.6129	.7902	.7757	1.2892	.2
.9	.6143	.7891	.7785	1.2846	52.1
38.0	0.6157	0.7880	0.7813	1.2799	52.0
.1	.6170	.7869	.7841	1.2753	51.9
.2	.6184	.7859	.7869	1.2708	.8
.3	.6198	.7848	.7898	1.2662	.7
.4	.6211	.7837	.7926	1.2617	.6
.5	.6225	.7826	.7954	1.2572	.5
.6	.6239	.7815	.7983	1.2527	.4
.7	.6252	.7804	.8012	1.2482	.3
.8	.6266	.7793	.8040	1.2437	.2
.9	.6280	.7782	.8069	1.2393	51.1
39.0	0.6293	0.7771	0.8098	1.2349	51.0
.1	.6307	.7760	.8127	1.2305	50.9
.2	.6320	.7749	.8156	1.2261	.8
.3	.6334	.7738	.8185	1.2218	.7
.4	.6347	.7727	.8214	1.2174	.6
.5	.6361	.7716	.8243	1.2131	.5
.6	.6374	.7705	.8273	1.2088	.4
.7	.6388	.7694	.8302	1.2045	.3
.8	.6401	.7683	.8332	1.2002	.2
.9	.6414	.7672	.8361	1.1960	50.1
40.0	0.6428	0.7660	0.8391	1.1918	50.0
	cos	sin	cot	tan	←↑

TABLE II (continued)

	sin	cos	tan	cot	
40.0	0.6428	0.7660	0.8391	1.1918	50.0
.1	.6441	.7649	.8421	1.1875	49.9
.2	.6455	.7638	.8451	1.1833	.8
.3	.6468	.7627	.8481	1.1792	.7
.4	.6481	.7615	.8511	1.1750	.6
.5	.6494	.7604	.8541	1.1708	.5
.6	.6508	.7593	.8571	1.1667	.4
.7	.6521	.7581	.8601	1.1626	.3
.8	.6534	.7570	.8632	1.1585	.2
.9	.6547	.7559	.8662	1.1544	49.1
41.0	0.6561	0.7547	0.8693	1.1504	49.0
.1	.6574	.7536	.8724	1.1463	48.9
.2	.6587	.7524	.8754	1.1423	.8
.3	.6600	.7513	.8785	1.1383	.7
.4	.6613	.7501	.8816	1.1343	.6
.5	.6626	.7490	.8847	1.1303	.5
.6	.6639	.7478	.8878	1.1263	.4
.7	.6652	.7466	.8910	1.1224	.3
.8	.6665	.7455	.8941	1.1184	.2
.9	.6678	.7443	.8972	1.1145	48.1
42.0	0.6691	0.7431	0.9004	1.1106	48.0
.1	.6704	.7420	.9036	1.1067	47.9
.2	.6717	.7408	.9067	1.1028	.8
.3	.6730	.7396	.9099	1.0990	.7
.4	.6743	.7385	.9131	1.0951	.6
.5	.6756	.7373	.9163	1.0913	.5
.6	.6769	.7361	.9195	1.0875	.4
.7	.6782	.7349	.9228	1.0837	.3
.8	.6794	.7337	.9260	1.0799	.2
.9	.6807	.7325	.9293	1.0761	47.1
43.0	0.6820	0.7314	0.9325	1.0724	47.0
.1	.6833	.7302	.9358	1.0686	46.9
.2	.6845	.7290	.9391	1.0649	.8
.3	.6858	.7278	.9424	1.0612	.7
.4	.6871	.7266	.9457	1.0575	.6
.5	.6884	.7254	.9490	1.0538	.5
.6	.6896	.7242	.9523	1.0501	.4
.7	.6909	.7230	.9556	1.0464	.3
.8	.6921	.7218	.9590	1.0428	.2
.9	.6934	.7206	.9623	1.0392	46.1
44.0	0.6947	0.7193	0.9657	1.0355	46.0
	cos	sin	cot	tan	

TABLE II   Trigonometric Functions—Decimal Degrees
(csc θ = 1/sin θ; sec θ = 1/cos θ) (continued)

↓⟶	sin	cos	tan	cot	
44.0	0.6947	0.7193	0.9657	1.0355	46.0
.1	.6959	.7181	.9691	1.0319	45.9
.2	.6972	.7169	.9725	1.0283	.8
.3	.6984	.7157	.9759	1.0247	.7
.4	.6997	.7145	.9793	1.0212	.6
.5	.7009	.7133	.9827	1.0176	.5
.6	.7022	.7120	.9861	1.0141	.4
.7	.7034	.7108	.9896	1.0105	.3
.8	.7046	.7096	.9930	1.0070	.2
.9	.7059	.7083	.9965	1.0035	45.1
45.0	0.7071	0.7071	1.0000	1.0000	45.0
	cos	sin	cot	tan	⟵↑

TABLE III Common Logarithms

N	0	1	2	3	4	5	6	7	8	9
1.0	.0000	.0043	.0086	.0128	.0170	.0212	.0253	.0294	.0334	.0374
1.1	.0414	.0453	.0492	.0531	.0569	.0607	.0645	.0682	.0719	.0755
1.2	.0792	.0828	.0864	.0899	.0934	.0969	.1004	.1038	.1072	.1106
1.3	.1139	.1173	.1206	.1239	.1271	.1303	.1335	.1367	.1399	.1430
1.4	.1461	.1492	.1523	.1553	.1584	.1614	.1644	.1673	.1703	.1732
1.5	.1761	.1790	.1818	.1847	.1875	.1903	.1931	.1959	.1987	.2014
1.6	.2041	.2068	.2095	.2122	.2148	.2175	.2201	.2227	.2253	.2279
1.7	.2304	.2330	.2355	.2380	.2405	.2430	.2455	.2480	.2504	.2529
1.8	.2553	.2577	.2601	.2625	.2648	.2672	.2695	.2718	.2742	.2765
1.9	.2788	.2810	.2833	.2856	.2878	.2900	.2923	.2945	.2967	.2989
2.0	.3010	.3032	.3054	.3075	.3096	.3118	.3139	.3160	.3181	.3201
2.1	.3222	.3243	.3263	.3284	.3304	.3324	.3345	.3365	.3385	.3404
2.2	.3424	.3444	.3464	.3483	.3502	.3522	.3541	.3560	.3579	.3598
2.3	.3617	.3636	.3655	.3674	.3692	.3711	.3729	.3747	.3766	.3784
2.4	.3802	.3820	.3838	.3856	.3874	.3892	.3909	.3927	.3945	.3962
2.5	.3979	.3997	.4014	.4031	.4048	.4065	.4082	.4099	.4116	.4133
2.6	.4150	.4166	.4183	.4200	.4216	.4232	.4249	.4265	.4281	.4298
2.7	.4314	.4330	.4346	.4362	.4378	.4393	.4409	.4425	.4440	.4456
2.8	.4472	.4487	.4502	.4518	.4533	.4548	.4564	.4579	.4594	.4609
2.9	.4624	.4639	.4654	.4669	.4683	.4698	.4713	.4728	.4742	.4757
3.0	.4771	.4786	.4800	.4814	.4829	.4843	.4857	.4871	.4886	.4900
3.1	.4914	.4928	.4942	.4955	.4969	.4983	.4997	.5011	.5024	.5038
3.2	.5051	.5065	.5079	.5092	.5105	.5119	.5132	.5145	.5159	.5172
3.3	.5185	.5198	.5211	.5224	.5237	.5250	.5263	.5276	.5289	.5302
3.4	.5315	.5328	.5340	.5353	.5366	.5378	.5391	.5403	.5416	.5428
3.5	.5441	.5453	.5465	.5478	.5490	.5502	.5514	.5527	.5539	.5551
3.6	.5563	.5575	.5587	.5599	.5611	.5623	.5635	.5647	.5658	.5670
3.7	.5682	.5694	.5705	.5717	.5729	.5740	.5752	.5763	.5775	.5786
3.8	.5798	.5809	.5821	.5832	.5843	.5855	.5866	.5877	.5888	.5899
3.9	.5911	.5922	.5933	.5944	.5955	.5966	.5977	.5988	.5999	.6010
4.0	.6021	.6031	.6042	.6053	.6064	.6075	.6085	.6096	.6107	.6117
4.1	.6128	.6138	.6149	.6160	.6170	.6180	.6191	.6201	.6212	.6222
4.2	.6232	.6243	.6253	.6263	.6274	.6284	.6294	.6304	.6314	.6325
4.3	.6335	.6345	.6355	.6365	.6375	.6385	.6395	.6405	.6415	.6425
4.4	.6435	.6444	.6454	.6464	.6474	.6484	.6493	.6503	.6513	.6522
4.5	.6532	.6542	.6551	.6561	.6571	.6580	.6590	.6599	.6609	.6618
4.6	.6628	.6637	.6646	.6656	.6665	.6675	.6684	.6693	.6702	.6712
4.7	.6721	.6730	.6739	.6749	.6758	.6767	.6776	.6785	.6794	.6803
4.8	.6812	.6821	.6830	.6839	.6848	.6857	.6866	.6875	.6884	.6893
4.9	.6902	.6911	.6920	.6928	.6937	.6946	.6955	.6964	.6972	.6981
5.0	.6990	.6998	.7007	.7016	.7024	.7033	.7042	.7050	.7059	.7067
5.1	.7076	.7084	.7093	.7101	.7110	.7118	.7126	.7135	.7143	.7152
5.2	.7160	.7168	.7177	.7185	.7193	.7202	.7210	.7218	.7226	.7235
5.3	.7243	.7251	.7259	.7267	.7275	.7284	.7292	.7300	.7308	.7316
5.4	.7324	.7332	.7340	.7348	.7356	.7364	.7372	.7380	.7388	.7396
N		1	2	3	4	5	6	7	8	9

TABLE III   Common Logarithms (continued)

N	0	1	2	3	4	5	6	7	8	9
5.5	.7404	.7412	.7419	.7427	.7435	.7443	.7451	.7459	.7466	.7474
5.6	.7482	.7490	.7497	.7505	.7513	.7520	.7528	.7536	.7543	.7551
5.7	.7559	.7566	.7574	.7582	.7589	.7597	.7604	.7612	.7619	.7627
5.8	.7634	.7642	.7649	.7657	.7664	.7672	.7679	.7686	.7694	.7701
5.9	.7709	.7716	.7723	.7731	.7738	.7745	.7752	.7760	.7767	.7774
6.0	.7782	.7789	.7796	.7803	.7810	.7818	.7825	.7832	.7839	.7846
6.1	.7853	.7860	.7868	.7875	.7882	.7889	.7896	.7903	.7910	.7917
6.2	.7924	.7931	.7938	.7945	.7952	.7959	.7966	.7973	.7980	.7987
6.3	.7993	.8000	.8007	.8014	.8021	.8028	.8035	.8041	.8048	.8055
6.4	.8062	.8069	.8075	.8082	.8089	.8096	.8102	.8109	.8116	.8122
6.5	.8129	.8136	.8142	.8149	.8156	.8162	.8169	.8176	.8182	.8189
6.6	.8195	.8202	.8209	.8215	.8222	.8228	.8235	.8241	.8248	.8254
6.7	.8261	.8267	.8274	.8280	.8287	.8293	.8299	.8306	.8312	.8319
6.8	.8325	.8331	.8338	.8344	.8351	.8357	.8363	.8370	.8376	.8382
6.9	.8388	.8395	.8401	.8407	.8414	.8420	.8426	.8432	.8439	.8445
7.0	.8451	.8457	.8463	.8470	.8476	.8482	.8488	.8494	.8500	.8506
7.1	.8513	.8519	.8525	.8531	.8537	.8543	.8549	.8555	.8561	.8567
7.2	.8573	.8579	.8585	.8591	.8597	.8603	.8609	.8615	.8621	.8627
7.3	.8633	.8639	.8645	.8651	.8657	.8663	.8669	.8675	.8681	.8686
7.4	.8692	.8698	.8704	.8710	.8716	.8722	.8727	.8733	.8739	.8745
7.5	.8751	.8756	.8762	.8768	.8774	.8779	.8785	.8791	.8797	.8802
7.6	.8808	.8814	.8820	.8825	.8831	.8837	.8842	.8848	.8854	.8859
7.7	.8865	.8871	.8876	.8882	.8887	.8893	.8899	.8904	.8910	.8915
7.8	.8921	.8927	.8932	.8938	.8943	.8949	.8954	.8960	.8965	.8971
7.9	.8976	.8982	.8987	.8993	.8998	.9004	.9009	.9015	.9020	.9025
8.0	.9031	.9036	.9042	.9047	.9053	.9058	.9063	.9069	.9074	.9079
8.1	.9085	.9090	.9096	.9101	.9106	.9112	.9117	.9122	.9128	.9133
8.2	.9138	.9143	.9149	.9154	.9159	.9165	.9170	.9175	.9180	.9186
8.3	.9191	.9196	.9201	.9206	.9212	.9217	.9222	.9227	.9232	.9238
8.4	.9243	.9248	.9253	.9258	.9263	.9269	.9274	.9279	.9284	.9289
8.5	.9294	.9299	.9304	.9309	.9315	.9320	.9325	.9330	.9335	.9340
8.6	.9345	.9350	.9355	.9360	.9365	.9370	.9375	.9380	.9385	.9390
8.7	.9395	.9400	.9405	.9410	.9415	.9420	.9425	.9430	.9435	.9440
8.8	.9445	.9450	.9455	.9460	.9465	.9469	.9474	.9479	.9484	.9489
8.9	.9494	.9499	.9504	.9509	.9513	.9518	.9523	.9528	.9533	.9538
9.0	.9542	.9547	.9552	.9557	.9562	.9566	.9571	.9576	.9581	.9586
9.1	.9590	.9595	.9600	.9605	.9609	.9614	.9619	.9624	.9628	.9633
9.2	.9638	.9643	.9647	.9652	.9657	.9661	.9666	.9671	.9675	.9680
9.3	.9685	.9689	.9694	.9699	.9703	.9708	.9713	.9717	.9722	.9727
9.4	.9731	.9736	.9741	.9745	.9750	.9754	.9759	.9763	.9768	.9773
9.5	.9777	.9782	.9786	.9791	.9795	.9800	.9805	.9809	.9814	.9818
9.6	.9823	.9827	.9832	.9836	.9841	.9845	.9850	.9854	.9859	.9863
9.7	.9868	.9872	.9877	.9881	.9886	.9890	.9894	.9899	.9903	.9908
9.8	.9912	.9917	.9921	.9926	.9930	.9934	.9939	.9943	.9948	.9952
9.9	.9956	.9961	.9965	.9969	.9974	.9978	.9983	.9987	.9991	.9996
N	0	1	2	3	4	5	6	7	8	9

Answers

Chapter 1

Exercises 1.1, pages 8–9

1. $[-7, -2)$

3. $(-4, 0]$

5. $(1, 3]$

7. $(-\infty, -1) \cup (1, \infty)$

9. $(-\infty, \infty)$

11. $(-2, 0] \cup (1, \infty)$

13. $(-\infty, -3) \cup (-1, 2]$

15. $(-\infty, -4) \cup [-1, 2]$

17. $y - 4$ **19.** $4 - y$ **21.** $10 - 2x$ **23.** $b - a$
25. $b - 2a$ **27.** $-x/5$ **29.** a^2 **31.** $-5 - 2a$
33. $3a - 14$ **35.** $\{-3, 9\}$ **37.** $\{2, -9\}$ **39.** $\{7/2\}$

41. $\varnothing$ **43.** $\{-3/2, -5\}$ **45.** $\{1, 7\}$ **47.** $\{9/4\}$
49. $\{9/4\}$ **51.** $\{-1\}$ **53.** $\{1/2\}$

55. $(-2, 2)$

57. $(-\infty, -3) \cup (3, \infty)$

59. $[-4, 4]$

61. $(1, 3)$

63. $[-2, 10]$

65. $(-3, 1)$

67. $(-\infty, -1] \cup [7, \infty)$

69. $(-\infty, -8] \cup [-4, \infty)$

71. $(1, 4)$

73. $[5, 9]$

75. $(-15/4, 5/2)$

77. $(-\infty, 1/2] \cup [7/2, \infty)$

79. $(-\infty, 1/4) \cup (5/4, \infty)$

81. $(-\infty, -2) \cup (4/5, \infty)$

83. $\varnothing$

85. $\{2\}$

87. $(-\infty, 1) \cup (1, \infty)$

89. $(-\infty, \infty)$

Exercises 1.2, pages 18–20

1. $64p^3q^6$ **3.** $4x^2/49$ **5.** $2/y^3$ **7.** $y^3/(2x^2)$
9. $1/(64x^7)$ **11.** $24/(p^4q^4)$ **13.** $-z^3/(64x^6y^6)$
15. b^6 **17.** $9x^{2m}/y^{2n}$ **19.** x^{n-1} **21.** 1.02×10^3
23. 4.161×10^7 **25.** 3.5×10^{-3} **27.** 8220
29. 0.66 **31.** 0.0000287 **33.** $4 - r^2 + 6r^4 - 2r^5$
35. $9 + 8z + 5z^2 + z^3$ **37.** $40q^2 + 19q - 3$
39. $10a^2 + 13ab - 3b^2$
41. $2x^4 - 5x^3 - 12x^2 - x + 4$
43. $2x^4 - x^3 + x^2 + 14x - 4$ **45.** $4r^2 - s^2$
47. $a^2 + 6abc^2 + 9b^2c^4$ **49.** $4x^6 - y^2$ **51.** $a^3 - 8b^3$
53. $x^3 - 6x^2y + 12xy^2 - 8y^3$ **55.** $a(x + 3y^2)(x - 3y^2)$
57. $(4x - 3y + 5)(4x - 3y - 5)$ **59.** $(2a^2 + 3b^2)^2$
61. $(3x - 2a)(x + 3a)$ **63.** $(2x + 3)(4x^2 - 6x + 9)$
65. $(a + b)(x^2 + d^2)$ **67.** $(5x + 1 + y)(5x + 1 - y)$
69. $(7 - z)/3$ **71.** $1/(w + 4)$
73. $(4w + 5)/(4w - 5)$
75. $(3x - 2y)(x^2 - xy + y^2)/2$
77. $[x(5 + 2x)]/[(x - 1)(3 + x)]$
79. $[a(2w + 5)]/[(w + 3)(w^2 + 4w + 16)]$
81. $1/[(y + 3)(2y + 5)]$ **83.** $1/(1 - u)$
85. $4d(5c - 4d)$ **87.** $(3y + 4)/[(y + 4)(y + 12)]$
89. $x(x + 1)/(x + 2)$ **91.** $1/(x - y)$
93. $(b - a)/(b + a)$ **95.** x **97.** $x/(x + 2)$
99. $xy/(y + x)$

Exercises 1.3, pages 26–27

1. $4xy^2$ **3.** $\sqrt[12]{y}$ **5.** xy^2 **7.** $2cd^2$ **9.** x^3y^3
11. $-p^2qr$ **13.** $5|a|$ **15.** $-|a - b|$ **17.** $|x - 1|$
19. $b^4|a^5|$ **21.** $5xy^2\sqrt{10y}$ **23.** $(r - s)\sqrt[3]{(r - s)^2}$
25. $\sqrt[4]{2}/a^2$ **27.** $a^4\sqrt{a/b}$ **29.** $4(x + y)^2\sqrt{3(x + y)}$
31. $x\sqrt{2}/2$ **33.** $\sqrt{6ab}/(2b^2)$ **35.** $\sqrt[3]{25a}/(5a^2)$
37. $b\sqrt[5]{3a^2bc^2}/(cd)$ **39.** $(\sqrt{x} + \sqrt{y})/(x - y)$
41. $-(\sqrt{2} + \sqrt{3})$ **43.** $\sqrt{2x(x^2 - 4)}/(2x)$
45. $-(1 - \sqrt{1 + x})/x$ **47.** $3\sqrt{3} - 4$ **49.** $1 - \sqrt{6}$
51. $x - 2\sqrt{x} + 1$ **53.** $x + \sqrt{xy} - 2y$ **55.** $10\sqrt{3}$
57. $-7b\sqrt{2a}$ **59.** $(9ab + 10b^2)\sqrt[3]{3ab^2}$ **61.** x^3
63. x^4y^{10} **65.** $a^{2/3}/b^2$ **67.** 1 **69.** $y^3/(4096x^2)$
71. $x - 1$ **73.** $2x^{1/6}$ **75.** $27x^{1/6}y^{1/3}/2$ **77.** x^2/y
79. $x^{1/4}y^{1/4}z^{7/12}$ **81.** $\sqrt[3]{x^2}$ **83.** $y\sqrt[3]{5x^2y}$ **85.** $\sqrt[3]{x}$
87. $\sqrt[3]{x^2}$ **89.** $x\sqrt[6]{x}$ **91.** $\sqrt[6]{b}$ **93.** 1.143×10^2
95. 9.487×10^{-1} **97.** 1.677×10^{-1}
99. 2.330×10^2

Exercises 1.4, pages 31–32

1. $4i$ **3.** $-7i$ **5.** -15 **7.** 3 **9.** $9 - i$
11. $15 + 67i$ **13.** $-4 - 2i$ **15.** $-5 + 3i$
17. $-1 - 17i$ **19.** $9 + 7i$ **21.** $-54 - 10i$
23. $27 - 36i$ **25.** $40i$ **27.** $14 + 5i$ **29.** $2 - \dfrac{7}{3}i$
31. $-2i$ **33.** $-\dfrac{4}{3}i$ **35.** $\dfrac{1}{3} + 2i$ **37.** $\dfrac{3}{25} - \dfrac{4}{25}i$
39. $\dfrac{5}{169} + \dfrac{12}{169}i$ **41.** $\dfrac{9}{5} + \dfrac{3}{5}i$ **43.** $\dfrac{27}{37} - \dfrac{23}{37}i$
45. $\dfrac{18}{5} + \dfrac{1}{5}i$ **47.** $-i$ **49.** $-\dfrac{13}{10} + \dfrac{1}{10}i$ **51.** 1
53. i **55.** $-i$ **57.** i **59.** -1 **61.** $-i$

Exercises 1.5, pages 38–40

1. $\pm5/2$ **3.** $-3, 0$ **5.** $-5, 2$ **7.** $1/3, 3$
9. $-2/7, 0, 1/7$ **11.** $-1, 1, 5/2$ **13.** $-7, 4$
15. $1/2, 3/2$ **17.** $-1, 0$ **19.** $-6, 2$ **21.** $\pm5/4$
23. $(-5 \pm \sqrt{5})/2$ **25.** $1 \pm i\sqrt{3}$ **27.** $-1.13,$
-2.79 **29.** $-0.58, -3.10$ **31.** $1/2, 7$
33. $-1 \pm \sqrt{2}$ **35.** $-2/9, 1/3$ **37.** $(-1 \pm \sqrt{11})/2$
39. $(2 \pm i)/2$ **41.** $16;$ 2 distinct real roots
43. $-31;$ 2 distinct complex numbers that are
conjugates   **45.** $0;$ 2 equal real roots
47. Approximately $48;$ 2 distinct real roots
49. Approximately $-1085;$ 2 distinct complex
numbers that are conjugates   **51.** $\{-3/4, -3\}$
53. $\{3, (-3 \pm 3i\sqrt{3})/2\}$
55. $\{0, 3/2, (-3 \pm 3i\sqrt{3})/4\}$

57. $\{-2, 1/3, (-1 \pm i\sqrt{3})/6, 1 \pm i\sqrt{3}\}$
59. $\{-1, 1/8\}$ **61.** $\{6\}$ **63.** $\{-10, -9\}$ **65.** $\varnothing$
67. $\{0, 3\}$ **69.** $21, 22$ or $-22, -21$ **71.** 30 meters by 60 meters   **73.** 14   **75.** 20 meters   **77.** 55 miles per hour   **79.** 160 miles per hour

Exercises 1.6, page 45

1. $(-\infty, -3) \cup (4, \infty)$ **3.** $[-1, 3]$
5. $(-\infty, 1] \cup [3, \infty)$ **7.** $(-\infty, -\sqrt{3}) \cup (\sqrt{3}, \infty)$
9. $(-\infty, 0) \cup (9, \infty)$ **11.** $[0, 4]$
13. $(-\infty, -\sqrt{2}) \cup (\sqrt{2}, \infty)$ **15.** $(-\infty, 0] \cup [2, \infty)$
17. $(-4, -2)$ **19.** $[-5, 3]$ **21.** $(-\infty, -5) \cup (2/3, \infty)$
23. $(2/3, 6/5)$ **25.** $(-\infty, 0) \cup (4/3, \infty)$
27. $(-7/3, 5/2)$ **29.** $(-7/4, 3)$
31. $[(-1 - \sqrt{2})/2, (-1 + \sqrt{2})/2]$
33. $(1, 2) \cup (3, \infty)$ **35.** $(1/2, 4/5)$ **37.** $(-2, 3/2)$
39. $[-5, -3)$ **41.** $(2, 3) \cup (7, \infty)$
43. $(-2, -1/2) \cup [0, \infty)$ **45.** $(-3/2, -1] \cup (0, \infty)$
47. w by $2w$, with $w \geq 30$ feet   **49.** $2 \leq t \leq 6$

Exercises 1.7, pages 51–52

1. $|x| - x = \begin{cases} 0, & \text{if } x \geq 0 \\ -2x, & \text{if } x < 0 \end{cases}$

3. $\dfrac{x}{|x|} = \begin{cases} 1, & \text{if } x > 0 \\ -1, & \text{if } x < 0 \\ \text{undefined}, & \text{if } x = 0 \end{cases}$

5. $\dfrac{|2x|}{x} = \begin{cases} 2, & \text{if } x > 0 \\ -2, & \text{if } x < 0 \\ \text{undefined}, & \text{if } x = 0 \end{cases}$

7. $\dfrac{|x - 2|}{2 - x} = \begin{cases} -1, & \text{if } x > 2 \\ 1, & \text{if } x < 2 \\ \text{undefined}, & \text{if } x = 2 \end{cases}$

9. $\dfrac{|4 - x|}{4 - x} = \begin{cases} -1, & \text{if } x > 4 \\ 1, & \text{if } x < 4 \\ \text{undefined}, & \text{if } x = 4 \end{cases}$

11. $\dfrac{|1 + x| - |1|}{x} = \begin{cases} 1, & \text{if } x \geq -1 \text{ and } x \neq 0 \\ -\dfrac{x + 2}{x}, & \text{if } x < -1 \\ \text{undefined}, & \text{if } x = 0 \end{cases}$

13. 9δ **15.** 2δ **17.** 4δ **19.** 7δ
21. $(x - \sqrt{3})(x + \sqrt{3})$ **23.** $(\sqrt{x} - \sqrt{2})(\sqrt{x} + \sqrt{2})$
25. $(x - \sqrt[3]{2})(x^2 + x\sqrt[3]{2} + \sqrt[3]{4})$
27. $(\sqrt[3]{x} + 1)(\sqrt[3]{x^2} - \sqrt[3]{x} + 1)$

29. $(\sqrt[3]{x} + \sqrt[3]{2})(\sqrt[3]{x^2} - \sqrt[3]{2x} + \sqrt[3]{4})$
31. $(\sqrt[3]{x} - \sqrt[3]{h})(\sqrt[3]{x^2} + \sqrt[3]{xh} + \sqrt[3]{h^2})$
33. $-(2x + 5)/(2x - 1)^3$
35. $-(7x + 2)/[x^3(x + 1)^6]$
37. $x^2(7x - 3)/(2x - 1)^{1/2}$
39. $[3x(3x + 8)]/[2(x + 2)^{3/2}]$
41. $4x^{1/2} - \dfrac{1}{x^{1/2}}$

43. $\dfrac{4x}{x^2 + x + 1} - \dfrac{3}{x^2 + x + 1}$ **45.** $\dfrac{x}{x^2 + 1} - \dfrac{3}{x^2 + 1}$

47. $\dfrac{x^2}{x^3 + 3x - 1} + \dfrac{x + 2}{x^3 + 3x - 1}$ **49.** $1/\sqrt{x}$

51. $1/(\sqrt{x - 2})$ **53.** $1/(\sqrt{x} + \sqrt{a})$
55. $1/(\sqrt{x + h} + \sqrt{x})$ **57.** $[x(3x - 8)]/[2(x - 2)^{3/2}]$
59. $5/[2(x + 3)^{3/2}(x - 2)^{1/2}]$ **61.** $(7x^2 - 1)/(4x^{5/4})$
63. $u^2 + a^2$, where $u = x + 3$ and $a = 1$
65. $3(u^2 - a^2)$, where $u = x + 2$ and $a = 3$
67. $2(u^2 - a^2)$, where $u = x - \dfrac{1}{4}$ and $a = \dfrac{\sqrt{33}}{4}$
69. $1/(u^2 + a^2)$, where $u = x - 2$ and $a = \sqrt{3}$
71. $1/[2(u^2 + a^2)]$, where $u = x + \dfrac{1}{2}$ and $a = \dfrac{1}{2}$

73. $3/[2(u^2 + a^2)]$, where $u = x + 1$ and $a = \sqrt{\dfrac{3}{2}}$

75. (a) $0, 2$; (b) -4 **77.** (a) $3, -3$; (b) -2
79. The expression is defined for all values of x, and it is never zero.   **81.** (a) 1; (b) 0 **83.** (a) $5/8$; (b) 1
85. (a) $\pm\sqrt{3}$; (b) $\pm\sqrt{5}$

Review Problems for Chapter 1, pages 54–55

1. $[-3, 2)$

2. $(-1, 4]$

3. $(-\infty, -2) \cup (0, 3]$

4. $(-\infty, 1] \cup [3, 5)$

5. $6 - 3x$ **6.** $8 - 2x$ **7.** $5x - 15$ **8.** $2x - 6$
9. $\{-6, -\frac{1}{2}\}$ **10.** $\{-10, 4/3\}$ **11.** $\{8, -14/3\}$
12. $\{4, -11/2\}$
13. $(-1, 3)$

14. $(1, 3)$

15. $(-\infty, -5] \cup [1, \infty)$ **16.** $(-\infty, -5] \cup [3, \infty)$

17. $(-\infty, 1) \cup (4, \infty)$ **18.** $(-\infty, -5/3) \cup (3, \infty)$

19. $\varnothing$ **20.** $\varnothing$ **21.** v^6/u^6 **22.** $1/(p^2q^2)$
23. $z^9/(x^6y^3)$ **24.** $(4w^4z^4)/(x^2y^2)$
25. $1 + 2x + x^2 - x^3 - x^4$ **26.** $1 + x + x^2 - x^4$
27. $y^4 - 4y^2 + 4$ **28.** $x^4 + 6x^2 + 9$ **29.** $p^3 - q^3$
30. $y^3 + 27$ **31.** $-x^3(x + 1)(x^2 - x + 1)$
32. $-8y(x + 2)(x^2 - 2x + 4)$
33. $(3x + y - z)(3x - y + z)$
34. $(4a + 3b - 1)(4a - 3b + 1)$
35. $[y(3x + 5)]/[(x + 2)(x^2 + 3x + 9)]$
36. $1/(3x + 2)$ **37.** $1/[(z - 3)(z + 2)]$
38. $3/[(b - 2)(b + 1)]$ **39.** $1/[(b - a)(b + a)]$
40. $(y + x)/(y - x)$ **41.** a/b^2 **42.** $3x^3/y^2$
43. $5x^2y^4$ **44.** $5a^3b$ **45.** $a^4b^{10}\sqrt{a}$ **46.** $r^3z\sqrt[3]{z}$
47. $(p\sqrt[4]{p^3q^3})/q^2$ **48.** $x\sqrt[4]{50xy^3}/(5y)$
49. $3(2x + \sqrt{x})/(4x - 1)$ **50.** $(2a - 3\sqrt{a})/(4a - 9)$
51. $6(x - 10y)\sqrt[3]{5xy}$ **52.** $(2s - r^2)\sqrt[5]{2rs^3}$ **53.** a^8/b^3
54. y^2/x **55.** $a - b$ **56.** $(b^3 - a^3)/(a^3b^3)$
57. $a^{1/4}b^{1/3}c^{1/5}$ **58.** $x^{2/3}yz^{1/2}$ **59.** $r^3\sqrt[6]{rs^5}$
60. $y\sqrt[15]{x^{11}y^8}$ **61.** $3i$ **62.** $5i$ **63.** $-2i\sqrt{2}$
64. $-4i\sqrt{2}$ **65.** $-5 - 4i$ **66.** $19 - 5i$
67. $-9 + 7i$ **68.** $-9 - 5i$ **69.** $-3 + 4i$
70. $-7 - 24i$ **71.** -16 **72.** -16
73. $-\dfrac{14}{17} - \dfrac{5}{17}i$ **74.** $-\dfrac{11}{29} - \dfrac{13}{29}i$ **75.** $-\dfrac{1}{2} + \dfrac{1}{2}i$
76. $-\dfrac{5}{3} - \dfrac{4}{3}i$ **77.** $-i$ **78.** i **79.** i **80.** -1
81. -3 **82.** $5/4$ **83.** $(-7 \pm \sqrt{17})/8$ **84.** $-4, 1/2$
85. $(-3 \pm \sqrt{5})/2$ **86.** $(1 \pm \sqrt{6})/5$ **87.** $-3/4, 2$
88. $-4/3, 3/2$ **89.** $(-5 \pm \sqrt{37})/6$
90. $(-3 \pm \sqrt{17})/4$ **91.** $-1, 3, (-3 \pm 3i\sqrt{3})/2$
92. $1, -1, (1 \pm i\sqrt{3})/2$ **93.** $\{9, 25\}$ **94.** $\{19, 84\}$
95. $\{1\}$ **96.** $\{7\}$ **97.** $\{14\}$ **98.** $\varnothing$ **99.** $\{-1\}$
100. $\{5\}$ **101.** $(-\infty, -4) \cup (1, \infty)$ **102.** $(-4, -2)$
103. $(-2, 0]$ **104.** $(-\infty, -1) \cup (3, \infty)$

105. $\dfrac{|2x| - 2x}{x} = \begin{cases} 0, & \text{if } x > 0 \\ -4, & \text{if } x < 0 \\ \text{undefined}, & \text{if } x = 0 \end{cases}$

106. $\dfrac{2x + |x|}{x} = \begin{cases} 3, & \text{if } x > 0 \\ 1, & \text{if } x < 0 \\ \text{undefined}, & \text{if } x = 0 \end{cases}$

107. $\dfrac{|x - 2| - |2|}{x} = \begin{cases} \dfrac{x - 4}{x}, & \text{if } x \geqslant 2 \\ -1, & \text{if } x < 2 \text{ and } x \neq 0 \\ \text{undefined}, & \text{if } x = 0 \end{cases}$

108. $\dfrac{|x - 3| - |3|}{x} = \begin{cases} \dfrac{x - 6}{x}, & \text{if } x \geqslant 3 \\ -1, & \text{if } x < 3 \text{ and } x \neq 0 \\ \text{undefined}, & \text{if } x = 0 \end{cases}$

109. $(\sqrt{x} - 2)(\sqrt{x} + 2)$ **110.** $(\sqrt{x} - 3)(\sqrt{x} + 3)$
111. $(x - \sqrt{a})(x + \sqrt{a})$ **112.** $(x - \sqrt{h})(x + \sqrt{h})$
113. $-(3x + 4)/(3x + 2)^3$
114. $(13 - 4x)/(2x + 1)^4$
115. $(x^2 + 1)(27x^2 - 4x + 3)/(6x - 1)^{1/2}$
116. $x^2(x - 1)^2(44x^2 - 2x - 9)/(8x + 3)^{3/2}$
117. $3/(\sqrt{3x + 3h} + \sqrt{3x})$
118. $2/(\sqrt{2x + 2h + 1} + \sqrt{2x + 1})$
119. $x(5x + 2)/(2x + 1)^2$
120. $2x(5x - 3)/(4x - 3)^2$

121. $2(u^2 - a^2)$, where $u = x + \dfrac{5}{4}, a = \dfrac{\sqrt{17}}{4}$

122. $4(u^2 - a^2)$, where $u = x - \dfrac{5}{4}, a = \dfrac{\sqrt{13}}{4}$

123. $5/[2(u^2 + a^2)]$, where $u = x - \dfrac{1}{2}, a = \dfrac{1}{2}$

124. $3/[2(u^2 - a^2)]$, where $u = x + 1, a = \sqrt{6}/2$
125. (a) $0, -1/7$; (b) $-1/6$ **126.** (a) $0, 4/7$; (b) $2/3$
127. (a) $0, -1$; (b) $-4/3$ **128.** (a) $0, 1/12$; (b) $1/9$

Critical Thinking Problems for Chapter 1, pages 56–58

1. Error

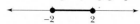

Correction

The inequality $|x| \leq 2$ is equivalent to $-2 \leq x \leq 2$. See accompanying graph.

2. Error

$4x - 1 \neq \pm 3$

Correction

$|4x - 1| = -3$ is impossible because an absolute value cannot be negative. The solution set is empty.

3. Errors

$$(x^{-2}y)^{-3} = \frac{1}{(\cancel{x^2}y)^3} = \frac{1}{\cancel{x^3}y^3}$$

Correction

$$(x^{-2}y)^{-3} = \frac{1}{(x^{-2}y)^3} = \frac{1}{x^{-6}y^3} = \frac{x^6}{y^3}$$

4. Error

$$r^2 - s^2 - r - s = (r^2 - s^2) \cancel{\times} (r - s)$$

Correction

$$r^2 - s^2 - r - s = (r^2 - s^2) - (r + s)$$
$$= (r - s)(r + s) - (r + s)$$
$$= (r + s)(r - s - 1)$$

5. Error

$$\frac{p - 3}{p + 1} - \frac{2p - 1}{p + 2} = \frac{\cancel{p - 3 - 2p + 1}}{(p + 1)(p + 2)}$$

Correction

$$\frac{p - 3}{p + 1} - \frac{2p - 1}{p + 2} = \frac{(p - 3)(p + 2)}{(p + 1)(p + 2)}$$
$$- \frac{(2p - 1)(p + 1)}{(p + 2)(p + 1)}$$
$$= \frac{p^2 - p - 6 - (2p^2 + p - 1)}{(p + 1)(p + 2)}$$
$$= \frac{p^2 - p - 6 - 2p^2 - p + 1}{(p + 1)(p + 2)}$$
$$= \frac{-p^2 - 2p - 5}{(p + 1)(p + 2)}$$
$$= -\frac{p^2 + 2p + 5}{(p + 1)(p + 2)}$$

6. Error

$$(-27)^{2/3} = [(-27)^{1/3}]^2 = (\cancel{\times 9})^2$$

Correction

$$(-27)^{2/3} = [(-27)^{1/3}]^2 = (-3)^2 = 9$$

7. Errors

$$2\sqrt{3} - 4\sqrt{2} = (2 \cancel{\times} 4)(\cancel{\sqrt{3} - \sqrt{2}})$$
$$= (-2)(\cancel{\sqrt{1}})$$

Correction

$2\sqrt{3} - 4\sqrt{2}$ cannot be combined by using the distributive property.

8. Error

$$\sqrt[3]{8x^2y^3} = \cancel{\sqrt{2 \cdot 4 \cdot x^2 \cdot y \cdot y^2}}$$

Correction

$$\sqrt[3]{8x^2y^3} = \sqrt[3]{2^3y^3x^2} = 2y\sqrt[3]{x^2}$$

9. Error

$$\sqrt{-25}\,\sqrt{-4} \cancel{\times} \sqrt{100}$$

Correction

$$\sqrt{-25}\,\sqrt{-4} = (5i)(2i) = -10$$

10. Error

$$\frac{1}{2 + i} = \frac{1}{2 + i} \cdot \frac{2 - i}{2 - i}$$
$$= \frac{2 - i}{4 \cancel{\times} 1}$$

Correction

$$\frac{1}{2 + i} \cdot \frac{2 - i}{2 - i} = \frac{2 - i}{4 + 1} = \frac{2}{5} - \frac{1}{5}i$$

11. Errors

$$x(6x - 1) = 1$$
$$x \cancel{\times} 1 \quad \text{or} \quad \cancel{6x - 1} = 1$$

Correction

$$6x^2 - x - 1 = 0$$
$$(3x + 1)(2x - 1) = 0$$
$$3x + 1 = 0 \quad \text{or} \quad 2x - 1 = 0$$
$$3x = -1 \qquad\qquad 2x = 1$$
$$x = -1/3 \qquad\qquad x = 1/2$$

Solution set: $\{-1/3, 1/2\}$.

12. Error

$$2 - \sqrt{1 - x} = 2x$$
$$\cancel{4 + 1 - x} = 4x^2$$

Correction

$$2 - \sqrt{1 - x} = 2x$$
$$2 - 2x = \sqrt{1 - x}$$
$$4 - 8x + 4x^2 = 1 - x$$
$$4x^2 - 7x + 3 = 0$$
$$(4x - 3)(x - 1) = 0$$
$$4x - 3 = 0 \quad \text{or} \quad x - 1 = 0$$
$$4x = 3 \qquad\qquad x = 1$$
$$x = 3/4$$

With $x = 3/4$, LHS $= 3/2 = 3/2 =$ RHS.
With $x = 1$, LHS $= 2 =$ RHS.
Solution set: $\{3/4, 1\}$.

13. Errors

$(x + 4)(x + 2) < 3$

~~$x > 4 < 3$~~ or ~~$x > 2 < 3$~~

Correction

$x^2 + 6x + 8 < 3$

$x^2 + 6x + 5 < 0$

$(x + 5)(x + 1) < 0$

$x + 5$	$- \quad - \quad 0 + + + + + +$
$x + 1$	$- - - - - - 0 + +$
$(x + 5)(x + 1)$	$+ + 0 - - - 0 + +$

$$\xleftarrow{\qquad} \underset{-5}{\circ} \rule{1cm}{0.4pt} \underset{-1}{\circ} \xrightarrow{\qquad}$$

Solution set: $(-5, -1)$.

14. Errors

$x - 3 = (\sqrt{x} - 3)(\sqrt{x} + 3)$

Correction

$x - 3 = (\sqrt{x} - \sqrt{3})(\sqrt{x} + \sqrt{3})$

15. Error

$$2x^2 + 6x - 7 = 2x^2 + 6x + (\ ) - 7 - (\ )$$
$$= 2x^2 + 6x + 9 - 7 - 9$$
$$= \cancel{(2x + 3)^2} - 16$$

Correction

$$2x^2 + 6x - 7 = 2\left[x^2 + 3x + (\ ) - \frac{7}{2} - (\ ) \right]$$

$$= 2\left[x^2 + 3x + \frac{9}{4} - \frac{7}{2} - \frac{9}{4} \right]$$

$$= 2\left[\left(x + \frac{3}{2} \right)^2 - \frac{23}{4} \right]$$

$$= 2(u^2 - a^2), \text{ where } u = x + \frac{3}{2}$$

$$\text{and } a = \frac{\sqrt{23}}{2}$$

Chapter 2

Exercises 2.1, pages 69–72

1. $D = \mathcal{R}, R = \mathcal{R}$ **3.** $D = [4, \infty), R = (-\infty, 0]$
5. $D = \mathcal{R}, R = [-3, \infty)$ **7.** $D = (-\infty, 1) \cup (1, \infty),$
$R = (-\infty, 0) \cup (0, \infty)$ **9.** $D = \mathcal{R}, R = [0, \infty)$
11. $D = \mathcal{R}, R = [3, \infty)$ **13.** Function   **15.** Not a
function   **17.** Not a function   **19.** Function
21. Function   **23.** 10   **25.** 16   **27.** $4b - 2$
29. $-4x - 2$ **31.** $(x + 1)^2$ **33.** 6   **35.** 12   **37.** 0
39. $-2x$ **41.** $2(x + h)$ **43.** $h(2x + h + 1)$ **45.** 3
47. $4x + 2h$ **49.** $6x + 3h - 2$
51. $6x^2 + 6xh + 2h^2$

53.

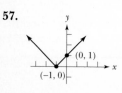

55.

(0, 1)

57.

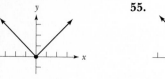

(0, 1)
(−1, 0)

59.

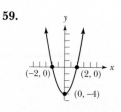

(−2, 0)   (2, 0)
(0, −4)

61.

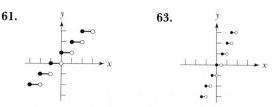

63.

65. $D = [-2, \infty), R = \mathcal{R}$, not a function
67. $D = [-3, 3], R = [-1, 1]$, function
69. $D = R = \mathcal{R}$, function
71. $D = (-\infty, -1] \cup [1, \infty), R = \mathcal{R}$, not a function

73.

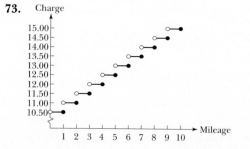

75. (a) $\ell = 60 - \frac{3}{2}w$; (b) $A = (60 - \frac{3}{2}w)(w)$
77. $V = 4x(9 - x)^2$ **79.** (a) $x = 16 - y$;
(b) $A = y^2/(4\pi)$; (c) $V = y^2(16 - y)/(4\pi)$

Exercises 2.2, pages 75–76

1. 8   **3.** 6   **5.** -9 **7.** 4   **9.** -3 **11.** 0   **13.** 5
15. 0   **17.** 5   **19.** 9   **21.** 3   **23.** 7
25. $(f + g)(x) = 3x - 1, x \in \mathcal{R};$
    $(f - g)(x) = -x - 3, x \in \mathcal{R};$
    $(f \cdot g)(x) = 2x^2 - 3x - 2, x \in \mathcal{R};$
    $(f/g)(x) = (x - 2)/(2x + 1), x \in \mathcal{R}$ and
    $x \neq -1/2$
27. $(f + g)(x) = x^2 + 3x - 1, x \in \mathcal{R};$
    $(f - g)(x) = 5x - 1 - x^2, x \in \mathcal{R};$
    $(f \cdot g)(x) = 4x^3 - 5x^2 + x, x \in \mathcal{R};$
    $(f/g)(x) = (4x - 1)/(x^2 - x), x \in \mathcal{R}$ and
    $x \neq 0, x \neq 1$
29. $(f + g)(x) = \sqrt{x} + 2x, x \in [0, \infty);$
    $(f - g)(x) = \sqrt{x} - 2x, x \in [0, \infty);$
    $(f \cdot g)(x) = 2x\sqrt{x}, x \in [0, \infty); (f/g)(x) = \sqrt{x}/(2x),$
    $x \in (0, \infty)$
31. $(f + g)(x) = 2x + 1 + \sqrt{x + 2}, x \in [-2, \infty);$
    $(f - g)(x) = 2x + 1 - \sqrt{x + 2}, x \in [-2, \infty);$
    $(f \cdot g)(x) = (2x + 1)\sqrt{x + 2}, x \in [-2, \infty);$
    $(f/g)(x) = (2x + 1)/\sqrt{x + 2}, x \in (-2, \infty)$
33. $(f + g)(x) = \sqrt{x - 2} + \sqrt{3 - x}, x \in [2, 3];$
    $(f - g)(x) = \sqrt{x - 2} - \sqrt{3 - x}, x \in [2, 3];$
    $(f \cdot g)(x) = \sqrt{5x - x^2 - 6}, x \in [2, 3];$
    $(f/g)(x) = \sqrt{x - 2}/\sqrt{3 - x}, x \in [2, 3)$
35. $(f + g)(x) = \sqrt{x + 1} + \sqrt{x + 6}, x \in [-1, \infty);$
    $(f - g)(x) = \sqrt{x + 1} - \sqrt{x + 6}, x \in [-1, \infty);$
    $(f \cdot g)(x) = \sqrt{x^2 + 7x + 6}, x \in [-1, \infty);$
    $(f/g)(x) = \sqrt{x + 1}/\sqrt{x + 6}, x \in [-1, \infty)$
37. $(f + g)(x) = x^2 - 4x + 2, x \in \mathcal{R};$
    $(f - g)(x) = 6x - 10 - x^2, x \in \mathcal{R};$
    $(f \cdot g)(x) = x^3 - 9x^2 + 26x - 24, x \in \mathcal{R};$
    $(f/g)(x) = (x - 4)/(x^2 - 5x + 6), x \in \mathcal{R}$ and
    $x \neq 2, x \neq 3.$
39. $(f + g)(x) = 7x^2 + 5x - 9, x \in \mathcal{R};$
    $(f - g)(x) = 5x^2 - x - 1, x \in \mathcal{R};$
    $(f \cdot g)(x) = 6x^4 + 20x^3 - 23x^2 - 23x + 20, x \in \mathcal{R};$
    $(f/g)(x) = (6x^2 + 2x - 5)/(x^2 + 3x - 4),$
    $x \in \mathcal{R}$ and $x \neq -4, x \neq 1$
41. $(f \circ g)(x) = x^2 + 1, x \in \mathcal{R}; (g \circ f)(x) = (x + 1)^2,$
    $x \in \mathcal{R}$
43. $(f \circ g)(x) = (x - 1)^{10}, x \in \mathcal{R}; (g \circ f)(x) = x^{10} - 1,$
    $x \in \mathcal{R}$
45. $(f \circ g)(x) = 2(x + 3)^2 + x + 3, x \in \mathcal{R};$
    $(g \circ f)(x) = 2x^2 + x + 3, x \in \mathcal{R}$
47. $(f \circ g)(x) = \sqrt{x - 3}, x \in [3, \infty);$
    $(g \circ f)(x) = \sqrt{x} - 3; x \in [0, \infty)$
49. $(f \circ g)(x) = 1/(1 - x), x \in \mathcal{R}$ and $x \neq 1;$
    $(g \circ f)(x) = 1 - (1/x), x \in \mathcal{R}$ and $x \neq 0$
51. $(f \circ g)(x) = \sqrt{x - 1}, x \in [1, \infty);$
    $(g \circ f)(x) = \sqrt{x + 2} - 3, x \in [-2, \infty)$
53. $(f \circ g)(x) = x, x \in \mathcal{R}; (g \circ f)(x) = x, x \in \mathcal{R}$

55. $(f \circ g)(x) = x, x \in \mathcal{R}$ and $x \neq 0; (g \circ f)(x) = x,$
    $x \in \mathcal{R}$ and $x \neq -1$
57. $(f \circ g)(x) = |x|, x \in \mathcal{R}; (g \circ f)(x) = x, x \in [1, \infty)$

The answers in Exercises 59–65 are not unique. One
correct solution is given.
59. $f(x) = x - 1, g(x) = x^3$
61. $f(x) = \sqrt{x}, g(x) = x + 3$
63. $f(x) = x^{50}, g(x) = 2x - 9$
65. $f(x) = x^3, g(x) = 1/(x + 3)$

Exercises 2.3, pages 84–86

1. x-intercept $\frac{3}{2}$
   y-intercept -3

3. x-intercept $\frac{5}{4}$
   no y-intercept

5. x-intercept $\frac{7}{2}$
   y-intercept $\frac{7}{5}$

7. x-intercept 3
   y-intercept 3

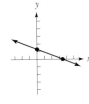

9. $d = 5, m = -4/3$ **11.** $d = 4, m$ undefined
13. $d = 6, m = 0$ **15.** $3/4$ **17.** Undefined
19. $7/2$ **21.** -2 **23.** 6
25. $x + 3y = 8$ **27.** $4x - y = 28$

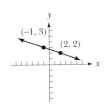

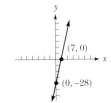

29. $4x - y = -3$ **31.** $y = -7$

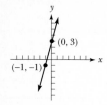   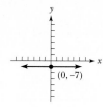

33. $6x + 5y = -45$ **5.** $5x - 7y = 7$

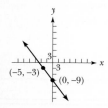   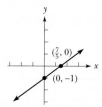

37. $x + y = 3$ **39.** $4x - y = 0$ **41.** $y = 4$
43.   **45.**

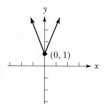   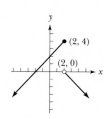

47. Yes   **49.** No   **51.** Yes   **53.** No
57. $f(x) = 20 + 18x$ **59.** $f(x) = 18{,}000 - 2500x$,
$0 \le x \le 6$ **61.** $9\frac{1}{3}$ inches   **63.** $53\sqrt{626} \approx 1326$ feet

Exercises 2.4, pages 91–92

1. V: $(0, -4)$; A: $x = 0$
Pts. $(-2, 0)$, $(2, 0)$

3. V: $(2, -1)$; A: $x = 2$
Pts. $(-1, 5)$, $(5, 5)$

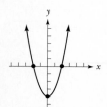   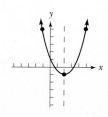

5. V: $(2, 4)$; A: $x = 2$
Pts. $(0, 0)$, $(4, 0)$

7. V: $(-1, 3)$; A: $x = -1$
Pts. $(-2, 4)$, $(0, 4)$

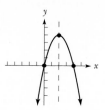   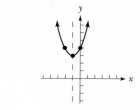

9. V: $(-4, -3)$; A: $x = -4$
Pts. $(-6, 1)$, $(-2, 1)$

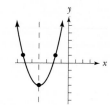

11. V: $(-1, -1)$; A: $x = -1$
Pts. $(-2, -2)$, $(0, -2)$

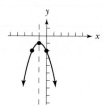

13. V: $(-2, 3)$; A: $x = -2$
Pts. $(0, -1)$, $(-4, -1)$

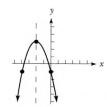

15. V: $(-2, 5)$; A: $x = -2$
Pts. $(-3, 8)$, $(-1, 8)$

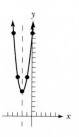

17. V: $(1, -3)$; A: $x = 1$
Pts. $(0, -5)$, $(2, -5)$

19. $f(4) = -11$ is the minimum.   **21.** $f(-5) = -45$ is the minimum.   **23.** $f(3) = 12$ is the maximum.
25. $f(1) = -1$ is the maximum.   **27.** $f(-3) = 0$ is the maximum.   **29.** $f(-3/2) = 29/4$ is the minimum.
31. The graph is shifted upward if $k > 0$ and downward if $k < 0$.   **33.** The parabolas open downward when $a < 0$.

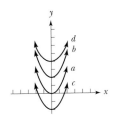

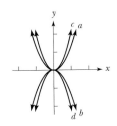

35. 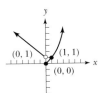   **37.**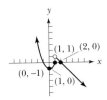

39.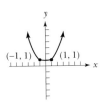

41. 27 and 27   **43.** 12 feet by 12 feet   **45.** The minimum cost is $C = \$737.50$ when $x = 250$.
47. The maximum height is 64 feet after 2 seconds.
49. 10 feet by 20 feet, with the long side parallel to the wall   **51.** $r = h = 20/(\pi + 4)$ feet

Exercises 2.5, pages 97–98

1. $g^{-1} = \{(3, 0), (2, 1), (2, -1)\}$; g is a function, g^{-1} is not a function.
3. $f^{-1} = \{(x, y): x = 3y - 2\}$; f and f^{-1} are both functions.
5. $p^{-1} = \{(x, y): x = |y|\}$; p is a function, p^{-1} is not a function.
7. $g^{-1} = \{(x, y): y = 2x^2\}$; g is not a function, g^{-1} is a function.
9. $f^{-1} = \{(x, y): x = y^2 + 2y\}$; f is a function, f^{-1} is not a function.

11. $g^{-1} = \{(x, y): x^2 + y^2 = 4\} = g$; g and g^{-1} are not functions.
13. $h^{-1} = \{(x, y): x = -\sqrt{4 - y^2}\}$; h is a function, h^{-1} is not a function.
15. (a) $y = x - 3$ **17.** (a) $4y - 3x = 12$
(b)   (b)

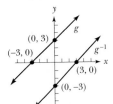   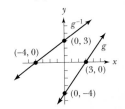

(c) g and g^{-1} are functions.   (c) g and g^{-1} are functions.

19. (a) $x = -1 - y^2$
(b)

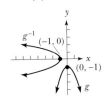

(c) g is a function, g^{-1} is not a function.
21. Not one-to-one   **23.** One-to-one
25. $f(x) = 2x - 4$, $f^{-1}(x) = \frac{1}{2}x + 2$;
$f^{-1}(f(x)) = \frac{1}{2}(2x - 4) + 2 = x - 2 + 2 = x$;
$f(f^{-1}(x)) = 2(\frac{1}{2}x + 2) - 4 = x + 4 - 4 = x$
27. $f(x) = x^2 + 2$, f is not one-to-one.
29. $f(x) = (x - 1)^2$, f is not one-to-one.
31. $f(x) = x^3 - 1$, $f^{-1}(x) = \sqrt[3]{x + 1}$;
$f^{-1}(f(x)) = \sqrt[3]{x^3 - 1 + 1} = \sqrt[3]{x^3} = x$;
$f(f^{-1}(x)) = (\sqrt[3]{x + 1})^3 - 1 = x + 1 - 1 = x$
33. f and g are inverse functions.   **35.** f and g are inverse functions.   **37.** f and g are inverse functions.
39. f and g are not inverse functions.   **41.** f and g are inverse functions.   **43.** f and g are not inverse functions.

Exercises 2.6, pages 105–107

1. a)   b) 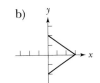   c)

3. a)   b)   c)

5. a)   b) 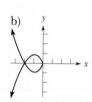   c)

7. a)   b)   c)

9. Symmetric with respect to the y-axis
11. Symmetric with respect to the x-axis, the y-axis, and the origin   **13.** Symmetric with respect to the x-axis   **15.** Symmetric with respect to the origin
17. Symmetric with respect to the x-axis, the y-axis, and the origin   **19.** Symmetric with respect to the origin   **21.** Symmetric with respect to the y-axis

23.   **25.**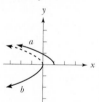

27.   **29.**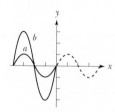

31. 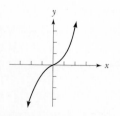   **33.**

35. 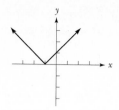   **37.**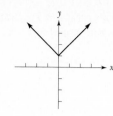

39. 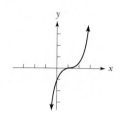   **41.**

43.

45. Circle with center $(0, 0)$ and radius 4   **47.** Circle with center $(3, -5)$ and radius 5   **49.** Circle with center $(-1, -3)$ and radius $\sqrt{10}$ **51.** Circle with center $(\frac{3}{2}, 0)$ and radius $\frac{3}{2}$ **53.** Not a circle
55. Circle with center $(2, -8)$ and radius 0 (point circle)   **57.** Circle with center $(-1, -2)$ and radius 3
59. $x^2 + (y - 1)^2 = 4$
61. $(x - 2)^2 + (y + 2)^2 = 18$
63. $(x + 3)^2 + (y + 4)^2 = 25$
65. $(x - 3)^2 + (y - 1)^2 = 18$
67. $(x + 1)^2 + (y + 1)^2 = 13$

69. 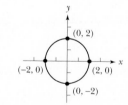   **71.**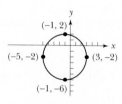

73. 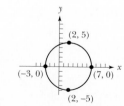   **75.**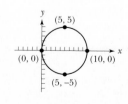

Exercises 2.7, pages 111–112

1.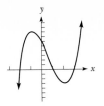

3.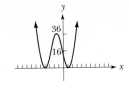

5.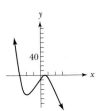

7.

9.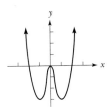

11.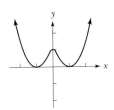

13.

15.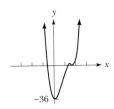

17.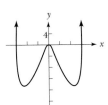

19.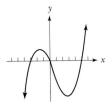

21.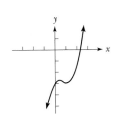

23.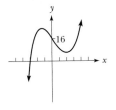

25.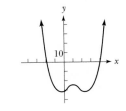

27.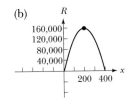

29.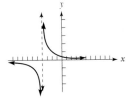

31. (a) $[0, 400]$ (b)

33. $[0, 4]$

Exercises 2.8, pages 119–120

1. ∞ **3.** $-\infty$ **5.** 0 **7.** 0 **9.** $-\infty$ **11.** ∞ **13.** 0
15. 0
17. $V: x = -4$ **19.** $V: x = 3$
    $H: y = 0$ $H: y = 2$

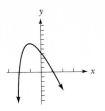
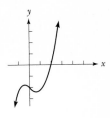

21. $V: x = -\frac{5}{3}$ **23.** $V: x = 0, x = 3$
    $H: y = -\frac{4}{3}$ $H: y = 0$

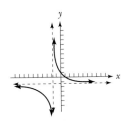

25. $V: x = -1$
  $H: y = 0$

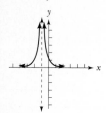

27. $H: y = 0$

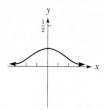

45. $V: x = 2$
  $O: y = x + 2$

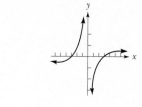

47. $V: x = 0$
  $H: y = 0$

29. $V: x = -2$
  $H: y = 0$

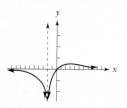

31. $V: x = 0, x = -1$
  $H: y = 0$

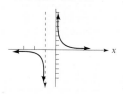

49. $V: x = 1, x = -1$
  $H: y = 1$

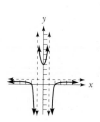

51.

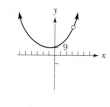

33. $V: x = -2, x = 3$
  $H: y = 0$

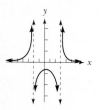

35. $V: x = 0, x = 1$
  $H: y = 1$

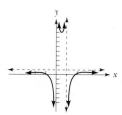

53.

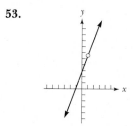

55.

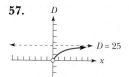

37. $V: x = -1, x = \frac{3}{2}$
  $H: y = \frac{1}{2}$

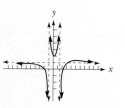

39. $V: x = -1$
  $O: y = 2x - 2$

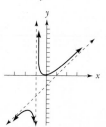

57.

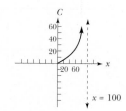

59.

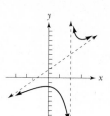

41. $V: x = 1, x = 2$
  $H: y = 0$

43. $V: x = 0, x = -2$
  $H: y = 0$

Review Problems for Chapter 2, pages 122–123

1. $D = [9, \infty), R = [0, \infty)$ **2.** (a) 8; (b) 24;
(c) $2\sqrt{2} + 3$ **3.** (a) $a + 4\sqrt{a} + 3$;
(b) $h(2x + h - 2)$

4.

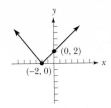

5.

6. $(f + g)(x) = \sqrt{x - 3} + x - 4, x \in [3, \infty)$
7. $(f - g)(x) = \sqrt{x - 3} - x + 4, x \in [3, \infty)$
8. $(f \cdot g)(x) = (x - 4)\sqrt{x - 3}, x \in [3, \infty)$
9. $(f/g)(x) = \sqrt{x - 3}/(x - 4), x \in [3, \infty)$ and $x \neq 4$
10. $(f \circ g)(x) = \sqrt{x - 7}, x \in [7, \infty)$

11. (a) x-intercept 4   (b) No x-intercept
   y-intercept -6 y-intercept 4

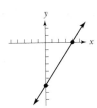

12. $m = 12/5, d = 13$ **13.** $8x + y = 23$
14. $3x - 4y = -18$ **15.** $2x + y = 2$ **16.** 9

17.

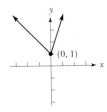

18. $V: (1, 2); A: x = 1$
   Pts. $(0, 5), (2, 5)$

19. $V: (2, 1); A: x = 2$
   Pts. $(1, 0), (3, 0)$

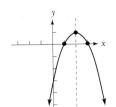

20. Minimum value of 5   **21.** Maximum value of 10
22. The maximum profit is $160 when $x = 400$.

23.

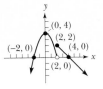

24. (a) $3y + 2x = 12$
   (b)

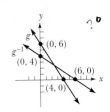

   (c) g and g^{-1} are
      functions.

25. (a) $y = x^2 + 1$
   (b)

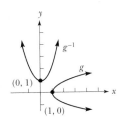

   (c) g is not a
      function, g^{-1} is a
      function.

26. f and g are not inverse functions   **27.** f and g
are inverse functions.   **28.** $f(x) = \frac{2}{3}x - 2,$
$f^{-1}(x) = \frac{3}{2}x + 3;$
$f^{-1}(f(x)) = \frac{3}{2}(\frac{2}{3}x - 2) + 3 = x - 3 + 3 = x;$
$f(f^{-1}(x)) = \frac{2}{3}(\frac{3}{2}x + 3) - 2 = x + 2 - 2 = x$
29. $f(x) = x^2 - 4, f$ is not one-to-one.

30–33.

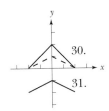

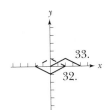

34. Symmetric with respect to the y-axis

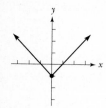

35. Symmetric with respect to the x-axis, the y-axis, and the origin

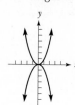

36. Not a circle   **37.** $(x - 2)^2 + (y + 3)^2 = 16$
38. $(x - 2)^2 + y^2 = 25$

39.

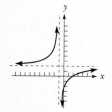

40.

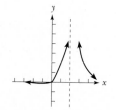

41. ∞ **42.** $-\infty$ **43.** 0
44. $V: x = -1; H: y = 2$ **45.** $V: x = 2; H: y = 0$

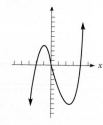

Critical Thinking Problems for Chapter 2, page 124

1. Error
$$f(g(7)) = \cancel{(7^2 - 1)}(\sqrt{7} - 3)$$

Correction
$$g(7) = \sqrt{7 - 3} = \sqrt{4} = 2$$
$$f(g(7)) = f(2) = 2^2 - 1 = 4 - 1 = 3$$

2. Errors
$$m = \frac{-2}{\cancel{3}} = -\frac{2}{3}$$
$$y - (-2) = -\frac{\cancel{2}}{\cancel{3}}(x - \cancel{3})$$

Correction
$$(x_1, y_1) = (3, 0) \quad \text{and} \quad (x_2, y_2) = (0, -2)$$
$$m = \frac{-2 - 0}{0 - 3} = \frac{2}{3}$$
$$y = \frac{2}{3}x + (-2)$$
$$3y = 2x - 6$$
$$2x - 3y = 6$$

3. Error
$$d = \sqrt{(4 - 1)^2 + (6 - 2)^2}$$
$$= \sqrt{3^2 + 4^2}$$
$$= \cancel{3 + 4}$$

Correction
$$\sqrt{3^2 + 4^2} = \sqrt{9 + 16} = \sqrt{25} = 5$$

4. Error
$$f = \{(x, y): y = \sqrt{x + 1}\},$$
$$f^{-1} = \left\{(x, y): y = \cancel{\frac{1}{\sqrt{x + 1}}}\right\}$$

Correction
$$f^{-1} = \{(x, y): x = \sqrt{y + 1}\}$$

5. Errors
$$x^2 + 2x + y^2 + 4y = 11$$
$$x^2 + 2x + 1 + y^2 + 4y + 4 = 11 + 1 + 4$$
$$(x + 1)^2 + (y + 2)^2 = 16$$
The center is at $\cancel{(1, 2)}$, and the radius is $\cancel{16}$.

Correction
The center is at $(-1, -2)$ and the radius is 4.

Chapter 3

Exercises 3.1, pages 131–132

1. $4i, -i$ **3.** $2i, \frac{1}{2}i$ **5.** $2i, -\frac{1}{2}i$ **7.** $\frac{1}{2} + \frac{1}{2}i, -1 - i$
9. $2x^2 + 5 + \dfrac{9}{x - 2}$ **11.** $-3x^2 + 9x - 25 + \dfrac{0}{x + 3}$
13. $x^4 + 2x^3 - x^2 - 3x - 9 - \dfrac{3}{x - 3}$
15. $x^3 - ax^2 + a^2x - a^3 + \dfrac{0}{x + a}$ **17.** 7
19. $4\sqrt{2}$ **21.** $-3i$ **23.** 9 **25.** $6 + 6\sqrt{2}$
27. $-16 + 3i$ **29.** $-16 - 22i$ **31.** No **33.** Yes
35. No **37.** Yes **39.** No **41.** Yes **43.** Yes
45. $x = 3, x = \pm i$ **47.** $k = 14/3$

Exercises 3.2, page 139

1. $P(x) = Q(x) = x^2 + 2x - 15$ **3.** $P(x) = x - 2i,$
$Q(x) = x^2 + 4$ **5.** $P(x) = x^2 + (-5 + i)x + 6 - 3i,$
$Q(x) = x^3 - 7x^2 + 17x - 15$
7. $P(x) = x^3 - (7 + i)x^2 + (17 + 2i)x - 15 + 3i,$
$Q(x) = x^5 - 11x^4 + 51x^3 - 119x^2 + 140x - 78$
9. 2 positive, 2 negative; 2 positive, 2 nonreal complex;
2 negative, 2 nonreal complex; 4 nonreal complex
11. 3 positive, 1 negative; 1 positive, 1 negative,
2 nonreal complex **13.** 2 negative, 4 nonreal complex;
6 nonreal complex **15.** 2 negative, 2 nonreal complex;
4 nonreal complex **17.** 4 nonreal complex
19. 6 nonreal complex **21.** $3i$ **23.** $1 - 2i$
25. $2i, -4i$ **27.** $2i, -4$ **29.** $2 - i, -2$
31. $3i, -\dfrac{1}{2} \pm \dfrac{\sqrt{3}}{2}i$ **33.** $1 - i, 1, -1$
35. $1 - i$ of multiplicity 2, 2

Exercises 3.3, pages 146–147

1. (a) 3; (b) -3 **3.** (a) 2; (b) -1 **5.** (a) 2;
(b) -3 **7.** (a) 2; (b) -3 **9.** $3, 2i, -2i$ **11.** $-3,$
$-1, \frac{1}{2}$ **13.** $2, (-1 \pm i\sqrt{23})/6$ **15.** $-2, 1, 1 \pm i$
17. $-1, \frac{3}{2}, \pm\sqrt{5}$ **19.** $-3, \frac{1}{2}, \pm\sqrt{2}$ **21.** No rational
zeros **23.** $1, -2$ **25.** $-2, -1, 1/3$ **27.** $-3, -5,$
$-1/2$ **29.** $2, -3/2$ **31.** $-3/2$ **33.** $2, -3/2$
35. No rational zeros **39.** $\frac{1}{2}$ meter **41.** 2
43. $5, (-5 + \sqrt{69})/2$ **45.** 3

Exercises 3.4, pages 149–150

1. 2.2 **3.** 0.5 **5.** 2.7 **7.** 0.4 **9.** 1.1 **11.** 1.3
13. 0.5 **15.** -0.8 **17.** 1.1 **19.** 1.2 **21.** $-1.5,$
$0.3, 2.2$ **23.** $-1.5, -0.4, 0.9$ **25.** $-2.3, -1.5, 4.8$
27. 0.9, 3.8

Review Problems for Chapter 3, pages 151–152

1. $(3 \pm \sqrt{33})i/4$ **2.** $i, -2i$ **3.** $-1 + i, (1 - i)/2$
4. $\dfrac{2x^3 - x^2 - 4x - 30}{x - 3} = 2x^2 + 5x + 11 + \dfrac{3}{x - 3}$
5. $\dfrac{3x^3 + 5x^2 + 7}{x + 2} = 3x^2 - x + 2 + \dfrac{3}{x + 2}$
6. $\dfrac{2x^4 + 6x^2 - 2x + 1}{x + 1} = 2x^3 - 2x^2 + 8x - 10 +$
$\dfrac{11}{x + 1}$
7. 6 **8.** 105 **9.** $-4 - 2i$ **10.** Yes **11.** Yes
12. No **13.** No **14.** Yes **15.** Yes
16. $x^2 - (4 + 2i)x + 8i$
17. $P(x) = x^4 + 3x^3 + 11x^2 + 27x + 18$
18. 2 positive, 1 negative; 1 negative, 2 nonreal
complex
19. 1 positive, 3 negative; 1 positive, 1 negative, 2
nonreal complex
20. 2 positive, 2 nonreal complex; 4 nonreal complex
21. 2 positive, 2 negative; 2 positive, 2 nonreal
complex; 2 negative, 2 nonreal complex; 4 nonreal
complex
22. $1 + i$ **23.** $2 + i, 2 - i$ **24.** $1 + i, -3$
25. $-1, 2, i$ **26.** (a) 2; (b) -3 **27.** -2 **28.** No
rational zeros **29.** $-2, 1, 1/3$ **30.** $2, 5, -1$
31. $3, -2, -4$ **32.** $2, (-1 \pm i\sqrt{3})/2$ **33.** $1, 4, -2$
34. $1/3, 1 + i, 1 - i$ **35.** $1/2, -1, -4/3$ **36.** $2/3,$
$-2, -5/3$ **37.** $-2, -4, -1/3$ **38.** -0.6 **39.** 0.8
40. -0.4

Critical Thinking Problems for Chapter 3, page 153

1. Error

$$
\begin{array}{r|rrr}
2 & 2 & -7 & 6 \\
  &   & 4 & -6 \\
\hline
  & 2 & -3 & 0
\end{array}
$$

Correction

$$
\begin{array}{r|rrrr}
2 & 2 & -7 & 0 & 6 \\
  &   & 4 & -6 & -12 \\
\hline
  & 2 & -3 & -6 & -6
\end{array}
$$

Since the remainder is -6, $x - 2$ is not a factor of
$P(x)$ and 2 is not a zero of $P(x)$.

2. Error

$P(x) = (x + 2)(x - 3i)$ ✗

Correction

$P(x) = (x + 2)(x - 3i)(x + 3i)$

$= x^3 + 2x^2 + 9x + 18$

3. Error

There are two variations of sign in $P(x) =$

$x^3 + x^2 - x + 15$ and one variation of sign in

$P(-x) = -x^3 + x^2 + x + 15$. Therefore $P(x)$ has two positive zeros and one negative zero. ✗

Correction

There are two possibilities. Either (a) $P(x)$ has two positive zeros and one negative zero, or (b) $P(x)$ has one negative zero and two nonreal complex zeros.

Chapter 4

Exercises 4.1, page 160

1. 3 **3.** -4 **5.** 5/2 **7.** $-2/3$ **9.** -2
11. $-5/2$ **13.** -1 **15.** $-3/2$ **17.** $-23/15$

19. 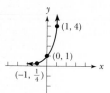 **21.**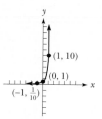

23. 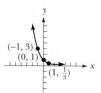 **25.**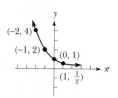

27. 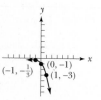 **29.**

31. 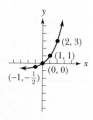 **33.**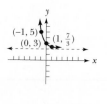

35. 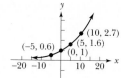 **37.**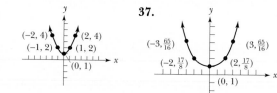

39. \$51,501.66 **41.** \$23,758.60 **43.** \$3280.50
45. 8,192,000 **47.** 33 (rounded)
49. $A(t) = 1000(\frac{1}{2})^t$

Exercises 4.2, page 165

1. 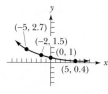 **3.**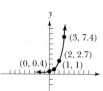

5. 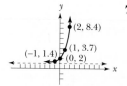 **7.**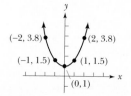

9.

11. 1 **13.** 0, -1 **15.** 1, -3 **17.** \$52,233.93
19. \$23,869.11 **21.** 92.85% **23.** 66.30 milligrams
25. 68.94% **27.** 43

Exercises 4.3, pages 170–171

1. $\log_2 8 = 3$ **3.** $\log_3 \frac{1}{81} = -4$ **5.** $\log_5 1 = 0$
7. $\log_{64} \frac{1}{16} = -\frac{2}{3}$ **9.** $5 = 5^1$ **11.** $81 = 3^4$
13. $\frac{1}{16} = 2^{-4}$ **15.** $\frac{25}{9} = \left(\frac{3}{5}\right)^{-2}$

17.
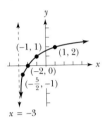
19.
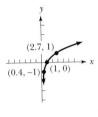

21.
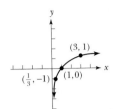
23.
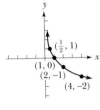

25. 0   **27.** 5   **29.** -2 **31.** 64   **33.** 1/9   **35.** 1/4
37. 4   **39.** 1/4   **41.** 1/27   **43.** 7   **45.** 1, -1
47. 4   **49.** 2/5   **51.** $3 \log_a x + 4 \log_a y$
53. $2 \log_a x + \log_a y + (1/2) \log_a z$
55. $\log_a x + \log_a y - \log_a z$
57. $2 \log_a x - \log_a y - (1/3) \log_a z$
59. $(4/3) \log_a x + \log_a y + 2 \log_a z$
61. $2 \log_a x - (5/3) \log_a y - \log_a z$ **63.** $\log_a xy^3$
65. $\log_a x^2yz^2$ **67.** $\log_a \dfrac{\sqrt{y}}{\sqrt[3]{z^2}}$ **69.** $\log_a \dfrac{x^3z^2}{64y^2}$

71. $\log_a \sqrt{\dfrac{x^3z^2}{y}}$ **73.** $\log_a \dfrac{27x^3y^9}{z^2}$

75. $x = \dfrac{\log_e (100/A)}{0.000411}$ **77.** $t = 2 \log_2 (N/2000)$

Exercises 4.4, pages 177–178

1. $\{1/3\}$ **3.** $\{9\}$ **5.** $\{7\}$ **7.** $\{8/7\}$ **9.** $\{10/3\}$
11. $\{1/2\}$ **13.** $\{2\}$ **15.** $\{2\}$ **17.** $\{-2, 2\}$ **19.** $\{2\}$
21. 2.58   **23.** -1.12 **25.** ± 1.02 **27.** -1.24
29. -23.2 **31.** 1.74   **33.** 0.461   **35.** 19.8
37. 1.77   **39.** 2.10   **41.** 4.84   **43.** 2.51   **45.** 1.44
47. -1.67 **49.** 3.21   **51.** 1.87   **53.** $1806
55. Approximately 5.12 years   **57.** $P = \$5528.75$
59. Approximately 6.93 years   **61.** Approximately
9.90 days   **63.** Approximately 6.10 days
65. Approximately 11,200 years   **67.** Approximately

1700 years   **69.** L increases by approximately 5
decibels.

Review Problems for Chapter 4, pages 178–179

1. (a) 9/4; (b) $\sqrt{6}/3$ **2.** $x = -3/4$ **3.** $x = -5/3$
4. $x = 31/8$ **5.** -9

6.
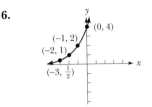
7.
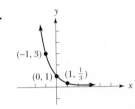

8.
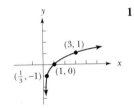
9.

10.
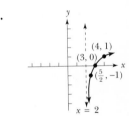
11.

12. 16   **13.** $-5/4$ **14.** 1/32   **15.** 9/2   **16.** e^3
17. -3 **18.** $2 \log_a x + \log_a z - 3 \log_a y$
19. $\dfrac{2}{3} \log_a x + \dfrac{1}{3} \log_a y - 4 \log_a z$ **20.** $\log_a (9x^5y/16z^4)$
21. $\log_a 2x^4y^3z^4$ **22.** 2.58   **23.** 1.55   **24.** 2.27
25. 3.24   **26.** $\{5/3\}$ **27.** $\{-1/3, -4/3\}$ **28.** $\{3\}$
29. 0, 3/2   **30.** 22.98   **31.** Approximately 49
months   **32.** Approximately 9.87 hours
33. Approximately 6.9 months   **34.** Approximately
25 minutes   **35.** Approximately 12.6 years
36. Approximately 6.74 years   **37.** Approximately
2009 days

Critical Thinking Problems for Chapter 4, page 180

1. Error

The error is the failure to check whether any of the functions in the original equation are undefined at $x = -\frac{8}{7}$.

Correction

Both $\log_2\left[-\frac{8}{7}\right]$ and $\log_2\left[-\frac{8}{7}+1\right]$ are undefined, since logarithms are not defined at negative numbers. The solution set is $\varnothing$.

2. Error

$\log[(2x+3)\not\times(3x-1)]=1$

Correction

$$\log(2x+3)(3x-1)=1$$
$$(2x+3)(3x-1)=10$$
$$6x^2+7x-13=0$$
$$(6x+13)(x-1)=0$$
$$6x+13=0 \quad\text{or}\quad x-1=0$$
$$x=-13/6 \quad\text{or}\quad x=1$$

The value $x=-\frac{13}{6}$ yields logarithms of negative

numbers in the original equation. The only solution is $x=1$.

3. Error

$x(2x+3)=\not{9}$

Correction

$$x(2x+3)=3^2$$
$$2x^2+3x-9=0$$
$$(2x-3)(x+3)=0$$
$$2x-3=0 \quad\text{or}\quad x+3=0$$
$$x=3/2 \quad\text{or}\quad x=-3$$

When $x=-3$, the original equation contains logarithms of negative numbers. The only solution is $x=\frac{3}{2}$.

Chapter 5

Exercises 5.1, pages 189–192

1. $48°$ **3.** $75°35'$ **5.** $71°17'46''$ **7.** $71.73°$
9. $14°16'12''$ **11.** $-729°30'20''$ **13.** $-18.83°$
15. $53.2406°$ **17.** $30°$ **19.** $-15°$ **21.** $-252°$
23. $495°$ **25.** $-\pi/2$ **27.** $2\pi/3$ **29.** $26\pi/9$
31. $5\pi/32$ **33.** $171.887°$ **35.** $29.393°$ **37.** 0.3229
39. 1.2608 **41.** 4π **43.** 3π **45.** $1/72$ revolution
47. $1/3$ revolution **49.** 355π meters **51.** 2.1π feet
53. $9/2$ centimeters **55.** $18\pi/25\approx 2.3$ feet
57. $\pi\approx 3.1$ inches **59.** 840 miles
61. $15\pi/11\approx 4.3$ miles per hour **63.** $3\pi\approx 9.4$
radians per second **65.** $20/(3\pi)\approx 2.1$ feet
67. $1000\pi/3\approx 1050$ miles per hour **69.** 80 feet
71. 37.5 feet **73.** 35 yards

Exercises 5.2, pages 200–202

In all these answers, the function values are given in this order: sine, cosine, tangent, cotangent, secant, cosecant.
1. $4/5, 3/5, 4/3, 3/4, 5/3, 5/4$ **3.** $24/25, -7/25,$
$-24/7, -7/24, -25/7, 25/24$ **5.** $-15/17, -8/17,$
$15/8, 8/15, -17/8, -17/15$ **7.** $-7\sqrt{51}/51,$
$-\sqrt{102}/51, 7\sqrt{2}/2, \sqrt{2}/7, -\sqrt{102}/2, -\sqrt{51}/7$
9. $1/\sqrt{1+x^2}, x/\sqrt{1+x^2}, 1/x, x,$
$\sqrt{1+x^2}/x, \sqrt{1+x^2}$ **11.** $\sqrt{1-x^2}, x,$
$\sqrt{1-x^2}/x, x/\sqrt{1-x^2}, 1/x, 1/\sqrt{1-x^2}$
13. $0.926, -0.377, -2.456, -0.407, -2.652, 1.080$
15. $-0.961, 0.276, -3.478, -0.288, 3.619, -1.041$

17. I **19.** I **21.** II **23.** III, IV **25.** II, IV
27. $3/5, 4/5, 3/4, 4/3, 5/4, 5/3$ **29.** $24/25, -7/25,$
$-24/7, -7/24, -25/7, 25/24$ **31.** $-\sqrt{10}/10,$
$-3\sqrt{10}/10, 1/3, 3, -\sqrt{10}/3, -\sqrt{10}$
33. $-2\sqrt{13}/13, 3\sqrt{13}/13, -2/3, -3/2, \sqrt{13}/3,$
$-\sqrt{13}/2$ **35.** $-\sqrt{21}/5, -2/5, \sqrt{21}/2, 2\sqrt{21}/21,$
$-5/2, -5\sqrt{21}/21$ **37.** $y, \sqrt{1-y^2}, y/\sqrt{1-y^2},$
$\sqrt{1-y^2}/y, 1/\sqrt{1-y^2}, 1/y$ **39.** $\sqrt{1-x^2},$
$x, \sqrt{1-x^2}/x, x/\sqrt{1-x^2}, 1/x, 1/\sqrt{1-x^2}$
41. $0, 1, 0$, undefined, 1, undefined **43.** $-1, 0,$
undefined, 0, undefined, -1 **45.** $67°$ **47.** $2\pi/5$
49. $\pi/9$ **51.** $-\sin(2\pi/9)$ **53.** $\cos 73°$
55. $-\tan(2\pi/9)$ **57.** $-\csc 20°30'$ **59.** $\sqrt{3}/2,$
$-1/2, -\sqrt{3}, -\sqrt{3}/3, -2, 2\sqrt{3}/3$
61. $1/2, -\sqrt{3}/2, -\sqrt{3}/3, -\sqrt{3}, -2\sqrt{3}/3, 2$
63. $-\sqrt{3}/2, 1/2, -\sqrt{3}, -\sqrt{3}/3, 2, -2\sqrt{3}/3$
65. $\sqrt{2}/2, -\sqrt{2}/2, -1, -1, -\sqrt{2}, \sqrt{2}$
67. $-1/2$ **69.** -1 **71.** $4\sqrt{3}/3$ **73.** False
75. False

Exercises 5.3, pages 206–207

1. $(-1, 0)$ **3.** $(0, 1)$ **5.** $(\sqrt{3}/2, 1/2)$
7. $(\sqrt{2}/2, -\sqrt{2}/2)$ **9.** $(-\sqrt{3}/2, 1/2)$
11. $(1/2, -\sqrt{3}/2)$ **13.** $(-\sqrt{2}/2, -\sqrt{2}/2)$
15. $(1/2, -\sqrt{3}/2)$ **17.** $(-\sqrt{2}/2, -\sqrt{2}/2)$
19. $(\sqrt{3}/2, -1/2)$ **21.** Quadrant III
23. Quadrant III **25.** Quadrant III
27. Quadrant IV **29.** 0

31. $-\sqrt{2}/2$ **33.** $-\sqrt{3}/3$
35. $-\sqrt{3}/2$ **37.** 1 **39.** 0 **41.** False **43.** True

Exercises 5.4, pages 210–211

1. 4.5600 **3.** 0.4679 **5.** -1.0951 **7.** -0.6861
9. 0.5357 **11.** -0.7989 **13.** 0.9883
15. -0.9926
17. $\tan\theta = -3/4$, $\cot\theta = -4/3$, $\sec\theta = 5/4$,
$\csc\theta = -5/3$
19. $\sin\theta = 12/13$, $\cos\theta = 5/13$, $\cot\theta = 5/12$,
$\csc\theta = 13/12$
21. $\sin\alpha = 8/17$, $\cos\alpha = -15/17$, $\cot\alpha = -15/8$,
$\sec\alpha = -17/15$
23. $\cos v = -24/25$, $\tan v = 7/24$, $\cot v = 24/7$,
$\csc v = -25/7$
25. $\sin t = 2\sqrt{2}/3$, $\cot t = -\sqrt{2}/4$, $\sec t = -3$,
$\csc t = 3\sqrt{2}/4$
27. $\sin s = -\sqrt{2}/2$, $\cos s = \sqrt{2}/2$, $\cot s = -1$,
$\sec s = \sqrt{2}$
29. $\tan\alpha = \sqrt{14}/2$, $\cot\alpha = \sqrt{14}/7$,
$\sec\alpha = -3\sqrt{2}/2$, $\csc\alpha = -3\sqrt{7}/7$
31. $\sin\theta = \sqrt{14}/7$, $\cos\theta = -\sqrt{35}/7$,
$\tan\theta = -\sqrt{10}/5$, $\sec\theta = -\sqrt{35}/5$
33. Impossible **35.** Possible **37.** Possible
39. Impossible **41.** Impossible **43.** Impossible
45. Impossible **47.** Impossible **49.** Impossible
51. Impossible

Exercises 5.5, pages 215–216

1. 0.5195 **3.** 0.5195 **5.** 1.980 **7.** -0.4791
9. 0.8450 **11.** -2.960 **13.** -2.577 **15.** -0.6626
17. 0.3007 **19.** 1.263 **21.** -0.3378 **23.** 0.4436
25. 32.0° **27.** 30.7° **29.** 54.6° **31.** 145.3°
33. 215.2° **35.** 333.4° **37.** 1.0558 **39.** 0.6226
41. 3.6507 **43.** 5.4774 **45.** 2.2020 **47.** 4.0753
49. 25°40′ **51.** 67°40′ **53.** 214°10′ **55.** 154°10′
57. 222°20′ **59.** 301°40′ **61.** 19°

Exercises 5.6, pages 223–226

1. $b = 9.4$, $c = 36$, $A = 75°$ **3.** $b = 0.80$, $c = 1.7$,
$B = 28°$ **5.** $a = 0.34$, $b = 0.51$, $A = 34°$
7. $c = 59$, $A = 42°$, $B = 48°$ **9.** $a = 0.11$, $b = 0.15$,
$B = 54°$ **11.** $b = 4.0$, $A = 37°$, $B = 53°$
13. $b = 207$, $c = 269$, $B = 50°20′$ **15.** $b = 191$,
$c = 201$, $A = 18.2°$ **17.** $a = 0.285$, $b = 0.148$,
$B = 27°30′$ **19.** $a = 187$, $b = 170$, $A = 47.8°$
21. $c = 13.4$, $A = 17.6°$, $B = 72.4°$ **23.** $b = 0.130$,

$A = 25.7°$, $B = 64.3°$ **25.** $b = 288.0$, $c = 321.9$,
$A = 26°32′$ **27.** $a = 28.01$, $b = 86.67$, $B = 72.09°$
29. $c = 13.01$, $A = 27.52°$, $B = 62.48°$
31. $a = 96.96$, $A = 79.92°$, $B = 10.08°$
33. 16 feet **35.** 26° **37.** 54.8 feet **39.** 67 feet
41. 350 feet **43.** 102 kilometers **45.** 248 feet
47. N 70° E **49.** 89 miles **51.** 120 feet,
0 not significant **53.** 73.0 feet

Exercises 5.7, pages 238–239

1. Amp = 4, $P = 2\pi/3$, no phase shift
3. Amp = 5, $P = \pi/2$, no phase shift
5. Amp = 1/2, $P = 2\pi$, phase shift is $\pi/4$ units to the
right. **7.** Amp = 1/3, $P = 2\pi/3$, phase shift is $\pi/3$
units to the left. **9.** Amp = 4, $P = 2\pi/3$, phase shift
is $\pi/12$ units to the right. **11.** Amp = 3, $P = \pi$,
phase shift is $\pi/8$ units to the left.

13.

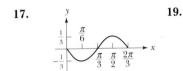

15.

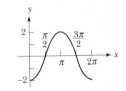

17.

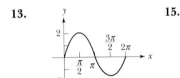

19.

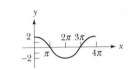

21.

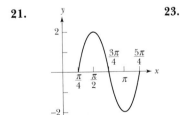

23.

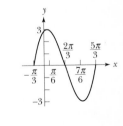

25.

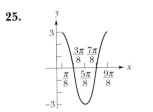

27.

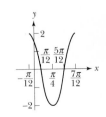

29.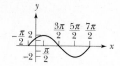

31.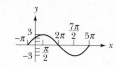

21.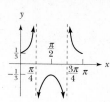

23.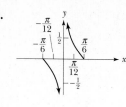

33.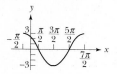

35.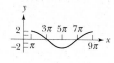

25.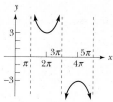

27.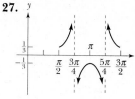

37.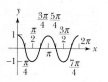

39.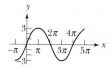

29.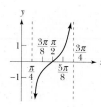

31.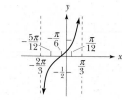

41. $y = 3 \sin \left(2x + \dfrac{\pi}{2} \right)$ **43.** $y = 4 \cos \left(3x - \dfrac{\pi}{2} \right)$

33.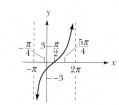

35.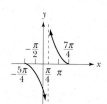

45.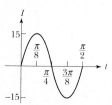

Exercises 5.8, pages 248–249

1. $P = \pi$, phase shift is $\pi/8$ units to the right.
3. $P = 2\pi/3$, phase shift is $\pi/9$ units to the left.
5. $P = \pi/2$, phase shift is $\pi/8$ units to the left.
7. $P = 2\pi$, phase shift is $\pi/2$ units to the right.
9. $P = 2\pi$, Amp $= 2$, no phase shift, no vertical
translation   **11.** $P = 4\pi$, Amp $= 2$, phase shift is $\pi/3$
units to the left, vertical translation is 1 unit upward.
13. $P = 6\pi$, amplitude is not defined, phase shift is
$3\pi/2$ units to the right, vertical translation is 2 units
upward.   **15.** $P = \pi/4$, amplitude is not defined,
phase shift is $\pi/4$ units to the right, vertical translation
is 1 unit upward.

37.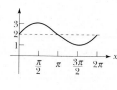

39.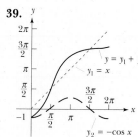

17.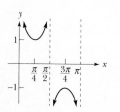

19.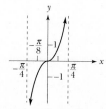

41.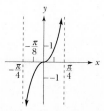

43.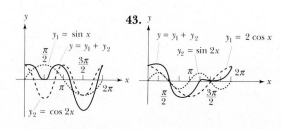

45.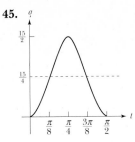

47.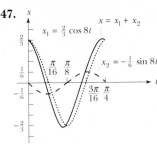

Review Problems for Chapter 5, pages 251–253

1. (a) Quadrant II, $483°$, $843°$, $-237°$, $-597°$;
(b) Quadrant IV, $316°$, $676°$, $-404°$, $-746°$ **2.** $6\frac{2}{3}$
feet   **3.** 6 feet   **4.** 65 feet   **5.** $\sin \theta = 3/7$,
$\cos \theta = -2\sqrt{10}/7$, $\tan \theta = -3\sqrt{10}/20$,
$\cot \theta = -2\sqrt{10}/3$, $\sec \theta = -7\sqrt{10}/20$, $\csc \theta = 7/3$
6. $\sin \alpha = -24/25$, $\cos \alpha = 7/25$, $\cot \alpha = -7/24$,
$\sec \alpha = 25/7$, $\csc \alpha = -25/24$ **7.** $\sin \beta = 2/3$,
$\cos \beta = \sqrt{5}/3$, $\tan \beta = 2\sqrt{5}/5$, $\sec \beta = 3\sqrt{5}/5$
8. 630 miles   **9.** 840 miles   **10.** $7\pi/40 \approx 0.55$ feet
per second   **11.** $7\pi/11 \approx 2.0$ miles per hour
12. $1056 \approx 1100$ radians per minute, two zeros not
significant   **13.** 88 radians per minute
14. (a) $-210°$ (b) $300°$ **15.** (a) $7\pi/2$; (b) $15\pi/64$
16. $3\pi \approx 9.4$ centimeters
17. 440 radians per minute
18. The function values are given in this order: sine,
cosine, tangent, cotangent, secant, cosecant.
(a) 1, 0, undefined, 0, undefined, 1
(b) $1/2$, $-\sqrt{3}/2$, $-\sqrt{3}/3$, $-\sqrt{3}$, $-2\sqrt{3}/3$, 2
(c) $-\sqrt{2}/2$, $\sqrt{2}/2$, -1, -1, $\sqrt{2}$, $-\sqrt{2}$
(d) -1, 0, undefined, 0, undefined, -1
(e) $-1/2$, $-\sqrt{3}/2$, $\sqrt{3}/3$, $\sqrt{3}$, $-2\sqrt{3}/3$, -2
(f) $-\sqrt{3}/2$, $-1/2$, $\sqrt{3}$, $\sqrt{3}/3$, -2, $-2\sqrt{3}/3$
19. (a) $(0, 1)$; (b) $(-1/2, \sqrt{3}/2)$ **20.** (a) Impossible;
(b) possible; (c) impossible; (d) possible;
(e) impossible   **21.** (a) 1.029; (b) -0.7490;
(c) 1.649; (d) -0.5793; (e) 0.9789   **22.** (a) 0.1363;
(b) -0.9217 **23.** (a) $24.7°$; (b) $64.3°$; (c) $152.0°$;
(d) $228.7°$; (e) $289.3°$ **24.** (a) 1.7017; (b) 3.4266;
(c) 1.9635; (d) 5.2971; (e) 4.4128   **25.** (a) 0.3200;
(b) 1.0996;   **26.** (a) $\pi/4$, $5\pi/4$; (b) $7\pi/6$, $11\pi/6$
27. $b = 310$ feet, $c = 320$ feet, zeros not significant,
$B = 75°$ **28.** 66 feet   **29.** 652 yards   **30.** 931
meters
31. (a) $P = \pi$, Amp $= 2$, phase shift is $\pi/2$ units to
the right, vertical translation is 1 unit upward.
(b) $P = \pi/2$, Amp $= 1/2$, phase shift is $\pi/4$ units
to the left, vertical translation is 2 units
downward.
(c) $P = \pi/3$, amplitude is not defined, no phase

shift, vertical translation is 4 units upward
(d) $P = 2\pi$, amplitude is not defined, phase shift
is $\pi/2$ units to the left, no vertical translation
(e) $P = 4\pi$, amplitude is not defined, phase shift
is $\pi/3$ units to the right, no vertical translation

32.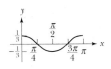

33.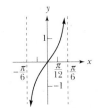

34.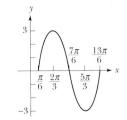

35.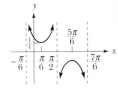

36.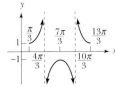

37.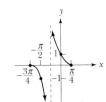

38.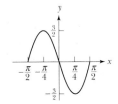

39.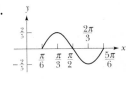

40.

41.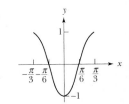

42.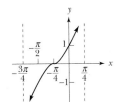

43.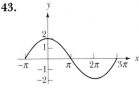

44.

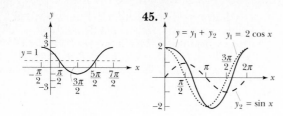

45.

$y = y_1 + y_2$ $y_1 = 2 \cos x$

$y_2 = \sin x$

46. $y = 3 \sin \left(x - \dfrac{\pi}{4}\right)$ **47.** $y = \dfrac{2}{3} \sin \left(3x + \dfrac{\pi}{2}\right)$

48. $y = \dfrac{1}{2} \cos(2x + \pi)$ **49.** $y = 2 \cos \left(x + \dfrac{\pi}{2}\right)$

50. Maximum value of $H = 14$ at $n = 171$; minimum value of $H = 31/3$ at $n = 353$.

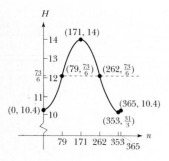

Critical Thinking Problems for Chapter 5, pages 254–256

1. Error

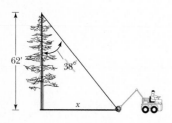

Correction

$\cot 38° = \dfrac{x}{62}$

$x = 62 \cot 38°$

$x = 79$ feet

38° = Angle of depression

2. Error

$48°24' = \cancel{48.24°}$

Correction

$48°24' = 48° + \dfrac{24°}{60} = 48° + 0.4° = 48.4°$

3. Error

$x = \cancel{3}$

Correction

$9 = x^2$

$x = -3$

$\cos \theta = \dfrac{x}{r} = -\dfrac{3}{5}$

4. Error

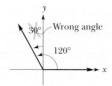

Correction

$\cos 120° = -\cos 60°$

$= -\dfrac{1}{2}$

5. Error

$\omega = \dfrac{v}{r}$

$= \dfrac{36}{9}$

Correction

$\omega = \dfrac{v}{r}$

$= \dfrac{36}{\frac{3}{4}}$

$= 48$ radians per second

6. Error

$y = 2 \sin \left(\dfrac{x}{2} - \dfrac{\pi}{4}\right)$

$= \sin 2 \left(\dfrac{x}{2} - \dfrac{\pi}{4}\right)$

Correction

The amplitude is $|a| = 2$, the period is $2\pi/(1/2) = 4\pi$, and a period starts where

$$\frac{x}{2} - \frac{\pi}{4} = 0$$

$$x = \frac{\pi}{2}.$$

The correct graph is shown in the following figure.

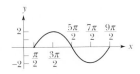

7. Errors

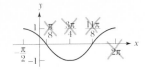

Correction

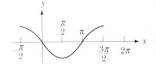

Chapter 6

Exercises 6.1, pages 264–266

15. $\sin \theta = \pm\sqrt{1 - \cos^2 \theta}$,
$\tan \theta = \pm\sqrt{1 - \cos^2 \theta}/\cos \theta$,
$\cot \theta = \pm\cos \theta/\sqrt{1 - \cos^2 \theta}$, $\sec \theta = 1/\cos \theta$,
$\csc \theta = \pm 1/\sqrt{1 - \cos^2 \theta}$
17. $\sin \theta = \pm\tan \theta/\sqrt{1 + \tan^2 \theta}$,
$\cos \theta = \pm 1/\sqrt{1 + \tan^2 \theta}$, $\cot \theta = 1/\tan \theta$,
$\sec \theta = \pm\sqrt{1 + \tan^2 \theta}$, $\csc \theta = \pm\sqrt{1 + \tan^2 \theta}/\tan \theta$
19. $\sec \theta$ **21.** $2 \cos \theta$ **23.** $1/(4 \cos u) = (\sec u)/4$
25. $27 \sec^3 u$ **27.** $\csc t$ **29.** $1/(a^2 \sec t \tan t)$

Exercises 6.2, pages 271–273

1. 0 **3.** 1 **5.** -1 **7.** $-\cos \theta$ **9.** $-\sin \theta$
11. $(\cos \theta - \sqrt{3} \sin \theta)/2$ **13.** $24/25, 0$
15. $-16/65, 56/65$ **17.** $7/25, -7/25$
19. $(5\sqrt{3} - 12)/26, (5\sqrt{3} + 12)/26$
21. $-(\sqrt{77} + \sqrt{10})/12, (\sqrt{10} - \sqrt{77})/12$
23. $\cos 3x$ **25.** $\cos (x/12)$

27. $y = \cos x$

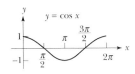

29. $y = -3 \sin x$

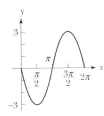

31. $y = -3 \cos 2x$

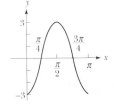

33. $\sqrt{2}(\sqrt{3} - 1)/4$ **35.** $\sqrt{2}(\sqrt{3} + 1)/4$
37. $\sqrt{2}(1 - \sqrt{3})/4$ **39.** $\sqrt{2}(1 + \sqrt{3})/4$

Exercises 6.3, pages 276–278

1. $\sqrt{3}/2$ **3.** 0 **5.** $-\sqrt{3}$ **7.** $-\sqrt{3}/3$
9. $180/163$ **11.** $5/301$ **13.** (a) 1; (b) $7/25$;
(c) undefined; (d) $7/24$ **15.** (a) $33/65$; (b) $-63/65$;
(c) $33/56$; (d) $63/16$ **17.** $\sqrt{2}(\sqrt{3} + 1)/4$
19. $\sqrt{2}(\sqrt{3} - 1)/4$
21. $(1 + \sqrt{3})/(1 - \sqrt{3}) = -(2 + \sqrt{3})$
23. $(1 - \sqrt{3})/(1 + \sqrt{3}) = \sqrt{3} - 2$
25. $\sin \theta$ **27.** $(-\sqrt{3}/2) \sin \theta + (1/2) \cos \theta$
29. $\tan \theta$ **31.** $(\tan \theta + 1)/(1 - \tan \theta)$

33.

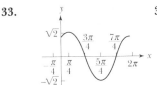

35.

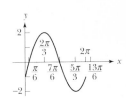

57. $\theta(t) = \dfrac{1}{2}\sin\left(4t + \dfrac{\pi}{6}\right)$, Amp $= \dfrac{1}{2}$, $P = \dfrac{\pi}{2}$

Exercises 6.4, pages 283–284

1. $\sin 44°$ **3.** $2\csc 210°$ **5.** $\cos(\pi/9)$ **7.** $\sec 2\gamma$
9. $\tan(2\pi/7)$ **11.** $\cot 2\alpha$ **13.** $\sin 121°$
15. $\tan 99°$ **17.** $\cos 189°$ **19.** $\tan 204°$
In answers 21, 23, 29, and 31, the function values are given in this order: sine, cosine, tangent, cotangent, secant, cosecant.
21. $-24/25$, $-7/25$, $24/7$, $7/24$, $-25/7$, $-25/24$
23. $4\sqrt{5}/9$, $-1/9$, $-4\sqrt{5}$, $-\sqrt{5}/20$, -9, $9\sqrt{5}/20$
25. $8\cos^4\theta - 8\cos^2\theta + 1$
27. $(3\tan\theta - \tan^3\theta)/(1 - 3\tan^2\theta)$
29. $2\sqrt{5}/5$, $\sqrt{5}/5$, 2, $1/2$, $\sqrt{5}$, $\sqrt{5}/2$
31. $\sqrt{30}/6$, $-\sqrt{6}/6$, $-\sqrt{5}$, $-\sqrt{5}/5$, $-\sqrt{6}$, $\sqrt{30}/5$
33. $\sqrt{2 - \sqrt{2}}/2$, $\sqrt{2 + \sqrt{2}}/2$, $\sqrt{2} - 1$
35. $\sqrt{2 + \sqrt{3}}/2$, $-\sqrt{2 - \sqrt{3}}/2$, $-(2 + \sqrt{3})$
37. $\sqrt{2 - \sqrt{2}}/2$, $-\sqrt{2 + \sqrt{2}}/2$, $1 - \sqrt{2}$
39. $-\sqrt{2 - \sqrt{3}}/2$, $-\sqrt{2 + \sqrt{3}}/2$, $2 - \sqrt{3}$

Exercises 6.5, pages 287–288

1. $\sin 111° + \sin 57°$ **3.** $4(\cos 25° + \cos 1°)$
5. $(\sin 11\alpha + \sin 5\alpha)/2$ **7.** $(\cos x - \cos 3x)/2$
9. $2\sin 49°\cos 7°$ **11.** $2\sin 55°\sin 38°$
13. $-2\cos 10\alpha\sin\alpha$ **15.** $2\cos 2x\cos x$

Exercises 6.6, pages 293–295

In answers 1-11, n represents an arbitrary integer.
1. $150° + n180°$ **3.** $30° + n360°$, $150° + n360°$
5. $60° + n120°$ **7.** $120° + n360°$, $240° + n360°$
9. $n360°$, $60° + n360°$, $300° + n360°$
11. $30° + n180°$, $150° + n180°$ **13.** $30°, 210°$
15. $135°, 225°$ **17.** $17.6°, 162.4°$ **19.** $66.5°, 127.5°$, $246.5°, 307.5°$ **21.** $0, \pi, \pi/4, 5\pi/4$ **23.** $0, \pi/4, 3\pi/4, \pi, 5\pi/4, 7\pi/4$ **25.** $\pi/6, 5\pi/6$ **27.** $\pi/3, \pi, 5\pi/3$ **29.** $7\pi/6, 3\pi/2, 11\pi/6$ **31.** $0, \pi/3, 5\pi/3$
33. $0, \pi$ **35.** $\pi/4, 3\pi/4, 5\pi/4, 7\pi/4$ **37.** $\pi/3, 3\pi/2, 5\pi/3$ **39.** $\pi/4, 5\pi/4$ **41.** $3\pi/2$ **43.** $4\pi/3$
45. $\pi/9, 4\pi/9, 7\pi/9, 10\pi/9, 13\pi/9, 16\pi/9$ **47.** $\pi/6$, $\pi/2, 5\pi/6, 7\pi/6, 3\pi/2, 11\pi/6$ **49.** $\pi/8, 5\pi/8, 9\pi/8$, $13\pi/8$ **51.** $\pi/12, 5\pi/12, 3\pi/4, 13\pi/12, 17\pi/12$, $7\pi/4$ **53.** $\pi/9, 5\pi/9, 7\pi/9, 11\pi/9, 13\pi/9, 17\pi/9$
55. $0, \pi/3, \pi, 5\pi/3$ **57.** 0 **59.** $\pi/4, \pi/2, 5\pi/4$,

$3\pi/2$ **61.** $\pi/3, 2\pi/3, 4\pi/3, 5\pi/3$ **63.** $\pi/3, 5\pi/3$
65. $2\pi/3, 4\pi/3$ **67.** $0, \pi/3, 2\pi/3, \pi, 7\pi/6, 4\pi/3$, $5\pi/3, 11\pi/6$ **69.** $\pi/4$ **71.** $\pi/3$ **73.** $t = n\pi/4$; $n = 0, 1, 2, \ldots$ **75.** $\pi/2, 7\pi/6, 3\pi/2, 11\pi/6$
77. $t = 5\pi/24 + n\pi/4$; $n = 0, 1, 2, \ldots$
79. $(\pi/2, 3), (7\pi/6, -3/2), (3\pi/2, -1), (11\pi/6, -3/2)$

Exercises 6.7, pages 302–303

1. $\pi/2$ **3.** $\pi/4$ **5.** $3\pi/4$ **7.** $-\pi/4$ **9.** $-\pi/3$
11. $5\pi/6$ **13.** Not defined   **15.** $\sqrt{3}/3$ **17.** -1.3
19. $5/6$ **21.** $4/5$ **23.** $-12/5$ **25.** 0.99
27. $5\pi/6$ **29.** 0.1978 **31.** 2.2108 **33.** 0.5250
35. -0.7046 **37.** $-17/49$ **39.** $-24/25$
41. $119/169$ **43.** $33/65$ **45.** $56/65$
47. $\sqrt{1 - x^2}/x$ **49.** $\sqrt{1 + x^2}/x$
51. $(1 - x^2)/(1 + x^2)$ **53.** $xy + \sqrt{(1 - x^2)(1 - y^2)}$
55. $-\sqrt{3}$ **57.** $-\sqrt{2}/2$ **59.** $-\sqrt{2}/2$ **61.** $\sqrt{2}/2$
63. $5/13$ **65.** 0 **67.** $1/2$ **69.** $-1/2$
71. No solution   **73.** $-1 \le x \le 1$

75.

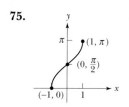

77.

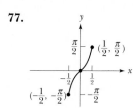

79.

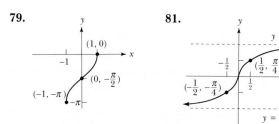

81.

Review Problems for Chapter 6, pages 305–306

4. $\sin\theta = 1/\csc\theta$, $\cos\theta = \pm\sqrt{\csc^2\theta - 1}/\csc\theta$, $\tan\theta = \pm 1/\sqrt{\csc^2\theta - 1}$, $\cot\theta = \pm\sqrt{\csc^2\theta - 1}$, $\sec\theta = \pm\csc\theta/\sqrt{\csc^2\theta - 1}$ **12.** (a) 2; (b) 0; (c) 1; (d) $\sqrt{3}/2$; (e) $\sqrt{2} + 1$ **13.** $\sqrt{2}(\cos\theta - \sin\theta)/2$
14. $-\sin\theta$ **15.** $\tan\theta$

16. $y = 2 \sin x$

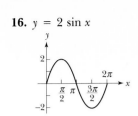

17. $y = 2 \sin \left(x + \dfrac{\pi}{3} \right)$

18. $\sin(\alpha - \beta) = 56/65$, $\cos(\alpha - \beta) = -33/65$, $\tan(\alpha - \beta) = -56/33$, $\cot(\alpha - \beta) = -33/56$, $\sec(\alpha - \beta) = -65/33$, $\csc(\alpha - \beta) = 65/56$
19. $\sin 2\theta = -24/25$ **20.** $\cos 2\theta = 7/25$
21. $\tan 2\theta = -24/7$ **22.** $\sin(\theta/2) = \sqrt{10}/10$
23. $\cos(\theta/2) = -3\sqrt{10}/10$ **24.** $\tan(\theta/2) = -1/3$
25. $1 - 8\sin^2\theta + 8\sin^4\theta$ **31.** (a) $\cos 8\theta + \cos 2\theta$;
(b) $\sin 10x + \sin 2x$; (c) $(\cos 2A + \cos 2B)/2$
32. (a) $2 \sin 3t \cos t$; (b) $2 \cos 6\alpha \cos \alpha$;
(c) $2 \sin 6t \sin 3t$ **33.** $210°, 330°$ **34.** $0°, 120°, 240°$
35. $45°, 135°$ **36.** $45°, 270°, 315°$ **37.** $0, \pi$
38. $\pi/6, 5\pi/6$ **39.** $\pi/2, 7\pi/6, 11\pi/6$
40. $\pi/2, 7\pi/6$ **41.** $4\pi/3$ **42.** $\pi/4, \pi/2, 3\pi/4,$
$5\pi/4, 3\pi/2, 7\pi/4$ **43.** 0 **44.** $3\pi/4$ **45.** $-\pi/4$
46. $\pi/4$ **47.** Not defined **48.** 0.2 **49.** $\sqrt{2}$
50. 0 **51.** $\frac{1}{4}$ **52.** $-\frac{4}{5}$ **53.** $-\frac{1}{8}$ **54.** $-\frac{24}{25}$ **55.** $\frac{16}{65}$
56. 1 **57.** $\frac{56}{65}$ **58.** $-\frac{16}{65}$ **59.** $-\frac{1}{2}$ **60.** $\frac{1}{2}$

61.

62.

63.

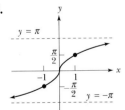

64.

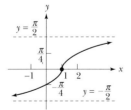

Critical Thinking Problems for Chapter 6, pages 307–308

1. Errors

$$\frac{\sin \theta}{\cos \theta} + \frac{\cos \theta}{\sin \theta} = \frac{\cancel{\sin \theta} + \cancel{\cos \theta}}{\cos \theta \sin \theta}$$

$$\frac{1}{\cos \theta} + \frac{1}{\sin \theta} = \frac{\cancel{\times}}{\cos \theta \sin \theta}$$

Corrections

$$\tan \theta + \cot \theta = \frac{\sin \theta}{\cos \theta} + \frac{\cos \theta}{\sin \theta}$$

$$= \frac{\sin^2 \theta + \cos^2 \theta}{\cos \theta \sin \theta}$$

$$= \frac{1}{\cos \theta \sin \theta}$$

$$= \frac{1}{\cos \theta} \cdot \frac{1}{\sin \theta}$$

$$= \sec \theta \csc \theta$$

2. Errors

$$\cancel{\cos \theta} = 1 \quad \text{or} \quad 2 \cos \cancel{\theta} \cancel{= 1}$$

Correction

$$2\cos^2 \theta - \cos \theta = 1$$

$$2\cos^2 \theta - \cos \theta - 1 = 0$$

$$(2\cos \theta + 1)(\cos \theta - 1) = 0$$

$$2\cos \theta + 1 = 0 \quad \text{or} \quad \cos \theta - 1 = 0$$

$$\cos \theta = -\frac{1}{2} \quad \text{or} \quad \cos \theta = 1$$

$$\theta = 120°, 240° \quad \text{or} \quad \theta = 0°$$

The solutions are $0°$, $120°$, and $240°$.

3. Errors

$$\frac{\cos 2\theta}{\cos^2 \theta} = \frac{\cos^2 \theta \cancel{\times} \sin^2 \theta}{\cos^2 \theta}$$

$$1 + \tan^2 \theta = 1 + (1 \cancel{\times} \sec^2 \theta)$$

Correction

$$\frac{\cos 2\theta}{\cos^2 \theta} = \frac{\cos^2 \theta - \sin^2 \theta}{\cos^2 \theta}$$

$$= \frac{\cos^2 \theta}{\cos^2 \theta} - \frac{\sin^2 \theta}{\cos^2 \theta}$$

$$= 1 - \tan^2 \theta$$

$$= 1 - (\sec^2 \theta - 1)$$

$$= 2 - \sec^2 \theta$$

4. Errors

$$(\cos \theta - \sin \theta)^2 = \cos^2 \theta \cancel{\times} \sin^2 \theta$$

$$\cos 2\theta = 1 - \cancel{\sin 2\theta}$$

Correction

$$(\cos \theta - \sin \theta)^2 = \cos^2 \theta - 2\cos \theta \sin \theta + \sin^2 \theta$$

$$= 1 - 2\sin \theta \cos \theta$$

$$= 1 - \sin 2\theta$$

5. Error

$$2 \sin x \cos x = \cos x$$
$$2 \sin x = 1 \, \times$$

Correction

$$2 \sin x \cos x = \cos x$$
$$2 \sin x \cos x - \cos x = 0$$
$$(2 \sin x - 1) \cos x = 0$$
$$2 \sin x - 1 = 0 \quad \text{or} \quad \cos x = 0$$
$$\sin x = \frac{1}{2}$$
$$x = \frac{\pi}{6}, \frac{5\pi}{6} \quad \text{or} \quad x = \frac{\pi}{2}, \frac{3\pi}{2}$$

The solutions are $\pi/6$, $\pi/2$, $5\pi/6$, and $3\pi/2$.

6. Error

$$\text{arccot} \, (-\sqrt{3}) = \arctan \left(\frac{1}{\sqrt{3}} \right) \, \times$$

Correction

$y = \text{arccot} \, (-\sqrt{3})$ is equivalent to $\cot y = -\sqrt{3}$ and $0 < y < \pi$. This is shown in the following figure. From the figure, we see that

$$\text{arccot} \, (-\sqrt{3}) = \frac{5\pi}{6}.$$

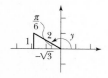

7. Error

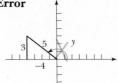

Correction

$y = \tan^{-1}(-3/4)$ is equivalent to $\tan y = -3/4$ and $-\frac{\pi}{2} < y < \frac{\pi}{2}$.

From the following figure, we see that

$$\sin (\tan^{-1}(-3/4)) = \sin y$$
$$= -\frac{3}{5}.$$

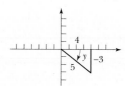

8. Error

$$\sin^{-1} \left(\sin \frac{5\pi}{6} \right) = \frac{5\pi}{6} \, \times$$

Correction

The range for $\sin^{-1} x$ is $-\frac{\pi}{2} \le y \le \frac{\pi}{2}$, and $\frac{5\pi}{6}$ is not in this range.

$$\sin^{-1} \left(\sin \frac{5\pi}{6} \right) = \sin^{-1}(1/2)$$
$$= \frac{\pi}{6}$$

Chapter 7

Exercises 7.1, pages 317–318

1. 1.7 **3.** 218 **5.** 19.16 **7.** 7° **9.** 90° **11.** 46°
13. No solution **15.** 66° or 114° **17.** C = 50°, $a = 58, b = 55$ **19.** A = 65.7°, $b = 12.2, c = 7.13$
21. No solution **23.** B = 16°, C = 37°, $b = 14$
25. B = 40°, C = 93°, $c = 110$ (zero not significant)
27. No solution **29.** A = 54°, C = 79°, $c = 24$; $A' = 126°, C' = 7°, c' = 3.0$ **31.** B = 49.2°,
$C = 68.5°, b = 152; B' = 6.2°, C' = 111.5°, b' = 21.7$
33. 35 feet, 41 feet **35.** 4.2 feet **37.** 16 feet
39. 110 miles (zero not significant) **41.** 4.3°

Exercises 7.2, pages 323–325

1. 17 **3.** 8.7 **5.** 8.7 **7.** 5.0 **9.** 120° **11.** 45°
13. No solution **15.** No solution **17.** B = 41°, C = 83°, $a = 25$ **19.** A = 10°, C = 58°, $b = 8.3$

21. $A = 26°$, $B = 37°$, $C = 117°$ **23.** $A = 20°$, $B = 149°$, $C = 11°$ **25.** $B = 27°30'$, $C = 125°20'$, $a = 304$ **27.** $A = 18.7°$, $C = 38.7°$, $b = 37.6$ **29.** 210 yards **31.** N 71°E **33.** 7.8 miles **35.** 60 feet **37.** 405 miles

Exercises 7.3, pages 329–330

1. 250, zero not significant **3.** 5.3 **5.** 0.73 **7.** 1200, zeros not significant **9.** 6.5 **11.** 600, last zero not significant **13.** 2800, zeros not significant **15.** 140, zero not significant **17.** No solution **19.** No solution **21.** 290, zero not significant **23.** 3.92 or 2.65 (two triangles) **25.** 220, zero not significant **27.** 18.6 square miles

Exercises 7.4, pages 338–340

1. **3.**

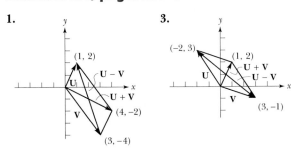

5. $\langle 11, -19 \rangle = 11i - 19j$ **7.** $\langle 29, -7 \rangle = 29i - 7j$ **9.** $\langle -1, 4 \rangle$, $\sqrt{17}$ **11.** $\langle 1, 4 \rangle$, $\sqrt{17}$ **13.** $4, 3\pi/2$ **15.** $8, 0$ **17.** $\sqrt{2}, 7\pi/4$ **19.** $4, 3\pi/4$ **21.** 39 pounds **23.** 35° **25.** 301 pounds **27.** 4.0 miles per hour, S 58°W **29.** 110 miles per hour, zero not significant; N 12°E **31.** 71.4 miles per hour, 173°30' **33.** 33.7 miles per hour **35.** 5.9° **37.** 10 miles per hour, N 89°E **39.** 0.39 tons

Exercises 7.5, pages 346–347

1. **3.**

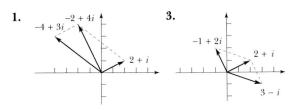

5. $\sqrt{2}(\cos 135° + i \sin 135°)$
7. (figure: $-1 + 3i$, $1 + 3i$, -2)
(figure: 5, $-3i$, $5 - 3i$)

9. $\sqrt{2}(\cos 135° + i \sin 135°)$
11. $3\sqrt{2}(\cos 315° + i \sin 315°)$
13. $4(\cos 180° + i \sin 180°)$
15. $7(\cos 90° + i \sin 90°)$
17. $8(\cos 210° + i \sin 210°)$
19. $\dfrac{\sqrt{2}}{4}(\cos 45° + i \sin 45°)$
21. $2(\cos 300° + i \sin 300°)$
23. $6(\cos 150° + i \sin 150°)$
25. $\cos 25.4° + i \sin 25.4°$
27. $\cos 136.1° + i \sin 136.1°$
29. $2\sqrt{2} + 2i\sqrt{2}$ **31.** $(-3/2) - (3\sqrt{3}/2)i$
33. $-9i$ **35.** $-1 - i$ **37.** $2.6730 + 1.3620i$
39. $4.6839 - 5.2020i$ **41.** $-12 + 12i\sqrt{3}$ **43.** -1
45. 6 **47.** $-8 + 8i\sqrt{3}$ **49.** $\sqrt{2} + i\sqrt{2}$ **51.** $2i$
53. $(\sqrt{2}/2) - (\sqrt{2}/2)i$ **55.** $(-\sqrt{3}/2) - (1/2)i$
57. $4[\cos (3\pi/4) + i \sin (3\pi/4)] = -2\sqrt{2} + 2i\sqrt{2}$, $4[\cos (-\pi/2) + i \sin (-\pi/2)] = -4i$
59. $6[\cos (2\pi/3) + i \sin (2\pi/3)] = -3 + 3i\sqrt{3}$, $(2/3)(\cos \pi + i \sin \pi) = -2/3$
61. $4(\cos 150° + i \sin 150°) = -2\sqrt{3} + 2i$, $\cos (-90°) + i \sin (-90°) = -i$
63. $4(\cos 180° + i \sin 180°) = -4$, $(1/2)[\cos (-90°) + i \sin (-90°)] = (-1/2)i$
65. $20(\cos 60° + i \sin 60°) = 10 + 10i\sqrt{3}$, $\dfrac{1}{5}[\cos (-60°) + i \sin (-60°)] = \dfrac{1}{10} - \dfrac{\sqrt{3}}{10}i$
67. $6\sqrt{2}(\cos 135° + i \sin 135°) = -6 + 6i$, $\dfrac{3\sqrt{2}}{4}(\cos 45° + i \sin 45°) = \dfrac{3}{4} + \dfrac{3}{4}i$

Exercises 7.6, pages 350–351

1. $\cos 126° + i \sin 126°$
3. $512(\cos 108° + i \sin 108°)$
5. $\cos (15\pi/8) + i \sin (15\pi/8)$
7. $9[\cos (8\pi/5) + i \sin (8\pi/5)]$
9. $(-\sqrt{3}/2) - (1/2)i$ **11.** 1 **13.** $-i$
15. $-128 - 128i\sqrt{3}$ **17.** $1024i$
19. $512 + 512i\sqrt{3}$ **21.** 16 **23.** $-250 + 250i$

25.

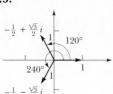

27.

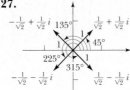

29.

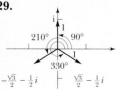

31.

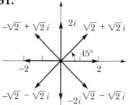

33.

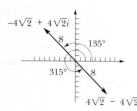

35.

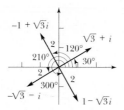

Review Problems for Chapter 7, pages 352–354

1. $A = 50°$, $b = 22$, $c = 15$ **2.** No solution
3. $A = 117.3°$, $B = 25.5°$, $a = 20.9$ **4.** $A = 84.6°$,
$C = 58.2°$, $a = 16.6$; $A' = 21.0°$, $C' = 121.8°$,
$a' = 5.99$ **5.** $A = 101.4°$, $B = 59.1°$, $C = 19.5°$
6. $B = 23°$, $C = 37°$, $a = 31$ **7.** $A = 54°$, $B = 18°$,
$c = 98$ **8.** $K = 130$, zero not significant
9. $K = 326$ **10.** $K = 130$, zero not significant
11. $-1 + i\sqrt{3}$ **12.** $-2\sqrt{3} - 2i$
13. $3(\cos 270° + i \sin 270°)$
14. $2\sqrt{2}(\cos 315° + i \sin 315°)$ **15.** $-12i$
16. $(-2\sqrt{3}/3) + (2/3)i$
17. $16(\cos 150° + i \sin 150°) = -8\sqrt{3} + 8i$,
$4(\cos 90° + i \sin 90°) = 4i$ **18.** $-9i$ **19.** $128i$
20. -64 **21.** $(1/2) + (\sqrt{3}/2)i$, $(-\sqrt{3}/2) + (1/2)i$,
$(-1/2) - (\sqrt{3}/2)i$, $(\sqrt{3}/2) - (1/2)i$
22. $2(\cos 15° + i \sin 15°)$, $2(\cos 75° + i \sin 75°)$,
$2(\cos 135° + i \sin 135°)$, $2(\cos 195° + i \sin 195°)$,
$2(\cos 225° + i \sin 225°)$, $2(\cos 315° + i \sin 315°)$
23. 2, $1 + i\sqrt{3}$, $-1 + i\sqrt{3}$, -2, $-1 - i\sqrt{3}$, $1 - i\sqrt{3}$
24. 84 feet   **25.** 16°   **26.** 38 inches
27. 254 miles per hour, 352.8°   **28.** 275°
29. N 67°E, 20 miles per hour   **30.** 21.1 pounds
31. 42°   **32.** 30.6°   **33.** 496 pounds
34. 66.1 pounds   **35.** 42.0 miles per hour
36. 55.3 miles per hour

37. $2(\cos 45° + i \sin 45°)$, $2(\cos 105° + i \sin 105°)$,
$2(\cos 165° + i \sin 165°)$, $2(\cos 225° + i \sin 225°)$,
$2(\cos 285° + i \sin 285°)$, $2(\cos 345° + i \sin 345°)$
39. $\cos 10° + i \sin 10°$, $\cos 130° + i \sin 130°$,
$\cos 250° + i \sin 250°$ **41.** $\cos 75° + i \sin 75°$,
$\cos 165° + i \sin 165°$, $\cos 255° + i \sin 255°$,
$\cos 345° + i \sin 345°$ **43.** $\cos 45° + i \sin 45°$,
$\cos 117° + i \sin 117°$, $\cos 189° + i \sin 189°$,
$\cos 261° + i \sin 261°$, $\cos 333° + i \sin 333°$
45. $2(\cos 45° + i \sin 45°)$, $2(\cos 117° + i \sin 117°)$,
$2(\cos 189° + i \sin 189°)$, $2(\cos 261° + i \sin 261°)$,
$2(\cos 333° + i \sin 333°)$ **47.** $2(\cos 105° + i \sin 105°)$,
$2(\cos 225° + i \sin 225°)$, $2(\cos 345° + i \sin 345°)$
49. $(3/2) + (3\sqrt{3}/2)i$, -3, $(3/2) - (3\sqrt{3}/2)i$
51. $(\sqrt{3}/2) + (1/2)i$, $(-\sqrt{3}/2) + (1/2)i$, $-i$
53. $\sqrt{3} + i$, $2i$, $-\sqrt{3} + i$, $-\sqrt{3} - i$, $-2i$, $\sqrt{3} - i$
55. $(\sqrt{3}/2) + (1/2)i$, $(-1/2) + (\sqrt{3}/2)i$,
$(-\sqrt{3}/2) - (1/2)i$, $(1/2) - (\sqrt{3}/2)i$
57. $2(\cos 18° + i \sin 18°)$, $2(\cos 90° + i \sin 90°)$,
$2(\cos 162° + i \sin 162°)$, $2(\cos 234° + i \sin 234°)$,
$2(\cos 306° + i \sin 306°)$ **59.** $\cos 20° + i \sin 20°$,
$\cos 140° + i \sin 140°$, $\cos 260° + i \sin 260°$

Critical Thinking Problems for Chapter 7, pages 355–356

1. Error

$$49 + 9 - 42(1/2) = \cancel{16}(1/2)$$

Correction

$$49 + 9 - 42(1/2) = 49 + 9 - 21$$
$$= 37$$
$$c = \sqrt{37}$$

2. Error

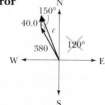

Correction

The heading of 120.0° is an angle to be measured clockwise from north, as shown in the following figure.

$$c^2 = (380)^2 + (40)^2 - 2(380)(40) \cos 60°$$
$$= 144,400 + 1600 - 30,400(1/2)$$
$$= 130,800$$
$$c = 362 \text{ miles per hour}$$

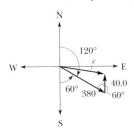

3. Error

$$-\sqrt{3} - i = 2(\cos \cancel{30}° + i \sin \cancel{30}°)$$

Correction

The *related angle* is 30°,
$$-\sqrt{3} - i = 2(\cos 210° + i \sin 210°).$$

4. Error

$$(1 + i\sqrt{3})^5 = \cancel{2}(\cos 60° + i \sin 60°)^5$$

Correction

$$(1 + i\sqrt{3})^5 = [2(\cos 60° + i \sin 60°)]^5$$
$$= 2^5(\cos 300° + i \sin 300°)$$
$$= 32 \left(\frac{1}{2} - \frac{\sqrt{3}}{2}i \right)$$
$$= 16 - 16i\sqrt{3}$$

Chapter 8

Exercises 8.1, pages 364–365

1. $\{(-2, 3)\}$

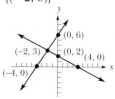

3. $\varnothing$

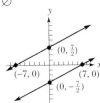

5. $\{(-2, 2), (6, 18)\}$

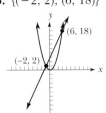

7. $\varnothing$

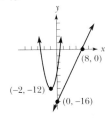

9. $\{(0, -2), (\frac{8}{5}, \frac{6}{5})\}$

11. $\{(8, 4), (\frac{8}{25}, -\frac{44}{25})\}$

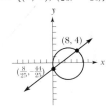

13. $\{(3, -1)\}$ **15.** $\{(3, 4)\}$ **17.** $\{(-3, 1)\}$
19. $\{(-1, -1), (-\frac{1}{7}, \frac{5}{7})\}$ **21.** $\varnothing$ **23.** $\{(0, 3), (0, -3)\}$
25. $\{(4, 3), (4, -3), (-4, 3), (-4, -3)\}$
27. $\{(1, 2\sqrt{2}), (1, -2\sqrt{2})\}$ **29.** $\{(\sqrt{3}, 1), (-\sqrt{3}, 1)\}$
31. $\{(\frac{1}{2}, 2), (7, \frac{1}{7})\}$ **33.** $\{(\frac{3}{2}, -\frac{4}{3}), (-\frac{3}{2}, \frac{4}{3})\}$
35. $\{(1, 2), (-1, -2), (\sqrt{2}, \sqrt{2}), (-\sqrt{2}, -\sqrt{2})\}$
37. $\{(2, 2), (-2, -2)\}$ **39.** $\{(1, 1), (-1, -1),$
$(5, -3), (-5, 3)\}$ **41.** $(20, 1000)$ **43.** $9500 at 9\%,
$5500 at 5\%$ **45.** 5.6 grams of the ingredient that is
50% protein, 22.4 grams of the other ingredient
47. 4 and 19, or -4 and -19 **49.** 4 and 7, or -4
and 7, or 4 and -7, or -4 and -7 **51.** $\pm\frac{4}{3}$
53. For 1200 items, both cost and revenue equal
$264,000. **55.** $s = 4$ feet, $\ell = 2\sqrt{3}$ feet

Exercises 8.2, pages 374–376

1. Not equal **3.** Equal **5.** Equal only if $x = 2$ and
$y = 3$ **7.** Equal only if $x = 2$ **9.** $\begin{bmatrix} 5 & 0 & 1 \\ 3 & 10 & -5 \end{bmatrix}$

11. Not possible **13.** $\begin{bmatrix} 2 & -8 \\ 4 & 7 \end{bmatrix}$ **15.** $\begin{bmatrix} -11 & 13 \\ 16 & -8 \\ -3 & 9 \end{bmatrix}$

17. Not possible **19.** $\begin{bmatrix} -1 \\ -8 \end{bmatrix}$ **21.** $\begin{bmatrix} 8 & 6 \\ 19 & 32 \\ 2 & 13 \end{bmatrix}$

23. $\begin{bmatrix} 50 & 6 & 13 & -23 \\ 34 & -21 & 24 & -14 \end{bmatrix}$ **25.** $\begin{bmatrix} 7 & -1 \\ 7 & -5 \end{bmatrix}$

27. $\begin{bmatrix} -1 & 2 \\ 5 & 20 \\ -11 & -13 \end{bmatrix}$ **29.** Not possible **31.** Not

possible **33.** $[22]$ **35.** $\begin{bmatrix} -8 & -22 & 4 \\ 4 & 11 & -2 \\ 4 & 11 & -2 \end{bmatrix}$

37. Not possible **39.** $\begin{bmatrix} 1 & 0 \\ 0 & 1 \end{bmatrix}$ **41.** $\begin{bmatrix} 1 & 0 & 0 \\ 0 & 1 & 0 \\ 0 & 0 & 1 \end{bmatrix}$

43. $\begin{bmatrix} 3 & 4 & 5 & 6 \\ 5 & 6 & 7 & 8 \end{bmatrix}$ **45.** $\begin{bmatrix} 1 & 8 & 11 \\ 1 & 1 & 13 \\ 1 & 1 & 1 \\ 1 & 1 & 1 \end{bmatrix}$

47. $A = \begin{bmatrix} 1 & 2 \\ 0 & 1 \\ 1 & 1 \end{bmatrix}$,

$B = \begin{bmatrix} 2 & 1 & 1 \\ -1 & 2 & 0 \end{bmatrix}$ is a possible answer.

49. $A = \begin{bmatrix} 1 & 0 \\ 1 & 2 \end{bmatrix}$, $B = \begin{bmatrix} 0 & 0 \\ 2 & 2 \end{bmatrix}$ is a possible answer.

51. $AB + AC = A(B + C) = \begin{bmatrix} -16 & -7 \\ -14 & -8 \\ -10 & -5 \end{bmatrix}$

53. $(A - B)(A + B) = \begin{bmatrix} 42 & -15 \\ 2 & 5 \end{bmatrix}$,

$A^2 - B^2 = \begin{bmatrix} 39 & -14 \\ -5 & 8 \end{bmatrix}$ **55.** \$3.44 at store A,

\$3.24 at store B, \$3.55 at store C

Exercises 8.3, pages 384–385

1. $\begin{bmatrix} 3 & -1 & | & 0 \\ -1 & 1 & | & 1 \end{bmatrix}$ **3.** $\begin{bmatrix} 3 & -2 & 5 & | & 0 \\ 4 & 7 & -1 & | & 0 \\ 1 & 0 & 1 & | & 0 \end{bmatrix}$

5. $\begin{bmatrix} 1 & -1 & 0 & 0 & | & 0 \\ 0 & 0 & 1 & 1 & | & 0 \\ 3 & 0 & 0 & 2 & | & 0 \\ 0 & 5 & -1 & 0 & | & 0 \end{bmatrix}$

7. $x + 2y = 5$
$\quad 3x + 4y = 6$
9. $x = a$
$\quad z = b$
$\quad y = c$
11. $x \qquad + z \qquad = 0$
$\qquad 2y + z + 3w = 7$
$\quad 3x + \ y + z + \ w = 1$
$\quad -3x + \ y - z + 2w = 5$
13. $x = 5, y = -4$ **15.** $a = -3, b = -2$ **17.** No
solution **19.** $x = -8, y = -5$ **21.** $x = 0, y = 0$
23. $x = 3, y = 1, z = -1$ **25.** $x = 4, y = 4, z = 1$
27. $x = 1, y = 3, z = -2$ **29.** $r = 4, s = -6, t = 3$
31. No solution **33.** $x = 1, y = -1, z = 2, t = -2$
35. There are many solutions of the form $a = 5 - 3r$,
$b = 10 - 7r, c = -4r, d = r$, where r is any real
number. **37.** (a) No solution exists for $c = -2$;
(b) exactly one solution for $c \neq 2$ and $c \neq -2$;
(c) many solutions for $c = 2$. **39.** (a) Solutions exist
for all values of c; (b) exactly one solution for $c \neq 0$;
(c) many solutions for $c = 0$.

Exercises 8.4, pages 389–390

1. $\begin{bmatrix} \frac{2}{3} & -\frac{1}{2} \\ \frac{1}{3} & 0 \end{bmatrix}$ **3.** $-\frac{1}{8}\begin{bmatrix} 2 & -3 \\ -2 & -1 \end{bmatrix}$ **5.** Does not exist

7. $-\frac{1}{29}\begin{bmatrix} 7 & -3 \\ -5 & -2 \end{bmatrix}$ **9.** $\begin{bmatrix} -7 & 1 & 1 \\ 1 & 0 & 0 \\ 3 & 0 & -1 \end{bmatrix}$

11. $\begin{bmatrix} 11 & -6 & 2 \\ 3 & -2 & 1 \\ 1 & -1 & 1 \end{bmatrix}$ **13.** $\begin{bmatrix} \frac{5}{4} & \frac{1}{4} & -\frac{1}{2} \\ \frac{3}{4} & \frac{3}{4} & -\frac{1}{2} \\ -\frac{1}{4} & -\frac{1}{4} & \frac{1}{2} \end{bmatrix}$

15. Does not exist **17.** $\begin{bmatrix} -23 & 4 & 5 & 2 \\ -97 & 15 & 22 & 8 \\ -50 & 8 & 11 & 4 \\ -13 & 2 & 3 & 1 \end{bmatrix}$

19. $x = 5, y = -3$ **21.** $x_1 = 3, x_2 = 3$ **23.** $x = 6$,
$y = 2$ **25.** $x = -\frac{3}{4}, y = \frac{7}{4}$ **27.** $x_1 = -4, x_2 = 0$,
$x_3 = -4$ **29.** $x = 2, y = 1, z = -3$ **31.** $x = 5$,
$y = 3, z = 1$ **33.** $x = 3, y = 1, z = -1$
35. $x = 1, y = 3, z = -2, w = 0$
37. $\begin{bmatrix} 1/a & 0 & 0 \\ 0 & 1/b & 0 \\ 0 & 0 & 1/c \end{bmatrix}$ **43.** $X = A^{-1}BC^{-1}$

Exercises 8.5, pages 395–396

1. $\dfrac{11}{x-3} - \dfrac{8}{x-2}$ **3.** $\dfrac{4}{x+2} - \dfrac{1}{x}$

5. $\dfrac{1}{x-1} + \dfrac{1}{x-2} - \dfrac{2}{x}$ **7.** $\dfrac{\frac12}{x} + \dfrac{\frac16}{x-2} - \dfrac{\frac23}{x+1}$

9. $\dfrac{7}{3x+2} - \dfrac{2}{x+1} + \dfrac{1}{x+2}$ **11.** $\dfrac{1}{x-2} - \dfrac{1}{(x-2)^2}$

13. $\dfrac{1}{x-1} + \dfrac{2}{(x-1)^2} + \dfrac{2}{(x-1)^3}$

15. $\dfrac{\frac12}{x+1} - \dfrac{1}{(x+1)^2} + \dfrac{\frac12}{x-1}$

17. $\dfrac{-\frac14}{x^2} + \dfrac{\frac1{16}}{x-2} - \dfrac{\frac1{16}}{x+2}$

19. $\dfrac{1}{x-1} + \dfrac{1}{(x-1)^2} - \dfrac{1}{x} - \dfrac{2}{x^2}$

21. $\dfrac{2}{x-2} + \dfrac{9}{(x-2)^2} + \dfrac{2}{3x-1}$

23. $\dfrac{x+2}{x^2+x+2} - \dfrac{1}{x-1}$ **25.** $\dfrac{4}{x-3} + \dfrac{x-1}{x^2+x-1}$

27. $\dfrac{2}{x-1} - \dfrac{1}{x+1} + \dfrac{4x}{x^2+1}$

29. $\dfrac{x}{x^2+1} - \dfrac{x+1}{(x^2+1)^2}$

31. $\dfrac{2x+1}{2x^2-x+1} - \dfrac{1}{(2x^2-x+1)^2}$

33. $x + \dfrac{6}{x-3} + \dfrac{2}{x+1}$

35. $2x - 1 - \dfrac{3}{x} + \dfrac{1}{(x-1)^2}$

Exercises 8.6, pages 401–402

1. -5 **3.** -1 **5.** -21 **7.** 0 **9.** $-1/9$ **11.** $5x$
13. 12 **15.** 7 **17.** 0 **19.** -1 **21.** 4 **23.** -3
25. -1 **27.** 0 **29.** -2 **31.** 1 **33.** -1
35. -2 **37.** $2, -3$ **39.** $-1, -2$ **41.** $5, -3, 2$
43. $0, 4, -3$

Exercises 8.7, pages 406–407

1. $x=-4$ **3.** $a=-1$ **5.** $x=1$ **7.** $x=-6$
9. $x=2$ **11.** $x=-2$ **13.** 15 **15.** 10 **17.** -20
19. -1 **21.** 23 **23.** 4 **25.** 11

Exercises 8.8, page 411

1. $x=-3, y=1$ **3.** $x=1, y=-2$ **5.** $a=-2$, $b=3$ **7.** Dependent system **9.** Inconsistent
11. $x=2, y=1, z=-1$ **13.** $x=5, y=-10$,

$z=4$ **15.** $x=-21, y=-12, z=39$ **17.** $a=1$, $b=2, c=3$ **19.** Dependent system **21.** $x_1=\frac{28}{13}$, $x_2=-\frac{6}{13}, x_3=\frac{53}{13}$ **23.** $x=-\frac12, y=\frac{15}{14}, z=-\frac{25}{14}$
25. $x=-1, y=-1, z=1$ **27.** $x=2, y=-1$, $z=0, w=1$ **29.** $x=w=1, y=z=0$
31. $m=2, b=-4$ **33.** $a=2, b=3, c=-4$

Exercises 8.9, pages 415–417

1. 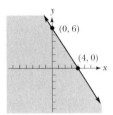 **3.**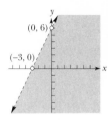

5. 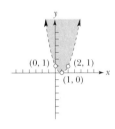 **7.**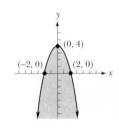

9. 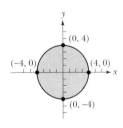 **11.**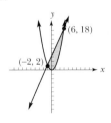

13. No solution **15.**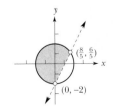

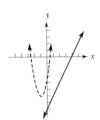

17.

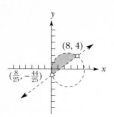

19.

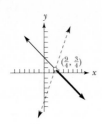

21.

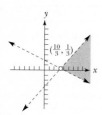

23.

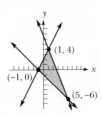

25.

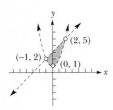

27.

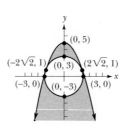

29.

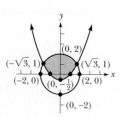

31.

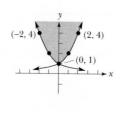

33. $x + 2y \geqslant 4$
$x - y \leqslant 1$

35. $y < -x^2 - 2x + 3$
$x \leqslant 0$

37. $y \geqslant x^2 - 1$
$x + y \leqslant 3$

39. $y \geqslant 2^x$
$x + y \leqslant 1$
$x \geqslant -2$

Exercises 8.10, pages 421–423

1. Maximum value of 14 at (6, 2)   **3.** Maximum value of 9 at (0, 2)   **5.** Minimum value of 11 at (1, 4) **7.** Maximum value of 14 at (6, 2)   **9.** Minimum value of 0 at (4, 1)   **11.** Maximum value of 4 at (3, 3) **13.** Make 30 cans of the first mixture and 12 cans of the second mixture for a maximum profit of \$180. **15.** Make 40 desks of the first type and 100 desks of the second type for a maximum profit of \$4800. **17.** Manufacture 8 chairs and 14 stools for a maximum profit of \$520.   **19.** Production of 14 cold cut trays and 9 seafood trays yields a maximum profit of \$59.50.

Review Problems for Chapter 8, pages 424–426

1. $\{(3, \frac{3}{2})\}$ **2.** $\{(6, 2)\}$ **3.** $\{(5, 2), (2, -1)\}$ **4.** $\{(3, 0), (8, 5)\}$ **5.** $\{(4, 5), (2, 1)\}$ **6.** $\{(\sqrt{6}, \sqrt{6}/6),$ $(-\sqrt{6}, -\sqrt{6}/6)\}$ **7.** $\begin{bmatrix} 4 & -3 \\ -10 & 17 \\ -2 & 2 \end{bmatrix}$ **8.** Not possible

9. $\begin{bmatrix} -7 & -19 & 12 \\ 2 & 7 & 17 \end{bmatrix}$ **10.** Not possible

11. $\begin{bmatrix} -7 & -8 \\ 9 & 10 \\ -5 & -1 \end{bmatrix}$ **12.** Not possible

13. $\begin{bmatrix} -3 & 23 & 35 \\ 4 & -12 & -20 \end{bmatrix}$ **14.** $x = 4, y = 6$ **15.** No solution   **16.** $x = 0, y = 2, z = -5$ **17.** $x = -1,$ $y = 2, z = -2$ **18.** $x = 4, y = -2, z = 1$

19. $x = -1, y = -2, z = 4$ **20.** $\begin{bmatrix} 3/2 & 2 \\ 1/2 & 1 \end{bmatrix}$

21. Does not exist   **22.** $\begin{bmatrix} 1 & 0 & 0 \\ 2 & 2 & -1 \\ -1 & -1 & 1 \end{bmatrix}$

23. $\begin{bmatrix} -3 & 2 & -2 \\ -4 & 2 & -1 \\ 2 & -1 & 1 \end{bmatrix}$ **24.** $\begin{bmatrix} -3 & 2 & 4 \\ 2 & -1 & -1 \\ 2 & -1 & -2 \end{bmatrix}$

25. Does not exist   **26.** $r = -9, s = -5$ **27.** $x = 1, y = 2$ **28.** $x = -2, y = 1, z = -4$ **29.** $x = 5, y = 0, z = 4$ **30.** $\dfrac{2}{x + 2} - \dfrac{1}{x - 3}$

31. $\dfrac{1}{x + 1} - \dfrac{2}{(x + 1)^2} + \dfrac{4}{x - 2}$

32. $\dfrac{4}{x-4} + \dfrac{-x+4}{x^2+2}$

33. $\dfrac{2}{x^2+x+1} + \dfrac{3x-1}{(x^2+x+1)^2}$ **34.** -14 **35.** 73

36. 39 **37.** -15 **38.** 0 **39.** -1
40. -1 **41.** 8 **42.** 6 **43.** Dependent
44. $x=2, y=-3$ **45.** $x=-2, y=3, z=1$
46. Dependent

47.

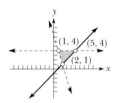

48.

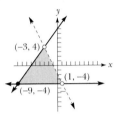

49.

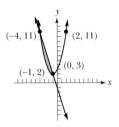

50.

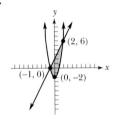

51. Maximum value of 29 at $(4, 3)$ **52.** Maximum value of 39 at $(3, \frac{3}{2})$ **53.** Preparation of 200 pounds of the \$6 mixture and 400 pounds of the \$5 mixture yields a maximum revenue of \$3200.

Critical Thinking Problems for Chapter 8, pages 427–428

1. (a) Error

$$[2 \quad -5 \quad 0] + [1 \quad 3] = \cancel{[3 \quad -2 \quad 0]}$$

Correction

Not possible

(b) Error

$$\begin{bmatrix} 0 & 1 \\ 2 & 3 \end{bmatrix}\begin{bmatrix} 4 & 5 \\ -1 & -2 \end{bmatrix} = \cancel{\begin{bmatrix} 0 & 5 \\ -2 & -6 \end{bmatrix}}$$

Correction

$$\begin{bmatrix} 0 & 1 \\ 2 & 3 \end{bmatrix}\begin{bmatrix} 4 & 5 \\ -1 & -2 \end{bmatrix} = \begin{bmatrix} -1 & -2 \\ 5 & 4 \end{bmatrix}$$

2. Errors

$$\begin{bmatrix} 1 & 1 & 2 & \vdots & 2 \\ 0 & 1 & 3 & \vdots & 1 \\ 1 & 0 & -2 & \vdots & 1 \end{bmatrix} \to \begin{bmatrix} 1 & 1 & 2 & \vdots & 2 \\ 0 & 1 & 3 & \vdots & 1 \\ 0 & -1 & -4 & \vdots & \cancel{*} \end{bmatrix} \to$$

$$\begin{bmatrix} 1 & 1 & 2 & \vdots & 2 \\ 0 & 1 & 3 & \vdots & 1 \\ 0 & 0 & -1 & \vdots & \cancel{*} \end{bmatrix} \to \begin{bmatrix} 1 & 0 & -1 & \vdots & \cancel{*} \\ 0 & 1 & 3 & \vdots & 1 \\ 0 & 0 & 1 & \vdots & \cancel{*} \end{bmatrix} \to$$

$$\begin{bmatrix} 1 & 0 & 0 & \vdots & \cancel{*} \\ 0 & 1 & 0 & \vdots & \cancel{*} \\ 0 & 0 & 1 & \vdots & 1 \end{bmatrix}$$

Correction

$$\begin{bmatrix} 1 & 1 & 2 & \vdots & 2 \\ 0 & 1 & 3 & \vdots & 1 \\ 1 & 0 & -2 & \vdots & 1 \end{bmatrix} \to \begin{bmatrix} 1 & 1 & 2 & \vdots & 2 \\ 0 & 1 & 3 & \vdots & 1 \\ 0 & -1 & -4 & \vdots & -1 \end{bmatrix} \to$$

$$\begin{bmatrix} 1 & 1 & 2 & \vdots & 2 \\ 0 & 1 & 3 & \vdots & 1 \\ 0 & 0 & -1 & \vdots & 0 \end{bmatrix} \to \begin{bmatrix} 1 & 0 & -1 & \vdots & 1 \\ 0 & 1 & 3 & \vdots & 1 \\ 0 & 0 & 1 & \vdots & 0 \end{bmatrix} \to$$

$$\begin{bmatrix} 1 & 0 & 0 & \vdots & 1 \\ 0 & 1 & 0 & \vdots & 1 \\ 0 & 0 & 1 & \vdots & 0 \end{bmatrix}$$

The solution is $x=1, y=1, z=0$.

3. Error

Since the denominator contains linear factors, we write

$$\frac{2x-4}{(x-1)^2(x+1)} = \frac{A}{x-1} \cancel{\times} \frac{B}{x+1}$$

$$2x-4 = A\cancel{(x+1)} + B\cancel{(x-1)}.$$

Correction

$$\frac{2x-4}{(x-1)^2(x+1)} = \frac{A}{x-1} + \frac{B}{(x-1)^2} + \frac{C}{x+1}$$

$$2x-4 = A(x-1)(x+1)$$
$$+ B(x+1) + C(x-1)^2$$

If $x=1$, then $-2=2B$, and $B=-1$.
If $x=-1$, then $-6=C(-2)^2$, and $C=-3/2$.
If $x=0$, then $-4=-A+B+C$, and $A=3/2$.
Thus

$$\frac{2x-4}{(x-1)^2(x+1)} = \frac{3/2}{x-1} - \frac{1}{(x-1)^2} - \frac{3/2}{x+1}.$$

4. Error

$$\begin{vmatrix} 3 & 1 & -1 \\ -1 & 1 & 2 \\ 1 & 2 & 1 \end{vmatrix} = 3\begin{vmatrix} 1 & 2 \\ 2 & 1 \end{vmatrix} - \cancel{1}\begin{vmatrix} 1 & -1 \\ 2 & 1 \end{vmatrix} + 1\begin{vmatrix} 1 & -1 \\ 1 & 2 \end{vmatrix}$$

Correction

$$\begin{vmatrix} 3 & 1 & -1 \\ -1 & 1 & 2 \\ 1 & 2 & 1 \end{vmatrix} = 3\begin{vmatrix} 1 & 2 \\ 2 & 1 \end{vmatrix} - (-1)\begin{vmatrix} 1 & -1 \\ 2 & 1 \end{vmatrix}$$

$$+ 1\begin{vmatrix} 1 & -1 \\ 1 & 2 \end{vmatrix}$$

$$= 3(1 - 4) + 1(1 + 2) + 1(2 + 1)$$
$$= 3(-3) + 1(3) + 1(3)$$
$$= -3$$

5. Error

$$\begin{vmatrix} 0 & 1 & 4 \\ -1 & 2 & 3 \\ 1 & -2 & 0 \end{vmatrix} \begin{array}{c} \cancel{R_2 \times 1} \\ = \end{array} \begin{vmatrix} 0 & 1 & 4 \\ \cancel{0} & \cancel{0} & \cancel{0} \\ 1 & -2 & 0 \end{vmatrix}$$

Correction

$$\begin{vmatrix} 0 & 1 & 4 \\ -1 & 2 & 3 \\ 1 & -2 & 0 \end{vmatrix} \begin{array}{c} R_3 + R_2 \\ = \end{array} \begin{vmatrix} 0 & 1 & 4 \\ 0 & 0 & 3 \\ 1 & -2 & 0 \end{vmatrix} =$$

$$1\begin{vmatrix} 1 & 4 \\ 0 & 3 \end{vmatrix} = 3$$

6. Errors

$$D_y = \begin{vmatrix} 2 & -8 & 2 \\ 0 & 4 & 2 \\ 3 & 2 & 1 \end{vmatrix} \begin{array}{c} \cancel{\tfrac{1}{2}R_2} \\ = \end{array} \begin{vmatrix} 2 & -8 & 2 \\ 0 & \cancel{2} & \cancel{1} \\ 3 & 2 & 1 \end{vmatrix} \begin{array}{c} -2\,C_3 + C_2 \\ = \end{array} \begin{vmatrix} 2 & -12 & 2 \\ 0 & 0 & 1 \\ 3 & 0 & 1 \end{vmatrix} = -1\begin{vmatrix} 2 & -12 \\ 3 & 0 \end{vmatrix} = -36$$

Thus $y = \cancel{-36}$

Correction

$$D_y = \begin{vmatrix} 2 & -8 & 2 \\ 0 & 4 & 2 \\ 3 & 2 & 1 \end{vmatrix} \begin{array}{c} -2C_3 + C_2 \\ = \end{array} \begin{vmatrix} 2 & -12 & 2 \\ 0 & 0 & 2 \\ 3 & 0 & 1 \end{vmatrix} = -2\begin{vmatrix} 2 & -12 \\ 3 & 0 \end{vmatrix} = -2(36) = -72$$

$$D = \begin{vmatrix} 2 & -4 & 2 \\ 0 & 3 & 2 \\ 3 & 0 & 1 \end{vmatrix} \begin{array}{c} -3C_3 + C_1 \\ = \end{array} \begin{vmatrix} -4 & -4 & 2 \\ -6 & 3 & 2 \\ 0 & 0 & 1 \end{vmatrix} = \begin{vmatrix} -4 & -4 \\ -6 & 3 \end{vmatrix} = -12 - 24 = -36$$

Thus $y = D_y/D = -72/(-36) = 2$.

Chapter 9

Exercises 9.1, pages 433–434

1. $\frac{1}{2}, \frac{1}{3}, \frac{1}{4}, \frac{1}{5}, \frac{1}{6}$　**3.** $0, \frac{1}{2}, \frac{2}{3}, \frac{3}{4}, \frac{4}{5}$　**5.** $-2, 4, -8, 16, -32$
7. $2, 2, 2, 2, 2$　**9.** $-1, \frac{1}{4}, -\frac{1}{9}, \frac{1}{16}, -\frac{1}{25}$　**11.** $1, \frac{2}{3}, \frac{3}{5}, \frac{4}{7}, \frac{5}{9}$
13. $1, 4, 7, 10, 13$　**15.** $x - 1, (x - 1)^2, (x - 1)^3,$
$(x - 1)^4, (x - 1)^5$　**17.** $-x^2, x^4, -x^6, x^8, -x^{10}$
19. $1, 2, 4, 8, 16, 32$　**21.** $-5, 5, 5, -5, -5, 5$
23. $2, 3, 8, 19, 46, 111$
25. $2 + 4 + 8 + 16 + 32 = 62$
27. $\frac{1}{4} + \frac{2}{5} + \frac{3}{6} + \frac{4}{7} + \frac{5}{8} + \frac{6}{9} + \frac{7}{10} = \frac{3119}{840}$
29. $0 + 1 + 3 + 6 + 10 + 15 + 21 = 56$
31. $-1 + 1 - 1 + 1 - 1 + 1 - 1 = -1$
33. $2 + 2 + 2 + 2 + 2 + 2 = 12$
35. $4 - 9 + 16 - 25 + 36 - 49 = -27$

37. $-1 + 0 + 1 + 8 + 27 + 64 + 125 = 224$
39. $(1 - \frac{1}{2}) + (\frac{1}{2} - \frac{1}{3}) + (\frac{1}{3} - \frac{1}{4}) + (\frac{1}{4} - \frac{1}{5}) +$
$(\frac{1}{5} - \frac{1}{6}) + (\frac{1}{6} - \frac{1}{7}) = \frac{6}{7}$
41. $\frac{1}{4} - \frac{1}{8} + \frac{1}{16} - \frac{1}{32} = \frac{5}{32}$　**43.** $\sum_{k=1}^{17} k$　**45.** $\sum_{j=1}^{7} 2^j$

47. $\sum_{n=1}^{7} (-1)^{n+1}(2n - 1)$　**49.** $\sum_{i=1}^{5} \frac{i}{i + 1}$

51. $\sum_{j=0}^{5} (-1)^j \frac{2j + 1}{2^{j+1}}$　**53.** $\sum_{k=1}^{43} \frac{1}{(2k - 1)(2k + 1)}$

55. $\sum_{i=0}^{n} \frac{x^i}{i + 1}$　**57.** $15/4$　**59.** $\sum_{k=1}^{4} \left(\frac{k}{2}\right)^2 \left(\frac{1}{2}\right)$

Exercises 9.2, pages 439–440

1. 1, 2, 3, 4, 5, 6   **3.** 13, 11, 9, 7, 5, 3   **5.** 3, -1, -5, -9, -13, -17 **7.** 2, 6, 10, 14, 18, 22
9. $d = 5$, $a_n = 5n - 22$ **11.** $d = -3$, $a_n = -3(n + 1)$ **13.** Not arithmetic   **15.** $d = -14$, $a_n = 21 - 14n$ **17.** $d = y$, $a_n = x - y + ny$
19. Not arithmetic unless $x = 0$ or $x = 1$. If $x = 0$, $d = 0$ and $a_n = 0$. If $x = 1$, $d = 0$ and $a_n = 1$.
21. $d = x$, $a_n = nx$ **23.** $a_{11} = -25$, $S_{11} = -165$
25. $a_{11} = 7$, $S_{11} = -88$ **27.** $a_{11} = -13$, $S_{11} = 22$
29. $a_{11} = m - 10x$, $S_{11} = 11(m - 5x)$
31. $a_{11} = x - 10k$, $S_{11} = 11x$ **33.** 53   **35.** -60
37. 23   **39.** $-\frac{3933}{4}$ **41.** $a_1 = -7$, $d = \frac{1}{2}$
43. $a_1 = 17$, $d = -5$ **45.** 91   **47.** $\frac{110}{3}$ **49.** 2550
51. -117 **55.** $-20, -25, -30, -35$
57. 4510   **59.** $n = 36$ **61.** 39   **63.** 9

Exercises 9.3, pages 448–449

1. $-1, 2, -4, 8, -16$ **3.** $4, 1, \frac{1}{4}, \frac{1}{16}$ **5.** $\frac{3}{4}, 3, 12, 48$
7. $\frac{1}{4}, -\frac{1}{2}, 1, -2, 4$ **9.** $-3, -6, -12$ or $-3, 6, -12$
11. $a_5 = 4$, $a_n = 2^n/8$, $S_5 = \frac{31}{4}$ **13.** $a_5 = 27$,
$a_n = -(-3)^n/9$, $S_5 = \frac{61}{3}$ **15.** $a_5 = -8$, $a_n = (-2)^n/4$,
$S_5 = -\frac{11}{2}$ **17.** $r = \frac{1}{2}$, $a_n = 14/2^n$ **19.** $r = \sqrt{2}/2$,
$a_n = (\sqrt{2})^{5-n}$ **21.** Not geometric   **23.** Not
geometric   **25.** $r = -\frac{1}{2}$, $a_n = (-4)(-\frac{1}{2})^{n-1}$
27. $r = -\frac{1}{7}$, $a_n = (-1)^{n-1}/7^{n-4}$ **29.** 31   **31.** $\frac{4323}{3125}$
33. $\frac{275}{81}$ **35.** $\frac{2101}{4}$ **37.** $r = \pm\sqrt{3}$, $a_1 = -\frac{1}{3}$
39. $r = 1.01$, $a_1 = 4/1.01$, or $r = -1.01$,
$a_1 = -4/1.01$ **41.** -2 **43.** $\frac{27}{8}$ **45.** Sum does not
exist.   **47.** Sum does not exist.   **49.** $2(2 + \sqrt{2})$
51. $\frac{50}{9}$ **53.** 30   **55.** 3   **57.** Sum does not exist.
Geometric series with $|r| = |6| > 1$.   **59.** 10
61. Sum does not exist. Geometric series with
$|r| = |-3| > 1$.   **63.** $\frac{13}{3}$ **65.** 1   **67.** $\frac{1}{99}$ **69.** $\frac{28}{9}$
71. $-\frac{4579}{1998}$ **73.** $255   **75.** $1,536,000   **77.** 10 feet
79. 256,000

Exercises 9.5, pages 460–461

1. $x^7 + 7x^6y + 21x^5y^2 + 35x^4y^3 + 35x^3y^4 + 21x^2y^5 + 7xy^6 + y^7$
3. $16x^4 + 32x^3y + 24x^2y^2 + 8xy^3 + y^4$
5. $x^{10} + 10x^8y + 40x^6y^2 + 80x^4y^3 + 80x^2y^4 + 32y^5$
7. $x^8 - 4x^6y^2 + 6x^4y^4 - 4x^2y^6 + y^8$
9. $16x^4 - 32x^3/y + 24x^2/y^2 - 8x/y^3 + 1/y^4$
11. $81x^4 + 36x^3y + 6x^2y^2 + 4xy^3/9 + y^4/81$
13. $x^{18} - 3x^{17} + 15x^{16}/4 - 5x^{15}/2 + 15x^{14}/16 - 3x^{13}/16 + x^{12}/64$
15. $a^5/32 - 15a^4b^2/16 + 45a^3b^4/4 - 135a^2b^6/2 + 405ab^8/2 - 243b^{10}$

17. 45   **19.** 80   **21.** $1760x^3y^9$ **23.** $-29{,}120x^{12}y^6$
25. $3003p^{12}q^{24}$ **27.** $8192y^{13}$
29. $1 + x/4 - 3x^2/32 + 7x^3/128 - 77x^4/2048 + \cdots$, $-1 < x < 1$
31. $1 - 2y/3 - y^2/9 - 4y^3/81 - 7y^4/243 + \cdots$, $-1 < y < 1$
33. $1 - x + 3x^2/2 - 5x^3/2 + 35x^4/8 - \cdots$, $-\frac{1}{2} < x < \frac{1}{2}$
35. $1 - 2x^2 + 3x^4 - 4x^6 + 5x^8 - \cdots$, $-1 < x < 1$
37. $1 + a + 3a^2/4 + a^3/2 + 5a^4/16 + \cdots$, $-2 < a < 2$
39. $x^4 + 4x^3y + 6x^2y^2 + 4xy^3 + y^4 + 4x^3 + 12x^2y + 12xy^2 + 4y^3 + 6x^2 + 12xy + 6y^2 + 4x + 4y + 1$
41. $x^3 - y^3 + z^3 - w^3 - 3x^2y + 3xy^2 + 3x^2z + 3xz^2 + 3y^2z - 3yz^2 - 3x^2w + 3xw^2 - 3y^2w - 3yw^2 - 3z^2w + 3zw^2 - 6xyz + 6xyw - 6xzw + 6yzw$
43. 1.061520   **45.** 0.922744   **47.** 27.270901
49. 1.0099505

Review Problems for Chapter 9, page 462

1. $0, \frac{1}{3}, \frac{1}{2}, \frac{3}{5}, \frac{2}{3}$ **2.** $-2, 2, -\frac{4}{3}, \frac{2}{3}, -\frac{4}{15}$ **3.** 1, 3, 5, 7, 9, 11
4. $2, 1, -1, -5, -13, -29$
5. $(-2) + (-2)^2 + (-2)^3 + (-2)^4 + (-2)^5 = -22$
6. $1 + \dfrac{1}{2} + \dfrac{1}{2^2} + \dfrac{1}{2^3} + \dfrac{1}{2^4} + \dfrac{1}{2^5} = \dfrac{63}{32}$
7. $5 + 5 + 5 + 5 + 5 + 5 + 5 = 35$
8. $\dfrac{2!}{2} + \dfrac{4!}{2^2} + \dfrac{6!}{2^3} + \dfrac{8!}{2^4} = 2617$
9. $\dfrac{1!}{-1} + \dfrac{2!}{(-1)^2} + \dfrac{3!}{(-1)^3} + \dfrac{4!}{(-1)^4} + \dfrac{5!}{(-1)^5} = -101$
10. $\dfrac{2!}{1!} + \dfrac{4!}{2!} + \dfrac{6!}{3!} + \dfrac{8!}{4!} = 1814$ **11.** $\displaystyle\sum_{k=1}^{7} 2k$
12. $\displaystyle\sum_{j=1}^{5} 2^{j-1}(2j)$ **13.** $\displaystyle\sum_{n=1}^{6} (-1)^{n+1}(2n - 1)$
14. $\displaystyle\sum_{i=1}^{7} \dfrac{i^2}{i + 1}$ **15.** $\displaystyle\sum_{j=1}^{6} \dfrac{j}{j + 2}$ **16.** $\displaystyle\sum_{k=1}^{6} \dfrac{k!}{2^{k-1}}$
17. $a_{11} = -8$, $S_{11} = 22$ **18.** $a_1 = -2$, $d = 2$
19. $5, 3, 1, -1, -3$ **20.** $a_{19} = -33$, $a_n = 5 - 2n$
21. $a_5 = -\frac{9}{2}$, $S_5 = -\frac{55}{18}$ **22.** (a) Geometric
sequence, $r = -\frac{3}{2}$, $a_n = -16(-\frac{3}{2})^n$; (b) Not a
geometric sequence   **23.** $r = -\frac{2}{3}$, $a_1 = \frac{729}{4}$
24. $a_6 = -\dfrac{1}{8}$, $a_n = 4\left(-\dfrac{1}{2}\right)^{n-1} = \dfrac{(-1)^{n-1}}{2^{n-3}}$
25. (a) The sum does not exist; (b) $\frac{27}{5}$.
26. -10 **27.** $\frac{106}{33}$ **28.** $x^6 - 12x^5y + 60x^4y^2 - 160x^3y^3 + 240x^2y^4 - 192xy^5 + 64y^6$ **29.** $-2016x^4y^{10}$
30. 1.14868

Critical Thinking Problems for Chapter 9, page 463

1. Errors

$$\sum_{n=2}^{7} (2n - 1) = \cancel{1} + 3 + 5 + 7\cancel{}$$

Correction

$$\sum_{n=2}^{7} (2n - 1) = 3 + 5 + 7 + 9 + 11 + 13$$

$$= 48$$

2. Error

$$a_7 = \cancel{a_1 r^6}$$

Correction

$$a_7 = a_1 + (7 - 1)d = 2 + 6(-3) = -16$$

3. Error

Since $a = 2/3$ and $r = -3/2$, the value of the sum is

$$\cancel{\frac{a}{1 - r}} \quad \cancel{\frac{\frac{2}{3}}{1 - (-\frac{3}{2})}} = \frac{\frac{2}{3}}{\frac{5}{2}} = \frac{4}{15}.$$

Correction

Since $|r| > 1$, the sum does not exist.

4. Errors

The third term is

$$\binom{9}{\cancel{3}} \cancel{x^6 (-2y)^3} = \frac{9 \cdot 8 \cdot 7 \cdot 6!}{3 \cdot 2 \cdot 1 \cdot 6!} x^6 8y^3 = 672 x^6 y^3.$$

Correction

The third term is

$$\binom{9}{2} x^7 (-2y)^2 = \frac{9 \cdot 8 \cdot 7!}{2! \cdot 7!} x^7 4y^2 = 144 x^7 y^2.$$

Chapter 10

Exercises 10.1, pages 471–472

1. $(x + 1)^2 = 4(y - 4)$　**3.** $(x - 1)^2 = -4y$
5. $(y + 2)^2 = 12(x + 1)$　**7.** $(y - 3)^2 = -8(x + 5)$
9. $(y - 1)^2 = -16(x - 1)$　**11.** $(y - 3)^2 = 8(x + 1)$
13. $(x - 1)^2 = 3(y - 2)$　**15.** $(y + 2)^2 = (\frac{4}{3})(x + 1)$

17. V: $(0, 1)$; F: $(0, 3)$
　　D: $y = -1$

19. V: $(-1, 0)$;
　　F: $(-1, -1)$
　　D: $y = 1$

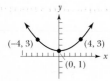

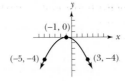

21. V: $(-1, 2)$; F: $(-\frac{3}{4}, 2)$
　　D: $x = -\frac{5}{4}$

23. V: $(2, -1)$; F: $(0, -1)$
　　D: $x = 4$

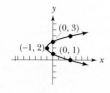

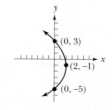

25. V: $(12, -2)$;
　　F: $(\frac{143}{12}, -2)$
　　D: $x = \frac{145}{12}$

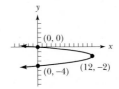

27. V: $(0, 5)$; F: $(\frac{1}{4}, 5)$
　　D: $x = -\frac{1}{4}$

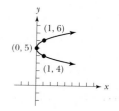

29. V: $(-3, 2)$;
　　F: $(-\frac{23}{8}, 2)$
　　D: $x = -\frac{25}{8}$

31. V: $(-3, 2)$;
　　F: $(-\frac{25}{8}, 2)$
　　D: $x = -\frac{23}{8}$

33.

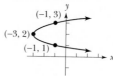

35.

37.

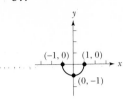

39.

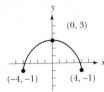

41.

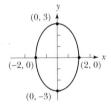

43. $(x - 1)^2 = y + 4$ **45.** $x^2 = -y + 4$
47. 9 meters **49.** $\frac{45}{4}$ centimeters

Exercises 10.2, pages 477–478

1. $V: (0, \pm 3)$;
  $F: (0, \pm \sqrt{5})$

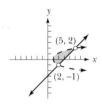

3. $V: (0, \pm 5)$;
  $F: (0, \pm \sqrt{21})$

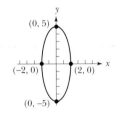

5. $V: (\pm 4, 0)$;
  $F: (\pm 2\sqrt{3}, 0)$

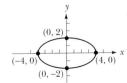

7. $V: \left(\pm \dfrac{5}{2}, 0 \right)$;
  $F: \left(\pm \dfrac{\sqrt{21}}{2}, 0 \right)$

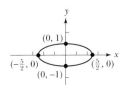

9. $V: \left(\pm \dfrac{3}{2}, 0 \right)$;
  $F: \left(\pm \dfrac{\sqrt{17}}{6}, 0 \right)$

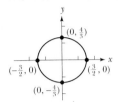

11.

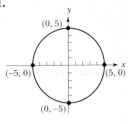

13. $V: \left(0, \pm \dfrac{5}{3} \right)$;
  $F: \left(0, \pm \dfrac{5\sqrt{7}}{12} \right)$

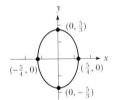

15. $V: (0, \pm 2\sqrt{2})$;
  $F: (0, \pm 2)$

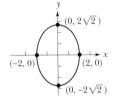

17. $V: (\pm 2\sqrt{3}, 0)$;
  $F: (\pm 2, 0)$

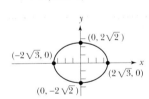

19.

21. $V: (2, 1 \pm 3)$;
  $F: (2, 1 \pm \sqrt{5})$

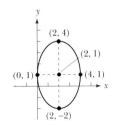

23. $V: (-3 \pm 4, 1)$
  $F: (-3 \pm 2\sqrt{3}, 1)$

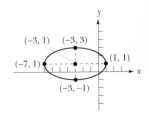

25. V: $(-1, 3 \pm 5)$
F: $(-1, 3 \pm \sqrt{21})$

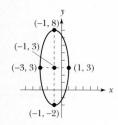

27. V: $(-2, 3 \pm 9)$
F: $(-2, 3 \pm 6\sqrt{2})$

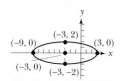

29. V: $(1, -3 \pm 4)$
F: $(1, -3 \pm 2\sqrt{3})$

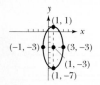

31. V: $(-3 \pm 6, 0)$
F: $(-3 \pm 4\sqrt{2}, 0)$

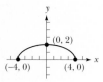

33.

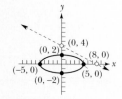

35.

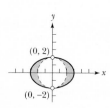

37. $\dfrac{x^2}{9} + \dfrac{y^2}{4} = 1$ **39.** $x^2 + \dfrac{y^2}{16} = 1$

41. $\dfrac{(x-1)^2}{4} + y^2 = 1$ **43.** $\dfrac{(x-2)^2}{4} + \dfrac{(y-3)^2}{9} = 1$

45. $\dfrac{(x+2)^2}{25} + \dfrac{(y-1)^2}{9} = 1$ **47.** $\dfrac{x^2}{9} + \dfrac{y^2}{5} = 1$

49. No solution   **51.**

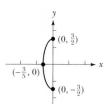
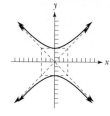

53. $\dfrac{x^2}{16} + \dfrac{y^2}{4} = 1$ **55.** $\dfrac{x^2}{4} + \dfrac{y^2}{36} = 1$
57. $9\sqrt{3}/2$ feet   **59.** $3\sqrt{3}$ feet

Exercises 10.3, pages 484–486

1. C: $(0, 0)$; V: $(\pm 3, 0)$
F: $(\pm 5, 0)$;
A: $y = \pm \frac{4}{3}x$

3. C: $(0, 0)$; V: $(0, \pm 2)$
F: $(0, \pm \sqrt{29})$;
A: $y = \pm \frac{2}{5}x$

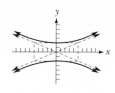
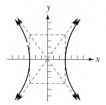

5. C: $(0, 0)$; V: $(0, \pm 3)$
F: $(0, \pm 3\sqrt{2})$;
A: $y = \pm x$

7. C: $(0, 0)$; V: $(\pm 3, 0)$
F: $(\pm \sqrt{13}, 0)$;
A: $y = \pm \frac{2}{3}x$

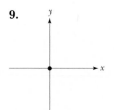
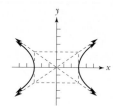

9.

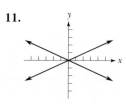

11.

13. C: $(0, 0)$; V: $(\pm 2, 0)$
F: $(\pm \sqrt{5}, 0)$;
A: $y = \pm \frac{1}{2}x$

15.

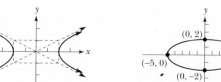

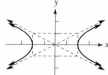

17. C: $(0, 0)$; V: $(0, \pm\frac{5}{4})$
F: $(0, \pm\frac{17}{12})$;
A: $y = \pm\frac{15}{8} x$

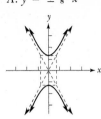

19.

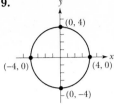

29. C: $(0, 1)$; V: $(0, 1 \pm 1)$
F: $(0, 1 \pm \sqrt{10})$; A: $y - 1 = \pm\frac{x}{3}$

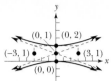

21. C: $(0, 0)$; V: $(\pm 2, 0)$
F: $(\pm 2\sqrt{3}, 0)$;
A: $y = \pm\sqrt{2}x$

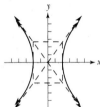

23. C: $(1, 2)$;
V: $(1 \pm 3, 2)$
A: $y - 2 =$
$\pm\frac{5}{3}(x - 1)$
F: $(1 \pm \sqrt{34}, 2)$

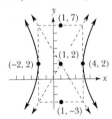

31. C: $(0, 1)$; V: $(\pm 2, 1)$
F: $(\pm 2\sqrt{5}, 1)$; A: $y - 1 = \pm 2x$

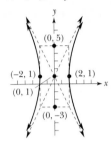

33. C: $(-2, -1)$; V: $(-2, -1 \pm 3)$
F: $(-2, -1 \pm \sqrt{13})$; A: $y + 1 = \pm\frac{3}{2}(x + 2)$

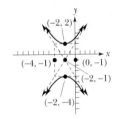

25. C: $(2, -3)$; V: $(2, -3 \pm 2)$
F: $(2, -3 \pm \sqrt{5})$; A: $y + 3 = \pm 2(x - 2)$

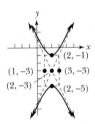

35.

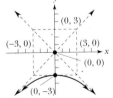

37.

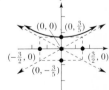

27. C: $(-2, -2)$; V: $(-2 \pm 3, -2)$
F: $(-2 \pm \sqrt{13}, -2)$; A: $y + 2 = \pm\frac{2}{3}(x + 2)$

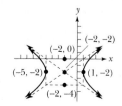

39. $\dfrac{x^2}{16} - \dfrac{y^2}{9} = 1$ **41.** $\dfrac{y^2}{4} - \dfrac{x^2}{12} = 1$

43. $\dfrac{x^2}{16} - \dfrac{y^2}{3} = 1$ **45.** $\dfrac{x^2}{64} - \dfrac{y^2}{36} = 1$

47. $A: \left(\dfrac{3\sqrt{65}}{13}, \dfrac{4\sqrt{26}}{13}\right)$; $B: \left(\dfrac{-3\sqrt{65}}{13}, \dfrac{4\sqrt{26}}{13}\right)$;

$C: \left(-\dfrac{3\sqrt{65}}{13}, \dfrac{-4\sqrt{26}}{13}\right)$; $D: \left(\dfrac{3\sqrt{65}}{13}, -\dfrac{4\sqrt{26}}{13}\right)$

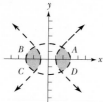

49. $A: (4\sqrt{2}, 4)$; $B: (-4\sqrt{2}, 4)$
$C: (-4\sqrt{2}, -4)$; $D: (4\sqrt{2}, -4)$

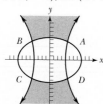

51. $\dfrac{x^2}{4} - \dfrac{y^2}{9} = 1$　　**53.** $\dfrac{y^2}{4} - \dfrac{x^2}{4} = 1$

Exercises 10.4, page 493

1. Circle　**3.** Ellipse　**5.** Parabola　**7.** Hyperbola
9. Hyperbola　**11.** Circle　**13.** Hyperbola
15. Parabola　**17.** Hyperbola　**19.** Ellipse
21. $\dfrac{(x')^2}{4} - \dfrac{(y')^2}{4} = 1$　　　**23.** $y' = 2(x')^2$

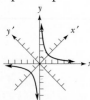　　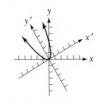

25. $\dfrac{(x')^2}{7} + \dfrac{(y')^2}{1} = 1$　　**27.** $\dfrac{(x')^2}{4} + \dfrac{(y')^2}{9} = 1$

29. $y' = -(x')^2$　　　　**31.** $(x')^2 - (y')^2 = 1$

33. $\dfrac{(x' - 1)^2}{4} - \dfrac{(y' - 1)^2}{4} = 1$

35. $y' = 2(x')^2 + 1$

37. $\dfrac{(x' + 2)^2}{4} + \dfrac{(y' - 1)^2}{3} = 1$

Exercises 10.5, pages 499–500

1. $(2\sqrt{3}, 2)$　**3.** $(-\sqrt{2}/2, \sqrt{2}/2)$　**5.** $(-1, \sqrt{3})$
7. $(-3\sqrt{3}/2, -3/2)$　**9.** $(-\sqrt{2}, -\sqrt{2})$　**11.** $(0, 5)$
13. $(-1, \sqrt{3})$　**15.** $(3, 0)$　**17.** $(2, 135°)$
19. $(2, 210°)$　**21.** $(1, 270°)$　**23.** $(4\sqrt{2}, 315°)$
25. $r \cos \theta = 3$　**27.** $\theta = 45°$　**29.** $r^2 = 1$, or $r = 1$,
or $r = -1$　**31.** $r^2 - 4r \cos \theta = 0$, or $r = 4 \cos \theta$
33. $y = 2$　**35.** $x^2 + y^2 + x - y = 0$
37. $x + y = 1$　**39.** $x^2 + y^2 - x = 1$
41. $x^2 + 4y = 4$　**43.** $3x^2 + 4y^2 - 12x = 36$

45.

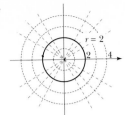

47.

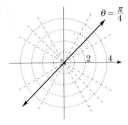

65.

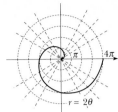

67.

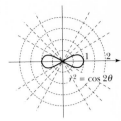

Exercises 10.6, page 506

1. $e = 1$, parabola, $r = 2 \csc \theta$ **3.** $e = 1$, parabola, $r = -\frac{3}{2} \sec \theta$ **5.** $e = \frac{1}{2}$, ellipse, $r = 6 \csc \theta$
7. $e = \frac{3}{2}$, hyperbola, $r = \frac{4}{3} \sec \theta$ **9.** $e = \frac{3}{2}$, hyperbola, $r = -2 \sec \theta$ **11.** $e = \sqrt{2}/2$, ellipse, $r = -2 \csc \theta$

13. Ellipse, $e = \frac{1}{2}$

r	θ
1/2	0
3/4	$\pi/2$
3/2	π
3/4	$3\pi/2$
1/2	2π

49.

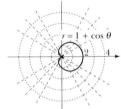

51.

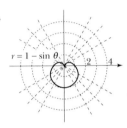

15. Ellipse, $e = \frac{1}{2}$

r	θ
3/4	0
1/2	$\pi/2$
3/4	π
3/2	$3\pi/2$
3/4	2π

53.

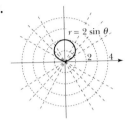

55.

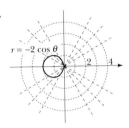

17. Parabola, $e = 1$

r	θ
3/2	0
Undef.	$\pi/2$
3/2	π
3/4	$3\pi/2$
3/2	2π

57.

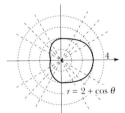

59.

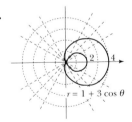

19. Hyperbola, $e = 2$

r	θ
-3	0
3	$\pi/2$
1	π
3	$3\pi/2$
-3	2π

61.

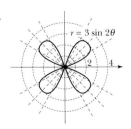

63.

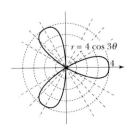

ANSWERS

A-78

21. Ellipse, $e = \frac{1}{2}$

r	θ
6	0
3	$\pi/2$
2	π
3	$3\pi/2$
6	2π

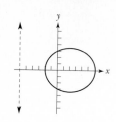

23. Parabola, $e = 1$

r	θ
Undef.	0
-2	$\pi/2$
-1	π
-2	$3\pi/2$
Undef.	2π

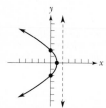

25. $r = \dfrac{12}{1 - 3\cos\theta}$ **27.** $r = \dfrac{2}{1 + \sin\theta}$

29. $r = \dfrac{5}{3 + 2\cos\theta}$ **31.** $r = \dfrac{6}{1 + \cos\theta}$

33. $r = \dfrac{3}{1 + 2\cos\theta}$ **35.** $r = \dfrac{30}{3 + 2\sin\theta}$ or

$r = \dfrac{6}{3 - 2\sin\theta}$

Exercises 10.7, pages 511–512

1. (a) $y = x - 6$
(b)

3. (a) $x + 2y = 2$
(b)
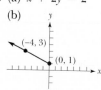

5. (a) $y = 2(x + 1)^2 + 1$
(b)
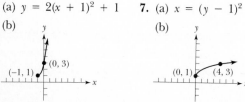

7. (a) $x = (y - 1)^2$
(b)

9. (a) $y = (x + 1)^2 + 2$
(b)
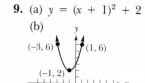

11. (a) $y = x^2$
(b)
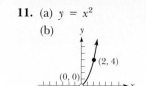

13. (a) $y = 6 - 2x$
(b)
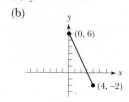

15. (a) $y = x^2$
(b)
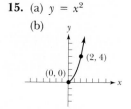

17. (a) $y = \sqrt{4 - x^2}$
(b)
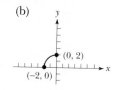

19. (a) $y = 1/x$
(b)
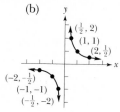

21. (a) $y = 1/x$
(b)
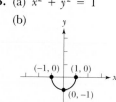

23. (a) $x^2 + y^2 = 1$
(b)
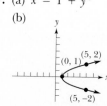

25. (a) $y = 1 + x^2$
(b)
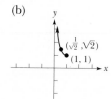

27. (a) $x = 1 + y^2$
(b)
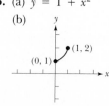

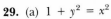
29. (a) $1 + y^2 = x^2$

(b)

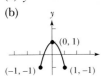

31. (a) $(x - 2)^2 + (y + 1)^2 = 1$

(b)

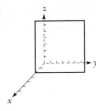

33.

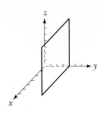

35.

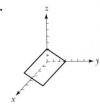

33. (a) $y = 1 - 2x^2$

(b)

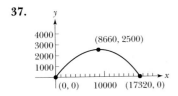

37.

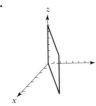

39.

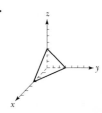

35. $10{,}000\sqrt{3} \approx 17{,}321$ feet

37.

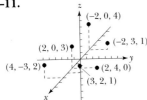

41.

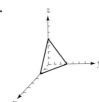

43.

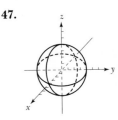

39. $y = (\tan \theta)x - (16 \sec^2\theta/v_0^2)x^2$

Exercises 10.8, page 522

1–11.

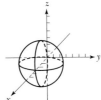

13. $3, (\frac{1}{2}, 1, 0)$ **15.** $3, (2, -\frac{1}{2}, 3)$ **17.** $2\sqrt{17}$, $(-4, 3, 3)$ **19.** $\sqrt{62}, (-\frac{1}{2}, 0, 7/2)$
21. $(x - 2)^2 + (y + 3)^2 + (z - 1)^2 = 1$
23. $(x - 1)^2 + (y + 2)^2 + (z - 2)^2 = 16$
25. $(x + 2)^2 + (y - 4)^2 + z^2 = 11$
27. $(-4, 2, -3), 4$
29. $(4, -1, 2), 5$ **31.** Not a sphere

45.

47.

49.

51. $x^2 + y^2 + z^2 = 26$
53. $(x + 2)^2 + (y + 3)^2 + (z - 1)^2 = 14$
55. The graph consists of the single point $(1, -5, 2)$.

Review Problems for Chapter 10, pages 524–525

1. $(x + 2)^2 = 12(y - 1)$　**2.** $(y + 1)^2 = -12(x - 4)$
3. $V: (-6, -1);$　　　　**4.** $V: (2, 2); F: (3/2, 2)$
　　$F: (-11/2, -1)$

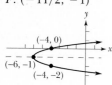

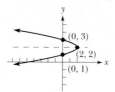

5. $C: (2, -1);$　　　　**6.** $C: (0, 0); V: (0, \pm 2)$
　　$V: (2 \pm 3, -1)$　　　　$F: (0, \pm\sqrt{29});$
　　$F: (2 \pm \sqrt{5}, -1)$　　　$A: y = \pm 2x/5$

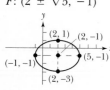

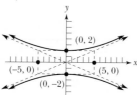

7. $C: (-3, 2); V: (-3 \pm 4, 2)$
　　$F: (-3 \pm 2\sqrt{5}, 2); A: y - 2 = \pm(x + 3)/2$

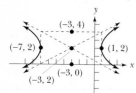

8.

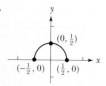

9.

10.

11.

12. $\dfrac{(x - 1)^2}{16} + \dfrac{(y - 2)^2}{9} = 1$

13. $\dfrac{(x - 1)^2}{9} + \dfrac{(y - 1)^2}{25} = 1$　**14.** $\dfrac{y^2}{36} - \dfrac{x^2}{4} = 1$

15. $\dfrac{x^2}{9} - \dfrac{y^2}{16} = 1$　**16.** Ellipse　**17.** Ellipse

18. Parabola　**19.** Hyperbola　**20.** Ellipse
21. Hyperbola　**22.** Hyperbola　**23.** Hyperbola

24. $\dfrac{(x')^2}{9} + \dfrac{(y')^2}{1} = 1$　　**25.** $\dfrac{(x')^2}{4} + \dfrac{(y')^2}{1} = 1$

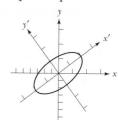

26. $y' = -2(x')^2$　　　**27.** $(x')^2 - (y' - 2)^2 = 1$

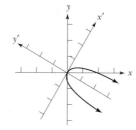

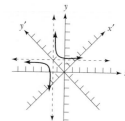

28. $r = 2 \cos \theta$　**29.** $r(3 \cos \theta + 4 \sin \theta) = 12$
30. $y - 2x = 4$　**31.** $xy = 1$

32. 　　　　**33.**

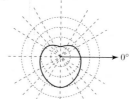

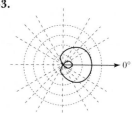

34.　　　　　**35.**

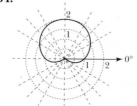

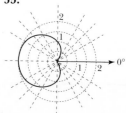

36. Parabola, $e = 1$

$(3, 0)$

37. Parabola, $e = 1$

$(1, 0)$

38. Ellipse, $e = \frac{1}{2}$

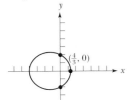

$\left(\frac{4}{3}, 0\right)$

39. Ellipse, $e = \frac{1}{2}$

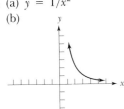

$(3, 0)$

40. $r = \dfrac{2}{1 + \cos \theta}$ **41.** $r = \dfrac{8}{3 + 2 \sin \theta}$

42. $r = \dfrac{8}{1 + 3 \cos \theta}$ **43.** $r = \dfrac{6}{2 + \cos \theta}$

44. (a) $y = 1/x^2$
(b)

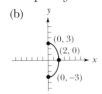

45. (a) $x = (y + 5)^2$
(b)

$(25, 0)$
$(0, -5)$
$(25, -10)$

46. (a) $\dfrac{x^2}{4} + \dfrac{y^2}{9} = 1$
(b)

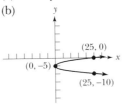

$(0, 3)$
$(2, 0)$
$(0, -3)$

47. (a) $\dfrac{x^2}{9} + \dfrac{y^2}{16} = 1$
(b)

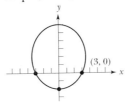

$(0, 4)$
$(3, 0)$

48. (a) $y = 2x^2$
(b)

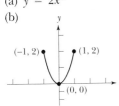

$(-1, 2)$ $(1, 2)$
$(0, 0)$

49. (a) $y = 1 - \dfrac{x^2}{4}$
(b)

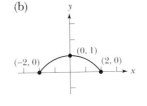

$(0, 1)$
$(-2, 0)$ $(2, 0)$

50. $2\sqrt{11}$, $(0, 4, 7)$ **51.** $2\sqrt{22}$, $(-1, -1, 3)$
52. 3, $(5/2, -1, 2)$ **53.** $\sqrt{69}$, $(2, -2, \frac{1}{2})$
54. $(x - 2)^2 + (y + 3)^2 + (z - 1)^2 = 16$
55. $(x + 3)^2 + y^2 + (z - 4)^2 = 1$
56. $(x + 2)^2 + (y - 3)^2 + (z - 5)^2 = 4$
57. $(x + 3)^2 + (y - 6)^2 + (z + 1)^2 = 9$
58. $C(6, -2, 3)$, $r = 7$ **59.** $C(0, 4, -3)$, $r = 5$
60. Not a sphere   **61.** Not a sphere

62.

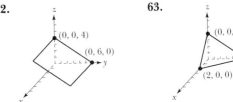

$(0, 0, 4)$
$(0, 6, 0)$

63.

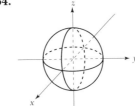

$(0, 0, 4)$
$(0, 6, 0)$
$(2, 0, 0)$

64.

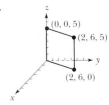

65.

$(0, 0, 5)$
$(2, 6, 5)$
$(2, 6, 0)$

Critical Thinking Problems for Chapter 10, pages 526–527

1. Error

$2\theta = 60°$ and
$x = x' \cos \cancel{60°} - y' \sin \cancel{60°}$
$y = x' \sin \cancel{60°} + y' \cos \cancel{60°}$

Correction

$2\theta = 60°$, so $\theta = 30°$ and
$x = x' \cos 30° - y' \sin 30° = \frac{1}{2}(\sqrt{3}x' - y')$
$y = x' \sin 30° + y' \cos 30° = \frac{1}{2}(x' + \sqrt{3}y')$

2. Error

The midpoint is given by
$$\left(\frac{5 \cancel{\times} 1}{2}, \frac{5 \cancel{\times} 3}{2}, \frac{10 \cancel{\times} 4}{2} \right).$$

Correction

The midpoint is given by
$$\left(\frac{5 + 1}{2}, \frac{5 + 3}{2}, \frac{10 + 4}{2} \right) = (3, 4, 7).$$

3. Error

$$x^2 + 6x + y^2 - 8y + z^2 - 2z = 25$$
$$x^2 + 6x + 9 + y^2 - 8y + 16 + z^2 - 2z + 1 = \cancel{25}$$

Correction

We add $9 + 16 + 1$ to *both* sides of the equation.

$$x^2 + 6x + 9 + y^2 - 8y + 16 + z^2 - 2z + 1 = 25 + 9 + 16 + 1$$

$$(x + 3)^2 + (y - 4)^2 + (z - 1)^2 = 51$$

The center is at $(-3, 4, 1)$, and the radius is $\sqrt{51}$.

4. Error

The graph is a ~~line~~ in the *yz*-plane. . . .

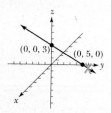

Correction

The graph is a plane, and we sketch the plane by drawing a parallelogram.

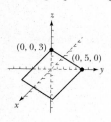

Exercises A.1, pages A-6–A-7

1. $1.6191 - 10$ **3.** 1.4871 **5.** 0.6542 **7.** 1
9. 6.0294 **11.** $7.0294 - 10$ **13.** 7.18
15. $283,000,000$ **17.** 0.00476 **19.** 0.001
21. 1.0047 **23.** 0.6202 **25.** 5.6210
27. $7.8875 - 10$ **29.** 1.292 **31.** $87,760$
33. 0.004954 **35.** 0.2445

Exercises A.2, page A-11

1. 65.3 **3.** 0.190 **5.** 2130 **7.** 1.96 **9.** 2.04
11. 0.00280 **13.** 0.04425 **15.** 0.8474 **17.** 0.7992
19. 557.1 **21.** $4,285,000,000$ **23.** 0.4601
25. 20.2 **27.** 8260

Index

A

Abscissa, 65
Absolute value, 5, 342
  equations, 6
  inequalities, 7
Acute angle, 183
Addition:
  of complex numbers, 28
  of fractions, 16–17
  of functions, 72
  of matrices, 368
  of ordinates, 247
  of polynomials, 12
  of vectors, 331
Additive identity, 369
Additive inverse, 369
Adjacent side, 216
Airspeed, 337
Algebraic expression, 11
Algebraic method of solution of
    an inequality, 42
Ambiguous case, 312
Amplitude:
  of a complex number,
    342
  of a function, 228
Angle, 182
  acute, 183
  central, 185
  complementary, 183
  composite, 266

coterminal, 193
degree measure, 183, 185
of depression, 220
of elevation, 220
of incidence, 214
initial side, 182
measure, 182, 185–187
negative, 182
obtuse, 183
positive, 182
quadrantal, 195
radian measure, 185
reference, 197
of refraction, 214
related, 197
right, 183
of rotation, 490
in standard position, 192
straight, 183
supplementary, 183
terminal side, 182
vertex, 182
Angular speed, 188
Antilogarithm, A-4
Approximate equality, 4
Arc length formula, 187
Arcsine function, 296
Area of a triangle, 326–327
Argument:
  of a complex number, 342
  of a function, 228

Arithmetic sequence, 434
Associative property:
  of addition of matrices, 369
  of multiplication of matrices,
    372
Asymptotes:
  horizontal, 115
  of a hyperbola, 480
  oblique, 118
  vertical, 115
Augmented matrix, 377
Axis, 64
  conjugate, 480
  imaginary, 341
  major, 474
  minor, 474
  polar, 493
  real, 341
  of symmetry, 88, 467
  transverse, 480

B

Back substitution, 378
Base, 10, 156–157
Base vectors, 335
Basic properties of radicals, 22
Basic properties of vector opera-
    tions, 334
Bearing, 221
Binomial, 11

Binomial coefficients, 456, 459
Binomial theorem, 456
Biorhythm, 228
Boundary, 412
Bounds for zeros, 143
Broken-line graphs, 82

C

Calculation rules for approximate
     numbers, 217–218
Cardioid, 497
Cartesian coordinate system, 64,
     514
Central angle, 185
Change-of-base formula, 176
Characteristic of a logarithm, A-3
Characteristic values, 402
Circle, 101, 202, 466
Circular function, 205
Coefficient matrix, 377, 388
Cofactor, 398
Cofactor expansion, 399–400
Cofunction, 217, A-13
Cofunction identities, 269
Column matrix, 367
Column operations, 403
Common difference, 434
Common logarithm, 172
Common ratio, 440
Commutative property, 369
Complementary angles, 183,
     A-12
Complementary functions, 217,
     A-13
Completing the square, 49
Complex fraction, 16
Complex numbers, 27
   absolute value, 342
   addition, 28
   amplitude, 342
   argument, 342
   conjugate, 29
   difference of, 28
   division of, 29
   equality, 28
   imaginary part, 27
   modulus, 342
   multiplication, 28, 345
   polar form, 342
   powers, 348
   quotients, 29, 345
   real part, 27
   roots, 349
   standard form, 28

   subtraction, 28
   trigonometric form, 342
Components:
   of an ordered pair, 61
   of a vector, 332
Composition function, 74
Compound amount, 158, 163
Compound inequality, 4
Conditional equation, 263
Cone, 466
Conformable matrices, 371
Conic sections, 466, 500
Conjugate of a binomial, 23
Conjugate of a complex number,
     29
Conjugate pairs of zeros, 137
Constant matrix, 377, 388
Constraints, 417
Conversion of degrees to radians,
     185
Conversion of radians to degrees,
     185
Convex region, 417
Coordinate planes, 513
Coordinates, 65, 493, 514
Cosecant function, 193, 205
Cosine function, 193, 205
Cosine:
   of the difference of two angles,
     267
   of half an angle, 281
   of the sum of two angles, 270
   of twice an angle, 278
Cosines, law of, 320, 321
Cotangent function, 193, 205
Coterminal angles, 193
Counterexample, 263
Course, 336
Cramer's Rule, 408, 410
Cube root, 20
Curve, 507
Cycloid, 510

D

Decibels, 174
Degree, 183
   of a term, 11
   of a polynomial, 11
De Moivre's Theorem, 348
Dependent system, 409, 410
Depreciation, 83
Descartes' rule of signs, 135
Determinant, 396–397
   cofactors of, 398

   expansion of, 399–400
   order of, 396
   properties of, 403–404
Diagonal matrix, 367
Diameter of a sphere, 516
Difference:
   common, 434
   of complex numbers, 28
   of matrices, 369
Dimension of a matrix, 367
Direction angle, 336
Directrix, 467, 500
Discriminant, 33, 487
Distance formula, 78, 515
Division:
   by detached coefficients, 128
   of functions, 72
   of polynomials, 128
   synthetic, 128
Divisor, 128
Domain, 60, 61
Dot product, 340
Double-angle identities, 278

E

Eccentricity, 479, 500
Eigenvalues, 402
Element:
   of a matrix, 367
   of a set, 2
Elimination method, 360, 379–380
Ellipse, 466, 472
   center of, 474
   foci of, 472
   major axis of, 474
   minor axis of, 474
   standard equations for, 474
   vertices of, 474
Empty set, 2
Equality:
   of complex numbers, 28
   of matrices, 368
   of sets, 2
   of vectors, 331
Equations:
   absolute value, 6
   conditional, 263
   exponential, 173
   inverse trigonometric, 301
   linear, 76
   logarithmic, 172
   polar, 497
   quadratic, 33
   trigonometric, 288

Equilateral triangle, 184
Exponent:
    definition of, 10, 24, 156
    laws of, 10
Exponential equation, 173
Exponential form, 24, 166
Exponential function, 157
Expressions:
    algebraic, 11
    rational, 15
    trigonometric, 258
Extraneous solutions, 35

F

Factor, 13, 107
    of multiplicity k, 134
Factorial, 455
Factoring by grouping, 14
Factorization formulas, 13
Factor theorem, 130
Feasible solutions, 417
Fibonacci sequence, 431
Finite sequence, 430
Focus:
    of a conic section, 500
    of an ellipse, 472
    of a hyperbola, 479
    of a parabola, 467
Four-leaved rose, 499
Function, 60, 61
    circular, 205
    composition, 74
    complementary, 217,
        A-13
    exponential, 157
    greatest integer, 68
    inverse, 95
    linear, 76, 417
    logarithmic, 166
    one-to-one, 95
    periodic, 227
    quadratic, 86
    rational, 112
    sinusoidal, 228
    step, 68
    trigonometric, 193
    wrapping, 204
Function value notation, 60
Fundamental identities,
    259
Fundamental principle of frac-
    tions, 15
Fundamental theorem of algebra,
    133

G

Gaussian elimination, 379
Gauss-Jordan elimination:
    to find an inverse, 386–387
    to solve a system, 381
General quadratic equation, 487
General term:
    of an arithmetic sequence, 437
    of a geometric sequence, 443
    of a sequence, 430
Geometric sequence, 440
Graph:
    of a cosecant function, 240
    of a cosine function, 235
    of a cotangent function, 246
    of an equation, 68, 516
    of an exponential function, 158
    of a function, 66
    of an inequality, 412
    of an interval, 5
    of a logarithmic function, 167
    of a polynomial function, 108
    of a rational function, 116
    of a relation, 66
    of a secant function, 242
    of a sine function, 229, 233
    of a tangent function, 245
Greater than, 4
Greatest integer function, 68
Ground speed, 337

H

Half-angle identities, 281
Half-line, 182
Half-plane, 412
Head of a vector, 330
Heading, 337
Heron's formula, 327
Horizontal asymptote, 115
Horizontal line, 77
Horizontal line test, 95
Horizontal translation, 103, 232
Hyperbola, 466, 479
    asymptotes of, 480
    axes of, 480
    center of, 480
    foci of, 479
    standard equations for, 481–482
    vertices of, 480

I

Identities, 207
    cofunction, 269
    double-angle, 278

    fundamental, 259
    half-angle, 281
    negative angle, 270
    product, 285
    Pythagorean, 259
    quotient, 208, 259
    reciprocal, 208, 259
    reduction, 276
Identity matrix, 373
Imaginary axis, 341
Imaginary number, 28
Imaginary part of a complex num-
    ber, 27
Inconsistent system, 409, 410
Index:
    of a radical, 21
    of refraction, 214
    of summation, 432
Induction hypothesis, 451
Induction principle, 450
Inequalities, 4
    absolute value, 7
    compound, 4
    linear, 412
    quadratic, 40
    systems of, 413
Infinite sequence, 430
Infinity, 4
Initial point of vector, 330
Initial side, 182
Inner product, 340
Intercepts of a graph, 76
Interpolation, A-4
Intersection, 2, 358
Interval notation, 4
Inverse:
    of a function, 95
    of a matrix, 385
    of a relation, 92
Inverse sine function, 296
Inverse trigonometric equations,
    301
Inverse trigonometric functions,
    297
Isosceles triangle, 184

K

Key points on a graph, 229, 235

L

Law of Cosines, 320, 321
Laws of Exponents, 10, 156–157
Law of Sines, 311

Leading coefficient, 11
Least common denominator, 16
Left-handed system, 513
Length:
    of an arc, 187
    of a vector, 330
Less than, 4
Linear depreciation, 83
Linear equation, 76
    point-slope form of, 81
    slope-intercept form of, 81
    standard form of, 80
Linear function, 76, 417
Linear inequality, 412
Linear interpolation, A-4
Linear programming problem, 417
Linear relation, 76
Linear speed, 188
Lines:
    equations of, 80–81
    parallel, 82
    perpendicular, 82
    straight, 76
Location theorem, 147
Logarithmic equation, 172
Logarithmic function, 166
Logarithms, 166
    characteristic, A-3
    common, 172
    mantissa, A-3
    natural, 175
    properties of, 168
Lower bound, 143
Lowest terms, 15

M

Magnitude of a vector, 330
Main diagonal elements, 367
Major axis of an ellipse, 474
Mantissa, A-3
Mathematical induction, 450
Matrix, 367
    addition of, 368
    augmented, 377
    coefficient, 377, 388
    column, 367
    constant, 377, 388
    determinant of, 396, 397
    diagonal, 367
    dimension of, 367
    equality of, 368
    identity, 373
    multiplication, 370

multiplicative inverse of, 385
    row, 367
    square, 367
    sum, 368
    unknown, 388
    zero, 369
Maximize, 419
Measure of an angle, 183, 185
Method:
    of elimination, 360, 378
    of substitution, 358
Midpoint formula, 516
Minimize, 419
Minor, 398
Minor axis of an ellipse, 474
Minute, 183
Modulus, 342
Monomial, 11
Multiplication:
    of complex numbers, 28, 345
    of functions, 72
    of matrices, 370
Multiplicative inverse of a matrix, 385
Multiplicity of zeros, 107, 134

N

nth root, 20
nth term:
    of an arithmetic sequence, 437
    of a geometric sequence, 443
    of a sequence, 430
Natural logarithm, 175
Natural number e, 161
Negative angle identities, 270
Negative infinity, 4
Newton's law of cooling, 179
Number line, 4
Numbers:
    complex, 27
    counting, 3
    irrational, 3
    natural, 3
    rational, 3
    real, 3

O

Oblique asymptote, 118
Oblique triangle, 184, 310
Obtuse angle, 183
Octant, 514
One-to-one function, 95
Operations with radicals, 21–26
Opposite side, 216

Order:
    of a determinant, 396
    of a matrix, 367
    of a radical, 21
Ordered pair, 61
Ordering of real numbers, 4
Ordinate, 65
Origin, 3, 64, 513

P

Parabola, 87, 466, 467
Parallel lines, 82
Parallelogram rule, 331, 341
Parameter, 507
Parametric equations, 507
Partial fractions, 390
Pascal's triangle, 460
Period, 227
Periodic function, 227
Perpendicular lines, 82
Phase shift, 232
Plane curve, 507
Point-circle, 104
Points of intersection, 358
Point-slope form, 81
Polar axis, 493
Polar coordinates, 493, 494
Polar equations, 497
    for conic sections, 502
Polar form, 342
Pole, 493
Polynomial, 11
    addition of, 12
    completely factored, 13
    degree of, 11
    division of, 128
    in simplest form, 12
    irreducible, 14
    monic, 11
    prime, 14
    rational zeros of, 140
    zeros of, 107, 140
Positive integers, 3
Powers of i, 31
Principal, 158
Principal nth root, 21
Principal square root, 20, 30
Principle of Mathematical Induction, 450
Product:
    of complex numbers, 28, 345
    of matrices, 370
    of vectors, 340
Product identities, 285

Properties:
  of determinants, 403–404
  of logarithms, 168, A-7
  of radicals, 22
Pythagorean identities, 259
Pythagorean Theorem, 78

Q

Quadrantal angles, 195
Quadrants, 65
Quadratic equation, 33
Quadratic formula, 33, 126
Quadratic function, 86
Quadratic inequality, 40
Quotient identities, 208, 259
Quotients:
  of complex numbers, 29, 345
  of polynomials, 128

R

Radian, 185
Radical, 21
  form, 24
  index of, 21
  order of, 21
Radicand, 21
Radius:
  of a circle, 101
  of a sphere, 516
Range, 61
  of a projectile, 512
  of the trigonometric functions, 209–210
Rational exponents, 24
Rational expression, 15
Rational function, 112
Rationalizing the denominator, 23
Rationalizing the numerator, 48
Rational number, 3
Rational zeros, 140
Ray, 182
Real axis, 341
Real part, 27
Reciprocal identities, 208, 259
Rectangular coordinate system, 64, 514
Reduction identity, 276
Reference angle, 197
Reference triangle, 198
Reflection, 102
Related angle, 197
Related angle theorem, 199
Related number, A-14
Related triangle, 198

Relations, 61
  linear, 76
  quadratic, 86, 469
Remainder theorem, 127
Resultant, 331
Revolution, 183
Right angle, 183
Right-handed system, 513
Root, nth, 20
Root, principal square, 20, 30
Roots of complex numbers, 349
Roots of an equation, 33–34
Rose, 499
Rotation of axes, 487
Rotation formulas, 488
Row matrix, 367
Row operations:
  on a determinant, 402
  on a matrix, 379
rth term of a binomial expansion, 458
Rules for calculations with approximate numbers, 217–218

S

Scalar, 333, 370
  multiplication, 333, 370
Scalar quantity, 330
Scientific notation, 18, A-2
Secant function, 193, 205
Second, 183
Sequence:
  arithmetic, 434
  geometric, 440
  Fibonacci, 431
  finite, 430
  infinite, 430
  terms of, 430
Series, 431
Set-builder notation, 2
Sets, 2
  empty, 2
  equal, 2
  intersection of, 2
  union of, 2
Side adjacent, 216
Side opposite, 216
Sides of an angle, 182
Sigma notation, 432
Sign graph, 41
Significant digits, 217
Signs of the trigonometric functions, 194
Similar triangles, 184

Simplest form of a polynomial, 12
Simplest radical form, 23
Sine, 193, 205
  of half an angle, 281
  of twice an angle, 278
Sine of the sum and difference of two angles, 273
Sines, Law of, 311
Sinusoidal function, 228
Slope of a line, 79
Slope-intercept form, 81
Snell's law, 214
Solutions:
  extraneous, 35
  feasible, 417
  of inequalities, 40, 412
  of nonlinear inequalities, 41
  of quadratic equations, 33
  of quadratic inequalities, 40
  of systems of equations, 358
  of systems of inequalities, 413
  of systems of linear equations, 378–379, 389, 409, 410
Special products, 12
Speed, 188, 337
Sphere, 516, 517
Square matrices, 367
Square root, 20
Standard form:
  for a complex number, 28
  for the equation of a circle, 103
  for the equation of an ellipse, 474
  for the equation of a hyperbola, 481–482
  for the equation of a line, 80
Standard position, 192
Step function, 68
Straight angle, 183
Straight line, 76
  depreciation, 83
Stretching, 102
Subset, 2
Substitution method, 358
Subtraction:
  of complex numbers, 28
  of functions, 72
  of matrices, 369
Sum:
  of an arithmetic sequence, 437
  of complex numbers, 28
  of an infinite geometric sequence, 446
  of matrices, 368

of *n* terms of a geometric sequence, 443
Sum identities, 286
Summation notation, 432
Supplementary angles, 183
Surface, 517
Symmetry, 87–88, 98
Synthetic division, 128
Systems of equations, 358
Systems of inequalities, 413
Systems of linear equations:
   dependent, 409, 410
   inconsistent, 409, 410

T
Tail of a vector, 330
Tangent function, 193, 205
Tangent of the sum and difference of two angles, 273
Terminal point of a vector, 330
Terminal side of an angle, 182
Terms of a sequence, 430
Tests for symmetry, 99
Trace, 520
Translation, 102–103
Triangle:
   equilateral, 184
   isosceles, 184
   oblique, 184, 310
   similar, 184
Trigonometric equations, 288
Trigonometric expressions, 258

Trigonometric form, 342
Trigonometric functions:
   of an angle, 193, 217
   of a real number, 205
Trinomial, 11
Typical shape:
   for a cosine curve, 235
   for a cotangent curve, 246
   for a sine curve, 229
   for a tangent curve, 245

U
Union of sets, 2
Unit circle, 202
Unit vector, 336
Unknown matrix, 388
Upper bound, 144

V
Variable:
   dependent, 61
   dummy, 432
   independent, 61
Variation of sign, 135
Vector, 330
   addition, 331
   quantity, 330
Velocity, 337
Vertex, 417
   of an angle, 182
   of a parabola, 88, 467

Vertical asymptote, 115
Vertical line, 77
Vertical line test, 67
Vertical translation, 102
Vertices:
   of an ellipse, 474
   of a hyperbola, 480
   of parabolas, 88, 467

W
Wrapping function, 204

X
x-axis, 64, 513
x-component, 332
x-coordinate, 65
x-intercept, 76

Y
y-axis, 64, 513
y-component, 332
y-coordinate, 65
y-intercept, 76

Z
z-axis, 513
Zero of a function, 164
Zero matrix, 369
Zero of a polynomial, 107, 132
   rational, 140
   real, 132
Zero vector, 331

Inverse Trigonometric Functions

$y = \sin^{-1} x = \arcsin x$ $\qquad -\dfrac{\pi}{2} \le y \le \dfrac{\pi}{2}$

$y = \csc^{-1} x = \operatorname{arccsc} x$ $\qquad -\dfrac{\pi}{2} \le y \le \dfrac{\pi}{2}, \ y \ne 0$

$y = \tan^{-1} x = \arctan x$ $\qquad -\dfrac{\pi}{2} < y < \dfrac{\pi}{2}$

$y = \cot^{-1} x = \operatorname{arccot} x$ $\qquad 0 < y < \pi$

$y = \cos^{-1} x = \arccos x$ $\qquad 0 \le y \le \pi$

$y = \sec^{-1} x = \operatorname{arcsec} x$ $\qquad 0 \le y \le \pi, \ y \ne \dfrac{\pi}{2}$

Law of Sines

$$\frac{a}{\sin A} = \frac{b}{\sin B} = \frac{c}{\sin C}$$

Law of Cosines

$a^2 = b^2 + c^2 - 2bc \cos A$ $\qquad \cos A = \dfrac{b^2 + c^2 - a^2}{2bc}$

$b^2 = a^2 + c^2 - 2ac \cos B$ $\qquad \cos B = \dfrac{a^2 + c^2 - b^2}{2ac}$

$c^2 = a^2 + b^2 - 2ab \cos C$ $\qquad \cos C = \dfrac{a^2 + b^2 - c^2}{2ab}$

Area of a Triangle

$$K = \tfrac{1}{2}bc \sin A = \tfrac{1}{2}ac \sin B = \tfrac{1}{2}ab \sin C$$

$$K = \frac{a^2 \sin B \sin C}{2 \sin A} = \frac{b^2 \sin A \sin C}{2 \sin B}$$

$$= \frac{c^2 \sin A \sin B}{2 \sin C}$$

$$K = \sqrt{s(s - a)(s - b)(s - c)},$$

where $s = \tfrac{1}{2}(a + b + c)$

Formulas for Geometric Sequences

$a_n = a_1 r^{n-1}$

$S_n = a_1 \dfrac{1 - r^n}{1 - r}$ if $r \ne 1$.

If $|r| < 1$, $\displaystyle\sum_{n=1}^{\infty} a_1 r^{n-1} = \frac{a_1}{1 - r}$.

If $|r| \ge 1$ and $a_1 \ne 0$, $\displaystyle\sum_{n=1}^{\infty} a_1 r^{n-1}$ does not exist.

Values of the Inverse Trigonometric Functions

$\sec^{-1} x = \cos^{-1} \dfrac{1}{x}$ $\qquad$ if $x \le -1$ or $x \ge 1$

$\csc^{-1} x = \sin^{-1} \dfrac{1}{x}$ $\qquad$ if $x \le -1$ or $x \ge 1$

$$\cot^{-1} x = \begin{cases} \tan^{-1} \dfrac{1}{x} & \text{if } x > 0 \\[2mm] \dfrac{\pi}{2} & \text{if } x = 0 \\[2mm] \pi + \tan^{-1} \dfrac{1}{x} & \text{if } x < 0 \end{cases}$$

Product and Quotient Rules

If $z_1 = r_1(\cos \theta_1 + i \sin \theta_1)$ and
$z_2 = r_2(\cos \theta_2 + i \sin \theta_2)$,

then

$$z_1 z_2 = r_1 r_2 [\cos (\theta_1 + \theta_2) + i \sin (\theta_1 + \theta_2)]$$

$$\frac{z_1}{z_2} = \frac{r_1}{r_2} [\cos (\theta_1 - \theta_2) + i \sin (\theta_1 - \theta_2)].$$

De Moivre's Theorem

If z has the trigonometric form $z = r(\cos \theta + i \sin \theta)$, then

$$z^n = r^n(\cos n\theta + i \sin n\theta).$$

Roots of Complex Numbers

If z is a nonzero complex number with trigonometric form

$$z = r(\cos \theta + i \sin \theta),$$

then the n nth roots of z are given by

$$\sqrt[n]{r} \left[\cos \left(\frac{\theta + k \cdot 360°}{n} \right) + i \sin \left(\frac{\theta + k \cdot 360°}{n} \right) \right]$$

for $k = 0, 1, 2, \ldots, n - 1$.

Formulas for Arithmetic Sequences

$$a_n = a_1 + (n - 1)d$$

$$S_n = \frac{n}{2} (a_1 + a_n)$$

$$S_n = \frac{n}{2} [2a_1 + (n - 1)d]$$